PLACE IN RETURN BOX to remove this checkout from your record.
TO AVOID FINES return on or before date due.

DATE DUE	DATE DUE	DATE DUE

MSU Is An Affirmative Action/Equal Opportunity Institution
c:\circ\datedue.pm3-p.1

INTRODUCTION TO GEOCHEMISTRY

Also Available from McGraw-Hill

Schaum's Outline Series in Science

Each outline includes basic theory, definitions and hundreds of example problems solved in step-by-step detail, and supplementary problems with answers.

Related titles on the current list include:

Acoustics
Analytical Chemistry
Applied Physics
Beginning Chemistry
Beginning Physics I
Biochemistry
Biology
Chemistry Fundamentals
College Physics
General, Organic, and Biological Chemistry
Genetics

Human Anatomy & Physiology
Introductory Geology
Lagrangian Dynamics
Modern Physics
Optics
Organic Chemistry
Physical Chemistry
Physical Science
Physics for Engineering & Science
Theoretical Mechanics
Zoology

Schaum's Solved Problems Books

Each title in this series is a complete and expert source of solved problems with solutions worked out in step-by-step detail.

Related titles on the current list include:

3000 Solved Problems in Biology
2500 Solved Problems in Chemistry
3000 Solved Problems in Organic Chemistry
2000 Solved Problems in Physical Chemistry
3000 Solved Problems in Physics

Available at most college bookstores, or for a complete list of titles and prices, write to:

Schaum Division
McGraw-Hill, Inc.
1221 Avenue of the Americas
New York, NY 10020

INTRODUCTION TO GEOCHEMISTRY

Third Edition

Konrad B. Krauskopf

Professor of Geochemistry Emeritus
Stanford University

Dennis K. Bird

Associate Professor of Geology
Stanford University

McGraw-Hill, Inc.

New York   St. Louis   San Francisco   Auckland   Bogotá
Caracas   Lisbon   London   Madrid   Mexico City   Milan
Montreal   New Delhi   San Juan   Singapore   Sydney   Tokyo   Toronto

This book was set in Times Roman.
The editors were Anne C. Duffy and John M. Morriss;
the production supervisor was Louise Karam.
The cover was designed by Rafael Hernandez.
Project supervision was done by Keyword Publishing Services Ltd.
R. R. Donnelley & Sons Company was printer and binder.

INTRODUCTION TO GEOCHEMISTRY

This book is printed on acid-free paper.

2 3 4 5 6 7 8 9 0 DOC DOC 9 0 9 8 7 6 5

ISBN 0-07-035820-6

Library of Congress Cataloging-in-Publication Data

Krauskopf, Konrad Bates, (date).
    Introduction to geochemistry / Konrad B. Krauskopf and Dennis K.
  Bird.—3rd ed.
      p.   cm.
    Includes index.
    ISBN 0-07-035820-6
    1. Geochemistry.   I. Bird, Dennis.   II. Title.
QE515.K7   1995
551.9—dc20   94-10069

ABOUT THE AUTHORS

Konrad B. Krauskopf is a Professor of Geochemistry Emeritus at Stanford University. He was born and raised in Madison, Wisconsin, and was given a BA degree in chemistry by the University of Wisconsin. Graduate work brought a PhD in chemistry from the University of California at Berkeley, then a PhD in geology from Stanford. He has taught geology and geochemistry at Stanford, and has conducted research on the geology of granitic and metamorphic rocks and on the geochemistry of ore deposits. His papers have been published in several professional journals, and he is the author of books on general physical science and on the geochemistry of radioactive waste disposal. He has seved as consultant to corporations and government agencies on subjects related to geochemistry, particularly on the disposal of hazardous wastes. He is a member of several professional organizations and has served as president of the Geochemical Society and the Geological Society of America.

Dennis K. Bird is an Associate Professor of Geology at Stanford University where he has taught geochemistry for 12 years. He was raised in a small town in the Sierra Nevada of eastern California and received his BS and PhD degrees in geology from the University of California. Dr. Bird is an associate editor for the *American Journal of Science*. He has published a number of papers on the thermodynamic properties of minerals, water-rock reactions, fluid flow, and fracture porosity in magma-hydrothermal systems. He has led 10 expeditions to East Greenland where he and his students study ancient geothermal systems formed during the opening of the North Atlantic Ocean.

CONTENTS

PREFACE

Revising a geochemistry textbook means a continual making of choices among subjects that should be emphasized, deemphasized, or newly introduced. So rapidly has knowledge of the Earth's chemistry expanded in the last couple of decades that a great wealth of material is at hand which seemingly deserves attention. But to include all of it, or even a large fraction, would inflate the book's volume beyond reason. Difficult choices must be attempted, and one cannot hope that the choices made will please all potential readers. Nor can one avoid the disquieting thought that regardless of the choices, research in some parts of the field is active enough to ensure that the book will be obsolete before it is printed.

Because geochemistry has become more quantitative, and because most students of the present generation have a more quantitative background in chemistry and mathematics than their predecessors, one of our choices has been to increase somewhat the rigor of treatment in several parts of the book. Elementary thermodynamics, at least derivation of the simpler formulas, has been introduced early rather than relegated to an appendix, and is used frequently in subsequent discussions. Effects of high temperature and pressure on geochemical processes are incorporated in the early chapters, not left as a new subject for later treatment. Topics that in our opinion particularly needed expansion include kinetics, surface chemistry, the geochemistry of isotopes, and details of element distribution and transport. The increasing attention given to environmental issues in recent years has prompted us to put new emphasis in many places on geochemical processes as a factor in environmental change.

The basic purpose of the book remains unaltered. It is intended as an introduction to an increasingly important part of earth science, for undergraduate students who have taken the usual pattern of elementary courses in earth sciences, chemistry, and mathematics. Emphasized throughout is the application of geochemistry to geologic problems, with selected examples from actual field and laboratory studies. Our hope is to make the subject seem an exciting area of research for students who are chemically inclined, and to give students of all camps a feeling for the usefulness of geochemistry as a background for many kinds of earth science inquiry.

The order of presentation of subjects should not be considered as fixed, but is adjustable for students of different backgrounds. The existing order begins with a review of very simple chemical ideas with immediate application to some familiar

xiii

geologic situations, designed for a student group whose exposure to chemistry may be some years in the past and for whom a start with thermodynamic formulas might seem unpleasantly abstract. A different group, say one with a good physical chemistry course fresh in mind, could well find it more profitable to start with the classical thermodynamics presented in Chapters 7 and 8, and to use the preceding chapters simply as illustrative material suitable for quick review. For those preferring to begin with a broad overview of the earth's chemical composition and history — and again who have had a recent brush with physical chemistry — a third alternative would put Chapter 21 (Historical Geochemistry) or Chapter 20 (Distribution of the Elements), or a combination of the two, at the beginning. Any deviation from the existing sequence will mean some awkwardness with references, of course, but the difficulty should not be serious.

In keeping with the pedagogical purpose of the book, most chapters end with a list of questions. These, it should be emphasized, are an integral part of the book. For a subject like geochemistry, no real understanding is gained by mere reading. Ideas become part of a student's useful mental equipment only when they are reflected on, digested, and manipulated, and problems are a necessary goad to such activity. The questions include not only those designed for practice in getting numerical results from equations, but also many requiring students to think about specific geologic problems, or to evaluate geologic hypotheses, in terms of the quantitative ideas just presented. Attention is continually directed, both in problems and in the text, to the relations between theoretical formulas and results of controlled laboratory experiments, on the one hand, and geologic field observations on the other. Only through the incentive that problems give for thinking about such relations can a student develop geochemical judgment, the feel for the kinds of geologic situations in which chemistry can be usefully employed and those in which the precise answers of chemistry are misleading or meaningless. A selected number of fully worked-out problems now appear in Appendix XI of the text.

At the end of each chapter appears also a list of books and articles, including those to which reference is made in the chapter and a few others intended as an introduction to current geochemical literature. Articles and books were picked more with an eye to intrinsic interest and suitability for undergraduate readers than for completeness of coverage.

To students, Stanford colleagues, and many elsewhere who have used the book we are grateful for pointing out errors and making suggestions for improvement. A few in particular should be named, who very kindly read drafts of entire chapters of the new edition and made comments that helped greatly in correcting mistakes and smoothing the language: Marco Einaudi, the chapter on ore-forming solutions, Simon Brassell, the chapter on organic geochemistry, George Parks and Peggy O'Day, the chapters on surface chemistry and clay minerals. To them we owe a great debt, but if errors remain the responsibility is solely ours. It is a pleasure to acknowledge also the assistance of many who helped us with the word processing and the figures, and to recognize the understanding perseverance of the editorial staff at McGraw-Hill in getting the book printed despite our tardiness in supplying them with a manuscript.

McGraw-Hill and the authors would like to thank the following, who reviewed selected parts of this revision: Charles E. Bickel, San Francisco State University; Carl Bowser, University of Wisconsin, Madison; William R. Brice, University of Pittsburgh at Johnstown; David R. Hickey, Formerly, Lansing Community College; Clark M. Johnson, University of Wisconsin, Madison; Teh-Lung Ku, University of Southern California; Tony Lasaga, Yale University; Craig E. Manning, University of California at Los Angeles; Stanley A. Mertzman, Franklin and Marshall College; Kula C. Misra, University of Tennessee, Knoxville; Philip Neuhoff, Stanford University; J. Donald Rimstidt, Virginia Technical University; Joaquin Ruiz, University of Arizona, Tucson; Peter Schlosser, Columbia University; William Seyfried, University of Minnesota; and Gregory Wheeler, California State University, Sacramento.

Konrad B. Krauskopf
Dennis K. Bird

LIST OF SYMBOLS

Parentheses after the name of a symbol indicate the customary units employed in its measurement and the number of the equation in which it first appears in the text

a activity (pure number) (2-72)

(aq) subscript: in water solution

atm atmosphere

c component, number of components (8-81)

C constant of integration (units variable) (10-2)

°C degrees Celsius

C heat capacity (joules degree^{-1}) (7-31)

    C_P heat capacity at constant pressure (7-33)

    C_V heat capacity at constant volume (7-32)

D_i diffusion coefficient of component i (cm^2 sec^{-1}) (11-42)

cm centimeter

e electron (9-4)

E electrode potential (volt) (9-14)

E° standard electrode potential (volt) (9-8)

Eh redox potential (volt) (9-22)

E internal energy (joule) (7-8)

E_a activation energy (joule)

f degrees of freedom (pure number) (8-81)

f as subscript: formation, e.g., H_f = heat of formation

f fugacity (bar) (8-23)

f Faraday constant (joules volt^{-1}) (9-14)

F function

g gram

g acceleration of gravity (cm sec^{-2})

(g) subscript: gas

G Gibbs free energy (kilojoule) (7-46)

$\Delta G°$ standard free energy change (kilojoules mol^{-1})

h height or vertical distance (meter)

H enthalpy (kilojoule) (7-21)

ΔH^o standard enthalpy change (kilojoules mol^{-1})

i component of a solution

I ionic strength (mols kg^{-1}) (2-70)

J joule

$\mathbf{J}$ flux (grams cm^{-2} sec^{-1}) (11-39)

$\mathbf{k}$ partition coefficient (pure number) (20-3)

k rate constant (units variable) (11-5)

K degrees kelvin

K equilibrium constant (units variable) (1-6)

K_D distribution coefficient (pure number) (5-5)

kbar, kg, km kilobar, kilogram, kilometer

l length or horizontal distance (meter)

(l) subscript: liquid

L liter

M mass (gram)

m molal concentration (mols kg^{-1}) (1-21)

$m_{i,\text{total}}$ total concentration of component i (2-12)

M molar concentration (mols L^{-1})

mg, ml, mm milligram, milliliter, millimeter

n number of mols

$\mathbf{n}$ number of electrons

ng, nm nanogram, nanometer

N normal concentration (equivalents kg^{-1})

p number of phases (8-81)

P pressure (bar, pascal, atm, kg cm^{-2})

pe electron activity (pure number) (9-31)

psi pounds per square inch

Q heat (joule) (7-8)

R dissolution rate (units variable, usually mols cm^{-2} sec^{-1}) (11-22)

R universal gas-law constant (J mol^{-1} deg^{-1}) (7-2)

$\mathbf{R}$ atom percent oxygen contributed by rock (11-37)

r subscript: reaction

ref subscript: reference

(s) subscript: solid

S entropy (joules degree^{-1}) (7-11)

ΔS^o standard change in entropy (joules mol^{-1} deg^{-1})

sec second

t time (second)

T temperature, kelvin (°C + 273.15)

v velocity (cm sec^{-1})

V volume (cm^3)

W work (joule, grams cm^{-3} sec^{-2}) (7-8)

W atom percent oxygen contributed by water (11-37)

X mol fraction (pure number)

z ionic charge (pure number) (2-70)

α fractionation factor for isotope separation (pure number) (10-25)

β beta particle or electron (10-11)

γ activity coefficient (mol^{-1}) (2-72)

δ measure of isotope separation compared to a standard (pure number) (10-26)

Δ_B^A isotope fractionation between phases A and B (10-33)

ε dielectric constant

ε relation of isotope ratio to ratio in primitive chondritic material (Fig. 10-5)

κ^{-1} distance from mineral surface to center of charge (nanometers) (6-6)

λ decay constant for radioactive decay (yr^{-1}) (10-1)

μ chemical potential (joules mol^{-1}) (8-3)

μm micrometer (micron)

ρ density (grams cm^{-3})

ρ_i partial mass density (gm cm^{-3}) (11-38)

σ charge on a mineral surface (mols of charge m^{-2}) (6-5)

υ stoichiometric reaction coefficient (pure number) (7-30)

ϑ kinematic viscosity ($cm^2\ sec^{-1}$) (11-40)

χ fugacity coefficient (pure number) (8-30)

INTRODUCTION TO GEOCHEMISTRY

CHAPTER
1

CHEMICAL EQUILIBRIUM

The Earth is a huge chemical machine, powered by its interior heat and less concentrated heat from the Sun on its surface. The inner heat drives convection in the Earth's mantle, the vast interior zone between iron core and silica-rich crust. Mantle material is mostly magnesium silicate, admixed with iron and minor amounts of other elements. Convection brings currents of this material up near the surface under mid-ocean ridges. Here the lower temperatures and lower pressures lead to partial chemical separation, or differentiation, of the minor constituents to form basaltic lava. The lava, with its higher content of silicon, aluminum, calcium, and the alkali metals, solidifies into rock of the oceanic crust, and moves away from the mid-ocean ridges as part of big lithospheric plates riding on horizontally moving segments of the convectional circulation. Ultimately the plates move down again into the mantle at subduction zones, carrying with them the newly differentiated basalt. Accompanying this movement is a further chemical separation, giving liquids that rise into the crust and solidify as the silica-rich granitic rocks that are so prominent a part of the continents.

Rocks of the crust, exposed to heat from the Sun and permeated by water and air, are the site of the second part of the Earth's chemical machinery. Water in all its forms, aided by active gases in the atmosphere and by organisms, reacts with crustal rocks in yet another stage of chemical differentiation. This time some of the separations are especially clean, silica going into quartz sand and chert, aluminum into clay minerals, calcium into limestone, some of the heavy metals into ores. The two parts of the chemical machine, convection in the interior and weathering near the surface, function semi-independently. But they are obviously at some points related, as new mantle material is brought up and exposed to surface agents, as differentiated rocks are carried down once more into the convecting mantle, as energy from both sources sets up circulation of fluids deep into rocks of the lithosphere.

This two-fold planetary machinery bristles with physical and chemical questions. First would be a query about convection: convection currents form only in fluids; how, then, can the mantle convect, when we know that the Earth's interior is largely solid? Then the details of differentiation: how does basalt form from peridotite in the upper mantle, and how does basalt (often plus other materials) generate granite? And how do water and air attack the minerals of granite? Or we could think of deeper questions: what is the end result of convection and differentiation? Do the processes of separating crust from mantle, and then the return of crustal material into the mantle by subduction, approach a state of balance? Or is the overall amount of crust steadily increasing? Also questions about time: how long in the Earth's history have convection and differentiation and subduction been going on, and at what rates? When and how did they start in the first place? And then questions of more immediate practical consequence: how do all these processes affect the surface of the continents, on which humanity makes its home? How do parts of the chemical machinery concentrate metals like gold, copper, nickel in some kinds of rock, and liquid hydrocarbons in other kinds? As humanity grows in number and in its use of Earth materials, will the planet's resources be sufficient for a long future? And will human activity have any influence on the Earth's chemical machinery, and, hence, on the environment we live in? These are but a sampling of the questions that could be posed.

To look for answers to many of them is the purpose of this book. But attacking them directly would hardly be profitable, just because of their complexity. We need first a background of chemical principles and geologic observations, on which an understanding of the Earth's chemistry can be built. To make a start on such a background is the goal of this chapter.

1-1 EQUILIBRIUM CONSTANTS

We begin with a basic question in chemistry: if two substances are mixed, under particular conditions of temperature and pressure, will a reaction take place? And if so, how far and how fast will it go? Applied to natural materials, this is a central theme of geochemistry. For particular substances, qualitative answers often come from everyday experience: we know that acids react readily with carbonates, that organic materials burn in oxygen, that salt will dissolve in water until the solution is saturated, that most reactions go faster if the temperature is raised. Such predictions become more useful if we attach numbers to them, numbers like equilibrium constants, solubility products, adsorption coefficients, rate constants. And among these numbers, we choose equilibrium constants for a first look.

Familiarity with equilibrium and equilibrium constants is a major help in understanding the processes responsible for the formation of minerals and mineral assemblages and their interaction with silicate liquids and aqueous solutions. It provides a background for attacking problems related to weathering, sedimentation, diagenesis, metamorphism, and hydrothermal and magmatic processes. Two variables of major importance in the study of equilibrium are temperature and pressure. An indication of their role in Earth processes is suggested by Fig. 1-1, which shows in a general way the temperature and pressure conditions characteristic of geologic environments in the Earth's crust. Much of this book will be concerned with reactions that take place in these environments, and the changes in composition of minerals and solutions that occur as the equilibria shift in response to changes in pressure–temperature conditions (Fig. 1-2). The relations among composition, temperature, and pressure are a dominant theme throughout.

The quantitative treatment of chemical equilibrium goes back, in a curious way, to the work of two Norwegian investigators of the mid-19th century, Cato Maxmilian Guldberg (1836–1902) and Peter Waage (1833–1900). They were concerned, as we are today, with the problem of what makes a reaction go, and they were thinking in terms of then-prevalent ideas about vaguely defined "chemical forces" and "chemical affinities." As a conclusion from studies of many reactions, they proposed that the "driving force" of a reaction is primarily dependent on the "mass" of each substance taking part. The meaning they attached to "driving force" is not very clear; they speculated that it was somehow related to reaction rate but did not make the connection explicit. The "mass" in their statement, from the context, would be better labeled "concentration." This rather obscure formulation has come down to us as "the law of mass action," which in more recent times has generally been stated, "The rate of a reaction is directly proportional to the concentration of each reacting substance."

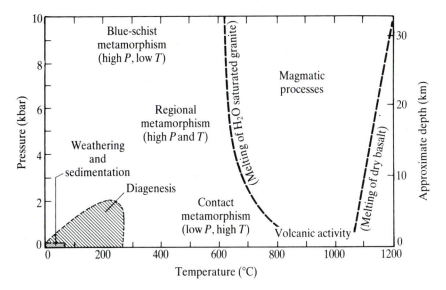

FIGURE 1-1
Pressure–temperature diagram illustrating the approximate conditions for weathering, diagenesis, metamorphism, and magmatism in the Earth's crust. Approximate depth, shown on the right side of the diagram, is computed from the equation, pressure = (density) × (acceleration due to gravity) × (depth), and an average rock density of 3.0 gm cm^{-3}.

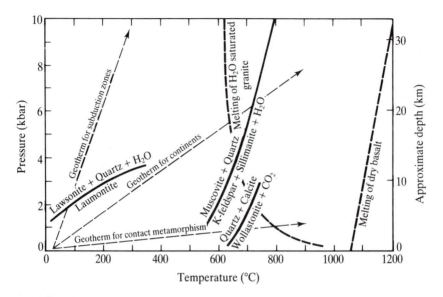

FIGURE 1-2

The same pressure–temperature grid as in Fig. 1-1, with added lines showing three typical geotherms and the pressure–temperature conditions for equilibrium in three metamorphic reactions. (*a*) In subducting lithosphere, pressure increases rapidly with depth while temperature remains fairly low; as pressure rises, the zeolite mineral laumontite (in altered basalts) becomes unstable with respect to lawsonite and quartz. (*b*) Under continents, where pressure and temperature increase together, a typical metamorphic reaction in pelitic rocks is the conversion, with rising temperature, of a muscovite–quartz mixture to sillimanite and K-feldspar. (*c*) Contact metamorphism in a siliceous limestone near a granitic intrusion, where pressures are low and temperatures high, results in a shift of equilibrium from quartz + calcite to wollastonite as the rock is heated. Fluid pressure is assumed to be equal to the total pressure for the three metamorphic reactions.

Even this formulation is beset with difficulties. For a reaction that takes place as a result of simple molecular collisions, with some fraction of the total number of collisions leading to reaction in a given time, the statement would be reasonable and accurate. But most reactions are not this simple. Commonly a reaction takes place in steps, and while each step may follow the "law," the sum of steps going at different rates does not. Hence the law of mass action probably does not merit the importance often attached to it. As a rule that holds for ideally simple reactions, and for postulated simple steps in reaction mechanisms, it has some usefulness, but as a general law for reaction rates it is misleading.

A more important result of Guldberg and Waage's work was the recognition that many reactions are incomplete because they are *reversible*, and that the same final mixture of substances is obtained when a reversible reaction is started from either end. In symbols, a reversible reaction can be represented

$$A + B \rightleftharpoons Y + Z, \tag{1-1}$$

where A, B, Y, and Z represent substances in the reaction.

We can mix equal amounts of A and B in one container, and equal amounts of Y and Z in another; when reaction has stopped in each, the same mixture of all four substances is present. In Guldberg and Waage's language, this meant that both reactions, forward and reverse, go until their "driving forces" become equal. If each driving force is shown as proportional to concentrations,

$$\text{Driving force of forward reaction} = k_1[A][B] \qquad (1\text{-}2)$$

$$\text{Driving force of reverse reaction} = k_2[Y][Z] \qquad (1\text{-}3)$$

then equality of the driving forces requires that $k_1[A][B] = k_2[Y][Z]$, or

$$\frac{[Y][Z]}{[A][B]} = \frac{k_1}{k_2} = K. \qquad (1\text{-}4)$$

Brackets indicate concentrations, k_1 and k_2 are proportionality constants, and K is another constant obtained by dividing k_1 by k_2. In other words, when a reversible reaction has gone as far as it can, the quotient obtained by multiplying the concentrations of the products and dividing by the concentrations of reactants is a constant quantity. This relation turns out to be experimentally verifiable, quite apart from any meaning we attach to "driving force" or any complexities of reaction mechanism.

In modern language we would say that a reversible reaction stops when it has reached a state of chemical equilibrium, that the stopping is only apparent because the two reactions are still going on, and that the state of equilibrium is merely a balance between opposing processes going at equal rates. The K is the *equilibrium constant*, a number that defines the equilibrium condition for a particular reaction.

The constancy of K is apparently "explained" by the fact that it is the quotient of two other constants, each of which describes a reaction rate. If one of the rates is changed, say by increasing the concentration of A, the rate of the forward reaction is temporarily increased, and the final equilibrium mixture will have more Y and Z and less B than before; but the rate constants, and of course their quotient also, remain the same. This reasoning is useful but only qualitatively correct, because most reaction rates are not accurately described as proportional to the simple first powers of concentrations. A more rigorous explanation of the constancy of K can be developed by starting from energy relations rather than reaction rates, as we will see in a later chapter (Sec. 8-4).

The history of the law of mass action is an illuminating episode in the development of scientific ideas. The law as originally stated was so vague as to have little meaning, and attempts to reframe the law in terms of reaction rates have caused much confusion. Yet it had the germ of an idea that led to the correct formulation of equilibrium constants, a discovery that marks a milestone in the development of physical chemistry. Normally we think of science as progressing by logical steps from one clearly defined concept to another, but this bit of history is clear proof that the path of progress is often more devious.

A more general equation for a chemical reaction is:

$$aA + bB + \ldots \rightleftharpoons yY + zZ + \ldots \qquad (1\text{-}5)$$

where A, B, Y, Z represent chemical formulas, a, b, y, z represent coefficients, the dots stand for additional reactants or products, and the double arrow indicates that the reaction is reversible. The corresponding equilibrium constant is[1]

$$K = \frac{[Y]^y[Z]^z \cdots}{[A]^a[B]^b \cdots}. \tag{1-6}$$

Note that the concentration of each substance is raised to a power given by its coefficient. For the present we may take this as a generalization from experimental results, without trying to justify it on the basis of reaction mechanism or thermodynamic principles.

1-2 AN EXAMPLE OF EQUILIBRIUM: HYDROGEN CHLORIDE

We apply Eq. (1-6) first to mixtures of hydrogen, chlorine, and hydrogen chloride, three gases often detectable at volcanic vents (Sec. 18-6). From laboratory study it is well-known that mixtures of hydrogen and chlorine can be made to react explosively to form hydrogen chloride, and likewise that hydrogen chloride raised to a high temperature decomposes into its elements. In other words, the reaction represented by the equation

$$H_2 + Cl_2 \rightleftharpoons 2HCl \tag{1-7}$$

is known to go both forward and backward, so that equilibrium among the three gases would be shown as

$$K = \frac{[HCl]^2}{[H_2][Cl_2]}. \tag{1-8}$$

The brackets in this equation indicate concentrations of the three gaseous species.

In what units should concentrations of gases be expressed? Obviously there is a wide choice of possible units: grams per liter, mols per liter, mols per kilogram, pounds per cubic foot, and many others. By convention, the unit ordinarily used is none of these, but a quantity proportional to them, the *partial pressure* in bars. (In a mixture of several gases, each gas behaves approximately as if it were alone in the space, and exerts a pressure independently of the others; this is its partial pressure, and the total pressure of the mixture is the sum of all the partial pressures—which is *Dalton's law of partial pressures*.) The constancy of the quotient in Eq. (1-8) can be

---

[1] A K defined in this manner, as a quotient of concentrations, is only approximately constant. When *activities* are used in place of concentrations, the K is a true constant. Because concentrations are easier to visualize than activities, the approximate concentration constants will be used at first, and activities will be introduced in Chapter 2 with a full explanation of their meaning in Chapter 8. Activities of substances in a reaction are numbers that are proportional to concentration, partial pressure, or composition of the substance; they are unitless, so that the true equilibrium constant is also a dimensionless number.

expressed equally well with any units, but to compare different constants and to standardize numerical values in tables, some arbitrary choice must be made—and for most purposes the partial pressure is the most convenient (see Sec. 1-9). Except where otherwise stated, concentrations of gases will be expressed as partial pressure in bars (symbol P)[1].

The measured value of K in Eq. (1-8) at 1000°C is 2.6×10^8, or $10^{8.4}$. In an equilibrium mixture at this temperature, suppose that H_2 and Cl_2 are present each at 1 bar pressure. What would be the partial pressure of HCl? Substitution gives

$$K = \frac{P_{HCl}^2}{1 \times 1} = 10^{8.4} \quad \text{or} \quad P_{HCl} = 10^{4.2} \text{ bar.} \tag{1-9}$$

Or suppose that the concentration of HCl is 0.1 bar and that of Cl_2 $10^{-5.0}$ bar. What is the equilibrium pressure of H_2?

$$\frac{(0.1)^2}{P_{H_2} \times 10^{-5}} = 2.6 \times 10^8 \quad \text{and} \quad P_{H_2} = 3.8 \times 10^{-6} \text{ bar.} \tag{1-10}$$

Thus, whatever the pressure of two of the gases may be, the third must have a value such that K will remain constant.

Once equilibrium is established, a change in concentration of one constituent must lead to changes in the others. For example, if equilibrium exists among H_2 and Cl_2, each at 1 bar, and HCl at $10^{4.2}$ bar, and if enough H_2 is added to make its total pressure momentarily 10 bar, then the three gases must react to bring K back to its original value. This means that more HCl must be formed by reaction between Cl_2 and some of the added H_2. If x is the amount of H_2 that reacts, then x bars of Cl_2 must also react and $2x$ bars of new HCl must be produced. Hence

$$\frac{(10^{4.2} + 2x)^2}{(10 - x)(1 - x)} = 10^{8.4}. \tag{1-11}$$

As often happens with equilibrium constants, some simple assumptions have led to a complicated algebraic equation. This one could be solved straightforwardly as a quadratic, but time is generally saved by hunting for a shortcut. Note here that x must be less than 1, so that $2x$ is negligible in comparison with $10^{4.2}$. Omitting $2x$ from the numerator simplifies the equation a great deal:

$$(10 - x)(1 - x) = 1. \tag{1-12}$$

Now x is fairly small in comparison with 10; if it is neglected in the first parenthesis, then $1 - x$ is approximately equal to 0.1, and x must be about 0.9. This is sufficiently accurate for most geologic purposes. If x is not neglected in the $10 - x$

---

[1] Three units of pressure are in common use, the *bar*, the *atmosphere*, and the *pascal*: 1 bar = 0.987 atm = 10^5 Pa. In this book the bar will be most frequently used. The difference between the bar and the atmosphere is so small that for most purposes in geochemistry the two may be considered roughly equivalent.

term, a slightly better value is obtained: $x = 0.89$. Hence a new equilibrium mixture is established among 9.11 bar of H_2, 0.11 bar of Cl_2, and $10^{4.2}$ bar of HCl.

It is important to sense how this sort of equilibrium reacts to any disturbance, like the addition of H_2 in the last example. Whatever changes of concentration are imposed, the equilibrium automatically adjusts itself so that K remains constant. One way to see why this happens is to recall that a chemical equilibrium is dynamic, maintained by a balance between opposing reactions that go on continuously. If more H_2 is added to an equilibrium mixture, the forward reaction

$$H_2 + Cl_2 \rightarrow 2HCl \tag{1-13}$$

is favored, while the reverse reaction is unaffected:

$$2HCl \rightarrow H_2 + Cl_2. \tag{1-14}$$

Hence for a brief time there is imbalance, until enough new HCl is built up to make its rate of decomposition equal to its rate of formation. Recall that a general property of equilibrium is to respond to disturbance in such a way as to counteract the disturbance as far as possible. When H_2 is added, the equilibrium responds by "trying" to decrease its concentration, hence by favoring for a time the forward reaction over the reverse reaction. From either point of view the equilibrium is pictured as consisting of active opposing processes, always compensating for changes imposed from without.

The formula for the equilibrium constant depends on how the chemical equation is stated. The HCl reaction, for example, could equally well be written backward:

$$2HCl \rightleftharpoons H_2 + Cl_2. \tag{1-15}$$

for which the equilibrium constant would be

$$\frac{P_{H_2} P_{Cl_2}}{P_{HCl}^2} = K'. \tag{1-16}$$

The new constant K' is the reciprocal of K in Eq. (1-8). Or we could write

$$\tfrac{1}{2}H_2 + \tfrac{1}{2}Cl_2 \rightleftharpoons HCl \quad \text{and} \quad \frac{[HCl]}{[H_2]^{1/2}[Cl_2]^{1/2}} = K'', \tag{1-17a, b}$$

where this K'' is the square root of the original K. Either K' or K'' could be used just as easily as K in calculating how concentrations change when one of the substances is added or removed. *An equilibrium constant has no meaning except in terms of a specified chemical equation.* One cannot speak of "the equilibrium constant of HCl," but only of the equilibrium constant for Eq. (1-7), for Eq. (1-15), for Eq. (1-17a), or for some other way of writing the reaction.

1-3 THE EFFECT OF TEMPERATURE

At low temperatures hydrogen burns readily in chlorine, and if the chlorine is in excess, no detectable H_2 is left over after the burning stops. Hydrogen chloride is stable at low temperatures, showing no tendency to decompose into its elements. These observations mean that the equilibrium constant for Eq. (1-7) under such conditions is a very large number; the forward reaction is practically complete, the backward reaction goes scarcely at all. On the other hand, at 2000°C hydrogen and chlorine react only slightly and HCl is rapidly decomposed, so that the equilibrium constant for Eq. (1-7) is a small number. Other equilibria show similar changes with temperature. An equilibrium constant, therefore, must be defined for a particular temperature as well as pressure, and has no meaning unless the temperature and pressure are specified. By convention, equilibrium constants are most commonly given for 25°C and 1 bar pressure, and these conditions are understood unless other values of temperature and pressure are specified.

At room temperature, then, an equilibrium mixture of the three gases we have been discussing might consist almost entirely of HCl alone, or it might contain HCl plus an excess of either element, with the second element present only in undetectable traces. A mixture of H_2 and Cl_2 at room temperature should be far from equilibrium and should react until at least one of the elements is nearly used up. Now laboratory experience seems to contradict this prediction, for a mixture of the two elements at room temperature appears to be inert as long as it is not touched with a flame or exposed to strong illumination. What is the difficulty?

We can say only that the reaction between H_2 and Cl_2 at ordinary temperatures is very slow. In terms of molecular structure, a partial explanation is that the elements can react only when the diatomic molecules H_2 and Cl_2 are broken up, and molecular collisions at room temperatures are, for the most part, not sufficiently violent to accomplish the splitting. Thus equilibrium is indeed displaced far toward HCl, but attainment of equilibrium may be indefinitely delayed because the reactions are sluggish. There is nothing in either the chemical equation or the equilibrium constant to warn us that slow reactions may make true equilibrium difficult to reach: the concept of equilibrium refers only to the final result, not to the means of attaining it.

This poses a real problem. If nonequilibrium mixtures can be seemingly stable, how do we recognize true equilibrium mixtures? How would we know *experimentally*, for example, that a mixture of hydrogen and chlorine with only a trace of HCl at 25°C is not actually in equilibrium? For this particular case the answer is easy, because all we need do is introduce a fairly minor disturbance—heat the mixture, touch it with a flame, expose it to light—and the resulting loud explosion will tell us emphatically that the mixture was not at equilibrium. A less spectacular but more rigorous test for equilibrium is to see whether the same mixture can be obtained by starting from the other end of the reaction, in this case by starting with pure HCl. We would find, of course, that HCl showed no tendency to decompose into our original mixture, which tells us immediately that one or the other cannot be in equilibrium. In general, *to see if a mixture of compounds is an equilibrium assemblage, we set up experiments to see whether the same mixture can be obtained by starting with pure components at each end of the reaction.* If no reaction takes place in either direction, we are probably dealing with slow processes, and we must find some way to speed them up. Until we do get reactions to take place, we have no way of being certain whether our original mixture is at equilibrium or is apparently inert because possible reactions may be very slow.

Slow reactions are always speeded up by a rise in temperature. The factors determining reaction rates are complex, and a quantitative treatment will be deferred to Chapter 11. A useful qualitative rule is that most reactions have their rates doubled or tripled by a 10°C rise in temperature. This means that equilibrium is always more easily attained at high temperatures. Even at the temperatures of molten rocks, however, we cannot assume that all chemical reactions reach equilibrium. In future discussions we shall often be faced with the question whether a given assemblage of minerals, liquids, and gases may be considered an equilibrium mixture.

To go back to the HCl example, hot gases from volcanic vents and fumaroles can sometimes be shown to contain the three gases H_2, Cl_2, and HCl. Furthermore, the relative amounts of the three are reported to be widely different even from fumaroles of approximately the same temperature. This is good evidence that not all the mixtures can be equilibrium assemblages—in other words, that during the violent escape of gas from lava beneath the surface, not enough time has elapsed for the gas mixture to adjust itself to changing conditions of temperature and pressure. This observation, together with the fact that the reaction of hydrogen with chlorine is *exothermic* (heat-producing), has suggested a possible explanation for temperature measurements showing that the surface of a lava pool may be hotter than its interior.

We shall find that most reactions involving polyatomic molecules are similar to the H_2–Cl_2 reaction in that they are slow at room temperature and reach equilibrium readily only at temperatures of a few hundred degrees. Reactions involving ions in water solution, however, are generally almost instantaneous even at low temperatures, because the reaction process requires no preliminary dissociation of stable molecules. Reactions of ions in silicate melts, on the other hand, may be slow, because movements of the ions are impeded by the extreme viscosity of the liquid. These generalizations have many exceptions, and we shall have to examine geologic processes individually to see whether or not equilibrium conditions can be expected in any particular circumstance.

1-4 A SECOND EXAMPLE: CARBON DIOXIDE IN WATER

The hydrogen chloride reaction has only minor interest in geology but serves as a beautifully simple illustration of the principles of equilibrium because all the substances involved are gases. More technically, we say that all mixtures of H_2, Cl_2, and HCl constitute a single *phase* (unless, of course, the temperature is low enough for chlorine or hydrogen chloride to liquefy), the word phase meaning any part of a system which is homogeneous and which is separated from other parts of the system by sharp boundaries. The equilibria of greatest concern to geologists are those involving more than one phase—liquid and gas, liquid and solid, or more complicated systems containing both liquid and gas together with several solid phases. Such equilibria are described as *heterogeneous*, in contrast to *homogeneous* equilibria like the HCl reaction which involves reaction among components of a single phase.

A simple two-phase system of everyday experience and of great importance geologically is the equilibrium between carbon dioxide and water. For a preliminary study of this system, we need only put a stopper in a bottle half full of water. Gaseous CO_2 from the air in the top of the bottle dissolves in the water, and dissolved CO_2 in the water escapes into the air. Soon a balance is established between these two processes:

$$CO_2 \rightleftharpoons CO_2 . \qquad (1\text{-}18)$$
$$\text{air} \qquad \text{water}$$

The importance of this reaction lies in the fact that the dissolved CO_2 reacts with water to form an acid, H_2CO_3, carbonic acid. Thus we can rewrite the equation for the equilibrium,

$$CO_2 + H_2O \rightleftharpoons H_2CO_3 . \qquad (1\text{-}19)$$
$$\text{air} \qquad \text{water} \qquad \text{water}$$

(We assume here that all the dissolved CO_2 forms H_2CO_3. This is not strictly correct; to be more accurate, we should consider also the equilibrium $CO_2(\text{dissolved}) + H_2O \rightleftharpoons H_2CO_3$, which actually is displaced far to the left. To simplify this and future discussions, we omit this step. As long as the assumption is used consistently, the omission does not affect geologic arguments.)

The equilibrium responds predictably to changes in conditions: if we increase the amount of CO_2 by adding some from a tank, we speed up the forward reaction and more CO_2 dissolves; if we decrease the pressure of CO_2 by attaching a vacuum pump, the forward reaction is slowed and the equilibrium shifts to the left; heating the bottle decreases the solubility of the gas (or the stability of H_2CO_3), and again the equilibrium shifts to the left. Or we could disturb the reaction chemically by adding a base to neutralize the acid: if a little NaOH solution is poured into the bottle, some of the H_2CO_3 is destroyed, thereby slowing the reverse reaction and permitting more CO_2 to dissolve.

A numerical value of the equilibrium constant for reaction (1-19),

$$K = \frac{[H_2CO_3]}{[CO_2][H_2O]}, \qquad (1\text{-}20)$$

can be found by looking up the solubility of CO_2. This is given in tables as 0.76 liter of carbon dioxide per liter of water at 25°C when the pressure of CO_2 is maintained at 1 bar. To express concentrations in solution, a common unit is mols per kilogram of water or *molality*; the 0.76 liter of CO_2 under these conditions would represent 0.76/24.5 or 0.031 mol, so that the concentration of H_2CO_3 in Eq. (1-20) may be written $0.031m$. (The symbol *m* means mols of solute per kilogram of water, and the number 24.5 is the volume in liters occupied by 1 mol of any gas at 25°C and 1 bar pressure.) For the concentration of CO_2 gas we use bars of pressure, as we did for the gases in the HCl equilibrium; in this case its concentration will be 1 bar. For the concentration of H_2O we could find mols per liter ($1000/18.016 = 55.5m$), but the standard convention is to represent its composition by the mol fraction of H_2O in solution. In all dilute solutions this mol fraction is approximately one. Hence we write

$$K = \frac{m_{H_2CO_3}}{P_{CO_2}} = \frac{0.031}{1} = 0.031 = 10^{-1.5}. \tag{1-21}$$

Having found the constant for the CO_2 reaction, we can now use it to calculate how much H_2CO_3 is present in water exposed to ordinary air. Air contains 0.03% CO_2 by volume; this means a volume fraction of 0.0003 and therefore a partial pressure of 0.0003 bar (since partial pressure is approximately proportional to mol fraction, and this in turn to volume fraction). Hence we substitute in Eq. (1-21):

$$K = 0.031 = \frac{m_{H_2CO_3}}{0.0003}, \tag{1-22}$$

from which

$$m_{H_2CO_3} = 0.031 \times 0.0003 = 10^{-1.5} \times 10^{-3.5} = 10^{-5}m. \tag{1-23}$$

This seems a very small concentration of acid, but it is sufficient to make natural waters much better weathering agents than they could be without it.

1-5   A THIRD EXAMPLE: CALCIUM SULFATE

Calcium sulfate (anhydrite) stirred in water dissolves to a slight extent, producing the free ions Ca^{2+} and SO_4^{2-}:

$$CaSO_4 \rightarrow Ca^{2+} + SO_4^{2-}. \tag{1-24}$$

anhydrite

Soon the solution becomes *saturated*, after which no further $CaSO_4$ will dissolve, however long the stirring is continued or however much solid $CaSO_4$ is added. This behavior suggests that an equilibrium has been reached—a *heterogeneous* equilibrium, since it includes the two phases, solution and solid $CaSO_4$. To prove that equilibrium exists, we try to make the reaction go backward, by mixing (in another container) separate solutions containing equal amounts of Ca^{2+} and SO_4^{2-}. A precipitate of $CaSO_4$ forms immediately, and we find that the concentrations of Ca^{2+} and SO_4^{2-} left in solution when the precipitate settles are the same as their concentrations in the original saturated solution. Hence we have shown that the reverse reaction

$$Ca^{2+} + SO_4^{2-} \rightarrow CaSO_4 \tag{1-25}$$

anhydrite

leads to the same result as the forward reaction, so that the existence of equilibrium is proved. We combine the equations into

$$CaSO_4 \rightleftharpoons Ca^{2+} + SO_4^{2-}. \tag{1-26}$$

anhydrite

(Actually the situation is not quite so simple. The precipitate formed when Ca^{2+} and SO_4^{2-} are mixed at room temperature is not anhydrous $CaSO_4$ but gypsum, $CaSO_4 \cdot 2H_2O$. This complication will be addressed later, but for the present discussion it is not important.)

The equilibrium constant for Eq. (1-26) is

$$K = \frac{[Ca^{2+}][SO_4^{2-}]}{[CaSO_4]}. \tag{1-27}$$

The concentrations of Ca^{2+} and SO_4^{2-} may be expressed as mols per kilogram of water, $m_{Ca^{2+}}$ and $m_{SO_4^{2-}}$. But what meaning can we give to the expression $[CaSO_4]$, referring to the "concentration" of solid calcium sulfate in equilibrium with its saturated solution? We might guess at first that the effect of $CaSO_4$ on the equilibrium, as determined by its rate of ionization, would depend on the amount of solid surface exposed to the solution. Actually, provided the mixture is kept well stirred, the amount of surface makes no difference; both reverse and forward reactions take place only at the solid surface, and one is no more affected than the other by the amount of surface exposed. (This argument disregards the effect of very minute crystals of the solid, which will be discussed later.) Hence the "concentration" of the pure solid is effectively constant, whether the amount present is small or large. To assign a value to this constant concentration we use the mol fraction of $CaSO_4$ in the solid anhydrite ($[CaSO_4] = X_{CaSO_4}$), so for the pure mineral $X_{CaSO_4} = 1$, and it does not appear in the expression for the equilibrium constant:

$$K = m_{Ca^{2+}} \, m_{SO_4^{2-}}. \tag{1-28}$$

Strictly speaking, the K here would be related to the K in Eq. (1-27) by the factor X_{CaSO_4}.

To generalize: *the concentration of any pure solid (or pure liquid) taking part in an equilibrium is assumed equal to 1, so that it need not appear explicitly in the* expression for the constant. For example, if solid $CaSO_4$ is dropped into a solution of barium chloride, the following equilibrium is quickly established:

$$Ba^{2+} + CaSO_4 \rightleftharpoons Ca^{2+} + BaSO_4. \tag{1-29}$$
$$\text{anhydrite} \qquad\qquad \text{barite}$$

Both the $CaSO_4$ and $BaSO_4$ in this equation are solid precipitates, so that the equilibrium constant is simply

$$K = \frac{m_{Ca^{2+}}}{m_{Ba^{2+}}}. \tag{1-30}$$

Another example is the formation of hematite from magnetite by heating in oxygen:

$$2Fe_3O_4 + \tfrac{1}{2}O_2 \rightleftharpoons 3Fe_2O_3. \tag{1-31}$$
$$\text{magnetite} \qquad\qquad \text{hematite}$$

If both Fe_3O_4 and Fe_2O_3 are pure solids, the constant must be

$$K = \frac{1}{P_{O_2}^{1/2}}, \tag{1-32}$$

meaning that the two oxides can exist together at equilibrium only at a single fixed pressure of oxygen (for a given temperature), and that this pressure is not at all dependent on the relative amounts of the solid oxides present.

A number like the K in Eq. (1-28) is an important kind of equilibrium constant called a *solubility product*. For any slightly soluble salt, a similar constant product of ionic concentrations can be set up. For example,

$$AgCl \rightleftharpoons Ag^+ + Cl^- \qquad \text{Solubility product} = K = m_{Ag^+}m_{Cl^-}. \qquad (1\text{-}33)$$

$$CaF_2 \rightleftharpoons Ca^{2+} + 2F^- \qquad \text{Solubility product} = K = m_{Ca^{2+}}m_{F^-}^2. \qquad (1\text{-}34)$$

and

$$As_2S_3 \rightleftharpoons 2As^{3+} + 3S^{2-} \qquad \text{Solubility product} = K = m_{As^{3+}}^2 \, m_{S^{2-}}^3. \qquad (1\text{-}35)$$

Numerical values of solubility products are given in the Appendix, Table VII-1.

The usefulness of solubility products can be illustrated by a few simple examples. For $CaSO_4$, the experimentally determined value of K is 3.4×10^{-5} at 25°C and 1 bar pressure. From this we can readily calculate the solubility of $CaSO_4$ in pure water. The solubility in mols per kilogram is equal to the concentration of Ca^{2+} (or of SO_4^{2-}) in the saturated solution, since every mol of $CaSO_4$ that dissolves gives 1 mol of each ion in solution. Hence

$$\text{Solubility} = m_{Ca^{2+}} = m_{SO_4^{2-}} \qquad (1\text{-}36)$$

and

$$(\text{Solubility})^2 = m_{Ca^{2+}}m_{SO_4^{2-}} = K = 3.4 \times 10^{-5}, \qquad (1\text{-}37)$$

whence

$$\text{Solubility} = 5.8 \times 10^{-3}m. \qquad (1\text{-}38)$$

For this ideally simple case, solubility is just the square root of the solubility product. *In general, this is not true for geological environments*, since most natural waters contain Ca^{2+} and SO_4^{2-} from other sources.

Solubilities as small as this are often expressed in a different kind of unit, *parts per million* (ppm). This means simply weight of solute in a million parts of solution, in any units—grams per million grams, tons per million tons, etc. To convert $5.8 \times 10^{-3}m_{CaSO_4}$ to ppm, for example, we first multiply by the molecular weight:

$5.8 \times 10^{-3} \times 136 = 0.79$ gram per kilogram of water

$\qquad\qquad = $ approximately 0.79 g of $CaSO_4$ per 1000 g of solution,

and this is equal to 0.79×1000 or 790 ppm. An equivalent expression often used in geochemistry is milligrams per kilogram (mg/kg).

The ideal relation between solubility and solubility product is less simple for a salt like CaF_2, because the ions are formed in unequal concentrations. Every mol of the solid that dissolves gives 1 mol of Ca^{2+} and 2 mols of F^-. Hence the solubility may be set equal to the concentration of Ca^{2+}, and the concentration of F^- may be expressed as $2m_{Ca^{2+}}$. The measured value of K (from Table VII-1) is $10^{-10.4}$ at 25°C. Then

$$K = 10^{-10.4} = m_{Ca^{2+}}m_{F^-}^2 = 4m_{Ca^{2+}}^3 \qquad (1\text{-}39)$$

and

$$m_{Ca^{2+}} = (0.25 \times 10^{-10.4})^{1/3} = 10^{-3.7}m = \text{solubility of } CaF_2. \qquad (1\text{-}40)$$

Note that this result holds only for the case of solid CaF_2 in contact with water containing no Ca^{2+} or F^- from other sources.

Returning to $CaSO_4$, let us see how the solubility is affected by other ions in solution, particularly by an excess of Ca^{2+} or SO_4^{2-}. What is the solubility, for example, of $CaSO_4$ in a solution of $0.1m$ $CaCl_2$? Let x be the solubility; then the concentration of Ca^{2+} in solution will be $(x + 0.1)$, since the $CaSO_4$ contributes x mols/kg and the $CaCl_2$ contributes 0.1. The concentration of SO_4^{2-} will be x, and the equilibrium constant is

$$K = m_{Ca^{2+}}m_{SO_4^{2-}} = (0.1 + x)(x) = 3.4 \times 10^{-5}, \qquad (1\text{-}41)$$

which can be rearranged to give

$$0.1x + x^2 = 3.4 \times 10^{-5}. \qquad (1\text{-}42)$$

This equation could be solved by the familiar quadratic formula, but again we look for a possible shortcut. We might guess that x will not be larger than the solubility we calculated for pure water, $5.8 \times 10^{-3}m$; hence the term x^2 should be small in comparison with $0.1x$. If x^2 is neglected, the equation gives at once

$$x = 3.4 \times 10^{-4}m. \qquad (1\text{-}43)$$

Then we check the validity of our assumption that x^2 is small by back-substitution:

$$0.1x + x^2 = 3.4 \times 10^{-5} + 11.6 \times 10^{-8} \approx 3.4 \times 10^{-5}. \qquad (1\text{-}44)$$

This statement is true to an accuracy of about 0.3%, which justifies the assumption.

Thus the solubility of $CaSO_4$ in $0.1m$ $CaCl_2$ solution is $3.4 \times 10^{-4}m$, compared with $5.8 \times 10^{-3}m$ in pure water. The decrease in solubility is a result we might have anticipated qualitatively, by noting that the excess Ca^{2+} would speed up the reverse reaction of the equilibrium

$$CaSO_4 \rightleftharpoons Ca^{2+} + SO_4^{2-} \qquad (1\text{-}26)$$

but would not affect the forward reaction. Clearly, excess of SO_4^{2-} would also lower the solubility; and in general, for any salt in equilibrium with its saturated solution, the solubility is less if an excess of one of its ions is present. This decrease in the solubility of a salt due to the presence of one of its own ions in solution is called the *common-ion effect*. The presence of ions *different* from those furnished by the salt itself generally makes the salt *more* soluble, a fact we could not predict from simple equilibrium reasoning (Sec. 2-6).

As a final example, let us calculate the value of K for the reaction of a barium salt with $CaSO_4$ [Eq. (1-29)]. Here two cations are in equilibrium with two salts, and the concentrations must adjust themselves so that both solubility products are maintained:

$$K' = m_{Ca^{2+}} m_{SO_4^{2-}} = 3.4 \times 10^{-5}, \tag{1-45}$$

$$K'' = m_{Ba^{2+}} m_{SO_4^{2-}} = 1.0 \times 10^{-10}. \tag{1-46}$$

The $m_{SO_4^{2-}}$ can be eliminated by dividing one equation by the other:

$$\frac{m_{Ca^{2+}}}{m_{Ba^{2+}}} = K = 3.4 \times 10^{5}. \tag{1-47}$$

In other words, $BaSO_4$ is so much more insoluble than $CaSO_4$ that the equilibrium concentration of Ca^{2+} is 340,000 times that of Ba^{2+}.

In all this discussion the temperature has been assumed fixed at 25°C and pressure at 1 bar. Solubilities change with a change in temperature and pressure, and the solubility products must increase or decrease accordingly. For most salts the solubility increases as the temperature rises, but the rate of increase is different for different salts and cannot be predicted from elementary rules presented here. (In later chapters we will see that prediction of changes in mineral solubility with increasing temperature and pressure is possible using fundamental energy and volume relations.)

The salt we have used as our principal example, calcium sulfate, is a geologically well-known compound, occurring as the two minerals gypsum and anhydrite (and also the very rare bassanite $CaSO_4 \cdot \frac{1}{2}H_2O$). Gypsum differs from anhydrite chemically in that it contains water (gypsum is $CaSO_4 \cdot 2H_2O$, anhydrite $CaSO_4$), and this leads to a slight difference in solubility. At ordinary temperatures gypsum is slightly less soluble (solubility product 2.0×10^{-5} instead of 3.4×10^{-5}), but at higher temperatures anhydrite is less soluble. The temperature at which their solubilities become equal is approximately 50°C in freshwater and 20°C in seawater. This makes it apparent that solubility products are by no means an adequate description of all the complications that may arise in the behavior of slightly soluble salts, but they are often useful in making approximate predictions for reasonably simple systems.

1-6   LE CHATELIER'S RULE

In a previous discussion (Sec. 1-2), it was mentioned as a general property of systems in equilibrium that the equilibrium responds to any disturbance by "trying to undo" or reverse the effects that the disturbance causes. The example cited was the effect on the HCl equilibrium of the addition of H_2: the equilibrium shifts so as to use up part of the added H_2 by reaction with Cl_2. The common-ion effect provides another simple illustration: if equilibrium between solid $CaSO_4$ and its saturated solution is disturbed by the addition of SO_4^{2-}, the equilibrium responds by trying to use up the added SO_4^{2-} by increased precipitation of $CaSO_4$. This property of equilibria goes by the name of *Le Chatelier's rule*, after the French physical chemist who first gave it explicit statement (Henri Louis Le Chatelier, 1850–1936).

The rule contains nothing very profound. In a sense it is hardly more than a restatement of the definition of dynamic equilibrium, for a system cannot be in equilibrium unless it seeks to maintain the "status quo," in other words, to counteract the effect of an outside disturbance. Still, Le Chatelier's statement is a handy device for predicting qualitatively which way an equilibrium will shift in various circumstances.

Let us apply it, for example, to the behavior of equilibria when temperature changes. We note first that the energy change in the two reactions of an equilibrium must be equal and opposite. If one reaction gives out heat (or is *exothermic*), the reverse reaction must absorb heat (or is *endothermic*), and according to the law of conservation of energy the amounts of heat must be the same. Now if equilibrium has been established, and the temperature is raised, Le Chatelier's rule tells us that the equilibrium must respond by trying to counteract the temperature increase. Thus in the example of high-grade metamorphism of pelitic rocks given in Fig. 1-2, the forward reaction of the equilibrium

$$KAl_2(AlSi_3O_{10})(OH)_2 + SiO_2 \rightleftharpoons KAlSi_3O_8 + Al_2SiO_5 + H_2O \qquad (1\text{-}48)$$

muscovite $\quad$ quartz $\quad$ K-feldspar $\quad$ sillimanite $\quad$ water

is strongly endothermic and the reverse reaction is exothermic; hence the dehydration of muscovite in the presence of quartz is favored by high temperatures, and the reaction will proceed from left to right as temperature rises. For another example, the ionic dissociation of most salts on dissolving in water is an endothermic reaction (as is shown convincingly by the cooling effect when a very soluble salt like sodium thiosulfate is stirred in water); hence when solubility equilibrium is established at one temperature, a rise in temperature will cause more of the salt to go into solution.

In general, therefore, the effect of temperature on an equilibrium can be qualitatively predicted by the simple rule that *endothermic reactions are favored by a rise in temperature, exothermic reactions by a fall in temperature.* It would also seem reasonable to suppose that the magnitude of the temperature effect would depend on the amount of the energy change, but the quantitative relationship we had best defer to a later chapter.

Another simple prediction from Le Chatelier's rule relates to the effect of total pressure on an equilibrium. Hitherto we have spoken of increasing or decreasing the partial pressure of a single gas, and this pressure has been treated simply as a measure of concentration. But now suppose that an equilibrium has been established and we raise the pressure on the entire system. How will the equilibrium respond to this disturbance? According to the rule, that reaction must be favored which will lead to a lowering of pressure. This will evidently be the reaction that gives the smaller total volume. Thus, if equilibrium is established between water and ice at the melting point,

$$H_2O \quad \rightleftharpoons \quad H_2O \qquad (1\text{-}49)$$

liquid $\qquad\qquad$ solid
density 1.0 $\qquad$ density 0.92 g/cm^3

and if pressure is increased, the reverse reaction is favored because liquid water has a lower specific volume (higher density) than ice. This is a long-winded way of stating the familiar fact that ice may be melted by the application of pressure. In a similar way we can see in Fig. 1-2 that an increase in pressure at low temperatures causes the zeolite laumontite to react to form lawsonite, quartz, and water, because the latter assemblage of phases has a lower specific volume.

Equilibria involving gases are especially sensitive to pressure changes. As an example, limestone at many granite contacts is partly converted to wollastonite by a reaction involving three solids and the single gas CO_2:

$$CaCO_3 + SiO_2 \rightleftharpoons CaSiO_3 + CO_2. \qquad (1\text{-}50)$$

calcite   quartz   wollastonite   gas

The forward reaction produces a large increase in volume, and high pressures would therefore tend to prevent the formation of wollastonite.

When both reactions of an equilibrium involve gases, increase in pressure favors the reaction that leads to the gas or gas mixture having the smaller volume. This can be readily predicted from the number of gas molecules shown in the equation (as a consequence of Avogadro's law that *equal numbers of gas molecules of any kind occupy the same volume, provided that the volumes are measured at similar temperatures and pressures*). For example, one hypothesis regarding the formation of cassiterite deposits from magmatic gases is represented by the equation

$$SnCl_4(g) + 2H_2O(g) \rightleftharpoons SnO_2(s) + 4HCl(g). \qquad (1\text{-}51)$$

cassiterite

At the high temperatures of cassiterite deposition, both $SnCl_4$ and H_2O would be gases, so that the equation shows three gas molecules on one side and four on the other. Hence the forward reaction would lead to an increase in volume and would be favored by low pressure.

These examples should make it clear that Le Chatelier's rule is a useful means for qualitatively predicting the effects on equilibria of changes in temperature and total pressure as well as changes of concentration. We have seen how concentration changes can be handled quantitatively also, by means of the equation for equilibrium constants. We shall return later (Chap. 8) to the problem of formulating quantitative rules for changes in temperature and total pressure.

1-7 STABILITY

One other concept needs qualitative mention now and quantitative refinement later. This is the idea of chemical stability, widely but often ambiguously used in geological discussions.

A stable substance is one that does not react readily in a particular environment. Gold under ordinary conditions is stable, because it is not noticeably attacked by air or most acids; metallic iron is unstable because it rusts when exposed to air. Water at usual temperatures is a stable compound, while TNT is an unstable compound. Sanidine is a form of potassium feldspar stable at high temperatures, adularia a form stable at low temperatures. Liquid water supercooled to $-5°C$ is unstable because it freezes quickly when disturbed. A supersaturated solution of potassium chloride is unstable because addition of a single grain of the salt causes rapid crystallization throughout the solution. These are all familiar examples of the stability concept.

Two precautions are needed to make usage of the term precise. First, stability has no meaning except when referred to particular external conditions. To say "iron is unstable" makes no sense; it is only mixtures of iron and air, or iron and acid, that are unstable. Water is stable enough at ordinary temperatures, but at 4000°C decomposes spontaneously into its elements. We often speak of olivine as a relatively unstable mineral, on the grounds that it weathers readily when exposed to air and that it reacts readily with silica at high temperatures; but there is nothing inherently unstable about olivine itself. Metamorphic rocks represent assemblages of minerals stable under certain temperature–pressure conditions but unstable when these conditions change. In using the term stability, we must keep in mind that a qualifying phrase is always implied—stable *with respect to* particular conditions of temperature and pressure, and particular kinds of associated substances.

The second precaution is to note that stability may be defined *by reference to either equilibrium or reaction rate.* In discussing mixtures of H_2, Cl_2, and HCl, we can say that a mixture rich in HCl is stable at ordinary temperatures, whereas a combination of H_2 and Cl_2 is stable above 2000°C, on the grounds that equilibrium is known to shift in these directions as the temperature changes. On the other hand, a mixture of H_2 and Cl_2 is apparently stable in the dark at room temperature, in the sense that no visible reaction occurs, but here the stability is due to a very slow rate of reaction. A similar example is galena exposed to air. The mineral is apparently perfectly stable, for museum specimens in contact with air persist indefinitely without visible change. Yet outcrops of veins containing galena commonly have anglesite entirely replacing the sulfide or surrounding remnants of the sulfide, suggesting that galena is slowly oxidized. Experimental work confirms this deduction by showing that equilibrium in the reaction

$$PbS + 2O_2 \rightleftharpoons PbSO_4 \qquad (1\text{-}52)$$
$$\text{galena} \qquad\qquad \text{anglesite}$$

is displaced far in the direction of the sulfate at ordinary temperatures. Hence the apparent stability of galena is a result of the slowness of the oxidation, and in terms of equilibrium anglesite is the more stable mineral. This double usage of the word stability is a frequent source of confusion.

To avoid ambiguity in this book, "stability" will be used only in the sense of "stable with respect to equilibrium." Substances or mixtures of substances which seem stable because they react very slowly will be referred to as "metastable", or "apparently stable." Both stability and metastability will become clearer after we have discussed energy changes in equilibrium reactions (Chap. 8).

1-8 CONVENTIONS

Several arbitrary conventions have been introduced in this chapter for the handling of equilibrium constants. These conventions and a few important definitions are repeated here for easy reference.

1. For a reversible reaction $aA + bB + \ldots \rightleftharpoons yY + zZ + \ldots$, the equilibrium constant ("concentration constant") is defined as

$$K = \frac{[Y]^y [Z]^z \ldots}{[A]^a [B]^b \ldots} \tag{1-6}$$

where the brackets indicate concentrations. We shall see in the next chaper that more accurate treatment of equilibrium constants requires the use of *activities* rather than concentrations. Activities are proportional to concentration, partial pressure, or composition. They are numbers without units, so the true equilibrium constant is also a dimensionless number.

2. K has a definite value for a particular chemical equation, not for a particular substance or a particular process.

3. Values of K are customarily given for 25°C and 1 bar total pressure, unless other temperatures and pressures are specified.

4. Concentrations of gases in the equilibrium-constant formula are generally expressed as partial pressures in *bars*. Alternatives often used in geochemical literature are *atmospheres* and *pascals*.

5. Concentrations of solutes in aqueous solution are given as mols of solute per kilogram of water (*molality* or *molal* concentrations, abbreviated *m*). Alternatively they may be expressed as mols of solute per liter of solution (*molarity* or *molar* concentration, abbreviated *M*). In dilute solutions the difference between the two is unimportant for geological calculations. A third concentration unit, often used for non-aqueous and high-temperature mixtures, is the *mol fraction* (X), the ratio of mols of solute to total mols present in the system.

6. Concentrations of solids and liquids, and of water in liquid aqueous solutions, are represented in terms of mol fractions, so in the equilibrium constant they do not appear in the formula if they are pure phases.

7. An equilibrium is called *homogeneous* if all the reacting substances are in a single phase, *heterogeneous* if they are in two or more phases. Most equilibria of interest in geology are heterogeneous.

8. A mixture is chemically *stable* either because it is at equilibrium or because reactions are slow. The first meaning is generally understood unless the second is specified.

1-9 A WORD ABOUT PROBLEM SOLVING

The solving of mathematical problems is not a main purpose of this book, but some facility in handling equilibrium constants is necessary for an understanding of geochemical arguments. The following remarks about some common kinds of problems may be helpful:

1. Equilibrium constants and the concentrations of ions in geologically important solutions are generally small numbers, best represented by expressions with negative exponents. Such expressions may be written with integral exponents (for example, 2.5×10^{-5}) or with fractional exponents (for example, $10^{-4.6}$). Sometimes one form is more convenient, sometimes the other. To convert from one to the other, find logarithms and then antilogarithms. For example, 2.5×10^{-5} may be written with a fractional exponent as $10^{-4.6}$, because

$$\log 2.5 \times 10^{-5} = \log 2.5 + \log 10^{-5} = 0.40 - 5 = -4.6$$

and antilog $-4.6 = 10^{-4.6}$. For a second example, $10^{-8.2}$ is equivalent to 6.3×10^{-9}, because $\log 10^{-8.2} = -8.2 = 0.8 - 9$, of which the antilog is 6.3×10^{-9}. With a little practice these conversions become almost automatic. In examples in future chapters the two modes of expression will be used interchangeably.

2. Since equilibrium problems are based on chemical equations, the first requirement in setting up a problem is to be sure that the equation or equations are correctly written. This means that each equation (1) must be balanced with respect to both numbers of atoms and numbers of charges, and (2) must be chemically reasonable for the environment specified in the problem. Unbalanced or unrealistic chemical equations are a common source of error in problem-solving.

3. For a problem involving a single equilibrium, generally the only equation needed is the expression for the equilibrium constant. This must be set up carefully, following the conventions summarized in the last section. The statement of the problem then gives enough relations among the concentrations so that a single one can be picked out as the unknown and labeled x. For example, when $CaSO_4$ is stirred in $0.1m$ $CaCl_2$ solution (Sec. 1-5), a convenient x is the solubility; from the stoichiometry this is equal to $m_{SO_4^{2-}}$, and the concentration of Ca^{2+} is x plus the 0.1 mol/kg already present from the $CaCl_2$.

4. For a problem involving several equilibria, it is often more convenient to write down several equations in several unknowns and solve them simultaneously. Such problems will be illustrated in Chapter 2.

5. It is characteristic of this kind of problem that the equation ultimately obtained for solution is difficult to handle mathematically, even when the original problem seems fairly simple. Practically always the equation is at least of second degree, and cubic and quartic equations are common. Algebraic methods of solving such equations exist, but they are tedious and cumbersome. Generally, however, approximate solutions can be found quickly by noting that some terms in an equation are necessarily smaller than others and so can be neglected during addition and subtraction. For many geologic purposes, approximate solutions are sufficiently accurate. Computer programs are available for solving the more intractable problems and problems where great accuracy is needed.

6. The solution to a problem cannot be more accurate than the data that go into it. If, for example, the solubility product of anhydrite is 3.4×10^{-5}, mathematics unrestrained by common sense would give for the solubility in pure water a figure $5.832 \times 10^{-3}m$. Since the equilibrium constant is given only to two significant figures, all digits beyond the first two in the answer are meaningless. Not only must spurious mathematical accuracy be avoided; in geochemical work, it is fruitless also to seek accuracy beyond what natural conditions warrant. Suppose, for example, that the solubility product of anhydrite has been measured very accurately, so that the solubility in pure water at 25°C can be confidently fixed at 5.82×10^{-3} rather than $5.80 \times 10^{-3}m$. This would have no geologic significance, because any natural solution shows enough variation in temperature and contains enough other ions to change the solubility by more than this difference.

7. Since answers are often arrived at by approximation and since the answers are expected to fit primarily geological rather than laboratory situations, it is important in problems of this kind that answers be checked for reasonableness. Once a problem has been solved, three questions should always be asked: (1) Is the answer grossly reasonable? (2) Does the solution fit the mathematical equations within a tolerable limit of accuracy? (3) Would the solution be significant in natural environments, where some uncontrolled variables are always present?

8. Problems in this book are approached on a basis of expediency rather than strict logic. Methods of attack are adapted to individual problems, and emphasis is kept as far as possible on physical relationships rather than mathematical elegance. Individuals differ in the kind of mathematical approach they can follow most easily, and some will find a more rigorous logical development preferable. For such readers two references from the list at the end of this chapter are particularly recommended: Garrels and Christ (1965) and Stumm and Morgan (1981).

PROBLEMS

Answers to some of the numerical problems are given in Appendix XII.

1. A mixture of hydrogen and oxygen at room temperature does not react appreciably but, when ignited by a flame or an electric spark, explodes violently as the two gases unite to form water. Water vapor is stable even at high temperatures, dissociating into its elements to an appreciable extent only above 1500°C. At 2000°C and 1 bar, it is about 0.4% dissociated.

   (a) Set up an expression for the equilibrium constant for the reaction

   $$2H_2 + O_2 \rightleftharpoons 2H_2O.$$

   (b) Calculate the value of the constant at 2000°C.

   (c) If additional hydrogen is added to an equilibrium mixture at 2000°C, how is the equilibrium affected?

   (d) If the total pressure on an equilibrium mixture at 2000°C is increased, how is the equilibrium affected?

(e) Which reaction, forward or reverse, is exothermic? How, then, is the composition of an equilibrium mixture affected by a rise in temperature?

(f) In items (c), (d), and (e), how is the equilibrium constant affected by the suggested changes?

(g) Is a mixture of hydrogen and oxygen at room temperature stable?

2. Is the equilibrium established when excess CaF_2 is stirred in water heterogeneous or homogeneous? What phases are present?

3. The solubility product of calcium carbonate (calcite) at 25°C is 4.5×10^{-9}.

(a) Calculate the solubility of calcite in pure water at 25°C. Express the solubility as (1) mols per kg of water, (2) grams of $CaCO_3$ per 100 ml of solution, (3) parts per million (ppm) of Ca in the solution.

(b) Calculate the solubility of $CaCO_3$ in a solution of $0.05m$ $CaCl_2$ at 25°C.

(c) What is the ratio of $m_{SO_4^{2-}}$ to $m_{CO_3^{2-}}$ in a solution at equilibrium with both $CaSO_4$ and $CaCO_3$?

4. The solubility of Ag_2SO_4 at 25°C is 0.8 g/100 g H_2O. Calculate the solubility product.

5. The equilibrium constant for Eq. (1-31) is 5×10^{43}. Calculate the pressure of oxygen in equilibrium with a mixture of hematite and magnetite at 25°C. How many *molecules* per liter does this represent?

6. The equilibrium constant for the reaction

$$2Fe^{3+} + 2Cl^- \rightleftharpoons 2Fe^{2+} + Cl_2$$

is approximately 10^{-20}. Calculate the equilibrium ratio $m_{Fe^{2+}}/m_{Fe^{3+}}$ for (a) $P_{Cl_2} = 1$ bar and $m_{Cl^-} = 1m$, and (b) $P_{Cl_2} = 10^{-10}$ bar and $m_{Cl^-} = 1m$. Would you expect chlorine to oxidize Fe^{2+} appreciably at ordinary temperatures? Would you expect to be able to smell Cl_2 over a solution of $FeCl_3$?

7. The solubility product of PbS is $10^{-27.5}$ and that of ZnS is $10^{-24.7}$. What is the ratio of $m_{Pb^{2+}}$ to $m_{Zn^{2+}}$ in a solution at equilibrium with both galena and sphalerite? If a solution containing 100 times as much Zn^{2+} as Pb^{2+} percolates through a mixture of the sulfides, would galena be replaced by sphalerite or sphalerite by galena?

8. The solubility of H_2S in water is 2.3 liters per liter of solution at 25°C and 1 bar. Using this figure, calculate the equilibrium constant for the reaction

$$H_2S(gas) \rightleftharpoons H_2S(aq).$$

9. From your general knowledge of the chemical behavior and geologic occurrence of the following substances, indicate which are stable and which are metastable at ordinary temperatures.

(a) Quartz exposed to air.

(b) Magnetite exposed to air.

(c) A mixture of olivine [$(Mg,Fe)_2SiO_4$] and silica.

(d) A mixture of kaolinite [$Al_2Si_2O_5(OH)_4$] and calcite.

(e) Petroleum exposed to air.

(f) Cassiterite (SnO_2) exposed to air.

(g) Calcium oxide exposed to air.

10. In Sec. 1-5, the ratio $m_{Ca^{2+}}/m_{Ba^{2+}}$ in equilibrium with both $CaSO_4$ and $BaSO_4$ was found to be 3.4×10^5. What are the actual concentrations of Ca^{2+} and Ba^{2+} in the solution? What is the concentration of SO_4^{2-}?

11. Suppose that a solution is just saturated with calcium sulfate, but no solid calcium sulfate is present. If a small amount of a more soluble calcium salt in fairly concentrated solution (say, $1m$ $CaCl_2$) is added, would you expect a precipitate to form? Explain.

REFERENCES AND SUGGESTIONS FOR FURTHER READING

Faure, G: *Principles and Applications of Inorganic Geochemistry*, Macmillan, New York, 1991. Chapter 11 is a good elementary introduction to chemical equilibrium and the handling of equilibrium constants.

Garrels, R. M., and C. L. Christ: *Minerals, Solutions, and Equilibria*, Harper and Row, New York, 1965. The mathematical handling of equilibrium constants for geologically important reactions is treated in great detail, with more attention to formal rigor than was attempted in this chapter. A good background in physical chemistry is assumed.

Petrucci, R. H.: *General Chemistry*, 5th ed., Macmillan, New York, 1989. The material covered in this book, or a similar textbook of modern chemistry, should be in the reader's background.

Stumm, W., and J. J. Morgan: *Aquatic Chemistry*, 2d ed., Wiley–Interscience, John Wiley & Sons, New York, 1981. A sophisticated treatment of equilibrium; requires a good knowledge of physical chemistry.

CHAPTER
2

AQUEOUS
SOLUTIONS

Water is a subtle but important agent of geologic change, abundant on the Earth's surface and present in the cracks and pore spaces of most rocks in the upper crust. It exerts a major influence on the course of weathering, diagenesis, metamorphism, even on magmatic processes and rock deformation. At the Earth's surface water is a transporting agent during erosion and sedimentation, and in its movement through the pore spaces of rocks it is effective in carrying both dissolved materials and thermal energy. Water is also the medium in which many chemical reactions occur at or near mineral surfaces. Because it expands on heating, water can be a source of mechanical energy for propagation of rock fractures, for explosive volcanic eruptions, and for phreatic discharges from active geothermal systems. Examining the complex role that water plays in a multitude of geologic systems will be a major theme of this book.

Although the most familiar of common liquids, water is also a very unusual one. For example, all other common liquids like alcohol, acetone, gasoline, and carbon tetrachloride increase in density regularly as temperature falls, and freeze to solids denser than the liquids. Water, by contrast, has its greatest density at 4°C above its melting point, and freezes to a solid that will float on the liquid. Another major peculiarity of water, the most important one for the present discussion, is its very high *dielectric constant*, higher than that of any other common liquid. This means that water has an extraordinary ability to hold apart, or prevent the neutralization of, dissolved electrically charged particles. (Quantitatively, the dielectric constant of a substance is the reciprocal of the constant in the Coulomb's-law expression for the force between electric charges contained in the substance.) The high value of the constant gives water an unusual capacity to dissolve substances made up of charged particles (ions). Such substances are called *electrolytes*, a term meaning that when in solution they can conduct an electric current and are thereby decomposed. Common electrolytes are acids, bases, and salts, and the first two of these are the subject of this chapter.

The ability of water to dissolve a great variety of substances means that pure water is practically never encountered on, within, or near the Earth. Even rainwater contains minute amounts of dissolved material, and water that has been in contact with rocks and soils and organic matter shows a wide range of chemical composition. To illustrate the range, Table 2-1 is a sampling of common types of natural aqueous solutions, including seawater, rainwater, shallow groundwater, and geothermal water. Note especially the analyses of two groundwaters from basaltic rocks in Iceland, one from a low-temperature spring and the other from a high-temperature geothermal system. Although these groundwaters were derived from rainwater, their compositions are quite distinct from that of the local rain, because the rainwater has reacted with the basalt through which it moved.

So we turn now to look in detail at some of the dissolved substances encountered in natural waters, particularly the substances called acids and bases.

TABLE 2-1

Representative concentrations of the major chemical components in natural waters including seawater, and rain, spring, and geothermal waters from Iceland. Analyses reported in mg/kg.

	Seawater[1]	Rain water[2]	Meteoric[2] spring water	Meteoric[3] geothermal water
Temperature	20°C	10°C	6°C	300°C
pH	8.2	5.6	8.82	7.09
Chloride	19,350	1.14	3.15	21.1
Sodium	10,760	1.02	22.6	187
Sulfate	2,710	1.48	10.9	102.7
Magnesium	1,290	0.24	2.39	0.004
Calcium	411	0.4	5.06	0.5
Potassium	399	0.51	1.01	26.6
Bicarbonate	142		44.0	245
Silica	0.5–10	1.2	17.4	777

[1] Average composition of seawater reported by Drever, 1982. Reprinted by permission.

[2] Rain water and meteoric spring water from the Krafla area of northern Iceland reported by Gislason, S. R., and H. P. Eugster: "Meteoric water–basalt interactions. II. A field study in Iceland," *Geochim. Cosmochim. Acta*, vol. 51, pp. 2841–2855, 1987. Reprinted by permission.

[3] Meteoric geothermal water from the Krafla geothermal system, well 9, northern Iceland reported by Arnórsson, S., E. Gunnlaugsson, and H. Svavarsson: "The chemistry of geothermal waters in Iceland. II. Mineral equilibria and independent variables controlling water composition," *Geochim. Cosmochim. Acta,* vol. 47, pp. 547–566, 1983. Reprinted by permission.

Note: *Meteoric water* refers to water in or derived from the atmosphere; as used in geochemistry it usually means groundwater derived from rain but modified by contact with rocks and/or soil. It has nothing to do with meteors or meteorites.

Throughout this inquiry we will be applying and expanding the principles of equilibrium introduced in the first chapter.

2-1 ACIDS AND BASES

The three words acid, base, and alkali are old ones, commonly used in geologic and chemical writing even before the sciences took modern form at the end of the eighteenth century. As knowledge about rock origins increased on the one hand and chemical relationships on the other, the meanings of the three words have undergone gradual changes and are not always quite the same as used by chemists and geologists.

Chemical Definitions

Acids, from the time of Robert Boyle (1663), have been described as substances with a sour taste and possessing the ability to dissolve many substances, to change the color of vegetable dyes like litmus, and to react with bases to form salts. Towards the end of the nineteenth century, as part of his ionic theory, Svante August Arrhenius (1859–1927) proposed that acids differ from other hydrogen-containing compounds in that they partially dissociate when dissolved in water to set free hydrogen ions, H^+. Early in the twentieth century, J. N. Brønsted (1879–1947) and others pointed out that this idea cannot be correct, because H^+ represents nothing more than an isolated proton, which could not possibly exist by itself in the presence of water. It would necessarily be hydrated, forming the ion H_3O^+ (hydronium ion), so that the dissociation of an acid would be represented by an equation like

$$HCl + H_2O \rightarrow H_3O^+ + Cl^- \tag{2-1}$$

rather than the simple formulation of Arrhenius:

$$HCl \rightarrow H^+ + Cl^-. \tag{2-2}$$

According to Brønsted, an acid is a molecule or ion capable of giving H^+ to another molecule or ion [as HCl gives it to H_2O in Eq. (2- 1)]; in other words, an acid is a proton donor. Other chemists have objected that this is still not quite correct, because water forms associated molecules like $(H_2O)_3$, so that the hydronium ion includes $H_5O_2^+$ and $H_7O_3^+$ as well as H_3O^+.

Strictly, then, there is no really correct chemical definition of acid that is simple enough for everyday use. Fortunately, despite their apparent differences, the various definitions give consistent and not very different interpretations of most common reactions, so we can choose the formulation that is most convenient for geologic purposes. This is the type of reaction favored by Arrhenius [Eq. (2-2)]: we define an acid as *a substance containing hydrogen which gives free hydrogen ions when dissolved in water*, and we describe the characteristic properties as the properties of the hydrogen ion.

The term *base* has a similar history. For many years it was used loosely to describe substances whose water solutions have a soapy feel, a bitter taste, the ability to neutralize acids, and the property of reversing the color changes that acids produce in vegetable dyes. Such properties are possessed by a variety of materials, seemingly not closely related—ammonia, metal oxides, carbonates, hydroxides, and several others. In the language of Arrhenius, the properties common to all bases are ascribed to hydroxide ion, OH^-, and the term is generally restricted to compounds like NaOH and $Ca(OH)_2$ which dissociate to give this ion directly:

$$Ca(OH)_2 \rightarrow Ca^{2+} + 2OH^-. \tag{2-3}$$

Brønsted refers to OH^- itself as a base and broadens the term *to include all ions and molecules which, like OH^-, are capable of uniting with H^+* ("proton acceptors"). As with acids, the two modes of description are less different than they seem, and we pick again the simple Arrhenius formulation. We define a base as *a substance containing the OH group that yields OH^- on dissolving in water*, and we describe the characteristic properties of bases as the properties of the hydroxide ion.

Alkali is an Arabic word originally used for the bitter extract produced by leaching the ashes of a desert plant. It was extended to the bitter-tasting salts that sometimes collect in desert lakes (chiefly the substance we call sodium carbonate), and ultimately to other compounds prepared from these salts and from extracts of plant ashes. Today, as used in chemistry, it refers generally to the soluble strong bases like NaOH, KOH, and $Ba(OH)_2$, but usage is not entirely consistent. The adjective *alkaline* is practically a synonym for *basic*, referring to any solution containing appreciable OH^- or any substance capable of forming such a solution. The derived term "alkali metal" means any metal of the group containing sodium, potassium, lithium, rubidium, cesium, and the term "alkaline-earth metal" means one of the group calcium, strontium, barium, (often including also magnesium, beryllium, and radium).

Geologic Usage

The various shades of meaning given to acid, base, and alkali by chemists are confusing enough, but in geology the terms acquire even wider meanings.

Many oxides of nonmetals dissolve in water to form acids (CO_2 and SO_3 are familiar examples), and in geologic writing the term "acid oxide" or "acid anhydride" is often extended to any nonmetallic oxide whether or not it is appreciably soluble. Hence SiO_2 becomes an "acid oxide," and rocks containing a high percentage of SiO_2 are "acid" rocks. Similarly, "basic oxide" is used for any metal oxide, regardless of whether it dissolves enough to give appreciable OH^-, and "basic" rocks are those containing an abundance of metal oxides (especially MgO, FeO, and CaO). The oxides of sodium and potassium, in geologic parlance, are generally called "alkaline" rather than basic, and rocks containing minerals with unusually large amounts of these oxides are described as alkaline or alkalic.

The terms base and basic are often used in geology with even broader significance. In rock analyses, "base" may refer to either oxides or metals; "base exchange" and "basic front" refer to movement of metal ions; and it is seldom quite clear whether "base" is restricted to iron and magnesium or includes also calcium, sodium, and potassium. Thus when a geologist speaks of an "alkaline solution", a "zone of acid alteration", or a "basic environment", the statement is not precise unless the terms are further defined.

In this book we adhere strictly to the chemical definitions as given above, using "acid" only with reference to hydrogen ion and "base" only with reference to hydroxide ion. "Alkali" will mean strong base, and "alkaline" will be a synonym for "basic," except that we will retain the common names "alkali metal" and "alkaline-earth metal". Instead of calling a quartz-rich rock "acidic," we will describe it as *felsic* or *silicic*, and "basic" rocks we will call *mafic*. Rocks rich in sodium and potassium we continue to call *alkalic* or *alkaline*, because no suitable alternative name has been proposed.

The pH

The reaction between an acid and a base, called *neutralization*, is illustrated in its simplest form when dilute solutions of HCl and NaOH are mixed:

$$HCl + NaOH \rightarrow H_2O + NaCl. \tag{2-4}$$

Because the acid, base, and salt are almost completely dissociated into ions, the reaction is more realistically expressed as:

$$H^+ + Cl^- + Na^+ + OH^- \rightarrow H_2O + Na^+ + Cl^-. \tag{2-5}$$

Or more simply, since Na^+ and Cl^- are unaffected,

$$OH^- + H^+ \rightarrow H_2O. \tag{2-6}$$

This same equation represents the neutralization of any strong acid by any strong base.

What happens if a solution containing exactly 1 equivalent[1] of acid is carefully neutralized with exactly 1 equivalent of base? Offhand it looks as if all the H^+ and OH^- would be used up, so that the neutral solution would contain none of these ions at all. Experimentally we find that this does not happen, for water itself is slightly dissociated. The reaction shown by Eq. (2-6) proceeds until H^+ and OH^- have equal concentrations of $10^{-7}m$, at which point the solution, by definition, is neutral. In other words, the dissociation constant for water is

$$H_2O \rightleftharpoons H^+ + OH^-, \qquad K_{water} = m_{H^+} m_{OH^-} = 10^{-14}. \tag{2-7}$$

It is fortunate for simplifying calculations that the exponent, shown in Eq. (2-7) rounded off to -14, is actually so nearly a whole number at 25°C. A more precise value is -13.998, but for all ordinary purposes at usual temperatures the value of this important constant is taken as 10^{-14}. The variation of the constant with temperature and pressure is shown in Fig. 2-1.

---

[1] An equivalent of an acid (or base) is its gram-molecular weight divided by the number of H's (or OH's) in its formula. Thus an equivalent of HCl is 36.5 g and an equivalent of H_2SO_4 is 49 g.

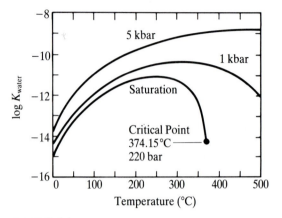

FIGURE 2-1
The logarithm of the equilibrium constant for the ionic dissociation of H_2O [Eq. (2-7)] as a function of pressure and temperature. The three curves show the change in the constant with temperature for pressures at liquid–vapor equilibrium for H_2O (the boiling-point or saturation curve), and for two fixed pressures, 1 kbar and 5 kbar. Note that the constant varies about six orders of magnitude over the 5 kbar and 500°C pressure–temperature range shown on the diagram. (Reprinted by permission from Helgeson, H. C., and D. H. Kirkham: "Theoretical prediction of the thermodynamic behavior of aqueous electrolytes at high pressures and temperatures: I. Summary of the thermodynamic/electrostatic properties of the solvent," *American Journal of Science*, vol. 274, p. 1089–1198, 1974).

The product $m_{H^+} m_{OH^-}$ is constant for all water solutions, not for neutral ones only. This means that H^+ is present even in strongly basic solutions and OH^- in strong acids—in very minute concentrations, of course, but enough to keep the product of m_{H^+} and m_{OH^-} equal to 10^{-14}. It follows that the acidity or alkalinity of a solution can be specified by giving the concentration of either H^+ or OH^- alone. Thus a $1N$ solution of a strong acid has $1 m_{H^+}$, a neutral solution has $10^{-7} m_{H^+}$, and a $1N$ solution of a strong base has $10^{-14} m_{H^+}$. (The symbol N, read "normal," means equivalents of acid or base per liter of solution.) The whole range from strong acids to strong bases can thus be expressed in terms of m_{H^+}, as a series of powers of 10. We simplify the representation even further by discarding the 10's and using only the exponents, and by changing the sign of the exponents from negative to positive. The numbers so obtained are called pH values. In more formal language, the pH of a solution is *the negative logarithm of the hydrogen-ion concentration*. A $1N$ solution of strong acid has a pH of 0, a neutral solution a pH of 7, a $1N$ solution of a strong base a pH of 14. A solution containing $3 \times 10^{-4} m_{H^+}$ has a pH of 3.5, one containing $10^{-5.3} m_{OH^-}$ a pH of 8.7.

In nature, observed pH's lie mostly in the range 4 to 9. Streams in humid regions generally show values between 5 and 6.5, in arid regions between 7 and 8. Soil water, especially if decaying vegetation is abundant, may have pH's down to 4 or a little lower. Seawater near the surface normally shows a pH of 8.1 to 8.3. Soil water and playa-lake water in deserts may have pH's of 9 or even higher. The lowest recorded pH's in nature, down even to negative values, are found in solutions in contact with oxidizing pyrite.

Dissociation Constants of Weak Acids

If solutions of hydrochloric and carbonic acids of similar molal concentration are compared, the HCl solution has a sourer taste, a more pronounced effect on indicator dyes, and a greater solvent action on metals. Hydrochloric acid is described as a *strong* acid, carbonic acid as a *weak* acid. The difference in behavior is explained by a difference in extent of dissociation as illustrated in Fig. 2-2(*a*), HCl in dilute solution being almost completely broken down into ions and H_2CO_3 only slightly so.

Weak acids containing more than one H atom per molecule dissociate in steps:

$$H_2CO_3 \rightleftharpoons H^+ + HCO_3^- \tag{2-8}$$

and

$$HCO_3^- \rightleftharpoons H^+ + CO_3^{2-}. \tag{2-9}$$

Like other equilibria, these reactions can be formulated in terms of constants representing quotients of concentrations:

$$\frac{m_{H^+} m_{HCO_3^-}}{m_{H_2CO_3}} = K_1 = 10^{-6.4} = 4.2 \times 10^{-7} \tag{2-10}$$

and

$$\frac{m_{H^+} m_{CO_3^{2-}}}{m_{HCO_3^-}} = K_2 = 10^{-10.3} = 5.0 \times 10^{-11}. \tag{2-11}$$

This particular kind of equilibrium constant is called a *dissociation constant*. Note that here we cannot disregard the concentration of the undissociated substance, as we could disregard the undissolved solid in formulating a solubility product (Sec. 1-5). The undissociated molecules [H_2CO_3 in Eq. (2-10) and HCO_3^- in Eq. (2-11)] are present *in the solution*, and obviously their concentration affects the concentration of H^+.

To illustrate the use of dissociation constants in finding hydrogen-ion concentrations, we calculate the pH of a solution of $0.01m$ H_2CO_3. The figure for concentration, $0.01m$, in this sort of expression refers to *total* carbonate (given the symbol, $m_{CO_3, \text{total}}$), i.e.,

$$m_{CO_3, \text{total}} = m_{H_2CO_3} + m_{HCO_3^-} + m_{CO_3^{2-}} = 0.01 \text{ mols/kg } H_2O. \tag{2-12}$$

In other words, "$0.01m$ H_2CO_3" means "a solution obtained by dissolving 0.01 mol of CO_2 in a kilogram of water." Some of the acid will be present as ions, but the designation $0.01m$ H_2CO_3 refers to the total amount dissolved. The expression $m_{H_2CO_3}$ in Eq. (2-12), by contrast, is the concentration of undissociated acid molecules alone.

Another equation can be set up to express the fact that the solution is electrically neutral, i.e., that the concentration of positive charges must equal the concentration of negative charges:

$$m_{H^+} = m_{OH^-} + m_{HCO_3^-} + 2m_{CO_3^{2-}}. \tag{2-13}$$

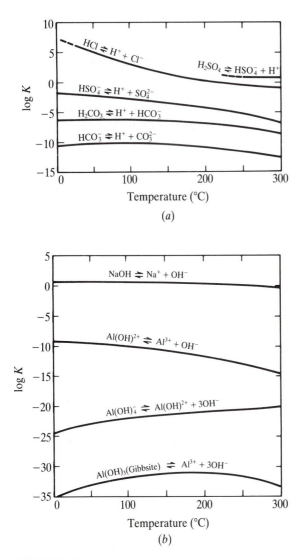

FIGURE 2-2
Equilibrium constants for the dissociation of acids (a) and bases (b) as a function of temperature. Strong acids and bases like HCl and NaOH exist in solution at ordinary temperatures and pressures chiefly as dissociated ions, but the dissociation is not complete so that equilibria are established in the solution. Note that at 25°C and 1 bar the concentration of undissociated HCl is about 6 orders of magnitude less than the product of $m_{H^+} m_{Cl^-}$, and that of undissociated NaOH is about one order of magnitude less than $m_{Na^+} m_{OH^-}$. Sulfuric acid is less completely dissociated than HCl, but still is considered a strong acid. Weak acids and bases, represented by carbonic acid and aluminum hydroxide and their ions, are less dissociated by several orders of magnitude.

(The coefficient before $m_{CO_3^{2-}}$ is necessary because each mol of a divalent anion is equivalent to 2 mols of negative charge. Thus if x mols of CO_3^{2-} are present, they must be balanced by $2x$ mols of H^+.) Still another relation that must be satisfied by any aqueous solution is Eq. (2-7):

$$m_{H^+} \times m_{OH^-} = 10^{-14}. \tag{2-14}$$

We now have a set of five equations [(2-10)–(2-14)] relating five unknowns (the concentrations of H^+, OH^-, H_2CO_3, HCO_3^-, CO_3^{2-}). To find the pH, we need only solve the five equations for m_{H^+} by the ordinary rules of algebra.

A glance at the equations, however, shows that the necessary algebra is far from simple. As noted before (Sec. 1-9), this is a common situation in problems based on equilibrium constants: we set up simple equations relating the concentrations of substances in an equilibrium and then find that the mathematics required to solve the equations is rather formidable. Sometimes a complete formal solution is necessary, and can be obtained most easily by using an appropriate computer program. For geological purposes, where great accuracy is seldom required, satisfactory answers can often be found quickly by making approximations.

Thus, in the present example, we note first that K_2 is almost 10,000 times smaller than K_1. This means that H^+ comes almost entirely from Eq. (2-8), that $m_{CO_3^{2-}}$ is very small, and that roughly $m_{H^+} = m_{HCO_3^-}$. If we assume that $m_{CO_3^{2-}}$ and m_{OH^-} are small enough to be neglected, Eqs. (2-10) and (2-12) become

$$\frac{m_{H^+}^2}{m_{H_2CO_3}} = 10^{-6.4} \tag{2-15}$$

and

$$m_{H_2CO_3} + m_{H^+} = 0.01. \tag{2-16}$$

The problem is thus reduced to two equations in two unknowns, which can be readily solved to give $m_{H^+} = 10^{-4.2}$, or a pH of 4.2. Concentrations of other ions would then be, approximately,

$$m_{HCO_3^-} = m_{H^+} = 10^{-4.2} m,$$

$$m_{OH^-} = \frac{10^{-14}}{m_{H^+}} = 10^{-9.8} m,$$

and

$$m_{CO_3^{2-}} = \frac{10^{-10.3} m_{HCO_3^-}}{m_{H^+}} = 10^{-10.3} m.$$

An answer obtained by making so many approximations must be checked by substituting in all the original equations. The error is found to be less than 1% in this example, which for most geological purposes (and many chemical purposes as well) is inconsequential. The assumptions are therefore justified, and a more exact solution obtained by solving the five equations algebraically would give an illusory accuracy. Thus a solution of $0.01m$ H_2CO_3 has a pH of 4.2, or a hydrogen-ion concentration of $10^{-4.2}m$, meaning that less than 1% of the acid molecules are broken up into ions.

Since it is a general rule that the first dissociation constant of a weak acid is greater than the second by a factor of 10^4 to 10^6, we may use the same reasoning for any weak acid, H_nA:

$$H_nA \rightleftharpoons H^+ + H_{n-1}A^-, \tag{2-17}$$

$$m_{H^+} = (m_{A, total} \times K_1)^{1/2}, \tag{2-18}$$

the symbol $m_{A, total}$ denoting the total concentration of A. Thus for H_2S, with $K_1 = 10^{-7}$ and $K_2 = 10^{-13}$, the hydrogen-ion concentration of a $0.001m$ solution is

$$m_{H^+} = (0.001 \times 10^{-7})^{1/2} = 10^{-5}m$$

and the pH would be 5.0.

A similar calculation gives an answer to a question suggested in Chapter 1: what is the acidity of a solution obtained by leaving water in contact with CO_2 in ordinary air? The amount of CO_2 dissolved at equilibrium is about $10^{-5}m$, so that the problem is identical to the preceding one except that $m_{CO_3, total}$ is now 10^{-5} instead of $10^{-2}m$. The result of the calculation is a pH of about 5.7 for any sample of water exposed to atmospheric CO_2 at 25°C—provided, of course, that the water is not in contact with other soluble material which could give a different pH. It is worth emphasizing that, although pure water has a theoretical neutral pH of 7.0, this pH is never encountered at the Earth's surface unless the water contains dissolved material that neutralizes the dissolved CO_2. All water that we ordinarily consider pure, either in nature or in the laboratory, acquires a pH near 5.7 if it stands for a short time exposed to air.

Dissociation Constants of Hydroxides

By analogy with acids, strong and weak bases can be distinguished on the basis of their degree of dissociation into ions. A strong base like NaOH is largely dissociated as shown in Fig. 2-2(b). At cation concentrations likely to be encountered in geology, the hydroxides of Li, K, Rb, Cs, Sr, and Ba may also be considered soluble strong bases. An example of a soluble weak base is ammonia:

$$NH_3 + H_2O \text{ (or } NH_4OH) \rightleftharpoons NH_4^+ + OH^-, \qquad K = 2 \times 10^{-5}. \tag{2-19}$$

Many weak bases have very low solubilities, as illustrated by $Al(OH)_3$ in Fig. 2-2(b).

Bases dissociate in steps, just as acids do, and the unraveling of the various dissociation reactions can be a complicated exercise. Consider, for example, the behavior of copper hydroxide. If a base is added to a solution of $CuCl_2$, a blue precipitate of $Cu(OH)_2$ appears. Detailed study of copper concentrations at various pH's shows that this compound, on dissolving, dissociates in steps:

$$Cu(OH)_2(s) \rightleftharpoons CuOH^+ + OH^-, \qquad K = m_{CuOH^+} m_{OH^-} = 10^{-13.0}, \qquad (2\text{-}20)$$

$$CuOH^+ \rightleftharpoons Cu^{2+} + OH^-, \qquad K = \frac{m_{Cu^{2+}} m_{OH^-}}{m_{CuOH^+}} = 10^{-6.3}. \qquad (2\text{-}21)$$

Furthermore, the solid hydroxide can react with OH^- to a slight extent (i.e., it is slightly amphoteric):

$$Cu(OH)_2(s) + OH^- \rightleftharpoons Cu(OH)_3^-, \qquad K = \frac{m_{Cu(OH)_3^-}}{m_{OH^-}} = 10^{-2.9}. \qquad (2\text{-}22)$$

This does not end the story, for copper can form additional ions in solution containing more than one copper atom per ion, for example, $Cu_2(OH)_2^{2+}$. And if certain common anions are present, copper forms precipitates like $Cu_4(OH)_6SO_4$ and $Cu_2(OH)_3Cl$ rather than the simple $Cu(OH)_2$. As if this were not enough, the precipitated $Cu(OH)_2$ is not stable in contact with the solution, but changes over a period of a few days into the slightly less soluble oxide, CuO. Equilibrium constants are known, at least approximately, for all the pertinent reactions, so that concentrations of the various copper species can be calculated at given pH's, but the algebra becomes complex and tedious.

2-2 SOLUTE HYDROLYSIS

Some salts dissolve in water to give neutral solutions, for example, NaCl, K_2SO_4, $Ba(NO_3)_2$. All of them have this in common, that they consist of the cation of a strong base and the anion of a strong acid. Other salts give distinctly acid or basic solutions, acid if the cation forms a weak base [for example, NH_4Cl, $Fe_2(SO_4)_3$] and basic if the anion forms a weak acid (for example, K_2CO_3, Na_2S). The reaction leading to an excess of H^+ or OH^- in these solutions is called *hydrolysis*. We can formulate hydrolysis reactions and predict how far they will go by a further application of equilibrium reasoning.

Hydrolysis is an old word meaning literally "break up by means of water." It is inherited from an earlier period in chemistry, when water was supposed to split a salt into an acid and a base:

$$K_2CO_3 + 2H_2O \rightarrow 2KOH + H_2CO_3. \qquad (2\text{-}23)$$

We would now describe the reaction differently, first mentally dissecting the salt into ions, then noting that K^+ is indifferent to the water or to anything else in solution, and so focusing our attention on the CO_3^{2-}:

$$CO_3^{2-} + H_2O \rightarrow HCO_3^- + OH^-. \qquad (2\text{-}24)$$

This is the essential reaction in the hydrolysis of carbonates. We no longer think of the water as breaking up the salt, but as uniting with one of the ions of the salt. This reaction does not go to completion but quickly establishes equilibrium; the position of equilibrium is determined by the extent of dissociation of the two substances H_2O

and HCO_3^-. In effect, there is competition for H^+ between CO_3^{2-} and OH^-:

$$H^+ + OH^- \rightarrow H_2O \tag{2-25}$$

and

$$H^+ + CO_3^{2-} \rightarrow HCO_3^-. \tag{2-26}$$

In the competition OH^- has a great advantage, since K for the first reaction is 10^{14} and that for the second is only $10^{10.3}$ [reciprocals of the K's in Eqs. (2-7) and (2-11)]. The CO_3^{2-} nevertheless gets enough H^+ away from OH^- to make the solution distinctly basic.

The hydrolysis of a salt like $Fe_2(SO_4)_3$ may be written

$$Fe^{3+} + H_2O \rightarrow FeOH^{2+} + H^+. \tag{2-27}$$

The SO_4^{2-} can be ignored because it is the anion of a strong acid. We can visualize the reaction as a partitioning of OH^- between Fe^{3+} and H^+, the H^+ having considerable advantage but the Fe^{3+} being able to attract enough OH^- to leave the solution markedly acidic. Thus in modern language, *hydrolysis is the reaction between water and the ion of a weak acid or a weak base*. If only one ion of a pair hydrolyzes, as in the above examples, the solution becomes basic or acidic; if both ions hydrolyze, their effects may partly or completely cancel each other and leave the solution nearly neutral.

The quantitative handling of hydrolysis is a simple extension of our work with dissociation constants. To find, for example, the approximate pH of a $0.01m$ solution of Na_2CO_3, we write for the equilibrium constant of Eq. (2-24),

$$K_{(2-24)} = \frac{m_{HCO_3^-} m_{OH^-}}{m_{CO_3^{2-}}}. \tag{2-28}$$

To evaluate this hydrolysis constant, we resort to a trick. If both numerator and denominator are multiplied by m_{H^+}, we obtain

$$K_{(2-24)} = \frac{m_{HCO_3^-} m_{OH^-} \cdot m_{H^+}}{m_{CO_3^{2-}} \cdot m_{H^+}} = \frac{m_{HCO_3^-}}{m_{CO_3^{2-}} \cdot m_{H^+}} m_{OH^-} \cdot m_{H^+}. \tag{2-29}$$

This is the product of two constants we already know; the first is the reciprocal of the constant in Eq. (2-11), and the second is the constant in Eq. (2-7b). Hence

$$K_{(2-24)} = \frac{K_{(2-7)}}{K_{(2-11)}} = \frac{10^{-14}}{5.0 \times 10^{-11}} = 2.0 \times 10^{-4}. \tag{2-30}$$

Then if the solution contains no other carbonate besides Na_2CO_3 and no other HCO_3^- except that produced by hydrolysis, we can set m_{OH^-} equal to $m_{HCO_3^-}$ and $m_{CO_3^{2-}}$ equal to $0.01 - m_{HCO_3^-}$ or $0.01 - m_{OH^-}$. Hence, from Eq. (2-29),

$$K_{(2-24)} = \frac{m_{OH^-}^2}{0.01 - m_{OH^-}} = 2.0 \times 10^{-4} \tag{2-31}$$

or

$$m_{OH^-}^2 + 2.0 \times 10^{-4} m_{OH^-} - 2.0 \times 10^{-6} = 0. \tag{2-32}$$

Solving this quadratic equation[1] gives $m_{OH^-} = 1.3 \times 10^{-3} m = 10^{-2.9} m$, and the pH is therefore $-\log m_{H^+} = -\log K_{water} + \log m_{OH^-}$ or $14 - 2.9 = 11.1$. Thus a dilute solution of a carbonate is fairly alkaline.

A similar equation can be set up for hydrolysis of a cation, for example the Fe^{3+} in Eq. (2-27):

$$K_{(2-27)} = \frac{m_{FeOH^{2+}} m_{H^+}}{m_{Fe^{3+}}}$$

$$= \frac{m_{FeOH^{2+}} m_{H^+}}{m_{Fe^{3+}}} \frac{m_{OH^-}}{m_{OH^-}} \qquad (2\text{-}33)$$

$$= \frac{m_{H^+} m_{OH^-}}{\left(\dfrac{m_{Fe^{3+}} m_{OH^-}}{m_{FeOH^{2+}}} \right)}$$

which could be evaluated by dividing 10^{-14} [the equilibrium constant for the dissociation of water, Eq. (2-7)] by the dissociation constant for $FeOH^{2+}$:

$$K_{hydrolysis} = \frac{K_{water}}{K_{dissociation}}. \qquad (2\text{-}34)$$

Eq. (2-34) is a general expression for finding hydrolysis constants, the $K_{dissociation}$ referring to dissociation of the weak acid or weak base that is formed in the hydrolysis reaction. We may generalize that simple solutions of any soluble carbonate or sulfide will necessarily be alkaline, and that solutions of simple salts of the common heavy metals [for example, $ZnCl_2$, $Pb(NO_3)_2$, $NiSO_4$] will necessarily be acidic, because of hydrolysis reactions analogous to Eqs. (2-24) and (2-27). We shall find these rules useful in discussing the many geologic processes in which hydrolysis plays an important role.

2-3 ESTIMATING IONIC CONCENTRATIONS

In a solution of a given pH, if dissolved carbonate is known to be present, would it exist chiefly as H_2CO_3, HCO_3^-, or CO_3^{2-}? If the solution contains iron in the trivalent state, is it mainly Fe^{3+}, $FeOH^{2+}$, or $Fe(OH)_2^+$? Would dissolved zinc be present as positive ions (Zn^{2+} and $ZnOH^+$) or as negative ions [$Zn(OH)_3^-$ and $Zn(OH)_4^{2-}$]? This is a kind of question often encountered in geochemistry when evaluating mineral solubilities and the transport of dissolved substances.

The distribution of carbonate species will serve as a convenient example. We know to begin with, in a general qualitative way, that dissolved carbonate must exist chiefly as H_2CO_3 in acid solutions, as CO_3^{2-} in basic solutions, and as HCO_3^- in some intermediate range. To fix the limits, we write equations for the two

---

[1] For the general quadratic equation $ax^2 + bx + c$, the solution is $x = \dfrac{-b \pm \sqrt{b^2 - 4ac}}{2a}$.

dissociation constants [Eqs. (2-10) and (2-11)] in the form

$$\frac{m_{HCO_3^-}}{m_{H_2CO_3}} = \frac{10^{-6.4}}{m_{H^+}} \quad \text{and} \quad \frac{m_{CO_3^{2-}}}{m_{HCO_3^-}} = \frac{10^{-10.3}}{m_{H^+}}. \tag{2-35a, b}$$

From these expressions, the concentrations of HCO_3^- and H_2CO_3 must be equal when m_{H^+} has a numerical value equal to $K_{(2-8)}$, and the concentrations of CO_3^{2-} and HCO_3^- are equal when $m_{H^+} = K_{(2-9)}$. Hence we can say immediately that H_2CO_3 is the dominant carbonate species in all solutions with pH less than 6.4 (or m_{H^+} greater than $10^{-6.4}m$), HCO_3^- is dominant in the pH range 6.4 to 10.3, and CO_3^{2-} is dominant at pH's above 10.3. These rules hold for any solution, regardless of how dilute or how concentrated it may be or what other solutes may be present.

Suppose now that we have given also a total analytical concentration of carbonate, say $0.001m$. In a solution whose pH is 6.4 the concentrations of both H_2CO_3 and HCO_3^- must be half this number, or $0.0005m$, and the concentration of CO_3^{2-} is very small; at pH 10.3 the concentrations of CO_3^{2-} and HCO_3^- are both $0.0005m$ and $m_{H_2CO_3}$ is very small. At pH's well below 6.4, $m_{H_2CO_3}$ is effectively $0.001m$; at pH's well above 10.3, $m_{CO_3^{2-}}$ is $0.001m$; and over a good part of the intermediate range, $m_{HCO_3^-}$ must be $0.001m$.

It is often desirable to know the concentrations of all three carbonate species in a given solution, even though one or two may be very minor. For example, what are the concentrations of CO_3^{2-} and HCO_3^- in a solution containing $0.001m$ total dissolved carbonate and having a pH of 8.0? This is the intermediate range, where most of the dissolved carbonate exists as HCO_3^-, so that this ion may be assigned a concentration of approximately $0.001m$. Then the equilibrium-constant equation for reaction (2-9) becomes

$$\frac{m_{CO_3^{2-}}}{0.001} = \frac{10^{-10.3}}{m_{H^+}} \tag{2-36}$$

and for reaction (2-8)

$$\frac{0.001}{m_{H_2CO_3}} = \frac{10^{-6.4}}{m_{H^+}}. \tag{2-37}$$

If m_{H^+} is 10^{-8}, these equations give $m_{CO_3^{2-}} = 10^{-5.3}$ and $m_{H_2CO_3} = 10^{-4.6}$.

To generalize these results, it is convenient to rewrite the two equations in logarithmic form:

$$\log m_{CO_3^{2-}} - (-3) = -10.3 - \log m_{H^+} \tag{2-38}$$

and

$$(-3) - \log m_{H_2CO_3} = -6.4 - \log m_{H^+}. \tag{2-39}$$

These may be simplified to

$$\log m_{CO_3^{2-}} = -13.3 + \text{pH}, \tag{2-40}$$

$$\log m_{H_2CO_3} = 3.4 - \text{pH}. \tag{2-41}$$

If now $\log m_{CO_3^{2-}}$ and $\log m_{H_2CO_3}$ are plotted against pH, both give straight lines with slopes of $+1$ and -1, respectively. These relations hold over the pH range in which HCO_3^- has the approximate concentration $0.001m$, in other words from roughly 7.0 to 9.5. Similar equations can be set up for other pH ranges, giving a combined plot shown in Fig. 2-3(a). This diagram is drawn for a total carbonate concentration of $0.001m$, but it can be used for any desired concentration by simply shifting the vertical scale up or down. From the diagram the concentration of each of the three carbonate species at any pH can be read as intersections of the appropriate lines with a vertical line through the pH value. In Fig. 2-3(b) the temperature dependence of the solution pH is shown where $m_{H_2CO_3}$ equals $m_{HCO_3^-}$ and where $m_{HCO_3^-}$ equals $m_{CO_3^{2-}}$. Similar diagrams for sulfide solutions are given in Fig. 2-4.

The two acids H_2CO_3 and H_2S are the most important weak acids in geological environments, and an understanding of the relations of the two acids and their ions, as summarized in Figs. 2-3 and 2-4, is essential to geochemical work with natural solutions. Similar diagrams may be constructed for other weak acids,

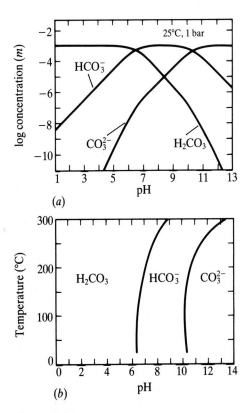

FIGURE 2-3

(a) Concentrations of carbonate species in solutions with total dissolved carbonate equal to $0.001m$ at 25°C and 1 bar. (b) The temperature dependence of the pH where the concentrations of H_2CO_3 and HCO_3^- are equal, and where the concentrations of HCO_3^- and CO_3^{2-} are equal.

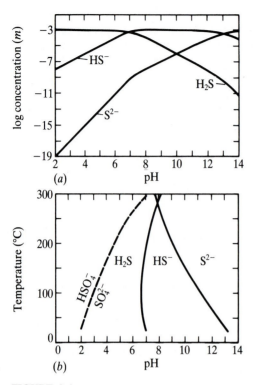

FIGURE 2-4

(*a*) Concentrations of sulfide species in solutions with total dissolved sulfide equal to $0.001m$ at 25°C and 1 bar. (*b*) The temperature dependence of the pH where the concentrations of H_2S and HS^- are equal, and where the concentrations of HS^- and S^{2-} are equal. The dashed line denotes equal concentrations of HSO_4^- and SO_4^{2-} for sulfuric acid.

and similar reasoning may be applied to other solutes that can exist in different species.

2-4 BUFFERS

The pH of ocean water sampled near the surface is almost always between the narrow limits 8.1 and 8.3. Locally and temporarily it may deviate from this range, but by and large the pH stays surprisingly constant. In the laboratory, a sample of seawater appears to resist a change in pH when acid or base is deliberately added. If, say, a liter of seawater and a liter of distilled water are set side by side and a few drops of HCl are added to each, the distilled water will show a pH of 2 or 3 while that of the seawater remains close to 8. Whence comes this capacity for maintaining a nearly constant concentration of hydrogen and hydroxide ions?

For an answer we look once more to equilibrium reactions. We start, however, with a system much simpler than seawater, a solution containing roughly $0.01m$ H_2CO_3 and $0.01m$ $NaHCO_3$. This mixture, we would find experimentally, behaves like seawater in that it shows little tendency to change its pH when a few drops of acid or base are added. What are the principal substances present in the solution? One is Na^+, but it needs no attention because it undergoes no reaction with common acids and bases. The other two principal substances are HCO_3^- and undissociated H_2CO_3, both present at concentrations of about $0.01m$ because their dissociation and hydrolysis reactions are too slight to affect their concentrations appreciably. From Fig. 2-3(a) we note that a solution with equal amounts of H_2CO_3 and HCO_3^- should have a pH of 6.4. We could get the same number from Eq. (2-10):

$$m_{H^+} = \frac{m_{H_2CO_3}}{m_{HCO_3^-}} K_{(2-8)} = \frac{0.01}{0.01} K_{(2-8)} = 10^{-6.4}m. \tag{2-42}$$

Now when acid is added, the H^+ reacts with HCO_3^-:

$$H^+ + HCO_3^- \rightarrow H_2CO_3. \tag{2-43}$$

This of course, will change the ratio of $m_{H_2CO_3}$ to $m_{HCO_3^-}$ in Eq. (2-42), and thereby the m_{H^+} also; but actually the pH is not very greatly affected unless the ratio changes a great deal. Suppose, for example, that enough acid is added to change the ratio from approximately 1 to approximately 2 (which could be accomplished by adding 3 ml of $1m$ HCl to a liter of solution); this would give $m_{H^+} = 2 \times 10^{-6.4}m$ and pH = 6.1 instead of 6.4. By way of contrast, if distilled water were used instead of the H_2CO_3–$NaHCO_3$ solution, the same amount of acid would give a pH of 2.5. The HCO_3^-, in effect, takes up enough of the added H^+ so that the amount of free H^+ in the solution is only slightly increased.

Now note what happens if a strong base is added. The OH^- reacts with the H_2CO_3:

$$H_2CO_3 + OH^- \rightarrow HCO_3^- + H_2O. \tag{2-44}$$

The ratio $m_{H_2CO_3}/m_{HCO_3^-}$ this time is decreased, and m_{H^+} is correspondingly decreased, but again the effect on the pH is relatively slight. Addition of 3 ml of $1m$ NaOH, for example, would make the pH about 6.7 instead of 6.4, whereas the same amount of base added to distilled water would give 11.5. The H_2CO_3 uses up most of the added OH^-, so that free hydroxide ion in the solution is increased only a trifle.

Solutions of this kind, capable of absorbing considerable H^+ or OH^- without showing much change in pH, are called *buffers*. In general, a buffer consists of approximately equal amounts of weak acid (like H_2CO_3) and a salt of the acid (like $NaHCO_3$). The pH maintained by any particular buffer is determined by the dissociation constant of the acid. Maintenance of pH is, of course, possible only within limits. When enough strong acid or base has been added to a buffer to change the ratio of acid to ion (H_2CO_3/HCO_3^-) by a factor of more than 10, the buffer loses its effectiveness, and the pH responds thereafter directly to increments of H^+ or OH^-.

Buffering action is not limited to acids. Mixtures of bases and the corresponding cations (say NH_3 and NH_4^+) would obviously be just as efficient. Or we could employ a slightly soluble salt as part of the buffer. A solution of calcium bicarbonate in contact with calcium carbonate (calcite), for example, tends to maintain its pH because H^+ is absorbed in the reaction

$$CaCO_3 + H^+ \rightarrow Ca^{2+} + HCO_3^- \qquad (2\text{-}45)$$
calcite

and OH^- in the reaction

$$Ca^{2+} + HCO_3^- + OH^- \rightarrow CaCO_3 + H_2O. \qquad (2\text{-}46)$$
calcite

The intermediate ion of a dihydrogen acid has some ability to control changes in pH all by itself, for it undergoes the two reactions,

$$HCO_3^- + H^+ \rightarrow H_2CO_3 \qquad (2\text{-}47)$$

and

$$HCO_3^- + OH^- \rightarrow CO_3^{2-} + H_2O. \qquad (2\text{-}48)$$

Many natural solutions have the right combinations of solutes to serve as effective buffers. Seawater is an excellent example, but its buffering action is complex. Any attempt to make seawater more acid is countered by the reactions

$$H^+ + CO_3^{2-} \rightarrow HCO_3^- \quad \text{and} \quad H^+ + HCO_3^- \rightarrow H_2CO_3. \qquad (2\text{-}49)$$

If solid calcite or aragonite ($CaCO_3$) is present, as it always is in many parts of the sea, excess H^+ is also reduced by the reaction $CaCO_3 + H^+ \rightarrow Ca^{2+} + HCO_3^-$. The addition of OH^- leads to the counterreactions

$$OH^- + HCO_3^- \rightarrow CO_3^{2-} + H_2O, \qquad (2\text{-}50)$$

$$OH^- + H_2CO_3 \rightarrow HCO_3^- + H_2O, \qquad (2\text{-}51)$$

and

$$Ca^{2+} + HCO_3^- + OH^- \rightarrow CaCO_3 + H_2O. \qquad (2\text{-}52)$$

The H_2CO_3 for the second reaction is present only in minute quantities, but the CO_2 of the atmosphere forms a reservoir of additional H_2CO_3 if the amount of base to be handled is large. In addition to all these reactions involving carbonates and carbonate ions, seawater contains enough boron so that reactions like

$$H^+ + H_2BO_3^- \rightarrow H_3BO_3 \qquad (2\text{-}53)$$

and

$$OH^- + H_3BO_3 \rightarrow H_2BO_3^- + H_2O \qquad (2\text{-}54)$$

play a significant secondary role.

These various processes serve to hold the pH of seawater in the neighborhood of 8 and probably have so held it for a long time in the geologic past. The ultimate controls, it should be noted, are CO_2 in the atmosphere and $CaCO_3$ in the bottom sediments. Any long-continued addition of acid, say as a result of large-scale production of HCl and CO_2 by volcanic activity, would lead to marked dissolution of the $CaCO_3$; and long-continued addition of base would mean a depletion of atmospheric CO_2. Since abundant calcium carbonate and carbon dioxide have been in contact with the oceans at least since the beginning of the Paleozoic era and probably in earlier times also, it seems unlikely that the pH of seawater has changed greatly during the latter part of geologic time. It may have been a few tenths of a unit lower during times when CO_2 in the air was abnormally high, for example during much of the Cretaceous period. One concern regarding the present-day slow increase in atmospheric CO_2 (Sec. 12-1) is the possible resulting change in seawater pH.

Although the carbonate system is thus the probable immediate control on the pH of seawater, a complete explanation would have to go deeper. We could ask, for example, why the oceans at some time in the past acquired a pH suitable for the precipitation of $CaCO_3$. At least theoretically the water might be much more acid and still exist in equilibrium with atmospheric CO_2. The ultimate explanation would go back to the properties of silicate minerals, whose reactions are slower than those of carbonates but which, over long periods, must determine the inorganic character of seawater. We shall return to this question later in Chapters 12 and 21.

2-5 ELECTROLYTE SOLUTIONS AND ION–ION ASSOCIATIONS

In electrolyte solutions like those we have been discussing, some of the ions and molecules exist in the form of *complexes*. This word, in its most general definition, refers to any combination of two or more atoms that can exist in solution, so on this basis such familiar species as OH^-, SO_4^{2-}, and H_2CO_3 would be complexes. In common usage, however, the word refers to more unusual atom combinations, for example $PbCl^+$, $NaCl(aq)$, HgS_2^{2-}, and $UO_2(CO_3)_2^{2-}$.

Other atom combinations, not usually considered as complexes, form by the reaction of ions with water molecules. Because the distribution of positive and negative charges in the water molecule is not symmetrical, water molecules are electric dipoles that can align themselves in an external electric field [Fig. 2-5(a)]. Thus, in the field around a charged ion in an aqueous solution, the H_2O dipoles orient themselves, positive ends facing a negatively charged ion and negative ends facing a positively charged ion [Fig. 2-5(b)]. The oriented dipoles form a protective "solvation shell" around each ion that helps to prevent contact with other ions. The solvation shell can be included in a formula [e.g., $Na(H_2O)_6^+$ is more accurate than the simple Na^+], but of course is usually omitted. In effect, these combinations of ions-plus-solvation-shells are a form of complex, and any electrolyte solution is made up of such complexes even though they are not indicated by the simple ionic formulas.

Chloride Complexes

The chloride ion, one of the most common anions in natural waters, forms a wide range of complexes with cations of alkali, alkaline-earth, and transition metals. Over

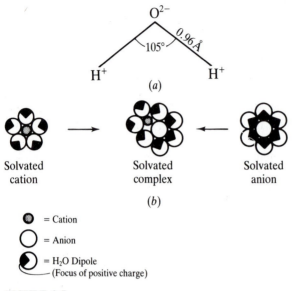

(a)

Solvated
cation

Solvated
complex

Solvated
anion

(b)

- = Cation
- = Anion
- = H₂O Dipole
  (Focus of positive charge)

FIGURE 2-5

(a) Diagram to illustrate the polarity of water molecules. Because the H₂O molecule is not symmetrical but has an angle of about 105°C between the H-O bonds, the center of positive charge does not coincide with the center of negative charge. Thus the H₂O molecule is an electrical dipole that can align itself in an electric field or orient itself about charged ions. (b) Schematic illustrations of a solvated cation, anion, and complex in an idealized aqueous solution. Arrows represent formation of a complex by collision of the two solvated ions. (Modified after Helgeson, H. C., D. H. Kirkham, and G. C. Flowers: "Theoretical prediction of the thermodynamic behavior of aqueous electrolytes at high pressures and temperatures: IV. Calculation of activity coefficients, osmotic coefficients, and apparent molal properties to 600°C and 5 kbar," *American Journal of Science*, vol. 281, pp. 1249–1516, 1981.)

the range of temperatures, pressures, and fluid compositions of geologic interest the relative stability of chloride complexes has an important influence on the solubilities of rock-forming minerals and on the ability of aqueous solutions to transport metals.

Dissolution of a salt like NaCl results in a decrease in the molal volume of the solution due to the collapse of the open structure of water and the alignment of water dipoles to form solvation shells in the vicinity of the dissociated Na⁺ and Cl⁻ ions. At low temperatures and pressures, where the density and dielectric constant for water are relatively large (Table 2-2 and Fig. 2-6), the solvation shells are effective in reducing the probability that a collision between the solvated Na⁺ and Cl⁻ ions will lead to the formation of a stable NaCl complex. Thus most salt solutions are largely dissociated at low temperatures and pressures. At high temperatures and moderate pressures, for example 500 bar and 500°C, the density and dielectric constant of water are both small, and the ability of an ion to form protective solvation shells is greatly reduced. This increases the probability that a collision will lead to formation of the NaCl complex.

These processes are reflected in the temperature–pressure variations of the equilibrium constant for the dissociation of aqueous NaCl(aq)

$$NaCl(aq) \rightleftharpoons Na^+ + Cl^-, \tag{2-55}$$

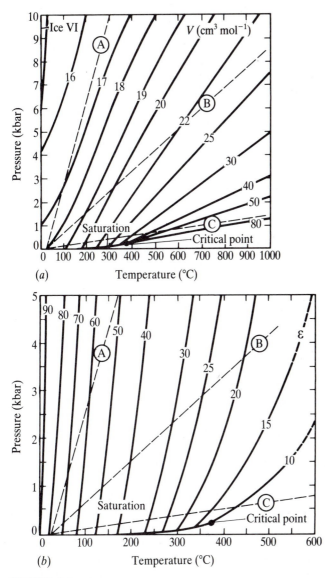

FIGURE 2-6
(a) Isochores of the molal volume ($cm^3 mol^{-1}$) of H_2O. The mathematical formula expressing the change of volume with temperature and pressure, $V = F(P,T)$, is called an equation of state. For a perfect gas this function is the familiar $V = nRT/P$ (where n is number of mols and R is the gas constant), but for water over a range of pressures and temperatures it becomes a complex polynomial. (b) The dielectric constant (ε) of water as a function of temperature and pressure. The constant is the ratio of the capacity of a condenser to store a charge when H_2O occupies the space between the plates to the capacity of the same condenser with a vacuum between the plates; it is a measure of the ability of water to hold apart ions of opposite charge in solution, hence of its ability to dissolve ionic solids. In both diagrams the saturation curve denotes liquid–vapor equilibrium (boiling-point curve), and the dashed lines represent typical geotherms for a subducting plate (A), regional metamorphism (B), and contact metamorphism (C). The stability of ice VI is shown in (a). (Computed from equations and data reported by Helgeson, H. C., and D. H. Kirkham: "Theoretical prediction of the thermodynamic behavior of aqueous electrolytes at high pressures and temperatures: I. Summary of the thermodynamic/electrostatic properties of the solvent," *American Journal of Science*, vol. 274, pp. 1089–1198, 1974.)

TABLE 2-2

The dissociation constant for the reaction $NaCl(aq) \rightleftharpoons Na^+ + Cl^-$, and the molal volume ($V$, cm^3 mol^{-1}), density (ρ, g cm^{-3}), and dielectric constant (ε) for water at pressures (bars) and temperatures (°C) corresponding to conditions typical of weathering (1 bar, 25°C), regional metamorphism (5000 bar, 500°C), and contact metamorphism (500 bar, 500°C).

P (bar)	T (°C)	log K_{NaCl}	V_{H_2O}	ρ_{H_2O}	ε_{H_2O}
1	25	0.93	18.07	1.00	78.47
5000	500	−0.65	20.74	0.87	18.56
500	500	−5.07	70.12	0.26	3.94

(Reprinted by permission from Helgeson, H. C., D. H. Kirkham, and G. C. Flowers: "Theoretical prediction of the thermodynamic behavior of aqueous electrolytes at high pressures and temperatures: IV. Calculation of activity coefficients, osmotic coefficients, and apparent molal properties to 600°C and 5 kbar, "*American Journal of Science*, vol. 281, pp. 1249–1516, 1981; and Helgeson. H. C., and D. H. Kirkham: "Theoretical prediction of the thermodynamic behavior of aqueous electrolytes at high pressures and temperatures: I. Summary of the thermodynamic/electrostatic properties of the solvent," *American Journal of Science*, vol. 274, pp. 1089–1198, 1974.)

as shown in Table 2-2. At low pressures and temperatures, near 1 bar and 25°C, the dissociation constant for NaCl(aq) is about 10; thus the product of the concentrations of the Na^+ and Cl^- ions is nearly an order of magnitude greater than the concentration of the NaCl(aq) complex in the solution. At high temperatures and moderate pressures, such as 500°C and 500 bar, the concentration of NaCl(aq) is about 5 orders of magnitude greater than the product $m_{Na^+} m_{Cl^-}$. The latter temperature–pressure condition is typical of contact metamorphism, as illustrated in Figs. 1-1 and 2-6. For pressure and temperature conditions typical of regional metamorphism, such as 5000 bar and 500°C (Table 2-2), the equilibrium constant for NaCl dissociation is slightly less than unity, meaning that here again NaCl(aq) predominates over the ionic species Na^+ and Cl^- but only slightly so.

Some metals, such as Ag, Au, Cu, Pb, Zn, Fe, and Hg, form a variety of chloride complexes in aqueous solutions. For example, the important chloride complexes of zinc are $ZnCl^+$, $ZnCl_2(aq)$, $ZnCl_3^-$, and $ZnCl_4^{2-}$. Dissociation constants for these complexes are given in Fig. 2-7 as a function of temperature. Unless the concentration of chloride ion is unusually high, however (greater than about $0.01m$), these constants show that Zn^{2+} is the most abundant zinc species in solution at low temperatures.

Let us consider, as an example, the effect of Zn-chloride complexes on the solubility of the ore mineral sphalerite (ZnS) and the ability of aqueous chloride solutions to transport zinc in shallow hydrothermal environments. Equilibrium

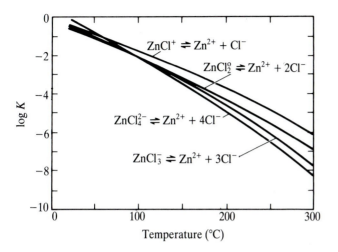

FIGURE 2-7
Equilibrium constants for the dissociation of the complexes $ZnCl^+$, $ZnCl_2(aq)$, $ZnCl_3^-$, and $ZnCl_4^{2-}$ as a function of temperature at pressures equal to the liquid–vapor equilibrium pressures for H_2O. (Reprinted by permission from Helgeson, H. C.: "Thermodynamics of hydrothermal systems at elevated temperatures and pressures," *American Journal of Science*, vol. 267, pp. 729–804, 1969).

between sphalerite and an aqueous solution can be represented by the reaction:

$$\underset{\text{sphalerite}}{ZnS} \rightleftharpoons Zn^{2+} + S^{2-}. \tag{2-56}$$

At 250°C and pressures equal to liquid–vapor equilibrium for H_2O the equilibrium constant is:

$$K_{(2\text{-}56)} = m_{Zn^{2+}} m_{S^{2-}} = 10^{-18.86}. \tag{2-57}$$

In a solution containing a small amount of S^{2-} ions, say $10^{-10} m$, the concentration of Zn^{2+} ion will be:

$$m_{Zn^{2+}} = \frac{K_{(2\text{-}56)}}{m_{S^{2-}}} = \frac{10^{-18.86}}{10^{-10}} = 10^{-8.86} m. \tag{2-58}$$

If Zn is not complexed with any anions, this represents the total concentration of Zn. The figure is so low that the ability of such a solution to transport enough zinc to form an ore deposit is questionable (see also Sec. 19-1).

Solutions associated with zinc deposits, however, generally contain substantial amounts of chloride, as is shown by analyses of fluid-filled inclusions in minerals from such deposits (Sec. 19-2). The concentration of dissolved zinc, therefore, would be augmented by contributions from the chloride complexes:

$$m_{Zn, total} = m_{Zn^{2+}} + m_{ZnCl^+} + m_{ZnCl_{2(aq)}} + m_{ZnCL_3^-} + m_{ZnCl_4^{2-}}. \tag{2-59}$$

We can estimate this total concentration by using the solubility product of sphalerite together with constants for the dissociation reactions given in Fig. 2-7. These equations can be rewritten as:

$$m_{ZnCl^+} = \frac{m_{Zn^{2+}} m_{Cl^-}}{K_{ZnCl^+}}, \tag{2-60}$$

$$m_{ZnCl_2(aq)} = \frac{m_{Zn^{2+}} m_{Cl^-}^2}{K_{ZnCl_2(aq)}}, \tag{2-61}$$

$$m_{ZnCl_3^-} = \frac{m_{Zn^{2+}} m_{Cl^-}^3}{K_{ZnCl_3^-}}, \tag{2-62}$$

and

$$m_{ZnCl_4^{2-}} = \frac{m_{Zn^{2+}} m_{Cl^-}^4}{K_{ZnCl_4^{2-}}}. \tag{2-63}$$

Eq. (2-60) through (2-63) define homogeneous equilibrium constants for the concentrations of the Zn-chloride complexes relative to the concentrations of Zn^{2+} and Cl^- ions. In this example, we will estimate the total concentration of zinc in a solution at equilibrium with sphalerite that contains $0.1m$ total Cl^- and $10^{-10} m_{S^{2-}}$.

Because the solution is in equilibrium with sphalerite, the concentration of Zn^{2+} in Eqs. (2-60) through (2-63) will be determined by the solubility product represented by Eq. (2-57) and the concentration of S^{2-}: thus $Zn^{2+} = 10^{-8.86} m$. We can substitute this value, together with values for the dissociation constants of the Zn-chloride complexes (Fig. 2-7), into Eqs. (2-60) through (2-63) and obtain the following relationships between the concentrations of the Zn-chloride complexes and the concentration of Cl^- ion:

$$m_{ZnCl^+} = \frac{10^{-8.86} m_{Cl^-}}{10^{-4.8}} = 10^{-4.06} m_{Cl^-}, \tag{2-64}$$

$$m_{ZnCl_2(aq)} = \frac{10^{-8.86} m_{Cl^-}^2}{10^{-5.5}} = 10^{-3.36} m_{Cl^-}^2, \tag{2-65}$$

$$m_{ZnCl_3^-} = \frac{10^{-8.86} m_{Cl^-}^3}{10^{-6.0}} = 10^{-2.86} m_{Cl^-}^3, \tag{2-66}$$

and

$$m_{ZnCl_4^{2-}} = \frac{10^{-8.86} m_{Cl^-}^4}{10^{-6.4}} = 10^{-2.46} m_{Cl^-}^4. \tag{2-67}$$

From these four equations we see that the concentration of Cl^- is several orders of magnitude greater than the concentration of any of the Zn-chloride complexes. Therefore, the mass-balance constraint on the total chloride concentration, which is written as

$$0.1m = m_{Cl^-,\ total} = m_{Cl^-} + m_{ZnCl^+} + 2m_{ZnCl_2(aq)} + 3m_{ZnCl_3^-} + 4m_{ZnCl_4^{2-}}, \tag{2-68}$$

can be simplified by the approximation that

$$m_{Cl^-} = m_{Cl^-,\,total} = 0.1m.$$

The total concentration of zinc in the solution is calculated using this approximation for m_{Cl^-} in Eqs. (2-64) through (2-67); the result is $m_{Zn(total)} = 10^{-4.83}m$. Thus the total amount of zinc dissolved in a solution in equilibrium with sphalerite is increased from $10^{-8.86}m$ to $10^{-4.83}m$, an increase of more than three orders of magnitude, with the addition of only $0.1mCl^-$ to the solution.

This example illustrates the importance of chloride complexes on the potential for aqueous solutions to transport metals in groundwater systems in sedimentary basins and subvolcanic environments where chloride concentrations often exceed $0.1m$. Chloride is not the only anion that forms substantial complexes in natural electrolyte solutions; other common ones are SO_4^{2-}, S^-, HS^-, CO_3^{2-}, HCO_3^-, and NO_3^{2-}. In the next section we consider in more detail how complex-forming anions affect the solubilities of rock-forming minerals.

2-6 ACTIVITY AND ACTIVITY COEFFICIENTS

Still another influence on the concentration of zinc in equilibrium with sphalerite is the presence of other ions in solution. We consider first the general effect on solubility exerted by other solute ions, and then return to a calculation of this effect for the specific example of zinc solubility in a chloride solution that we have been considering.

We have spoken earlier (Sec. 1-5) about the effect of a common ion on the solubility of minerals. For example, the addition of either Ca^{2+} or CO_3^{2-} to a saturated solution of $CaCO_3$ will decrease the solubility of the mineral calcite by an amount that is easily computed from the solubility product for the reaction,

$$CaCO_3 \rightleftharpoons Ca^{2+} + CO_3^{2-}. \tag{2-69}$$
$$\text{calcite}$$

Electrolytes which do not supply a common ion generally have the opposite effect, making the solubility greater. To account for this observation, let us consider some details of the precipitation process. Ca^{2+} ions in water solution, as noted above, are surrounded by water molecules with their negative ends close to the ions and their positive ends facing outward [Fig. 2-5(b)]. The CO_3^{2-} ions would be similarly hydrated, with the water molecules in their vicinity showing opposite orientations. These loosely held retinues of molecules around ions serve to shield Ca^{2+} and CO_3^{2-} from coming close to each other and precipitating. The shield is evidently not very efficient, because most of the ions precipitate out of the solution in spite of it.

Now suppose an electrolyte, say NaCl, is added. The solution will contain abundant Na^+ and Cl^- ions, and we might expect these ions as well as water molecules would be attracted to Ca^{2+} and CO_3^{2-}. Each Ca^{2+} we can picture as the center of a cluster of water molecules and Cl^- ions, each CO_3^{2-} as the center of a cluster of water molecules and Na^+ ions. The added ions will increase the shielding action of the water, and it will be harder for the Ca^{2+} and CO_3^{2-} to find each other and precipitate. The solubility is therefore greater, and the amount of increase should depend, within limits, on the concentration of added salt.

We can carry this naive qualitative reasoning a bit further. If increased solubility in electrolyte solutions results from protective shields of opposite charge around each ion, we could guess that the effect would be greater in solutions containing divalent ions than in those with only univalent ions, since the attraction for multiply charged ions should be stronger. Thus the solubility of $CaCO_3$ should be higher in $0.1m$ $MgSO_4$ than in $0.1m$ NaCl solution. By the same argument, the increase in solubility should be greater for salts made up of multivalent ions than for those with only univalent ions; for example, the solubility of $CaCO_3$ should be increased more than the solubility of AgCl if both are put in $NaNO_3$ solution. We could predict also that the effect of electrolytes on the solubility of undissociated solutes like H_2CO_3 or H_4SiO_4 should be small, since ions would have little tendency to cluster around uncharged molecules. In general, predictions of this sort are borne out by experiment.

Somewhat greater precision can be given to these qualitative conclusions if the concentration of an electrolyte solution is expressed as ionic strength rather than simple molality. The *ionic strength* of a solution is defined as

$$I = \frac{1}{2} \sum_i m_i z_i^2,$$ (2-70)

where m_i is the concentration of an ion in mols per kilogram of water, z_i is its charge, and the sum is taken over all the ions in the solution. For a solution containing $0.1m$ $BaCl_2$ and $0.04m$ $NaNO_3$, for example, the ionic strength would be

$$I = \frac{1}{2}(0.1 \times 4 + 0.2 \times 1 + 0.04 \times 1 + 0.04 \times 1) = 0.34.$$ (2-71)

The ionic strength differs from the sum of the molal concentrations in a complex solution in that it emphasizes the effect of higher charges of multivalent ions. Empirically it is found that the ionic strength gives a good measure of the effect of electrolytes on solubility. For insoluble salts of a given type (univalent like AgCl; didivalent like $CaSO_4$; unidivalent like Ag_2SO_4; and so on), the effect on solubility of any solution of given ionic strength, regardless of its composition, is roughly the same. Furthermore, experiments show that the solubility of a given salt generally shows a rough proportionality to the square root of ionic strength, at least for fairly dilute solutions. These are useful rules, but they are only approximate and have many exceptions.

To attach numbers to this effect of electrolytes on solubility, we introduce a new quantity called *activity*. Suppose that the solubility of a salt is measured in solutions of different ionic strengths. By plotting the solubility against ionic strength (or better, against its square root), we can extrapolate back to hypothetical solubility at zero ionic strength (Fig. 2-8). Generally this will be nearly the same as the solubility measured when the salt is stirred in pure water. The value of the solubility at this point is called the *activity* of the salt in a saturated solution, and the

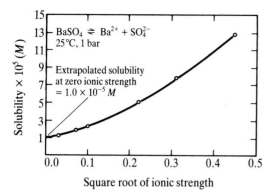

FIGURE 2-8
Change in the solubility of $BaSO_4$ (in mols per liter) as a function of the square root of ionic strength at 25°C and 1 kbar. The equilibrium constant for barite-solution equilibrium is 10^{-10} (Appendix VII), thus the solubility of $BaSO_4$ in pure water $(I = 0)$ is $10^{-5}M$.

solubility product is called the *activity product*, or the product of the activities of the separate ions.

By definition, the activity product is constant for any saturated solution at a given temperature. In a solution with ionic strength higher than zero, then, the increased solubility may be expressed by saying that the measured concentration of $CaCO_3(aq)$ is greater than its activity. The relation between the two quantities is expressed by means of a factor called the *activity coefficient*:

$$a_{CaCO_3(aq)} = \gamma_{CaCO_3(aq)} m_{CaCO_3(aq)}, \tag{2-72}$$

where a is the activity and γ is the activity coefficient. Similar expressions can be set up for activities and activity coefficients of the separate ions:

$$a_{Ca^{2+}} = \gamma_{Ca^{2+}} m_{Ca^{2+}} \quad \text{and} \quad a_{CO_3^{2-}} = \gamma_{CO_3^{2-}} m_{CO_3^{2-}}. \tag{2-73a, b}$$

In very dilute solutions (ionic strength less than, say, 0.001) the activity coefficients are close to 1; this means that activities are practically equal to concentrations, and the activity product is numerically practically equal to the solubility product which we have previously defined in terms of concentrations (Sec. 1-5):

$$a_{Ca^{2+}} a_{CO_3^{2-}} = K_{(2-69)} \approx m_{Ca^{2+}} m_{CO_3^{2-}}. \tag{2-74}$$

As the ionic strength of the solution increases, the activity coefficients decrease. For practically all ions the activity coefficient remains less than 1 for ionic strengths up to about 2. Some salts and some ions in very concentrated solutions, however, have activity coefficients much greater than 1, which means that, as a solution becomes concentrated, the solubility of a slightly soluble salt may pass through a maximum and then decrease rapidly.

Activities have a more general significance than the above definition would indicate. They need not be defined only for a saturated solution, but may be used in any solution. The activity of an ion, so to speak, is its *effective concentration*, the part of its analytical concentration that determines its behavior toward other ions with which it may react. At a given ionic strength the activity is proportional to the concentration, and the proportionality constant is the activity coefficient. In other words, Eqs. (2-72) and (2-73) remain true, whether the solution is saturated with the mineral calcite or not.

What units should we use to express activities? If activity is defined as a sort of "corrected" concentration, as indicated in the last few paragraphs, the appropriate units would seem to be the same as those used for concentration, mols per kilogram of water in aqueous solutions. (And activity coefficients would be simply ratios of corrected to measured concentrations, or dimensionless numbers.) Alternatively activities can be defined as ratios of corrected concentrations to concentrations in some *standard state*—pure substances, or an ideal solution of some specified molality, temperature, and pressure. In this case, activities would be dimensionless numbers, and activity coefficients would have dimensions of reciprocal molalities (m^{-1}). Actually, as we shall see in Chapter 8, activities have a broader significance than we have given them here, and in this broader context the second definition is preferable. So we shall treat activities as dimensionless numbers, but in simple calculations of solubility it is often convenient to visualize them as modified molalities.

Now, how can we measure activities and activity coefficients? The subject is too complicated for detailed treatment here, but a few general remarks are in order. We have mentioned activity coefficients for both salts and separate ions; actually only those for the salts are directly measurable, and the coefficients for ions are estimated from those for salts by making reasonable assumptions. One useful method of determining activities experimentally was described above: measurement of solubilities of a salt in solutions of different ionic strengths, and extrapolation of the solubilities to infinite dilution ($I = 0$). Other experimental methods include measurements of electromotive force, of vapor pressures, and of distribution of solutes between solvents. Such measurements are described in detail in some of the references at the end of this chapter.

For work with reasonably dilute solutions where great accuracy is not important, activity coefficients can be estimated on the basis of a theoretical treatment formulated by Peter Debye and Erich Hückel in the early nineteen hundreds. In this theory, ions are regarded as point centers of charge in a medium with a dielectric constant equal to that of the pure solvent. These are obviously idealizations, but for solutions with ionic strengths up to about 0.1 they give results in reasonable agreement with experiment. The simplest form of the Debye–Hückel theory gives as an expression for the activity coefficient of an ion

$$-\log \gamma = Az^2 I^{1/2}, \tag{2-75}$$

where z is the charge on the ion, I is the ionic strength of the solution, and A is a constant depending on temperature and the dielectric constant of the solvent, equal approximately to 0.51 in water at 25°C. The ionic strength, whose importance was recognized empirically before the Debye–Hückel theory was proposed, appears in

TABLE 2-3
Activity coefficients calculated from Davies' equation

Ionic strength	Ionic charge		
	±1	±2	±3
0.001	0.97	0.87	0.73
0.005	0.93	0.74	0.51
0.01	0.90	0.66	0.40
0.05	0.82	0.45	0.16
0.1	0.78	0.36	0.10
0.2	0.73	0.28	0.06
0.5	0.69	0.23	0.04
0.7	0.69	0.23	0.04

the theoretical derivation as an essential factor in determining the size of γ. A more satisfactory expression for log γ, derived in part from an extension of the theory and in part empirically, has been given by Davies (1962):

$$-\log \gamma = Az^2 \left(\frac{I^{1/2}}{1 + I^{1/2}} - 0.2I \right). \tag{2-76}$$

Values of γ calculated from this equation are given in Table 2-3. The values can be considerably refined by using more complicated expressions for particular kinds of ions, but for many geological purposes the numbers in Table 2-3 are sufficiently accurate.

Some of the basic assumptions of the Debye–Hückel theory break down at ionic strengths greater than 0.1, and extension of Davies' equation beyond this point is purely empirical. At ionic strengths above about 0.5, the equation gives only order-of-magnitude estimates. Such estimates may be useful in geochemical speculations, but their possible gross inaccuracy should be kept in mind.

From Table 2-3 the approximate quantitative effect of electrolytes on the solubility of $CaCO_3$ can be easily read. Even at an ionic strength as low as 0.001 (for example, in a $0.001m$ NaCl solution), the activity of Ca^{2+} and CO_3^{2-} is reduced to 87% of their concentration. This means that the solubility is $\frac{100}{87}$, or 1.15 times as great as its value at infinite dilution. The solubility product (K_{sp}), using concentrations, would be greater than the equilibrium activity product (K) by the square of this number, since from Eqs. (2-73) and (2-74),

$$K_{sp} = m_{Ca^{2+}} m_{CO_3^{2-}} = \frac{a_{Ca^{2+}} a_{CO_3^{2-}}}{\gamma_{Ca^{2+}} \gamma_{CO_3^{2-}}} = \frac{K}{0.87^2} = 1.3K. \tag{2-77}$$

If the ionic strength of the solution is 0.5, the solubility is increased to $\frac{100}{23}$, or more than 4 times the value in a very dilute solution, and the concentration product is nearly 200 times the activity product. Clearly, the effect of other electrolytes in solution is one of the major reasons why the measured concentration of $CaCO_3$ in some natural waters can be much larger than would be predicted from the solubility product alone.

More generally, the numbers in Table 2-3 show the deviations to be expected in the solubilities of any slightly soluble compounds due to the presence of electrolytes in solution. The numbers we have used heretofore for solubility products, and the numbers given in Appendix VII, are actually activity products. We have considered them equivalent to concentration products, and we have used them to calculate concentrations in solutions of fairly high ionic strength. This procedure gives reasonably accurate results in solutions whose electrolyte concentration is less than $0.001m$ and gives an order-of-magnitude accuracy at concentrations up to $0.1m$. For geologic purposes this kind of accuracy is often sufficient, and we shall continue to assume in many later discussions that activity coefficients can be set equal to 1 and disregarded. But for precise work, even in fairly dilute solutions, the activity corrections cannot be neglected.

2-7 CONVENTIONS REGARDING ACTIVITIES

Introduction of activities to account for the increase in mineral solubility with increasing ionic strength makes necessary a few additions to the list of conventions given in Sec. 1-8.

1. Activities are defined formally as dimensionless numbers. They are measured relative to a standard state of reference chosen for each substance at a specified temperature, pressure, and composition. The reasoning behind this convention is fully developed in Chapter 8. For many purposes it is convenient to visualize the activities of aqueous species as modified molal concentrations—modified to take account of the influence on the concentration of a given solute species by other ions in solution. Numerical values of activity and concentration are related by the activity coefficient, $a_i/m_i = \gamma_i$.

2. Equilibrium constants expressed in terms of activities are true constants. They vary as a function of temperature and pressure, but are independent of composition. For example, values of the equilibrium constant for Eq. (2-55),

$$K_{(2-55)} = \frac{a_{Na^+}a_{Cl^-}}{a_{NaCl}} = \left(\frac{m_{Na^+}m_{Cl^-}}{m_{NaCl}}\right) \cdot \left(\frac{\gamma_{Na^+}\gamma_{Cl^-}}{\gamma_{NaCl}}\right), \qquad (2-78)$$

are given in Table 2-2 at several different temperatures and pressures. Homogeneous equilibrium between NaCl and the ions Na^+ and Cl^- exists when the ratio of the product $a_{Na^+} \cdot a_{Cl^-}$ to a_{NaCl} is equal to this constant. For many examples used in this chapter we have assumed a unit value for activity coefficients, or that the ratio of activity coefficients shown in the last term of Eq. (2-78) is close to unity, which allows us to characterize equilibrium with simple products and ratios of molal concentrations.

3. In dilute solutions the difference in numerical values for activities and molal concentrations is small, but in concentrated solutions may be an order of magnitude or more.

4. Activities of pure solids and pure liquids are equal to 1, by convention. Activities of gases will be addressed later (Chap. 8), but for reactions that take place at low to moderate pressures, numerical values for gas activities are approximately equal to partial pressures.

5. Water in dilute solutions has an activity close to 1.

6. Unionized solutes [for example, H_4SiO_4, $SiO_2(aq)$, H_2CO_3] generally have activity coefficients close to or equal to 1.

7. Empirical and theoretical expressions used for estimating activity coefficients for ionized solutes are functions of ionic strength, a number defined as half the sum of concentrations of ions multiplied by the squares of their charge.

2-8 ACTIVITY CORRECTIONS APPLIED TO SPHALERITE SOLUBILITY

As a final example, we investigate the effect of ionic strength on the solubility of the mineral sphalerite (ZnS) in a $NaCl-ZnCl_2(aq)-H_2S-H_2O$ solution at 250°C and 50 bar. In this problem both ion-ion interaction as represented by individual ion activity coefficients and ion–ion association that forms aqueous complexes will be considered. Substances that may be present in the solution include: Zn^{2+}, $ZnOH^+$, $ZnCl^+$, $ZnCl_2(aq)$, $ZnCl_3^{2-}$, $ZnCl_4^{2-}$, H_2S, HS^-, S^{2-}, $NaCl(aq)$, $NaOH(aq)$, Na^+, Cl^-, H^+, OH^-, and $HCl(aq)$. To calculate the concentrations of these species in equilibrium with sphalerite we must know the equilibrium constants for reactions describing the solubility of sphalerite and the dissociation of the aqueous species, the activity coefficients for all the aqueous species, the ionic strength of the solution, and the compositional constraints on the fluid phase as determined by chemical analysis and/or by local equilibrium with a mineral or an assemblage of minerals.

The molalities and activity coefficients for the 15 aqueous species together with the ionic strength of the solution give a total of 31 unknowns. We can solve these unknowns by evaluating 31 independent equations, which include: the equilibrium constant for the dissociation of water [Eq. (2-7)] and the dissociation constants of the 9 complexes [for example, Eqs. (2-60)–(2-63)], an equation that defines electrical neutrality in the solution [similar to Eq. (2-13)], 4 equations that describe the total concentration of Na, Cl, Zn, and S in the fluid phase [for example, Eq. (2-59)], equations for the activity coefficients of each of the aqueous species [Eq. (2-76)], and an equation for calculating the ionic strength of the solution [Eq. (2-70)]. A mathematical solution to a set of equations like these is best obtained with high-speed digital computers, using computer programs of the sort described in the end-of-chapter references. Such computer programs provide a ready means for conducting numerical experiments to determine the distribution of species in aqueous solutions that are in heterogeneous equilibrium with ore minerals or rock-forming minerals.

In the example considered here we specify a solution pH at 5.54, which is neutral pH at 250°C (see Fig. 2-1), and a total concentration of S equal to $10^{-5}m$. The total chloride concentration is varied incrementally in a series of computer experiments from 0.001 up to $1.2m$, and Na^+ is added to the solution in each experiment to maintain electrical neutrality. Our final constraint is maintenance of equilibrium with sphalerite, which requires:

$$K_{(2-56)} = a_{Zn^{2+}}a_{S^{2-}} = m_{Zn^{2+}}\gamma_{Zn^{2+}}m_{S^{2-}}\gamma_{S^{2-}}. \tag{2-79}$$

With these input constraints the distribution of aqueous species was calculated with the computer program EQ3NR (Wolery, 1992). The calculation procedure used by this program assumes unit activity coefficients for neutral species and calculates individual ion activity coefficients using an extended version of the Debye–Hückel equation:

$$-\log \gamma_i = \frac{Az_i^2 I^{1/2}}{1 + \mathring{a}BI^{1/2}} + bI, \tag{2-80}$$

where A and B are functions of temperature, density and the dielectric constant of water, å is the effective diameter of the hydrated ion, and b is a parameter derived from empirical observations (tabulated values of A, B, and å are given in Nordstrom and Munoz, 1994, Chapter 7). The results of these computer experiments are graphically shown in Fig. 2-9 and are discussed below.

The calculated total concentration of zinc [Eq. (2-59)] in equilibrium with sphalerite increases many orders of magnitude with increasing ionic strength up to $I = 1.0$ [Fig. 2-9(a)]. This is a consequence of two factors previously discussed: the decrease in individual ion activity coefficients with increasing ionic strength, and the increasing degree of formation of Zn-chloride complexes as chloride concentration increases.

Activity coefficients of selected species in this solution are given in Fig. 2-9(b). Note that the activity coefficients for monovalent species are all similar, differing only by the effect of the term representing the effective diameter of the hydrated ion [å in Eq. (2-80)]. Coefficients for the divalent ions are also similar among themselves, but their absolute values are substantially lower than those for the monovalent species, as qualitatively predicted above.

Because $\gamma_{Zn^{2+}}$ and $\gamma_{S^{2-}}$ decrease with increasing ionic strength [Fig. 2-9(b)], the product of the molalities of Zn^{2+} and S^{2-} must increase in order to satisfy the constraints of heterogeneous equilibrium between sphalerite and the solution as defined by Eq. (2-56). For example, the molality of Zn^{2+} at an ionic strength of 0.001 is $\approx 10^{-9}$, and at an ionic strength of 0.5 about an order of magnitude larger, $\approx 10^{-8}$. In contrast to this variation for the Zn^{2+} ion, the total concentration of zinc in the solution with an ionic strength of 0.5 is $\approx 10^{-4}m$ [Fig. 2-9(a)], representing an increase of about 5 orders of magnitude relative to the dilute solutions with $I < 0.001$. This large increase in sphalerite solubility is due to formation of Zn-chloride complexes, the cumulative percentages of which are shown as a function of ionic strength in Fig. 2-9(c). In this figure we see that the Zn^{2+} ion represents a significant portion of the total zinc in solution only at very low ionic strengths. At

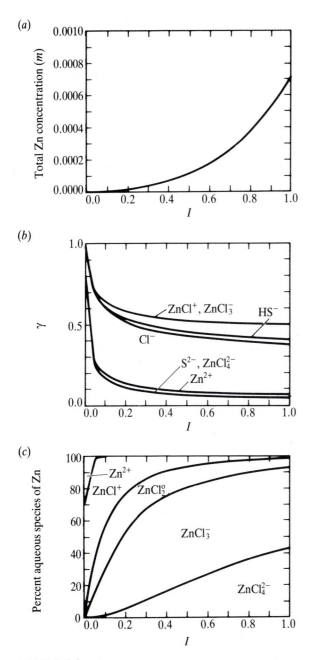

FIGURE 2-9
Calculated properties of an aqueous electrolyte solution in equilibrium with sphalerite at 250°C and 50 bar as a function of ionic strength. Ionic strength was increased by adding NaCl to a series of computer experiments where pH = 5.54 and the total molality of sulfur is 10^{-5}. (a) Total concentration of zinc [Eq. (2-59)]. (b) Individual ion activity coefficients [Eq. (2-80)]. (c) Cumulative percentages of aqueous species of zinc.

an ionic strength of 0.5, where the total zinc in solution is about $10^{-4}m$, 58.7% of the zinc is in the form of $ZnCl_3^-$, 23.5% as $ZnCl_4^{2-}$, 12.1% as $ZnCl_2(aq)$, 5.6% as $ZnCl^+$ and <0.1 as Zn^{2+}.

SUMMARY

Aqueous solutions are common throughout the Earth's surface and upper crust. The physical and chemical properties of these solutions have an important influence on weathering, diagenesis, and metamorphism, and under certain circumstances on igneous processes. Natural waters exhibit a wide range of chemical compositions that are a consequence of chemical interactions with the geologic environment. Because water at low to moderate temperatures has a high dielectric constant, much of the dissolved material in natural solutions is in the form of ions. Within the solutions a wide variety of chemical reactions can occur: important ones we have considered in this chapter include reactions of acids and bases, solute hydrolysis, and the formation of metal complexes. In a later chapter we will extend the discussion to reactions that involve oxidation and reduction.

For equilibrium at a given temperature, pressure, and bulk fluid composition, the distribution of aqueous species must satisfy equilibrium relations for all possible chemical reactions in the system. Calculation of the conditions of homogeneous equilibrium usually requires solving a series of equations that describe equilibrium for these reactions, the conditions of electrical neutrality, and the mass-balance constraints. Examples presented in this chapter illustrate the dramatic effect on the solubility of minerals caused by the formation of complexes and ion–ion interaction within the solution. An understanding of the homogeneous equilibrium constraints in an aqueous solution is essential to evaluating heterogeneous equilibrium between these solutions and rock-forming minerals. In the next chapter we consider further examples of these heterogeneous reactions.

PROBLEMS

1. List the substances (ions and/or molecules) present in a solution made by stirring sodium sulfide (Na_2S) in water to give a concentration of $0.01m$. Which substances are present in large concentrations and which in very minute concentrations? How does the concentration of each substance change if HCl is added to the solution? Is the solution alkaline to start with? Does it remain alkaline as acid is added?

2. Nepheline basalt is a rock containing the minerals nepheline (approximately $NaAlSiO_4$), plagioclase ($NaAlSi_3O_8 + CaAl_2Si_2O_8$), and pyroxene (approx. $CaMgSi_2O_6$). In what sense is this rock alkaline? In what sense is it basic?

3. Analysis of a groundwater sample shows 44.9 g/liter of Na^+, 6.6 g/liter of Ca^{2+}, 81.9 g/liter of Cl^-, and 1.0 g/liter of SO_4^{2-}. No other ions are present in appreciable amounts except H^+ or OH^-. Is the solution acid or alkaline, in the chemical sense? Approximately what is its pH?

4. Calculate the H^+ concentration and the pH of a solution of $0.0001m$ H_2CO_3, and calculate also what fraction of H_2CO_3 has dissociated. Compare this fraction with the

fraction dissociated when the concentration is $0.01m$ (Sec. 2-1). What generalization can you make about the effect of dilution on degree of dissociation?

5. The solubility of amorphous silica in pure water is about 120 ppm of SiO_2 at 25°C and 1 bar. The solution contains silicic acid, H_4SiO_4, whose first dissociation constant is $10^{-9.9}$. What is the pH of a solution saturated with amorphous silica?

6. What is the activity coefficient of Mg^{2+} in a solution at 25°C with $m_{MgCl_2} = 0.01$ and $m_{NaCl} = 0.05$? Find $\gamma_{Mg^{2+}}$ from both Eqs. (2-75) and (2-76).

7. Find the pH of a solution of K_2CO_3 containing 20 g of the salt per kilogram of water.

8. If a solution having a concentration of $0.001m$ each of Al^{3+}, Cu^{2+}, Fe^{2+}, Fe^{3+}, Mg^{2+}, and Zn^{2+} is gradually made alkaline, in what order would you expect the hydroxides to precipitate? (Use data in Appendix VII.)

9. If a solution containing Ca^{2+}, Mg^{2+}, Mn^{2+}, Ba^{2+}, and Zn^{2+}, each in $0.001m$ concentration, comes in contact with a solution containing CO_3^{2-}, in what order will precipitates form? Answer the same question for a solution containing SO_4^{2-}.

10. In the normal pH range of surface waters (4 to 9), what would be the dominant dissolved species (ions or molecules or both) containing boron? Silicon? Fluorine? Divalent selenium? Quinquevalent phosphorus?

11. In what part of the normal pH range would the dominant carbon-containing species be undissociated H_2CO_3? In what part would it be HCO_3^-? In what sort of geologic environment would CO_3^{2-} be dominant?

12. A solution contains $0.1m$ total carbonate (H_2CO_3 + HCO_3^- + CO_3^{2-}). If the pH is 6, find the concentration of each carbonate species present. Compare your answers with Fig. 2-3. If the solution contains also $10^{-4}m$ Ca^{2+}, is it saturated, unsaturated, or supersaturated with $CaCO_3$?

13. Is the solution in Prob. 12 a buffer? If 1 ml of $6m$ HCl is added to a liter of this solution, what is the principal reaction that takes place? What is the final pH?

14. In Fig. 2-4(a), what determines the location of the point where the lines for H_2S and HS^- cross? The lines for HS^- and S^{2-}? Can you frame a generalization for the form of such a diagram for any dihydrogen acid, H_2A? From Fig. 2-4(a), read the concentration of S^{2-} (a) in a solution with $m_{S,total} = 10^{-3}m$ and a pH of 4.5 (b) in a solution with $m_{S,total} = 10^{-5}m$ and a pH of 7.0.

15. Suppose that a solution with pH 11 stands in contact with air. Over a period of a day or so, before evaporation has reduced its volume appreciably, would you expect its pH to remain constant? Answer the same question for a solution with pH 4.

REFERENCES AND SUGGESTIONS FOR FURTHER READING

Barnes, I.: *Field measurement of alkalinity and pH,* U.S. Geological Survey Water-Supply Paper 1535-H, 1964. Description of the precautions needed to obtain accurate measurements of pH and carbonate-ion concentration.

Drever, J. I.: *The Geochemistry of Natural Waters,* Prentice-Hall, Englewood Cliffs, New Jersey, 1982. Chapter 2 has a clear and concise treatment of activity–concentration relationships, and Chapter 3 gives details of the carbonic acid and carbonate systems.

Faure, G.: *Principles and Applications of Inorganic Geochemistry,* Macmillan, New York, 1991. Chapters 11 and 12 discuss in simple terms the dissociation of acids and bases, hydrolysis, activities and concentrations, and the solubility of calcium carbonate.

Langmuir, D.: "The geochemistry of some carbonate groundwaters in central Pennsylvania," *Geochim. et Cosmochim. Acta,* vol. 35, pp. 1023–1045, 1971. Details of the composition of spring and well waters issuing from limestones and dolomites.

Nordstrom, D. K., and J. L. Munoz: *Geochemical Thermodynamics,* 2d ed., Blackwell Scientific Publications, 1994. Chapter 13 describes many of the more popular and more sophisticated computer programs used in geochemical modelling.

Stumm, W., and J. J. Morgan: *Aquatic Chemistry,* 2d ed., Wiley– Interscience, New York, 1981. Chapters 3 and 4 give an advanced treatment of acid–base reactions and the carbonic acid system.

Wolery, T. J.: EQ3NR, a computer program for geochemical aqueous specification-solubility calculations, version 7.0. UCRL-MA-110662 PT III. Lawrence Livermore Laboratory, Livermore, California, 1992.

CHAPTER 3

SOLUTION—MINERAL EQUILIBRIA
PART 1: CARBONATES

Carbonic acid and the carbonate minerals provide another good illustration of the use of equilibrium reasoning in geochemistry. Interactions among these compounds determine the conditions under which limestones and dolomites are formed or dissolved, and likewise the conditions of formation of carbonate minerals as cements in soils and sandstones and as vein fillings. We start with some qualitative remarks about the most common of these substances, the carbonate of calcium, then go on to a more quantitative treatment and to the reactions of other common carbonate minerals.

3-1 SOLUBILITY OF CALCITE

Calcium carbonate occurs in nature as the two common minerals calcite and aragonite (Sec. 3-3). A third crystal form (vaterite) can be prepared artificially, and is known as a very rare mineral in nature. Under usual conditions near the Earth's surface calcite is the most stable and most abundant of the three forms, and for the discussion in this section and the next it will be the center of attention. Aragonite and vaterite may be assumed to show similar chemical behavior, but in general to react more rapidly and to have greater solubility.

A strong acid dissolves calcite by the familiar reaction

$$CaCO_3 + 2H^+ \rightarrow Ca^{2+} + H_2O + CO_2. \tag{3-1}$$
calcite

At low acid concentrations a more accurate equation would be

$$CaCO_3 + H^+ \rightarrow Ca^{2+} + HCO_3^- \tag{3-2}$$
calcite

showing that H^+ takes CO_3^{2-} away from Ca^{2+} to form the very weak (little dissociated) acid HCO_3^-. (Still greater accuracy would require consideration of the complex ion $CaHCO_3^+$, but under usual conditions its concentration is small and for the present can be neglected.) These reactions would take place in nature, for example, where acid solutions from the weathering of pyrite encounter limestone. The reactions can be reversed by any process that uses up H^+. For example, if a base is added,

$$Ca^{2+} + HCO_3^- + OH^- \rightarrow CaCO_3 + H_2O. \tag{3-3}$$
calcite

Quite evidently, the solubility of $CaCO_3$ is determined in large part by the pH of its environment. By using equilibrium constants for the above reactions we could express this dependence quantitatively, but we look first at some qualitative relationships.

Under natural conditions the dissolving of calcium carbonate is somewhat more complicated, because the acids involved are usually weak rather than strong. When limestone dissolves in carbonic acid, for example, the overall process may be summarized by the reaction

$$CaCO_3 + H_2CO_3 \rightarrow Ca^{2+} + 2HCO_3^-. \tag{3-4}$$
calcite

Note that the two HCO_3^- ions come from different sources: one is simply left over from the dissociation of H_2CO_3, and the other is formed by the reaction of H^+ from the acid with $CaCO_3$ [as shown by Eq. (3-2)]. Eq. (3-4) is the essential reaction for an understanding of carbonate behavior in nature. The forward reaction shows what happens when limestone weathers, when limestone is dissolved to form caves, or when marble is dissolved by ore-bearing solutions in the walls of a fissure. The reverse of Eq. (3-4) represents the precipitation of calcium carbonate in the sea, as a cementing material in sedimentary rocks, or where droplets evaporate at the tip of a stalactite.

The effect of pH on solubility is shown as well by Eq. (3-4) as by the simpler equations preceding it. At low pH, where most dissolved carbonate exists as H_2CO_3 [Fig. 2-3(a)], the forward reaction is favored; at high pH, the reverse reaction leads to precipitation. Eq. (3-4) shows also that the solubility depends on the partial pressure of CO_2 above the solution, since this pressure helps to determine the concentration of dissolved H_2CO_3 by the reaction,

$$H_2O(l) + CO_2(g) \rightleftharpoons H_2CO_3(aq). \tag{3-5}$$

Any process that increases the amount of CO_2 available to the solution makes more $CaCO_3$ dissolve; anything that decreases the amount of CO_2 causes $CaCO_3$ to precipitate.

Reactions (3-1)–(3-4) have been written as one-way processes ($\rightarrow$) resulting in the dissolving or precipitation of $CaCO_3$. They could, of course, equally well represent heterogeneous equilibria ($\rightleftharpoons$), where calcite is in contact with a solution and with air, and where forward and reverse reactions are going at the same rate. The principal ions present in such a solution are shown in Fig. 3-1, and some of the ways in which equilibria among them respond to changes in conditions are described in the following paragraphs.

Temperature. The solubility of calcite and other crystal forms of $CaCO_3$ in pure water decreases somewhat as the temperature rises (Fig. 3-2). This is opposite to the behavior of most rock-forming minerals. An increase in temperature generally results in higher solubilities, but a number of carbonates and sulfates are exceptions. In addition to this effect, the solubility of $CaCO_3$ in natural waters decreases at higher temperatures because CO_2, like any other gas, is less soluble in hot water than in cold water. Generally the solubility of carbonates is influenced much more by this change in solubility of CO_2 than by the temperature dependence of the solubility itself. As an illustration of the effect of temperature, calcite dissolves at great depths in the ocean, where the water is perennially cold, but precipitates near the surface, especially in the tropics, where the water is warm.

Pressure. The effect of pressure by itself, independent of its effect on CO_2, is to increase the solubility of calcite slightly. Where pressure is very large, its effect can be substantial; in the deep parts of the ocean, for example, pressure alone increases the solubility to about twice its surface value. The main reason for an influence of pressure in near-surface environments, however, is the amount of dissolved CO_2 when the pressure of the gas changes in the surrounding atmosphere. Theoretically even day-to-day barometric changes should have a detectable effect on solubility,

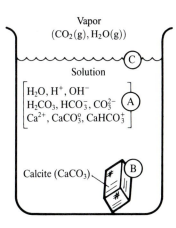

FIGURE 3-1
Conditions for heterogeneous equilibrium between calcite and a solution of carbonic acid, and between solution and carbon dioxide in the gas. The principal ions present in the solution are indicated. Examples of heterogeneous equilibria would be:

$$CaCO_3 \text{(calcite)} \rightleftharpoons Ca^{2+} + CO_3^{2-},$$
$$CaCO_3 \text{(calcite)} + H^+ \rightleftharpoons Ca^{2+} + HCO_3^-,$$
$$H_2O \text{(l)} + CO_2 \text{(g)} \rightleftharpoons H_2CO_3 \text{(aq)},$$

and examples of homogeneous equilibria would be:

$$CaCO_3 \text{(aq)} \rightleftharpoons Ca^{2+} + CO_3^{2-},$$
$$CaHCO_3^+ \rightleftharpoons Ca^{2+} + HCO_3^-.$$

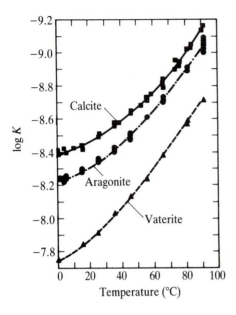

FIGURE 3-2

Change with temperature of the solubility products of three polymorphs of $CaCO_3$. The symbols represent experimental data. (Reprinted by permission from Plummer and Busenberg, 1982.)

and the local production of CO_2 in abnormal amounts, say by a forest fire, an industrial plant, or a volcanic eruption, could cause a marked increase temporarily. But circulation of the atmosphere is so effective in keeping the partial pressure of CO_2 uniform that this factor is probably less important than others.

Organic Activity. Many organisms use calcium carbonate in the construction of their shells. Just how they accomplish this is not certain, but they flourish in greatest numbers in water approximately saturated with calcite, where only a minor change in pH is needed to cause precipitation. Green plants may cause precipitation of calcium carbonate indirectly, by removing CO_2 from water in the process of photosynthesis. Abundant green algae in the warm waters of the Bahama Banks, for example, aid in the precipitation of the limy mud and sand with which the banks are covered.

Decay. Decay of organic matter in the presence of air or aerated water produces CO_2 in large amounts and hence makes calcium carbonate in the vicinity more soluble. If access of air is restricted or cut off entirely, the processes of decay are more complicated and the effect on solubility of carbonate minerals is not predictable. Any CO_2 or H_2S produced would make the water acid and lead to increased solubility, but ammonia is a common product of anoxic decay which would have the opposite effect. Measurements of pH in stagnant waters suggest that anaerobic decay most often causes increased acidity.

Carbonic acid, as these many examples show, is important in controlling the solubility of carbonate minerals in most natural environments, but carbonates are somewhat soluble even in water containing no CO_2:

$$CaCO_3 + H_2O \rightleftharpoons Ca^{2+} + HCO_3^- + OH^-. \tag{3-6}$$

calcite

This is a hydrolysis reaction, possible because HCO_3^- is a weak acid (meaning that the H-CO$_3$ bond is strong). Even the small amount of CO_3^{2-} produced by the dissolving calcite can take a little H^+ away from the OH^- of water. The reaction, of course, cannot go far in the forward direction, not nearly so far as the corresponding reaction for the CO_3^{2-} ion from a soluble carbonate [Eq. (2-24)], because here the reverse reaction is aided by the insolubility of calcite. But the reaction does go far enough to make water in contact with carbonates appreciably basic. In nature most water solutions are exposed to CO_2 from the atmosphere, and this complicates the hydrolysis. Experimentally it is found that water containing suspended calcite and left exposed to the air acquires a pH of approximately 8.

In summary, the solubility of calcite (and other carbonates) in nature is controlled by fairly simple equilibria involving H_2CO_3, HCO_3^-, CO_3^{2-}, and water. The principal equilibrium [Eq. (3-4)] is very sensitive to changes in the amount of dissolved CO_2, and this is dependent on a variety of influences. Much surface water and groundwater is approximately saturated with calcite, and such water can either dissolve or precipitate the carbonate, depending on slight alterations in external conditions. Hydrolysis of even fairly insoluble carbonates is sufficient to make solutions in contact with them slightly basic.

3-2 CALCIUM CARBONATE: SOLUBILITY CALCULATIONS

We set out now to apply the quantitative reasoning of the last chapter to these reactions involving calcite. We will try, for example, to answer such questions as: How much calcium carbonate can be dissolved in a solution of known composition and pH? What is the equilibrium pH of water standing in contact with limestone? If a base is added slowly to a solution containing calcium ion and carbonic acid, at what pH will calcium carbonate begin to precipitate?

For the solubility of calcite ($CaCO_3 \rightleftharpoons Ca^{2+} + CO_3^{2-}$) at 25°C we find from Appendix VII that the equilibrium constant is

$$K = \frac{a_{Ca^{2+}} a_{CO_3^{2-}}}{a_{CaCO_3}^{calcite}} = 10^{-8.35} = 4.5 \times 10^{-9}. \tag{3-7}$$

If the calcite is pure and the solution dilute the equilibrium constant is approximately equal to the product of $m_{Ca^{2+}}$ times $m_{CO_3^{2-}}$. (Here and elsewhere we will state equations in terms of the formally correct activities, but for applications to dilute solutions we will often substitute as approximations the more familiar molal concentrations. Whenever greater exactness is needed, or when more concentrated solutions are being considered, activities will of course be required.) The solubility in pure water, according to the methods of Chapter 1, should then be the square root of this number, or about $6.8 \times 10^{-5} m$. Because of the hydrolysis of CO_3^{2-}, however, the process of dissolution is somewhat more complicated.

First we consider the calcite hydrolysis reaction, Eq. (3-6). The equilibrium constant can be evaluated by combining other constants as follows:

$$K_{(3-6)} = a_{Ca^{2+}} a_{OH^-} a_{HCO_3^-}. \tag{3-8}$$

Multiply the right-hand side by the ratio $(a_{H^+} a_{CO_3^{2-}})/(a_{H^+} a_{CO_3^{2-}})$ and rearrange to obtain

$$K_{(3-6)} = a_{Ca^{2+}} a_{CO_3^{2-}} \times a_{OH^-} a_{H^+} \times \frac{a_{HCO_3^-}}{a_{H^+} a_{CO_3^{2-}}},$$

which expresses $K_{(3-6)}$ as the product of equilibrium constants we have seen in Eqs. (2-7), (2-9), and (3-7),

$$K_{(3-6)} = K_{(3-7)} \times K_{(2-7)} \times \frac{1}{K_{(2-9)}}.$$

Substituting numerical values for these constants gives

$$K_{(3-6)} = 10^{-8.3} \times 10^{-14} \times \frac{1}{10^{-10.3}} = 10^{-12}.$$

In pure water, where the activity coefficients are approximately unity, the three ions Ca^{2+}, OH^-, and HCO_3^- should be formed in equal amounts [Eq. (3-6)], so that

$$m_{Ca^{2+}} = m_{OH^-} = m_{HCO_3^-} \tag{3-9}$$

and from Eq. (3-8) we solve for the value of $m_{Ca^{2+}}$,

$$m_{Ca^{2+}} = \sqrt[3]{10^{-12}} = 10^{-4} \text{ mol/kg } H_2O.$$

Hence calcite should dissolve in CO_2-free water to a concentration of $0.0001 m$ or about 0.01 g/kg H_2O, and the solution should acquire a pH of 10 (because $m_{OH^-} = 10^{-4}$).

This is not quite the end of the story, because at such a high pH the acid HCO_3^- is appreciably dissociated. From Fig. 2-3(a), in fact, we can read that the ratio $m_{CO_3^{2-}}/m_{HCO_3^-}$ should be about 0.5. This means that our assumption about the equality of the three ions is not valid and that more $CaCO_3$ must dissolve to maintain the ion product at 10^{-12}. Using the rough figure 0.5 for the carbonate–bicarbonate ratio, we guess that the concentrations of HCO_3^- and OH^- would be about two-thirds of the concentration of Ca^{2+}, instead of equal to it, so that the last equation should be

$$m_{Ca^{2+}} = \tfrac{3}{2} m_{OH^-} = \tfrac{3}{2} m_{HCO_3^-}, \tag{3-10}$$

and from Eq. (3-8)

$$m_{Ca^{2+}} = \sqrt[3]{\frac{9}{4}} \times 10^{-12} = 1.3 \times 10^{-4} m.$$

Thus the solubility is raised to a figure about twice as large as the $6.8 \times 10^{-5}m$ obtained by simply taking the square root of the solubility product [Eq. (3-7)]. At the same time the OH^- concentration is lowered slightly, so that the pH is approximately 9.9 instead of 10.0.

This problem of calculating concentrations when calcite is in equilibrium with pure water may be set up more formally as follows. We note that the reactions contributing to the equilibrium involve a total of six concentrations: m_{H^+}, m_{OH^-}, $m_{Ca^{2+}}$, $m_{CO_3^{2-}}$, $m_{HCO_3^-}$, and $m_{H_2CO_3}$. To solve a problem with six unknowns requires that we have six equations. Three of these equations are supplied by the equilibrium constants for the dissociation of H_2O, H_2CO_3, and HCO_3^- [Eqs. (2-7), (2-10), and (2-11)], as required by homogeneous equilibrium in the solution, and a fourth equation comes from the reaction describing heterogeneous equilibrium of this solution with calcite [Eq. (3-4)]. Another equation expresses the fact that all carbonate in solution is supplied by dissolution of calcite, so that the sum of all carbonate species must equal the concentration of calcium:

$$m_{Ca^{2+}} = m_{CO_3^{2-}} + m_{HCO_3^-} + m_{H_2CO_3}, \qquad (3\text{-}11)$$

The sixth equation comes from the requirement that the solution remains electrically neutral: the total concentration of positively charged ions must equal the total for negatively charged ions, with doubly charged ions counted twice:

$$2m_{Ca^{2+}} + m_{H^+} = 2m_{CO_3^{2-}} + m_{HCO_3^-} + m_{OH^-}. \qquad (3\text{-}12)$$

The array of six equations can then be solved in various ways, most easily by using a computer program. An advantage of the formal approach is that results of greater accuracy can be obtained, but a simpler calculation like that in the last paragraph often provides all the accuracy that is justified by the quality of the original data.

The corresponding calculation for water containing CO_2 is based on Eq. (3-4), for which the equilibrium constant is [again following the procedure for Eq. (3-8) and assuming activity coefficients of unity],

$$K_{(3\text{-}4)} = \frac{m_{Ca^{2+}} m_{HCO_3^-}^2}{m_{H_2CO_3}}, \qquad (3\text{-}13)$$

$$K_{(3\text{-}4)} = m_{Ca^{2+}} m_{CO_3^{2-}} \times \frac{m_{HCO_3^-} m_{H^+}}{m_{H_2CO_3}} \times \frac{m_{HCO_3^-}}{m_{CO_3^{2-}} m_{H^+}},$$

$$K_{(3\text{-}4)} = K_{(3\text{-}7)} K_{(2\text{-}8)} \times \frac{1}{K_{(2\text{-}7)}},$$

and

$$K_{(3\text{-}4)} = 10^{-8.3} \times 10^{-6.4} \times 10^{-10.3} = 10^{-4.4}.$$

In the reaction of Eq. (3-4), 2 mols of HCO_3^- are produced for every mol of Ca^{2+}. Hence, if there is no other source of these ions,

$$m_{HCO_3^-} = 2m_{Ca^{2+}}. \qquad (3\text{-}14)$$

For equilibrium with CO_2 in the atmosphere, the concentration of dissolved CO_2 or H_2CO_3 is approximately $10^{-5}m$ (Sec. 1-4). We can then substitute in the expression for the equilibrium constant:

$$\frac{m_{Ca^{2+}} m_{HCO_3^-}^2}{m_{H_2CO_3}} = \frac{4m_{Ca^{2+}}^3}{10^{-5}} = 10^{-4.4}, \tag{3-15}$$

whence

$$m_{Ca^{2+}} = 10^{-3.3} = 5.0 \times 10^{-4}, \tag{3-16}$$

which is almost four times the concentration in pure water. To find the pH of the solution, we substitute values for $m_{HCO_3^-}$ [Eqs. (3-14) and (3-16)] and $m_{H_2CO_3}$ in the expression for the dissociation constant of carbonic acid [Eq. (2-10)]:

$$\frac{m_{H^+} m_{HCO_3^-}}{m_{H_2CO_3}} = \frac{m_{H^+} \times 10^{-3.0}}{10^{-5}} = 10^{-6.4}, \tag{3-17}$$

whence $m_{H^+} = 10^{-8.4}$, and pH $= 8.4$.

This calculation, it should be noted, is only approximate because $m_{CO_3^{2-}}$ is assumed to be negligibly small. If the concentration of this ion were appreciable (in other words, if the simple dissolving of calcite to form Ca^{2+} and CO_3^{2-} is significant in comparison with the reaction of H_2CO_3), then it would no longer be true that $m_{Ca^{2+}} = \frac{1}{2} m_{HCO_3^-}$, and a more complicated procedure would be necessary. The magnitude of $m_{CO_3^{2-}}$ can be estimated from the relation [derived from Eq. (2-11)]

$$m_{CO_3^{2-}} = 10^{-10.3} \frac{m_{HCO_3^-}}{m_{H^+}} = 10^{-10.3} \frac{10^{-3.0}}{10^{-8.4}} = 10^{-4.9}m. \tag{3-18}$$

This is only about 1/100 of the concentration of HCO_3^-, so for all ordinary purposes the assumption is justified. Alternatively, relations among the variables pH, $m_{Ca^{2+}}$, and P_{CO_2} can be usefully derived from the equilibrium for Eq. (3-1), for which the expression for the equilibrium constant can be written in logarithmic form as

$$\log \frac{a_{Ca^{2+}}}{a_{H^+}^2} = -\log P_{CO_2} + \log K_{(3-1)} \tag{3-19}$$

(on the assumption that a_{H_2O} is 1). This is an equation for the straight line shown in Fig. 3-3(a) at 25°C and 1 bar total pressure where the value of $K_{(3-1)}$ is $10^{9.6}$. Then if we fix the partial pressure of CO_2, for example $P_{CO_2} = 0.1$ bar, the ratio of activities in Eq. (3-19) will be a constant equal to

$$\log \frac{a_{Ca^{2+}}}{a_{H^+}^2} = -(-1.0) + 9.6 = 10.6. \tag{3-20}$$

Rearranging gives

$$\log \frac{a_{Ca^{2+}}}{a_{H^+}^2} = \log a_{Ca^{2+}} + 2pH = 10.6$$

or

$$\log a_{Ca^{2+}} = -2pH + 10.6, \tag{3-21}$$

which defines an inverse linear relationship between the activity of Ca^{2+} and pH in a solution at equilibrium with calcite at a fixed partial pressure of CO_2 [Fig. 3-3(b)]. Results of similar calculations for pH and concentration of Ca^{2+} as functions of the partial pressure of CO_2 in Figs. 3-3(c) and (d) show that the solubility of calcite increases and pH decreases with increasing P_{CO_2}.

In dilute surface waters at temperatures near 25°C, then, the solubility of calcite ranges from about $1.3 \times 10^{-4}m$ [Eqs. (3-8) and (3-10)] to $5.0 \times 10^{-4}m$ [Eq. (3-16)], or 0.01 to 0.05 g/kg, depending on the degree of saturation with CO_2. The higher figure may be exceeded in colder water because CO_2 becomes more soluble; it will be exceeded also in places where CO_2 is unusually abundant or where some other source of acid keeps the pH low. In soils, for example, the decomposition of organic matter gives local concentrations of CO_2 often on the order of 0.1 bar, and sometimes as high as 1 bar. The pH's of solutions that have

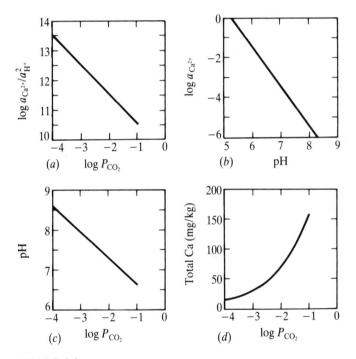

FIGURE 3-3
Relations among the variables pH, $a_{Ca^{2+}}$, and P_{CO_2} for heterogeneous equilibrium, at 25°C and 1 bar total pressure, of calcite in contact with solution [diagram (a), Eq. (3-19)], and in a particular solution containing carbon dioxide at a partial pressure of 0.1 bar [diagram (b), Eq. (3-21)]. The lines in the figures show conditions for calcite saturation. Further constraints may be placed on solution compositions by considering the total concentrations of solute components, the ionic strength, individual ion activity coefficients, and charge balance. Diagrams (c) and (d) show the computed increase in calcite solubility and the decrease in pH with increasing P_{CO_2}, using these variables with the equations and data in the computer program EQ3NR (see Sec. 2-8 and the reference to Nordstrom and Munoz on p. 56).

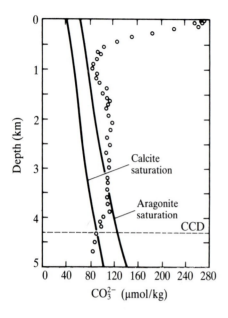

FIGURE 3-4

Measured carbonate-ion concentrations plotted against water depth for a station in the western South Atlantic Ocean. The lines show concentrations of carbonate ions in seawater at saturation with calcite and aragonite. The increase in concentration with depth is largely due to the change of solubility with pressure. The dashed line marked CCD is the calcite compensation depth; below this depth seawater is undersaturated with respect to calcite and above this depth seawater is usually supersaturated with respect to calcite. (Reprinted by permission from Broecker and Peng, 1982.)

come to equilibrium with calcite should be in the range 8 to 10—close to the lower figure at the Earth's surface, and close to the upper figure at a depth not influenced by the atmosphere.

These numbers give us a quantitative expression of the variation in solubility of $CaCO_3$ that we predicted from qualitative arguments at the beginning of this chapter. Our next move obviously should be to compare the theoretically derived numbers with actual measured concentrations of $CaCO_3$ in natural solutions. The comparison is easily made, but it turns out to be most disillusioning. Concentrations of $CaCO_3$ in natural waters are extremely variable and only rarely come close to the numbers predicted in the last few paragraphs. Low concentrations can be plausibly explained as the result of failure of solutions to reach equilibrium with solid carbonate. In many natural waters, such as seawater (Fig. 3-4), however, the discrepancy is in the opposite direction: concentrations are embarrassingly high, much higher than can be accounted for even with generous assumptions about temperature, CO_2 pressure, and acidity.

Possible reactions for these higher concentrations are discussed in the following six sections. They are worth examining in some detail, because they illustrate beautifully the difficulties in trying to make theoretical predictions adaptable to complex natural environments.

3-3 CALCITE AND ARAGONITE

One complication affecting solubility is the fact mentioned above that $CaCO_3$ is polymorphous. Measured solubilities of the three crystal forms are different, as is shown by the solubility products in Fig. 3-2. A similar difference in solubility was noted earlier (Sec. 1-5) for the two common forms of calcium sulfate, gypsum and anhydrite.

The existence of different solubilities for the same chemical compound suggests an awkward question. Suppose that equilibrium has been established between solid aragonite and its saturated solution. The product of the concentrations of the two ions is larger than the equilibrium product for calcite. Why, then, does calcite not precipitate? Theoretically, it would seem that the excess Ca^{2+} and CO_3^{2-} should combine to form calcite; then more aragonite should dissolve to reestablish its own equilibrium, more calcite should precipitate, and so on—the net result being the slow replacement of aragonite by calcite.

Practically, this does not happen, or at least it does not happen rapidly enough to be observed. A plausible explanation is that nuclei are not present for calcite to precipitate around. The solution is supersaturated with respect to calcite, and like any supersaturated solution it is essentially unstable; but when supersaturation is very slight, as in this case, crystallization is extremely slow unless seed crystals are present. If finely ground calcite is added to the equilibrium solution, and if the mixture is warmed to speed up reactions, the change becomes appreciable.

Aragonite is therefore unstable with respect to calcite under ordinary conditions. It becomes the stable form of calcium carbonate at high pressures, as might be anticipated from its greater density (2.9 g/cm^3, in contrast to 2.7 for calcite). The necessary pressure is so high, however, that it cannot be a factor in environments of sedimentation near the Earth's surface. This leads to another set of awkward questions: Why does aragonite ever appear in sedimentary rocks? When conditions are right for calcium carbonate to precipitate, why doesn't the more stable calcite form in all cases?

There is no entirely satisfactory answer to this question. It is an empirical fact that many polymorphous substances show this same tendency to precipitate first in metastable forms, which change only slowly to the stable varieties. We can guess that the manner of crystallization depends at least in part on reaction rates, the metastable forms being able to crystallize more quickly. In the case of calcium carbonate, experiments show that calcite and aragonite often precipitate together, the proportion of aragonite being greater if the reaction is carried out at high temperatures. The presence of certain other ions in solution, notably Mg^{2+} and Sr^{2+}, also aids the formation of aragonite. Living creatures that use calcium carbonate in their shells may precipitate either polymorph, some species favoring one and some the other; many pelecypods precipitate both in alternate layers. In whatever way aragonite may form, during the course of geologic time it gradually alters to calcite, so that in old rocks and shells aragonite is a rare mineral.

3-4 SUPERSATURATION

The fact that calcite does not precipitate immediately from solutions at equilibrium with aragonite is proof that calcium carbonate can exist for extended periods in supersaturated solution. Why do we not use this simple fact to explain all the anomalously high concentrations of calcium carbonate in natural solutions?

The question is a tricky one to handle, because the phenomenon of supersaturation is imperfectly understood. Some compounds, it is well-known, can remain in solution almost indefinitely at concentrations far above saturation equilibrium; others precipitate whenever the saturation value is exceeded by a very small amount; and there is little theoretical basis for predicting which way a given compound will behave. Furthermore, the onset of precipitation is influenced by a variety of factors that are difficult to control or predict: mechanical disturbances, dust particles in the solution, the material of the container, unevenness of surfaces in contact with the solution. Probably in any precipitation process at least momentary supersaturation exists before precipitation begins, but the persistence of supersaturation depends unpredictably on these factors as well as on the nature of the solute and on its concentration. In any given case predictions are possible only on the basis of empirical evidence.

For $CaCO_3$, laboratory experiments show that this compound ordinarily precipitates rapidly, without any lag ascribable to supersaturation. If the experiment is done carefully, so as to minimize external disturbance and to build up the concentrations of Ca^{2+} and CO_3^{2-} slowly, supersaturation up to about twice the normal solubility can be demonstrated and can be maintained for times that are long by laboratory standards. In natural waters, especially where CO_2 is being slowly removed by aquatic plants, apparent temporary supersaturation up to five times the solubility has been recorded. But such cases are rare, and in general the high concentrations of $CaCO_3$ sometimes found in nature can be more plausibly explained in other ways.

The suggestion that supersaturation might account for abnormal carbonate concentrations in nature is an example of a kind of explanation that often looks superficially attractive. It depends on a nonequilibrium phenomenon—on the hypothesis that carbonate solubilities in natural environments are determined by processes to which the laboratory-derived rules of chemical equilibrium do not apply. In effect, it appeals to a supposed slow rate of change from a metastable to a stable mixture. Now there is nothing basically wrong with this kind of assumption; certainly slow rates and nonequilibrium mixtures are commonplace in the laboratory as well as in nature. Here, however, the hypothesis of large and widespread deviation from equilibrium in carbonate solutions is not supported by empirical data, so that it becomes a pure assumption. The assumption cannot really be proved wrong, because quantitative information on reaction rates is still meager, but it gives no insight into the geochemistry of carbonates, since no specific prediction can be made from it. This is the kind of hypothesis that at first glance seems promising, but on closer examination turns out to be so vague and general that it explains nothing. If equilibrium does not exist, or is not closely approached, then the field is wide open for almost any kind of speculation. In some cases this supposition is necessary, but ordinarily it is good policy to turn to nonequilibrium hypotheses only as a last resort, after explanations based on equilibrium have been tried and found wanting.

3-5 EFFECT OF GRAIN SIZE ON SOLUBILITY

Another factor that modifies the solubility relations of $CaCO_3$ is the size of crystals exposed to a saturated solution. Experimentally, very tiny grains show a greater solubility (and, of course, a greater solubility product) than do large crystals. The mechanism of dissolving furnishes a ready explanation: ions escape from a crystal most easily on exposed corners and edges, which are more numerous on small particles than on large ones. The effect is noticeable only when particle dimensions are extremely small, say of the order of a few millimicrons. Because of the greater solubility of small particles, ions in equilibrium with them constitute a super-saturated solution with respect to large particles. This means that the small ones should eventually dissolve, and the large ones should grow at their expense. A precipitate on first forming necessarily consists of very tiny grains; on standing the grains grow in size, at first rapidly and then more slowly, so that with most substances the crystals will have reached stable dimensions within a few seconds or minutes. Thus the measured solubility is the equilibrium solubility of larger crystals, and ordinarily the higher solubility of tiny particles is only a temporary phenomenon.

The growth of large grains at the expense of small ones in contact with a saturated solution is a familiar process both in geologic environments and in chemical laboratories. The chemist, faced with the problem of filtering a very fine-grained precipitate like $BaSO_4$ or $Fe(OH)_3$, often "digests" the precipitate by holding it near the boiling point for a few minutes. At the high temperatures the growth of particles is accelerated, and the crystals soon become large enough for a filter paper to hold them. A geologist sees in this same phenomenon an explanation for the recrystallization and increase in grain size of many rocks formed originally as fine chemical precipitates. The coarse grains of some limestones and the development of chalcedony or quartz from opal are common examples.

3-6 EFFECT OF ION ASSOCIATION ON SOLUBILITY

Another factor that might conceivably enhance the solubility of $CaCO_3$ is the possible association of its ions to form complexes. Thus far we have assumed that Ca^{2+} and CO_3^{2-} have no tendency to associate in solution, except to form a solid precipitate when the product of their concentrations exceeds the solubility product. This is nearly correct, but not quite. The two ions join to a slight extent into uncharged dissolved $CaCO_3(aq)$ molecules:

$$Ca^{2+} + CO_3^{2-} \rightleftharpoons CaCO_3(aq), \tag{3-22}$$

where

$$K = \frac{a_{CaCO_3(aq)}}{a_{Ca^{2+}} a_{CO_3^{2-}}} = 10^{3.2}. \tag{3-23}$$

As long as $a_{CO_3^{2-}}$ remains less than $10^{-3.2}$ (approximately $0.0006m$), most of the dissolved calcium remains as the free ion Ca^{2+}. Solutions with $a_{CO_3^{2-}}$ greater than this can exist in nature only in unusual, highly alkaline environments, so that ordinarily $CaCO_3(aq)$ plays a minor role. In most situations of geologic interest, the solubility of solid $CaCO_3$ would be increased no more than a percent or so by this factor.

More important than $CaCO_3(aq)$ in enhancing the solubility of $CaCO_3$ are other complexes that Ca^{2+} can form with ions that may be abundant in natural solutions, for example $CaHCO_3^+$, $CaCl^+$, $CaOH^+$, and $CaSO_4(aq)$. Of these the most stable is the neutral sulfate complex $CaSO_4(aq)$, which in seawater can increase the solubility of $CaCO_3$ minerals by about 10%. The others, except in solutions with unusually high concentrations of the complexing ion, have much smaller effects.

Neutral complexes like $CaCO_3(aq)$ and $CaSO_4(aq)$ are common in the chemistry of didivalent salts. The K's for the association reactions, so far as they are known, lie in the range 10^2 to 10^4. This means that the complexes have a significant effect on solubility only if the anion concentration is high, and in the dilute solutions found in most geologic situations the possibility of association can be safely neglected. Some of the charged complexes, however, particularly those formed by heavy metals like Cu, Pb, U, or Au, can influence solubility profoundly, as we have noted in Sec. 2-5 and will see again later when we consider the transport of these metals in ore-forming solutions (Chap. 19).

3-7 EFFECT OF ORGANISMS ON SOLUBILITY

Organic activity, as noted briefly in Sec. 3-1, affects solubility in various ways. Some inorganic compounds are used for the construction of shells and skeletons; the compound that is now the center of attention, calcium carbonate, is an excellent example. Some organisms in nearly every phylum of invertebrates have found calcium carbonate a useful structural material, and its use goes back in geologic time at least to the beginning of the Cambrian period. The mechanism of its precipitation in shells is not known, but apparently it can form only in water that is saturated or nearly so with respect to calcium carbonate. In part, at least, the precipitated shell substance is covered with a thin film of organic material that protects it from dissolving if the water becomes temporarily unsaturated, either by cooling or by dilution. Whether calcium carbonate can be made to precipitate by organisms in a solution not saturated with its ions, so that its apparent solubility is reduced below the equilibrium value, is an unanswered question. Certainly there is no evidence that any organism can cause its precipitation from solutions very far below the saturation concentration. Thus organisms can prevent supersaturation, and perhaps can lower the equilibrium solubility slightly, but at least in their shell-building activities they do not influence the solubility very greatly. Their role is chiefly to make use of a process which, in their absence, would take place inorganically.

Another way in which organisms can affect precipitation is by using the precipitation reaction to provide energy for their life processes. This applies chiefly to slow reactions that are far from equilibrium. Compounds of Fe(II) in solution, for example, react slowly with atmospheric oxygen to precipitate Fe(III) oxide, and some kinds of bacteria take advantage of this reaction to obtain energy. The ultimate solubility of Fe(III) oxide is not affected by the bacteria; they simply promote a reaction which would take place slowly in their absence, in this case a reaction that gives out considerable energy. They serve, so to speak, as an organic catalyst. The precipitation of calcium carbonate would probably not be used in this manner for obtaining energy, because the reaction is rapid and in natural solutions is seldom far from equilibrium, so that the amount of energy available is small.

A third sort of influence exerted by organisms is the causing of precipitation simply as an incidental consequence of their life processes. Water plants, as we have noted earlier, use up dissolved CO_2 during photosynthesis, thereby reducing the amount of H_2CO_3 in solution and leading to precipitation of $CaCO_3$ [the reverse of Eq. (3-4)]. Precipitation of $CaCO_3$ may also be favored by the decay of organisms, if the products of decay are alkaline, or impeded if the products are acid, but unfortunately the acidity or alkalinity to be expected from decay in particular circumstances is not predictable from present data.

Thus the effect of organisms on the formation of insoluble compounds is generally indirect, primarily by creating conditions favorable for their precipitation or dissolution.

3-8 PRECIPITATION OF CaCO₃ IN SEAWATER

As a way of summarizing this long discussion of the factors that influence solubility, we survey briefly the behavior of calcium carbonate in the ocean.

Seawater is an exceedingly complex solution (Table 2-1 and Appendix III), containing electrolytes in great variety plus an abundance of living and dead organic material. The ordinary laws of dilute solution cannot be applied, or at best need great modification. The chemistry of seawater can be described fairly satisfactorily in general terms, but details about the behavior of even so simple a substance as calcium carbonate remain obscure.

To evaluate the effects of individual ion activity coefficients and the formation of complexes on the solubility of calcite in seawater, we employ the computer program (EQ3NR), discussed in Sec. 2-8, to calculate the equilibrium distribution of aqueous species in seawater (see Table 2-1). In Table 3-1 we list the most abundant species, their molal concentrations, activity coefficients, and activities. The table shows that approximately 85% of the total calcium in seawater occurs as the Ca^{2+} ion, the rest as complexes with sulfate and chloride ions. Of the total carbonate in solution, most is in complexes, only about 2% as CO_3^{2-}. These complexes, together with the individual ion activity coefficients of 0.227 for Ca^{2+} and 0.223 for CO_3^{2-}, have the effect of increasing the apparent solubility of calcite in

TABLE 3-1

Concentration, activity coefficients (γ), and activities of aqueous species in seawater[1]

Aqueous Species	Molality ($\times 10^3$)	log γ	log Activity
Cl^-	523.600	−0.187	−0.468
Na^+	451.800	−0.190	−0.535
Mg^{2+}	47.390	−0.535	−1.860
SO_4^{2-}	26.700	−0.744	−2.318
NaCl(aq)	16.570	0.000	−1.781
K^+	9.879	−0.220	−2.225
Ca^{2+}	8.734	−0.644	−2.703
$MgCl^+$	5.398	−0.190	−2.458
HCO_3^-	1.540	−0.158	−2.970
$CaSO_4$(aq)	1.220	0.000	−2.914
KSO_4^-	0.312	−0.158	−3.664
$MgHCO_3^+$	0.250	−0.190	−3.793
$CaCl^+$	0.210	−0.190	−3.869
$MgCO_3$(aq)	0.110	0.000	−3.960
$CaCl_2$(aq)	0.052	0.000	−4.284
CO_3^{2-}	0.038	−0.652	−5.079
$CaHCO_3^+$	0.037	−0.190	−4.623
$CaCO_3$(aq)	0.035	0.000	−4.456
KCl(aq)	0.017	0.000	−4.774
CO_2(aq)	0.013	0.054	−4.835
$MgOH^+$	0.006	−0.190	−5.435
OH^-	0.003	−0.187	−5.785

[1] Calculated using computer program EQ3NR (see Sec. 2-8) and major element concentrations reported in Table 2-1 at 25°C and 1 bar pressure. Solution pH is 8.22, the calculated ionic strength is 0.66, and the activity of water is 0.98.

seawater relative to that in pure water. Thus the activity product for calcium carbonate in seawater is (from Table 3-1)

$$a_{Ca^{2+}} a_{CO_3^{2-}} = 10^{-2.7} \times 10^{-5.1} = 10^{-7.6} = 1.6 \times 10^{-8}, \qquad (3\text{-}24)$$

approximately 5 times greater than the equilibrium constant for calcite solubility and approximately 3 times greater than the constant for aragonite solubility. From a comparison of this number with the solubility curves and seawater compositions reported as a function of depth in Fig. 3-4, we can conclude that much of seawater is approximately saturated (or slightly supersaturated) with calcium carbonate. This seems reasonable, inasmuch as calcium carbonate is observed to be dissolving in some parts of the ocean and precipitating in other parts.

The slight shifts in the marine environment that exercise major control over precipitation and dissolution of calcite are so obvious that they hardly need pointing out. Precipitation is favored where the water is warm and where CO_2 is being lost through evaporation or photosynthesis, as on the shallow banks off Florida and the

Bahamas. Cold water and abundance of CO_2 promote dissolution; the scarcity of calcareous shells in parts of the deep ocean below about 4000 meters, for example, can be plausibly explained by the presence at these depths of cold masses of CO_2-saturated water moving slowly equatorward from the polar regions (cf. Sec. 12-3). Tide pools in temperate regions show evidence of precipitation of calcium carbonate when the trapped seawater warms up during the day, and of dissolution at night when the water becomes cold.

The calcium carbonate that precipitates from seawater may be either calcite or aragonite, the circumstances that favor one or the other being still not entirely clear. With time, either during or after diagenesis of the carbonate sediments, precipitated aragonite changes to the more stable calcite. Precipitation may occur either inorganically or through the agency of organisms; organisms may accomplish the precipitation either in the building of their shells, or incidentally as they remove CO_2 from the water, or during decay if they supply alkaline materials to the water. The old argument as to the relative amounts of limestone formed by inorganic precipitation on the one hand and by organic processes on the other has never been settled, but it is probably not a matter of great importance. In modern seas the places where abundant inorganic precipitation might be expected are precisely those places where organisms flourish most luxuriantly. Investigators disagree as to whether the fine-grained carbonate precipitated indirectly by organisms can be distinguished from an inorganic precipitate. If the distinction is difficult in modern sediments, it is doubtless impossible in older rocks where later recrystallization would have obscured whatever minute textural differences may have been present originally.

Once deposited as bottom sediment, calcium carbonate gradually hardens into limestone, through a series of processes that collectively come under the heading of *diagenesis*: growth of large crystals at the expense of small ones, conversion of aragonite into calcite, replacement of other materials by calcite, and deposition of calcite between the grains by circulating solutions. The ease with which $CaCO_3$ recrystallizes even at low temperatures is attested by the coarse, intergrowing grains of many older limestones. An unexplained detail in carbonate geochemistry is the fact that some limestones from far back in the Paleozoic remain very fine-grained, contrasting sharply with the recrystallized textures that are common even in much more recent limestones.

Thus the chemistry of calcium carbonate in geologic processes is fairly well understood. Much of it is embodied in simple displacements of the equilibrium with carbonic acid and bicarbonate. Additional concepts are needed to account for the relations of the two polymorphs, the behavior of very tiny grains, the effect of dissolved electrolytes, and the influence of organisms. Quantitative treatment is feasible only for relatively simple freshwater solutions, but numbers that show fair agreement with observation can be obtained even for solutions as concentrated as seawater. We shall find that the rules of solubility developed in this discussion apply to many other chemical precipitates of geologic interest. Other carbonates in particular are governed by equilibria similar to the ones we have used for calcium carbonate, as we note in the next section.

3-9 OTHER SIMPLE CARBONATES

All the carbonates that precipitate as sediments or form during metamorphism and hydrothermal processes are somewhat soluble in carbonic acid, and all of them crystallize out of solution when the concentration of carbonic acid diminishes. This means that other carbonates enter into equilibria like the major calcium carbonate equilibrium [Eq. (3-4)], according to reactions that can be symbolized by

$$MCO_3 + H_2CO_3 \rightleftharpoons M^{2+} + 2HCO_3^-, \tag{3-25}$$

where the M^{2+} represents Fe^{2+}, Sr^{2+}, Mn^{2+}, and so on. Representative solubility products of carbonate minerals are given in Table 3-2. Like $CaCO_3$, the other carbonates are more soluble when the concentration of dissolved CO_2 is high, whether because of low temperature or high pressure or decay of organic matter, and less soluble when CO_2 is removed from solution by heating, by decrease in pressure, by addition of alkali, or by the photosynthetic activity of plants. The chemistry of carbonates is essentially simple, and their solubilities can be described in terms of a few well-understood equilibria.

Replacement of one carbonate by another is often observed in metamorphic and sedimentary rocks and in hydrothermal veins. The conditions under which replacement can occur may be partly reconstructed from the data of Table 3-2. For example, conditions for the replacement of calcite by siderite can be evaluated using the reaction

$$\underset{\text{calcite}}{CaCO_3} + Fe^{2+} \rightleftharpoons Ca^{2+} + \underset{\text{siderite}}{FeCO_3}, \tag{3-26}$$

for which the equilibrium constant is (assuming unit activities of pure $CaCO_3$ and $FeCO_3$ and unit activity coefficients for aqueous species)

$$K = \frac{m_{Ca^{2+}}}{m_{Fe^{2+}}} = \frac{m_{Ca^{2+}} m_{CO_3^{2-}}}{m_{Fe^{2+}} m_{CO_3^{2-}}} = \frac{4.5 \times 10^{-9}}{2.0 \times 10^{-11}} = 225. \tag{3-27}$$

This means that at ordinary temperatures calcite should be replaced by siderite if it is in contact with a solution containing Fe^{2+} at a concentration more than $\frac{1}{225}$ that of

TABLE 3-2
Solubility products of carbonates

$MgCO_3$	3.2×10^{-8}
$CaCO_3$(calcite)	4.5×10^{-9}
$SrCO_3$	1.0×10^{-9}
$BaCO_3$	5.0×10^{-9}
$MnCO_3$	5.0×10^{-10}
$FeCO_3$	2.0×10^{-11}
$NiCO_3$	1.3×10^{-7}
$ZnCO_3$	1.0×10^{-10}
$PbCO_3$	8.0×10^{-14}

From Appendix VII-1.

Ca^{2+}, and likewise that siderite can be replaced by calcite only if the solution has more than 225 times as much Ca^{2+} as Fe^{2+}.

Because the concentration of Fe^{2+} in natural solutions is seldom greater than $\frac{1}{225}$ that of Ca^{2+}, replacement of calcium carbonate minerals by siderite is not often observed. Similarly the cations of other simple carbonates (Mn^{2+}, Sr^{2+}, Zn^{2+}, etc.) generally have such low concentrations in groundwaters and surface waters that replacement is not often to be expected. An apparent exception is Mg^{2+}, which forms the well-known mineral magnesite and exists in natural solutions at concentrations similar to those of Ca^{2+}. Rather than replacing calcite or aragonite with magnesite, however, Mg^{2+} in its reaction with $CaCO_3$ commonly undergoes a more complicated reaction—which is the subject of the next section.

3-10 THE DOLOMITE PROBLEM

The most stubborn question in carbonate geochemistry is the origin of the double carbonate dolomite [$CaMg(CO_3)_2$)]. The "dolomite problem" may be stated very simply: Dolomite rock is one of the commonest sedimentary materials, appearing as thick and extensive beds in strata of all ages from the Precambrian to the Cenozoic; there is no geologic evidence to indicate that its formation took place under unusual conditions of temperature or pressure; yet efforts to prepare dolomite in the laboratory under simulated sedimentary conditions have failed, and very little dolomite is observed to be forming in nature in ordinary sedimentary environments. It is true that laboratory precipitation of dolomite at low temperatures is possible, from solutions at pH's greater than 9.5 and containing high concentrations of SO_4^{2-} and NO_3^-; it is also true that formation of dolomite as a primary precipitate in nature has been reported in a number of unusual environments: hot springs, in sediments of salt lakes, and in muds from salt lagoons undergoing strong solar evaporation. But the extreme conditions represented by both the laboratory experiments and the field occurrences, compared with usual sedimentary environments, make the abundance of dolomite in the sedimentary record seem all the more mysterious. Why should this one common sediment be such a conspicuous anomaly?

For an answer we look first to crystallography. The mineral dolomite is a double carbonate of magnesium and calcium, $CaMg(CO_3)_2$, with a structure that may be visualized as a distorted NaCl framework (Sec. 5-1) in which the anions are CO_3^{2-} groups and the cations are regularly alternating Ca^{2+} and Mg^{2+}. The regular alternation is important. This is a special, highly ordered crystal structure, which perhaps takes a long time to grow. When attempts are made to precipitate dolomite in the laboratory, the usual result is a mixture of calcite and hydromagnesite, simple compounds which can form rapidly in solution. The magnesium-rich calcium carbonates in the shells of some marine organisms are not dolomite but structures with a random distribution of Ca and Mg—again the sort of structure that might be expected to form fairly rapidly. Dolomite is readily prepared artificially at temperatures somewhat over 100°C, the function of temperature probably being to speed up the movement of ions so that Ca and Mg can find their places in the ordered structure within a reasonable time. Attempts to prepare the compound at successively lower temperatures give precipitates whose x-ray diffraction patterns show the characteristic lines of dolomite becoming progressively less numerous and more fuzzy. A reasonable inference from these facts is that dolomite forms at ordinary temperatures so very slowly that we have no chance to observe the process in nature or to duplicate it in the laboratory. This slowness of formation, due to the necessity of attaining a highly ordered structure, seems a convincing answer to at least part of the dolomite riddle.

One might expect to gain information about the origin of dolomite from solubility data, as one can for other carbonates. The difficulty lies in the uncertainty about attainment of solution–mineral equilibrium. The amount of dolomite that goes into solution can be measured readily enough, but we cannot be sure that the solution has reached saturation because the ions will not recombine to form solid dolomite. An approximation to the solubility can be obtained by stirring dolomite in water and following the changes in concentration of one of its ions, or of the pH, until the concentrations no longer change; if reproducible results are obtained in such experiments, it seems likely that a condition approaching equilibrium has been reached. Experiments of this sort give an apparent solubility product,

$$K = a_{Ca^{2+}} a_{Mg^{2+}} a^2_{CO_3^{2-}}, \tag{3-28}$$

that is slightly less than the same product of ions obtained from a mixture of calcite and magnesite ($MgCO_3$) treated in the same manner. In other words, reactions like

$$\underset{\text{calcite}}{CaCO_3} + Mg^{2+} + 2HCO_3^- \rightleftharpoons \underset{\text{dolomite}}{CaMg(CO_3)_2} + H_2CO_3 \tag{3-29}$$

and

$$2\underset{\text{calcite}}{CaCO_3} + Mg^{2+} \rightleftharpoons \underset{\text{dolomite}}{CaMg(CO_3)_2} + Ca^{2+} \tag{3-30}$$

are displaced in the direction of dolomite at usual concentrations of Ca^{2+} and Mg^{2+}, but the difference in solubilities is so slight that there is little drive to make the two ions assume the ordered structure of dolomite.

Regarding the numerical value of K in Eq. (3-28) there is still wide disagreement. Published values range from 10^{-17} to 10^{-19}, most of them near the higher figure. Measurements of $a_{Ca^{2+}}$, $a_{Mg^{2+}}$, and $a_{CO_3^{2-}}$ in cave waters by Holland et al. (1964) give an activity product higher than 10^{-15}, with no precipitation of dolomite; whatever the true solubility product may be, this indicates that supersaturation with respect to dolomite can persist for long periods.

Another difficulty in measuring the solubility of dolomite, or of any similar double salt, stems from a question as to just how the dissolving takes place. Does the salt dissolve as a whole (*congruent* dissolution), to give equimolal quantities of Mg^{2+} and Ca^{2+}, or does the more soluble part of the salt, the $MgCO_3$, dissolve in greater amount (*incongruent* dissolution)? The two possibilities may be symbolized

$$\underset{\text{dolomite}}{CaMg(CO_3)_2} \rightarrow Ca^{2+} + Mg^{2+} + 2CO_3^{2-} \text{ (congruent dissolution)} \tag{3-31}$$

and

$$CaMg(CO_3)_2 \rightarrow CaCO_3 + Mg^{2+} + CO_3^{2-} \quad \text{(incongruent dissolution).} \quad (3\text{-}32)$$

$\quad$ dolomite $\qquad$ calcite

Experiments show that dolomite dissolves congruently at ordinary temperatures, but at higher temperatures the dissolution is at least partly incongruent.

$\quad$ Geologically there is not much evidence that dolomites in older strata formed as primary precipitates, except possibly the dolomites associated with evaporite deposits. Many dolomites contain structures, particularly fossils, which originally must have been calcium carbonate, so that these dolomites certainly were formed by a reaction between Mg^{2+} and a $CaCO_3$ sediment. The characteristic poor preservation of fossils in dolomite, the coarseness of grain, and the commonly observed cavities and pore spaces are all indications that dolomite forms by reactions like Eqs. (3-29) and (3-30). The Mg^{2+} may come from seawater in contact with the limy sediment or buried with it, from ions taken up in the original $CaCO_3$ structure (particularly in shells), or from later solutions moving through the sediment; the reaction may represent an addition of bicarbonate [Eq. (3-29)], a replacement of original Ca^{2+} [Eq. (3-30)], or even in part a precipitation of small amounts of original dolomite. In general, for any particular dolomite, it is impossible to sort out the effects of these various reactions. The conversion of calcium carbonate to dolomite commonly takes place shortly after deposition of the original sediment (during diagenesis), as is shown by the replacement of entire beds by dolomite and by the lack of influence on dolomitization of later structures in the rock. On the other hand, partial dolomitization of some limestones along networks of veinlets must represent the work of later solutions acting on solid rock.

$\quad$ Thus details of the formation of dolomite remain obscure, but recent experimental studies plus geologic observations have furnished fairly convincing evidence that most dolomite is not a primary precipitate but forms rather as a product of slow reactions altering originally deposited calcium carbonate. Solutions that accomplish the alterations are most effective if they have a fairly high salinity and pH, a low Ca^{2+}/Mg^{2+} ratio, and a somewhat elevated temperature.

$\quad$ The effects of Ca/Mg ratios and temperature on the replacement of calcite by dolomite can be illustrated with the phase diagrams in Fig. 3-5, which show calculated phase relations among the carbonate minerals calcite, dolomite, and magnesite in contact with solutions. The three lines in Fig. 3-5(a) represent conditions for saturation of a fluid with respect to calcite, dolomite, or magnesite at 1 bar total pressure, 0.1 bar partial pressure of CO_2, and 25°C. The reactions for these equilibria are:

$$CaCO_3 + 2H^+ \rightleftharpoons Ca^{2+} + H_2O + CO_2, \quad (3\text{-}1)$$

$\quad$ calcite

$$CaMg(CO_3)_2 + 4H^+ \rightleftharpoons Ca^{2+} + Mg^{2+} + 2H_2O + 2CO_2, \quad (3\text{-}33)$$

$\quad$ dolomite

and

$$MgCO_3 + 2H^+ \rightleftharpoons Mg^{2+} + H_2O + CO_2. \quad (3\text{-}34)$$

$\quad$ magnesite

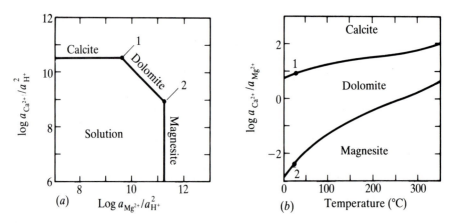

FIGURE 3-5
Calculated phase relations among aqueous solutions and the carbonate minerals calcite, dolomite, and magnesite. The lines in diagram (a) show conditions for saturation with respect to the three minerals at 1 bar total pressure, 0.1 bar partial pressure of CO_2, and 25°C. Diagram (b) shows changes with temperature of the equilibrium ratios $a_{Ca^{2+}}/a_{Mg^{2+}}$ in solutions at equilibrium with the pairs calcite–dolomite and dolomite–magnesite. Pressure corresponds to liquid–vapor equilibrium for H_2O. The points 1 and 2 are the same in the two diagrams. (Diagrams computed using equilibrium constants listed in Bowers, T. S., K. J. Jackson, and H. C. Helgeson: *Equilibrium Activity Diagrams*, Springer Verlag, 396 pp., 1984.)

The saturation lines in Fig. 3-5(a) are calculated using equations written by rearranging the logarithms of the equilibrium constants for Eqs. (3-1), (3-33), and (3-34). Assuming that the activity of water is approximately unity, we write for these equations,

$$\log \frac{a_{Ca^{2+}}}{a_{H^+}^2} = -\log P_{CO_2} + \log K_{(3\text{-}1)}, \tag{3-35}$$

$$\log \frac{a_{Ca^{2+}}}{a_{H^+}^2} = -\log \frac{a_{Mg^{2+}}}{a_{H^+}^2} - 2\log P_{CO_2} + \log K_{(3\text{-}33)}, \tag{3-36}$$

and

$$\log \frac{a_{Mg^{2+}}}{a_{H^+}^2} = \log P_{CO_2} + \log K_{(3\text{-}34)}. \tag{3-37}$$

Under the conditions of fixed temperature, total pressure, and partial pressure of CO_2 these equations form straight lines on a plot of log $(a_{Ca^{2+}}/a_{H^+}^2)$ against log $(a_{Mg^{2+}}/a_{H^+}^2)$, as illustrated in Fig. 3-5(a) (cf. Fig. 3-3).

The intersections of the saturation lines in Figure 3-5(a) denote the conditions for equilibrium among calcite, dolomite, and a solution at point 1, and among dolomite, magnesite, and a solution at point 2. Point 1 can be represented by a chemical reaction balanced by combining reactions (3-1) and (3-33) to give

$$2CaCO_3 + Mg^{2+} \rightleftharpoons CaMg(CO_3)_2 + Ca^{2+}, \tag{3-38}$$
$$\text{calcite} \qquad\qquad\qquad \text{dolomite}$$

where

$$\log \frac{a_{Ca^{2+}}}{a_{Mg^{2+}}} = \log K_{(3-38)}. \tag{3-39}$$

Note that the activity ratio of Ca^{2+} to Mg^{2+} in a solution in equilibrium with calcite and dolomite is a function only of the equilibrium constant and not a function of P_{CO_2}. This ratio is ~ 8 at 1 bar total pressure and 25°C, and it increases with rising temperature as shown in Fig. 3-5(b). We conclude from the phase relations plotted in Fig. 3-5 that dolomite is favored over calcite by low activity ratios of Ca^{2+} to Mg^{2+} and by an increase in temperature.

SUMMARY

The theoretical ideas about chemical equilibrium developed in the first two chapters have given us here a basis for understanding the geochemistry of carbonates in sedimentary environments. The influence of acids and bases, of changes in temperature and pressure, and of other ions in solution can be related to simple equilibrium processes. Given a deposit of calcium carbonate, we can set limits on the geologic conditions under which it could have been deposited. Given a solution of known composition, we can predict the conditions under which calcium carbonate would precipitate from it. If data are available on the geology of an area, we can suggest where calcium carbonate should be in process of solution and where in process of precipitation, and how much calcium the groundwater would contain at various points. From the properties of other carbonates, we can describe the conditions under which they might replace, or be replaced by, the carbonate of calcium. In these various senses, chemistry provides an "understanding" of carbonates.

For geochemical questions relating to the carbonates we can find particularly specific and clean-cut answers. The compounds have relatively simple structures, and processes of solution and precipitation (except for dolomite) involve rapid reactions in which ions in dilute solution play an important role. This is the kind of situation where the quantitative rules of chemical equilibrium are particularly effective as a basis for predictions. As we turn our attention to more complex, slower-reacting compounds, we shall find that our explanations necessarily become more general and less quantitative.

PROBLEMS

1. Explain why underground water in contact with limestone is alkaline. Why is the pH higher if the water is out of contact with air?
2. Would you expect calcite to be appreciably soluble in a solution whose pH is maintained at 4? A solution at pH 11? A solution containing $0.1 m H_2S$? Where might you find such solutions in nature?
3. If calcite is in equilibrium with a solution containing $0.001 m$ total dissolved carbonate (see Fig. 2-3), what is the concentration of dissolved Ca^{2+} in the solution as a function of pH? Make a graph of your results.

4. Calculate the solubility of calcite in water standing in contact with a soil atmosphere containing CO_2 at a partial pressure of 0.1 bar, and find the pH of the saturated solution.

5. (a) Given that the solubility product of rhodochrosite, $MnCO_3$, is $10^{-9.3}$, calculate the equilibrium constant for the reaction

$$MnCO_3 + H_2CO_3 \rightleftharpoons Mn^{2+} + 2HCO_3^-.$$

   (b) Qualitatively, how is the equilibrium constant affected by a change in pH from 6 to 8?

   (c) How is the solubility of $MnCO_3$ affected by this change?

6. Suggest possible explanations for the facts that (a) calcium carbonate is the most abundant carbonate in sedimentary rocks, and (b) calcium in sedimentary environments appears chiefly in the form of carbonate, rather than as calcium sulfide, chloride, oxide, fluoride, or phosphate.

7. Equilibrium constants for the ionization of phosphoric acid at 25°C and 1 bar are

$$H_3PO_4(aq) \rightleftharpoons H^+ + H_2PO_4^- \quad K_1 = 10^{-2.1},$$
$$H_2PO_4^- \rightleftharpoons H^+ + HPO_4^{2-} \quad K_2 = 10^{-7.2},$$
$$HPO_4^{2-} \rightleftharpoons H^+ + PO_4^{3-} \quad K_3 = 10^{-12.2}.$$

   From these activity products, would you estimate that phosphoric acid is stronger or weaker than carbonic acid? What would be the principal phosphate ion or ions in a solution at pH 4? At pH 10? If calcium phosphate, $Ca_3(PO_4)_2$ ($K = 10^{-28.7}$), were placed in contact with each of these solutions, how much would dissolve?

8. The equilibrium constant for fluorite (CaF_2) solubility is $10^{-10.4}$ at 25°C. The concentration of Ca^{2+} in seawater is about 400 ppm, and of F^- about 1.3 ppm, and the activity coefficients for Ca^{2+} and F^- in seawater are 0.23 and 0.65, respectively. Using this information, show that these figures are in agreement with the observation that fluorite practically never occurs as a primary precipitate in marine sediments. What common chemical sediment does contain notable amounts of fluorine?

9. From the solubility products for CaF_2 and $CaCO_3$(calcite), find the equilibrium constant for the reaction

$$CaCO_3 + 2F^- \rightleftharpoons CaF_2 + CO_3^{2-}.$$

   This number should enable you to make predictions about the kind of solutions that might cause replacement of calcite by fluorite, or of fluorite by calcite. For example, suppose that a solution containing $0.0001m$ Ca^{2+}, $0.001m$ CO_3^{2-}, and $0.001m$ F^- percolates through limestone. Would calcite be replaced by fluorite? If the same solution moves through a deposit of fluorite in a vein, would calcite replace fluorite?

10. Quantitatively, how would a slight change in pH affect the ability of a solution to replace calcite or fluorite? (See Table VII-2 in the Appendix for dissociation constants of HF and HCO_3^-.)

11. The solubility products of strontianite ($SrCO_3$) and celestite ($SrSO_4$) are 1.0×10^{-9} and 3.0×10^{-7}, respectively. Using the data given in Table 3-1, predict whether celestite ($SrSO_4$) or strontianite ($SrCO_3$) would be the first to precipitate in places where solutions bring large amounts of Sr^{2+} into the sea.

12. Is seawater saturated, unsaturated, or supersaturated with respect to $BaSO_4$? Use the solubility product of barite given in Appendix VII-1, the activity of sulfate ion given in Table 3-1, and for Ba^{2+} use the following values: $Ba^{2+} = 0.015$ ppm and $\gamma_{Ba^{2+}} = 0.23$.

REFERENCES AND SUGGESTIONS FOR FURTHER READING

Back, W., B. B. Hanshaw, L. N. Plummer, P. H. Rahn, C. T. Rightmire, and M. Rubin: "Process and rate of dedolomitization: Mass transfer and ^{14}C dating in a regional carbonate aquifer," *Bull. Geol. Soc. America*, vol. 94, pp. 1415–1429, 1983. Study of reactions among carbonate minerals, sulfate minerals, and groundwater in a limestone aquifer draining the Black Hills.

Broecker, W. S., and T.-H. Peng: *Tracers in the Sea*, Eldigio Press, New York, 1982. Chapter 2 describes the distribution of calcite in marine sediments and the spatial variations of $CaCO_3$ saturation in seawater.

Garrels, R. M., and C. Christ: *Minerals, Solutions, and Equilibria,* Harper and Row, New York, 1965. Chapter 3 covers the calculation of carbonate equilibria in great detail, for a variety of situations of geologic importance.

Holland, H. D., T. V. Kirsipu, J. S. Huebner, and U. M. Oxburgh: "On some aspects of the chemical evolution of cave waters," *Jour. Geology*, vol. 72, pp. 36–37, 1964. An interesting study of changes in the composition of cave waters, and of the solubility in such waters of calcite, aragonite, and dolomite.

Langmuir, D.: "Geochemistry of some carbonate waters in central Pennsylvania," *Geochim. et Cosmochim. Acta*, vol. 35, pp. 1023–1046, 1971. Analyses of spring and well waters in a limestone–dolomite terrain show that many waters are approximately saturated. Good discussion of precautions needed for laboratory study of natural waters.

Machel, H. G., and E. W. Mountjoy: "Chemistry and environments of dolomitization—a reappraisal," *Earth Science Reviews*, vol. 23, pp. 175–222, 1986. A review of current ideas about the formation of dolomite in sedimentary environments.

Mucci, A.: "The solubility of calcite and aragonite in seawater at various salinities, temperatures, and one atmosphere of total pressure," *American Jour. Sci.*, vol. 283, pp. 780–799, 1983. A good example of recent work on the experimental measurement of the solubility of carbonate minerals in seawater.

Plummer, L. N., and E. Busenberg: "The solubilities of calcite, aragonite, and vaterite in CO_2–H_2O solutions between 0 and 90°C," *Geochim. et Cosmochim. Acta*, vol. 46, pp. 1011–1040, 1982. A careful experimental study of the relation between solubility and temperature.

Shatkay, M., and M. Magaritz: "Dolomitization and sulfate reduction in the mixing zone between brine and meteoric water in the newly exposed shores of the Dead Sea," *Geochim. et Cosmochim. Acta*, vol. 51, pp. 1135–1142, 1987. Detailed study of one area where dolomite is observed to be forming at present, and a good review of recent ideas about the origin of dolomite in general.

CHAPTER
4

SOLUTION—MINERAL EQUILIBRIA PART II: SILICATES

Compounds containing silicon and oxygen—silica and the silicates—are the most abundant materials in most rocks, soils, and sediments. Altogether the silicate minerals (including forms of silica, which is not strictly a silicate but is similar in its chemical behavior) make up some 90% of the Earth's crust. Because they are so abundant and form in so many different ways, an acquaintance with the geochemical behavior of silicates is essential to an understanding of many geologic processes. To begin the study of these compounds, we look first at reactions that describe heterogeneous equilibrium between solutions and common rock-forming silicates. The ideas of chemical equilibrium from preceding chapters are applied to the solubility of the simple minerals quartz and kaolinite, then to more complex silicates like the feldspars. In later chapters we will find that phase relationships and parageneses of silicate minerals make possible inferences about the origin and history of many kinds of rocks, including changes in their composition and mineralogy and the temperatures and pressures under which these chemical changes take place.

4-1 QUARTZ SOLUBILITY

Quartz crystals, amethyst, agate, and opal are all familiar forms of silicon dioxide or silica (SiO_2). It is hard to think of these substances as formed from aqueous solutions, because at ordinary temperatures and pressures they are so very insoluble. Yet their common occurrence as crystals or amorphous masses lining cavities and veins in igneous and metamorphic rocks, as sinter around hot springs, and as cementing material in sedimentary rocks is good evidence that silica has indeed been transported by and deposited from an aqueous solution. This apparent contradiction of ordinary experience means, of course, that the solubility must be different under other conditions of temperature, pressure, and solution composition.

The different polymorphs (crystalline forms) of silica found in nature include α-quartz, β-quartz, tridymite, cristobalite, stishovite, moganite, and coesite, plus various amorphous forms. Chert is a mixture of microcrystalline α-quartz and moganite, often called chalcedony; opal is amorphous silica containing some water. Together these minerals form about 12% of the Earth's crust. Pressures and temperatures at which the various polymorphs are stable are shown in Fig. 4-1. Here we focus on the mineral α-quartz, the most common SiO_2 polymorph in the upper crust.

Because the formation of α-quartz in many kinds of rock so obviously involves the transport of silica in solution, the solubility of this mineral has been the object of numerous experimental studies. Such experiments are not easy, because the reactions leading to equilibrium are slow and because silica tends to form colloidal suspensions, so that it is hard to be sure when equilibrium has been attained. Some of the results are shown in Fig. 4-2, together with similar data for amorphous silica. Note that the solubility of both forms of silica increases rapidly with a rise in temperature, that the solubilities differ by more than an order of magnitude in the lower part of the temperature range, and that the solubilities reported by different investigators are not entirely consistent. The discrepancies can probably be attributed to difficulties in establishing equilibrium, but the size of the dissolving particles and analytical uncertainties may also play a role. Such

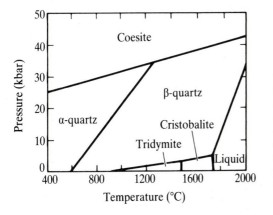

FIGURE 4-1
Pressure and temperature stability of several polymorphs of SiO_2. (Reprinted by permission from Putnis, A.: *Introduction to Mineral Sciences*, Cambridge University Press, 457 pp., 1992, Fig. 6-45, p. 171.) The polymorph stishovite is stable at pressures greater than 80 kbar, hence would be represented by a field well above coesite.

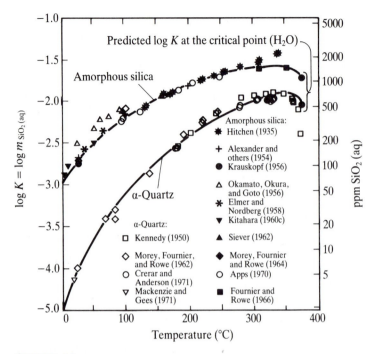

FIGURE 4-2

Experimental (symbols) and theoretical (curves) values of the solubility of α-quartz and amorphous SiO_2 as a function of temperature at pressures equal to liquid–vapor equilibrium for H_2O and at solution pH's <9. (Modified from Walther and Helgeson, 1977.) The names and dates following the symbols in the figure refer to articles where the experimental studies were published. The complete references for these articles are given in Walther and Helgeson's paper.

discrepancies are common, of course, in measurements of solubility, but for most substances are not this large.

The solubility difference between α-quartz and amorphous silica is similar to that between calcite and aragonite (Chap. 3) in that both are examples of the general rule that polymorphs of a given compound show different solubilities, the most stable form having the lowest solubility and the least stable form the highest. If solubility curves for other forms of silica were plotted on Fig. 4-2, they would lie between the two that are shown.

When equilibrium has been established between α-quartz and a silica solution at pH's less than about 9, the reaction can be represented by the simple equation,

$$\underset{\text{quartz}}{SiO_2} + 2H_2O \rightleftharpoons H_4SiO_4. \tag{4-1}$$

This implies that $SiO_2(aq)$ in the solution exists chiefly as an uncharged monomeric species combined with two H_2O dipoles. This is not true for high-pH solutions (see below) or near the critical point of H_2O. If quartz and H_2O are nearly pure, and therefore have unit activities, the equilibrium constant for Eq. (4-1) is

$$K_{(4-1)} = a_{H_4SiO_4} \approx m_{H_4SiO_4}. \qquad (4-2)$$

We are assuming that the activity coefficient for H_4SiO_4 is unity, a reasonable assumption for such dilute solutions. The value of $K_{(4-1)}$, and hence the concentration of aqueous silicic acid in equilibrium with quartz, is about $10^{-4}m$ at 25°C and 1 bar.

H_4SiO_4 is a weak acid, and its stepwise dissociation is similar to that of H_2CO_3 [Eqs. (2-8) and (2-9)]:

$$H_4SiO_4 \rightleftharpoons H_3SiO_4^- + H^+ \qquad (4-3)$$

and

$$H_3SiO_4^- \rightleftharpoons H_2SiO_4^{2-} + H^+, \qquad (4-4)$$

for which the equilibrium constants $K_{(4-3)}$ and $K_{(4-4)}$ are $10^{-9.9}$ and $10^{-11.7}$, respectively, at 25°C and 1 bar (Appendix VII). Measurements of the total concentration of aqueous silica in the solution may thus be represented as the sum of all the silica species,

$$m_{SiO_2(aq)_{total}} = m_{H_4SiO_4} + m_{H_3SiO_4^-} + m_{H_2SiO_4^{2-}}. \qquad (4-5)$$

If the solution is in equilibrium with quartz, its total silica concentration is a function of hydrogen-ion concentration and the equilibrium constants for reactions (4-1), (4-3), and (4-4). Mathematically this is expressed by combining the equilibrium-constant expressions for these three reactions with Eq. (4-5) to obtain:

$$m_{SiO_2(aq)_{total}} = K_{(4-1)} + \frac{K_{(4-1)}K_{(4-3)}}{m_{H^+}} + \frac{K_{(4-1)}K_{(4-3)}K_{(4-4)}}{m_{H^+}^2}. \qquad (4-6)$$

In this equation we have made the approximation that the activity coefficients for all aqueous species (γ_i) are equal to unity. When the molality of H^+ is less than about 10^{-9} (pH > 9), the two right-hand terms in Eq. (4-6) become important, thus requiring the concentration of aqueous silica in equilibrium with quartz to increase as shown in Fig. (4-3). The thin lines in Fig. (4-3) denote fluid compositions consistent with reaction (4-1) and the reactions,

$$\underset{\text{quartz}}{SiO_2} + 2H_2O \rightleftharpoons H_3SiO_4^- + H^+ \qquad (4-7)$$

and

$$\underset{\text{quartz}}{SiO_2} + 2H_2O \rightleftharpoons H_2SiO_4^{2-} + 2H^+. \qquad (4-8)$$

The thick solid curve in the figure is the total concentration of silica required for equilibrium with quartz [Eq. (4-5)]. Positions of the lines in this figure are calculated by procedures similar to that used in constructing Fig. 2-3(a) in Sec. 2-3.

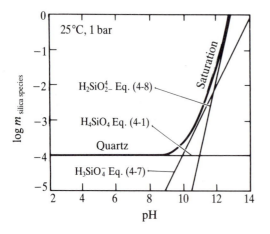

FIGURE 4-3
Solubility of quartz as a function of pH at 1 bar and 25°C. Lines marked H_4SiO_4, $H_3SiO_4^-$, and $H_2SiO_4^{2-}$ show the concentrations of these species in equilibrium with quartz, as indicated by Eqs. (4-1), (4-7) and (4-8). The thick solid line is the total solubility of aqueous silica computed from Eq. (4-6).

The effect of temperature and pressure on the solubility of quartz, as shown by numerous experimental studies, is summarized in the curves of Fig. 4-4(a). These curves were calculated from analysis of experimental data in the pH-independent region of quartz solubility (i.e., at pH < 9 at 25°C and 1 bar pressure; see Fig. 4-3). The curves show that the solubility of α-quartz increases with increasing pressure at constant temperature. It also increases with increasing temperature at constant pressure, except at high temperatures and at pressures less than 1000 bars, where the solubility at constant pressure decreases with increasing temperature. This latter trend is accentuated near the critical point of water [see the 250-bar solubility curve in Fig. 4-4(a)]. These temperature and pressure variations of quartz solubility will be expressed in terms of heat of reaction and volume relations in Chapter 8.

Calculated solubilities of α-quartz for geotherms representative of subduction zone, continental, and subvolcanic terrains are shown in Fig. 4-4(b). This diagram forms the basis for qualitative analysis of transport and deposition of silica in these geologic environments. For example, consider pore fluids that undergo decompression and cooling as they rise through the upper portions of the Earth's crust. If the starting temperature is less than about 400°C, under any of the geothermal gradients the solubility of quartz decreases as the fluid moves upward. This means that the fluid will become supersaturated and quartz may precipitate, forming quartz veins along the channels of movement. The same precipitation may occur at higher temperatures if the gradient is typical of that in subduction-zone or continental environments. But in regions with subvolcanic temperature gradients the situation is more complicated: above about 400°C, as the diagram shows, the solubility of quartz increases rather than decreases, and supersaturation caused by cooling can occur only below this temperature. In accordance with this prediction from Fig. 4-4(b), measured temperatures of formation of the numerous quartz veins in subvolcanic environments are uniformly below 400°C (methods of measuring such temperatures, using fluid-filled inclusions in quartz, are described in Chap. 19).

Hot springs in regions of volcanic activity provide a natural example of the

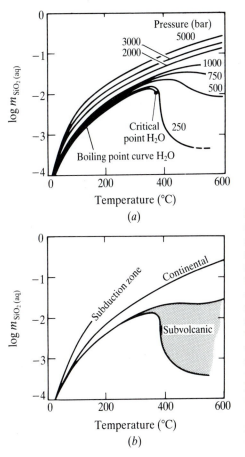

FIGURE 4-4

(*a*) Isobaric temperature dependence of quartz solubility in water calculated from equations and data reported by Walther and Helgeson (1977). (*b*) The temperature dependence of the concentration of aqueous silica in local equilibrium with quartz for the three geothermal gradients shown in Fig. 1-2. The hatched area for the subvolcanic geotherm at temperatures >400°C shows the probable range in silica concentrations for pressures between approximately 250 to 750 bar, which are typical pressures for the deep portions of subvolcanic hydrothermal systems. The solubility curve for the subduction zone geotherm ends at 5000 bar and 150°C, which is the upper pressure limit for the equation of state of aqueous silica given by Walther and Helgeson.

transport and deposition of silica by aqueous solutions. These springs discharge groundwaters that have been heated in a subsurface geothermal reservoir and that by reaction with adjacent rocks have reached local equilibrium (become saturated) with quartz. Movement of the quartz-saturated fluid to the surface commonly leads to precipitation of metastable amorphous silica-rich sinters about the spring orifice. In some springs the fluid rises so fast that the equilibrium does not have time to adjust to the conditions of changing temperature and pressure. If boiling or mixing with cooler dilute waters does not occur, the concentration of silica in the supersaturated discharge fluids may be the same, or nearly the same, as in the subsurface reservoir from which the fluids came. Measurement of silica concentrations in such springs, together with the solubility relations shown in Fig. 4-2, provides a means of estimating temperatures in the geothermal reservoir. This simple technique of geothermometry has been used successfully in the exploration of geothermal reservoirs worldwide. Such reservoirs of hot groundwaters in the Earth's crust are used in many places to provide thermal energy for space heating and for generating electricity.

4-2 KAOLINITE SOLUBILITY

The clay mineral kaolinite is formed by weathering and hydrothermal alteration of other aluminosilicates, typically the feldspars. At some localities, for example Cornwall in southwestern England, altered granitic rocks are mined for huge quantities of kaolinite needed for the industrial production of paper, ceramics, plastics, paint, and rubber. The hydrolysis of feldspars to form kaolinite is complex, and will be deferred to the next section. To help in understanding the hydrolysis reactions, we look first at (1) the stability of kaolinite relative to other minerals in the chemical system Al_2O_3–SiO_2–H_2O, and (2) heterogeneous equilibria among kaolinite, quartz, and a solution.

The composition of kaolinite $[Al_2Si_2O_5(OH)_4]$ is graphically represented on the ternary diagram of Fig. 4-5, together with other rock-forming minerals in the system Al_2O_3–SiO_2–H_2O. Besides kaolinite, two other aluminosilicates are shown: the clay mineral pyrophyllite $[Al_2Si_4O_{10}(OH)_2]$ and the three polymorphs of Al_2SiO_5 called andalusite, kyanite, and sillimanite. Gibbsite $[Al(OH)_3]$ and diaspore $[AlO(OH)]$ are hydroxides of Al, and corundum is the oxide. The lines in the figure connecting minerals and H_2O represent possible combinations of phases that may occur in nature. For example, the triangle A-B-C denotes an assemblage of kaolinite + quartz + water. Such lines are called *tie lines*, and the triangles they form are referred to as *compatibility triangles*. Both are used to represent stable assemblages of minerals and solutions.

As with minerals in the system SiO_2–H_2O (Sec. 4-1), each mineral shown in Fig. 4-5 is stable over a specific range of temperature, pressure, and fluid composition. Fig. 4-6 illustrates the stability of these minerals as a function of temperature and concentration of $SiO_2(aq)$ in the fluid phase at a fixed pressure of 1 kbar. The symbols denote fluid compositions determined by experiments involving mixtures of two minerals and a small amount of water that are allowed to

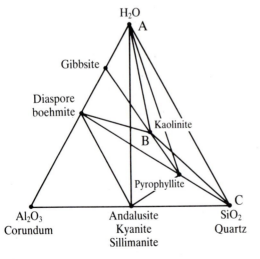

FIGURE 4-5
Composition of minerals in the system Al_2O_3–SiO_2–H_2O.

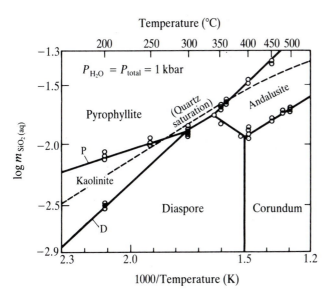

FIGURE 4-6

Experimental (symbols) and calculated (lines) phase relations in the system Al_2O_3–SiO_2–H_2O at 1 kbar fluid pressure. Lines in the diagram are calculated from experimental data using techniques that will be described in Chapters 7 and 8. (Modified from Hemley, J. J., J. W. Montoya, J. W. Marinenko, and R. W. Luce: "General equilibria in the system Al_2O_3–SiO_2–H_2O and some implications for alteration/mineralization processes." *Economic Geology*, vol. 75, pp. 210–228, 1980.)

equilibrate at a specified temperature and pressure. The scatter in the data for a particular phase boundary at one temperature probably represents varying degrees of disequilibrium in the final experiment, due to slow reaction rates and to supersaturation caused by the very fine grain-size of the solids used in the experiments.

The lines marked D and P in Fig. 4-6 outline the stability field of kaolinite in the presence of water. These two curves represent equilibrium for the reactions

$$Al_2Si_2O_5(OH)_4 \rightleftharpoons 2AlO(OH) + 2SiO_2(aq) + H_2O \qquad (4-9)$$
$$\text{kaolinite} \qquad\qquad \text{diaspore}$$

and

$$Al_2Si_4O_{10}(OH)_2 + H_2O \rightleftharpoons Al_2Si_2O_5(OH)_4 + 2SiO_2(aq). \qquad (4-10)$$
$$\text{pyrophyllite} \qquad\qquad\qquad \text{kaolinite}$$

At 1 kbar pressure kaolinite is stable at temperatures $< 300°C$. For temperatures below 275°C, kaolinite is the only mineral in the system Al_2O_3–SiO_2–H_2O that can be in equilibrium with quartz and water. At temperatures greater than about 375°C andalusite is the only phase stable with quartz. Diaspore and corundum are stable only in solutions undersaturated with quartz, and with the exception of a narrow range in temperature and fluid compositions, pyrophyllite is stable only with solutions supersaturated with respect to quartz.

Experiments summarized in Fig. 4-6 show that kaolinite can coexist with solutions having a wide range of silica concentrations. Not shown in this diagram is the dependence of kaolinite-solution equilibria on the solution pH, and on the concentration and extent of complexing of aqueous Al. To show this dependence, we first consider the equilibrium of kaolinite with dilute acid solutions (where $a_{H_2O} = 1$), an equilibrium that can be described by the reaction,

$$Al_2Si_2O_5(OH)_4 + 6H^+ \rightleftharpoons 2Al^{3+} + 2SiO_2(aq) + 5H_2O. \qquad (4\text{-}11)$$
$$\text{kaolinite}$$

The logarithmic equation for the equilibrium constant is

$$2 \log a_{Al^{3+}} + 2 \log a_{SiO_2(aq)} - 6 \log a_{H^+} = \log K_{(4\text{-}11)}. \qquad (4\text{-}12)$$

To represent this relationship graphically, Eq. (4-12) is re-written as

$$\log \frac{a_{Al^{3+}}}{a_{H^+}^3} = -\log a_{SiO_2(aq)} + 0.5 \log K_{(4\text{-}11)}. \qquad (4\text{-}13)$$

Eq. (4-13) defines a straight line with a slope of -1 and a y-intercept equal to $0.5 \log K_{(4\text{-}11)}$ on a plot of $\log a_{Al^{3+}}/a_{H^+}^3$ versus $\log a_{SiO_2(aq)}$, as shown in the graph of Fig. 4-7. If the activity coefficients of aqueous species (γ_i) are close to unity, as would be true for dilute solutions, the axes of this graph can be expressed as,

$$\log \frac{a_{Al^{3+}}}{a_{H^+}^3} = \log m_{Al^{3+}} + 3pH \qquad (4\text{-}14)$$

and

$$\log a_{SiO_2(aq)} = \log m_{SiO_2(aq)}. \qquad (4\text{-}15)$$

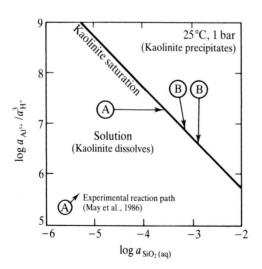

FIGURE 4-7

Activity–activity phase diagram showing calculated equilibrium (line) for kaolinite in dilute acid solution at 1 bar and 25°C, and the experimental approach to equilibrium solubility (arrows). The arrows illustrate the changes reported in solution compositions over a period of 1237 days, for reactions involving solutions that are undersaturated with respect to kaolinite (arrow A) and solutions supersaturated with respect to kaolinite (arrows B). (Figure simplified from data presented by May, H. M., D. G. Kinniburgh, P. A. Heimki, and M. L. Jackson: "Aqueous dissolution, solubilities, and thermodynamic stabilities of common aluminosilicate clay minerals: kaolinite and smectites," *Geochim. Cosmochim. Acta,* vol. 50, pp. 1667–1707, 1986.)

This allows measured compositions of solutions to be easily plotted on the diagram. (The ratio of the activity of a cation to the activity of hydrogen ion, as represented by the y-axis in Fig. 4-7, is a commonly used variable to illustrate phase relations among minerals and solutions.)

The line in Fig. 4-7 shows the composition of solutions in equilibrium with kaolinite. Fluids that plot above this line are supersaturated with kaolinite, those that plot below are undersaturated. This type of phase diagram, called an *activity–activity diagram*, provides a basis for representing equilibrium among minerals and solutions, and for showing changes in solution properties during chemical reactions between these solutions and rock-forming minerals (see Sec. 4-4). For example, the arrows in the figure illustrate changes in solution composition measured during experiments where kaolinite was allowed to react with solutions that were either undersaturated (arrow A) or supersaturated (arrows B) with respect to kaolinite.

Calculations for aluminosilicates like kaolinite are complicated because Al^{3+} ions hydrolyze to form complexes with OH^-. The most important of these complexes are $Al(OH)_2^+$ and $Al(OH)_4^-$:

$$Al^{3+} + 2H_2O \rightleftharpoons Al(OH)_2^+ + 2H^+ \qquad (4\text{-}16)$$

and

$$Al(OH)_2^+ + 2H_2O \rightleftharpoons Al(OH)_4^- + 2H^+. \qquad (4\text{-}17)$$

The total measured concentration of Al ($m_{Al,total}$) in the solution is now expressed as

$$m_{Al,\,total} = m_{Al^{3+}} + m_{Al(OH)_2^+} + m_{Al(OH)_4^-}. \qquad (4\text{-}18)$$

To illustrate the effect of Al-speciation and pH on the solubilities of aluminosilicate minerals, we consider heterogeneous equilibria among kaolinite, quartz, and a dilute solution at 1 bar and 25°C. Because kaolinite and quartz are pure or nearly pure phases, we assume that $a_{kaolinite} = a_{quartz} = 1.0$ in the equations below. In addition, we limit the discussion to dilute solutions ($a_{H_2O} = 1.0$), in which activity coefficients of aqueous species (γ_i) are close to unity so that $a_i \cong m_i$. Three reactions describe these equilibria, one for each Al-species:

$$\underset{\text{kaolinite}}{Al_2Si_2O_5(OH)_4} + 6H^+ \rightleftharpoons 2Al^{3+} + \underset{\text{quartz}}{2SiO_2} + 5H_2O, \qquad (4\text{-}19)$$

$$\underset{\text{kaolinite}}{Al_2Si_2O_5(OH)_4} + 2H^+ \rightleftharpoons 2Al(OH)_2^+ + \underset{\text{quartz}}{2SiO_2} + H_2O, \qquad (4\text{-}20)$$

and

$$\underset{\text{kaolinite}}{Al_2Si_2O_5(OH)_4} + 3H_2O \rightleftharpoons 2Al(OH)_4^- + 2H^+ + \underset{\text{quartz}}{2SiO_2}. \qquad (4\text{-}21)$$

Concentrations of the Al-species may be expressed (in logarithmic form) in terms of the equilibrium constants for these reactions:

$$\log m_{Al^{3+}} = -3pH + 0.5 \log K_{(4\text{-}19)}, \qquad (4\text{-}22)$$

$$\log m_{Al(OH)_2^+} = -pH + 0.5 \log K_{(4\text{-}20)}, \qquad (4\text{-}23)$$

and

$$\log m_{Al(OH)_4^-} = pH + 0.5 \log K_{(4\text{-}21)}. \qquad (4\text{-}24)$$

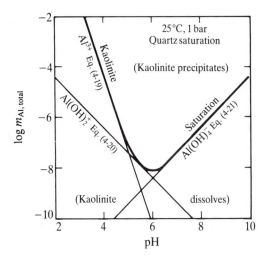

FIGURE 4-8

The calculated concentrations of Al-species (thin lines) and total Al [thick curve, Eq. (4-18)] in a solution in equilibrium with kaolinite and quartz as a function of pH at 1 bar and 25°C.

Each equation is shown by a straight line on a graph of the logarithm of molality of Al-species versus pH (Fig. 4-8). Note that the thin lines representing these equations have different slopes: the concentrations of Al^{3+} and $Al(OH)_2^+$ decrease with increasing pH, but the concentration of $Al(OH)_4^-$ increases. Slopes of the lines are determined by the stoichiometry of the reactions describing solution–mineral equilibria [Eqs. (4-19) to (4-21)], and positions of the lines are determined by the equilibrium constants for the reactions. The linear relationships are those required by heterogeneous equilibrium among kaolinite, quartz, and solution.

The total concentration of Al in the solution is calculated by substituting the exponential forms of Eqs. (4-22) through (4-24) into Eq. (4-18) to obtain

$$m_{Al,\,total} = m_{H^+}^3 K_{(4\text{-}19)}^{0.5} + m_{H^+} K_{(4\text{-}20)}^{0.5} + \frac{K_{(4\text{-}21)}^{0.5}}{m_{H^+}}, \qquad (4\text{-}25)$$

and is shown by the thick curve in Fig. 4-8. Kaolinite exhibits a minimum in its solubility at the pH where there are nearly equal concentrations of positively and negatively charged Al-complexes [for example, where the lines marked $Al(OH)_2^+$ and $Al(OH)_4^-$ intersect in Fig. 4-8]. Also note that the total Al-concentration in equilibrium with kaolinite and quartz is approximately represented by the molality of a single one of the Al-species [as given by one of the Eqs. (4-22) to (4-24)], except at pH values where there are nearly equal concentrations of Al^{3+} and $Al(OH)_2^+$, or of $Al(OH)_2^+$ and $Al(OH)_4^-$.

Similar solubility relations can be derived for the other aluminosilicate and alumino-hydroxide minerals shown in Fig. 4-5. For example, curves in Fig. 4-9(*a*) illustrate calculated Al-concentrations in equilibrium with pyrophyllite + quartz and gibbsite + quartz, together with the curve for kaolinite + quartz equilibrium shown in Fig. 4-8. At the pressure and temperature conditions of this solubility diagram (1 bar and 25°C), the stable mineral assemblage is kaolinite + quartz. This assemblage is characterized by the minimum solubility in Fig. 4-9(*a*). Note that

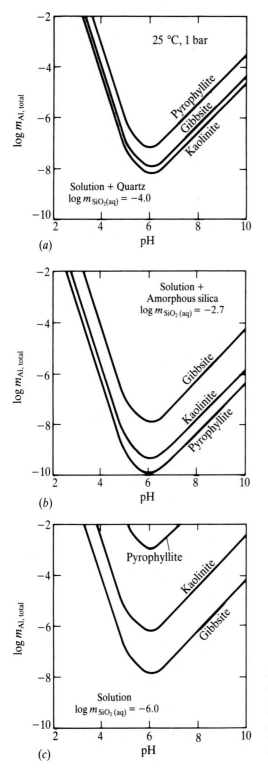

(a)

(b)

(c)

FIGURE 4-9
Calculated solubilities of pyrophyllite, gibbsite, and kaolinite in aqueous solutions in equilibrium with quartz (a), with amorphous silica (b), and in solutions undersaturated with respect to both quartz and amorphous silica (c).

fluids plotted on the curves representing equilibrium of pyrophyllite + quartz or gibbsite + quartz would be supersaturated with respect to kaolinite. Compare these relationships with the curves in Fig. 4-9(b),which are calculated for fluids in equilibrium with amorphous silica (see Fig. 4-2), and with the curves in Fig. 4-9(c), which are constructed for fluids that are undersaturated with respect to both quartz and amorphous silica. In Fig. 4-9(b) pyrophyllite in the presence of amorphous silica is stable relative to gibbsite and kaolinite, and in Fig. 4-9(c) gibbsite is the stable phase. Such diagrams provide a background for the discussion in the next section of the more complicated reactions involving feldspar and water.

4-3   FELDSPAR–SOLUTION EQUILIBRIA

Feldspars are the most abundant minerals in the Earth's crust, occurring in many kinds of sedimentary, metamorphic, and igneous rocks. They show much variety in chemical composition and complexities in their crystal structures, some of which will be discussed in the next chapter. Here we focus on solution–mineral equilibria involving the feldspars, and for this purpose we choose the common alkali feldspar with the formula $KAlSi_3O_8$. This formula represents the approximate composition of several familiar minerals (orthoclase, microcline, sanidine, adularia) that have slight differences of stability under different conditions of temperature and pressure, but for present purposes we will simply lump them as *K-feldspar.* The K-feldspar minerals are formed during diagenesis and metamorphism of a wide variety of rock types, and during crystallization from certain kinds of magmas. Because of their great abundance and the diversity of geologic environments in which they are found, the changes they undergo during weathering, diagenesis, and hydrothermal alteration are an important field of geochemical study.

Measuring the low-temperature solubility of K-feldspar, as well as other common silicate minerals, cannot be accomplished by the relatively simple procedures we have used for the carbonate minerals. Seemingly, determining the solubility of K-feldspar at 25°C should be a simple matter of finding total concentrations of potassium, aluminum, and silica in a solution allowed to stand in contact with the feldspar, but this ignores complications from incongruent dissolution of the solid, slow reactions, and the formation of secondary minerals (Sec. 11-2). At higher temperatures reactions are faster and their products more definite, so we look first at the experimental behavior of K-feldspar in water at temperatures between 250 and 500°C.

From qualitative experiments it has long been known that feldspar sealed in a bomb with water and heated to temperatures of a few hundred degrees will react with the solution to form mica or various types of clay minerals. The nature of the minerals produced depends on the temperature, amount of water, and amounts of other materials that may be present, such as HCl and metal oxides or chlorides. Kaolinite forms at temperatures less than about 275°C, with or without HCl, provided that metal compounds are low or absent; the mica muscovite $[KAl_2(AlSi_3O_{10})(OH)_2]$ appears at higher temperatures and at higher ratios of KCl to HCl; the complex clay mineral montmorillonite is favored by oxides or chlorides of sodium and the alkaline-earth metals. These results confirm, in general, observations of the weathering of feldspars near the Earth's surface, and alteration of feldspars in subvolcanic hydrothermal systems.

Qualitative observations of this sort have been refined in laboratory experiments by Hemley and Jones (1964), the results of which are in part summarized in Fig. 4-10. Instead of leaving the pressure of water vapor to be determined by the amount of H_2O placed in a sealed bomb at the beginning of the experiment, these workers arranged to have the bomb's interior connected with an outside source of water vapor, so that the pressure could be delicately controlled at all times. Solutions in their experiments were saturated with respect to quartz. Amounts of HCl and KCl were systematically varied, and the products were identified at various temperatures. Figure 4-10 shows the products obtained in the system K_2O–Al_2O_3–SiO_2–H_2O under a pressure of 1000 bars of water pressure.

At temperatures less than about 350°C there are two principal equilibria in this system:

$$3KAlSi_3O_8 + 2H^+ \rightleftharpoons KAl_2(AlSi_3O_{10})(OH)_2 + 6SiO_2 + 2K^+, \qquad (4\text{-}26)$$

$$\text{K-feldspar} \qquad\qquad\qquad \text{muscovite} \qquad\qquad \text{quartz}$$

$$2KAl_2(AlSi_3O_{10})(OH)_2 + 2H^+ + 3H_2O \rightleftharpoons 3Al_2Si_2O_5(OH)_4 + 2K^+. \quad (4\text{-}27)$$

$$\text{muscovite} \qquad\qquad\qquad\qquad \text{kaolinite}$$

These hydrolysis reactions show that equilibrium in both cases depends on the ratio of K^+ to H^+. The equilibrium ratio at a given temperature is higher for reaction (4-26) than for reaction (4-27), and for each reaction the ratio becomes larger as the temperature falls. We can infer from the phase diagram that K-feldspar is stable relative to mica and clay minerals at higher temperatures and at higher KCl to HCl

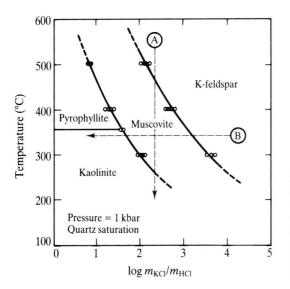

FIGURE 4-10

Experimental (symbols) and predicted (curves) phase relations among K-feldspar, muscovite, pyrophyllite, and kaolinite as a function of temperature and the ratio of concentrations of KCl to HCl in the solution. (Modified from Hemley and Jones, 1964.)

ratios in the fluid. Increasing HCl relative to KCl causes K-feldspar to react with the solution to form first muscovite and then kaolinite or pyrophyllite.

The phase diagram in Fig. 4-10 is useful for predicting processes by which feldspar is formed or altered in geologic systems. For example, in many igneous and metamorphic rocks feldspars near veins are replaced first by mica, then by clays. From the phase diagram we see that this mineral paragenesis could be produced by a decrease in temperature at constant fluid composition in the vein (arrow A in Fig. 4-10), or by a decrease in the ratio of KCl to HCl in the vein fluids at constant temperature (arrow B in Fig. 4-10). These two possibilities, as well as combinations of the two, have been proposed as explanations for mineral zoning involving the replacement of feldspars by hydrous minerals in many subvolcanic hydrothermal systems. This process, commonly called hydrogen-ion metasomatism because H^+ is added to the rock during the alteration of feldspar, will be considered in more detail in Chapters 18 and 19.

4-4 THEORETICAL ACTIVITY–ACTIVITY PHASE DIAGRAMS

A convenient method for representing graphically the complex equilibria established when K-feldspar is in contact with water is to set up phase diagrams in which logarithms of cation-to-hydrogen-ion activity ratios (for example, $\log a_{K^+}/a_{H^+}$) are used as axes. We have used this method before to show carbonate–solution equilibria [Fig. 3-3(b)] and kaolinite–solution equilibria (Fig. 4-7).

Equilibrium between K-feldspar and an aqueous solution can be represented by a number of chemical reactions, of which one example is

$$\underset{\text{K-feldspar}}{KAlSi_3O_8} + 4H^+ \rightleftharpoons K^+ + Al^{3+} + 3SiO_2(aq) + 2H_2O. \qquad (4\text{-}28)$$

On the assumption of unit activity for H_2O and K-feldspar, the equilibrium constant for this reaction is

$$K_{(4\text{-}28)} = \frac{a_{K^+}a_{Al^{3+}}a_{SiO_2(aq)}^3}{a_{H^+}^4}, \qquad (4\text{-}29)$$

or, in logarithmic form written in terms of cation to hydrogen-ion activity ratios,

$$\log K_{(4\text{-}28)} = \log \frac{a_{K^+}}{a_{H^+}} + \log \frac{a_{Al^{3+}}}{a_{H^+}^3} + \log a_{SiO_2(aq)}. \qquad (4\text{-}30)$$

The three variables on the right-hand side of Eq. (4-30) define a surface in a three-dimensional diagram, Fig. 4-11. Fluids with values of $\log a_{K^+}/a_{H^+}$, $\log a_{Al^{3+}}/a_{H^+}^3$ and $\log a_{SiO_2(aq)}$ that plot on this surface are saturated with respect to K-feldspar; in other words, the surface represents solutions in equilibrium with K-feldspar. Fluids that plot below the surface are undersaturated, fluids that plot above the surface supersaturated, with respect to K-feldspar.

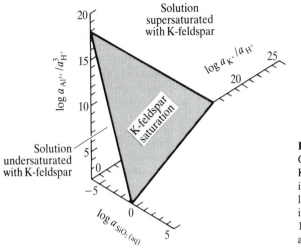

FIGURE 4-11
Calculated saturation surface for K-feldspar-solution equilibrium in terms of $\log a_{K^+}/a_{H^+}$, $\log a_{Al^{3+}}/a_{H^+}^3$, and $\log a_{SiO_2(aq)}$ in the fluid phase at 25°C and 1 bar for unit activities of H_2O and K-feldspar.

The molalities of K^+, Al^{3+}, H^+, and $SiO_2(aq)$ used to represent fluid compositions on this diagram are those that satisfy homogeneous equilibrium in the fluid phase as described in Sec. 2-6. Concentrations of these species may not represent the total dissolved concentrations of potassium, aluminum, and silica if these substances form complexes of appreciable concentrations [for example, Eqs. (4-5) and (4-25)]. As a consequence, phase diagrams of this kind do not provide information about the absolute solubilities of minerals. They are useful, however, for (1) determining whether a particular solution is in equilibrium with a mineral or an assemblage of minerals, and (2) predicting the chemical characteristics of solutions from petrographic analysis of mineralogic phase relations and paragen-eses.

The diagram of Fig. 4-11 can be expanded to display also equilibrium relationships among minerals that are often noted as products of alteration of K-feldspar by acid solutions—mica, clay minerals, and gibbsite. In a sense, this means extrapolating the curves of Fig. 4-10 to lower temperatures and putting them in a different geometric arrangement. As an example, we consider the expanded three-dimensional diagram (Fig. 4-12) for a temperature of 25°C and 1 bar pressure, assuming unit activity for H_2O and all minerals.

The diagram has the same axes as Fig. 4-11, and the planar surface shown in this figure is now labeled Ksp. The other planes in Fig. 4-12 are saturation surfaces for the alteration minerals: muscovite, kaolinite, pyrophyllite, and gibbsite. These surfaces are located, just as was Ksp, by calculations from equations for the equilibrium constants for reactions representing solution–mineral equilibria. As an example, the plane for muscovite (Mus) is derived from the equilibrium constant for the reaction

$$KAl_2(AlSi_3O_{10})(OH)_2 + 10H^+ \rightleftharpoons K^+ + 3Al^{3+} + 3SiO_2(aq) + 6H_2O, \quad (4\text{-}31)$$

muscovite

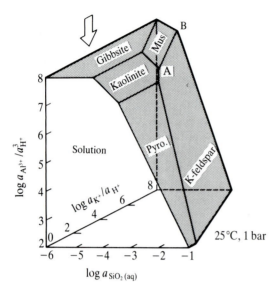

FIGURE 4-12
Theoretical phase diagram illustrating saturation surfaces for K-feldspar [Ksp, Eq. (4-28)], muscovite [Mus, Eq. (4-32)], kaolinite [Kaol, Eq. (4-10)], pyrophyllite (Pyro), and gibbsite at 25°C and 1 bar where activities of H_2O and all minerals are unity. Quartz saturation is not shown, but would be represented by a vertical plane with log $a_{SiO_2(aq)}$ equal to -4.

which in logarithmic form for unit activities of water and muscovite is

$$\log K_{(4\text{-}31)} = \log \frac{a_{K^+}}{a_{H^+}} + 3 \log \frac{a_{Al^{3+}}}{a_{H^+}^3} + 3 \log a_{SiO_2(aq)}. \qquad (4\text{-}32)$$

The intersection of two mineral saturation surfaces in Fig. 4-12 represents equilibrium among the two minerals and the solution. Thus the line marked A-B in the figure, which is the intersection of planes defined by Eqs. (4-30) and (4-32), represents the constraints on fluid composition imposed by local equilibrium among K-feldspar, muscovite, and a solution. As in Fig. 4-11, fluids that plot below the saturation surfaces are undersaturated with respect to the minerals, and those that plot above the planes are supersaturated.

Because the concentration of total Al or of the Al^{3+} ion is generally small in natural solutions and is difficult to measure accurately, this concentration is often eliminated as a variable by graphically projecting the phase relations through the log $a_{Al^{3+}}/a_{H^+}^3$ axis in Fig. 4-12 (indicated by the vertical arrow), giving the two-dimensional plot shown in Fig. 4-13. Lines in this figure are projections of the intersections of the saturation surfaces in Fig. 4-12. They may be calculated from equations relating the various pairs of minerals, equations in which Al^{3+} does not appear. For example, the line separating the fields of K-feldspar and muscovite (line marked A-B in Figs. 4-12 and 4-13) is represented by the equilibrium constant expression for the reaction

$$\underset{\text{K-feldspar}}{3KAlSi_3O_8} + 2H^+ \rightleftharpoons \underset{\text{muscovite}}{KAl_2(AlSi_3O_{10})(OH)_2} + 2K^+ + 6SiO_2(aq), \qquad (4\text{-}33)$$

for which the logarithmic form of the equilibrium constant equation is

$$\log \frac{a_{K^+}}{a_{H^+}} = -3 \log a_{SiO_2(aq)} + 0.5 \log K_{(4\text{-}33)}. \qquad (4\text{-}34)$$

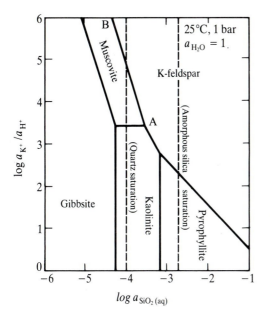

FIGURE 4-13
Theoretical activity–activity phase diagram showing stability relations among the minerals K-feldspar, muscovite, kaolinite, pyrophyllite, gibbsite, quartz, amorphous silica, and an aqueous solution in terms of the variables $\log a_{K^+}/a_{H^+}$ and $\log a_{SiO_2(aq)}$ at 1 bar and 25°C.

The dashed lines on Fig. 4-13 represent saturation of the solution with quartz and amorphous silica according to the equilibria:

$$SiO_2 \rightleftharpoons SiO_2(aq) \qquad (4\text{-}35)$$
$$\text{quartz}$$

and

$$SiO_2 \rightleftharpoons SiO_2(aq). \qquad (4\text{-}36)$$
$$\text{amorphous} \atop \text{silica}$$

Equilibrium constants for these two reactions are equal to the activities of dissolved silica in equilibrium with the two forms of solid SiO_2 .

Thus each solid line in Fig. 4-13 represents equilibrium between two minerals and the solution, and areas bounded by the lines (as projections of the saturation surfaces in Fig. 4-12) represent the range in solution variables a_{K^+}/a_{H^+} and $a_{SiO_2(aq)}$ where one of the minerals is stable.

Theoretical diagrams like those illustrated by Figs. 4-12 and 4-13 are widely used to facilitate description and interpretation of water–rock reactions for many kinds of mineral assemblages and many kinds of geologic environments. The diagrams, it should be emphasized, are only in part based on experimental results or field observations at the particular temperatures and pressures indicated. The positions of lines and surfaces are calculated from thermodynamic properties of the minerals and solutions using methods we will examine in Chapter 8. Even without knowledge of details of the thermodynamic background, the diagrams provide convenient graphical representation of observed relationships in nature.

SUMMARY

Silica and the silicates are the most abundant minerals in the Earth's crust. Their formation and alteration during water–rock reactions are a major part of the processes of weathering, diagenesis, and metamorphism. Such reactions determine the make-up of mineral assemblages and the extent of metasomatic mass transfer between aqueous solutions and their mineralogic environment. Some of the reactions can be duplicated experimentally, but their laboratory study is difficult because of slow reaction rates and the frequent appearance of metastable phases.

As an introduction to the study of silicate reactions, we have examined in this chapter some aspects of equilibria established between a water solution and the common minerals quartz, kaolinite, and K-feldspar. The solubilities and relative stabilities of these minerals are complex functions of temperature, pressure, and solution composition. Methods of studying and graphically displaying such relationships, as suggested in this chapter, will have many applications in our future exploration of water–rock interactions in a variety of geologic environments.

PROBLEMS

In answering questions **1, 2**, and **3**, assume that the activities of minerals and H_2O are equal to unity, and that activity coefficients for all dissolved species are also unity.

1. Using the equations and data given below, calculate the concentration of Al^{3+} and K^+ ions in a solution with a pH of 8.0 in equilibrium with muscovite, quartz, and K-feldspar, at 1 bar pressure and 25°C.

$$KAlSi_3O_8 + 4H^+ \rightleftharpoons K^+ + Al^{3+} + 3SiO_2(aq) + 2H_2O$$

K-feldspar

$$\log K = 0.085$$

$$KAl_2(AlSi_3O_{10})(OH)_2 + 10H^+ \rightleftharpoons K^+ + 3Al^{3+} + 3SiO_2(aq) + 6H_2O$$

muscovite

$$\log K = 14.57$$

$$SiO_2 \rightleftharpoons SiO_2(aq)$$

quartz

$$\log K = -4.00$$

2. What are the concentrations of Al^{3+}, $Al(OH)^{2+}$, and $Al(OH)_4^-$ as a function of pH in an aqueous solution containing 0.00001 m total Al? Present your results on a diagram showing the logarithm of Al-concentration plotted against pH at 25°C, 1 bar, using the following equilibrium constants for the reactions:

$$Al(OH)_4^- \rightleftharpoons Al^{3+} + 4OH^-$$

$$\log K = -32.73$$

$$Al(OH)^{2+} \rightleftharpoons Al^3 + OH^-$$

$$\log K = -9.25$$

$$H_2O \rightleftharpoons H^+ + OH^-$$

$$\log K = -14.00$$

3. Using the equations and data given in questions **1** and **2**, construct a diagram showing the relationship between log m_{K^+} and pH for equilibrium among muscovite, quartz, and a solution with a total Al-concentration of $0.00001m$.

4. Write balanced chemical equations for the phase boundaries in the diagram of Fig. 4-13, then use equilibrium constants for these reactions given in Appendix VII to construct a graph of this phase diagram at 25°C and 1 bar. On the diagram, project the compositions of seawater and rain water using the data and information provided in Tables 2-1 and 3-1. Use the diagram to answer the following questions: (*a*) Can gibbsite be stable in acid solutions? (*b*) Can gibbsite and K-feldspar be stable together at equilibrium with a solution? (*c*) Can pyrophyllite be stable in contact with a solution that is not super-saturated with quartz?

5. Construct two-dimensional activity–activity phase diagrams for projections of Fig. 4-12 through (*a*) the log $a_{SiO_2(aq)}$ axis and (*b*) the log a_{K^+}/a_{H^+} axis at 25°C and 1 bar and at 500°C and 500 bar using equilibrium constants from Appendix VII. In diagrams (*a*) the axes will be log a_{K^+}/a_{H^+} and log $a_{Al^{3+}}/a_{H^+}^3$, and all chemical reactions will be balanced conserving Si in the solid phases. In diagrams (*b*) the axes will be log $a_{Al^{3+}}/a_{H^+}^3$ and log $a_{SiO_2(aq)}$, and chemical reactions will be balanced conserving K^+ in the solids. Write a one page discussion of the phase diagrams you have constructed. Include in the essay a description and interpretation of the phase relations among minerals and solutions and a comparison of the differences in predicted phase relations at the two different temperatures and pressures.

6. Answer the following questions concerning the reaction of a granite, consisting chiefly of quartz and K-feldspar, with dilute solutions. Use Fig. 4-13 and the phase diagrams generated in question **5** to address these questions.

   (*a*) A granite is in contact with a solution that has a pH of 6 and contains $0.1m_{K^+}$. Describe the sequence of reactions you would expect to take place (theoretically) if the pH of this solution were to decrease gradually to 4.

   (*b*) Granite adjacent to a quartz–sulfide vein is commonly altered, showing a thin layer of sericite (fine-grained muscovite) next to the vein and then a strip a few centimeters wide of whitish rock in which the feldspars have been largely changed to kaolinite. If the vein was formed by a warm, slightly acid solution moving through the granite, suggest an explanation for this pattern of alteration.

   (*c*) In a granite that has been altered by hydrothermal solutions, would you expect to find gibbsite as an alteration product directly in contact with K-feldspar? Why or why not?

   (*d*) In soils formed from granitic rocks in a humid temperate climate, the chief clay mineral is usually kaolinite. Remembering that solutions in such soils may have high CO_2 contents from decaying vegetation, what range of concentrations of K^+ would you expect to find in the solutions?

7. Partial chemical analysis of the solution from a geothermal well at 200°C gives:

$$m_{Na,\,total} = 1 = m_{Na^+} + m_{NaCl(aq)}$$

$$m_{Cl,\,total} = 1 = m_{Cl^-} + m_{NaCl(aq)}$$

$$m_{Ca,\,total} = 0.0016m = m_{Ca^{2+}}$$

$$pH = 6.4$$

The symbols $m_{Na,total}$ and $m_{Ca,total}$ denote the total analytical concentration of Na and Ca in the solution. Using these data and the activity–activity phase diagram given below, decide what mineral or minerals in the system $CaO–Na_2O–Al_2O_3–SiO_2–H_2O–HCl$ might be in equilibrium with the fluid. Assume unit activity of H_2O and unity for all activity coefficients, and consider only complexing of Na^+ and Cl^- to form $NaCl(aq)$, where

$$NaCl(aq) \rightleftharpoons Na^+ + Cl^-$$

$$K_{200°C, 15\,bar} = \frac{a_{Na^+} \cdot a_{Cl^-}}{a_{NaCl(aq)}} = 2.63$$

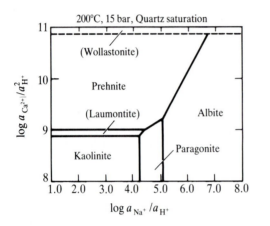

REFERENCES AND SUGGESTIONS FOR FURTHER READING

Bowers, T. S., K. J. Jackson, and H. C. Helgeson: *Equilibrium Activity Diagrams*, Springer Verlag, Berlin, Heidelberg, New York, Tokyo, 1984. A compilation of theoretical activity–activity phase diagrams for coexisting minerals and aqueous solutions at pressures and temperatures to 5 kbar and 600°C. Also includes extensive tables of equilibrium constants for mineral hydrolysis reactions.

Drever, J. I.: *The Geochemistry of Natural Waters*, 2d ed., Prentice-Hall, Englewood Cliffs, N.J., 1988. Chapter 6 provides an excellent discussion of the details for computing activity–activity phase diagrams. It includes techniques for locating metastable phase boundaries, and for graphical derivation of the topology of phase boundaries based on compositions of coexisting minerals.

Faure, G.: *Inorganic Geochemistry*, Macmillan, New York, 1991. Chapter 14 gives details for calculating the phase diagram in Fig. 4-13.

Garrels, R. M., and C. L. Christ: *Solutions, Minerals, and Equilibria*, Harper and Row, New York, 1965. Chapter 10 presents a variety of diagrams used to represent phase relations among silicates.

Hemley, J. J., and W. R. Jones: "Chemical aspects of hydrothermal alteration with emphasis on hydrogen metasomatism," *Economic Geology* , vol. 59, pp. 538–569, 1964.

Walther, J. V. and H. C. Helgeson: "Calculation of the thermodynamic properties of aqueous silica and the solubility of quartz and its polymorphs at high pressures and temperatures," *American Journal of Science*, vol. 277, pp. 1315–1351, 1977. Demonstration that the solubility of quartz at pressures and temperatures up to 5 kbar and 600°C, as calculated from its thermodynamic properties, agrees with experimental measurements.

CHAPTER 5

CRYSTAL CHEMISTRY

Equilibrium reactions of carbonate and silicate minerals with aqueous solutions have been the center of attention in the last two chapters. We turn now to a closer examination of minerals themselves. Minerals, by definition, are naturally-occurring homogeneous solids, for the most part inorganically formed, with definite chemical compositions and specific arrangements of atoms. More than 3800 mineral species have been identified, but we limit discussion here to the crystal chemistry of the small number that are major constituents of rocks in the Earth's crust. Only about a dozen kinds of minerals make up the greater part of common rocks, and these dozen are composed of only a small number of elements—only 8 (O, Si, Al, Fe, Mg, Ca, Na, K) out of the more than 100 elements that are known. These eight (Fig. 5-1) account for more than 99% of the mass of the Earth's crust. The two mineral species quartz and feldspar make up almost two-thirds of the crustal volume. If pyroxenes, olivine, and hydrated silicates are added to these two, over 90% of the volume is included in the sum. The study of crystal chemistry is greatly simplified by the small number of elements and the small number of minerals that we need be concerned with in discussing most geological processes and geologic environments.

The mineral structures of most interest in geology are those of crystalline solids composed of ions. A few materials found in nature, for example diamond and sulfur, are constructed of un-ionized atoms. A few others, like asphalt, are best regarded as made up of molecules; in these the structure of the molecule itself may be complex and may largely determine the properties of the substance. But the rocks

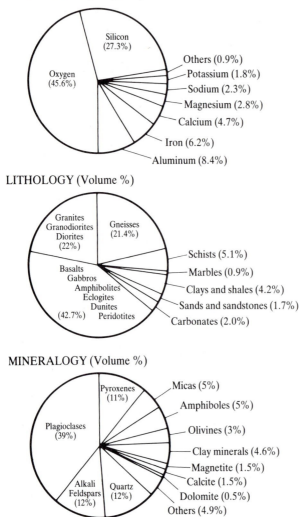

COMPOSITION (Weight %)

Silicon (27.3%)

Oxygen (45.6%)

Others (0.9%)
Potassium (1.8%)
Sodium (2.3%)
Magnesium (2.8%)
Calcium (4.7%)
Iron (6.2%)
Aluminum (8.4%)

LITHOLOGY (Volume %)

Granites
Granodiorites
Diorites
(22%)

Gneisses (21.4%)

Basalts
Gabbros
Amphibolites
Eclogites
Dunites
(42.7%) Peridotites

Schists (5.1%)
Marbles (0.9%)
Clays and shales (4.2%)
Sands and sandstones (1.7%)
Carbonates (2.0%)

MINERALOGY (Volume %)

Pyroxenes (11%)

Plagioclases (39%)

Alkali
Feldspars
(12%)

Quartz (12%)

Micas (5%)
Amphiboles (5%)
Olivines (3%)
Clay minerals (4.6%)
Magnetite (1.5%)
Calcite (1.5%)
Dolomite (0.5%)
Others (4.9%)

FIGURE 5-1
Composition, lithology, and mineralogy of the Earth's crust. (Modified from data in Gribble, C. D. *Rutley's Elements of Mineralogy*, 27th ed., Unwin Hyman, London, p. 482, 1988; and Wyllie, P. J.: *The Dynamic Earth*, Wiley, New York, p. 416, 1971.)

and soils with which a geologist commonly deals are composed almost wholly of ionic solids, and it is the structure of these that will be the subject of our inquiry here.

5-1 STRUCTURE OF NaCl

A familiar crystalline solid, ordinary table salt, has an especially simple structure that will serve as a good starting point. Salt occurs in nature as the mineral *halite*, which has the composition of NaCl and is a major constituent of evaporite deposits.

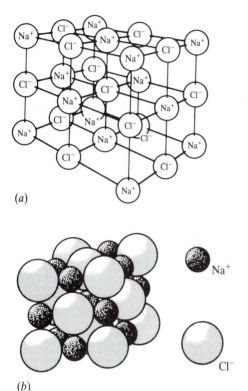

(a)

(b)

FIGURE 5-2
Arrangement of ions in the structure of NaCl.
(a) the relative sizes and distances are not to
scale. (b) Relative sizes drawn to scale.

The arrangement of ions in the structure of halite has a cubic pattern, as shown in Fig. 5-2. Note that Na^+ ions alternate with Cl^-, that every Na^+ is surrounded by six Cl^- ions at the corners of an octahedron and every Cl^- by six Na^+ ions. Figure 5-2(a) shows only the arrangement, not the sizes; actually the ions should be regarded as touching their nearest neighbors, so that the entire volume of the crystal is occupied except for spaces between the spheres, as illustrated in Fig. 5-2(b).

Now what are these objects called ions, which we write as Na^+ and Cl^- and which we represent in diagrams as tiny spheres? In the familiar model, a simple ion like Na^+ or Cl^- consists of a compact positive nucleus made up of protons and neutrons, surrounded by negative electrons. Both nucleus and electrons are tiny compared with their distance apart. The positive charge of Na^+ means that its nucleus contains one positive charge (one proton) in excess of the number of surrounding electrons, and the negative charge of Cl^- means that it contains one electron in excess. All this is simple enough, but difficulties arise if we try to make the picture more definite by specifying the arrangement and movement of the electrons. For some purposes the electrons can be thought of as minute particles revolving around the nucleus in orbits, like planets around the sun, but this model is too simple to describe all aspects of electronic behavior. Actually the positions and motions of the electrons cannot be specified precisely; according to quantum theory, we can only indicate the probability of finding electrons at particular places within an ion or atom. The region of probability for a given electron is described as its *orbital*, rather than an orbit. This gives us a rather fuzzy and elusive sort of unit out of which to build our crystalline structures.

Fortunately the deeper questions about the nature of the electron are not essential for considering the elementary relations of ions in crystals. The important thing here is to remember that an ion consists essentially of widely spaced electric charges, so that its boundaries are actually boundaries of electric fields. We cannot expect, then, that ions will behave entirely like rigid spheres, nor can we expect to assign precise values to their radii or volumes.

The size of an ion, despite the fact that it cannot be defined or measured with great exactness, is a useful concept in discussing crystal structures. Some idea of the magnitude of ionic sizes can be gained by a simple calculation. The molecular weight of sodium chloride tells us that 58.5 g of NaCl must contain 6.02×10^{23} (Avogadro's number) molecules of NaCl, or 6.02×10^{23} Na^+ ions and the same number of Cl^- ions. The density of NaCl is 2.16 g/cm^3, so that these ions occupy a volume of 58.5/2.16, or about 27 cm^3. If we assume that the two ions have roughly the same size, the volume of a single ion would be

$$\frac{27}{2 \times 6.02 \times 10^{23}} = 22 \times 12^{-24} \ cm^3.$$

The diameter of an ion would then be approximately the cube root of this figure or 2.8×10^{-8} cm, and the radius 1.4×10^{-8} cm (generally written 1.4 Å, = 1.4 angstroms). Actually the radii are not equal, Cl^- having a radius of about 1.8 Å and Na^+ about 1.0 Å.

Better values of ionic sizes can be obtained by measurements of spacings of planes in crystal structures by means of x-rays, coupled with measurements of molar refractivities and with calculations from quantum mechanics. Methods of finding ionic sizes differ in detail, so that currently several sets of radii are available. Some of the differences are real, in that ionic radii for some purposes are defined differently, and therefore measured differently, than for other purposes. Some of the differences can be ascribed to experimental uncertainties and to differences in radii selected as standards. The discrepancies are mostly in the second decimal place and are not of great concern here.

The most widely used set of radii is given in Appendix VI, and values for some of the common ions are shown pictorially in Fig. 5-3. Most of the values shown are the so-called octahedral radii, calculated on the assumption that the ion has octahedral coordination—in other words, that its closest neighbors in crystal structures are six other ions at the corners of an octahedron, like the six Cl^- ions around each Na^+ in the NaCl structure. For other types of coordination (for example, tetrahedral, as in quartz) the radii would be slightly different; thus the octahedral radius of aluminum is 0.54 Å and the tetrahedral radius about 0.41 Å. Octahedral coordination is the commonest type for the cations of geologic interest and is a good average between the extremes found in crystals, hence the octahedral radii are the most generally useful.

Cations	Radii	Coordination with O^{2-}	Anions

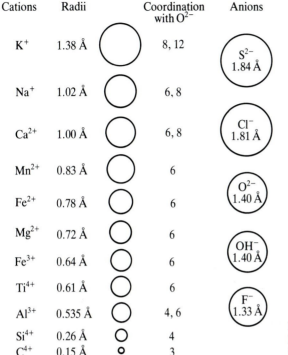

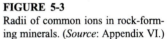

FIGURE 5-3
Radii of common ions in rock-forming minerals. (*Source*: Appendix VI.)

An arrangement like the NaCl structure of Fig. 5-2 represents, for these particular ions, the most stable possible pattern, i.e., the arrangement in which the forces between adjacent Na^+ and Cl^- particles are a maximum, or for which the potential energy is a minimum. If we focus attention on the forces, we may speak of an *ionic bond* between Na^+ and Cl^-, and since NaCl is particularly stable, we would describe this bond as stronger than that, say, in a crystal of AgCl. Or if we think in terms of energy, we could say that the *crystal energy*—the amount of energy needed to tear apart a mol of NaCl into widely separated individual ions—of NaCl is larger than the crystal energy of AgCl.

The arrangement of ions in a crystal would be expected to have a close relation to some of the ordinary large-scale properties of the crystal, particularly its symmetry, cleavage, and effect on polarized light. The cubic pattern of ions in sodium chloride, for example, is reflected in the cubic habit of halite crystals, in the three mutually perpendicular planes of cleavage, and in the ability of light to traverse the crystal with the same speed in any direction. Historically these megascopic properties were studied in great detail long before anything was definitely known about the particles and the geometric patterns of crystal structures. It is one of the great triumphs of twentieth-century science that the inner structures elucidated by x-rays have been found to correlate with and to explain so elegantly the intricate relationships among geometric and optical properties discovered by nineteenth-century crystallographers.

5-2   OTHER IONIC SOLIDS

The pattern of ions in sodium chloride is very simple, and the correlation with elementary crystal properties is transparently obvious. Can we find similar patterns and similar correlations in other crystalline solids? To what extent can the properties of solids be related to geometric arrangements of ions treated simply as small, hard spheres?

We look first at compounds chemically similar to NaCl, consisting merely of an alkali metal joined to a halogen. These show a gratifying similarity, both in megascopic properties and in patterns of ions, except for a few salts of lithium, cesium, and fluorine. These three elements have ions of extreme sizes, small for lithium and fluorine and large for cesium; we might guess, therefore, that an explanation for the exceptions is somehow connected with different ionic volumes. The simplest assumption is that any given combination of ions will form a structure in which the ions are packed as close together as possible, since this would make the forces between them a maximum. The packing arrangement to secure this closest approach would necessarily depend on the relative sizes of the anions and the cations. On this assumption our problem becomes the purely geometric one of how to arrange spheres of different sizes so as to give the densest possible packing.

From solid geometry the following rules can be worked out for three-dimensional structures, which are illustrated in Fig. 5-4:

1. If the radius ratio (radius of cation divided by radius of anion) is unity, the pattern of closest packing has each ion surrounded by twelve ions of opposite charge that are located at the midpoint of the edges of a cube. This is the structure of cesium oxide (see Table 5-1).
2. If the radius ratio lies between 0.73 and 1, closest packing is secured when each ion has eight ions of opposite sign as closest neighbors that are located at the corners of a cube. This is often described as the cesium chloride structure, consistent with the radius ratio $Cs^+/Cl^- = 0.92$.
3. If the radius ratio is between 0.41 and 0.73, the pattern of closest packing has each ion surrounded by six ions of opposite sign located at the corners of an octahedron. This is the sodium chloride structure (radius ratio for $Na^+/Cl^- = 0.56$).
4. If the radius ratio is in the range 0.22 to 0.41, each cation can surround itself with only four anions located at the corners of a tetrahedron. This kind of structure is found in sphalerite, ZnS, with a ratio $Zn^{2+}/S^{2-} = 0.41$, and is often called the sphalerite structure.

For a great many simple compounds these rules hold remarkably well, and crystal structures can be in large measure predicted from geometry.

A further word is needed about nomenclature. In sodium chloride structures, with a pattern of six ions around each ion of opposite sign, the cation is said to have a *coordination number* of 6, or to be 6-coordinated. The anion could equally well be regarded as having a coordination number of 6, but conventionally the term is more

Radius ratio	Coordination number	Atomic structure
1	12	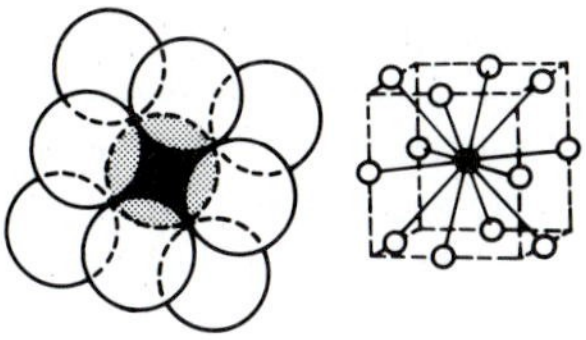
0.73–1	8	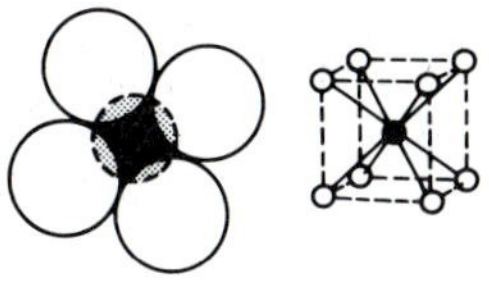
0.41–0.73	6	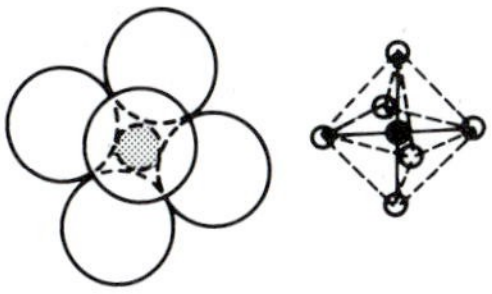
0.22–0.41	4	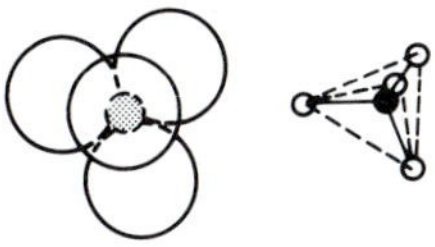
0.15–0.22	3	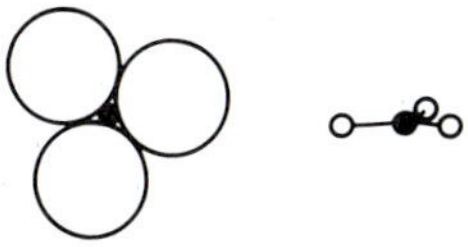

FIGURE 5-4

Relation between radius ratio, coordination number, and the geometric arrangement of anions (open spheres) about a central cation (solid spheres).

commonly applied to the cation. The six ions surrounding a central ion are arranged at the corners of an octahedron (see Fig. 5-4), so that the sodium chloride structure is often spoken of as *octahedral*. Similarly the cesium chloride structure may be referred to as *cubic*, and the cesium ion may be assigned a coordination number of 8. The sphalerite structure is *tetrahedral*, and the zinc ion in sphalerite is 4-coordinated. In mineral structures the commonest anion by far is O^{2-}, with an ionic radius of 1.40 Å; most of the common cations have radii in the range 0.60 to 1.10 Å, giving radius ratios with oxygen of 0.43 to 0.79; hence the most frequent coordination number in minerals is 6. This is why ionic radii are generally tabulated as octahedral radii rather than tetrahedral or cubic.

TABLE 5-1

Comparison of observed coordination numbers with numbers predicted from geometric radius ratios[1]

Ion	Radius ratio: $\left(\dfrac{\text{Ionic radius}}{\text{radius of } O^{2-}}\right)$	Coordination predicted from ratio	Observed coordination numbers	Theoretical limiting radius ratios
Cs^+	1.19	12	12	
Rb^+	1.08	12	8–12	
				———— 1.00
K^+	0.98	8	8–12	
Sr^{2+}	0.84	8	8	
Na^{2+}	0.73	8	6,8	
				———— 0.73
Ca^{2+}	0.71	6	6,8	
Fe^{2+}	0.56	6	6	
Mg^{2+}	0.51	6	6	
				———— 0.41
Al^{3+}	0.38	4	4,6	
				———— 0.22
Si^{4+}	0.18	3	4	

[1] Data from Appendix VI. Radius of O^{2-} assumed = 1.40 Å.

The same sort of geometric argument can be used to make predictions about the crystal structures of more complex compounds, and for many of these the predictions fit observed facts just as admirably as they do for simple ones.

Is crystallography, then, no more than an extension of solid geometry? The answer, of course, is "No." A great deal more than geometry is involved, but we have so far looked only at specially selected examples for which geometric reasoning works fairly well. In view of the complex electronic structure of ions, it is cause for wonder that the simple geometric model of ions as hard round spheres is actually so successful for a large number of substances.

The failure of geometric reasoning as a complete explanation is illustrated by Table 5-1, which shows a comparison between predicted and observed coordination numbers for several cations in compounds with oxygen. For cations whose radius ratios fall well within the theoretical limits, like Sr^{2+} and Mg^{2+}, the agreement of observed and predicted values is excellent. But cations like Ca^{2+}, Na^+, and Al^{3+}, whose ratios are close to the limiting figures, show variable coordination which the simple theory could not predict. A similar table for compounds containing sulfur would bring to light more serious discrepancies. Geometric reasoning breaks down also for complex anion groups like SO_4^{2-}, CO_3^{2-}, AsS_2^-; these groups show smaller interionic distances, and correspondingly greater stability, than would be expected from the sizes of the S^{6+}, C^{4+}, and As^{3+} ions alone. Our next step must be to inquire into the reasons for these deviations from strict geometric rules.

5-3 COVALENT BONDS

The nature of the deviation from radius-ratio predictions may be illustrated with some compounds of calcium and cadmium. These two metals have ions of almost identical radius: 1.00 Å for Ca^{2+} and 0.95 Å for Cd^{2+}. In their compounds with oxygen (ionic radius 1.40 Å) the interionic distances are 2.40 Å for CaO and 2.34 Å for CdO. There is nothing new here: the interionic distance for CdO is a trifle less than the sum of the radii, but the difference is hardly more than the uncertainty in the radii themselves. For the sulfides (ionic radius of $S^{2-} = 1.84$ Å), however, the interionic distances are 2.80 Å for CaS and 2.51 Å for CdS. Obviously something is amiss. It looks as if the ions in cadmium sulfide are squashed against each other, so that the interionic distance is considerably less than the sum of the radii. The crystal structures suggest the same thing: both compounds would be expected to have the NaCl structure (radius ratios 0.54 and 0.52), but the prediction is fulfilled only for CaS; CdS has the sphalerite structure instead, showing that the ions are so deformed that only four rather than six can be grouped around an ion of opposite sign.

To describe the apparent deformation of ions in some crystal structures, we look at the nature of the bond between atoms. In compounds of simple ionic character like NaCl, the formation of a bond between two atoms (an *ionic bond*) involves the complete removal of the outermost electron from one atom (Na) and its incorporation into the structure of the other (Cl), giving the latter a negative charge and leaving a positive charge on Na. Compounds like CdS have bonds of a different sort, in that electrons are partly *shared* between the atoms rather than completely transferred from one atom to another. The shared electrons are commonly grouped in pairs, which may be thought of as occupying positions about halfway between the atoms.

Chemical bonds of this latter kind, consisting of electron pairs held jointly between adjacent atoms, go by the name of *covalent* (or homopolar) bonds. Compounds with covalent bonds differ from ionic compounds, as a rule, in their slight solubilities in water and in their failure to conduct an electric current when melted. The formation of covalent bonds may lead to very stable molecules with little attraction for one another (for example, Cl_2, CH_4, SO_2), in which case the compounds have low melting and boiling points and form solids with little strength or hardness. Alternatively such bonds may give continuous three-dimensional structures in which atoms are linked to others on all sides, as in diamond; in this case the substances have high melting points and exceptional hardness.

The ionic bonds in crystalline NaCl and the covalent bonds in Cl_2 or diamond are extreme types. In most substances the bonds are neither purely ionic nor purely covalent, but have an intermediate character; in other words, the electron pair between adjacent atoms may be regarded as somewhat displaced toward one of the atoms, but still attached to both. In crude symbols, with : representing the electron pair,

Na : Cl	H : Cl	Cl : Cl
ionic bond, NaCl	polar bond, HCl	covalent bond, Cl_2

Bonds of intermediate type are called *polar-covalent* or simply *polar*, the term polar meaning that one end of such a bond is relatively more positive than the other. The partial separation of charge may give polar molecules (like HCl), or it may be compensated by oppositely directed bonds in the same molecule (for example, CCl_4). Water is an interesting case: although we might expect the polarity of the two bonds to compensate each other (H-O-H), the molecule is not linear but bent, as we noted before [Fig. 2-5(*a*)]. The four electron pairs around the oxygen nucleus are approximately at the corners of a tetrahedron; the angle between the O-H bonds is 104.5°, close to the theoretical tetrahedral angle 109°28′. Because of the bending, one side of the molecule has a net negative charge and the other side a positive charge.

Many polar compounds are soluble in water, but they generally do not dissociate appreciably (for example, sugar and alcohol). Some polar compounds, however, react with water to produce ions:

$$HCl + H_2O \rightleftharpoons H_3O^+ + Cl^-. \tag{5-1}$$

This is a more accurate picture of the reaction we usually symbolize as $HCl \rightleftharpoons H^+ + Cl^-$ (Sec. 2-2).

Bonds in minerals, which cover the whole range from ionic to covalent, may be described as having so much "ionic character" or "covalent character." In CaS, for example, the bond is largely ionic, whereas the bond in CdS has considerable covalent character. The geometric radius-ratio rules of the last section apply strictly only to ionic compounds, and show increasing deviations as bonds become more covalent. Fortunately the bonds in many common minerals are sufficiently ionic that the rules hold fairly well, and the effects of covalency can be treated as minor corrections. In other words, to answer a question posed several paragraphs back, mineralogical crystallography can indeed be regarded as largely an exercise in solid geometry, but with many complications arising from the partial covalent character of bonds.

The bonds in radicals like SO_4^{2-} and CO_3^{2-} are largely covalent and particularly strong, so that these groups are stable and remain intact through many chemical reactions. In the structures of crystals, such groups are so compact that to a first approximation they may be regarded as simple anionic units. For example, the structure of calcite may be pictured as a distorted NaCl structure, the Ca^{2+} and CO_3^{2-} ions playing the roles of Na^+ and Cl^-.

5-4 ELECTRONEGATIVITY

So far the discussion of bond type has been entirely qualitative. We have merely noted that bonds in some minerals have a partly covalent character, so that predictions about crystal structures from the radius-ratio rules are not wholly accurate. Obviously it would be desirable to make the concept quantitative—to express the amount of covalent character by numbers, and to relate these numbers to other properties of elements so that the covalent character of a given bond could be predicted. This program has proved difficult to carry out in detail, but several semiempirical schemes have been suggested by which the amount of covalent character can at least be estimated.

The most successful of these schemes is based on a concept called *electronegativity.* The general, qualitative meaning of this word is familiar from elementary chemistry: chlorine is an electronegative element because it readily forms negative ions in solution; sodium is an electropositive element because it forms positive ions; and copper is more electronegative (or less electropositive) than iron because Cu^{2+} will take electrons away from iron metal. The electronegativity of an element is clearly related to its ionization potential (for cations) and its electron affinity (for anions), but the relationship is not simple. Two examples of sets of numbers expressing electronegativity are given in Appendix VI and in abbreviated form in Table 5-2. One set (columns headed "Electronegativity") consists of values calculated by Pauling (1960) from bond strengths as measured by heats of formation, with an arbitrary range from 0.7 for Cs to 4.0 for F. The other set (Percent ionic character) is based on electronegativities estimated by Povarennykh (1956) from ionization potentials and electron affinities; the numbers are not electronegativities as such, but percentages of ionic character of bonds with oxygen calculated from the electronegativities by Smith (1963).

Appendix VI gives an alphabetical listing for easy reference, but Table 5-2 is arranged so as to bring out more clearly the relations of the two sets of numbers with each other and with chemical properties of the elements. From Pauling's numbers we can generalize that a bond formed between any two atoms in the table is almost purely covalent if the electronegativities are similar, and largely ionic if the electronegativities are very different. Note, for example, that the electronegativity difference for NaCl is 2.1, for CaS 1.5, for CuS 0.5, and for CS_2 0.0, in agreement with the increasing covalent character of the bonds in this series of compounds. Smith's percentages express similar electronegativity differences for compounds of

TABLE 5-2

Partial list of electronegativities and percentages of ionic character of bonds with oxygen[1]

Ion	Electro-negativity	% ionic character	Ion	Electro-negativity	% ionic character	Ion	Electro-negativity	% ionic character
Cs^+	0.7	89	Mn^{2+}	1.5	72	Si^{4+}	1.8	48
K^+	0.8	87	Zn^{2+}	1.7	63	C^{4+}	2.5	23
Na^+	0.9	83	Sn^{2+}	1.8	73	P^{5+}	2.1	35
Li^+	1.0	82	Pb^{2+}	1.8	72	N^{5+}	3.0	9
Ba^{2+}	0.9	84	Fe^{2+}	1.8	69	Se	2.4	
Ca^{2+}	1.0	79	Fe^{3+}	1.9	54	S	2.5	
Mg^{2+}	1.2	71	Ag^+	1.9	71	O	3.5	
Be^{2+}	1.5	63	Cu^+	1.9	71	I	2.5	
Al^{3+}	1.5	60	Cu^{2+}	2.0	57	Cl	3.0	
B^{3+}	2.0	43	Au^+	2.4	62	F	4.0	

[1] Data from Appendix VI.

TABLE 5-3
Periodic classification of the elements

Period ↓ / Group →	IA	IIA	IIIB	IVB	VB	VIB	VIIB	VIII	VIII	VIII	IB	IIB	IIIA	IVA	VA	VIA	VIIA	0
1	H 1																	He 2
2	Li 3	Be 4											B 5	C 6	N 7	O 8	F 9	Ne 10
3	Na 11	Mg 12											Al 13	Si 14	P 15	S 16	Cl 17	Ar 18
4	K 19	Ca 20	Sc 21	Ti 22	V 23	Cr 24	Mn 25	Fe 26	Co 27	Ni 28	Cu 29	Zn 30	Ga 31	Ge 32	As 33	Se 34	Br 35	Kr 36
5	Rb 37	Sr 38	Y 39	Zr 40	Nb 41	Mo 42	(Tc) 43	Ru 44	Rh 45	Pd 46	Ag 47	Cd 48	In 49	Sn 50	Sb 51	Te 52	I 53	Xe 54
6	Cs 55	Ba 56	57-71[1]	Hf 72	Ta 73	W 74	Re 75	Os 76	Ir 77	Pt 78	Au 79	Hg 80	Tl 81	Pb 82	Bi 83	Po 84	(At) 85	Rn 86
7	(Fr) 87	Ra 88	89[2]															

Transition Metals

Rare-earth metals[1] (lanthanides)	La 57	Ce 58	Pr 59	Nd 60	(Pm) 61	Sm 62	Eu 63	Gd 64	Tb 65	Dy 66	Ho 67	Er 68	Tm 69	Yb 70	Lu 71
Actinide metals[2]	Ac 89	Th 90	Pa 91	U 92	(Np) 93	(Pu) 94	(Am) 95	(Cm) 96	(Bk) 97	(Cf) 98	(Es) 99	(Fm) 100	(Md) 101	(No) 102	(Lr) 103

Elements whose symbols are enclosed in parentheses do not occur in nature in appreciable amounts, but have been prepared artificially by nuclear reactions. The number for each element is its atomic number.

each cation with oxygen; thus the bonds Na-O, Ca-O, Cu-O, C-O have, respectively, 83, 79, 57, and 23% ionic character. In general, as would be expected, Pauling's numbers are low and Smith's high for the active metals at the left side of Table 5-2, and Pauling's are high and Smith's low for the nonmetals at the right side. Since the two sets of numbers represent two different ways of calculating electronegativity, however, the agreement is far from perfect. Discrepancies are particularly evident among the metals from the middle of the periodic system (*transition metals*) shown in the center of Table 5-2; as a single example, Pauling's numbers would make the Sn-O bond less ionic than the Zn-O bond, while Smith's numbers make it more ionic.

It is a general rule, illustrated by these two sets of numbers and also by several alternative sets that have been proposed as a measure of electronegativity, that numbers can be made to express very nicely the chemical properties of elements at the ends of the periods in the periodic table (Table 5-3), but that unresolvable difficulties arise in trying to express the subtle and complicated relationships among the transition metals in the interior of the table. Electronegativity is a useful concept, but it cannot be depended on for wholly accurate predictions about character of bonding and coordination numbers in all kinds of compounds.

5-5 GENERAL RULES ABOUT BOND TYPE

The ionic radii and electronegativities in Appendix VI permit the formulation of a few useful rules about chemical bonds:

1. For a given cation and two different anions, the bond with the larger anion is more covalent (MgS is more covalent than MgO).
2. For a given anion and two different cations, the bond with the smaller cation is more covalent (MgO is more covalent than BaO).
3. For ions of similar size and different charge, the one with the highest charge forms the most covalent bonds (Ca-O is more covalent than Na-O in Na_2O).
4. Ions of metals in the middle of the long periods of the periodic table form more covalent bonds with anions than do ions of similar size and charge in the first two or three groups of the table (CdS is more covalent than CaS, and FeO is more covalent than MgO).

These rules have many exceptions, but they often help in making qualitative predictions about crystal structures and about the distribution of rare elements in geologic materials.

Such rules, it should be emphasized once again, are based on a crude model of electron-pair bonds linking spherical, somewhat deformable ions. The number of exceptions to the rules is a measure of the crudity of the model. Considerable refinement in predictions is possible by calculating bond energies and bond angles from quantum mechanics (see, for example, Fyfe, 1964), but for most purposes in geochemical arguments the simpler picture is adequate.

Covalent and ionic bonds are the only bond types of much interest in geology, but one other kind needs brief mention. The *metallic* bond, characteristic of all metals, is formed typically in substances whose atoms do not have sufficient valence electrons to fill completely a given stable shell of 8, either by sharing or by transfer. In this case the electrons may be pictured as largely free to wander from atom to atom, hence capable of moving along a metal under the influence of an electric potential difference. Some of the sulfide minerals, especially those with metallic luster, contain partly free electrons and so exhibit to some degree the characteristics of the metallic bond.

5-6   SILICATE STRUCTURES

The crystal structures of most interest in geology are also the most complicated— the structures of the multitudinous compounds called silicates.

We note first that the element silicon is a nonmetal of intermediate electronegativity, that it has an oxidation number of 4 and also a coordination number of 4, and that therefore the fundamental unit of silicate structures ought to be the anion SiO_4^{4-}, in which the silicon ion is surrounded by four oxygen ions at the corners of a tetrahedron. This expectation is borne out in nature, for silicon practically always occurs in SiO_4 tetrahedra. A few silicate minerals contain this group in the form of simple ions; for example, the common mineral olivine (Mg_2SiO_4) has a structure consisting of alternate Mg^{2+} and SiO_4^{4-} ions, much as magnesite consists of alternating Mg^{2+} and CO_3^{2-} ions. In such compounds silicon acts like a typical nonmetal, and its structural chemistry is no more complicated than that of sulfur or phosphorus or the carbon of carbonates. But compared with these other nonmetals, silicon displays astonishing versatility, for its simple ionic compounds are only the first step of an elaborate structural chemistry that includes rings, chains, sheets, and solid frameworks of interconnected silicon–oxygen groups. Why should this particular nonmetal form compounds of such enormous variety?

For an answer we recall that silicon, although a nonmetal, has properties in some measure intermediate between those of nonmetals and metals. A more metallic element like aluminum or magnesium would form a structure with oxygen in which metal ions and oxygen ions are linked together in a strong three-dimensional framework. Silicon exhibits this kind of behavior in quartz and the other silica minerals (tridymite, cristobalite, coesite, stishovite): here silicon remains at the center of SiO_4 tetrahedra, but each oxygen ion is linked with an adjacent silicon ion (Fig. 5-5). A more nonmetallic element like carbon or sulfur would form a volatile oxide consisting of self-contained molecules (like CO_2 or SO_3) having only slight attraction for one another, and would also form ionic compounds with metals in whose structure the nonmetal is part of a compact anion (like CO_3^{2-} or SO_4^{2-}). Silicon forms no volatile oxide, but it does follow the behavior of nonmetals, as we saw in the last paragraph, by entering structures like that of olivine in the form of simple anions. Thus the behavior of silicon straddles the roles of metal and nonmetal. It forms not only structures typical of the extremes, but intermediate structures in which it is part of large silicon–oxygen units that are at the same time anions and more or less continuous frameworks. This dual capacity of silicon, the ability to play the role of metal or nonmetal or anything in between, accounts for the diversity of silicate structures.

This explanation is largely in terms of chemical properties. We could say the same thing in more structural language by noting that the size of the Si^{4+} ion is intermediate between that of the smallest common metal ions (Ti^{4+} and Al^{3+}) and the largest multivalent nonmetal ions (P^{5+}, S^{6+}). If the silicon ion were smaller, it could polarize oxygen ions more effectively, deforming them to fit around it in compact, self-contained molecules or anions. If it were larger, its attraction for oxygen ions would be less, so that Si^{4+} and O^{2-} would simply be independent units in a framework like that of metal oxide crystals. The in-between size gives silicon the ability to perform either function in a crystal structure.

In silicates, then, the silicon ions occur always surrounded by oxygen ions at the corners of a tetrahedron. The tetrahedral groups may be independent anions, or they may be linked together in a variety of ways. The linkage simply means that one or more oxygens in a given tetrahedron are also part of adjacent tetrahedra. The possible varieties of structure permitted by these linkages are illustrated in Fig. 5-5 and described below:

Independent tetrahedral groups (*nesosilicates*). Silicon–oxygen tetrahedra as independent anions; the most familiar example is olivine.

Multiple tetrahedral groups (*sorosilicates*). Two to six tetrahedra linked together to form larger independent anions. A typical example is hemimorphite, $Zn_4Si_2O_7(OH)_2 \cdot H_2O$.

Ring structures of linked tetrahedra (*cyclosilicates*). The commonest example is beryl, $Be_3Al_2Si_6O_{18}$, with rings consisting of six tetrahedra.

Chain structures (*inosilicates*). Tetrahedra linked to form linear chains of indefinite length. Two kinds of chains are found: single chains with a silicon–oxygen ratio of $1:3$, characteristic of the pyroxenes, and cross-linked double chains with a silicon–oxygen ratio of $4:11$, characteristic of the amphiboles. The chains are bonded to one another by metal ions.

Sheet structures (*phyllosilicates* or *layer silicates*). Three oxygens of each tetrahedron are linked with adjacent tetrahedra, forming flat sheets of indefinite extent. In effect this is the double-chain inosilicate structure extended in two dimensions instead of one. The silicon–oxygen ratio is $2:5$. This structure is found in the micas and clay minerals; the sheet structures with a hexagonal pattern are reflected in the perfect basal cleavage and pseudohexagonal habit of these minerals.

Framework structures (*tectosilicates*). Three-dimensional networks, each tetrahedron sharing all its oxygens with adjacent tetrahedra, thus giving a structure with a silicon–oxygen ratio of $1:2$. Quartz and the other silica minerals are good examples (Fig. 5-5). Other minerals may have this structure provided that some of the silicon ions are replaced by ions of lower charge; the commonest substitution is Al^{3+} for Si^{4+}, giving a negative charge to the framework which is balanced by positive metal ions. The feldspars and zeolites are familiar examples of this kind of structure.

Clasification	Structural arrangement

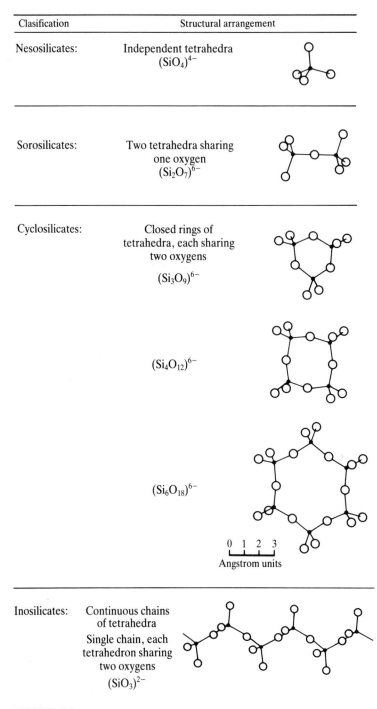

Nesosilicates: Independent tetrahedra
(SiO$_4$)$^{4-}$

Sorosilicates: Two tetrahedra sharing
one oxygen
(Si$_2$O$_7$)$^{6-}$

Cyclosilicates: Closed rings of
tetrahedra, each sharing
two oxygens

(Si$_3$O$_9$)$^{6-}$

(Si$_4$O$_{12}$)$^{6-}$

(Si$_6$O$_{18}$)$^{6-}$

0 1 2 3
Angstrom units

Inosilicates: Continuous chains
of tetrahedra

Single chain, each
tetrahedron sharing
two oxygens
(SiO$_3$)$^{2-}$

FIGURE 5-5
Structural characteristics of the silicates. (Reprinted by permission from Berry *et al.*, 1983.)

Classification	Structural arrangement
Inosilicates:	

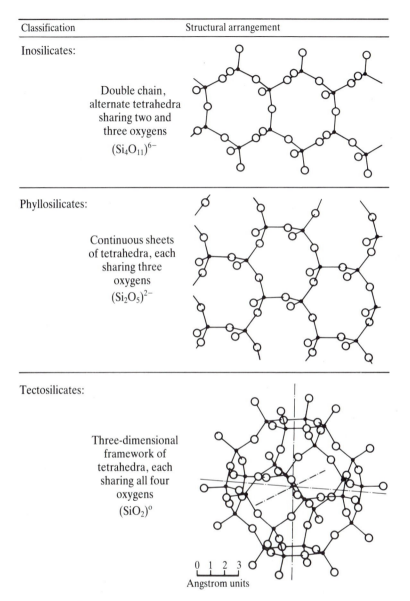

Double chain,
alternate tetrahedra
sharing two and
three oxygens
$(Si_4O_{11})^{6-}$

Phyllosilicates:

Continuous sheets
of tetrahedra, each
sharing three
oxygens
$(Si_2O_5)^{2-}$

Tectosilicates:

Three-dimensional
framework of
tetrahedra, each
sharing all four
oxygens
$(SiO_2)^{0}$

0 1 2 3
Angstrom units

FIGURE 5-5 (contd)

5-7 POLYMORPHISM

Two substances are *isomorphous* if they have similar crystal structures but different chemical formulas, *polymorphous* if they have similar formulas but different structures. Calcite and aragonite, for example, are polymorphs of $CaCO_3$; α-quartz, β-quartz, tridymite, and cristobalite are polymorphs of SiO_2; andalusite, kyanite, and sillimanite are polymorphs of Al_2SiO_5. If a compound is known in only two crystalline modifications, it is often called *dimorphous* rather than polymorphous; thus pyrite is a dimorph of marcasite.

From a structural standpoint, polymorphism means that the ions making up a compound can be arranged in two or more different patterns. For example, rubidium ion, Rb^+, with an ionic radius intermediate between the radii of Cs^+ and Na^+, forms a chloride (RbCl) with the 8-coordinated cesium chloride structure at high pressures and the more open 6-coordinated structure of sodium chloride at ordinary pressures. Another example is the series of carbonates $MgCO_3$, $CaCO_3$, $SrCO_3$, $BaCO_3$, arranged in the order of increasing size of the cation; the compounds of the large ions Sr^{2+} and Ba^{2+} show the aragonite structure and that of the small ion Mg^{2+} the calcite structure, and only $CaCO_3$ with its cation of intermediate size is capable of crystallizing in either structure.

Polymorphism, however, is much more complicated than such a simple rule would indicate. Ideally we can refer it to geometry, but the factors influencing the geometry are numerous and inadequately understood. Temperature, of course, is one such factor, and pressure is another; in general, a particular structure is stable only over a certain range of temperature and pressure. Rhombic sulfur, for example, is stable at temperatures below 96°C, monoclinic above; and the temperature of the change, or *transition temperature*, is raised if the pressure is increased. Speaking still in generalities, we can say that high-temperature polymorphs have more open structures with higher symmetry than their low-temperature equivalents, and that high-pressure polymorphs have more closely packed structures with higher coordination numbers than low-pressure forms. The form in which a substance crystallizes, however, may be changed by impurities, even though these are sometimes present in very minor amounts. The form also depends on rate of crystallization: high-temperature forms sometimes appear in low-temperature environments, apparently because crystallization took place so rapidly that arrangement of ions into the more stable and more ordered arrangement of the low-temperature polymorph was impossible. Cristobalite, for example, although stable only at temperatures above 1470°C, is commonly found encrusting cavities in lava which could not have been at temperatures over a few hundred degrees, either because it contains enough impurities to modify its stability range profoundly or because it crystallized very rapidly.

The influence of temperature and pressure on polymorph stability is illustrated by the three forms of aluminum silicate, Al_2SiO_5, shown in Fig. 5-6. The lines representing solid–solid transformations for the reactions kyanite $\rightleftharpoons$ andalusite, kyanite $\rightleftharpoons$ sillimanite, and andalusite $\rightleftharpoons$ sillimanite were derived from carefully conducted experiments. There are large uncertainties in the positions of the lines for these reactions because of difficulties in establishing equilibrium in laboratory experiments on reactions that involve only solids. Nevertheless, these three minerals are common constituents of metamorphic rocks, and the relative positions of their transformations provide one means of estimating pressure and temperature conditions of metamorphic processes.

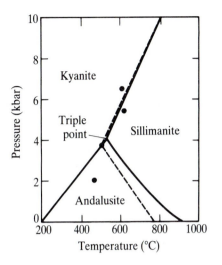

FIGURE 5-6

Pressure and temperature stabilities of Al_2SiO_5 polymorphs kyanite, sillimanite, and andalusite. Solid lines represent equilibrium for the solid–solid transformations as deduced from experimental observations reported by Bohlen, S. R., A. Montana, and D. M. Kerrick: "Precise determinations of the equilibria kyanite–sillimanite and kyanite–andalusite and a revised triple point for Al_2SiO_5 polymorphs," *American Mineralogist,* vol. 76, pp. 677–680, 1991. The dashed lines are those reported by Holdaway, M. J., and B. Mukhopadhyay: "A reevaluation of the stability relations of andalusite: thermochemical data and phase diagram for the aluminum silicates," *American Mineralogist,* vol. 78, pp. 298–315, 1993. The "triple point" shows the pressure and temperature where the three polymorphs are in equilibrium. Experimental values of pressure and temperature for the triple point obtained by others are indicated by the solid symbols.

5-8 ISOMORPHISM AND SUBSTITUTIONAL ORDER–DISORDER

The formula of olivine is customarily written $(Mg,Fe)_2SiO_4$. This means that olivine contains both magnesium and iron, that the ratio of the two metals is variable from one specimen to another but the ratio of total metal to silicon (in atoms or gram-atoms) remains constant, and that magnesium ordinarily is present in greater amount than iron. For this mineral all ratios of magnesium to iron are possible, and the two *end members* Mg_2SiO_4 (forsterite) and Fe_2SiO_4 (fayalite) have a similar crystal form and crystal structure. This relationship is described as *isomorphism*, and olivine is said to be an *isomorphous mixture* of its two end members. Alternatively olivine may be called a *solid solution* of fayalite and forsterite, the term referring to the fact that this solid resembles a liquid solution in that it remains homogeneous when its components are varied over a certain range of compositions. Still another way of describing the relationship of the two metals is to say that Mg^{2+} *substitutes* for Fe^{2+} in the crystal structure.

Isomorphism is a very common phenomenon, and its explanation seems obvious: when two compounds differ in only a single constituent and when the two kinds of ions that play the role of this constituent have similar sizes, we might expect that the two could substitute for one another in crystallographic sites. Thus the ions Mg^{2+} and Fe^{2+} have roughly the same radii (0.72 Å and 0.78 Å, respectively), and the olivine structure can accommodate either or both. A similar isomorphous relationship between magnesium and iron is common in other silicate minerals; many of the pyroxenes [for example, diopside–hedenbergite, $Ca(Mg,Fe)Si_2O_6$] and amphiboles [for example, actinolite, $Ca_2(Mg,Fe)_5(Si_4O_{11})_2(OH)_2$] are familiar examples.

But it does not follow that all pairs of magnesium and iron compounds are isomorphous. Very little magnesium is found in pyrite (FeS_2), and very little iron in epsomite, $MgSO_4 \cdot 7H_2O$, for example. We might guess that the slight difference in ionic size prevents isomorphism in these compounds, or alternatively we recall that iron (a transition metal in the middle of the first long period of the periodic system) should form bonds of more covalent character than magnesium (near the beginning of a period), so that despite the similarity in size the ions can play somewhat different roles in crystal structures. As a general rule, ionic size is the most important determinant of isomorphism, but similarity of bond character is another major factor.

In olivine the two metals can substitute for one another in any proportions; in other words, the two end members of the solid solution are completely miscible. More commonly the amount of substitution is limited, especially if the difference in ionic size between two ions is greater than about 15%. For example, K-feldspar generally has a little Na^+ substituting for K^+, but the sizes are so different (radii 1.02 Å and 1.38 Å, respectively) that only a little can be accommodated without altering the crystal structure. In general, the extent of substitution of one ion for another depends on the properties of the ions and the nature of the crystal framework; it also depends on temperature, since at high temperature the added thermal energy serves to increase the vibratory motion of ions, expanding the structure and making it more tolerant of foreign particles. To continue with the feldspar example, increasing amounts of sodium substitute for potassium as the temperature rises, until at about 650°C the two compounds $KAlSi_3O_8$ and $NaAlSi_3O_8$ become isomorphous and form a complete solid-solution series. On cooling, the isomorphous mixture separates (or *exsolves*) into two phases—the familiar perthite and antiperthite intergrowths of K-rich and Na-rich feldspars. This relationship is shown in the phase diagram of Fig. 5-7. The solid line in the figure, showing temperatures and compositions where Na-rich and K-rich feldspars coexist as two separate phases, is called a *solvus*.

The actual location of the solvus in Fig. 5-7 is uncertain because of slowness of reaction rates in laboratory experiments and because of complications arising from exchanges of Al and Si in the crystal structure. For other common minerals that exhibit unmixing of high-temperature solid solutions (e.g., Ca–Mg carbonates, calcite and dolomite; pyroxenes, Ca-rich and Ca-poor pyroxenes; and ternary feldspars, K-feldspar and plagioclase) experimental difficulties are less serious. For

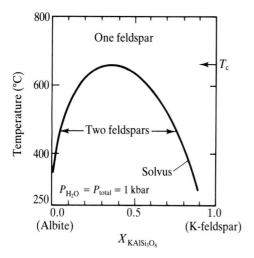

FIGURE 5-7
The alkali-feldspar solvus as a function of temperature and mol fraction of the $KAlSi_3O_8$ component, at a water pressure of 1 kbar. At temperatures above T_c (called the critical temperature), Na^+ and K^+ show complete substitution in the alkali-feldspar structure. At temperatures below T_c, Na-rich and K-rich feldspars coexist, with compositions that are defined by the curve (solvus). (Reprinted by permission from Smith, P., and I. Parsons: "The alkali-feldspar solvus at 1 kilobar water-vapour pressure," *Mineralogical Magazine*, vol. 39, pp. 747–767, 1974.)

these minerals the solvi have been accurately determined, thus permitting compositions of coexisting mineral pairs to be used in estimating the temperature of equilibration during geologic processes.

The alkali feldspars provide an example of a more complicated kind of substitutional relationship. The two ions Al^{3+} and Si^{4+} occupy four tetrahedral sites in the feldspar structure that differ slightly in size and nearest neighbor configuration. In feldspars formed at relatively low temperatures, Al^{3+} ions occupy one kind of site and three Si^{4+} ions the other three, with the sizes of the ions well adapted to the differing properties of the sites. At higher temperatures, say in K-feldspar formed by rapid cooling of magma, the distribution of ions is more random, some Si^{4+} occupying sites better adapted to Al^{3+} and some Al^{3+} in sites appropriate for Si^{4+}. The low-temperature distribution is said to be *ordered*, the high-temperature distribution *disordered*. A disordered distribution is typical of K-feldspar in volcanic rocks, where the random pattern was frozen in by rapid quenching of the lava; this is the feldspar called *high-sanidine*. In contrast, K-feldspars formed in slow-cooling plutons, such as granite, are highly ordered with Al partitioned into the largest and most distorted tetrahedral sites, producing the ordered feldspar called *maximum microcline*.

Similarly, in pyroxenes the cations Mg^{2+} and Fe^{2+} are systematically partitioned between two distinct octahedral sites, designated M1 and M2. The M2 site is slightly larger than the M1 site and thus can accommodate large cations. In an ordered distribution of the two cations, then, we would expect to find the larger cation (Fe^{2+}) concentrated in the M2 sites and the smaller cations (Mg^{2+}) in M1. At higher temperatures we would expect increasing disorder between the two sites. Expectations like this can be summarized by writing the formula for orthopyroxene in the form:

$$(Mg, Fe^{2+})^{M1}(Fe^{2+}, Mg)^{M2}Si_2O_6, \qquad (5-2)$$

which indicates that the composition of the mineral is variable, with Fe^{2+} exchanging for Mg^{2+}, and that for any specified composition Mg^{2+} is preferentially partitioned into the M1 site and Fe^{2+} into M2. These relations are confirmed by the data shown in Fig. 5-8, data obtained by spectroscopic analysis of orthopyroxenes that had been heated at temperatures between 500° and 800°C for several weeks. Clearly the upper curve in the figure shows only small amounts of Fe^{2+} in M1 sites for the relatively ordered structure at 500°C, and a steady increase as temperature rises toward the state of complete disorder indicated by the straight diagonal line.

The curves in Fig. 5-8 are called *distribution isotherms*. They represent equilibrium partitioning of Fe^{2+} and Mg^{2+} between the two types of octahedral sites, which can be described by the intracrystalline reaction,

$$Fe^{2+}_{M2} + Mg^{2+}_{M1} \rightleftharpoons Fe^{2+}_{M1} + Mg^{2+}_{M2}. \tag{5-3}$$

This reaction represents a state of homogeneous equilibrium in pyroxene, and is analogous to chemical reactions we used in Chapter 2 to describe homogeneous equilibrium among ions and complexes in an aqueous solution. One might expect that the isotherms in Fig. 5-8 would provide a basis for predicting the temperature of pyroxene formation by plotting measured site distributions of igneous and metamorphic pyroxenes on the figure. With the exception of rapidly cooled volcanic rocks, temperatures estimated by this technique are unreasonably low relative to those computed using other methods, such as the intercrystalline partitioning of cations discussed below. It turns out that the state of atomic ordering in a mineral may change after formation of the mineral. This change is a complex function of intracrystalline reaction kinetics and of cooling rates. In principle, if the reaction kinetics are understood, the degree of atomic ordering in a mineral may be used to predict cooling rates of igneous and metamorphic rocks.

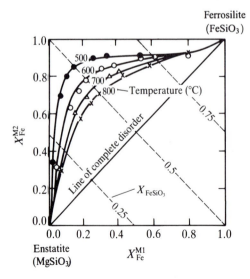

FIGURE 5-8
Distribution of Fe^{2+} and Mg^{2+} in the octahedral M1 and M2 sites in orthopyroxene. Symbols represent measured average site occupancies at various temperatures. (Reprinted by permission from Saxena, S., and G. Ghose: "Mg^{2+}–Fe^{2+} order–disorder and the thermodynamics of the orthopyroxene crystalline solution," *American Mineralogist.* vol. 56, pp. 532–559, 1971, Fig. 1, p. 535.) For reference, the solid diagonal line shows site occupancies for a completely disordered arrangement of octahedral cations. Dashed diagonal lines are constant compositions of orthopyroxene (mol fraction of $FeSiO_3$), ranging from enstatite at the lower left to ferrosilite in the upper right.

The equilibrium intracrystalline partitioning of elements between lattice sites in a mineral, as shown in Fig. 5-8, can also be applied to intercrystalline partitioning of cations between minerals of different structures. As an example, we consider partitioning of Fe^{2+} and Mg^{2+} between the nesosilicate garnet [represented by the end members $Fe_3Al_2Si_3O_{12}$ (almandine) and $Mg_3Al_2Si_3O_{12}$ (pyrope)] and the phyllosilicate biotite [$KFe_3AlSi_3O_{10}(OH)_2$ (annite) and $KMg_3AlSi_3O_{12}(OH)_2$ (phlogopite)]. This is a mineral assemblage commonly formed during amphibolite facies metamorphism of pelitic rocks. The intercrystalline exchange reaction is,

$$Fe_3Al_2Si_3O_{12} + KMg_3AlSi_3O_{10}(OH)_2 \rightleftharpoons$$

garnet   biotite

$$Mg_3Al_2Si_3O_{12} + KFe_3AlSi_3O_{10}(OH)_2 . \quad (5\text{-}4)$$

garnet   biotite

The partitioning represented by this reaction has been experimentally determined at temperatures between 550° and 800°C, and the change in extent of partitioning with temperature makes it possible to determine temperatures of equilibration during metamorphism from measured compositions of garnet and biotite. Values of the distribution coefficient (K_D) for this reaction, defined as

$$K_D = \frac{X_{Mg\text{-}gt}X_{Fe\text{-}bi}}{X_{Fe\text{-}gt}X_{Mg\text{-}bi}} \quad (5\text{-}5)$$

(where $X_{Mg\text{-}gt}$ means the mol fraction of the Mg end member of garnet, etc.), are computed from experiments where garnet and biotite were allowed to equilibrate at temperatures over 550°C. These values are shown in Fig. 5-9. The distribution coefficient, like the concentration constant presented in Sec. 1-8 and the solubility product discussed in Chapter 2, is equal to the thermodynamic equilibrium constant [$K_{(5\text{-}5)}$] only if the activities of the mineral components (a_i) are equal to the mol fractions of the end member components (X_i).

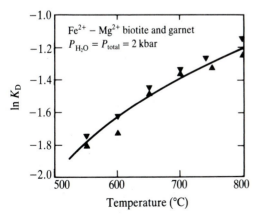

FIGURE 5-9

Temperature dependence of the distribution coefficient for the intercrystalline partitioning of Fe^{2+} and Mg^{2+} between garnet and biotite. Symbols represent experimental partition coefficients (K_D) computed using Eq. (5-5) for the compositions of garnet and biotite equilibrated in laboratory experiments at 2.07 kbar. The line is fitted to the experimental data, as represented by Eq. (5-6). (Reprinted by permission from Ferry, J. M., and F. S. Spear: "Experimental calibration of the partitioning of Fe and Mg between biotite and garnet." *Contributions to Mineralogy and Petrology*, vol. 66, pp. 113–117, 1978.)

The curve in Fig. 5-9 was drawn to fit the experimental data, and is represented by the equation,

$$T = \frac{2109}{0.782 - \ln K_D} - 273, \tag{5-6}$$

where T is in degrees kelvin. This equation allows temperature to be estimated from measured compositions of coexisting garnet and biotite in metamorphic rocks. A word of caution is appropriate, however, in that temperature calculated from Eq. (5-6) will be dependent on pressure and on the possible substitution of other minor components in the garnet and biotite solid solutions, as well as on the extent to which equilibrium partitioning of Fe^{2+} and Mg^{2+} was attained during metamorphism. Geothermometers like this are probably accurate to plus or minus 50°C.

5-9   THE PERIODIC LAW

The classification of the elements according to the periodic law has been mentioned several times in this chapter, and will appear frequently hereafter. It is time to give this basic generalization of chemistry a more formal introduction, and especially to indicate its strong relation to the chemistry of crystals.

Based originally on chemical properties of the elements as observed in ordinary laboratory reactions, especially on the property of valence, Mendeleev's (1869) classification has proved also to be an expression of regularities in the arrangement of electrons in atomic structures. Its importance in chemistry has increased through the years, and it is scarcely less important in geochemistry. Already we have noted the relevance of an element's position in Mendeleev's table to its electronegativity and to the kinds of bonds it forms with other elements, hence to the sorts of crystal structure into which its atoms will fit most readily. This is only the beginning of the many applications we shall consider in later chapters.

In the usual expression of the periodic law as shown in Table 5-3, the horizontal sequences of elements are the *rows* or *periods* of the table and the vertical sequences are the *columns* or *groups*. The number under the symbol for each element is the atomic number, which is equal to the total number of electrons in the neutral atom (and also to the total number of positive charges in the atomic nucleus). For the *main-group elements* (those at either side of the table), the Roman numerals at the top of the columns indicate the number of electrons in the outer shells of the atoms, or the *valence* electrons, the electrons which are shared or transferred during the formation of chemical bonds and which therefore are most important in determining the chemical behavior of an element. These same numbers are also used as designations of element groups: thus the elements N, P, As, Sb, Bi belong to Group V and have atoms containing 5 valence electrons.

The principal relationships of the main-group elements are clear immediately. Exclusive of the inert gases (which form a group at the extreme right), active metals stand at the left and active nonmetals at the right side of the table. Within each row properties show a gradual change with increasing atomic number, metallic activity decreasing and nonmetallic activity increasing. Down each column is a similar but less pronounced change toward elements of increasing metallic character. This means that the most active metals of all are in the lower left corner and the most active nonmetals in the upper right. Ionic radii also change in fairly regular fashion across the rows and down the columns of the table: generally in each row the biggest ions appear at the beginning and end, and in each column the biggest ions are at the bottom. Thus the largest metallic ions appear in the lower left corner, and the largest nonmetallic (negatively charged) ions in the lower right.

Relationships among the *transition metals* in the middle of the table are more complex. These elements have atoms with 1, 2, or 3 valence electrons, and with 8 to 18 electrons in the shell below the valence shell. Since the number of valence electrons is small, the elements are all metals. Many of them can assume several oxidation states in their compounds, because some electrons in the second shell of their atoms ("*d*" electrons) have nearly the same energy as the outer electrons and can act as additional valence electrons. Some of these elements have faint relationships with main-group elements, as indicated by the Roman numerals designating A and B "subgroups." In general, however, these elements show less regularity in their properties than elements in the main groups, and their properties are less readily predictable from their position in the table.

The *lanthanide*, or *rare-earth*, elements and the *actinide* elements, which form part of the transition group but for convenience are shown at the bottom of the table, present still further complications. All the rare-earth elements and most of the actinides have atoms with 3 valence electrons and 8 electrons in the shell below the valence shell. Each atom differs from the one preceding it by 1 electron in the third shell ("*f*" electrons), two shells below the valence shell. Since the outer electron structures are so similar, the elements in each of the two groups show very similar chemical properties, so similar that the elements are difficult to separate in the laboratory and generally occur intimately associated in nature. This generalization is less true for the actinides than for the rare earths, because some of the former have higher stable oxidation states (additional electrons coming from the shell under the valence shell) which make their separation easier.

The importance of the table for crystal chemistry lies in the information it gives about the ability of rare elements to substitute for abundant elements in mineral structures. This is a complex subject that we will explore in more detail later (Chap. 20). For the moment we look only at what the table can tell us about the ability of rare metals to replace major ones in the silicate minerals that make up the bulk of most ordinary rocks.

As we have noted, the really abundant metallic constituents of silicates are limited to six: Na, K, Ca, Mg, Fe, and Al, which exist as ions linked through oxygen to Si. Other metals appear in silicate structures chiefly as ions substituting in minor amounts for one or more of this group. The position of a rare metal in the periodic table generally makes possible a good guess as to which of the abundant metals it can most easily replace, and as to the probable extent of the replacement.

Note that the six abundant ions all belong to Groups IA, IIA, and IIIA, except for Fe; this element is part of the group probably just because of its extreme overall abundance in the Earth's materials. Rare metals whose ions readily substitute for the major ones likewise in large part belong to these same three groups, with some additions from Group IIIB. These are all, or nearly all, elements that are strongly electropositive and form bonds with oxygen of mostly ionic character. Specific substitutions depend chiefly on ionic size, and can be grouped according to size ranges of the major ions:

Ions that replace K^+, with radii well over 1.1 Å

Ions that replace Na^+ and Ca^{2+}, with radii 0.9–1.1 Å

Ions that replace Mg^{2+} and Fe^{2+}, with radii 0.7–0.9 Å

Ions that replace Al^{3+}, with radii less than 0.7 Å

As we have noted before, substitution of one ion for another is generally limited if their sizes differ by more than 15%, and if their numbers of valence electrons differ by more than one unit.

We expect, then, that rubidium (ionic radius 1.52 Å) and barium (1.35 Å) would be found as minor constituents of potassium minerals (radius of K^+, 1.38 Å), that lithium (0.76 Å) might replace magnesium, that yttrium and the rare earths (radii 0.86–1.01 Å) would substitute for calcium more extensively than for sodium, that gallium (0.62 Å) might be found in aluminum minerals, and so on. Predictions like this are not always valid, because in particular cases substitution of ions may be influenced by other factors, but generally they are borne out by silicate analyses.

Thus many of the rare elements found in silicate rocks, as well as the major elements, appear in the groups labeled IA, IIA, IIIA, and IIIB of Mendeleev's table. This clustering of elements simply reflects the fact that the properties of these elements make possible extensive substitution in silicate structures. Many other elements, notably those in Groups IVB, VB, VIB, and VIIB, are also found as common trace constituents of silicate rocks, but for different reasons that we will look into later on.

This brief look at the relation of mineral structures to one part of the periodic table is intended only as an introduction to a set of ideas that constitutes a large part of geochemistry. We shall hear much more about the table and its capacity to bring order into crystal chemistry in later chapters.

SUMMARY

The minerals that compose ordinary rocks are a small number out of several thousands that are known, and these common minerals in turn are made up of only a few abundant elements. The less common minerals and rarer elements are of great interest, and we shall have much to say about them later, but the rocks and soils that occupy most of the geologist's attention are built from surprisingly few constituents.

In most minerals the building blocks are ions, and the ions are put together in the geometrically regular structures called crystals. The patterns in these structures are determined by the sizes of the ions and the nature of the bonds formed between them. The patterns can change with temperature and pressure, since the ions are in vibratory motion and accommodate their geometric arrangements to the space available. In silicate minerals the patterns are especially complex, ranging from isolated SiO_4 groups to rings, chains, and three-dimensional networks of these groups.

For some chemical compositions more than one crystal structure is possible, and each polymorphous structure is stable over a particular range of temperature and pressure. Conversely, some structures can have different chemical compositions, since ions of different elements can occupy the same position in a crystal lattice provided their sizes and bonding properties are sufficiently similar. The substitution of one ion for another may range up to complete replacement, in which case the two structures are said to be isomorphous, or to form a complete solid-solution series. More commonly the substitution is only partial, its extent depending delicately on temperature and pressure. Analysis of the amount of substitution therefore provides a means of estimating temperatures (and sometimes pressures) of mineral formation. We shall find this technique particularly useful in discussions of metamorphic reactions and ore formation.

The ability of crystal structures to accommodate small amounts of foreign ions will also be important when we look into the geologic behavior of the rarer elements, since these often substitute for the ions of more abundant elements in common minerals rather than forming minerals of their own. Substitution of rare-element ions for the more abundant ions can be to a large extent predicted from their position in the periodic table of the elements.

PROBLEMS

1. Why does the zinc ion, Zn^{2+}, in many of its compounds show tetrahedral coordination, although its ionic radius is similar to that of ions like Mg^{2+} which show only octahedral coordination?

2. Why is the radius of Fe^{3+} smaller than the radius of Fe^{2+}? Would you expect, as a general rule, that the more highly charged ions of a multivalent element would have smaller radii than the ions of lower charge? Why?

3. The following substitutions are uncommon in minerals. Inspect each one, to see whether its infrequency of occurrence conforms to the general rules of isomorphous substitution:

$$Li^+ \text{ for } Na^+ \qquad C^{4+} \text{ for } Si^{4+} \qquad Cd^{2+} \text{ for } Na^+$$
$$Cu^+ \text{ for } Na^+ \qquad Sc^{3+} \text{ for } Li^+ \qquad Cl^- \text{ for } F^-$$

4. Why are intergrowth textures (perthite and antiperthite) common in alkali feldspars (albite and orthoclase) but not in the plagioclase feldspars?

5. A possible method of estimating the temperature of formation of sulfide veins is based on a determination of the iron content of sphalerite. Assuming that iron was present in excess in the solutions from which sphalerite crystallized, would you expect the iron content of sphalerite to be greater at high temperature or low? Why?

6. In each of the following pairs, choose the one in which the chemical bond would have more covalent character:

$$KCl \text{ and } KI \qquad Li_2S \text{ and } Cs_2S \qquad BaCl_2 \text{ and } HgCl_2$$
$$KCl \text{ and } BaCl_2 \qquad Cu_2O \text{ and } Cu_2S \qquad B_2O_3 \text{ and } Al_2O_3$$

7. Describe in general terms the crystal structure of diopside, forsterite, analcite, muscovite, tremolite.

8. Why do we not find a variety of complex carbonate minerals with structures analogous to silicate structures?

9. If a cation can show more than one type of coordination, would you expect the higher coordination to be found in minerals formed at high temperatures or at low temperatures? Why? A good example is Al^{3+}, which shows 4-coordination in orthoclase, $KAlSi_3O_8$, and 6-coordination in kaolinite, $Al_2Si_2O_5(OH)_4$.

10. Consider a series of cations having the same electronic structure, for example, the series: Na^+, Mg^{2+}, ..., Cl^{7+}. Do ionic radii show a regular change through this sequence? Can you suggest a reason for the pattern of change?

REFERENCES AND SUGGESTIONS FOR FURTHER READING

Berry, L. G., B. Mason, and R. V. Dietrich: *Mineralogy*, 2d ed., W. H. Freeman, San Francisco, 1983. A standard mineralogy textbook.

Bloss, F. D.: *Crystallography and Crystal Chemistry*, Holt, Rinehart and Winston, New York, 1971. A standard textbook.

Fyfe, W. S.: *Geochemistry of Solids*, McGraw-Hill, New York, 1964. A sophisticated but clearly written treatment of the nature of chemical bonds and their role in crystal structures from the point of view of quantum mechanics.

Gill, R.: *Chemical Fundamentals of Geology*, Unwin Hyman, London, 1989. Chapters 5 through 8 give a good summary of atomic theory, the periodic table, chemical bonding, and the physical and chemical properties of minerals.

Mason, B., and C. B. Moore: *Principles of Geochemistry*, 4th ed., Wiley, New York, 1982. A good brief reference book on all aspects of geochemistry. Chapter 4 gives an excellent elementary survey of crystal chemistry.

Pauling, L.: *The Nature of the Chemical Bond*, 3rd ed., Cornell University Press, Ithaca, New York, 1960. Methods of determining ionic radii are described in pages 511 to 519, and the development of an electronegativity scale in pages 88 to 102.

Shannon, R. D.: "Revised effective ionic radii," *Acta Crystallographica*, Sect. A, pp. 751–767, 1976. A detailed discussion of various influences on ionic radii, and a table of radii for many valence states and coordination numbers.

Smith, F. G.: *Physical Geochemistry*, Addison–Wesley, Reading, Massachusetts, 1963. Chapter 2 is an excellent brief account of the development of ideas about coordination and chemical bonding. It includes a table of electronegativities as estimated by Povarennykh and percentages of ionic character of bonds in oxides and sulfides as calculated from the electronegativities by Smith.

Whittaker, E. J. W., and R. Muntus: "Ionic radii for use in geochemistry," *Geochim. et Cosmochim. Acta*, vol. 34, pp. 945–956, 1970. Comprehensive table of ionic radii, and critical discussion of bases for estimating the radii.

CHAPTER

6

SURFACE CHEMISTRY:
THE SOLUTION–MINERAL
INTERFACE

In the preceding chapters we have discussed geochemical reactions from an overall, generalized point of view—the concentrations of ions and complexes formed when a silicate or carbonate mineral dissolves, the precipitates formed when sulfate is added to solutions of alkaline-earth ions, the change in solubility of an oxide when H^+ or OH^- is added. For some purposes a more detailed treatment is needed. All of these reactions—and, in fact, most geochemical reactions in nature—result from processes taking place at the contact of a mineral surface with a solution, and the nature of the surface may influence both the extent and rate of reaction. In this chapter we look briefly at fine details of mineral surfaces and solutions in immediate contact with them. Such details are often important in controlling the concentrations of substances in solution, especially when a solution is in contact with fine-grained material where the area of exposed surface is very large.

In a practical sense, surface reactions play a major role in determining the quality of groundwater moving slowly through rocks and soils. Changes in the nature and amount of surface exposed can serve as a control on the migration of toxic substances and the redistribution of plant nutrients and pesticides. From a broader point of view, surface reactions serve as an aid in the interpretation of many types of ore deposits and textures of metamorphic rocks.

6-1 NATURE OF THE INTERFACE

Water is in direct contact with minerals throughout a large part of the Earth's crust (Fig. 6-1). Of the rainwater that falls on continents, much seeps into the ground and percolates through soils and sediments where it comes in contact with minerals at the edges of intergranular pores and cavities formed by the dissolution of unstable minerals [Fig. 6-1(b)]. Deeper in the crust, solution-mineral interfaces are associated with fractures, microcracks, and grain-to-grain contacts [Fig. 6-1(c)].

The interface between a mineral and a solution is marked by a discontinuity in the density, composition, and structure of the two phases. The mineral is an organized three-dimensional arrangement of ions in a crystal structure, while the solution consists of freely-moving water dipoles together with solvated ions and complexes. Reactions at the interface cause atomic reorganization in both phases.

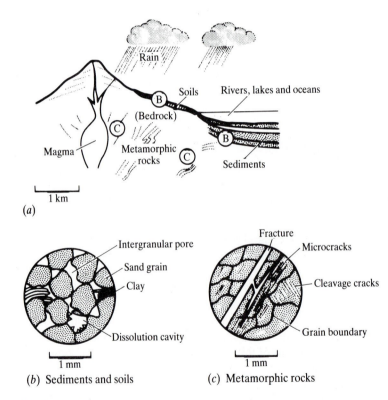

FIGURE 6-1
Schematic illustration of solution-mineral interfaces in the Earth's crust. (a) Rainwater falls on mountains, erodes the surface, and collects in rivers, lakes, and oceans. It percolates through soils, sediments, and into the underlying bedrock where it is involved in hydrothermal systems and metamorphic processes. (b) Thin-section drawing of sediments and soils, showing the pore spaces where solutions are in contact with minerals. (c) Thin-section drawing of metamorphic rocks illustrating fractures, microcracks, cleavage cracks, and grain boundaries where solutions are in contact with minerals.

A mineral surface can be thought of as a large crystallographic defect where ions in the mineral are exposed to the surrounding solution. When a mineral surface is formed, say by fracturing, the surface region may undergo a relaxation that changes the lengths and angles of bonds between ions in the crystal structure. In some cases atoms near the surface move to new equilibrium positions different from those in the bulk mineral. On a microscopic scale the surface is complex, consisting of terraces and numerous edges formed by steps, kinks, and cavities, as illustrated in Fig. 6-2(a). Similar topographic features of mineral surfaces probably occur on the atomic scale [Fig. 6-2(b)].

Mineral surfaces commonly have a non-zero electric charge. The existence of the charge is easily demonstrated by passing an electric current through a solution containing a suspension of very small mineral particles (a colloidal suspension). The particles migrate to one electrode or the other, the direction of motion being determined, of course, by the sign of their charge, which is produced by electrical imbalances within the particles and at their surface. The resulting electrostatic forces modify the arrangement of water dipoles and aqueous species near the interface as ions accumulate at the mineral surface to preserve net electrical neutrality.

One way in which an electrical imbalance may develop in a mineral is by changes in crystal structure near the mineral surface, say by the formation of vacancies or by isomorphic substitution such as Al^{3+} for Si^{4+} or Ca^{2+} for Na^+. A charge so produced is called the *permanent structural charge*. It is typically close to zero for the hydrous oxide minerals and simple layer silicates, such as kaolinite. More complex layer silicates, like montmorillonite, can have a significant permanent structural charge that is usually negative as a consequence of Al^{3+} substitution for Si^{4+}. For the common clay minerals, the number of mols of permanent structural charge per kilogram of mineral is between -0.7 and -1.7 for smectites and montmorillonites, between -1.9 and -2.8 for illites, and between -1.6 and -2.5 for vermiculites.

A net charge at the interface may also be produced by chemical reactions between ions in solution and those at the mineral surface. One important kind of reaction involves surface hydroxyl ions (OH^-) attached to metal cations in the mineral. In silicate, oxide, and hydroxide minerals, for example, surface hydroxyls are formed by reaction of low-coordinated metal cations near the mineral surface with water molecules in the solution, as illustrated in Fig. 6-3. The hydroxyl sites shown in Fig. 6-3(c) are very reactive, in that here a proton from the adjacent solution may be accepted or removed from the mineral surface. For example, the attachment of a hydrogen ion in solution to surface OH's coordinated to a single silicon (Si-OH) or aluminum (Al-OH) atom in the mineral can be represented by the chemical reactions

$$Si-OH + H^+ \rightarrow Si-OH_2^+ \tag{6-1}$$

and

$$Al-OH + H^+ \rightarrow Al-OH_2^+. \tag{6-2}$$

These reactions contribute to the total positive electric charge of the mineral, and are generally favored by interactions with acid solutions. In basic solutions a proton

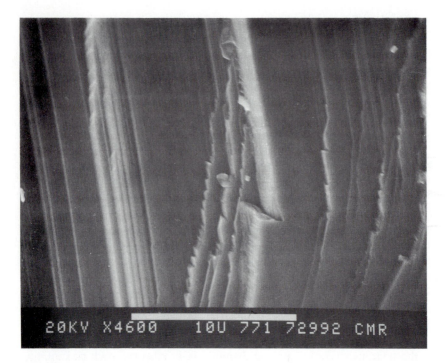

FIGURE 6-2(a)

Photographs of mineral surfaces. (*a*) Scanning electron microscope (SEM) image of a fractured surface of hematite showing the irregular microscopic topography typical of many mineral surfaces. Scale bar is 10 microns in length. Photograph by R. Nevle, Stanford University.

may be lost from the surface OH by

$$Si-OH + OH^- \rightarrow Si-O^- + H_2O \tag{6-3}$$

and

$$Al-OH + OH^- \rightarrow Al-O^- + H_2O, \tag{6-4}$$

reactions that provide a negative contribution to the mineral's electric charge.

Because of such reactions, a mineral's net charge can change sign from positive to negative with increasing pH. The pH at which the mineral's charge becomes zero is called a *point of zero charge* (PZC) or pH_{PZC}. At pH's lower than pH_{PZC} a mineral has a net positive charge due to reactions such as Eqs. (6-1) and (6-2), and for higher pH's the mineral has a net negative charge [Eqs. (6-3) and (6-4)].

The PZC can be measured by noting the pH at which small particles of suspended minerals (colloids) do not move in an external electric field. Measured values of the PZC for some common minerals in dilute solutions are listed in Table 6-1. Note that quartz, albite, and the clay minerals have negative surface charges in all but the most acid of solutions. These minerals attract cations to their surfaces to neutralize the negative charge. Other minerals, for example, the oxides and

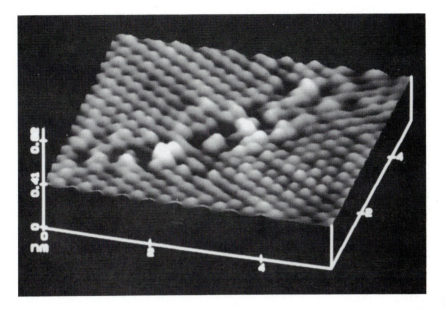

FIGURE 6-2(b)

(b) Scanning tunneling microscope (STM) image of the {001} surface of hematite. The image is tilted back from the viewer to give perspective to the surface. Each peak in the portion of the image showing the hexagonal pattern represents a single oxygen atom on the hematite surface. The oxygen atoms are roughly 3 Å apart. The disturbance running diagonally through the center of the image is most likely a line of atomic defects. Iron atoms do not appear in this image because the microscope conditions are not appropriate for electronic tunneling to the metal. Image collected by C. Eggleston, Stanford University. Reference: Eggleston, C. M., and M. F. Hochella, Jr.: "The structure of hematite {001} surfaces by scanning tunneling microscopy: Image interpretation, surface relaxation, and step structure," *American Mineralogist,* vol. 77, pp. 911–922, 1992.

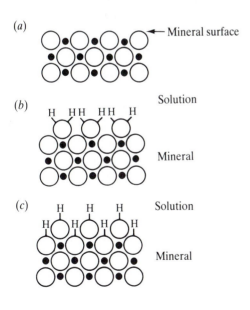

FIGURE 6-3
Schematic cross section of the surface portion of a metal oxide mineral showing the arrangement of metal ions (solid circles) and oxygen ions (open circles). The surface ions have low coordination at the mineral surface (a) and in the presence of water the metal ions coordinate H_2O molecules at the mineral surface (b). These water molecules may dissociate to form a hydroxylated surface of OH ions as shown in diagram (c). (Reprinted by permission from Schindler, P.: "Cation adsorption by hydrous metal oxides and clay," in Anderson, N., and A. Rubin, (eds.): *Adsorption of Inorganics at the Solid/Liquid Interface*, Ann Arbor Science, Ann Arbor, Michigan, pp. 91–141, 1981.

TABLE 6-1
The point of zero charge (PZC:) for some common rock-forming minerals in dilute aqueous solutions. PZC is the pH where the mineral has zero charge [$\sigma_{mineral} = 0$ in Eq. (6-5)]. At pH values less than the PZC the mineral has a net positive charge, at values greater than the PZC a net negative charge.

Mineral	PZC
Quartz	2–3
Albite	2.0
Montmorillonite	2–3
Kaolinite	4.6
Hematite	5–9
Goethite	7.3–7.8
Gibbsite	9.0
Corundum	8.5–9.5

After Stumm and Morgan (1981); Davis, J. A., and D. B. Kent: "Surface complexation modeling in aqueous geochemistry," in Hochella, M. F., Jr., and A. F. White (eds.), "Mineral–Water Interface Geochemistry," *Reviews in Mineralogy*, vol. 23, Mineralogical Society of America, 1990; Parks, G. A.: "The isoelectric points of solid oxides, solid hydroxides, and aqueous hydroxo complex systems," *Chem. Rev.*, vol. 65, pp. 177–198, 1965.

hydroxides of Al (corundum and gibbsite), have positive surface charges in solutions with a pH less than 9, and attract negatively charged aqueous species to their surfaces. The surface charge for oxides and hydroxides of Fe can be either negative or positive in the pH range of most natural solutions. The difference in the PZC between quartz (pH = 2–3) and corundum (pH = 9.0–9.3) reflects the ability of Si-OH and Al-OH surface sites to gain or lose protons over a range of pH as represented by Eqs. (6-1) through (6-4).

Ions other than H^+ can also be attached to or released from mineral surfaces. These ions contribute to the overall mineral charge, and are thus important in determining the kinds of solute species that are attracted to mineral surfaces in groundwater systems. The nature of these surface reactions is a complex function of the reactivity of different surface sites and of the composition and pH of the adjacent solution.

The total net charge of a mineral ($\sigma_{mineral}$, measured in mols of charge per square meter of mineral surface) is the sum of the permanent structural charge (σ_{psc}) and the charge due to surface reactions with the solution (σ_{rexn}):

$$\sigma_{mineral} = \sigma_{psc} + \sigma_{rexn}. \tag{6-5}$$

To neutralize this charge, ions (called *counter-ions*) accumulate near the solution–mineral interface. The arrangement of these ions near the interface forms what is commonly called an *electric double layer* (Fig. 6-4). This can be visualized as two layers of ions, an inner layer attached to the mineral surface by electrostatic forces or by the formation of complexes with atoms at the mineral surface (sometimes called the Stern layer), and a more diffuse outer layer (the Gouy layer) where ions are free to move in the solution. In the Gouy layer the local electrical imbalance of anions and cations (σ_{diffuse}) decreases exponentially away from the mineral [Fig. 6-4(b)]. The distance between the mineral–solution interface and the center of charge within this diffuse outer layer of ions [given the symbol κ^{-1} in Fig. 6-4(b)] is proportional to the square root of the ionic strength of the solution [Eq. (2-70)]. For water at 20°C this distance is

$$\kappa^{-1} \approx 2.8 \times 10^{-8}(I^{-1/2}) \text{ cm}. \tag{6-6}$$

Thus in dilute waters the thickness of the diffuse outer layer of ions is much larger (κ^{-1} is usually 5 to 20 nm) than in saline waters ($\kappa^{-1}_{\text{seawater}} \approx 0.4$ nm). Because the total net charge of the mineral and solution must be zero, an electrical balance exists

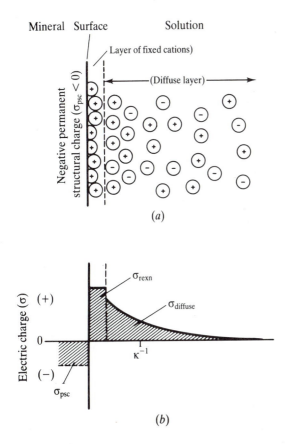

(a)

(b)

FIGURE 6-4
(a) Schematic illustration of the accumulation of ions near the surface of a negatively charged mineral, showing the formation consisting of a fixed layer of cations (the Stern layer) and a diffuse layer (the Gouy layer). Cations and anions in solution are represented by + and −, respectively, and the circles about these ions denote solvation by water molecules. (b) The net charge at the solution–mineral interface. Shaded areas marked σ_{psc}, σ_{rexn}, and σ_{diffuse} schematically denote the permanent structural charge of the mineral, the net electric charge of fixed ions at the solution–mineral interface, including both inner-sphere and outer-sphere complexes, and ions in the diffuse layer, respectively [see Eqs. (6-5) and (6-7)]. The symbol κ^{-1} [Eq. (6-6)] is the distance from the mineral–solution interface to the center of charge in the diffuse layer.

between the mineral charge [Eq. (6-5)] and the net charge of the diffuse layer in the surrounding solution,

$$\sigma_{mineral} + \sigma_{diffuse} = 0. \qquad (6\text{-}7)$$

The diffuse layer of ions may influence the movement of anions and cations in groundwaters flowing through fine-grained sediments. For example, when clay minerals are compacted during sediment burial (perhaps at depths >1 km), the diffuse layer of ions about one clay particle overlaps a similar layer of an adjacent particle. Because clays have a negative net surface charge ($\sigma_{mineral} < 0$), cations accumulate in these diffuse layers to balance the mineral's net charge [Eq. (6-7)]. If a significant portion of the sediment's porosity is occupied by solutions that are part of the diffuse outer layer of ions, water molecules and cations will move through the small pore networks but anions will be partially excluded. Thus some clay layers may act as semipermeable membranes that modify the composition of pore fluids in deep sedimentary basins.

Electrical imbalances within a mineral or at its surface cause a rearrangement of ions in solution and can lead to chemical reactions at the interface. The nature of the atomic sites exposed at the mineral surface and the charge on the surface control the types of aqueous species that accumulate on the mineral surface and the ability of these ions to exchange with other ions in the solution. In the following sections we explore these kinds of processes in detail.

6-2 SORPTION

When a mineral is placed in an aqueous solution, solute ions, complexes, or molecules accumulate onto the surface of the solid material. This process, called *sorption*, has a strong influence on the mobility and uptake of metals and organic compounds in groundwaters, and can ultimately lead to the formation of new minerals from solution.

Sorption occurs by three principal processes. One is by formation of a *surface precipitate* that has a structure or composition different from that of the host mineral [Fig. 6-5(a); in petrology this is referred to as an epitaxial overgrowth]. Another is by incorporating solute species into the mineral structure either by diffusion or by dissolution of the mineral and reprecipitation, a mechanism called *absorption* or *coprecipitation*. A third process, the one we will emphasize in this section, is the accumulation of aqueous species on the mineral surface without formation of a three-dimensional molecular arrangement typical of a mineral, a process referred to as *adsorption*. It is often difficult in both experimental and natural systems to determine the extent to which each of these kinds of sorption occurs. For many geologic problems such details are unimportant.

We might expect that the amount of solute adsorbed on a mineral surface would increase with an increase in its concentration in solution, a prediction that is confirmed by experimental results like those shown in Fig. 6-6. The experimental data are plotted as the amount of Sr^{2+} sorbed on solid MnO_2 as a function of Sr^{2+} concentration in solution at constant pressure, temperature, pH, and solution ionic

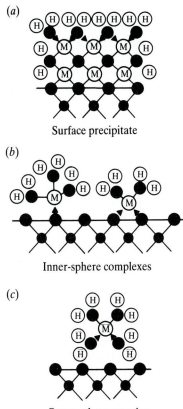

(a)

Surface precipitate

(b)

Inner-sphere complexes

(c)

Outer-sphere complex

FIGURE 6-5
Schematic illustration of a surface precipitate (a), two types of inner-sphere complexes (b), and an outer-sphere complex (c). The horizontal line denotes the boundary between the mineral below and the solution. Small solid circles in the mineral represent metal ions and the large solid circles are oxygens. M represents sorbate cations, and H denotes hydrogens in water molecules solvated about the surface complexes. (Reprinted by permission from Brown, G. E. Jr.: "Spectroscopic studies of chemisorption reaction mechanism at oxide–water interfaces," in Hochella, M. F. Jr., and A. F. White, (eds): *Mineral–Water Interface Geochemistry, Reviews in Mineralogy*, vol. 23, pp. 309–364, 1990.)

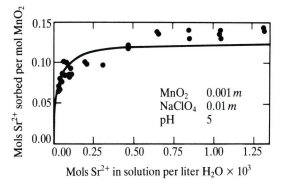

FIGURE 6-6
The symbols are experimentally determined values for sorption of Sr^{2+} on solid MnO_2 as a function of the concentration of Sr^{2+} in solution (25°C, 1 bar). The curve represents the computed Langmuir isotherm [Eq. (6-10)]. (Reprinted by permission from Posselt, H. S., F. J. Anderson, and W. J. Weber Jr.: "Cation adsorption on colloidal hydrous manganese dioxide," *Environmental Science and Technology*, vol. 2, pp. 1087–1093, 1968.)

strength. A line through these data points is called an *adsorption isotherm*, and the one shown in the figure represents one type of isotherm computed for a simplified type of surface reaction. Here adsorption of Sr^{2+} is represented by the reaction

$$Sr^{2+}_{(in\ solution)} + Surface\ sites \rightleftharpoons Sr^{2+}_{(adsorbed)}. \tag{6-8}$$

The equilibrium constant for the reaction is

$$K = \frac{[Sr^{2+}_{(adsorbed)}]}{[Sr^{2+}_{(in\ solution)}][Surface\ sites]}, \tag{6-9}$$

where K is the numerical constant of equilibrium and the brackets represent concentrations: mols per surface area or mols per unit mass of mineral for surface sites and adsorbed Sr^{2+}, and mols per kg water for the aqueous Sr^{2+}. If the maximum amount of Sr^{2+} that can be adsorbed on MnO_2 is defined as the constant $[Sr^{2+}_{(adsorbed)}]^{max}$, which is equal to the sum of [Surface sites] plus $[Sr^{2+}_{(adsorbed)}]$, Eq. (6-9) can be solved for the amount of Sr^{2+} adsorbed:

$$[Sr^{2+}_{(adsorbed)}] = \frac{[Sr^{2+}_{(adsorbed)}]^{max}[Sr^{2+}_{(in\ solution)}]K}{1 + [Sr^{2+}_{(in\ solution)}]K} \tag{6-10}$$

Eq. (6-10) is referred to as the *Langmuir equation*, and the curve for the equation on a graph like Fig. 6-6 is called a *Langmuir isotherm*. At low concentrations of the adsorbing species $[Sr^{2+}_{(in\ solution)}]$, the denominator in Eq. (6-10) is close to unity $(1+[Sr^{2+}_{(in\ solution)}]K \approx 1)$, so the amount of adsorption increases linearly with increasing concentration of $Sr^{2+}_{(in\ solution)}$. For high concentrations of the adsorbing species, where $1 + [Sr_{(in\ solution)}]K \approx [Sr_{(in\ solution)}]K$, the amount adsorbed equals the constant $[Sr_{(adsorbed)}]^{max}$. Some experimental data, like those shown in Fig. 6-6, are closely approximated by the Langmuir equation, but deviations from this idealized behavior are common. Thus to thoroughly explore the nature of sorption in complex geologic systems, it is best to start with conceptual atomistic models of surface processes at the solution–mineral interface.

Adsorption takes place by reactions between a solution and a mineral in which the solute is removed from the bulk solution and attached to the mineral surface. The stable molecular unit that forms is called a *surface complex*, of which there are two common types based on atomic arrangement and bonds formed between the solute and the mineral. In one type, called an *inner-sphere complex*, ionic or covalent bonds are formed between the solute species and a specific crystallographic site on the mineral surface [Fig. 6-5(*b*)]. This kind of complex has no water molecules positioned between the mineral surface and adsorbed species. The other common type is an *outer-sphere complex*, where one or more water molecules are located between the mineral and the solute species [Fig. 6-5(*c*)]. Outer-sphere complexes are loosely attached to the mineral surface by electrostatic bonding, and are not constrained to any specific crystallographic site on the mineral surface. A third kind of adsorption refers to ions in the diffuse layer farther from the mineral surface, ions shown schematically in Fig. 6-4 that contribute to the net charge of

σ_{diffuse} in Eq. (6-7). These *diffuse-ion swarms* do not form discrete complexes attached to the mineral surface, but are solvated ions weakly attracted by electrostatic forces to charged mineral surfaces.

The amount of adsorption depends, in general, on the relative and absolute concentrations of the adsorbing solute (*adsorbate*) and on the surface area of the mineral (*adsorbent*). In experiments conducted at constant temperature, pressure, and initial concentration of adsorbate, pH is the most important variable controlling adsorption. This reflects the influence of reactions such as Eqs. (6-1) through (6-4) on the nature of surface complexation and the total net charge of the mineral. Other factors that can affect the adsorption process are the ionic strength of the solution and the extent of hydrolysis or complexation of the adsorbate ion.

To illustrate some of these features, we consider an example of adsorption on the mineral goethite [FeO(OH)]. This hydrated iron oxide commonly forms a characteristic reddish-yellow surface coating on many kinds of minerals during weathering and hydrothermal alteration. Its surface chemistry is thus important in evaluating the transport and uptake (sorption) of solute components in soils, rivers, lakes, and groundwater systems.

When goethite is placed in a solution that contains small amounts of Pb^{2+}, almost no Pb is adsorbed onto the mineral surface if the solution has a low pH, but nearly all of it is adsorbed if the pH is high. The transition from approximately 0% to >90% uptake occurs over a narrow range of pH. This is illustrated by the data in Fig. 6-7(*a*), which show results of adsorption experiments using three different initial concentrations of Pb in solution.

The equilibrium distribution of aqueous species in the Pb-bearing solution used in one of the experiments in Fig. 6-7(*a*) is shown by the curves in i .g. 6-7(*b*). These curves were calculated using a procedure similar to that used to predict the distribution of Zn-species in NaCl-solutions in Chapter 2. Comparison of the two diagrams in Fig. 6-7 shows that: (1) Pb is adsorbed on goethite over a pH range where the solution is undersaturated with Pb minerals [such as solid lead oxide in Fig. 6-7(*b*)], and (2) there is no direct correlation between the adsorption isotherm and the formation of Pb-complexes in the solution. For the case of Pb on goethite, the experimental studies indicate that the dependence of adsorption on pH is a consequence of surface reactions that lead to the formation of inner-sphere surface complexes. One such reaction can be written as

$$Fe-OH + Pb^{2+} \rightarrow Fe-OPb^+ + H^+, \qquad (6-11)$$

which represents the formation of a Pb-surface complex ($Fe-OPb^+$) where Pb bonds directly to a surface oxygen coordinated to a single Fe^{3+} in goethite.

Varying the total ionic strength of the solution in a series of adsorption experiments provides interesting insights into atomic-scale surface processes. For transition metals, such as Pb on goethite, the percent of metal adsorption is independent of the ionic strength of the solution as illustrated by the data shown in Fig. 6-8(*a*). This suggests that the nature of mineral surface sites and their reactions with adsorbate ions, as illustrated by Eq. (6-11), are most important in controlling adsorption. In contrast, experiments using alkaline-earth metals as adsorbates [such

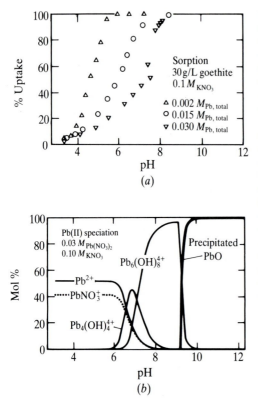

FIGURE 6-7
(*a*) Experimentally determined adsorption of
Pb^{2+} on goethite. The experiments were conducted with 30 grams of goethite per liter of
0.1 M_{KNO_3} solution for three initial concentrations of Pb at 25°C and 1 bar ($M_{Pb,total}$,
denotes the total molarity of dissolved Pb).
(Experimental data (symbols) are from Roe,
A. L., K. F. Hayes, C. Chisholm-Brause, G. E.
Brown, G. A. Parks, K. O. Hodgson, and J. O.
Leckie: "X-ray adsorption study of lead
complexes at α-FeOOH/water interfaces."
Langmuir, vol. 7, pp. 367–373, 1991.)
(*b*) Calculated distribution of aqueous species
in the 0.03 $M_{Pb,total}$ solution used in one of the
experiments shown in diagram (*a*). The thin
curves represent the mol percent of Pb species
as a function of pH calculated for homogeneous equilibrium in a solution that is not in
contact with goethite, using equations and
data summarized in Chapter 2. The thick
curve denotes the pH where the solution is
saturated with solid PbO (Reprinted by permission from Parks, G. A.: "Surface energy
and adsorption at mineral/water interfaces: an
introduction," in Hochella, M. F. Jr., and A. F.
White, (eds): *Mineral–Water Interface Geochemistry, Reviews in Mineralogy,* vol. 23, pp.
133–176, 1990.)

as Ba^{2+} in Fig. 6-8(*a*)] show a strong dependence between adsorption and ionic
strength. Note in the figure that at constant pH the amount of Ba^{2+} adsorbed
decreases with an increase in ionic strength, indicating that electrostatic forces and
competition with other ions in solution for positions near the mineral surface are
important features controlling adsorption.

Recent spectroscopic studies suggest that the difference in adsorption
behavior between transition metals and metals of the alkali and alkaline-earth
groups is explainable by the kinds of surface complexes they form. The transition
metals characteristically form inner-sphere complexes, so that their adsorption is
largely insensitive to the mineral surface charge and the properties of the solution.
They are tightly bound to the mineral surface and are not readily exchanged with
other solute species. The nature of the inner-sphere complexes appears to change
with the amount of sorption. For example, investigations of Pb on goethite, and of
Co on Al-oxide, rutile, quartz, and kaolinite, have shown that at low concentrations
the adsorbates form mononuclear inner-sphere complexes represented by reactions
like Eq. (6-11) and schematically illustrated in Fig. 6-5(*b*). With increasing
adsorbate concentration, clusters of multinuclear surface complexes are formed
(Fig. 6-9) which may act as nucleation sites for the growth of metal-hydroxide
surface precipitates as their saturation in the fluid phase is approached. In contrast to

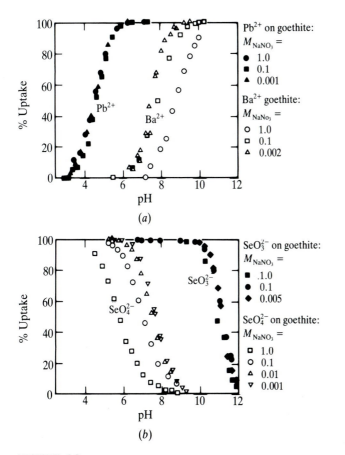

FIGURE 6-8
Adsorption of Pb^{2+}, Ba^{2+}, SeO_4^{2-}, and SeO_3^{2-} on goethite in solutions of varying ionic strength. The ionic strength of solutions used in these experiments is changed by varying the total amount of $NaNO_3$ in solution from $0.001M$ to $1M$ [see Eq. (2-70) for calculation of ionic strength]. (Experimental data from Hayes, K. F.: "Equilibrium, spectroscopic, and kinetic studies of ion adsorption at the oxide/aqueous interface," Ph.D. dissertation, Stanford University, 1987.)

the transition metals, alkali and alkaline-earth ions are considered to be adsorbed as outer-sphere complexes and as diffuse-ion swarms, loosely bonded to the mineral surface and readily exchanged with other ions in solution. Adsorption of these ions is sensitive to changes in the mineral's surface charge and to properties of the solution like ionic strength.

Sorption of anions on oxide and hydroxide minerals shows features similar to those described above for cation adsorption on goethite, except that the pH dependence is just the opposite, as shown in Fig. 6-8(b). At low pH's there is nearly complete anion adsorption, and at high pH's very little. The data in the figure illustrate a further complication of adsorption processes, in that the oxidation state of the adsorbate may be important in determining its interaction with the mineral

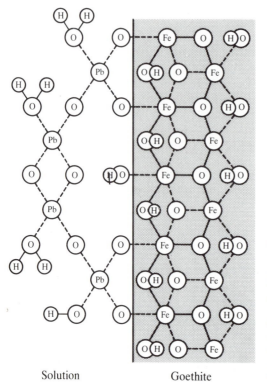

Solution   Goethite

FIGURE 6-9
Schematic two-dimensional representation of a multinuclear surface complex of Pb bound by two inner-sphere surface complexes to goethite. (Reprinted by permission from Roe, A. L., K. E. Hayes, C. Chisholm-Brause, G. E. Brown, G. A. Parks, K. O. Hodgson, and J. O. Leckie: "X-ray adsorption study of lead complexes at α-FeOOH/water interfaces," *Langmuir*, vol. 7, pp. 367–373, 1991.)

surface. Note that adsorption of selenite (SeO_3^{2-}) is essentially independent of the ionic strength of the solution, but that adsorption of selenate (SeO_4^{2-}) appears to be a function of the solution's ionic strength. The selenite probably adsorbs in the form of inner-sphere complexes, selenate as outer-sphere complexes. Similar to selenite in this respect are phosphate, borate, arsenate, and carboxylate, while chloride, nitrate, chromate, bisulfide, sulfate, bicarbonate, and carbonate resemble selenate in adsorbing largely as diffuse-ion swarms or outer-sphere complexes.

A practical example illustrating the importance of adsorption and its dependence on solution pH is found in the leaching of heavy metals from waste material produced during mining of sulfide-rich ore deposits. Consider the geologic setting illustrated in Fig. 6-10, showing an ore deposit rich in the sulfide minerals galena (PbS) and pyrite (FeS_2). The ore deposit was formed at the contact where a granite pluton intruded a limestone. Uplift and erosion have exposed the ore deposit, which was mined for its metals. Tailings of waste rock from the mine were dumped in a nearby ravine exposing numerous fresh grains of sulfides to reactions with rainwater as it percolates through the tailings. Dissolution and oxidation of the pyrite have reduced the solution pH in the tailings to very low values (possibly < 2), according to a reaction that may be symbolized,

$$2FeS_2 + 7.5O_2 + H_2O \rightarrow 2Fe^{3+} + 4SO_4^{2-} + 2H^+. \tag{6-12}$$
pyrite

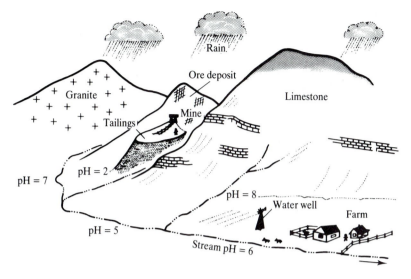

FIGURE 6-10
A schematic geologic setting illustrating heavy metal transport and deposition in an acid mine drainage (see text for details).

The Fe^{3+} produced during pyrite alteration is transported in stream and ground-waters down the drainage system where it precipitates as a coating of fine-grained goethite on sediments and soils. Precipitation results from downstream increase in the stream's pH as the water mixes with runoff waters that have reacted with the nearby limestone and granite in the bedrock. A similar sequence of events— oxidation of pyrite, movement of Fe^{3+} in highly acid solutions, precipitation of the hydrated oxide as acidity diminishes—can be followed on a large scale in the waste heaps from open pit mining of pyrite-bearing coal, and has become a major environmental problem where coal is being or has been mined in large amounts.

From the scenario in Fig. 6-10 we can make some simple predictions concerning the transport and deposition of Pb in the drainage system. Galena reacts with H_2CO_3 and O_2 in rainwater that percolates through the tailings, releasing Pb^{2+} into solution:

$$PbS + 2H_2CO_3 + 2O_2 \rightarrow Pb^{2+} + 2HCO_3^- + SO_4^{2-} + 2H^+. \qquad (6\text{-}13)$$
galena

The Pb concentration of the solution will be determined by such factors as the dissolution rate of galena, the solubility of Pb minerals such as anglesite ($PbSO_4$) or cerussite ($PbCO_3$), and the flux of rainwater through the tailings. Because the pH is low (< 2) we predict from Fig. 6-7(a) that very little of the dissolved Pb will be adsorbed onto goethite that coats minerals in the mine tailings. The Pb is transported downstream until the solution pH increases to > 6, where nearly 100% of the Pb in solution is adsorbed onto the surface of goethite in the sediments and soils [see Fig.

6-7(a)]. Because Pb forms tightly bound inner-sphere complexes on goethite, it cannot be readily removed from these sediments and soils unless the pH is lowered significantly. It is through processes of this sort that toxic heavy metals are introduced into and concentrated within groundwaters, sediments, and soils, where they may be harmful to biological communities. Such concentrations of sorbed metals, although sometimes a health hazard, have been found useful in geochemical prospecting for ore deposits.

6-3 ION EXCHANGE

Ions that accumulate on mineral surfaces are held by forces that range from weak to very strong, depending on the kind of mineral and the kind of ion. Those adsorbed as outer-sphere complexes or diffuse ion swarms (see Figs. 6-4 and 6-5) are easily replaced by other ions in solution. If a mineral with one kind of ion adsorbed on its surface is added to an electrolyte solution containing different ions, some of the original ions will be set free in the solution and some of the new ions will be adsorbed in their place. This phenomenon goes by the name of *ion exchange.*

Ion exchange is not limited to ions adsorbed on mineral surfaces, but may be applied also to the substitution of one ion for another in crystals. The most familiar example of this kind of ion exchange is the process commonly used for softening water. If hard water containing Ca^{2+} ions is passed over crystals of a zeolite (usually an artificial zeolite) with Na^+ as part of its crystal structure, the Ca^{2+} takes the place of the Na^+ in the crystals and the relatively harmless Na^+ is added to the water. When most of the Na^+ has been replaced, the zeolite must be regenerated, which is accomplished by passing over it a concentrated NaCl solution. The ion exchange is now reversed because of the high concentration of Na^+, and the crystals become once more a Na-rich zeolite. In symbols the reaction may be thought of as an equilibrium,

$$Na-zeolite + Ca^{2+} \rightleftharpoons Ca-zeolite + 2Na^+, \qquad (6\text{-}14)$$

and the equilibrium may be displaced in one direction or the other by changing the relative concentrations of Na^+ and Ca^{2+}. This is a particularly clear and simple example of ion exchange, but more complicated cases can always be broken down into similar exchange equilibria.

Two other important examples of exchange between aqueous ions and ions in a crystal lattice are provided by carbonate minerals and alkali feldspars. Heterogeneous equilibrium between dilute water solutions and pure (stoichiometric) calcite, albite, and K-feldspar was discussed in Chapters 3 and 4. These minerals all exhibit isomorphic substitution of ions in their crystal structures (see Sec. 5-8), which changes their composition and affects their equilibria with the solution. In calcite ($CaCO_3$), for example, substitution of x mols of Mg^{2+} for Ca^{2+} forms magnesian calcite ($Ca_{1-x}Mg_xCO_3$), which is the composition of the shell material of many marine organisms. This material is a solid solution of x mols of $MgCO_3$ and $1-x$ mols of $CaCO_3$. Equilibrium between a magnesian calcite and a solution can be represented by an ion-exchange reaction similar to Eq. (6-14),

$$CaCO_3 + Mg^{2+} \rightleftharpoons MgCO_3 + Ca^{2+}. \tag{6-15}$$

If we assume, for simplicity, that magnesian calcite and the fluid are ideal solutions (that is, all activity coefficients are equal to unity) the equilibrium constant,

$$K = \frac{m_{Ca^{2+}} X_{MgCO_3}}{m_{Mg^{2+}} X_{CaCO_3}}, \tag{6-16}$$

defines the relationship between the ratio of Ca^{2+} to Mg^{2+} in the fluid and the composition of the coexisting magnesian calcite. In a similar way we can write for an alkali feldspar, where K^+ substitutes for Na^+,

$$NaAlSi_3O_8 + K^+ \rightleftharpoons KAlSi_3O_8 + Na^+. \tag{6-17}$$

The ability of an ion in solution to exchange with an ion in the crystal structure of a mineral is a complex function of its bonding in the crystal lattice and the mechanism of exchange. Ions loosely held in the crystal structure, like the inner-layer cations of clay minerals, are more readily exchanged than cations in octahedral or tetrahedral coordination in the sheet silicate structure. Although the mechanisms of ion exchange are poorly understood, they probably involve solid-state diffusion and selective mineral dissolution followed by reprecipitation incorporating new ions from the solution. These processes are collectively referred to as absorption (Sec. 6-1).

The exchangeability of adsorbed ions on mineral surfaces depends on how they are attached. The weakest bonds to mineral surfaces are those formed by ions in diffuse-ion swarms and outer-sphere complexes [Figs. 6-4 and 6-5(c)]. For the more tightly held inner-sphere and multinuclear surface complexes [Figs. 6-5(b) and 6-9], the kind of bonding is determined by the nature of the various atomic sites exposed on the surfaces, corners, and edges of a mineral. Ion attachment at the edges of mineral particles is not very different from actual substitution in the crystal structure itself, the sort of substitution that is involved in water-softening by zeolites. Thus we encounter all gradations, from ions so loosely held to surfaces that they are readily exchanged for any others, to ions that are integral parts of a crystal structure. These latter are readily exchanged in minerals with particularly open structures like zeolites, but more commonly are exchangeable only when they are on corners or edges of crystal fragments.

In experiments on ion exchange it is seldom possible or even desirable to determine just what kinds of attachment of ions to surfaces are involved. Usually the important consideration is that some surfaces or some kinds of minerals can be shown to hold adsorbed ions much more strongly, and to hold them in much greater amounts, than do others. For example, most cations are more effectively removed from solution by shaking the solution with montmorillonite than with kaolinite, when the two clays are added in equal amount and have comparable grain size.

For any given kind of particle or surface, different ions show a wide range of sorption. The behavior of zeolites in water softening is a good example: Ca^{2+} is removed by Na-zeolite even from a very dilute solution, but to regenerate the Na-zeolite requires that the Ca-zeolite be treated with a very concentrated Na^+ solution. In other words, equilibrium is established for Eq. (6-14) when the ratio m_{Na^+} to $m_{Ca^{2+}}$ is much greater than 1. Experiments of this kind, involving measurements of the amount of one ion released from a sorbent by a given concentration of another, provide us with a measure of the relative degree of ion exchange. Unfortunately, most experiments do not lead to simple rules about the behavior of ions toward sorbents, apparently because sorption depends on too many factors: the nature of the sorbent (chemical composition, structure, particle size, method of preparation, aging), the temperature, and the kind and amounts of other ions present in the solution.

When more than one type of sorbate ion is present in solution, the different ions compete for positions near the solution–mineral interface. A list of ions in the order of the strength of their bonds with a mineral surface is called a *selectivity sequence*. It is important for determining the extent of uptake of a particular sorbate and the effect of one ion on the uptake of another in complex geologic systems. As a first guess, we might suppose that the relative sorbability of two ions would depend on such properties as their sizes, their charges, and their ability to form covalent bonds, the same properties that determine their roles in crystal structures. It would seem natural, for example, that (1) ions that form inner-sphere complexes would have a greater affinity for a mineral surface than those forming only outer-sphere or diffuse-ion swarms, (2) the smaller of two ions would be more firmly held to a surface than the larger, (3) a multivalent ion would be more firmly attached than a univalent ion, and (4) an ion whose bonds have strong covalent character would be more readily adsorbed than one whose bonds are dominantly ionic. In some experiments these expectations are fulfilled, but exceptions are so numerous and so serious that the generalizations are not very useful. The second one, for example, is often violated. We would predict, for example, that Na^+ should be more readily adsorbed from solution than its larger relative K^+, but experimental results are usually just the reverse. The entire series of alkali metals, in fact, in most adsorption experiments shows a progressive *decrease* in sorbability from Cs^+ to Li^+, so that:

alkali cation selectivity sequence

$$Cs^+ > Rb^+ > K^+ > Na^+ > Li^+$$

decrease in the relative degree of sorption $\Rightarrow$

This is in direct contradiction to the second rule above. One can explain away the discrepancy by noting that the smallest ion, Li^+, is the most highly hydrated and therefore effectively larger than Cs^+ in water solution, but this introduces yet another variable into adsorption experiments.

In addition to solute properties, the nature of mineral surface sites and the charge on mineral surfaces are also important factors in determining selectivity sequences of groups of ions on different minerals. Consider, for example, transition metal cations that commonly form inner-sphere complexes on oxide and hydroxide minerals. Because the formation of these surface complexes is a sensitive function of pH-dependent surface reactions involving different types of atomic sites on the mineral, there is little consistency in the experimentally determined selectivity sequences for these cations.

Ion-exchange involving clay minerals plays an important role in the chemistry of soils and surface waters. Clay minerals occur as flat plates like mica flakes, much smaller in one dimension than in the other two. In a given sample the grains may have an enormous range in size and surface area, from those easily visible with an optical microscope to those whose dimensions can be determined only by electron microscopy. The flattened shapes of clay particles, together with peculiarities of the crystal structure discussed in Chapter 5, give opportunity for sorbing ions in a variety of ways. Some ions are held to broken bonds on the edges of the flakes, some to the flat surfaces; some make their way into spaces between the layers of the crystal structure, and some enter into the structure of the crystal itself to take the place of one of its constituents. Ions in these different positions vary greatly in their exchangeability; some are replaced by others even from dilute solutions, but some are so firmly held that exchange is slight even with concentrated solutions. Exchangeability also depends on the kind of ions; some ions by reason of charge or size fit into the clay structure better than others. Still another factor in determining ion exchange is the kind of clay mineral or minerals present, since the clays differ greatly among themselves in their capacity for picking up and holding foreign ions. In general the ion exchange capacity of clay minerals is very high for vermiculites and smectites and relatively low for illites and kaolinites. The processes of ion exchange in clays are thus extremely complicated and only partly understood. This is a subject of much current research interest, for the properties of clays as raw materials for agriculture, ceramics, and engineering purposes depend in large part on the quantity and nature of their adsorbed ions.

The ability of clay minerals to adsorb water is another important surface property affecting the physical characteristics of clays and soils. Different clays show enormous variation in the amount of water they can absorb. Kaolinite clays are relatively nonplastic, because their particles hold only a little water. A pure montmorillonite clay, in contrast, adsorbs water until the clay swells to several times its original volume and acquires a gel-like consistency. Most clay samples show behavior between these extremes.

The geochemistry of ion exchange is important for two principal reasons. The more obvious one is the mechanism it provides for redistribution of metal ions between solution and sediments. Thus the high Na^+/K^+ ratio in seawater can be partly explained by the greater ease of sorption of K^+ on clays, organic matter, and chemical precipitates; and the high concentrations of metals like Co^{2+} and Pb^{2+} often found in manganese ores can be accounted for by sorption of the ions on manganese dioxide minerals. The second important effect of ion exchange is the influence of different ions on the properties of the adsorbing substance. This is especially significant for clays; the outstanding example is the difference in properties between clays containing chiefly adsorbed Na^+ and clays containing chiefly Ca^{2+} ions, the former being sticky and impervious while the latter are granular, easily worked, and readily permeable. For agricultural purposes the calcium clays are more desirable, and gypsum is often added to convert soil clays into the calcium variety. For water reservoirs, on the other hand, it is sometimes necessary to convert calcium clays to sodium clays by adding brine to make the bottom sediments less permeable.

Despite the importance of sorption and ion exchange, our knowledge regarding them is still largely empirical. So many variables are involved, and some are so difficult to control, that experiments all too often lead only to broad generalizations rather than to specific rules. In discussions to follow we shall often mention sorption and ion exchange as important geologic processes, but we shall seldom be able to predict just what their effects should be in particular situations.

6-4   COLLOIDS

The term "colloid" goes back to a period in the history of chemistry when a distinction was attempted between "crystalloid" matter on the one hand and noncrystalline, or "colloid" matter on the other—the noncrystalline matter including substances like glue, soap, and gelatin which dissolved in water to give viscous solutions and which could not be obtained in the form of crystals. The distinction proved to be artificial. The characteristics associated with colloids turned out to be simply the properties of matter in a finely divided state, and any substance that could be finely divided would exhibit these properties. Hence there is no such thing as a "colloid" in the original sense of the word. In modern chemistry a substance is considered colloidal, or in a colloidal state, if it consists of very tiny particles dispersed in another substance. Just how small the particles must be is not universally agreed upon, but a widely used definition puts the range of colloidal diameters at 0.01 to 10 µm.

In geology we are concerned almost exclusively with water as the dispersion medium and with crystalline or amorphous solid material as the dispersed particles. The most notable exception is a colloidal system consisting of water suspended in petroleum, or petroleum in water; such a dispersion of one liquid in another is called an *emulsion*. In the geochemistry of the atmosphere, dispersions of fine solid particles and liquid droplets in the atmospheric gases, called *aerosols*, are sometimes important. Some naturally occurring crystals may contain inclusions of solid, liquid, or gas in the colloidal size range, thus forming systems with a solid dispersion medium, but the colloidal properties of such occurrences have only minor geologic significance.

Colloidal suspensions in flowing streams, groundwaters, and hydrothermal solutions are an efficient means of element mass transfer. The enormous surface area of the colloidal particles makes reactions at the solution–particle interface a critical factor in determining both the migration and fixation of elements. Thus the geochemistry of colloidal phenomena is a part of surface chemistry essential to studies of groundwater and soil contaminants, weathering, diagenesis, and the formation of some types of ore deposits.

Properties of Colloids

Many solid-in-liquid colloidal systems may be prepared in two forms. If gelatin, for example, is dissolved in warm water, it forms a clear, transparent liquid which to all appearances is an ordinary solution. To specify such a liquid, actually a colloidal system but resembling a solution in its transparency and fluidity, we use the term *sol*. If the gelatin sol is cooled and allowed to stand, its character changes: we say it "sets," forming a translucent or transparent solid called a *gel* where the colloidal particles form a network of fibers or chains in which water is trapped. Many colloids, like gelatin, may be obtained in either a sol or gel form, depending on the concentration, methods of preparation, and time of standing.

Gelatin is an example of a large group of colloidal substances that dissolve directly to form disperse systems. The resulting systems are stable indefinitely, the suspended particles showing no tendency to settle out. Such colloids, provided the concentration is high enough, readily set to gels, and the process of gelation is reversible, in that the gel disperses to form a sol if more water is added. The ease with which the gels form and disperse, together with other less direct evidence, suggests that particles of these substances are accompanied in the sol by much adsorbed water. An appropriate name for such colloids is *hydrophilic*, which in Greek means "water-loving."

In contrast, a great many substances do not disperse themselves spontaneously but may be obtained by indirect means. If a reducing agent is added to a solution of a gold compound, for example, the resulting metallic gold may not precipitate, but may instead remain suspended in the form of colloidal particles, which color the solution bright red ("ruby gold") because of their ability to scatter light. Similarly, if one attempts to precipitate arsenic sulfide by bubbling hydrogen sulfide through an arsenic solution, the metal sulfide often remains suspended in colloidal form. Many other common precipitates form as colloids if the concentrations are in certain ranges, often to the embarrassment of the analyst who is trying to use the precipitation to separate one substance from another. These types of colloids are generally less stable than a colloid like gelatin, settling out partly or completely on standing or when disturbed. Particles of these colloids are known to have much less adsorbed water, and the colloids are accordingly called *hydrophobic*, or "water-fearing."

A few colloids are difficult to classify as either hydrophobic or hydrophilic. Silica, for example, behaves like a hydrophilic colloid in that a dilute silica sol is stable indefinitely and in that it readily forms a gel. On the other hand, silica does not spontaneously disperse itself in water, in this respect exhibiting the characteristic behavior of a hydrophobic colloid. Despite exceptions of this sort, the distinction between colloids with much adsorbed water and with little adsorbed water is a useful one.

Colloidal particles lie in the size range between that of true solutions (particle diameters of the order of 10^{-3} μm) and that of suspensions whose particles quickly settle out (particle diameters greater than 1 to 10 μm). Colloids are distinguished from suspensions by their apparent homogeneity and their stability, but the

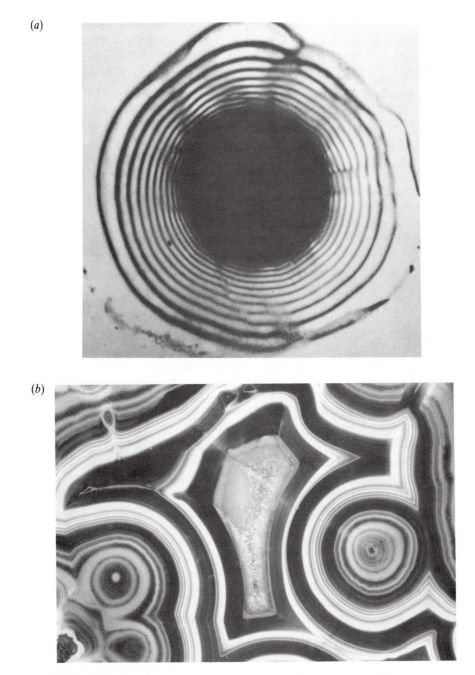

FIGURE 6-11
Photographs of Liesegang bands. (*a*) Silver chromate bands produced by dropping a small crystal of silver nitrate into a dilute solution of potassium dichromate in gelatin. (Photograph reprinted by permission from McBirney, A. R., and Noyes, R. M.: "Crystallization and layering of the Skaergaard Intrusion," *Journal of Petrology*, vol. 20, pp. 487–554, 1979.) (*b*) Banded agate. (Photograph by B. J. Skinner, reprinted by permission from Dietrich, R. V., and B. J. Skinner: *Gems, Granites, and Gravels*, Cambridge University Press, 173 pp., 1990, Plate 15c.)

distinction between colloids and true solutions is more difficult, because super-ficially they look alike. One indication that colloidal particles are present is the ability of a colloid to scatter light. This is readily demonstrated by passing a beam of light through a sol; the beam is outlined by a soft glow in the liquid, much as the path of light through a darkened room may be outlined by a scattering from dust particles, whereas a light beam through a true solution is invisible from the side. This scattering of light, called the *Tyndall effect*, in some colloids is so pronounced that the colloid shows *opalescence*—an apparent milkiness, or a difference in color when examined by reflected and transmitted light.

Another characteristic of colloids is the lack of predictable effect on vapor pressure, boiling point, and melting point of water. A substance in true solution lowers the vapor pressure, depresses the freezing point, and raises the boiling point; the effects on these properties follow simple, well-known rules that depend on the concentration and molecular weight of the dissolved material. Colloids, on the other hand, affect the properties of the liquid erratically. Hydrophobic colloids generally show very slight effects, whereas hydrophilic colloids may change the behavior of the liquid profoundly due to the adsorption of large amounts of water onto the surfaces of the colloidal particles.

Finally, a striking characteristic of colloids in the form of a gel is the development of colored diffusion bands, *Liesegang bands*, when an electrolyte is allowed to diffuse into the gel [Fig. 6-11(*a*)]. If a gel contains chromate ion, for example, and a silver salt diffuses into it, regularly spaced bands of precipitated silver chromate develop in the gel. This phenomenon, first studied by the German chemist R. E. Liesegang, is still not entirely understood, but in general it must depend on a slow movement of silver ions leading to a depletion of chromate ions in some parts of the gel and supersaturation with silver chromate in other parts. The diffusion bands are strikingly similar to the rhythmic banding found in agate [Fig. 6-11(*b*)] and in some ore deposits, suggesting that diffusion in gels may account for the origin of the natural color patterns.

Surface Chemistry of Colloidal Particles

Many of the important geochemical characteristics of colloids are related to the high ratio of surface area to volume of colloidal particles. The great surface area of colloids needs no more demonstration than an elementary calculation. If a cube 1 cm on a side is cut into eight equal parts, its surface area is increased from 6 to 12 cm^2; if each fragment is divided into eight smaller fragments 2.5 mm on a side, the area becomes 24 cm^2; and if this process of subdivision is continued until the diameter of each fragment is 0.01 μm, within the colloidal range, the surface area will have increased to 6 million cm^2, or 600 square meters. As a consequence of their high ratio of surface area to volume, particle surface charge is a determining factor in the stability of colloidal suspensions. The surface charge means also that sorption and ion exchange play an important role in the migration of many elements as ions attached to colloidal particles.

The electric charge of a colloidal particle is the sum of the permanent structural charge of the particle and the charge due to surface reactions with the solution [Eq. (6-5)]. Because of the enormous total surface area which the particles expose to the solution, pH dependent surface reactions and ion adsorption have a strong influence on the net surface charge. There is no simple rule by which the sign of this charge may be predicted, but empirically it is found that the colloids of most interest in geology develop charges consistent with the points of zero charge measurements given in Table 6-1. To this list we can add sulfide and organic particles, which typically have negative charges over the pH range of most natural water.

The electric charge on colloidal particles is the principal reason that they remain dispersed indefinitely. Particles in the colloidal size range would stay suspended for a time simply because of random bombardment by molecules of the solution (*Brownian movement*), but this bombardment would not be sufficient to prevent them from settling eventually. When the particles all have charges of the same sign, however, their mutual repulsion makes settling difficult. The stability of the suspension depends on the thickness of the diffuse layer of ions about the particles [Fig. 6-4(*a*)], a thickness determined by the electric potential at the boundary between the inner fixed layer and the outer diffuse layer [Fig. 6-4(*b*)], and on the ionic strength of the solution.

Now if the total particle surface charge is decreased due to surface reactions with the solution, or if the ionic strength of the solution is increased, the layer of diffuse ion swarms becomes compressed so that electrostatic forces are not strong enough to counteract the van der Waals forces of attraction between the particles. It follows that anything which will neutralize or diminish the charge should cause the particles to settle out or *flocculate*. This prediction is borne out by the fact that many colloids are readily flocculated simply by changing the solution pH to values close to the particles' PZC (see Table 6-1), or by adding an electrolyte. Flocculation can also occur by the addition of a colloid that is simultaneously adsorbed to the surfaces of adjacent particles linking them together to form a microgel. This latter process is schematically illustrated in Fig. 6-12 where colloidal particles of silica with a negative surface charge (Table 6-1) are joined together by the addition of a colloid with a positive surface charge.

Flocculation by Electrolytes

The precipitation of colloids by electrolytes is a complex phenomenon, for which theoretical explanations are meager. It seems contradictory to learn that some electrolyte must be present in the first place for the colloid to form and be stabilized—because if electrolytes are completely absent, the colloidal particles would find no ions to adsorb—and then to learn that addition of more electrolyte has the opposite effect and causes the colloid to coagulate. The important consideration, apparently, is the amount and kind of ions present. Generally sols are more stable in very dilute electrolyte solutions than in concentrated ones, and are more easily precipitated by addition of some kinds of ions than by

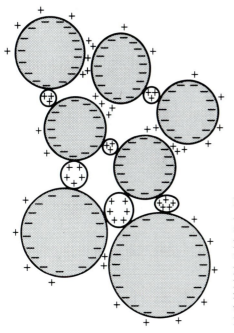

FIGURE 6-12

A schematic model illustrating flocculation of silica sol [large circles with a negative (−) particle charge] by colloidal particles of opposite ionic charge [irregular areas with a positive (+) particle charge]. The +'s outside the colloidal particles represent cations in solution that accumulate near the negatively charged silica. Note that these cations can act as binding agents linking adjacent particles of silica.

others. Two rules are useful in predicting the effect of electrolytes: (1) doubly and triply charged ions are more effective in coagulating colloids than singly charged ions, and (2) H^+ and OH^- are especially effective as coagulants due to the effect of surface reactions such as reactions (6-1) through (6-4) on the total charge of a colloidal particle.

These rules may be illustrated by experiments like the following: A stable arsenic sulfide sol is prepared by bubbling H_2S through a dilute solution of arsenious acid, H_3AsO_3. If solutions of NaCl and $CaCl_2$ of similar concentration are added to different portions of the sol, the calcium salt will prove much more effective in causing precipitation of the sulfide. This is because the sulfide particles are negatively charged, hence are flocculated by positive ions, and because the Ca^{2+} ions have a higher charge than Na^+ ions. If solutions of NaCl and Na_2SO_4 are used as electrolytes, no difference will be found in their effectiveness as coagulants because the positive ion in each is the same, Na^+. If HCl and NaOH are used, the former will be a better coagulant because H^+ reacts with the negative sol particles. If the same electrolyte solutions are now tried with a positive sol, say $Al(OH)_3$, the NaCl and $CaCl_2$ will show equal effectiveness as precipitants, Na_2SO_4 will prove more active than NaCl, and NaOH will be better than HCl. Note that experiments of this sort provide a means for determining the sign of the electric charge [Eq. (6-5)] of a colloid without recourse to an electric current.

In simple cases of this sort the logic is reasonably straightforward, but difficulties and exceptions are found when it is extended to other colloids and other electrolytes. For geologic applications the most important generalizations we can make are simply that different colloids may have charges of different sign, that many colloids are unstable in fairly concentrated electrolyte solutions, and that, for any given colloid, some electrolytes are more effective than others in causing precipitation. Rules of this kind provide possible explanations for the partial separation of substances during sedimentation and during ore deposition.

Stability and Transport of Colloids

One of the important geological questions about colloids may be phrased like this: for a given metal, is there a compound that forms a sol readily that incorporates the metal in the colloidal particle or on its surface, so that the sol can serve as a means of transporting the metal in surface or groundwaters? The effectiveness of the sol in carrying the metal must depend on its stability, for obviously a colloid that flocculates readily would not serve. The ultimate fate of the metal will depend on the conditions necessary to bring about flocculation. Iron, for example, in the Fe(III) state is known to form a very insoluble hydroxide, which often remains dispersed as a stable $Fe(OH)_3$ sol; the amount of this sol in stream water is greater than that in the sea, suggesting that the sol is flocculated by surface reactions involving electrolytes in seawater.

The stability of a given colloid is difficult to predict. Any colloidal dispersion is essentially unstable; its particles are large enough to settle out under the influence of gravity, but are prevented from doing so by the accident of adsorbing enough ions to maintain an electric charge. Many colloids demonstrate their instability by gradually precipitating on standing, but some are stable enough to remain suspended indefinitely. We have noted that colloids are precipitated by electrolytes, and that some electrolytes are better precipitants than others. Yet this rule is hedged around with difficulties. Nevertheless, the rule about colloids being precipitated by electrolytes has enough validity for us to state that particles remain suspended in freshwater longer than in seawater, and that the extraordinary clearness of the water in many salt lakes is due to the rapid flocculation of any suspended material brought into them.

The effect of heat on colloids is a problem on which chemical data are practically nonexistent, but which has great interest for geologists because of its bearing on the possible role of colloids in hydrothermal solutions. One might expect that a high temperature in general would favor flocculation of colloids, because increased agitation would give the particles more opportunity to come into contact and because the growth of large particles at the expense of small particles would be accelerated. Certainly for many colloids this generalization holds; a favorite trick in chemical analysis, for example, is to speed up the flocculation of a partly colloid precipitate by keeping it for a time near the boiling point of water. Yet iron hydroxide sols are best prepared in the laboratory in boiling solution, and experiments have shown that mercury sulfide sols are stable for long periods at 100 °C and gold sols at temperatures as high as 150 °C. Until more data become available, speculations about the behavior of colloids in hydrothermal solutions can have little guidance from physical chemistry.

One curious observation about the stability of colloids is the fact that a given colloid may be much more stable in the presence of a second colloid than it is by itself. When ammonium sulfide is added to a copper solution, for example, a precipitate of copper sulfide appears at once; if the solution contains a mere trace of gelatin, the copper sulfide does not precipitate but is stabilized as a sol. The sulfide is said to be "protected" by the gelatin, and the gelatin is called a *protective colloid*. In general, protective colloids are hydrophilic. Geologically, these types of colloids are of particular interest in surface waters containing an abundance of organic matter, for the hydrophilic organic colloids serve as efficient protectors of such inorganic material as iron hydroxide and manganese dioxide, making these colloids more stable and hence more easily transportable than they otherwise would be. The stabilizing of iron hydroxide by organic protective colloids adds to the environmental problems of acid mine drainage described in Sec. 6-3, because iron in the form of a protected colloid can be transported much farther and so cause more widespread damage than when it is carried as a simple ion or unprotected colloidal particles.

Geologic Evidences of Former Colloids

The importance of colloids in present-day geologic processes is plain to see. The transport of insoluble metal compounds by streams, the precipitation of fine-grained material in seawater, the adsorption of ions by clays and organic matter in the soil are only a few examples of the role played by finely dispersed particles. What evidence do we have that colloids have affected geologic processes in the past as well? The question is a difficult one because colloids are essentially unstable. In the course of geologic time they can be expected to flocculate and perhaps to crystallize, so that the materials we examine today may show little sign of their colloidal history.

Materials that are amorphous or very finely crystalline, that commonly occur with smooth rounded surfaces (botryoidal or mammillary surfaces), that show indistinct color banding similar to Liesegang rings, and that consist of compounds known to form colloids readily in either natural or artificial environments, are often assigned a colloidal origin. Chert is a good example: its texture, its banding, its rounded surfaces resemble so closely gelatinous silica produced in the laboratory that the inference of origin as a mass of silica gel seems natural. There is one pitfall here, in that artificial gel contains 90% or more water, whereas chert seldom has more than 5%; furthermore, if artificial gels are allowed to dehydrate, they lose their gelatinous appearance and smooth surfaces and disintegrate into flaky opaque material. The objection may not be insuperable, because chert in its process of formation may pick up silica as it loses water, but the entire sequence of events has not yet been duplicated in the laboratory. A colloidal origin is often inferred even for materials that do not readily form gels in the laboratory, particularly calcite and siderite, when these substances occur in discrete bodies with microcrystalline texture and smooth, hummocky surfaces. Septarian nodules are a good example: the crystal-lined cracks in these odd structures can be made plausible on the colloid hypothesis by supposing that they represent shrinkage cracks produced by partial dehydration of the gel as it hardened.

The same general criteria for colloidal origin are also extended to textures of metallic ores in hydrothermal veins, despite the lack of experimental evidence on colloidal behavior at high temperatures. Finely crystalline ores with rounded surfaces in open cavities, particularly if they show bands of different composition, are commonly cited, and the assumption of colloidal origin is sometimes stretched even to finely banded ores without rounded surfaces. With our present knowledge of colloidal chemistry it is hardly worthwhile to debate the pros and cons of this hypothesis. The assumption that these ores were deposited from colloidal solutions is not unreasonable, but there is little positive evidence to support it. The converse statement, however, that coarse crystalline ores cannot be deposited from colloidal solutions, has little justification. It is true enough that coarse crystals do not form from colloids at low temperatures in the short times of laboratory experiments, but about the capacity of such crystals to grow from high-temperature colloids in the course of geologic time we simply have no present basis for guessing.

Compared with other branches of physical chemistry, the geochemistry of colloids is still in a primitive state, necessarily so because of the enormous complexity of the surface chemistry of colloidal phenomena.

SUMMARY

Most geochemical reactions in nature occur at the interface between minerals and solution. The nature of the atomic sites exposed at the mineral surface and the charge on the surface control the types of aqueous species that accumulate on a mineral and the ability of these ions to exchange with other ions in the solution. Sorption and ion exchange are important in determining the extent of solute transport and fixation in surface and groundwaters, as well as in the solutions involved in deeper geologic processes such as diagenesis, hydrothermal ore deposition, and metamorphism. Particles with sizes in the colloidal range, because of their enormous surface area, are especially active in sorption and ion-exchange reactions. The formation, transport, and flocculation of colloids are important in many aspects of the geochemistry of surface and underground waters.

PROBLEMS

1. The following cation selectivity sequence has been experimentally observed for the relative adsorption of alkaline-earth metals on a clay-rich soil:

$$Ba^{2+} > Sr^{2+} > Ca^{2+} > Mg^{2+}$$

Explain, in terms of adsorption processes, why this particular ordering sequence occurs. The observed cation selectivity sequence for the relative adsorption of transition metals in this soil is:

$$Cu^{2+} > Ni^{2+} > Co^{2+} > Fe^{2+} > Mn^{2+}$$

Compare and contrast the adsorption processes and selectivity sequences between the transition metal cations and the alkaline-earth cations.

2. Explain why increasing ionic strength decreases the thickness of the layer of diffuse ions about a charged mineral surface.

3. Kesterson National Wildlife Refuge in California until recently received agricultural drainage water which was high in selenium. Excess selenium is toxic to the waterfowl which are abundant in the Kesterson area. Use your knowledge of the mechanisms of SeO_4^{2-} and SeO_3^{2-} sorption on surfaces to predict which ionic species is more mobile and thus more toxic to the waterfowl.

4. By what observations or experiments could you prove that a red liquid containing gold is a sol of the metal rather than a true solution?

5. Describe two kinds of experiments by which you could establish that the particles of a gold sol carry a positive charge.

6. What would you expect to happen when sols of Fe(III) oxide and arsenic sulfide are mixed?

7. One method of determining the molecular weight of a substance in solution is to measure the depression of the freezing point caused by known concentrations of the dissolved substance. Would this be a feasible method for finding the average weight of the particles of a sol? Why or why not?

8. On an artificial island in San Francisco Bay an attempt was made to construct a freshwater pond, but the clay fill of the island was too permeable for the pond to hold water. The difficulty was solved by filling the pond with seawater and keeping it full for several days, after which the seawater was pumped out and replaced by freshwater. The water shows little tendency to seep away. Suggest an explanation.

9. What is meant by the following terms?

(a) Hydrophilic colloid.

(b) Emulsion.

(c) Ion exchange.

(d) Sorption.

(e) Liesegang bands.

(f) Protective colloid.

10. Water taken from hot springs in Yellowstone Park at temperatures near 100°C has a pH of about 7 and contains dissolved silica at a concentration of approximately 350 ppm SiO_2. Hot-spring orifices are commonly surrounded by low mounds of opalite silica (geyserite). If a sample of the water is placed in a stoppered bottle and kept at room temperature, successive analyses over a period of weeks show diminishing amounts of dissolved silica, but the solution shows no visible change. After several years, the sample is still clear and colorless, and analyses give a constant value of about 140 ppm SiO_2; if salt is added to the water, gelatinous flocs appear. Explain these observations.

REFERENCES AND SUGGESTIONS FOR FURTHER READING

Hochella, M. F., Jr., and A. F. White: "Mineral–Water Interface Geochemistry," *Reviews in Mineralogy*, vol. 23, Mineralogical Society of America, 603 pp., 1990. This volume contains 14 articles reviewing recent research in the field of surface chemistry. It includes papers on atomic theory, sorption experiments, spectroscopic studies of mineral surfaces, and mechanisms of dissolution and mineral growth.

Iler, R. K.: *The Chemistry of Silica: Solubility, Polymerization, Colloid and Surface Properties, and Biochemistry*, Wiley, New York, 866 pp., 1979. An advanced textbook on the surface chemistry and colloidal phenomena of silica.

van Olphen, H.: *An Introduction to Clay Colloidal Chemistry*, 2d ed., Wiley, New York, 1977. A standard textbook on the surface chemistry and colloidal phenomena of clays.

Roedder, E.: "The noncolloidal origin of 'colloform' textures in sphalerite ores," *Econ. Geology*, vol. 63, pp. 451–471, 1968. A critical study of textures in metallic ores that are often thought to be evidence of colloidal origin.

Sposito, G.: *The Chemistry of Soils*, Oxford University Press, 277 pp., 1989. Excellent textbook reviewing the surface chemistry, sorption, ion exchange, and colloidal phenomena in soils.

Stumm, W.: *Chemistry of the Solid–Water Interface*, Wiley, New York, 428 pp., 1992. A recent textbook covering many aspects of the chemistry of the mineral–water interface.

Stumm, W., and J. J. Morgan: *Aquatic Chemistry, An Introduction Emphasizing Chemical Equilibria in Natural Waters*, 2d ed., Wiley, New York, 780 pp., 1981. Chapter 10, The Solid–Solution Interface, is an in-depth presentation of surface chemistry, sorption, ion exchange, and colloidal phenomena.

CHAPTER

7

CHEMICAL THERMODYNAMICS: FUNDAMENTAL PRINCIPLES

In this chapter we return to a formal study of chemical equilibrium, but the approach will be from a new direction. We look now at the energy evolved and consumed in chemical reactions and at the relation of this energy to equilibrium. This is a branch of physical chemistry called thermodynamics, a basic subject that permeates much of modern chemistry and geochemistry. As a predictive tool, thermodynamics has its greatest usefulness in the closed, carefully controlled systems of the chemical laboratory, but we shall find it an aid also in setting limits on what is possible and what is impossible in the more complex systems with which a geologist must deal.

7-1 THERMODYNAMIC SYSTEMS

In geochemistry we apply the principles of equilibrium thermodynamics to a well-defined macroscopic volume of the universe that is referred to as a *thermodynamic system* (Fig. 7-1). We are interested in what happens inside the system. That is, for a given set of conditions what chemical reactions will occur in the system? And how much reaction must take place to attain overall chemical equilibrium?

The dimensions of the system can be chosen to suit specific geochemical problems. We may choose systems that include the entire Earth and its atmosphere, or ones that are limited to a small portion of the Earth such as a hand specimen of rock. Another kind of system consists of experimental apparatus, for example a pressurized heating-vessel used in a laboratory. Thermodynamic systems are classified on the basis of whether or not energy (heat and work) and matter

165

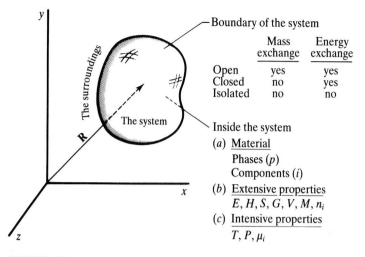

FIGURE 7-1

The thermodynamic system is located in space relative to a coordinate system by the vector (**R**). It is a well-defined volume separated from the surroundings by its boundaries, and classified in terms of its ability to exchange energy and mass with the surroundings. The interior, composed of phases whose composition is represented by chemical components, is described by a series of extensive and intensive properties.

can be transported across the boundaries outlining the system. There are three types: *isolated systems* cannot exchange energy or matter, *closed systems* can exchange energy but not matter, and *open systems* can exchange both energy and matter. Most thermodynamic systems of geologic interest belong to the last type, open to exchange of energy and matter. Some natural systems, however, have such slow rates of mass exchange with their surroundings that they can be treated as closed thermodynamic systems in a study of chemical equilibrium.

The material in a system is composed of *phases*, each phase having distinct physical and chemical properties. In geochemistry the phases are typically minerals that form rocks, plus fluids, including vapor, water, or magma, that are in contact with the minerals. The composition of each phase is described by a series of discrete chemical formula units, called *chemical components*. For example, we can represent the compositions of alkali feldspar minerals in Fig. 5-7 with the components $NaAlSi_3O_8$ and $KAlSi_3O_8$, or the minerals and aqueous solution of the phase diagram in Fig. 4-10 with the components K_2O, Al_2O_3, SiO_2, H_2O, and HCl.

Some properties of a system, like mass and volume, are defined by the system as a whole, and are called *extensive* properties. These properties are additive. For example, the mass of a system is equal to the sum of the masses of all the phases it contains. Internal energy (E), entropy (S), enthalpy (H), Gibbs free energy (G), and the number of mols of chemical components (n_i, where n is the number of mols of a specific component i) are all extensive properties that we will use in this chapter to describe chemical equilibrium. On the other hand, properties like pressure and temperature take well-defined values at each point within the system, and do not depend on amounts of material. These are called *intensive* properties. They are *not* additive; the temperature of a system is *not* equal to the sum of the temperatures of the individual phases. One example of an intensive property that we will use in the following chapter to describe equilibrium among the phases in a system is chemical potential (μ_i). Handling equilibrium problems in terms of intensive variables is advantageous because these properties are independent of the mass of the system.

The *thermodynamic state* of a system is described by some specified group of independent intensive and/or extensive variables. Any particular property of the system that can be expressed in terms of these variables is called a *function of the state of the system*. As an example, consider a system with one phase, a gas, confined inside a cylinder with a piston at one end that allows the volume of the gas to expand or contract. If we describe the state of the system by specifying temperature (T) and pressure (P), then the volume (V) of the gas will be a function of the state of the system represented by the equation

$$V = F(T, P). \tag{7-1}$$

For one mol of an ideal gas this functional relationship is the familiar gas law,

$$\bar{V} = \frac{RT}{P}, \tag{7-2}$$

which permits calculation of the molal gas volume ($\bar{V}$) from the variables temperature and pressure.

Functions like Eq. (7-1) are commonly used in thermodynamics to calculate how certain properties respond to changes in the independent variables used to describe the system. For example, how does the volume of the gas in our cylinder respond to changes in temperature and pressure? Such a relationship is represented mathematically by the total derivative of Eq. (7-1),

$$dV = \left(\frac{\partial V}{\partial T}\right)_P dT + \left(\frac{\partial V}{\partial P}\right)_T dP, \tag{7-3}$$

and is graphically illustrated in Fig. 7-2. The change in volume is equal to the sum of two terms: the first is the rate of change in V with respect to T at constant P multiplied by the total change in temperature (dT), and the second is the rate of change in V with respect to P at constant T multiplied by the total change in pressure (dP).

The relation expressed by Eq. (7-3) for one mol of an ideal gas is obtained by first differentiating Eq. (7-2) with respect to pressure and temperature,

$$\left(\frac{\partial \bar{V}}{\partial P}\right)_T = -\frac{RT}{P^2} \tag{7-4}$$

and

$$\left(\frac{\partial \bar{V}}{\partial T}\right)_P = \frac{R}{P}, \tag{7-5}$$

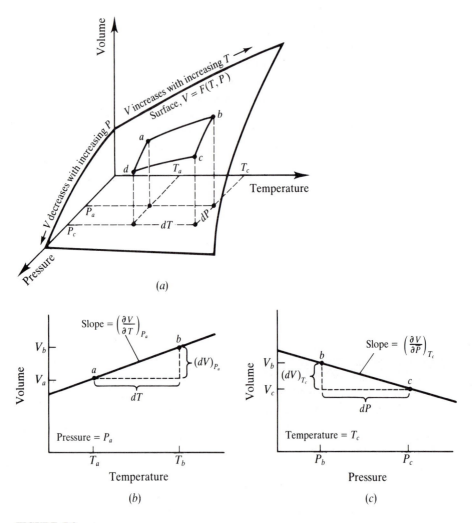

FIGURE 7-2
Schematic illustration of the differential change in volume (dV) when temperature and pressure change from T_a, P_a to T_c, P_c (diagram (a)). Because $V = F(T,P)$, the total derivative of volume [Eq. (7-3)] is composed of two parts: changes with T at constant P along path a to b as shown in diagram (b), and changes with P at constant T along path b to c in diagram (c). Note that the same total differential change in V is obtained along the alternate path a to d to c.

then substituting into Eq. (7-3):

$$dV = -\frac{RT}{P^2}\,dP + \frac{R}{P}\,dT. \tag{7-6}$$

To obtain the total change in molal volume this equation is integrated from the P and T conditions of the initial state ($P_{initial}$, $T_{initial}$) to that of the final state (P_{final}, T_{final}).

Note in Fig. 7-2 that it does not matter in which order we consider the differential operations involving temperature and pressure [Eq. (7-3)]. The change in volume (for volume in general, not only molal volumes) during a transformation that goes from an initial state to a final state is simply the difference in the volume written as $\Delta V = V_{final} - V_{initial}$. The change in volume with respect to changes in temperature and pressure will be the same because volume is a function of the state of the system and is independent of the path followed in changing the state of the system. This independence of path is characteristic of all extensive properties.

Chemical equilibrium will be described in this chapter in terms of the variables temperature, pressure, and composition, using a function of the state of a system called the Gibbs free energy (G),

$$G = F(T, P, n_i). \qquad (7\text{-}7)$$

To develop relationships between equilibrium and these variables T, P, and n_i, we begin with the laws of thermodynamics, which introduce two additional functions of the state of the system, internal energy (E) and entropy (S).

7-2 THE LAWS OF THERMODYNAMICS

The laws of thermodynamics cannot be proved mathematically. They are simply generalizations derived from laboratory experiments and from ordinary experience that provide practical and useful explanations for many natural phenomena.

First Law: Conservation of Energy

The first law of thermodynamics expresses a relation between heat and work, specifically the heat added to or taken from a system and the associated work done by or done to the system. To give the law a mathematical statement, we postulate the existence of a function of the state of a system called its *internal energy, E*, which can be thought of as the sum of the kinetic and potential energies of its particles. The absolute value of E would be difficult to determine, but for most purposes we are interested only in its changes. Suppose that a change in E is brought about by addition of heat (Q) to a system from its surroundings, and that the system does mechanical work (W) on the surroundings as a result. The first law states that the change in internal energy is equal to the difference between the heat added and the work done, $\Delta E = E_{final} - E_{initial} = Q - W$, or in differential form

$$dE = dQ - dW. \qquad (7\text{-}8)$$

In this expression dQ has a positive sign because it represents energy added to the system from its environment, while dW is given a negative sign because it shows energy lost by the system when it does work on the environment. The law, in effect, is a restatement of the law of conservation of energy: an amount of energy is supplied to the system in the form of heat, an amount is removed in the form of work, and the difference is the change of energy within the system. Energy has moved from one system to another, and a part of it has changed its form from heat to work, but the total amount of energy has remained constant.

To measure the work, suppose the system is arranged so that work is done by expansion against an external pressure. Specifically, suppose that the system is confined to a vertical cylinder and that expansion results in raising a mass (M) at the top of the cylinder against the force of gravity. If the vertical distance moved is Δh, the mechanical energy expended is $Mg\Delta h$, where $\mathbf{g}$ is the acceleration of gravity. This energy can also be expressed as $P\Delta V$, where P is pressure and ΔV is the total change in volume, since

$$P = \frac{\text{force}}{\text{area}} = \frac{M\mathbf{g}}{\text{area}} \quad \text{and} \quad \Delta V = \text{area} \times \Delta h.$$

If the expansion is performed reversibly, meaning that internal and external pressures never differ by more than infinitesimal amounts, P may be regarded not as the effect of an added mass, but as the prevailing external pressure acting on the system. Then we write in differential form

$$dW = PdV, \tag{7-9}$$

and Eq. (7-8) becomes

$$dE = dQ - PdV. \tag{7-10}$$

The energy E is a property of the system, determined by the nature and arrangements of its constituent particles. By contrast, Q and W are not properties of the system; their values are dependent on processes by which the state of the system is changed. By different experimental arrangements, the same change in energy, E, can be produced with different combinations of Q and W.

Second Law: Creation of Entropy

The second law of thermodynamics defines a function of the state of the system called *entropy* (S). Like energy, entropy is an extensive property of the system. The second law states that the change in S of a closed system due to a *reversible process* (that is, a process where infinitesimal changes can make the process reverse itself at any time) is equal to the heat absorbed during this process divided by the absolute temperature,

$$dS \equiv \frac{dQ}{T}. \tag{7-11}$$

Thus S is related to Q through the intensive variable T in much the same way as V is related to W by the intensive variable P [$dV = dW/P$, see Eq. (7-9)]. If an irreversible (nonequilibrium) process occurs in the system, the second law states that the total change in entropy will be greater than the heat adsorbed divided by T,

$$dS > \frac{dQ}{T}. \tag{7-12}$$

This statement is very succinct, and commonly fails to convey the significance of the second law except to those who have had long practice in its use. To help make clear the meaning of entropy and its importance in geochemistry, we look first at two hypothetical experiments.

Suppose that an ideal gas is confined under pressure in a cylinder by a piston that moves without friction, and suppose that an amount of heat Q is added to the gas. The gas responds by expanding against the pressure P exerted by the piston and thereby does an amount of work $P\Delta V$, where ΔV is the total change in volume during the experiment ($\Delta V = V_{final} - V_{initial}$). Suppose further that the addition of heat and the resulting expansion take place slowly, so that the pressure of the gas on the piston is never more than infinitesimally higher than the external pressure, and so that the temperature of the gas remains constant. At any time during the expansion it would be possible, by an infinitesimal increase in the external pressure, to reverse the operation—to do work on the gas by moving the piston inward, the work being converted into heat which would escape to the surroundings. Under these conditions we say that the expansion takes place reversibly. In thermodynamics reversible and equilibrium are equivalent terms. The entropy change during the expansion ($\Delta S = S_{final} - S_{initial}$), then, is equal to the total amount of heat absorbed divided by the absolute temperature [Eq. (7-11)].

Suppose, still using an ideal gas, we perform the experiment differently: we insert a shutter at the original position of the piston, move the piston back to its position at the end of the preceding experiment, evacuate the space between shutter and piston, then withdraw the shutter and let the gas expand freely to fill the enlarged space. The end result of the experiment is identical with that of the first experiment, for an ideal gas undergoes no temperature change during free expansion. This time we have added no heat and obtained no work, so that both Q and W are zero. The total entropy change (ΔS), however, must be the same as before, since the initial and final states of the gas are identical in the two experiments.

The second experiment is considered to be highly *irreversible* (nonequilibrium): there is no possible way, while the gas is expanding, by an infinitesimal change in external conditions to make the molecules move back to their former restricted volume. The total change in entropy during this experiment is equivalent to that transferred to the system by the heat absorbed *if* the process were carried out reversibly as in the first experiment. The amount of entropy that is supplied from the surroundings in the second experiment is zero; but the total entropy acquired by the gas is of course the same in both experiments, since it depends only on the state of the gas and not on how the expansion is carried out. The entropy acquired by the gas in the second experiment is therefore *new* entropy which was not present originally. In an irreversible process total entropy is not conserved but increases. Since all actual processes are to some extent irreversible, this means that, in any energy change, the total entropy of all systems involved increases. The amount of the net entropy increase measures the extent of irreversibility.

These experiments illustrate the dual role that entropy plays. In a reversible process, the change in entropy of a closed system is proportional to heat exchange with the surroundings. In an irreversible process the entropy change may have nothing to do with heat, but is defined by some change in the configuration of the system, in this case by the expansion of the gas to a greater volume. The *measure* of entropy change is still the heat that *would* be absorbed *if* the process were carried out reversibly, and the entropy change is the same whether the process is reversible or not. The differential change in entropy (dS) during any transformation in a closed system will thus be made up of two parts: dS_{ext}, which arises from the exchange of entropy between the system and its exterior,

$$dS_{ext} = \frac{dQ}{T},$$ (7-13)

and dS_{int} which arises from the creation of entropy inside the system. That is,

$$dS = dS_{ext} + dS_{int}.$$ (7-14)

Entropy created inside the system is due only to irreversible (nonequilibrium) processes; it is always greater than zero during such processes and equal to zero at equilibrium,

$$dS_{int} \geq 0.$$ (7-15)

As we shall see later, the equality in Eq. (7-15) will provide an important thermodynamic criterion for judging chemical equilibrium in geologic systems.

Difficulty in understanding entropy often arises because of confusion regarding the systems under consideration. In an individual system, entropy may increase or decrease or remain constant during various parts of a cycle of changes. In reversible processes the entropy change in one system, or one part of a system, must be balanced by opposite changes in another system; in irreversible processes the entropy increase in some system or systems is not completely balanced by entropy decreases elsewhere.

Because adding heat to a system always increases its entropy, the meaning of entropy will perhaps become clearer from a general consideration of the possible effects of the added heat. In our hypothetical experiment with a heated gas expanding against a movable piston, the added entropy causes the gas to become more attenuated. Other familiar effects of heat are the melting of a crystalline solid, the evaporation of a liquid, the dissolving of increasing amounts of a salt in water. Thus an expanded gas has greater entropy than compressed gas, a liquid has greater entropy than the corresponding solid, a gas has greater entropy than the liquid from which it comes, and the entropy of a salt-plus-water system increases as more of the salt dissolves. A common theme here is increase in disorder—from a compressed state to an expanded state, or from the ordered arrangement of particles in a crystal structure to the random dispersion in a liquid or gas or solution. Thus addition of heat leads to disorder, and entropy may be regarded as a measure of the disorder. Of course heat is not necessary to produce disorder; a system can become disordered either by the addition of heat from the outside (exchange of entropy from one system to another, or dS_{ext}) or from a change in the state of the system without addition of heat (free expansion, evaporation, in other words dS_{int}). In either case, entropy increases as disorder increases. Because all processes are to some extent irreversible, the entropy of the world is increasing—as an expression of the tendency of natural processes toward disorder.

Thus when we write a mathematical expression for the second law as

$$dS \geq \frac{dQ}{T}, \qquad (7\text{-}16)$$

the equality refers to the entropy added in a reversible change, and the inequality defines the additional increase from the natural irreversible tendency toward disorder. The inevitable change toward disorder can be expressed in versions of the second law that emphasize the unique characteristics of heat as a form of energy: (1) Heat always flows from hot objects to cold objects, never spontaneously from cold to hot. (2) It is impossible for a self-acting machine to transfer heat from a cold object to a hot object. (3) In any energy transformation, some of the original energy always appears in the form of heat energy which is no longer available for conversion into other forms of energy. Entropy provides a means of making such statements quantitative.

Combined Laws: Thermodynamic Potentials

Combining equations representing the first and second laws provides useful mathematical expressions relating energy to equilibrium. For any reversible energy transformation in a closed system we can substitute $dQ = TdS$ and $dW = PdV$ into Eq. (7-8) to obtain

$$dE = TdS - PdV. \qquad (7\text{-}17)$$

Changes in the internal energy are determined *only* by changes in entropy and volume, which we express in a form similar to Eq. (7-1) by writing $E = F(S, V)$.

We now consider a transformation in a closed system, where an irreversible chemical reaction occurs so that $dS_{int} > 0$. By combining Eqs. (7-13) and (7-14), we determine that dQ is equal to the difference between the total change in entropy (dS) and the entropy produced (dS_{int}) when both are multiplied by absolute temperature,

$$dQ = TdS - TdS_{int}. \qquad (7\text{-}18)$$

Substituting this relationship into Eq. (7-10) gives another expression for dE, written specifically for an irreversible process,

$$dE = TdS - PdV - TdS_{int}. \qquad (7\text{-}19)$$

Now imagine a reaction that occurs under constraints so that the total change in S and V is zero; that is $dS = dV = 0$ and $dE = -TdS_{int}$. Because internal entropy production, dS_{int}, is always greater than or equal to zero [Eq. (7-15)], we have

$$(dE)_{S,V} \leq 0. \qquad (7\text{-}20)$$

The inequality means that any irreversible processes will always be accompanied by a decrease in internal energy when S and V are constant. Furthermore, at equilibrium $dS_{int} = 0$, requiring that $(dE)_{S,V} = 0$; this corresponds to a minimum in the internal energy of the system. Thus internal energy (E) may be referred to as a *thermodynamic potential* associated with the variables S and V, in that it can be used to determine the direction of reaction and conditions of equilibrium.

For the geochemist it is not very practical to characterize equilibrium in terms of the variables S, V, and E. Other thermodynamic potentials called *enthalpy* and *free energy* will provide more practical information about chemical reactions and equilibria in both experimental and natural systems.

7-3   ENTHALPY

Definition

In thermodynamic terms, enthalpy (H) is an extensive property of a system that is closely related to heat energy. It is defined as

$$H = E + PV, \tag{7-21}$$

or

$$dH = dE + PdV + VdP. \tag{7-22}$$

For any process that occurs at constant pressure Eq. (7-22) becomes

$$dH = dE + PdV = dQ. \tag{7-23}$$

Thus differential changes in H are equal to the heat adsorbed. For this reason H is often called the "heat content" of a system. As it was just defined, however, H has a broader meaning, not limited to isobaric processes, so that the name "enthalpy" is preferable. If a reaction takes place with both pressure and volume constant, $dH = dE$, so that the entire energy change is represented by the enthalpy.

Heats of Reaction

The simple and obvious way to measure the energy of a chemical reaction is to put the ingredients in a calorimeter and record the amount of heat energy absorbed or liberated as the reaction takes place. To standardize the results so that one reaction can be compared with another, we specify that the process shall take place at constant temperature and pressure, or at least that the products of reaction shall be brought back to the initial temperature and pressure before the heat change is measured. Under these conditions we speak of the heat given out or taken up as the *heat of reaction*. It is measured in joules [1 joule (J) = 1 kg^2 m^2 sec^{-2} and 1 kilojoule (kJ) = 10^3 J], and may be written as a part of the equation:

$$H_2(g) + 0.5\ O_2(g) \rightarrow H_2O(l) + 285.8\ kJ\ mol^{-1}. \tag{7-24}$$

This means that 285.8 kJ of heat are evolved for each mol of liquid H_2O formed, when the reaction is carried out at 25°C and 1 bar.

From a slightly different point of view, we may think of the reactants as possessing a certain amount of chemical energy and the products a different amount. In the example just cited, the product H_2O has less energy than does the combination of H_2 and O_2, the excess being given up in the form of heat when the reaction takes place. In symbols,

$$\Delta H = H_{\text{products}} - H_{\text{reactants}} = -285.8 \text{ kJ}. \tag{7-25}$$

The symbol ΔH means the change in H of the system when the reactants are transformed into the products. Here the H for the system becomes less during the reaction, so ΔH must have a negative sign. In general, ΔH is negative for exothermic reactions like this one, and positive for all endothermic reactions (see Sec. 1-6).

As was true for E, the absolute enthalpy of a substance or a mixture of substances cannot be measured. The quantity that concerns us in discussing a reaction is not the individual H's but the ΔH, the enthalpy change. This is easily measured as the heat change when the reaction takes place at constant temperature and pressure and is obviously the same number as the heat of reaction with the sign reversed. This odd procedure—introducing hypothetical quantities that cannot be measured, and using a negative sign for heat evolved—at first seems to make a simple subject needlessly complicated, but we shall find that it fits well into a large logical framework.

For the hydrogen–oxygen reaction, then, we write,

$$H_2(g) + 0.5 \, O_2(g) \rightarrow H_2O(l) \qquad \Delta H = -285.8 \text{ kJ mol}^{-1}. \tag{7-26}$$

The enthalpy change is different at different temperatures and pressures, so we specify that -285.8 kJ mol^{-1} refers to the reaction when it is carried out at 25 °C and 1 bar (or when the resulting water is condensed and brought to these conditions). All enthalpy changes, unless otherwise specified, refer to these same conditions. Furthermore, ΔH obviously must be measured for a particular amount of substance; in our example the -285.8 kJ is the enthalpy change when 1 mol of hydrogen gas reacts with 0.5 mol of oxygen. We could write equally well,

$$2H_2(g) + O_2(g) \rightarrow 2H_2O(l) \qquad \Delta H = -571.5 \text{ kJ}. \tag{7-27}$$

In other words, a specific value of ΔH applies only to a particular formulation of the chemical equation. If the equation is multiplied by a number, the ΔH must be multiplied also, a relationship that is true for any extensive property of a chemical reaction. Obviously, if an equation is reversed, the sign of ΔH changes:

$$H_2O(l) \rightarrow H_2(g) + 0.5 \, O_2(g) \qquad \Delta H = +285.8 \text{ kJ}, \tag{7-28}$$

meaning simply that energy must be supplied to break water down into its elements. Thus the positive sign for ΔH indicates that the reaction in this direction is endothermic.

Computing Reaction Enthalpy

For any compound, the enthalpy change involved in forming it from its elements is called its *heat of formation*. Thus the heat of formation of water is -285.8 kJ mol^{-1} [Eq. (7-26)]. Heats of formation are useful for a variety of purposes, and have been measured for many compounds of geochemical interest. The numbers given in tables usually refer to 25°C and 1 bar and to a single mol of the compound, unless conditions are specified otherwise. Tabulated values also refer to particular states of reactants and products, states described as *standard states*. In general the standard state of a pure element or compound is taken as its most stable form at 25°C and 1 bar, but the definition is less straightforward at elevated temperatures and pressures and for mixtures and solutions. We will consider standard states in more detail later, but for the moment we need only note that heats of reaction referred to standard states are symbolized with a superscript, ΔH_f^o, and that some of the values of ΔH_f^o for compounds of most interest in geochemistry are given in Appendix VIII.

By adding and subtracting heats of formation, the enthalpy change for any reaction can be computed. For example, the burning of hydrogen sulfide may be written,

$$2H_2S(g) + 3O_2(g) \rightarrow 2H_2O(l) + 2SO_2(g). \tag{7-29}$$

The heats of formation are:

$$H_2(g) + 0.5\,S_2(g) \rightarrow H_2S(g) \qquad \Delta H_f^o = -20.6\,\text{kJ mol}^{-1}$$

$$H_2(g) + 0.5\,O_2(g) \rightarrow H_2O(l) \qquad \Delta H_f^o = -285.8\,\text{kJ mol}^{-1}$$

$$S(g) + O_2(g) \rightarrow SO_2(g) \qquad \Delta H_f^o = -296.8\,\text{kJ mol}^{-1}$$

Now if these three reactions are doubled, if the second and third are added together, and if the first is subtracted from the sum, we obtain the original reaction. Hence we find ΔH^o for the overall reaction by combining the ΔH_f^o for the reactants minus those of the products: $\Delta H^o = -1124.0$ kJ.

A more general formulation is expressed by the stoichiometric equation. This is a general mathematical expression used to represent the difference between the properties of products and reactants in a chemical reaction. For the enthalpy of reaction (ΔH_r^o) at 1 bar and 25°C the stoichiometric equation is

$$\Delta H_r^o = \sum_i v_i \Delta H_{f,i}^o, \tag{7-30}$$

where ΔH_r^o is the heat of a specific chemical reaction denoted by the subscript r, and v_i (the Greek letter nu) is the stoichiometric reaction coefficient for the ith species of the reaction, which by convention is positive for products and negative for reactants. Thus for Eq. (7-29) we have $v_{SO_2} = +2$, $v_{H_2O} = +2$, $v_{O_2} = -3$, and $v_{H_2S} = -2$. The term $\Delta H_{f,i}^o$ is the heat of formation of a particular species or compound (i) in the reaction. We now write Eq. (7-30) as

$$\Delta H_r = v_{SO_2}\Delta H_{f,\,SO_2}^o + v_{H_2O}\Delta H_{f,\,H_2O}^o + v_{O_2}\Delta H_{f,\,O_2}^o + v_{H_2S}\Delta H_{f,\,H_2S}^o,$$

or substituting values of v_i and $\Delta H_{f,i}^o$,

$$\Delta H_r = 2(-296.8\,\text{kJ}) + 2(-285.8\,\text{kJ}) + (-3)(0\,\text{kJ}) + (-2)(-20.6\,\text{kJ}) = -1124.0\,\text{kJ}.$$

This is a general rule for obtaining enthalpy changes of a chemical reaction at 1 bar and 25°C from tabulated values of heats of formation.

Enthalpy at Elevated Temperatures: The Heat Capacity

Enthalpies of formation for substances of geologic interest, as noted above, are usually tabulated at 25°C. Because many geochemical problems concern processes that occur at elevated temperatures, a means for calculating heats of reactions at temperatures $> 25°C$ is often necessary. For this purpose we introduce the property called *heat capacity*.

The heat capacity of a substance is the amount of heat required to raise the temperature of 1 mol by 1°C. More rigorously, it may be defined as the limit of the ratio of heat added to the temperature change produced, as the latter approaches zero:

$$C = \frac{dQ}{dT}. \tag{7-31}$$

If V is constant during the addition of heat, the heat absorbed equals the change in internal energy [see Eq. (7-10)]:

$$C_V = \left(\frac{\partial E}{\partial T}\right)_V, \tag{7-32}$$

where the term C_V is heat capacity at constant volume. If P is constant rather than V, heat absorbed goes into both temperature rise and expansion; hence, combining Eq. (7-23) and (7-31), we obtain

$$C_P = \left(\frac{\partial H}{\partial T}\right)_P, \tag{7-33}$$

which tells us that the change of H with respect to T is the heat capacity at constant pressure, C_P.

Measured values of C_P for the monoclinic feldspar microcline are shown in Fig. 7-3 as a function of temperature. The heat capacity curve for most minerals is

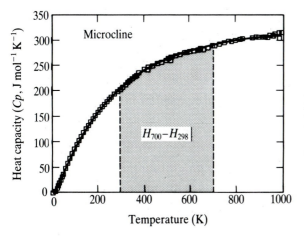

FIGURE 7-3
Heat capacity of microcline ($KAlSi_3O_8$) between 0 and 1000 K. (Symbols are data reported by Hemingway et al., 1981, and Openshaw et al., 1976). The shaded area under the C_p curve between $T = 700$ K and $T_{ref} = 298$ K represents the quantity $H_T - H_{T_{ref}}$ in Eq. (7-36).

similar to this one; that is, C_P increases from near 0 to approximately 400–600 K, then at higher temperatures the relationship between C_P and temperature becomes approximately linear. Near temperatures where polymorphic phase transformations or atomic disordering occur in a mineral, there will be steep changes or discontinuities in the temperature dependence of C_P. Relations between C_P and temperature are more complex for H_2O and for aqueous electrolyte solutions. Water has a maximum in C_P at high temperatures and low pressures near its critical point (374°C and 221 bar), and electrolytes are characterized by even more extreme temperature variations in C_P at low pressures.

To facilitate calculations of reaction enthalpy we use an equation that approximates experimental measurements of C_P, such as the data represented by the symbols in Fig. 7-3. One simple equation that provides a reasonable fit to experimental C_P data for many common minerals is

$$C_P = a + bT - cT^{-2}. \tag{7-34}$$

The a, b, and c terms are constants derived from fitting the line to the experimental data, and they have characteristic values for each substance. More complex equations would be required to describe the heat capacity of minerals with temperature-dependent atomic transformations, H_2O, and electrolytes.

Using an equation like (7-34), we can compute the effect of temperature on the enthalpy of a substance. First we rearrange Eq. (7-33) to give $dH = C_P dT$, then integrate from some reference temperature (T_{ref}, which is usually 25°C, 298 K), to some specified elevated temperature (T),

$$\int_{H_{T_{ref}}}^{H_T} dH = \int_{T_{ref}}^{T} C_P dT. \tag{7-35}$$

The left-hand side of this equation integrates to give $H_T - H_{T_{ref}}$, and for C_P on the right-hand side we substitute the value given by Eq. (7-34) to obtain,

$$H_T - H_{T_{ref}} = a(T - T_{ref}) + \frac{b}{2}(T^2 - T_{ref}^2) + c\left(\frac{1}{T} - \frac{1}{T_{ref}}\right). \tag{7-36}$$

This relationship is illustrated graphically by the shaded area in Fig. 7-3.

The enthalpy of reaction at some elevated temperature may be calculated in two different ways. One method is to compute values of $H_T - H_{T_{ref}}$ for every substance in the reaction, then apply the stoichiometric equation [see Eq. (7-30)] to obtain

$$\sum_i v_i(H_T - H_{T_{ref}}) = \sum_i v_i H_T - \sum_i v_i H_{T_{ref}} = \Delta H_T - \Delta H_{T_{ref}}. \tag{7-37}$$

The heat of reaction is computed by solving for ΔH_T,

$$\Delta H_T = \Delta H_{T_{ref}} + \sum_i v_i(H_T - H_{T_{ref}}). \tag{7-38}$$

In Eq. (7-38) the value of $\Delta H_{T_{ref}}$ is computed using heats of formation for the substances in the reaction [see Eq. (7-30)], and $H_T - H_{T_{ref}}$ is computed using Eq. (7-36).

Alternatively we may compute the change in heat capacity for the reaction (ΔC_P) using the stoichiometric equation,

$$\Delta C_P = \sum_i v_i C_{P_i}, \tag{7-39}$$

where C_{P_i} is the heat capacity of the ith substance in the reaction. The heat of reaction is then determined by solving the integral

$$\int_{H_{T_{\text{ref}}}}^{H_T} d\Delta H = \int_{T_{\text{ref}}}^{T} \Delta C_P dT \tag{7-40}$$

for ΔH_T:

$$\Delta H_T = \Delta H_{T_{\text{ref}}} + \int_{T_{\text{ref}}}^{T} \Delta C_P dT. \tag{7-41}$$

Enthalpy Change as a Measure of Reactivity

In Chapter 1 we set up as a major goal of geochemistry the ability to predict what will happen when one substance is mixed with another. Will a reaction take place, or will the substances exist side by side in chemical equilibrium? At first glance it looks as if enthalpy changes might be a key to answer such questions, for exothermic reactions often take place spontaneously while endothermic processes generally do not occur unless energy is supplied. Further, the more exothermic a reaction is (the higher its negative ΔH), the more energetically we expect it to proceed. The hydrogen–oxygen reaction [Eq. (7-26)], for example, is strongly exothermic and occurs with explosive violence if it is touched off with flame or spark. The reaction between water and carbon dioxide to form carbonic acid, on the other hand, is only mildly exothermic and reaches equilibrium with no obvious indication that a reaction is taking place at all.

To say that the energy available in a reaction mixture determines how readily and how violently the reaction will take place seems entirely natural. It conforms to our intuitive prejudices, derived from experience with mechanical systems, where, in general, energy available does determine whether a given process will go of its own accord and how much work or heat we can expect to get out. A little reflection, however, will suggest that the generalization cannot be strictly true for chemical processes. Some reactions take place spontaneously despite the fact that they are endothermic; of their own accord they extract the necessary heat from their surroundings, cooling adjacent objects below the general temperature level. A good example is the dissociation of potassium nitrate: when this salt is stirred in water, the container quickly becomes cold because the reaction

$$KNO_3(s) \rightarrow K^+ + NO_3^- \quad \Delta H^\circ = +35.1 \text{ kJ mol}^{-1} \tag{7-42}$$

absorbs heat from its surroundings. Many other common dissolution processes are similarly endothermic. The enthalpy change, therefore, is by no means an infallible measure of the tendency of a reaction to take place.

It remains true that for very many reactions the size of the enthalpy change provides at least a gross indication of reactivity. In the absence of other data, enthalpy changes can be used fairly satisfactorily to make predictions about the behavior of unfamiliar substances. We might expect, therefore, that a quantity related to enthalpy would give us the necessary refinement to make the predictions exactly. This quantity is called the *free energy*.

7-4 FREE ENERGY

Definition

To see why enthalpy is not the only factor of importance in making a reaction take place, let us imagine a process in which there is no enthalpy change at all. We have two gases in containers separated by a partition; we remove the partition and let the gases mix. If we symbolize one gas by A, the other by B, and the mixture by A–B, the "equation" for the process is

$$A + B \rightarrow A\text{-}B. \tag{7-43}$$

The experiment is imaginary because we assume that A and B behave as perfect gases, a reasonable assumption as long as the pressure is low. The mixing of two perfect gases involves no energy change, so we may write for the reaction $\Delta H = 0$. Despite the lack of any evolved energy, the reaction takes place readily of its own accord.

Why does it take place? Because the mixture represents a state of greater disorder or randomness, or a state of higher probability, than the two separate pure gases. From ordinary experience we know that natural processes tend to produce disordered arrangements from ordered arrangements; think, for example, how quickly a neat pile of papers is scattered by a gust of wind, how readily the ordered arrangement of particles in a crystal disappears when salt dissolves in water, how effectively the processes of decay destroy the complexly organized structures of a dead animal. The mixing of two gases is a simple example of a general pattern in nature. Quite apart from energy changes, natural processes go spontaneously from states of order to states of disorder, and this tendency toward disorder is the other factor besides enthalpy that makes chemical reactions take place.

Thus the tendency of a reaction to take place can be thought of as dependent on two factors, the change in enthalpy and the change in degree of disorder:

$$\textit{tendency to react} = \Delta H + \textit{change in disorder}. \tag{7-44}$$

Now in a previous section (Sec. 7-2) we have noted that entropy can be used as a measure of the degree of disorder. As an energy term, then, we can express "change in disorder" by multiplying the change in entropy by temperature:

$$\textit{change in disorder} = T\Delta S. \tag{7-45}$$

And if we now describe "tendency to react" as a new variable called "change in free energy," symbolized as ΔG, we can rewrite Eq. (7-44) in mathematical form:

$$\Delta G = \Delta H - T\Delta S. \tag{7-46}$$

The minus sign means that an entropy increase adds to the negative value of ΔH, in other words increases the tendency of a reaction to take place.

The new symbol we have introduced, G, stands for free energy, or more properly Gibbs free energy. This is an extensive property of a system, defined as

$$G \equiv H - TS \tag{7-47}$$

or

$$G = E - TS + PV. \tag{7-48}$$

Absolute values of G, as for the values H and E, would be difficult to obtain. Our chief concern is with *changes* in G during reactions:

$$\Delta G = G_{\text{products}} - G_{\text{reactants}}. \tag{7-49}$$

Values of free energy differences, ΔG, like similar values for ΔH and ΔE, are commonly expressed in joules or kilojoules.

Free Energy as a Criterion of Equilibrium

Because the quantity ΔG expresses the ability of substances to react and the extent to which reactions will go, it plays an important role in making predictions about a great variety of geochemical processes. To demonstrate this relationship between free energy and equilibrium for a reaction mixture, we begin by differentiating Eq. (7-48) to obtain

$$dG = dE - TdS - SdT + VdP + PdV. \tag{7-50}$$

From Eq. (7-19) we see that

$$dE - TdS + PdV = -TdS_{\text{int}}, \tag{7-51}$$

so that Eq. (7-50) becomes

$$dG = -SdT + VdP - TdS_{\text{int}}. \tag{7-52}$$

Now imagine a chemical reaction occurring in a closed system under conditions of constant T and P, so that $dT = dP = 0$ and Eq. (7-52) reduces to

$$(dG)_{T,P} = -TdS_{\text{int}}, \tag{7-53}$$

or from Eq. (7-15)

$$(dG)_{T,P} \leq 0. \tag{7-54}$$

Thus Gibbs free energy is a thermodynamic potential associated with the variables T and P. Any spontaneous irreversible process in a closed system at constant T and P will cause a decrease in the Gibbs free energy, and equilibrium will occur when $(dG)_{T,P} = 0$, corresponding to a minimum in the function G. These energy relations are graphically illustrated in Fig. 7-4 for a hypothetical chemical reaction between substances A and B, where equilibrium of A $\rightleftharpoons$ B occurs when G is a minimum, and the spontaneous irreversible reactions A $\rightarrow$ B and B $\rightarrow$ A are accompanied by a decrease in the free energy of the system as shown in the figure.

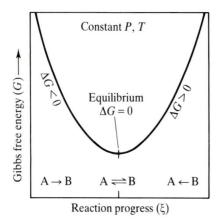

FIGURE 7-4
Gibbs free energy (G) is a thermodynamic potential that acts as an indicator of equilibrium for any process that occurs in a closed system at constant T and P [Eq. (7-54)]: any spontaneous irreversible process will cause a decrease in G, and equilibrium occurs when G is a minimum. This schematic diagram illustrates these relations for a reaction between two hypothetical substances, A and B, as a function of a reaction progress variable (ξ): positive changes ($d\xi > 0$) occur when $\Delta G < 0$ and A $\rightarrow$ B, and negative changes ($d\xi < 0$) occur when $\Delta G > 0$ and A $\leftarrow$ B. Equilibrium, A $\rightleftharpoons$ B, corresponds to $\Delta G = 0$.

Now in terms of a chemical reaction, where ΔG is expressed by Eq. (7-49), the rules are very simple:

1. If $\Delta G = 0$, the reaction mixture is at equilibrium.
2. If $\Delta G < 0$, the reaction will take place spontaneously in the direction of lower Gibbs free energy (although the rate may be so slow that no reaction is apparent).
3. If $\Delta G > 0$, the reaction cannot take place as written unless energy is supplied from an external source.

Similar rules cannot hold for enthalpy changes, because some endothermic reactions (for which $\Delta H > 0$) take place spontaneously. Such endothermic reactions would have a negative ΔG, but a positive $T\Delta S$ term [Eq. (7-46)] larger numerically than ΔG, so that ΔH becomes positive. These reactions are not common, and by and large a negative ΔG means a negative ΔH, especially if the ΔG is a large number, say more than 40 kJ. But ΔG has the advantage over ΔH in that it gives an exact measure of the position of a reaction with respect to equilibrium for the constraints of constant temperature and pressure, whereas ΔH gives only a rough indication.

Values of ΔG for simple reactions can be combined, just as values of ΔH are combined, to give values for more complex reactions. And again like ΔH, values of ΔG for the formation of compounds from their elements can be usefully tabulated, with the same provisos about temperature, pressure, and standard states (*standard free energies of formation*, ΔG_f°). A list of such values for many compounds of geochemical importance is given in Appendix VIII.

Two examples will illustrate the use of free energy as a criterion of equilibrium. The reaction between sulfur and oxygen has a large negative ΔG,

$$S + O_2 \rightarrow SO_2 \qquad \Delta G_f^\circ = -300.2 \text{ kJ mol}^{-1} \text{ at } 25°C. \qquad (7\text{-}55)$$

This means that a mixture of sulfur and oxygen is far from equilibrium: once ignited, the mixture will react until one or the other substance is almost entirely converted to SO_2. The reaction between hydrogen and iodine, on the other hand, has a small positive free energy change

$$\tfrac{1}{2}H_2 + \tfrac{1}{2}I_2 \leftarrow HI \qquad \Delta G_f^{\circ} = +1.3 \, \text{kJ mol}^{-1} \text{ at } 25^{\circ}C. \qquad (7\text{-}56)$$

Hence a mixture of hydrogen gas at 1 bar, hydrogen iodide gas at 1 bar, and solid iodine, would be approximately at equilibrium. A little of the hydrogen iodide would tend to break down into its elements, but much would still be present when true equilibrium was established.

Measurement of ΔG

The direct measurement of ΔG for most reactions is a difficult operation. One method, applicable to many oxidation–reduction processes, depends on measurements of electromotive force and use of the relation

$$\Delta G^{\circ} = -W' = n\mathbf{f}E^{\circ}. \qquad (7\text{-}57)$$

This equation expresses the fact that electromotive force (E°) multiplied by the amount of electric charge moving through a cell (the charge carried by a mol of electrons, $\mathbf{f}$, times the number of mols, n) gives the maximum electrical work (W') that the cell reaction can accomplish (Sec. 9-2). Another possible method involves measurement of activities of reactants and products when a reaction has reached equilibrium, and use of the equation

$$\Delta G^{\circ} = -RT \ln K, \qquad (7\text{-}58)$$

where K is the equilibrium constant (this is discussed in detail in the next chapter). But obviously it would be desirable to have a general method for determining free energy changes from heat measurements alone. Such a method might be based on the relation

$$\Delta G^{\circ} = \Delta H^{\circ} - T\Delta S^{\circ} \qquad (7\text{-}59)$$

if ΔH° and ΔS° can be determined independently. The enthalpy term is easily found by measuring the heat of reaction in a calorimeter. Now if we had a way of obtaining entropy change also, the problem of measuring ΔG° would be solved.

It turns out that values of ΔS can be found by determining absolute entropies from heat capacity measurements. Since $dQ = TdS$ [Eq. (7-11)] we can express the heat capacity at constant pressure in terms of the change in entropy,

$$C_P = \left(\frac{dQ}{dT}\right)_P = T\left(\frac{\partial S}{\partial T}\right)_P. \qquad (7\text{-}60)$$

Upon rearranging

$$dS = C_P \frac{dT}{T} = C_P \, d\ln T, \qquad (7\text{-}61)$$

or

$$\int_{S_{T_1}}^{S_{T_2}} dS = S_{T_2} - S_{T_1} = \int_{T_1}^{T_2} C_P \, d\ln T. \qquad (7\text{-}62)$$

Hence when C_P of a substance is known as a function of temperature, the change in S with temperature may be calculated.

If we knew an absolute value for S at one temperature, its value at any other temperature could be found by carrying out the integration of Eq. (7-62). Arbitrarily let us assign a value of zero to the entropy of some crystal form of each element at the absolute zero of temperature. Then the entropy of this crystal form at a temperature T can be determined by measuring its heat capacity over the temperature interval 0 to T (involving slight extrapolation at the lower limit, since absolute zero is experimentally unattainable) and integrating between these limits.

Now a striking experimental fact emerges if we then determine the entropy differences between various crystal forms of the same element: the differences become smaller and smaller as the temperature drops toward absolute zero. In other words, if we make the assumption of zero entropy at $T = 0$ for one crystal form, then the entropy of other crystal forms is also zero at this temperature. Even more striking is the experimental fact that entropy differences in reactions involving combinations of crystalline elements to form pure crystalline compounds also fall toward zero as T approaches zero. We can generalize from such experimental results: *if the entropy of each element in some crystal state is assumed to be zero at $T = 0$, the entropy of other pure crystalline solids is also zero at this temperature.* This statement is called the Third Law of Thermodynamics. Although formulated much later than the first and second laws and for a time seriously questioned, its validity has now been established by many different kinds of experiments.

The third law makes possible the determination of absolute entropies for most pure crystalline solids. For example, the standard third law entropy ($S°$) of a pure crystalline solid can be computed by integrating Eq. (7-62) from 0 K where $S = 0$ to 298.15 K,

$$S°_{298.15\,K} = \int_{T=0}^{T=298.15\,K} C°_P \, d\ln T. \tag{7-63}$$

This is graphically illustrated in Fig. 7-5 for the mineral phlogopite. Note that S is different from E and H, in that its absolute value can be found.

It should be noted that the relationship between C_P and T is complicated at very low temperatures and the integration is often best performed by graphical methods. In addition, purity and perfection of crystal form are essential in a careful evaluation of third law entropies; glassy solids, solid solutions, and imperfect crystals would all have finite positive entropies at absolute zero. If any phase transitions occur in the substance between 0 and 298.15 K, they must also be considered in evaluating the third law entropy. Entropies of liquids and gases are obtainable from those of the corresponding solids by adding the entropies of melting or vaporization (heats of fusion or vaporization divided by absolute temperature at the transition points). Absolute entropies of gases may be calculated also from spectroscopic data by the methods of statistical mechanics. Entropies of ions and undissociated molecules in solution may be found, with a few additional assumptions, from heats of solution. Thus absolute entropies are obtainable, at least

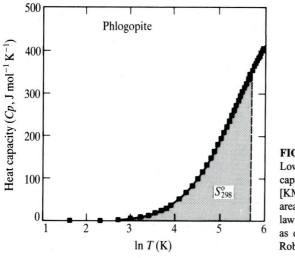

FIGURE 7-5
Low-temperature (5 to 400 K) heat capacity of phlogopite $[KMg_3(AlSi_3O_{10})(OH)_2]$. The shaded area schematically represents the third-law entropy of phlogopite at 298.15 K, as defined by Eq. (7-63). (Data from Robie and Hemingway, 1984.)

in principle, for most substances of geochemical interest. Some of these are listed in Appendix VIII.

The ΔS's for reactions are simply differences between the combined absolute entropies of reactants and products. These, multiplied by the absolute temperature and subtracted from ΔH's give values for ΔG. This is the most generally useful method of finding free energies of reaction.

Free Energy: Conventions and Qualifications

The conventions used in free energy calculations are summarized here for easy reference:

1. As we will use the term, free energy refers only to processes taking place at constant temperature and constant pressure (*Gibbs free energy*). The term may be defined equally well for constant temperature and volume processes (*Helmholtz free energy*), but because reactions of interest in geochemistry take place more commonly under conditions of constant pressure than under constant volume, Gibbs free energies are the more useful.

2. Numerical values of free energies, like those for enthalpies, are known only as relative quantities, never as absolute ones. All we ever get from experiment is a difference of enthalpy and a difference of free energy between the products and the reactants of a chemical process. We are concerned only with ΔG and ΔH, never with G and H themselves.

3. Free energies, like enthalpies, are expressed in joules (J) or kilojoules (kJ), or sometimes in calories or kilocalories. The convention of sign is similar: negative free energy means energy evolved; positive free energy means energy absorbed.

4. The free energy of formation of a compound (ΔG_f) is the free energy change accompanying the formation of 1 mol from its elements. The *standard* free energy of formation of a compound from its elements (ΔG_f°) is the free energy change when the reaction takes place with all substances in their specified *standard states*. A variety of standard states may be chosen for special purposes (this will be discussed further in Sec. 8-3), but the ones used most commonly in calculations of free energies are similar to those used for enthalpies of formation: a temperature of 25°C or 298.15 K, a pressure of 1 bar, and a provision that pure substances be in their most stable states under these conditions. For example, sulfur is assumed to be orthorhombic (rather than monoclinic), iron to have the crystal form called alpha-iron, and silica to be in the form of α-quartz, since these are the most stable states of these substances at the specified conditions. In reactions involving mercury or bromine, the standard state for the element would be the pure liquid. The standard state of gases is more difficult to define precisely, but for most purposes can be taken as the pure gas with an activity of 1 at a specified temperature and pressure (usually taken as 1 bar and any temperature)..

5. Substances in solution also may pose a difficult problem with regard to choice of standard states, but in the relatively dilute solutions encountered most commonly in geochemistry a convenient choice is again unit activity, for both solvent and solutes. Thus for H_2S as a pure gas and in aqueous solution, we write

$$S_{rh} + H_2(g) \rightarrow H_2S(g) \qquad \Delta G_f^\circ = -33.5 \text{ kJ} \qquad (7\text{-}64)$$

$$S_{rh} + H_2(g) \rightarrow H_2S(aq) \qquad \Delta G_f^\circ = -28.0 \text{ kJ}. \qquad (7\text{-}65)$$

The symbol S_{rh} means rhombic sulfur, $H_2(g)$ and $H_2S(g)$ mean hydrogen gas and hydrogen sulfide gas at 1 bar, and $H_2S(aq)$ means dissolved H_2S at an activity of 1.

6. For ions a further convention is needed, because ions are never produced singly. We can determine, for example, the free energy change in the reaction $Na + \frac{1}{2}Cl \rightarrow Na^+ + Cl^-$, but not for the separate processes Na going to Na^+ or $\frac{1}{2}Cl_2$ going to Cl^-. This means that one dissociation reaction must be arbitrarily assigned a free energy value, and the other dissociation processes can then be referred to it. The convention universally adopted is

$$\tfrac{1}{2}H_2(g) \rightleftharpoons H^+ + e^- \qquad \Delta G^\circ = 0. \qquad (7\text{-}66)$$

The symbol H^+ means hydrogen ion at an activity of 1 and the symbol e^- refers to the electron removed from the hydrogen atom.

7. Values of the standard Gibbs free energy of formation are important because they can be added and subtracted to determine free energy changes of reactions (ΔG°), just as heats of formation are used to calculate enthalpy changes using Eq. (7-30). Written in terms of the stoichiometric equation the free energy change of a reaction is

$$\Delta G_r^\circ = \sum_i v_i \Delta G_f^\circ. \qquad (7\text{-}67)$$

As in Eq. (7-30), v_i is the stoichiometric reaction coefficient, positive for products and negative for reactants, so that the free energy of the reaction is the difference between the free energies of all the products and the free energies of the reactants.

SUMMARY: USES AND LIMITATIONS OF FREE ENERGY

In effect, by introducing the concept of free energy change we have developed a second method of describing the tendency of a mixture to react. The first device was the equilibrium constant, from which we could deduce the concentration ratios that exist at equilibrium, and thus determine in what direction a given mixture must react to attain equilibrium. The standard free energy change gives us the same sort of information, for it shows, by its sign and its magnitude, whether a mixture with unit concentrations will react one way or the other and to what extent. We shall see in the next chapter that the equilibrium constant is actually defined in terms of standard free energy changes. So what have we gained by introducing this second method for describing equilibrium?

Free energies have two great advantages: they are additive, and they require less space in tables. In tables a separate equilibrium constant must be listed for each reaction, whereas free energies of formation in a relatively brief list can be used to compute free energy changes (and equilibrium constants as we shall see in the next chapter) for a large number of reactions.

Free energies give us a quantitative measure of *stability*, a concept we have discussed at some length earlier (Sec. 1-7). A mixture of substances all of whose possible reactions have zero free energy could undergo no change, hence would be stable; a mixture permitting reactions with negative free energy changes would be unstable, and the magnitude of the possible free energy changes would indicate the degree of instability. As in the earlier discussion, it should be emphasized that this kind of stability refers only to possible energy changes, not to the rate at which these energy changes actually occur. A piece of coal exposed to the air is apparently stable, but the stability is due only to the slowness with which coal reacts with oxygen at ordinary temperatures. In this book *stability* or *true stability* will always refer to stability with respect to energy changes, and the apparent stability of some mixtures due to slow reaction rates will be called *metastability*.

The inability to give information about reaction rates is one of the severe limitations on the application of free energy reasoning in geochemistry. A second serious limitation is the inability of free energy reasoning to deal with geologic processes that take place in open systems, systems in which matter is continually being added from the outside or is continually flowing away, so that real chemical equilibrium is not attained. Consider the movement of fluid through a vein, for example, or the rush of gases through a volcanic orifice. Even when flow is not an important factor, the slowness of reaction and incompleteness of mixing may prevent attainment of equilibrium, as is obvious for immature soils and for half-digested inclusions in granites. Despite these restrictions, we shall find that reasoning on the basis of free energy and equilibrium will help in setting limits to geologic processes, in telling us at least which reactions are possible and which are completely out of the question. In restricted areas of geology, particularly in chemical sedimentation, metamorphism, and the crystallization of magmas, equilibrium reasoning can be carried further and leads to exact predictions that can be checked against natural occurrences.

PROBLEMS

1. The area, a, of a rectangle is a function of the height, h, and the length, l; that is

$$a = F(h, l) \qquad \text{where} \qquad a = hl.$$

The independent variables are h and l; a is the dependent variable. Two other possible dependent variables are the perimeter, p,

$$p = F(h, l) \qquad \text{where} \qquad p = 2h + 2l$$

and the diagonal, d,

$$d = F(h, l) \qquad \text{where} \qquad d = (h^2 + l^2)^{0.5}.$$

Find the values of the following partial derivatives in terms of h and l, or find a numerical answer:

$$\left(\frac{\partial a}{\partial h}\right)_l; \quad \left(\frac{\partial a}{\partial l}\right)_h; \quad \left(\frac{\partial p}{\partial h}\right)_l; \quad \left(\frac{\partial l}{\partial h}\right)_p; \quad \left(\frac{\partial l}{\partial h}\right)_d.$$

2. Prove that the expression for work performed by an ideal gas is

$$dW = R\,dT - (RT/P)\,dP.$$

3. The coefficients of expansion and compression for a substance are defined as

$$\alpha = \frac{1}{V}\left(\frac{\partial V}{\partial T}\right)_P \qquad \text{and} \qquad \beta = -\frac{1}{V}\left(\frac{\partial V}{\partial P}\right)_T.$$

Derive expressions for α and β in terms of density (ρ). At the critical point, where $(\partial P/\partial V)_T = 0$, show that both α and β are infinite.

4. Using the data of Appendix VIII, find the standard enthalpy changes and free energy changes for the following reactions at 25°C:

$$\text{Sn} + \text{O}_2 \rightleftharpoons \text{SnO}_2,$$

$$\text{KCl(s)} \rightleftharpoons \text{K}^+ + \text{Cl}^-,$$

$$2\text{FeCO}_3 + \tfrac{1}{2}\text{O}_2 + 2\text{H}_2\text{O(l)} \rightleftharpoons \text{Fe}_2\text{O}_3 + 2\text{H}_2\text{CO}_3\text{(aq)}.$$

To reach equilibrium, which of these reactions would go as written, i.e., from left to right? Which is more stable at ordinary temperatures in contact with air, tin or tin oxide? Hematite or siderite?

5. Using the data of Appendix VIII, find the standard free energy change and enthalpy change for the reaction

$$\underset{\text{K-feldspar (microcline)}}{2\text{KAlSi}_3\text{O}_8} + 2\text{H}^+ + \text{H}_2\text{O} \rightleftharpoons \underset{\text{kaolinite}}{\text{Al}_2\text{Si}_2\text{O}_5(\text{OH})_4} + 2\text{K}^+ + \underset{\text{quartz}}{4\text{SiO}_2}.$$

From these numbers calculate the standard entropy change for the reaction, and compare your result with the value calculated directly from the standard entropies of formation listed in Appendix VIII. (Any discrepancy probably means that the different numbers were determined by different investigators at different times, and the difference is just expectable experimental error.) Is the reaction exothermic or endothermic? In what sort of environment would you expect to find it occurring in nature?

6. In each of the following pairs, which has the greater amount of entropy? Assume that all variables not mentioned are the same for both members of a pair.

(a) A mol of calcite at 0°C, (a') a mol of calcite at 100°C.

(b) A sample of air under a pressure of 1 bar, (b') the same sample under 10 bar.

(c) Crystalline salt plus water, (c') a solution formed by dissolving the salt in water.

(d) Crystalline basalt, (d') the magma from which the basalt crystallized.

7. Calcite is more soluble at low temperatures than at high temperatures. What does this tell you about the sign of $\Delta H°$ for the reaction $Ca^{2+} + CO_3^{2-} \rightarrow CaCO_3$? Verify your answer by calculating the standard change in enthalpy for this reaction.

8. Because the standard state adopted for minerals is unit activity of the pure phase at any temperature and pressure, we can define the standard molal Gibbs free energy of a mineral at some elevated temperature (T) and pressure (P) by the equation

$$\Delta G°_{\text{mineral}, T, P} = \Delta G°_{\text{f, mineral}} + (G°_{T, P} - G°_{T_{\text{ref}}, P_{\text{ref}}}).$$

The term $\Delta G°_{\text{mineral}, T, P}$ is called the *apparent standard molal Gibbs free energy* of the mineral at the specified T and P, $\Delta G°_{\text{f, mineral}}$ is the free energy of formation of the mineral from its elements at the reference conditions of 25°C and 1 bar, and the parenthetical term is the difference in the absolute value of the standard molal Gibbs free energy of the mineral at the temperature and pressure of interest and in the reference conditions (25°C and 1 bar).

(a) To compute the value of $\Delta G°_{\text{mineral}}$ at a temperature greater than 25°C and at the reference pressure of 1 bar, the parenthetical term in the above equation is written as

$$(G°_{T, P_{\text{ref}}} - G°_{T_{\text{ref}}, P_{\text{ref}}}) = -S°_{T_{\text{ref}}, P_{\text{ref}}}(T - T_{\text{ref}}) + \int_{T_{\text{ref}}}^{T} C_{P_{\text{ref}}}° dT - T \int_{T_{\text{ref}}}^{T} C_{P_{\text{ref}}}° d\ln T.$$

Derive this expression.

(b) Using the above equations and the data given in the table below, compute the temperature of equilibrium between kyanite and andalusite at a pressure of 1 bar. The letters a, b, and c in the table correspond to the coefficients in Eq. (7-34) for the temperature dependence of heat capacity. (Reference: Helgeson, H. C., J. M. Delany, H. W. Nesbitt, and D. K. Bird: "Summary and critique of the thermodynamic properties of rock-forming minerals," *American Journal of Science*, vol. 278A, 229 pp., 1978; see pages 28–29.)

Mineral	$\Delta G°_{\text{f, mineral}}$ J/mol	$S°_{T_{\text{ref}}, P_{\text{ref}}}$ J/mol K	a J/mol	b × 10^3 J/mol K^2	c × 10^{-5} J K/mol
Andalusite	−2,429,176	92.88	172.8444	26.3282	51.8485
Kyanite	−2,430,720	83.68	173.1887	28.5202	53.8987

REFERENCES AND SUGGESTIONS FOR FURTHER READING

Anderson, G. M., and D. A. Crerar: *Thermodynamics in Geochemistry*, Oxford University Press, 588 pp., 1993.

Denbigh, K.: *The Principles of Chemical Equilibrium, with Applications in Chemistry and Chemical Engineering*, Cambridge University Press, 3rd ed. 494 pp., 1971. A standard textbook in chemical thermodynamics.

Fletcher, P.: *Chemical Thermodynamics for Earth Scientists*, Longman, London and New York, 464 pp., 1993. An advanced textbook in thermodynamics of geologic substances.

Hemingway, B. S., K. M. Krupka, and R. A. Robie: "Heat capacities of the alkali feldspars between 0 and 1000 K," *American Mineralogist*, vol. 66, pp. 1202–1215, 1981.

Nordstrom, D. K., and J. L. Munoz: *Geochemical Thermodynamics*, Blackwell Scientific Publications, 2d ed. 493 pp., 1994. Chapter 3 provides a clear and concise treatment of the fundamental principles of chemical thermodynamics.

Openshaw, R. E., B. S. Hemingway, R. A. Robie, D. R. Waldbaum, and K. M. Krupka: "The heat capacities at low temperatures and entropies at 298.15 K of low albite, analbite, microcline, and high sanidine," *Journal of Research, U.S. Geol. Survey*, vol. 4, No. 2, pp. 195–204, 1976.

Robie, R. A., and B. S. Hemingway: "Heat capacities and entropies of phlogopite $[KMg_3(AlSi_3O_{10})(OH)_2]$ and paragonite $[NaAl_2(AlSi_3O_{10})(OH)_2]$ between 5 and 900 K and estimates of the enthalpies and Gibbs free energies of formation," *American Mineralogist*, vol. 69, pp. 858–868, 1984.

CHAPTER
8

CHEMICAL THERMODYNAMICS: PHASE EQUILIBRIA

The idea embodied in free energy as a measure of equilibrium may be expressed in a variety of ways. Here we develop a practical means for using free energy to describe heterogeneous equilibrium in complex geologic systems.

8-1 CHEMICAL POTENTIAL

We begin by considering a very simple kind of heterogeneous equilibrium, the equilibrium between liquid water and water vapor in a closed container,

$$H_2O(\text{vapor}) \rightleftharpoons H_2O(\text{liquid}). \tag{8-1}$$

The condition that equilibrium exists we can state by saying that, if a very small amount of liquid is vaporized and if a small amount of vapor is condensed, the change in the free energy is zero. One may think of the liquid water as capable of vaporizing, or as tending to vaporize, or as seeking to "escape" from the liquid into the vapor phase. Thus we can assign to the water a certain *escaping tendency* with respect to the vapor. Similarly we can describe the vapor as having a certain escaping tendency with respect to the liquid. The condition of equilibrium would then be simply that the two escaping tendencies are equal. If the temperature is raised, the escaping tendency of the liquid increases more than that of the vapor, so that additional liquid must vaporize before equilibrium is reestablished.

As a measure of the escaping tendency, we may use the free energy per mol, or *molal free energy*, of water and its vapor. We designate molal free energy with a bar over the letter symbol: $\overline{G} = G/n$, where G represents the free energy of n mols. The condition of equilibrium at constant temperature and pressure requires that the free energy of the reaction be zero so that $\overline{G}_{\text{liquid}} = \overline{G}_{\text{vapor}}$. Alternatively, we may express this condition by thinking of the vaporization of an infinitesimal amount of water— the escape of dn mols of water from the liquid phase into the gas phase. The free energy lost by the liquid is $\overline{G}_{\text{liquid}}\, dn$, and that gained by the vapor is $\overline{G}_{\text{vapor}}\, dn$. If equilibrium exists, the net free energy change must be zero:

$$dG = \overline{G}_{\text{liquid}}\, dn - \overline{G}_{\text{vapor}}\, dn = 0. \qquad (8\text{-}2)$$

This is clearly equivalent to $\overline{G}_{\text{vapor}} = \overline{G}_{\text{liquid}}$. Molal free energies so defined are not a very practical measure for evaluating equilibrium, since numerical values of absolute free energies are unknown. For actual use we would substitute free energies of formation assigned on the basis of conventions described in Chapter 7.

The molal free energy may be given another name, *chemical potential*, suggesting that this quantity represents a sort of energy level of a substance in one phase of a system. An actual transfer of energy attendant on vaporization or condensation is expressed by the chemical potential multiplied by the number of mols transferred—in other words, by the total free energy change. Analogously in mechanics an object suspended above the Earth can be said to have a mechanical potential, measured by its height times the acceleration of gravity, gh; energy released when the object falls is the product of this potential multiplied by its mass, Mgh. Or in electricity: an electric charge has a certain electric potential (measured in volts) with respect to another charged object, and electrical energy released when the charge moves is the product of electric potential and the amount of charge. Chemical potential is less easily visualized, but it plays the same role in chemical reactions that these more familiar potentials play in mechanical and electrical processes. We may rephrase our condition of chemical equilibrium [Eq. (8-2)] by saying that the chemical potentials of H_2O in the liquid and vapor phases must be equal.

The idea of chemical potential becomes more useful when we consider complicated systems involving solutions. Suppose we change our simple example by stirring salt into the liquid water: the equilibrium will readjust itself, and we inquire how the necessary condition for equilibrium between liquid and vapor can best be expressed. If we try to use equality of the molal free energies of H_2O in the two phases, we meet the difficulty that we have not yet defined the molal free energy of H_2O in the solution. It cannot be simply the free energy of the water present divided by the number of mols, because dissolving the salt involves an energy change. In other words, the free energy of H_2O is now a complex function of concentration as well as of temperature and pressure.

We can still set up a number to represent the molal free energy, however, if we consider the change in free energy of the solution produced by an infinitesimal change in the amount of water. Instead of a quotient of macroscopic quantities, G/n, which we can use for a pure substance, we set up a differential coefficient in terms of a specific chemical component represented by the subscript i,

$$\lim_{\Delta n_i \to 0} \left(\frac{\Delta G}{\Delta n_i} \right) = \left(\frac{\partial G}{\partial n_i} \right)_{P,T,n_j} \equiv \mu_i. \tag{8-3}$$

The subscript T, P, n_j means that differentiation is carried out for constant pressure, temperature, and the concentrations of all other components of the solution except the ith (represented by the term n_j). In mathematical language this is a partial derivative (see Fig. 7-2), and the quantity it represents is therefore called the *partial molal free energy.* The term chemical potential is a synonym for this expression, as well as for the molal free energy of pure substances; it is often designated by the Greek letter mu, μ, usually with a subscript denoting the chemical component and a superscript representing the phase, such as $\mu_{\mathrm{H_2O}}^{\mathrm{vapor}}$.

A general statement of the condition of equilibrium can now be proposed for any system, however complex: *the chemical potential of each component must be the same in all phases of the system.* To prove this we again consider two different types of experiments, each with identical initial and final states, so that changes in functions of the state of the system will be the same for both experiments.

The first experiment is the one we used to derive Eq. (7-19): an irreversible chemical reaction occurs in a closed system producing entropy within the system, so that the change in the internal energy is

$$dE = TdS - PdV - TdS_{\mathrm{int}}. \tag{7-19}$$

An example would be the irreversible hydrolysis of K-feldspar in an acid solution to produce the clay mineral kaolinite,

$$\underset{\text{K-feldspar}}{\mathrm{KAlSi_3O_8}} + \mathrm{H^+} + \tfrac{1}{2}\mathrm{H_2O} \rightarrow \mathrm{K^+} + \tfrac{1}{2}\underset{\text{kaolinite}}{\mathrm{Al_2Si_2O_5(OH)_4}} + 2\mathrm{SiO_2(aq)}. \tag{8-4}$$

The initial state consists of one mol of K-feldspar and an acid solution, and the final state has half a mol of kaolinite and a solution enriched in $\mathrm{K^+}$ and $\mathrm{SiO_2(aq)}$.

In the second experiment the initial and final states of the first experiment are exactly reproduced, but without any chemical reaction whatsoever. This is accomplished by making the boundaries of the system open to the transport of mass, so material inside the system can be exchanged with material from the surroundings. First we remove the one mol of K-feldspar and all the solution from our system. Next, we place inside the system half a mol of kaolinite and a solution of the exact composition produced by the first experiment.

The initial and final states of our two experiments are identical, so that changes in E, S, and V are equal. In the first experiment a portion of the energy change was also due to irreversible entropy production inside the system $(-TdS_{\mathrm{int}})$. But in the second experiment there is no entropy production. Instead the energy was altered by exchange of mass between the system and its surroundings. We represent this energy change by the number of mols of each component (dn_i) multiplied by the chemical potential of that component [see Eq. (8-2)]. So for the second experiment the change in internal energy is

$$dE = TdS - PdV + \sum_i \mu_i dn_i. \tag{8-5}$$

Since the differential changes in E, S, and V are the same in both experiments, Eqs. (7-19) and (8-5) are combined to give

$$-TdS_{int} = \sum_i \mu_i dn_i. \tag{8-6}$$

Thus the condition for chemical equilibrium defined by the second law [$dS_{int} = 0$, Eq. (7-15)] can be stated as

$$\sum_i \mu_i dn_i = 0. \tag{8-7}$$

This general equation is clearly equivalent to the specific equation [Eq. (8-2)] we deduced earlier for liquid–vapor equilibrium of H_2O based on escaping tendency alone.

We now assume that Eq. (8-7) can be used to express changes in internal energy within a closed system when the redistribution of elements among phases is due to chemical reactions within the system. The criterion for equilibrium of a chemical reaction is now written as

$$\sum_i \mu_i v_i = 0, \tag{8-8}$$

where the change in the number of mols of a reaction species [dn_i in Eq. (8-7)] is equal to the stoichiometric reaction coefficient, v_i [which is taken to be positive for products and negative for reactants as in Eq. (7-30)]. To illustrate this we write Eq. (8-8) for the liquid–vapor equilibrium of H_2O [Eq. (8-1)],

$$\mu_{H_2O}^{liquid}(+1) + \mu_{H_2O}^{vapor}(-1) = 0 \tag{8-9}$$

or

$$\mu_{H_2O}^{liquid} = \mu_{H_2O}^{vapor}, \tag{8-10}$$

and for equilibrium between quartz and an aqueous solution,

$$\underset{\text{quartz}}{SiO_2} \rightleftharpoons SiO_2(aq), \tag{8-11}$$

Eq. (8-8) becomes

$$\mu_{SiO_2}^{aqueous}(+1) + \mu_{SiO_2}^{quartz}(-1) = 0 \tag{8-12}$$

or

$$\mu_{SiO_2}^{aqueous} = \mu_{SiO_2}^{quartz}. \tag{8-13}$$

These relationships, developed from the basic principles of thermodynamics, demonstrate that: *at equilibrium the chemical potential of a thermodynamic component is the same in all phases.*

Other useful relationships between chemical potential and Gibbs free energy are derived by substituting the identity of Eq. (8-6) into Eq. (7-52),

$$dG = -SdT + VdP + \sum_i \mu_i dn_i. \tag{8-14}$$

This equation shows that changes in G are determined by changes in T, P, and n_i; that is, G is a function of temperature, pressure, and composition, as stated earlier in Eq. (7-7),

$$G = F(T, P, n_{i(i=1, 2, 3, ...)}). \tag{8-15}$$

The subscript $i = 1, 2, 3, \ldots$ represents chemical components used to describe the composition of all phases in our system. Differentiating Eq. (8-15) leads to an expression for the total derivative of G,

$$dG = \left(\frac{\partial G}{\partial T}\right)_{P,n_i} dT + \left(\frac{\partial G}{\partial P}\right)_{T,n_i} dP + \sum_i \left(\frac{\partial G}{\partial n_i}\right)_{P,T,n_j} dn_i, \tag{8-16}$$

and comparison with Eq. (8-14) gives the following identities,

$$S = -\left(\frac{\partial G}{\partial T}\right)_{P,n_i}; \quad V = \left(\frac{\partial G}{\partial P}\right)_{T,n_i}; \quad \mu_i = \left(\frac{\partial G}{\partial n_i}\right)_{P,T,n_j}. \tag{8-17a, b, c}$$

We now see that our function of the state of the system called G has the following properties: its partial change with respect to temperature is simply the negative of the entropy of the system, its partial change with respect to pressure is the volume of the system, and its partial change with respect to composition is the chemical potential.

The extensive parameters G, S, and V in Eqs. (8-17a, b) refer to the entire system. However, similar types of equations can also be written for a chemical reaction,

$$\Delta S = -\left(\frac{\partial \Delta G}{\partial T}\right)_P; \quad \Delta V = \left(\frac{\partial \Delta G}{\partial P}\right)_T, \tag{8-18a, b}$$

or for a specific phase (represented here by the subscript ψ,

$$S_\psi = -\left(\frac{\partial G_\psi}{\partial T}\right)_P; \quad V_\psi = \left(\frac{\partial G_\psi}{\partial P}\right)_T. \tag{8-19a, b}$$

These identities have many practical uses in geochemistry. They may be integrated or used graphically to predict the effects of temperature and pressure on the free energy of substances or chemical reactions.

As an example, consider what happens to the liquid–vapor reaction of water [Eq. (8-1)] when temperature is changed at constant pressure. From Eq. (8-19a) we see that the entropy of the two phases (S_{liquid} and S_{vapor}) will determine how the free energies (G_{liquid} and G_{vapor}) change with temperature. Because the entropy of vapor is greater than that of the liquid, the slope of a curve for G_{vapor} will be steeper than that for G_{liquid} when both are plotted against T, as shown schematically in Fig. 8-1(*a*). The curves intersect at the temperature where ΔG is zero and the liquid and vapor are in equilibrium. At lower temperatures, because $S_{vapor} > S_{liquid}$, the free energy curve for the vapor will be above that for the liquid. Here ΔG is negative and the liquid is the stable phase. For higher temperatures, where ΔG is positive, the vapor will be stable. These relationships are easily visualized in a graph like Fig. 8-1(*a*), where the stable phase is always the one with the lower free energy, and the phase with the larger entropy (the vapor in this case) is stable at high temperatures.

Similar relations can be deduced by qualitatively evaluating the pressure dependence of free energy using Eq. (8-19b). An example for the liquid–vapor reaction of water is shown in Fig. 8-1(*b*). Because water as a vapor has a higher

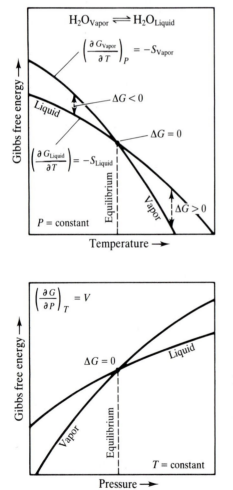

FIGURE 8-1
Schematic illustration of the difference in the Gibbs free energy of formation of H_2O liquid and H_2O vapor as a function of temperature at constant pressure (*a*) and as a function of pressure at constant temperature (*b*). In both diagrams the stable phase is the one with the lower free energy, and equilibrium between the two phases exists when $\Delta G = 0$. The stable phase at high temperatures is the one with the higher entropy (the vapor), and the stable phase at high pressures is the one with the higher density or smaller volume (the liquid).

molal volume than the liquid, its free energy curve will be steeper when plotted against pressure. As a consequence, at pressures greater than liquid–vapor equilibrium the phase with the smaller volume (the liquid) will have the lower free energy and thus be the stable phase.

Other important relations between chemical potential and free energy are obtained if we integrate Eq. (8-16) over the entire mass of the system. Noting that T and P are independent of mass, we find that G is uniquely defined by the sum of the chemical potentials multiplied by the number of mols of each chemical component (i),

$$G = \sum_i \mu_i n_i. \tag{8-20}$$

And upon differentiation we obtain

$$dG = \sum_i \mu_i dn_i + \sum_i n_i d\mu_i, \tag{8-21}$$

which can be combined with Eq. (8-14) to give

$$0 = SdT - VdP + \sum_i n_i d\mu_i. \tag{8-22}$$

The latter equation is called the Gibbs–Duhem equation. It defines relationships among changes in the intensive variables T, P, and μ_i for systems in a state of internal equilibrium. We shall see later that these three equations form a useful basis for describing equilibrium phase relations in complex geologic systems, but first we introduce the concepts of fugacity, activity, and standard states.

8-2 FUGACITY

Mathematically the partial molal free energy has the drawback that it approaches an infinite negative value as the concentration becomes indefinitely small. To circumvent this difficulty, it is often convenient to use still another measure of escaping tendency, called *fugacity*, defined by the equation

$$\left(\frac{\partial \ln f_i}{\partial \mu_i} \right)_{P,T} = \frac{1}{RT}, \tag{8-23}$$

where f_i is the fugacity and μ_i is the chemical potential as defined by Eq. (8-3). Alternatively this may be expressed in the integrated form

$$\mu_i - \mu_i' = RT \ln \frac{f_i}{f_i'}, \tag{8-24}$$

where μ_i and μ_i' are the partial molal free energies, and f_i and f_i' the fugacities, of one substance (i) at two different concentrations (or, for gases, at two different pressures). The fugacity, in effect, is a sort of idealized vapor pressure. It is measured in bars, and is equal to the vapor pressure when the vapor behaves as a perfect gas.

Fugacities may be used for solids and liquids as well as for gases. We may speak of the fugacity of a vapor above a liquid, or the fugacity of the liquid itself, just as the vapor pressure of a liquid may be considered a property of either the vapor or the liquid.

The relation between fugacity and vapor pressure can be made clear by an argument based on the equation for change in free energy with pressure (8-17b) for a perfect gas, where nRT/P may be substituted for V,

$$\left(\frac{\partial G}{\partial P}\right)_T = \frac{nRT}{P}.$$ (8-25)

Differentiation with respect to n at constant T and P gives

$$\left(\frac{\partial}{\partial n}\left(\frac{\partial G}{\partial P}\right)_T\right)_{T,P} = \left(\frac{\partial}{\partial P}\left(\frac{\partial G}{\partial n}\right)_{T,P}\right)_T = \left(\frac{\partial \mu}{\partial P}\right)_T = \frac{RT}{P}.$$ (8-26)

If dP is moved to the right side of the equation, integration gives

$$\int_{\mu'}^{\mu} d\mu = \int_{P'}^{P} \frac{RT}{P} dP,$$ (8-27)

and

$$\mu - \mu' = RT \ln \frac{P}{P'}.$$ (8-28)

For a specific chemical component (i) of the gas we substitute Dalton's law $(P_i = X_i P$, which states that the partial pressure of a gas, P_i, is equal to its mol fraction, X_i, times the total pressure, P, noting that $X_i \equiv X_i'$) so that,

$$\mu_i - \mu_i' = RT \ln \frac{P_i}{P_i'}.$$ (8-29)

This has the same form as Eq. (8-24), indicating that vapor pressure may be used as an approximation for fugacity to the same extent that a perfect gas approximates the behavior of a real gas.

Deviation from ideal gas behavior is measured by the ratio of the fugacity to the partial pressure of a chemical component (i) in the gas; this ratio is given the name *fugacity coefficient* (χ_i),

$$\frac{f_i}{P_i} = \chi_i.$$ (8-30)

Upon substituting Dalton's law we obtain the following relation between fugacity and composition for a nonideal gas,

$$f_i = X_i \chi_i P.$$ (8-31)

8-3   ACTIVITY AND STANDARD STATES

In very dilute solutions the fugacity of a volatile solute becomes equal to its vapor pressure, and this in turn is proportional to its concentration in the solution, a relationship called Henry's law. To express the deviation of more concentrated solutions from Henry's law, it is convenient to use *relative fugacities*, fugacities referred to the fugacity in some assumed standard state:

$$\mu_i - \mu_i^\circ = RT \ln \frac{f_i}{f_i^\circ}. \tag{8-32}$$

This equation is identical with Eq. (8-24) if μ_i' and f_i' are defined as the values of these quantities in a standard state. The standard state may be selected in different ways for different kinds of problems. For solids and solvent components of fluids the standard state usually chosen is one of unit activity of the pure substance at some specified T and P, or for some range of T and P. For solutes a convenient standard state is a hypothetical solution of unit concentration that obeys Henry's law (practically all real solutions deviate considerably from Henry's law at concentrations near $1m$).

The ratio f_i/f_i° in Eq. (8-32) is defined as the *activity* of the ith species or component (a term introduced in Sec. 2-6). Thus Eq. (8-32) can be rewritten as

$$\mu_i - \mu_i^\circ = RT \ln a_i. \tag{8-33}$$

In this way the chemical potential of a thermodynamic component in our system is considered to be the sum of two terms: the chemical potential in some specified standard standard state (μ_i°) and a term including the activity (a_i),

$$\mu_i = \mu_i^\circ + RT \ln a_i. \tag{8-34}$$

Note that the last term is simply a measure of the difference in the chemical potentials of a component in our system and the component in some chosen standard state, that is

$$a_i \equiv \frac{f_i}{f_i^\circ} = \exp\left(\frac{\mu_i - \mu_i^\circ}{RT}\right). \tag{8-35}$$

Like fugacity, activity is related to concentration. It is, in fact, a sort of idealized concentration, the concentration that would be expected if the solution obeyed Henry's law at all concentrations. The ratio of activity to concentration (m_i) or to mol fraction (X_i),

$$\frac{a_i}{m_i} = \gamma_i \tag{8-36}$$

and

$$\frac{a_i}{X_i} = \lambda_i, \tag{8-37}$$

expresses the extent to which solutions deviate from Henry's law; this ratio is given the name *activity coefficient*. The concepts of activity and activity coefficient described here may be generalized to apply to volatile as well as nonvolatile solutes and to any kind of solution. We have met the terms before in our discussion of solubility equilibria in Chapter 2.

The best units to use for activities are not agreed upon by all writers. If the activity of a solute in a saturated solution is defined as in Sec. 2-6, that is, as the equilibrium concentration extrapolated to zero ionic strength, the appropriate unit would be the same as the unit used to express concentration. With the more general definition given above [Eq. (8-35)], however, *activity is the ratio of two fugacities*, hence *it is a dimensionless quantity.* There is no discrepancy in the two definitions; they can easily be reconciled by adjusting the definitions of standard fugacity or of the activity coefficient. The way these related quantities are defined is largely a matter of choice, and practically it makes little difference, since numerical values of activity are the same with either usage.

8-4   THE EQUILIBRIUM CONSTANT

In Chapter 1 we introduced the equilibrium constant as a quotient of concentrations based on qualitative observations, and used it in later chapters to describe conditions of equilibrium between aqueous solutions and minerals. Here we derive an exact formulation of the equilibrium constant based on thermodynamic principles.

For constant temperature and pressure, where dT and dP are zero, the relationships among G, μ_i, and n_i defined by Eqs. (8-14), (8-21), and (8-22) reduce to

$$0 = \sum_i n_i d\mu_i \qquad (8\text{-}38)$$

and

$$dG = \sum_i \mu_i dn_i. \qquad (8\text{-}39)$$

The latter equation can be written for a chemical reaction as

$$\Delta G = \sum_i \mu_i \nu_i, \qquad (8\text{-}40)$$

an equation similar in form to stoichiometric equations derived earlier involving free energy, composition, and chemical potential [Eqs. (7-49), (7-67), (8-2), (8-7), and (8-8)]. Now if we substitute for μ_i the standard state and activity relations defined by Eq. (8-34), noting that

$$\Delta G^\circ = \sum_i \mu_i^\circ \nu_i, \qquad (8\text{-}41)$$

we obtain

$$\Delta G = \Delta G^\circ + RT \sum_i \nu_i \ln a_i. \qquad (8\text{-}42)$$

The superscript in the symbol ΔG° indicates that this is the *standard Gibbs free energy* of reaction, the free energy change for the reaction that would occur when all substances are present in their standard states (Sec. 7-4).

Equilibrium at constant T and P requires that the free energy of reaction, ΔG, must be equal to zero, so we can rewrite Eq. (8-42) for the condition of equilibrium (converting to the more familiar base 10 logarithm) as

$$-\Delta G^\circ = 2.303\, RT \sum_i \nu_i \log a_i. \tag{8-43}$$

Thus for equilibrium among the hypothetical components X, Y, Z, and Q represented by the reaction

$$x\text{X} + y\text{Y} \rightleftharpoons z\text{Z} + q\text{Q}, \tag{8-44}$$

where x, y, z, and q are stoichiometric reaction coefficients, we have

$$\Delta G^\circ = -2.303\, RT\, (z \log a_Z + q \log a_Q - x \log a_X - y \log a_Y) \tag{8-45}$$

or

$$\Delta G^\circ = -2.303\, RT\, (\log a_Z^z + \log a_Q^q - \log a_X^x - \log a_Y^y). \tag{8-46}$$

Rearrangement of Eq. (8-46) gives, for equilibrium,

$$\frac{-\Delta G^\circ}{2.303\, RT} = \log \left(\frac{a_Z^z \cdot a_Q^q}{a_X^x \cdot a_Y^y} \right). \tag{8-47}$$

The quotient on the right-hand side of the equation must be constant if ΔG is to remain zero. The separate activities may change, but any change is compensated by other changes so that the value of the quotient remains constant. We call this quotient the equilibrium constant, K, for the reaction of Eq. (8-44):

$$K = \left(\frac{a_Z^z \cdot a_Q^q}{a_X^x \cdot a_Y^y} \right). \tag{8-48}$$

The following general relations can now be written among K, the standard Gibbs free energy, and the activities of components in a chemical reaction:

$$\log K = \frac{-\Delta G^\circ}{2.303 RT} = \sum_i \nu_i \log a_i \tag{8-49}$$

and

$$K = 10^{\left(\frac{-\Delta G^\circ}{2.303 RT} \right)} = \prod_i a_i^{\nu_i}. \tag{8-50}$$

The right-hand expression in Eq. (8-50) is a general statement for the stoichiometric activity product of a chemical reaction; it is explicitly stated by the quotient in Eqs. (1-6) and (8-48).

From the derivations above we see that the equilibrium constant has a value that is determined by the standard states (μ_i°) chosen for all the substances in the reaction. It is a function of T and P, and is independent of the compositions of the phases involved in the actual reaction. In earlier chapters the equilibrium constant was treated as a product of concentrations [Eq. (1-6)]; actually the number refers to activities [Eqs. (8-48) to (8-50)], so that K is a dimensionless quantity (as noted earlier, Sec. 1-8).

As an example of the relationship between the equilibrium constant and standard molal Gibbs free energy we look again at the solubility of anhydrite previously considered in Sec. 1-5. At equilibrium $CaSO_4(\text{anhydrite}) \rightleftharpoons Ca^{2+} + SO_4^{2-}$ [Eq. (1-26)] and, if the anhydrite is pure so that $a_{\text{anhydrite}} = 1$, the constant is $K = a_{Ca^{2+}} a_{SO_4^{2-}}$. For a solution of $CaSO_4$ in pure water, the concentrations of Ca^{2+} and SO_4^{2-} are so small that the difference between concentrations and activities is unimportant, and the experimental value of K is $m_{Ca^{2+}} m_{SO_4^{2-}} = 3.4 \times 10^{-5}$. From this value the standard free energy change can be computed:

$$\Delta G^\circ = -2.303RT \log K = +25.5 \text{ kJ}. \tag{8-51}$$

The positive number means that the reaction tends to go to the left when all substances are at unit activities, that is, solid $CaSO_4$ and the two ions Ca^{2+} and SO_4^{2-} each at an activity of unity. This, of course, agrees with laboratory experience: in a solution that contains such a high concentration of Ca^{2+} and SO_4^{2-}, solid $CaSO_4$ (in the form of gypsum) will precipitate rapidly until equilibrium is established.

For another example, consider the oxidation of galena to anglesite:

$$\underset{\text{galena}}{PbS} + 2O_2 \rightarrow \underset{\text{anglesite}}{PbSO_4} . \tag{8-52}$$

The standard free energy of this reaction is computed from Eq. (7-67), which reduces to the difference in free energies of formation of anglesite and galena ($\Delta G^\circ = -714$ kJ). The equilibrium constant, computed from Eq. (8-50), is $K = 10^{125}$. Both the high negative ΔG° and the enormous exponent in K indicate that this reaction goes practically to completion toward the right. For the reaction

$$\underset{\text{galena}}{PbS} + SO_4^{2-} \leftarrow \underset{\text{anglesite}}{PbSO_4} + S^{2-} \qquad \Delta G^\circ = +115.9 \text{ kJ} \tag{8-53}$$

on the other hand, $K = 10^{-20.3}$; here the high positive ΔG° and the large negative exponent in K both mean that this reaction is displaced far to the left. For the reaction

$$\underset{\text{galena}}{PbS} + 2H^+ \rightleftharpoons H_2S + Pb^{2+} \qquad \Delta G^\circ = +46.4 \text{ kJ} \tag{8-54}$$

K is $10^{-8.1}$. Here the small positive ΔG° and the small exponent in K mean that the reaction is displaced toward the left, but that all four substances can exist together in appreciable amounts at equilibrium—as is shown by the experimental observations that lead sulfide is precipitated when H_2S is passed through an acid solution of a lead salt, and that a perceptible odor of H_2S is evident when a fairly strong acid is dropped on galena.

8-5 EQUILIBRIUM CONSTANTS: TEMPERATURE AND PRESSURE DEPENDENCE

Thermodynamic relations presented above allow us to express quantitatively how equilibrium constants change with temperature and pressure, and thus to determine conditions of equilibrium for the geologic processes summarized in Fig. 1-1.

The equilibrium constant is a function of T and P expressed mathematically as

$$K = F(T, P). \tag{8-55}$$

Differential changes in K are described by the total derivative of this equation (see Fig. 7-1), which in logarithmic form appears as

$$d \ln K = \left(\frac{\partial \ln K}{\partial T}\right)_P dT + \left(\frac{\partial \ln K}{\partial P}\right)_T dP. \tag{8-56}$$

To see how K changes with variations in T and P, we need equations for the two partial derivatives in Eq. (8-56) written in terms of measurable thermodynamic quantities, specifically T, P, $\Delta H°$, and $\Delta V°$.

Temperature Dependence

First we derive the change in $\ln K$ with respect to T at constant P. Combining Eq. (8-49) and Eq. (7-59) allows the equilibrium constant to be expressed in terms of $\Delta H°$ and $\Delta S°$:

$$\ln K = \frac{-\Delta G°}{RT} = \frac{-\Delta H°}{RT} + \frac{\Delta S°}{R}. \tag{8-57}$$

Differentiating with respect to T at constant P gives

$$\left(\frac{\partial \ln K}{\partial T}\right)_P = \frac{-1}{R} \left(\frac{\partial \Delta G° / T}{\partial T}\right)_P$$
$$= \frac{-\Delta H°}{R} \left(\frac{\partial T^{-1}}{\partial T}\right)_P - \frac{1}{RT} \left(\frac{\partial \Delta H°}{\partial T}\right)_P + \frac{1}{R} \left(\frac{\partial \Delta S°}{\partial T}\right)_P. \tag{8-58}$$

On the right-hand side of the equation we differentiate the first term, then substitute for the last two partial derivatives their definitions in terms of $\Delta C_P°$ [Eqs. (7-33) and (7-60)],

$$\left(\frac{\partial \ln K}{\partial T}\right)_P = \frac{\Delta H°}{RT^2} - \frac{\Delta C_P°}{RT} + \frac{\Delta C_P°}{RT}. \tag{8-59}$$

The last two terms cancel, leaving

$$\left(\frac{\partial \ln K}{\partial T}\right)_P = \frac{\Delta H°}{RT^2}. \tag{8-60}$$

Using the more familiar decimal logarithms, we have

$$\left(\frac{\partial \log K}{\partial T}\right)_P = \frac{\Delta H^\circ}{2.303RT^2}. \tag{8-61}$$

This equation, referred to as the van't Hoff equation, shows that differential changes in log K with respect to T are proportional to the standard heat of reaction and inversely proportional to the square of temperature.

The importance of the heat of reaction in determining how the equilibrium constant responds to changes in temperature can be summarized as follows: if ΔH° is zero, the value of K will not change when temperature is changed; if the reaction is endothermic, log K increases with increasing temperature, and the larger the value of ΔH° the greater the change with temperature; and if the reaction is exothermic, log K decreases with increasing temperature, the magnitude of change again depending on the value of ΔH°. This agrees with qualitative predictions from Le Chatelier's rule. For example, when ΔH° is positive, meaning that the reaction is endothermic, the differential coefficient of Eq. (8-61) is also positive; in other words, K increases as T increases, so that the equilibrium is displaced in the direction which shows absorption of heat.

We can now attach numbers to the temperature dependence of mineral solubilities and phase assemblages discussed in Chapter 4. Consider first the solubility of quartz presented in Fig. 4-4. The reaction is

$$\underset{\text{quartz}}{SiO_2} \rightleftharpoons SiO_2(aq), \tag{4-35}$$

and the equilibrium constant is merely the activity of aqueous silica, since the activity of pure quartz is chosen to be unity at any temperature and pressure, by convention. In dilute solutions the activity of aqueous silica may be taken as equal to its concentration, and this may be expressed as the measured solubility of quartz, that is, log $K = \log a_{SiO_2(aq)} \approx \log m_{SiO_2(aq)}$. The temperature dependence of quartz solubility can now be expressed by Eq. (8-61) as

$$\left(\frac{\partial \log m_{SiO_2(aq)}}{\partial T}\right)_P = \frac{\Delta H^\circ}{2.303RT^2}. \tag{8-62}$$

For pressures above 1 kbar quartz solubility increases with increasing temperature at constant pressure, as shown in Fig. 4-4; for these conditions the reaction must be endothermic [ΔH° is positive in Eq. (8-62)]. The maximum in quartz solubilities at pressures below 1 kbar in Fig. 4-4 corresponds to pressures and temperatures where ΔH° is equal to zero and Eq. (8-62) becomes

$$\left(\frac{\partial \log m_{SiO_2(aq)}}{\partial T}\right)_P = 0. \tag{8-63}$$

At lower temperatures the reaction is endothermic, but at temperatures higher than that for maximum solubility and at these low pressures we deduce that quartz solubility is exothermic because of the negative slopes of the curves in diagrams

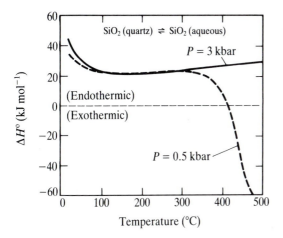

FIGURE 8-2
Computed values of the enthalpy of reaction for quartz solubility at 3 kbar and 0.5 kbar. Note that the enthalpy of reaction is essentially constant over the temperature range of 100°C to 300°C where it is nearly independent of changes in pressure. However, at high temperatures and low pressures (for example 0.5 kbar) the enthalpy of reaction changes rapidly with increasing temperature. (Curves computed from equations and data reported by Walther and Helgeson, 1977.)

where $\log K$ is plotted as a function of T. These predictions are borne out by the computed values of $\Delta H°$ for Eq. (4-35) shown in Fig. 8-2.

Applying the same type of reasoning to experimental phase relations shown in Fig. 4-6, we see that reactions of kaolinite going to diaspore [Eq. (4-9)], and of pyrophyllite to kaolinite [Eq. (4-10)], are endothermic reactions because these phase boundaries have positive slopes in a plot of $\log K$ as a function of temperature. On the other hand, reactions involving K-feldspar, muscovite, kaolinite, and quartz in Eqs. (4-26) and (4-27) are exothermic because: (1) the equilibrium constants are both equal to $\log K = 2 \log (a_{K^+}/a_{H^+})$, (2) their temperature derivative is

$$\left(\frac{\partial \log(a_{K^+}/a_{H^+})}{\partial T}\right)_P = \frac{\Delta H°}{4.606RT^2},\tag{8-64}$$

and (3) the measured ratio of KCl to HCl in solution decreases with increasing temperature (Fig. 4-10); thus $\Delta H°$ must be negative. We see from these examples that solubility and phase equilibrium experiments provide useful information about the heat energy of chemical reactions.

To evaluate the absolute change in $\log K$ with temperature we integrate Eq. (8-61) from some reference temperature (T_{ref}, which is usually 25°C (298 K) where values of $\log K_{T_{ref}}$ and $\Delta H°$ are known) to some specified temperature (T) of interest,

$$\int_{K_{T_{ref}}}^{K_T} d \log K = \int_{T_{ref}}^{T} \frac{\Delta H°}{2.303RT^2}\, dT\tag{8-65}$$

and

$$\log K_T = \log K_{T_{ref}} + \int_{T_{ref}}^{T} \frac{\Delta H°}{2.303T^2}\, dT.\tag{8-66}$$

If ΔC_P° of the reaction is zero, so that ΔH° is independent of temperature [Eq. (7-33)], the equilibrium constant will be a linear function of T^{-1}:

$$\log K_T = \log K_{T_{ref}} - \frac{\Delta H^\circ}{2.303R}\left(\frac{1}{T} - \frac{1}{T_{ref}}\right). \tag{8-67}$$

Otherwise we need equations like Eqs. (7-34) and (7-39), describing the temperature dependence of ΔC_P°, to solve the integral of Eq. (8-66).

As an example of how Eq. (8-67) is used, we consider the familiar equilibrium among K-feldspar, muscovite, quartz, and an aqueous solution represented by

$$3KAlSi_3O_8 + 2H^+ \rightleftharpoons 2K^+ + KAl_2(AlSi_3O_{10})(OH)_2 + 6SiO_2, \tag{4-26}$$

K-feldspar   muscovite   quartz

for which log K at 1 bar and 25°C is 9.58 and the heat of reaction is -26.3 kJ. On the assumption that ΔH° is independent of temperature, we can calculate values of log K at some higher temperature, say 250°C, by Eq. (8-67) (note $R = 8.314$ J mol^{-1} K^{-1}),

$$\log K_{T=523} = 9.58 - \frac{-26,300 \text{ J mol}^{-1}}{2.303 \times 8.314 \text{ J mol}^{-1}\text{ K}^{-1}}\left(\frac{1}{523K} - \frac{1}{298K}\right) = 7.6. \tag{8-68}$$

In other words, the equilibrium constant has decreased two orders of magnitude as the temperature rises from 25 to 250°C. This is the direction of change we would predict qualitatively from Le Chatelier's principle for a reaction with negative ΔH°. So if pure K-feldspar, muscovite, and quartz were in equilibrium with a dilute solution where $m_{K^+} = 10^{-3}$, the pH of the solution at 25°C would be

$$\log K = 2\log a_{K^+} - 2\log a_{H^+} = 2\log a_{K^+} + 2\text{pH},$$

$$\text{pH} = 0.5\log K - \log a_{K^+} = 0.5(9.58) - (-3.0) = 7.8,$$

and at 250°C it would be

$$\text{pH} = 0.5\log K - \log a_{K^+} = 0.5\,(7.6) - (-3.0) = 6.8.$$

The heat capacities of aqueous species and water change dramatically with increasing temperature at low pressures, so our assumption about the temperature independence of reaction enthalpy for Eq. (4-26) is not strictly valid. The equilibrium constant computed at 250°C above is only a first approximation. Accurate computation of the solution pH at 250°C requires additional information on the temperature dependence of ΔH° [see Eqs. (7-38) and (7-41)].

As a second example of the use of van't Hoff's equation, let us try the reverse problem: given values of the equilibrium constant over a range of temperature, to calculate the enthalpy change of the reaction. We use a simple reaction considered earlier, the dissolving of quartz in water represented by Eq. (4-35). Experimental values for the solubility of quartz over a temperature range of 70 to 250°C are shown in Fig. 8-3. Solubilities are plotted as logarithms, and temperatures as reciprocals of the absolute temperature times 1000. The reason for this procedure lies in the form

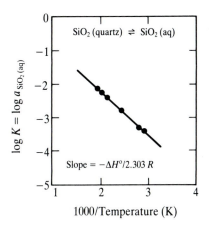

FIGURE 8-3
The logarithm of the equilibrium constant for quartz solubility [Eq. (4-35)] as a function of the reciprocal of temperature times 1000. The symbols represent experimental measurements reported by Morey et al., 1962. The line is a linear fit to the experimental data: log $K = -1.346$ $(1000T^{-1}) + 0.491$.

of Eq. (8-67), which can be rewritten as

$$\log K_T = -\frac{\Delta H^\circ}{2.303R}\frac{1}{T} + \left(-\log K_{T_{ref}} + \frac{\Delta H^\circ}{2.393R}\frac{1}{T_{ref}}\right). \qquad (8\text{-}69)$$

This is an equation for a straight line which has the form of

$$y = mx + b,$$

where y corresponds to the dependent variable (log K_T), m is the slope of the line $(-\Delta H^\circ/2.303R)$, x represents the independent variable $(1/T)$, and b is a constant equal to the parenthetical term in Eq. (8-69). The experimental data shown in Fig. 8-3 clearly fit this requirement over this limited temperature interval. (See for comparison Fig. 4-4.)

When the data are plotted and the best straight line is fitted to the points, the slope of the line is equal to $-\Delta H^\circ/2.303R$. Hence ΔH° in kJ mol^{-1} can be found by multiplying the slope by 2.303R (19.147 J mol^{-1} K^{-1}). In Fig. 8-3 the slope is -1.346 K, and ΔH° for dissolving quartz is 25.85 kJ. This number can be used to estimate the enthalpy of formation of aqueous silica because $\Delta H^\circ = \Delta H^\circ_{f,\,SiO_2(aq)} - \Delta H^\circ_{f,\,SiO_2(quartz)}$ [Eq. (7-30)]: from Appendix VIII, the value of $\Delta H_{f,quartz}$ is -910.65 kJ mol^{-1}; the figure for aqueous silica must then be 25.85 kJ more positive than this or approximately 884.8 kJ mol^{-1}. This is a widely used method for obtaining enthalpy changes for reactions and enthalpies of formation that would be difficult to measure directly.

Pressure Dependence

We can also differentiate Eq. (8-49) with respect to P at constant T to obtain an expression for the pressure dependence of K:

$$\left(\frac{\partial \log K}{\partial P}\right)_T = -\frac{1}{2.303RT}\left(\frac{\partial \Delta G^\circ}{\partial P}\right)_T. \qquad (8\text{-}70)$$

The right-hand partial derivative is equal to the standard volume of reaction [Eq. (8-17b)],

$$\left(\frac{\partial \log K}{\partial P}\right)_T = -\frac{\Delta V^\circ}{2.303RT}. \tag{8-71}$$

Thus the pressure dependence of the equilibrium constant will be determined by the standard volume of reaction and temperature. Qualitatively, if the volume change of the reaction is less than zero, log K will increase and ΔG° [Eq. (8-49)] will decrease with increasing pressure. In other words the reaction is displaced to the right, in the direction of decreasing volume, as Le Chatelier's rule requires.

Integration of Eq. (8-71) from a reference pressure (P_{ref}), where the value of $K_{P_{ref}}$ is known, to some specified pressure (P) gives

$$\log K_P = \log K_{P_{ref}} - \frac{1}{2.303RT}\int_{P_{ref}}^{P} \Delta V^\circ \, dP. \tag{8-72}$$

To use this equation we must express ΔV° as a function of P before we can integrate. If ΔV° is independent of P the integral becomes

$$\log K_P = \log K_{P_{ref}} - \frac{\Delta V^\circ}{2.303RT} (P - P_{ref}), \tag{8-73}$$

and log K will be a linear function of P at constant T. The slope of a plot of log K versus P will be $-\Delta V^\circ/2.303RT$.

As an example of the use of Eq. (8-73) we again consider equilibrium of K-feldspar, muscovite, quartz, and solution [Eq. (4-26)], for which we computed a value of log K at 250°C of 7.6 using numbers for K and ΔH° at 1 bar and 25°C. Now, how will this value for the equilibrium constant change if pressure is increased to 5000 bar and temperature is held constant at 250°C? To estimate this we use a value of ΔV° of -34.8 cm^3 mol^{-1} (computed from molal and partial molal volumes of reaction species at 250°C), assume that ΔV° is independent of pressure, and use a conversion factor of 10 cm^3 bar = 1 J:

$$\log K_{P=5000} = 7.59 - \frac{-34.8 \text{ cm}^3 \text{ mol}^{-1}}{2.303 \times 8.314 \text{ J mol}^{-1} \text{K}^{-1} \times 523 \text{ K}}\left(\frac{4999 \text{ bar}}{10 \text{ cm}^3 \text{ bar}}\right) = 9.3. \tag{8-74}$$

Because the volume change of reaction is negative, the differential coefficient for Eq. (8-71) is positive and log K increases with increasing pressure as shown in Fig. 8-4. This change with pressure is opposite to the change in K we computed earlier for a temperature increase from 25 to 250°C. The effects of increasing both T and P may, under circumstances when both ΔH° and ΔV° are of the same sign, cancel each other in Eq. (8-56); thus in the present example, log K at 1 bar and 25°C is nearly equal to that at 5000 bar and 250°C.

As our final example we use Eq. (8-73) to predict pressures for reactions deep in the Earth from assemblages and compositions of minerals found on the surface. This procedure, referred to as geobarometry, helps geologists evaluate physical conditions of igneous and metamorphic processes that cannot be directly observed

3 K-feldspar + 2H$^+$ $\rightleftharpoons$ 2K$^+$ + Muscovite + 6 Quartz

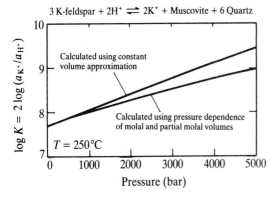

FIGURE 8-4
Computed values of the equilibrium constant for the reaction of K-feldspar, muscovite, quartz, and solution [Eq. (4-26)] as a function of pressure at a constant temperature of 250°C. The upper curve is computed assuming that the volume change of reaction is independent of pressure, while the lower curve accounts for the pressure dependence of the volume of aqueous species in the reaction.

or measured. Of the many reactions we could consider, the one we choose involves only three minerals, the familiar minerals quartz and albite and a Na–Al pyroxene called jadeite (NaAlSi$_2$O$_6$). Pyroxenes rich in jadeite are found with quartz and sometimes albite in blueschist metamorphic rocks, rocks that formed at high pressures and low temperatures in regions where plates of the lithosphere are being subducted into the mantle. In these rocks jadeitic pyroxene exhibits a wide range of isomorphic substitutions with the components diopside (CaMgSi$_2$O$_6$) and acmite (NaFe^{3+}Si$_2$O$_6$). In contrast, the quartz and albite are nearly pure phases. The equilibrium reaction is

$$\text{NaAlSi}_2\text{O}_6 + \text{SiO}_2 \rightleftharpoons \text{NaAlSi}_3\text{O}_8 .$$ (8-75)

jadeite   quartz   albite

Because quartz and albite are nearly pure their activities are taken as unity, so the equilibrium constant equals the reciprocal of the activity of the jadeite component in pyroxene. We assume, as a first approximation, that the jadeite is an ideal solid solution so that the activity of jadeite in the pyroxene is equal to the mol fraction of jadeite [X_{jadeite}, Eq. (8-37)] and the equilibrium constant for Eq. (8-75) becomes

$$K = a_{\text{jadeite}}^{-1} \approx X_{\text{jadeite}}^{-1}.$$ (8-76)

If we assume that the pyroxene has not changed in composition since it formed, the mol fraction of jadeite will be a measure of the equilibrium constant at the T and P of metamorphism. If an independent estimate of temperature can be made, the measured pyroxene compositions can be used to predict metamorphic pressures.

The pressure dependence of the pyroxene compositions is given by combining Eqs. (8-76) and (8-71),

$$\left(\frac{\partial \log X_{\text{jadeite}}}{\partial P}\right)_T = \frac{\Delta V^\circ}{2.303RT}.$$ (8-77)

The volume change for Eq. (8-75) is positive (approximately 17 cm^3 mol^{-1} at 1 bar and 25°C), so by analogy with Fig. 8-1(b) we see that the assemblage jadeite + quartz has the greater density and will be the mineral assemblage stable at high pressure. One can predict, then, that the mol fraction of jadeite in pyroxene in equilibrium with quartz and albite will increase with increasing pressure [the differential coefficient in Eq. (8-77) is positive].

To use the composition of jadeitic pyroxene as a geobarometer at some specified temperature (T), we integrate Eq. (8-77) (assuming that the reaction volume is independent of pressure). One of the integration limits (P_{ref}) will represent the pressure of equilibrium for pure jadeite, quartz, and albite, so that $\log K_{P_{ref}} = -\log X_{jadeite} = 0$; the other integration limit (P) will correspond to the pressure of equilibrium for albite, quartz, and a pyroxene of some measured composition represented by $X_{jadeite}$. The resulting equation is

$$\log X_{jadeite} = \frac{\Delta V^\circ}{2.303RT}(P - P_{ref}), \tag{8-78}$$

which is an equation for a straight line on a diagram of $\log X_{jadeite}$ plotted as a function of P; the line (Fig. 8-5) has a slope of $\Delta V^\circ/2.303RT$. Solving for pressure gives

$$P = \frac{2.303RT}{\Delta V^\circ}\log X_{jadeite} + P_{ref}, \tag{8-79}$$

an equation that allows us to estimate metamorphic pressures from measured compositions of pyroxene in equilibrium with albite and quartz.

For example, at a temperature of 600°C, where the equilibrium pressure (P_{ref}) for pure jadeite, quartz, and albite is experimentally determined to be 16.85 kbar (Fig. 8-5), Eq. (8-79) becomes

$$P = \frac{2.303 \times 8.314 \text{ J mol}^{-1} \text{ K}^{-1} \times 873 \text{ K}}{17 \text{ cm}^3 \text{ mol}^{-1} \times 0.1 \text{ J cm}^{-3} \text{ bar}^{-1}}\log X_{jadeite} + 16,850 \text{ bar}. \tag{8-80}$$

As with our previous examples the conversion factor of 1 J = 10 cm^3 bar is required to maintain the correct units. So if $X_{jadeite}$ is 0.5 and the temperature of metamorphism is about 600°C, the predicted pressure is approximately 14 kbar, as shown in Fig. 8-5.

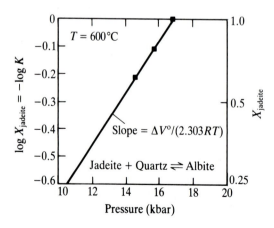

FIGURE 8-5

Calculated (line) and experimental (symbols) relations between pyroxene composition and pressure at 600°C for jadeitic–pyroxene, quartz, and albite equilibrium [Eq. (8-75)]. The symbols are unpublished experimental data of Jun Liu, Stanford University.

8-6 THE PHASE RULE

The equations developed in this chapter provide a thermodynamic basis for calculating equilibrium phase relations among minerals and solutions using the variables temperature, pressure, molal concentrations, and mol fractions. One important question remains for consideration: What determines the number of phases when a mixture of substances is in equilibrium? Or, in different phrasing, how many thermodynamic variables must be fixed to establish equilibrium in an assemblage of phases?

Consider, for example, the equilibrium of muscovite, quartz, K-feldspar, sillimanite, and water shown in Fig. 1-2, where the state of equilibrium is specified as a function of the variables temperature and pressure. How many of these variables must be fixed to ensure that equilibrium exists? Here the answer is *one*: if we fix pressure, say at 3 kbar, the temperature is automatically determined at about 630°C. We are free to choose one of the variables arbitrarily, but not two. For another example, take the equilibrium of quartz and water shown as a function of three intensive variables: temperature, pressure, and $m_{SiO_2(aq)}$ in Fig. 4-4(a). (Note that the third term, $m_{SiO_2(aq)}$, can be considered to represent the chemical potential of SiO_2 because of the relation $\mu_{SiO_2} = \mu_{SiO_2}^{\circ} + RT \ln a_{SiO_2} = \mu_{SiO_2}^{\circ} + RT \ln m_{SiO_2}$). This time the number of variables that can be chosen arbitrarily is *two*: if we pick only one, say a temperature of 200°C, we still do not know whether we are on a line representing equilibrium unless either pressure or concentration is specified also.

A general answer to such questions about the number of variables that must be fixed to establish equilibrium lies in a simple relationship called the *phase rule*. This is one more useful tool whose original formulation we owe to the American chemist J. Willard Gibbs toward the end of the 19th century. The rule tells us that the number of phases (p) and number of chemical components (c) needed to describe the composition of all the phases determines the number of intensive variables that must be specified at equilibrium. The number of variables to which arbitrary values can be assigned is called the *variance* or number of *degrees of freedom*. It is commonly given the symbol f.

In symbolic terms the phase rule is stated as

$$f = c + 2 - p. \tag{8-81}$$

To illustrate its use, we apply it to the two examples considered above. The muscovite–quartz–K-feldspar–sillimanite–water equilibrium has 5 phases made up of 4 components (K_2O, Al_2O_3, SiO_2, and H_2O), so that Eq. (8-81) becomes $f = 4 + 2 - 5 = 1$. This agrees with our conclusion from Fig. 1-2: *one* of the intensive variables, either temperature or pressure, can be arbitrarily specified, and the other is thereby determined. A reaction of this sort is described as *univariant*. The quartz–water equilibrium has 2 phases and 2 components (SiO_2, H_2O), so that the variance is $f = 2 + 2 - 2 = 2$; thus values of two variables, say temperature and pressure, can be fixed arbitrarily, and by fixing these we automatically determine the equilibrium value of the third, $m_{SiO_2(aq)}$ [Fig. 4-4(a)]. Such a reaction is said to be *divariant*.

The equation for the phase rule [Eq. (8-81)] can be arrived at in a number of ways. It can be deduced qualitatively by considering equilibrium in simple systems, like the ones discussed above, or it can be obtained from a series of equalities among the chemical potentials of thermodynamic components and constraints imposed by the mol fractions used to describe the composition of all phases in the system. Here we derive the rule using the fundamental thermodynamic relation of the Gibbs–Duhem equation [Eq. (8-22)].

Consider again the equilibrium of quartz and water in the system SiO_2–H_2O. An explicit statement of the Gibbs–Duhem equation can be written for each of the two phases: for quartz we have

$$S_{quartz}\, dT - V_{quartz}\, dP + n_{SiO_2,\,quartz}\, d\mu_{SiO_2} = 0, \tag{8-82}$$

and for the solution

$$S_{solution}\, dT - V_{solution}\, dP + n_{SiO_2,\,solution}\, d\mu_{SiO_2} + n_{H_2O,\,solution}\, d\mu_{H_2O} = 0. \tag{8-83}$$

These two equations determine the possible changes of intensive variables (T, P, and μ_i) in quartz and in the solution. At equilibrium the value of each intensive variable will be constant throughout the phases. Thus the state of the system is determined by the two variables, temperature and pressure, plus the chemical potential of each component in the system; that is, a total number of $c + 2$ variables. For our particular example the number is $2 + 2 = 4$. Since we have only 2 independent equations and 4 independent variables, we must arbitrarily specify two of the variables in order to have a unique solution to the equations; in other words, the number of unknowns must equal the total number of independent equations. The equality of Eq. (8-81) is now apparent: if we have p phases, then the number of degrees of freedom is the difference between the number of unknowns ($c + 2$) and the total number of independent statements of the Gibbs–Duhem equation (p), or $f = c + 2 - p$.

The phase rule has many uses in geochemistry, especially in evaluating experimental observations and in constructing theoretical phase diagrams. Its application to geologic systems is complicated by the large number of components needed to describe the composition of all the phases and by the fact that geologic systems are typically open to the transport of chemical components, especially those in the fluid phase. Nevertheless, the phase rule is another useful tool in geochemistry for which we shall find many applications in the study of metamorphic and magmatic processes in later chapters.

SUMMARY

The various concepts we have introduced here are treated at length in textbooks of physical chemistry and thermodynamics listed in the references at the end of this chapter. The important considerations are, in summary: the idea of chemical equilibrium may be visualized in terms of escaping tendencies, the tendency of each constituent to escape from one phase into another; escaping tendencies may be expressed quantitatively by chemical potentials (molal free energies for pure substances, partial molal free energies for constituents of solutions), by fugacities, or by activities and equilibrium constants once appropriate standard states are adopted; and the deviation of solutions from ideal behavior may be expressed by activity and fugacity coefficients. Variation of the equilibrium constant with temperature and pressure is determined by the heat and volume of reaction, a relation that provides a theoretical basis for estimating temperatures and pressures of geological processes. First order approximations of the temperature and pressure dependence of the equilibrium constant can often be made by using the assumption of constant heat and volume of reaction, which greatly simplifies the computations. It should be noted, however, that the validity of any thermodynamic calculation is determined by the extent to which the equations, data, and approximations used are consistent with reliable experimental and geologic observations. In subsequent chapters we will build upon the principles developed here to evaluate complex phase relations and reaction paths in a wide variety of geochemical environments.

PROBLEMS

1. For the reaction of water and carbon to make a combustible fuel (the "water–gas reaction"),

$$H_2O(g) + C \rightleftharpoons H_2 + CO,$$

$\Delta H°$ at 25°C is +131 kJ and the equilibrium constant K is $10^{-16.0}$. Calculate K for 100°C and for 200°C on the assumption that $\Delta H°$ is constant. What is the fugacity ratio of f_{H_2} to f_{H_2O} in equilibrium with 0.1 bar of CO at these two temperatures? Assume that fugacity is equal to partial pressure.

2. From the data of Appendix VIII, calculate the equilibrium constant at 25°C and at 250°C for the reaction

$$N_2 + 3H_2 \rightleftharpoons 2NH_3.$$

In volcanic gas escaping from a fumarole at 600°C, would you expect the nitrogen to exist largely in the form of ammonia or largely as the free element? Assume $\Delta H°$ for the reaction is independent of temperature.

3. From the data of Appendix VIII, calculate the solubility product at 25°C for the reaction,

$$PbS \rightleftharpoons Pb^{2+} + S^{2-}.$$
galena

What would be the solubility of PbS in pure water if this were the only reaction taking place? The measured solubility is 3×10^{-4} g/liter. Explain why the two values differ so markedly.

4. In theories of formation of sulfide veins, a critical factor is the amount of free S^{2-} present in the vein solutions. Would the concentration of this ion resulting from dissociation of H_2S be greater or less at 100°C than at 25°C? Calculate the change in the two dissociation constants of H_2S between these temperatures. Assume $\Delta H°$ for the reactions are independent of temperature.

5. It is often suggested that solutions which transport and deposit metallic sulfides are alkaline. At ordinary temperatures would galena dissolve appreciably in an alkaline solution to form $Pb(OH)_3^-$? In this and similar questions, assume that "appreciable solution" means a concentration of at least $10^{-5}m$, which for most common metals is roughly equal to 1 ppm.

6. A common reaction during high-grade metamorphism of pelitic rocks involves the breakdown of muscovite in the presence of quartz to form K-feldspar and sillimanite:

$$\text{muscovite} + \text{quartz} \rightarrow \text{K-feldspar} + \text{sillimanite} + H_2O. \qquad [\text{Eq.}(1\text{-}48)]$$

Experimental studies of the reaction show that equilibrium at 2 kbar occurs at 600°C. As in many other dehydration reactions, the entropy, enthalpy, volume, and heat capacity of reaction are all greater than zero. With this information answer the following questions:

(a) If we choose a standard state for minerals and water of unit activity of the pure phase at any temperature and pressure, then what is the value of log K and $\Delta G°$ for this reaction at 2 kbar and 600°C?

(b) With additional information that the isothermal compressibility of the reaction, $(\partial V/\partial P)_T$, is less than zero, sketch the following schematic diagrams: the free energy of reaction as a function of temperature at constant pressure, and the free energy of reaction as a function of pressure at constant temperature. Be sure that the curves in your diagrams satisfy the constraints imposed by the first and second derivatives of $\Delta G°$ for Eq. (1-48).

(c) If there is isomorphous substitution of paragonite $[NaAl(Al_2Si_3O_{10})(OH)_2]$ in the muscovite to form a solid solution mineral, how would this change the equilibrium temperature at 2 kbar? Assuming that $\Delta H°$ of the reaction is independent of temperature near the 2 kbar equilibrium and equal to 73 kJ, at what temperature are K-feldspar, sillimanite, quartz and water in equilibrium with a muscovite solid solution when

$X_{KAl(Al_2Si_3O_{10})(OH)_2} = 0.5$? Assume that $a_{KAl(Al_2Si_3O_{10})(OH)_2} = X_{KAl(Al_2Si_3O_{10})(OH)_2}$.

7. In a submarine hydrothermal system under a mid-ocean ridge, alteration of basalt above the magma chamber under the ridge forms albite ($NaAlSi_3O_8$), epidote (clinozoisite, $Ca_2Al_3Si_3O_{12}(OH)$), prehnite ($Ca_2Al_2Si_3O_{10}(OH)_2$), and quartz. Answer the following questions concerning phase relations and solution compositions in this geothermal system.

(a) List the thermodynamic criteria required for heterogeneous equilibrium among the minerals and the hydrothermal solution.

(b) How many degrees of freedom must be specified to describe the equilibrium?

(c) Evidence from freezing and heating experiments on fluid inclusions within these minerals suggests solution compositions between $0.1m$ and $0.5m$ NaCl and temperatures near 300°C. If we assume that these solutions were in equilibrium with the mineral assemblage, what would be the range in solution pH values? Use the equilibrium below and assume unit activity for minerals and H_2O and that $a_{Na^+} \approx m_{Na^+}$.

$$\text{prehnite} + \text{albite} + H^+ \rightleftharpoons \text{clinozoisite} + 3 \text{ quartz} + Na^+ + H_2O$$

$$\log K = 4.6 \text{ at } T = 300°C \text{ and } P = 250 \text{ bar}$$

(d) The graph below illustrates the computed temperature dependence of the enthalpy of reaction at 250 bar. Write a short essay (using equations with your text) describing the relative changes in pH of the fluid with decreasing temperature from 350°C to 100°C. Assume that the solution is buffered by equilibrium with albite, clinozoisite, prehnite, and quartz, and that the activity of Na^+ in the solution is *constant* during cooling of the basalts.

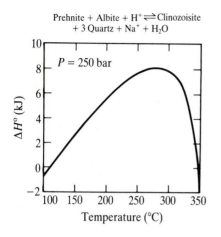

Prehnite + Albite + $H^+ \rightleftharpoons$ Clinozoisite + 3 Quartz + $Na^+ + H_2O$

8. Analysis of the chemical composition of fluids produced from deep drillholes in active geothermal systems shows that the ratio of m_{Na^+} to m_{K^+} decreases with increasing temperature. This relation is believed to be due to the temperature dependence of the reaction

$$\text{albite} + K^+ \rightleftharpoons \text{K-feldspar} + Na^+.$$

Estimate the enthalpy of this reaction using the temperatures and alkali cation ratios reported below from geothermal systems in New Zealand.

Temperature (°C)	log (m_{Na^+}/m_{K^+})
275	1.08
260	0.93
285	0.90
280	0.92
270	0.92
292	0.89
258	0.92
200	1.12
250	1.17
230	1.31
275	0.97
220	1.06
214	1.06
231	1.12

Data from Fournier, R. O., and A. H. Truesdell: "An empirical Na–K–Ca geothermometer for natural waters," *Geochimica et Cosmochimica Acta*, vol. 37, p. 1255–1275, 1978.

REFERENCES AND SUGGESTIONS FOR FURTHER READING

The references at the end of Chapter 7 by Denbigh, Nordstrom and Munoz, and Anderson and Crerar all provide a more in-depth study of the principles of chemical thermodynamics and their application to phase equilibrium. The references below are related to more specific areas of phase equilibrium in geochemical systems.

Helgeson, H. C.: "Description and interpretation of phase relations in geochemical processes involving aqueous solutions," *American Journal of Science*, vol. 268, pp. 415–438, 1970. Applications of the phase rule to solution–mineral equilibria, and the utility of activity–activity phase diagrams.

Morey, G. W., R. O. Fournier, and J. J. Rowe: "The solubility of quartz in water in the temperature interval from 25° to 300°C at pressures corresponding to liquid–vapor equilibrium for H_2O," *Geochim. Cosmochim. Acta*, vol. 26, pp. 1029–1043, 1962.

Nordstrom, D. K., and J. L. Munoz: *Geochemical Thermodynamics*, The Benjamin/Cummings Publishing Co., 1985. In addition to the chapters on thermodynamics, Chapter 4 on "Phase Equilibria in Simple Systems" is a good reference concerning the phase rule and its application to a variety of phase diagrams.

Rumble, D., III: "Gibbs phase rule and its application in geochemistry," *Jour. Washington Academy of Sciences*, vol. 64, pp. 199–208, 1974. A detailed review of the phase rule and how it can be used to extract information on the formation of metamorphic rocks.

Walther, J. V., and H. C. Helgeson: "Calculation of the thermodynamic properties of aqueous silica and the solubility of quartz and its polymorphs at high pressures and temperatures," *American Journal of Science*, vol. 277, pp. 1315–1351, 1977.

CHAPTER
9

OXIDATION
AND
REDUCTION

The reddish-brown coloration of many weathered rock surfaces is a familiar sight. The color is generally caused by fine-grained Fe(III) compounds, formed by reaction of Fe(II) minerals with oxidizing agents in rain and air. Reactions of this sort, involving a change in oxidation state, are referred to as oxidation–reduction reactions. They occur not only during rock weathering, but in diagenetic, metamorphic, and magmatic processes as well. Oxidation–reduction reactions have a controlling influence on the solubility and transport of some minor elements in seawater, groundwater, and hydrothermal solutions. The stability of many ore minerals and the sequence of mineral crystallization in magmas are sensitive functions of oxidation state. In this chapter we explore the chemistry of oxidation–reduction reactions and develop methods to represent their equilibria graphically.

Oxidation means an increase in oxidation state, reduction a decrease. For example, when zinc displaces copper from a solution of copper sulfate,

$$Zn + Cu^{2+} \rightarrow Zn^{2+} + Cu, \qquad (9\text{-}1)$$

the zinc is oxidized (change in oxidation state from 0 to +2) and the copper is reduced (+2 to 0). When chlorine displaces bromine from a solution of sodium bromide,

$$Cl_2 + 2Br^- \rightarrow 2Cl^- + Br_2, \qquad (9\text{-}2)$$

chlorine is reduced (0 to -1) and bromine is oxidized (-1 to 0). And when gold is dissolved by the action of MnO_2 in hydrochloric acid solution,

$$3MnO_2 + 2Au + 12H^+ + 8Cl^- \rightarrow 3Mn^{2+} + 2AuCl_4^- + 6H_2O, \tag{9-3}$$

manganese is reduced (+4 to +2) and gold is oxidized (0 to +3). Alternatively, oxidation may be described as a loss of electrons and reduction as a gain; in the zinc–copper reaction, for example, each Zn atom loses two electrons to a copper ion. Note that any reaction of this sort must involve both an oxidation and a reduction, and that the total changes in oxidation state must balance (see Appendix X for methods of balancing oxidation–reduction reactions).

Conditions of equilibrium for reactions like these can be described using the methods of the preceding chapter. But in addition to equilibrium constants and free energy, another device is available for describing the tendency of one substance to react with another when the reaction involves changes in oxidation state. This device, called the oxidation potential or oxidation–reduction potential, often provides the most convenient method of getting the desired quantitative information.

9-1 OXIDATION POTENTIALS

Any oxidation–reduction reaction, theoretically at least, can be set up so that the transfer of electrons from one element to another will take place along a wire. For the zinc–copper reaction the arrangement is very simple: pieces of the two metals are connected by a wire and submerged in copper sulfate solution. The piece of zinc slowly dissolves, fresh copper from the solution plates out on the copper metal, and a current flows through the wire. The process occurring at the zinc electrode may be symbolized

$$Zn \rightarrow Zn^{2+} + 2e^-, \tag{9-4}$$

where e^- indicates an electron. The liberated electrons move along the wire to the copper electrode, where they are used in the reaction

$$Cu^{2+} + 2e^- \rightarrow Cu. \tag{9-5}$$

Reactions of this kind, showing the processes that occur as electrons are produced or consumed at an electrode, are called *half-reactions*, or *electrode reactions*. Addition of two half-reactions gives the complete oxidation–reduction reaction; thus Eq. (9-5) added to Eq. (9-4) gives Eq. (9-1). For a more complicated illustration, the electrode reactions corresponding to Eq. (9-3) are:

$$3MnO_2 + 12H^+ + 6e^- \rightarrow 3Mn^{2+} - 6H_2O \tag{9-6}$$

and

$$2Au + 8Cl^- \rightarrow 2AuCl_4^- + 6e^-. \tag{9-7}$$

The potential difference between the electrodes of our zinc–copper cell can be measured by adding a galvanometer to the circuit. The amount of the potential difference depends on a great many variables, but we arrange to keep most of these constant. Thus we fix the activity of both Cu^{2+} and Zn^{2+} at 1; we make sure that the metal of the electrodes is pure and has a clean surface; we hold the temperature at 25°C and the pressure at 1 bar; and we arrange to have as small a flow of current as possible. Under such conditions the measurement of potential difference is reproducible and may be compared with potential differences measured similarly for other oxidation–reduction reactions. (We pass over the technical difficulties in such measurements, which for some reactions are very troublesome.)

If we set up a number of cells similar to the zinc–copper cell, using various metals in contact with solutions of metal ions, we find that the metals can be arranged in a series according to their ability to displace one another from solution and according to the size of the potential difference produced by different pairs. Thus zinc displaces copper, copper displaces silver, and silver displaces gold; and the potential difference of a zinc–silver or a zinc–gold cell is greater than the potential difference of a zinc–copper cell. On the other hand, the reverse reactions do not take place appreciably: silver placed in copper sulfate solution or in zinc sulfate solution causes no detectable reaction. Experiments of this sort give us the familiar *electromotive series* of metals, according to which we express the chemical reactivities of various metals with respect to one another.

The electromotive series can be made quantitative by assigning a potential difference to each half-reaction. This requires that one half-reaction be chosen as a standard and given an arbitrary potential of zero, so that other half-reactions can be measured against it. A convenient choice is the hydrogen couple,

$$\tfrac{1}{2}H_2 \rightarrow H^+ + e^- \qquad E° = 0.00 \text{ volt.} \tag{9-8}$$

If we arrange a cell with zinc as one electrode and hydrogen as the other (by letting hydrogen at 1 bar and 25°C bubble over a platinum rod), and use a solution containing H^+ and Zn^{2+} both at an activity of 1, we obtain the potential difference for the overall reaction,

$$Zn + 2H^+ \rightarrow Zn^{2+} + H_2 \qquad E° = -0.76 \text{ volt,} \tag{9-9}$$

and we use this number as the potential for the zinc electrode reaction:

$$Zn \rightarrow Zn^{2+} + 2e^- \qquad E° = -0.76 \text{ volt.} \tag{9-10}$$

There is no way of measuring potentials for half-reactions independently. We get them only as differences between pairs of half-reactions, so that the actual numbers are no more than relative voltages compared with the hydrogen electrode.

Electrode potentials and potential differences for complete reactions are both designated by the symbol E. For electrode reactions like Eq. (9-10) the potential E is also called the *electromotive force*. The symbol $E°$ refers to *standard* potentials or potential differences, for reactions that take place with all substances at unit activities; most commonly it refers to reactions at 25°C and 1 bar, but may be used for other specified temperatures or pressures. Standard electrode potentials for some reactions of geologic interest are tabulated in Appendix IX.

Several details about the arrangement in this table should be noted. Each half-reaction is given with the reduced form of the element on the left and the oxidized form on the right. Strong reducing agents appear toward the top of the table, strong oxidizing agents near the bottom. Some reactions follow different courses in acid and basic solutions, because certain precipitates and complex ions are stable in one kind of solution but not in the other; hence it is necessary to include an auxiliary table for those reactions that occur only in an alkaline environment. The $+$ and $-$ signs given to the voltages are purely arbitrary, and unfortunately are not uniform from one reference to another. In this book, we follow the usual practice in geochemical literature, making voltages more reducing than the hydrogen electrode negative and those more oxidizing than the hydrogen electrode positive.

Qualitatively, the table tells us at a glance what reactions are possible and what are not. The reduced form of any couple will react with the oxidized form of any couple *below* it, but not with the oxidized form of a couple above it. Thus Pb will reduce Ag^+ but not Al^{3+}. Two reactions close together in the table will reach equilibrium with all substances present in appreciable amounts; thus metallic lead and tin are stable in contact with a solution containing fairly large amounts of both Pb^{2+} and Sn^{2+}. Technically, of course, all possible reactions indicated by the table are equilibrium processes. When we say "Pb will not reduce Al^{3+}," we mean more precisely that Pb will react until a very small amount of Pb^{2+} is formed, but that the ratio $a_{Al^{3+}}/a_{Pb^{2+}}$ is very large at equilibrium.

To use the table quantitatively to find the potential difference for a particular reaction, we need only subtract one half-reaction from another and subtract the corresponding voltages. Each reaction must be multiplied by a coefficient that will make the electron changes the same, for no free electrons can appear in the overall reaction. *The voltages, however, are not multiplied by the coefficients*; unlike free energies and enthalpies, the voltages are measurements of potential only and do not change with amount of substance present. For example, to find the potential for the oxidation of Fe^{2+} by MnO_2 in acid solution, we look up the two electrode reactions

$$Fe^{2+} \rightarrow Fe^{3+} + e^- \qquad E° = +0.77 \text{ volt}, \tag{9-11}$$

$$Mn^{2+} + 2H_2O \rightarrow MnO_2 + 4H^+ + 2e^- \qquad E° = +1.23 \text{ volt}. \tag{9-12}$$

To make the electron changes balance, we multiply the iron half-reaction by 2, and then subtract the manganese half-reaction. The standard potential for the overall reaction is found by subtracting the half-reaction potentials without multiplication:

$$MnO_2 + 4H^+ + 2Fe^{2+} \rightleftharpoons Mn^{2+} + 2H_2O + 2Fe^{3+} \qquad E° = -0.46 \text{ volt}. \tag{9-13}$$

The convention regarding $+$ and $-$ signs means that a reaction which takes place spontaneously must have a negative voltage, and one that requires outside energy has a positive voltage—a convention similar to the one we have used for free energy and enthalpy.

9-2 RELATION OF OXIDATION POTENTIAL TO FREE ENERGY

Since the potential difference for a reaction computed from the data of Appendix IX is a measure of how far the reaction mixture is from equilibrium, it must clearly be related to the free energy change of reaction (ΔG). The relation is simple,

$$\Delta G = \mathbf{nf}\,E. \tag{9-14}$$

Here $\mathbf{n}$ is the number of electrons that the equation shows shifting from one kind of atom to another, E is the potential difference, and $\mathbf{f}$ is the Faraday constant. This constant is a number which, when multiplied by voltage, gives energy; the usual expression for it is 96,485 coulombs, the corresponding energies being expressed in volt-coulombs, or joules. For example, the standard free energy change in the manganese–iron reaction just discussed is

$$\Delta G^\circ = \mathbf{nf}\,E^\circ = 2 \times 96{,}485(-0.46) = -88{,}766\ \text{joules} = -88.8\ \text{kJ}. \tag{9-15}$$

This checks well with the figure -88.4 kJ which is obtained by adding up free energies of formation in the usual way. Note that a potential difference of half a volt corresponds to a fairly large free energy change.

It should be emphasized again that free energies depend on how the equation is written—i.e., on whether the coefficients are doubled or tripled or halved—while the potential does not. The difference is taken care of in Eq. (9-14) by the factor $\mathbf{n}$, which, of course, changes if the coefficients change. By convention, the equality of Eq. (9-14) refers to electrode reactions where electrons appear on the right-hand side, as in the generalized reaction,

$$reduced\ state \rightleftharpoons oxidized\ state + \mathbf{n}e^-. \tag{9-16}$$

So far we have talked exclusively about standard potentials, which are potentials for unit activities at a temperature of 25°C. To find the potential difference for a reaction under different conditions, we make use of the relations developed earlier for free energies. By combining Eqs. (9-15) and (8-42), we obtain for the hypothetical reaction of Eq. (8-44),

$$E = \frac{\Delta G}{\mathbf{nf}} = \frac{\Delta G^\circ}{\mathbf{nf}} + \frac{RT}{\mathbf{nf}}\ \ln\ \frac{a_Z^z a_Q^q}{a_X^x a_Y^y}, \tag{9-17}$$

$$= E^\circ + \frac{2.303 RT}{\mathbf{nf}}\ \log\ \frac{a_Z^z a_Q^q}{a_X^x a_Y^y}, \tag{9-18}$$

an expression called the Nernst equation. For reactions at 25°C the combination of constants before the logarithm is $0.059/\mathbf{n}$, so that

$$E = E^\circ + \frac{0.059}{\mathbf{n}}\ \log\ \frac{a_Z^z a_Q^q}{a_X^x a_Y^y}. \tag{9-19}$$

For example, to find the potential for the manganese–iron reaction [Eq. (9-13)] in a solution with pH 3 and unit concentrations of other ions:

$$E = -0.46 + \frac{0.059}{2}\ \log\ \frac{a_{\text{Mn}^{2+}}\,a_{\text{Fe}^{3+}}^2}{a_{\text{H}^+}^4\,a_{\text{Fe}^{2+}}^2},$$

$$= -0.46 + 0.03\log\ \frac{1}{(10^{-3})^4},$$

$$= -0.46 + 12 \times 0.03 = -0.10\ \text{volt}. \tag{9-20}$$

The smaller negative value for E means that the reaction would have less tendency to take place, as we would expect when the concentration of one of the reactants is reduced. (If the Nernst equation is used to calculate potentials at temperatures other than 25°C, the E° term refers to the standard potential at temperature T, not at 25°C.)

The relation of the oxidation potential to the equilibrium constant for a reaction may be formulated by combining Eqs. (9-15) and (8-51):

$$E° = \frac{\Delta G°}{nf} = -\frac{2.303RT \log K}{nf} = -\frac{0.059}{n} \log K. \tag{9-21}$$

To find oxidation potentials at elevated temperatures and pressures we simply use values of $\Delta G°$ or K computed for the conditions of interest using methods outlined in Chapter 8.

What advantages can be claimed for oxidation potentials as opposed to free energies in handling problems of equilibrium? Their chief merit is convenience: from a table of potentials like Appendix IX one can see at a glance, with no calculation except a mental note of how far apart two half-reactions are in the table, whether a given oxidation–reduction process can be expected to take place and approximately how far the reaction will go. If a more quantitative estimate is needed, the calculation involves only two figures rather than several.

9-3 REDOX POTENTIALS

One further advantage of oxidation potentials is their usefulness in treating problems which concern not only specific reactions but also the general oxidizing or reducing characteristics of a geologic environment. It is common knowledge, for example, that dissolved sulfur is largely in the form of SO_4^{2-} in water of the open sea, where conditions are oxidizing, but chiefly in the form of H_2S in the stagnant bottom waters of enclosed basins. Oxidation potentials make it possible to refine such qualitative statements and to estimate semi-quantitatively just what concentrations of the various ions and compounds of sulfur can exist in these environments.

The ability of a natural environment to oxidize sulfur, or bring about any other oxidation or reduction process, is measured by a quantity called its *redox potential.* Experimentally this is determined by immersing an inert electrode, usually platinum, in the environment—say in a sample of seawater, swamp muck, or soil–and determining the potential difference between the platinum and a hydrogen electrode or some other electrode of known potential. Measured redox potentials of seawater, for example, range between +0.3 volt for aerated water to −0.5 volt for water from bottom sediments containing organic matter. The term redox potential is used by some geochemists also as a synonym for oxidation potential, applicable to potentials of individual half-reactions as well as to potentials of environments. In either usage it is commonly given the symbol Eh to remind us that the potential difference is measured relative to the standard hydrogen electrode [Eq. (9-8)].

As an example, suppose we find the redox potential of a sample of water to be +0.5 volt, and inquire as to the dominant form of dissolved iron in this environment.

If the solution is acid, the choice lies between Fe^{2+} and Fe^{3+} (neglecting possible complexes). For the standard potential of the Fe^{2+}–Fe^{3+} couple [(Eq. (9-11)] we read from Appendix IX a value +0.77 volt. The measured potential is more reducing than this; hence we would expect qualitatively to find Fe^{2+} the chief ion. To get a quantitative value for the Fe^{2+}/Fe^{3+} ratio, we substitute in Eq. (9-19), using 0.77 as $E°$ and 0.5 as E (equivalent to Eh as defined above):

$$Eh = E = 0.5 = 0.77 + \frac{0.059}{1} \log \frac{a_{Fe^{3+}}}{a_{Fe^{2+}}},$$

$$\log \frac{a_{Fe^{3+}}}{a_{Fe^{2+}}} = -\frac{0.27}{0.059} = -4.58, \tag{9-22}$$

$$\frac{a_{Fe^{3+}}}{a_{Fe^{2+}}} = 10^{-4.58} = 2.6 \times 10^{-5}.$$

In this water, therefore, the equilibrium activity of Fe^{2+} is nearly 40,000 times that of Fe^{3+}. [Note that the activity of the electron which appears in Eq. (9-11) can be omitted from Eq. (9-22) because the electrode potential is measured against the hydrogen electrode, with activities of H_2 and H^+ assumed equal to 1; cf. Eqs. (9-9) and (9-10).]

Redox potential in many ways is analogous to pH. It measures the ability of an environment to supply electrons to an oxidizing agent, or to take up electrons from a reducing agent, just as the pH of an environment measures its ability to supply protons (hydrogen ions) to a base or to take up protons from an acid. In a complex solution like seawater or water in a soil the redox potential is determined by a number of reactions, just as pH is determined by the combined effects of the carbon dioxide system, the boric acid system, and various organic acids. The particular reactions are difficult to identify, and are less important than the overall ability of the environment to maintain its Eh and pH constant when small amounts of foreign material are added.

Although usually expressed, like $E°$, as a voltage with reference to a standard hydrogen electrode, redox potential may also be described by a number more explicitly analogous to pH, a number called *electron activity* and symbolized *pe*. This is the negative logarithm of an assumed activity of electrons in a solution, just as pH is the negative logarithm of the activity of protons in the solution. As an example, for the Eh of a solution containing the ions Fe^{2+} and Fe^{3+} we generally write

$$Eh = E° + \frac{2.303RT}{f} \log (a_{Fe^{3+}}/a_{Fe^{2+}}), \tag{9-23}$$

where $E°$ is the standard potential of the half-reaction of Eq. (9-11), and where the activity of the electron is omitted from the logarithmic term even though it appears in the expression for the reaction in Eq. (9-11). This is permissible, as noted above, because the potential is measured relative to the standard hydrogen electrode, so that Eq. (9-11) is actually

$$Fe^{2+} + H^+ \rightleftharpoons Fe^{3+} + \tfrac{1}{2}H_2. \tag{9-24}$$

Because $a_{H^+} = a_{H_2} = 1$ in the standard hydrogen electrode, the activity product of Eq. (9-24) is

$$\frac{a_{Fe^{3+}}a_{H_2}^{1/2}}{a_{Fe^{2+}}a_{H^+}} = \frac{a_{Fe^{3+}}}{a_{Fe^{2+}}}. \tag{9-25}$$

Alternatively we could focus attention on the simple expression of Eq. (9-11) without the hydrogen terms and regard e^- as a substance in solution, so that the equilibrium constant would be expressed

$$K = \frac{a_{Fe^{3+}}a_{e^-}}{a_{Fe^{2+}}}, \tag{9-26}$$

or in logarithmic form

$$\log K = \log (a_{Fe^{3+}}/a_{Fe^{2+}}) + \log a_{e^-}, \tag{9-27}$$

where a_{e^-} is the activity of the electrons.

The relationship between Eh and a_{e^-} is obtained by first rearranging Eq. (9-27):

$$\log (a_{Fe^{3+}}/a_{Fe^{2+}}) = \log K - \log a_{e^-}, \tag{9-28}$$

and combining Eqs. (9-15) and (8-49),

$$E° = \frac{2.303RT}{f} \log K. \tag{9-29}$$

Substitution of these equations [Eq. (9-29) and (9-28)] in Eq. (9-23) gives

$$Eh = \frac{-2.303RT}{f}\log K + \frac{2.303RT}{f}\log K - \frac{2.303RT}{f}\log a_{e^-}, \tag{9-30}$$

$$Eh = +\frac{2.303RT}{f}pe, \tag{9-31}$$

where the symbol $pe \equiv -\log a_{e^-}$. For a system at 25°C, then, Eh = $(2.303 \times 8.314 \times 298/96,485)pe = 0.059pe$. Thus the two numbers for expressing redox potential are proportional, but the proportionality constant changes with temperature. Either Eh or pe can be used to represent redox potential, and little advantage can be claimed for choosing one over the other. The main difference is that the slopes of lines representing redox reactions on an Eh–pH diagram are determined, in part, by temperature [Eq. (9-18)], but for diagrams using pe as a descriptive variable the slopes of reaction lines are solely a function of reaction stoichiometry.

Unfortunately, redox potentials in nature cannot be determined as simply and unambiguously as this discussion has implied. The difficulty is that some of the reactions that determine redox potentials are slow, so that instantaneous readings with a platinum electrode do not give valid equilibrium potential differences. This is particularly true for reactions involving oxygen, which, of course, include a great many of the most important oxidation reactions in nature. Most reactions in which oxygen plays a role take place by a series of steps, and one of the steps is very slow. Hence redox potentials measured in oxygen-containing environments are generally lower than equilibrium values, and there is no simple way to apply a correction factor. This means that most redox potential measurements in nature give us only qualitative or semi-quantitative information. It is useful nevertheless to make calculations based on the measured values and on theoretical potentials, since such calculations can at least set limits to the processes we may expect in natural environments.

9-4 LIMITS OF pH AND Eh IN NATURE

To make predictions about geologic processes, we need at least a rough idea as to the ranges of natural Eh and pH values. The ranges will obviously be more restricted than those with which the chemist is accustomed to deal in the laboratory.

The limits of pH we have mentioned in earlier discussions. The solutions of highest acidity found in nature are those formed by the dissolving of volcanic gases and by the weathering of ores containing pyrite. Locally such solutions may attain pH's less than zero ($m_{H+} > 1$). Acidities of this magnitude are quickly lowered by reaction with adjacent rocks, and the rocks are thereby drastically altered, as is evident in the bleached and porous zones commonly found near fumaroles, hot springs, and pyritic ore deposits. Given enough time, contact with ordinary silicate or carbonate rocks would neutralize the solutions. Complete neutralization is generally prevented by dissolution of carbon dioxide from the atmosphere and of organic acids formed by decaying organic matter; these two are the source of acidity in most near-surface waters, giving pH's commonly in the range 5 to 6. Lower pH's are found in the A horizons of pedalfer soils, especially podzols, where values as low as 3.5 are sometimes recorded (see Sec. 13-10 for soil classification). Disregarding the possible extremes, we can reasonably select a figure of 4 as the *usual* lower limit of pH's in natural environments.

At the other end of the scale, CO_2-free water in contact with carbonate rocks can acquire by hydrolysis a pH of about 10, and in contact with the silicates of ultramafic rocks a pH of nearly 12. Similar high values may be found in desert basins, where fractional crystallization and fractional solution have segregated alkaline salts like sodium carbonate and sodium borate. But most surface waters have sufficient contact with the atmosphere that such high alkalinities are not attained, and a reasonable upper limit of pH in *most* near-surface environments is about 9.

The strongest oxidizing agent commonly found in nature is the oxygen of the atmosphere. Stronger agents than this cannot persist, for the reason that they would react with water to liberate oxygen. Thus the upper limit of redox potentials is defined by the reaction

$$H_2O \rightleftharpoons \tfrac{1}{2}O_2 + 2H^+ + 2e^- \qquad E° = +1.23 \text{ volts.} \qquad (9\text{-}32)$$

The potential of this half-reaction clearly depends on the pH, as shown by the expression of Eq. (9-19) for Eq. (9-32):

$$Eh = 1.23 + 0.03 \log a_{O_2}^{0.5} \cdot a_{H+}^2. \qquad (9\text{-}33)$$

For the usual concentration of O_2 we may use 0.2 atm (or approximately 0.2 bar), since oxygen makes up about one-fifth of the atmosphere by volume. Hence

$$Eh = 1.23 + 0.03 \log (0.2)^{0.5} + 0.059 \log a_{H^+},$$
$$= +1.22 - 0.059 \text{ pH}. \tag{9-34}$$

Slowness of reaction ("overvoltage effects") should make it possible for stronger oxidizing agents to exist locally and temporarily. Actually, however, measured oxidizing potentials in nature are always well below this limit, so that the empirical equation

$$Eh = 1.04 - 0.059 \text{ pH} \tag{9-35}$$

is a more realistic upper boundary (Baas Becking et al., 1960). The discrepancy probably means that oxidation reactions involving O_2 have complicated mechanisms, possibly with a slow step in which traces of hydrogen peroxide act as an intermediate (Sato, 1960).

Reducing agents likewise are limited to substances that do not react with water, the reaction this time resulting in liberation of hydrogen. The limiting redox potential is that of the hydrogen electrode reaction,

$$H_2 \rightleftharpoons 2H^+ + 2e^- \qquad E° = 0.00 \text{ volt}, \tag{9-36}$$

for which

$$Eh = 0.00 + 0.03 \log a_{H^+}^2 - 0.03 \log a_{H_2},$$
$$= -0.059 \text{ pH} - 0.03 \log a_{H_2}. \tag{9-37}$$

Since the pressure of hydrogen gas in near-surface environments cannot exceed 1 atm, the maximum possible reducing potential in the presence of water would be

$$Eh = -0.059 \text{ pH} - 0.03 \log (1) = -0.059 \text{ pH}. \tag{9-38}$$

Conceivably local conditions might permit stronger reducing reactions, particularly within bodies of organic material (coal or petroleum) out of contact with water.

9-5 Eh–pH DIAGRAMS

The usual limits of Eh and pH that we have just discussed may be conveniently plotted on a graph with Eh as ordinate and pH as abscissa (Fig. 9-1). Fig. 9-1(a) shows the stability of water as described by the lines for Eqs. (9-34) and (9-38). The shaded area in Fig. 9-1(b) outlines the range of measurements in many kinds of surface and near-surface environments.

To show the conditions under which various oxidation–reduction processes may be expected to occur in nature, potentials for these processes are plotted on graphs like those in Fig. 9-1. For example, Fig. 9-2 shows one method of diagramming some of the oxidation reactions of iron in solution. Consider first the couple

$$Fe^{2+} \rightleftharpoons Fe^{3+} + e^- \qquad E° = +0.77 \text{ volt}. \tag{9-11}$$

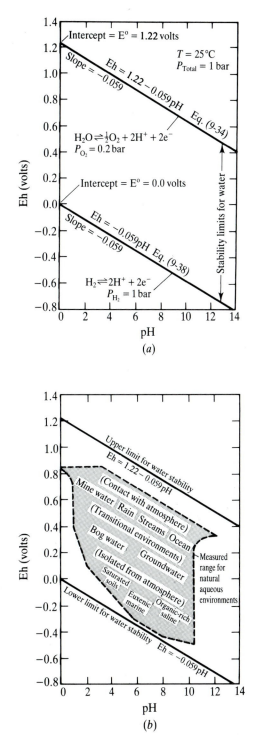

FIGURE 9-1
Framework of Eh–pH diagrams. Diagonal lines in (a) define the upper [Eq. (9-34)] and lower [Eq. (9-38)] stability limits for water at 25°C and 1 bar. The shaded area in (b) shows the measured limits of Eh and pH in natural environments. (Reprinted by permission from Baas Becking et al., 1960.)

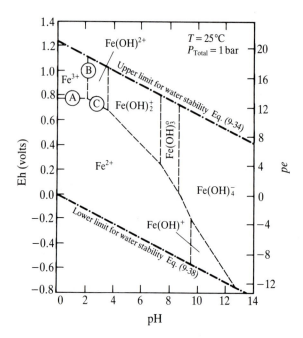

FIGURE 9-2
Eh-pH diagram for the simple ions of iron at 25°C and 1 bar. The dashed lines denote Eh and pH conditions for equal activities of the aqueous species. Lines marked A, B, and C represent equal activities of iron species determined from Eq. (9-11), (2-27), and (9-42), respectively. Redox potential is expressed as volts (Eh) on the left side and as pe on the right side. (Reprinted by permission from Nordstrom and Munoz, 1985.)

This reaction is independent of pH, since neither H^+ nor OH^- appears in the equation, so that its standard potential may be plotted as a horizontal line (A). This line corresponds to the Eh where the activity of Fe^{3+} equals the activity of Fe^{2+} in the solution, a requirement of Eq. (9-23) when Eh = E°. At any pair of Eh–pH values above the line the ratio $a_{Fe^{3+}}/a_{Fe^{2+}}$ is greater than 1, at any pair below the line less than 1. The line cannot be continued far into the diagram because Fe^{3+} reacts with water to form $Fe(OH)^{2+}$:

$$Fe^{3+} + H_2O \rightleftharpoons Fe(OH)^{2+} + H^+ \qquad \log K = -2.2. \qquad (2\text{-}27)$$

For unit activity of water the equilibrium constant is

$$\log K = \log \frac{a_{Fe(OH)^{2+}}}{a_{Fe^{3+}}} + \log a_{H^+}, \qquad (9\text{-}39)$$

$$= \log \frac{a_{Fe(OH)^{2+}}}{a_{Fe^{3+}}} - pH, \qquad (9\text{-}40)$$

and setting the ratio $a_{Fe(OH)^{2+}}/a_{Fe^{3+}}$ equal to one gives

$$\log K = -pH = -2.2. \qquad (9\text{-}41)$$

Eq. (9-41) plots as the vertical line marked B in Fig. 9-2. The area in the figure outlined by the upper stability limit for water and lines A and B is the range of Eh and pH values where Fe^{3+} is the predominant (most abundant) species of iron. In Chapter 2 we used similar methods for defining the range of pH for the predominant carbonate (Fig. 2-3) and sulfur (Fig. 2-4) species in solution.

The next step is to construct a line showing equal activities of Fe^{2+} and $Fe(OH)^{2+}$, by adding reactions (9-11) and (2-27):

$$Fe^{2+} + H_2O \rightleftharpoons Fe(OH)^{2+} + H^+ + e^-. \qquad (9\text{-}42)$$

The standard potential for this reaction, $E° = 0.900$, is determined by calculating the standard free energy of reaction (9-11) from its standard potential using Eq. (9-15), and by adding to this the standard free energy for Eq. (2-27) as computed from Eq. (8-49). The resultant standard free energy for Eq. (9-42) (86,853 J mol^{-1}) then gives the standard potential, by use of Eq. (9-15). A line representing equal activities of Fe^{2+} and $Fe(OH)^{2+}$ is now derived from the Nernst equation (9-19),

$$Eh = E° + 0.059 \left(\log \frac{a_{Fe(OH)^{2+}}}{a_{Fe^{2+}}} + \log a_{H^+} \right), \qquad (9\text{-}43)$$

by substituting $E° = 0.900$, $a_{Fe^{2+}} = a_{Fe(OH)^{2+}}$, and $-\log a_{H^+} = pH$:

$$Eh = 0.900 - 0.059\ pH. \qquad (9\text{-}44)$$

This equation, plotted as line C in Fig. 9-2, has a slope of -0.059 and an intercept of 0.900. The point where lines A, B, and C intersect represents Eh and pH values where the three species Fe^{3+}, Fe^{2+}, and $Fe(OH)^{2+}$ have equal activities. Using these methods, similar lines can be constructed for the other species of Fe(III) and Fe(II) in solution.

The completed diagram shows that the change from Fe(II) to Fe(III) falls approximately in the middle of the field representing conditions in nature [Fig. 9-1(b)], so that we would expect to find changes from one to the other very frequent, depending on slight shifts in the pH or Eh of the environment. This obviously corresponds with everyday experience. We find iron compounds reduced in the organic-rich surface layer of a soil and oxidized beneath, reduced in bottom sediments of the sea and oxidized in seawater itself, and so on. The diagram also indicates that oxidation of iron takes place much more completely in alkaline solution than in acid, which fits the observation that larger amounts of dissolved iron are commonly present in slightly acid stream waters than in the faintly alkaline water of the oceans because of the low solubility of hydrous Fe(III) compounds.

One other kind of information can be given on Eh–pH diagrams: the concentration of Fe^{2+} or of total Fe^{2+} plus Fe^{3+} in equilibrium with various minerals at different Eh–pH conditions. This is accomplished first by setting up equations for the electrode potentials of reactions between minerals such as magnetite and hematite,

$$\underset{\text{magnetite}}{2Fe_3O_4} + H_2O \rightleftharpoons \underset{\text{hematite}}{3Fe_2O_3} + 2H^+ + 2e^- \qquad E° = +0.22 \text{ volts}. \qquad (9\text{-}45)$$

For unit activity of water and solids the Nernst equation is

$$Eh = E° + \frac{0.059}{2} \log a_{H^+}^2 \qquad (9\text{-}46)$$

or

$$Eh = 0.22 - 0.059\ pH, \qquad (9\text{-}47)$$

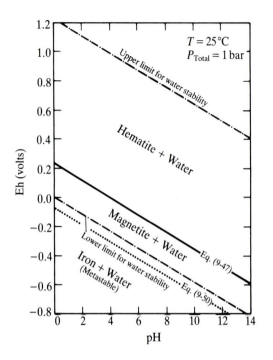

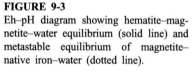

FIGURE 9-3
Eh–pH diagram showing hematite–magnetite–water equilibrium (solid line) and metastable equilibrium of magnetite–native iron–water (dotted line).

which plots as the solid straight line in Fig. 9-3. This line defines Eh–pH conditions for hematite–magnetite equilibrium. A line for equilibrium between magnetite and native iron is constructed in a similar manner:

$$\underset{\text{iron}}{3Fe} + 4H_2O \rightleftharpoons \underset{\text{magnetite}}{Fe_3O_4} + 8H^+ + 8e^- \qquad E° = -0.08 \text{ volts}, \qquad (9\text{-}48)$$

$$Eh = -0.08 + \frac{0.059}{8} \log a_{H^+}^8, \qquad (9\text{-}49)$$

and

$$Eh = -0.08 - 0.059 \text{ pH}. \qquad (9\text{-}50)$$

Eq. (9-50) plots below the lower stability limit of water in Fig. 9-3, meaning that metallic iron is not to be expected in sedimentary environments because its presence would require too low a redox potential.

To compute the concentration of iron species in equilibrium with hematite and magnetite we combine the relationships derived in Figs. 9-2 and 9-3. For example, the activity of Fe^{2+} in a solution saturated with hematite is derived from the reaction

$$2Fe^{2+} + 3H_2O \rightleftharpoons \underset{\text{hematite}}{Fe_2O_3} + 6H^+ + 2e^-, \qquad (9\text{-}51)$$

$$Eh = E° + 0.03 \log \frac{a_{H^+}^6}{a_{Fe^{2+}}^2}. \qquad (9\text{-}52)$$

The standard potential $E°$ given in tables is 0.65 volts. Hence

$$Eh = 0.65 - 0.18 \text{ pH} - 0.059 \log a_{Fe^{2+}}. \tag{9-53}$$

Rearrangement gives $\log a_{Fe}^{2+}$ as a function of pH and Eh:

$$\log a_{Fe^{2+}} = \frac{0.65 - 0.18 \text{ pH} - Eh}{0.059}. \tag{9-54}$$

In the field where hematite is the stable iron mineral, hematite cannot precipitate unless the activity of Fe^{2+} exceeds values given by this equation.

Eq. (9-54) defines a planar three-dimensional surface in terms of the variables $\log a_{Fe^{2+}}$, Eh, and pH. Any fluid that plots on this surface is saturated with hematite. Such a three-dimensional plot would be similar in form to the one shown in Fig. 4-11 for feldspar–solution equilibrium. To graphically represent the concentration of Fe^{2+} in equilibrium with hematite on a two-dimensional Eh–pH diagram, we first specify a concentration of Fe^{2+} (assuming that $a_{Fe^{2+}} = m_{Fe^{2+}}$) and use Eq. (9-54) to compute the equilibrium line in terms of Eh and pH. One such line is shown in Fig. 9-4 (dotted line marked D) for a concentration of $Fe^{2+} = 10^{-6}m$. A similar line may be derived for magnetite–solution equilibrium (dotted line marked E in Fig. 9-4). At Eh values greater than 0.77 volt, where Fe^{3+} is the predominant iron species, we can use the equation

$$2Fe^{3+} + 3H_2O \rightleftharpoons \underset{\text{hematite}}{Fe_2O_3} + 6H^+ \tag{9-55}$$

to compute the vertical line marked F in the figure to complete the diagram.

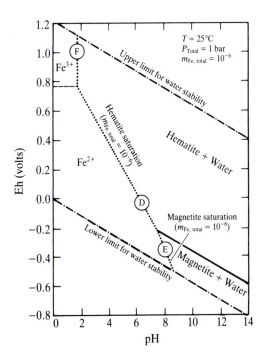

FIGURE 9-4

Eh–pH diagram showing equilibrium conditions for iron oxides and a solution containing $10^{-6}m$ total iron (dotted line marked $m_{Fe,total} = 10^{-6}$). The dashed line separates the predominance areas for aqueous Fe^{3+} and Fe^{2+} (Fig. 9-2), and the solid line denotes equilibrium between hematite and magnetite (Fig. 9-3).

The dotted line (F-D-E) in Fig. 9-4 represents the Eh and pH conditions for equilibrium of hematite or magnetite with a solution containing $10^{-6}m$ Fe^{2+} at Eh values less than +0.77 volt, and $10^{-6}m$ Fe^{3+} at Eh values greater than +0.77 volt. Because the dotted line lies within the predominance areas of these two iron species in Fig. 9-2, we can say that the dotted line in Fig. 9-4 represents iron oxide equilibria with a solution containing $10^{-6}m$ total iron ($m_{Fe,total} = 10^{-6}$). For a total iron concentration greater than this the computed line would plot to the left of the one shown in the figure, and for a lower concentration the line would plot to the right. In the latter case the calculations can be simplified by taking into account the predominance areas of other iron species in solution (Fig. 9-2).

Eh–pH diagrams may be made geologically more realistic by plotting fields of stability for other common rock-forming minerals. Suppose we inquire, for example, as to the Eh–pH conditions under which siderite would be deposited in preference to hematite. The equation relating the two minerals is

$$2FeCO_3 + 3H_2O \rightleftharpoons Fe_2O_3 + 2H_2CO_3 + 2H^+ + 2e^-. \qquad (9\text{-}56)$$
siderite   hematite

From free energies we calculate the E° value, +0.30 volt, and then set up the equation for Eh at various concentrations:

$$Eh = E° + 0.03 \log a^2_{H_2CO_3} a^2_{H^+}, \qquad (9\text{-}57)$$

$$Eh = 0.30 + 0.059 \log a_{H_2CO_3} - 0.059 \text{ pH}. \qquad (9\text{-}58)$$

For any given total concentration of dissolved CO_2, $a_{H_2CO_3}$ is a function of pH alone (Fig. 2-3). Hence Eh can be expressed as a function (albeit a rather complicated one) of pH, and the corresponding line can be drawn on an Eh–pH graph to express equilibrium between siderite and hematite (line A, Fig. 9-5). Other iron minerals can be included by setting up similar equations and making reasonable assumptions about total dissolved sulfur, total dissolved silica, and so on. Figure 9-5 includes hematite, siderite, magnetite, and pyrite, for assumed concentrations of total sulfur (H_2S, HS^-, S^{2-}, HSO_4^-, and SO_4^{2-}) equal to $10^{-6}m$ and of total carbonate (H_2CO_3, HCO_3^-, and CO_3^{2-}) equal to $1m$. For the siderite field the appropriate equation would be

$$FeCO_3 \rightleftharpoons Fe^{2+} + CO_3^{2-} \qquad K = 10^{-10.7}, \qquad (9\text{-}59)$$
siderite

whence

$$a_{Fe^{2+}} = \frac{10^{-10.7}}{a_{CO_3^{2-}}}. \qquad (9\text{-}60)$$

The concentration of CO_3^{2-} would then be expressed as a function of total carbonate and pH, and substitution in this equation gives values of $a_{Fe^{2+}}$. In Fig. 9-5 the dotted line showing siderite equilibrium with a solution where the Fe^{2+} activity is 10^{-6} is parallel to the Eh axis, since the equation involves no oxidation or reduction.

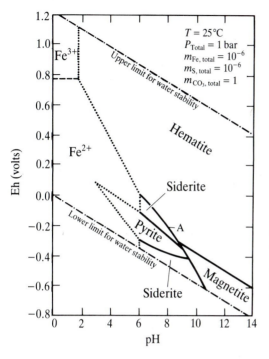

FIGURE 9-5
Eh–pH diagram showing fields of common iron minerals. Total concentration of dissolved carbonate $1m$, of dissolved sulfur $10^{-6}m$. The dotted boundaries on the left side of the diagram represent equilibrium of hematite, siderite, and pyrite with a solution containing $10^{-6}m$ total iron. (Reprinted by permission from Garrels and Christ, 1965.)

The Eh–pH relations in Fig. 9-5 may be interpreted as follows: Hematite is the stable mineral of iron in all moderately and strongly oxidizing environments. In reducing environments the stable mineral may be pyrite, siderite, or magnetite, depending on concentrations of sulfur and carbonate in the solution. For the conditions of high total carbonate ($1m$) and low total sulfur ($10^{-6}m$) shown in Fig. 9-5, siderite has two fields of stability separated by the field of pyrite, and magnetite is stable only in contact with strongly basic solutions. If dissolved carbonate is smaller and dissolved sulfur is higher, the field of pyrite expands until it fills nearly all the lower part of the diagram; a small field in which pyrrhotite is stable may appear at the extreme lower edge of the natural Eh range. If both carbonate and sulfur are very low, the field of magnetite extends into near-neutral environments. All the mineral transitions involving oxidation are favored by basic solutions, so that, for example, hematite may form from siderite in response to an increase in either pH or Eh. The occurrence of siderite is practically restricted to neutral and basic environments; it can precipitate from weakly acid solutions only if the concentration of dissolved iron is abnormally high. Within narrow ranges of Eh and pH most pairs of iron minerals are stable together—magnetite–hematite, hematite–siderite, magnetite–siderite, siderite–pyrite. Even the pair hematite–pyrite, although forbidden by the low total sulfur of Fig. 9-5, has a stable existence in environments with higher sulfur. Most of these conclusions are familiar enough as geological deductions based on field associations of iron minerals, but the diagram displays the underlying chemistry and makes the conclusions more quantitative.

9-6 FUGACITY DIAGRAMS

Eh and *pe* are the variables commonly employed to express redox conditions in sedimentary and groundwater environments, i.e., for equilibria under usual conditions of temperature and pressure. These variables can be used also at higher temperatures and/or pressures, by making the necessary corrections to Gibbs free energies and equilibrium constants by the methods described in Chapter 8, but other variables are often more convenient.

One example is the fugacity of oxygen, f_{O_2}, a gas that can be assumed present in contact with any redox equilibrium and whose fugacity will change as redox conditions change. Consider the equilibrium between hematite and magnetite

$$6Fe_2O_3 \rightleftharpoons 4Fe_3O_4 + O_2. \tag{9-61}$$
$$\underset{\text{hematite}}{} \quad \underset{\text{magnetite}}{}$$

For pure hematite and magnetite the equilibrium constant is [Eq. (8-49)]

$$\log K = \frac{-\Delta G°}{2.303RT} = \log a_{O_2}, \tag{9-62}$$

or expressed in terms of fugacity using the definition of Eq. (8-35),

$$\log K = \log \frac{f_{O_2}}{f°_{O_2}}. \tag{9-63}$$

If we choose a standard state for oxygen as unit fugacity of the pure gas at 1 bar and any temperature ($f°_{O_2} = 1$ bar), Eq. (9-63) reduces to

$$\log K = \log f_{O_2}, \tag{9-64}$$

where the term "f_{O_2}" is actually the ratio of the fugacity of oxygen (measured in bars) to its fugacity in the standard state (1 bar) and the equilibrium constant is a dimensionless number (see Sec. 8-4).

The hematite–magnetite reaction [Eq. (9-61)] is endothermic, so oxygen fugacity increases with increasing temperature [see Eq. (8-61)]. This increase is illustrated by the curve marked HM in Fig. 9-6. The fugacity of oxygen in equilibrium with hematite and magnetite is very small, about 10^{-80} bar at low temperatures, increasing to values of about 10^{-10} bar at magmatic temperatures. If H_2O is present, the equilibrium can also be represented by

$$3Fe_2O_3 + H_2 \rightleftharpoons 2Fe_3O_4 + H_2O, \tag{9-65}$$
$$\underset{\text{hematite}}{} \quad \underset{\text{magnetite}}{}$$

and the equilibrium constant is

$$\log K = \log (f_{H_2O}/f_{H_2}). \tag{9-66}$$

For this reaction the oxidation state of the system is expressed by the fugacity ratio of H_2O to H_2.

An equilibrium curve like the one for hematite and magnetite in Fig. 9-6 is sometimes referred to as an oxygen buffer because the presence of these minerals in a rock may control the oxidation state as temperature (and pressure) change. Also

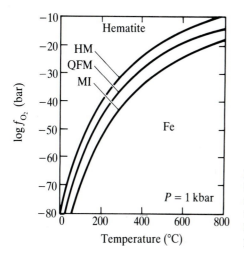

FIGURE 9-6
Oxygen fugacity for equilibria of hematite–magnetite [HM, Eq. (9-61)], quartz–fayalite–magnetite [QFM, Eq. (9-67)], and magnetite–iron [MI, Eq. (9-68)], as a function of temperature at a pressure of 1 kbar.

shown in Fig. 9-6 are equilibrium curves for the assemblage quartz, magnetite, and fayalite, and for magnetite and native iron, as represented by the reactions

$$\underset{\text{magnetite}}{2Fe_3O_4} + \underset{\text{quartz}}{3SiO_2} \rightleftharpoons \underset{\text{fayalite}}{3Fe_2SiO_4} + O_2 \tag{9-67}$$

and

$$\underset{\text{magnetite}}{\tfrac{1}{2}Fe_3O_4} \rightleftharpoons \underset{\text{iron}}{\tfrac{3}{2}Fe} + O_2. \tag{9-68}$$

The oxidation states of most geologic systems lie between the curves for hematite stability [Eq. (9-61), oxidizing conditions] and for native iron [Eq. (9-68), reducing conditions].

For further illustration, we look at oxidation–reduction relations among the simple compounds of sulfur. We start with a standard Eh–pH diagram for sulfur at 25°C and 1 bar [Fig. 9-7(a)], assuming a dilute solution with a total concentration of dissolved sulfur species of $0.001m$. The diagram shows fields of stability for the common oxidation states of sulfur in ordinary environments: -2 for sulfides, 0 for the native element, $+6$ for sulfates. The narrow wedge marked S_{rh} on the left-hand side of the diagram means that native sulfur (rhombic) is stable in contact with near-surface solutions over a small range of Eh and pH. The other fields are marked with the formulas of the principal ions or molecules present. Over most of the area of any one field, the designated species makes up more than 99% of the total concentration of sulfur in solution, hence it has a concentration of approximately $0.001m$. Near any boundary line the concentration diminishes; the line itself is drawn through points where the two species on either side have equal concentrations, approximately $0.0005m$. Note that the important sulfur-containing ion in oxidizing solutions in SO_4^{2-}, that reducing solutions have chiefly H_2S at pH's less than 7 and chiefly HS^- at pH's greater than 7, and that S^{2-} is not a major constituent of

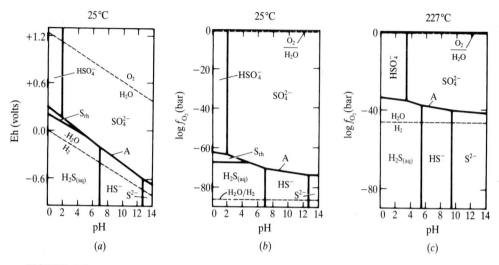

FIGURE 9-7

Eh–pH and f_{O_2}–pH diagrams for stable sulfur species. Total concentration of dissolved sulfur species 0.001m. (*a*): Eh–pH, 25°C, 1 bar. (*b*): f_{O_2}–pH, 25°C, 1 bar. (*c*): f_{O_2}–pH, 227°C, pressure equal to liquid–vapor equilibrium for H_2O.

any geologically important solution. These are simply quantitative expressions of our general knowledge about the chemical behavior of sulfur.

The diagram, like any Eh–pH diagram, has stringent limitations. The boundaries are calculated for equilibrium conditions, and take no account of slow reactions or metastable forms of sulfur. Actually the reduction of sulfate to sulfide is so slow as to be undetectable unless bacteria are present, so that SO_4^{2-} may have at least a temporary existence in strongly reducing solutions. Ions that are metastable under these conditions (for example, SO_3^{2-}, S_2^{2-}, $S_2O_3^{2-}$) do not appear at all. A number of heavy metals form complex ions with various sulfur species, and if these are present the diagram would be more complicated. Fig. 9-7(*a*) thus shows only the bare bones of sulfur chemistry, but for present purposes this is adequate.

Converting Fig. 9-7(*a*) to a f_{O_2}–pH diagram requires that the equation for each line be expressed as a reaction involving O_2 as an oxidizing agent. Take, for example, the line marked A on Fig. 9-7(*b*), separating fields of SO_4^{2-} and HS^-. The appropriate equation is

$$HS^- + 2O_2 \rightleftharpoons SO_4^{2-} + H^+. \tag{9-69}$$

Free energies of formation (Appendix VIII) are $+12.1$ kJ mol^{-1} for HS^-, -744.5 kJ mol^{-1} for SO_4^{2-}, and 0 for $2O_2$ and H^+, so that $\Delta G°$ for the reaction is -756.6 kJ mol^{-1}. From this figure the equilibrium constant can be calculated and expressed as a quotient of activities [Eq. (8-49)]:

$$\log K = -(-756.6/5.708) = +132.6,$$

$$= \log a_{SO_4^{2-}} - pH - \log a_{HS^-} - 2 \log f_{O_2}. \tag{9-70}$$

At the boundary, by definition, $a_{SO_4^{2-}} = a_{HS^-}$, so that the equation reduces to

$$\log f_{O_2} = -\tfrac{1}{2}pH - 66.3. \tag{9-71}$$

For comparison, the corresponding line on Fig. 9-7(a), also marked A, is drawn by calculating the standard Gibbs free energy of reaction and thence $E°$ and Eh for the reaction

$$HS^- + 4H_2O \rightleftharpoons SO_4^{2-} + 9H^+ + 8e^-. \tag{9-72}$$

Thus each line on one diagram has a matching line on the other. The stability fields have different shapes, but the relations among them are identical. It should be noted that the absurdly low values for oxygen fugacity indicated on Fig. 9-7(b), down to 10^{-80} bar, have no meaning as actual, measurable gas pressures, but are simply numerical indices of the oxidation state of the system.

To adapt the diagram to a higher temperature, we need only modify the values of the equilibrium constant obtained from the equation for each line, using the methods outlined in Chapter 8, and from these get values of $\log f_{O_2}$ as a function of pH. A diagram for the sulfur system at a temperature of 227°C, constructed in this manner, is shown as Fig. 9-7(c). Note that the field of native sulfur has vanished and that the field boundaries have shifted somewhat, but general relations among the sulfur species are still much the same.

As on the Eh–pH diagram of Fig. 9-5, we can also represent mineral stabilities on f_{O_2}–pH diagrams. One example is shown in Fig. 9-8 for iron sulfide and iron oxide minerals at 250°C, where the solution contains $0.1m$ total dissolved sulfur. Phase diagrams like the ones given in Figs. 9-5 and 9-8 provide a useful basis for description and interpretation of mineral stabilities in terms of pH and oxidation state. In later chapters we will use diagrams of this sort to estimate conditions for

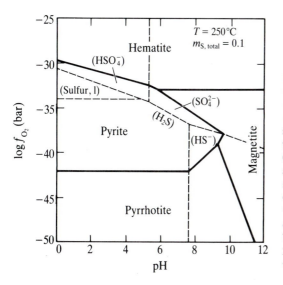

FIGURE 9-8
Oxygen fugacity–pH diagram for mineral stabilities in the system Fe–S–O (stability fields are shown by the thick solid lines) at 250°C for solutions with a total concentration of dissolved sulfur of $0.1m$. Dashed lines show relationships among the predominant sulfur species in solution. (Diagram simplified from Crerar, D. A., and H. L. Barnes: "Ore solution chemistry. V. Solubilities of chalcopyrite and chalcocite assemblages in hydrothermal solutions at 200°C to 350°C," *Economic Geology*, vol. 71, pp. 772–794, 1976.)

diagenesis (Chap. 14) and for the formation of hydrothermal ore deposits (Chap. 19).

This same procedure, obviously, can be used to show relations between any two variables that are sensitive to change in acidity or oxidation potential—or to any other significant properties of a system. For example, the phase relations given in Fig. 9-8 could also be represented by the variables of sulfur fugacity and oxygen fugacity, or sulfur fugacity and the fugacity ratio of H_2O to H_2. The best variables to use in any particular case depend on the conditions of an experiment, or of a natural environment, and on the ease with which the variables can be measured. At high temperatures, as illustrated by the examples above, fugacities of gases are often the variables of choice.

Whatever variables are used, the limitations of such stability diagrams must be kept in mind. They are drawn to show equilibrium relations, and equilibrium is often not attainable in reasonable time. Usually other variables play a role, besides the two that are selected; a three-dimensional figure can be tried, to accommodate a third, but is generally inconvenient; variables beyond three can be handled only by cumbersome partial sections. And there is always the limitation posed by uncertainty in the thermodynamic data. With all these qualifications, are diagrams of this sort worth the time and effort they take to construct? The same question can be asked about any quantitative representation of geologic relations based on laboratory data for simple compounds and solutions. The answer must come from experience: Do the diagrams in fact summarize geologic and experimental observations realistically, and do they lead to predictions that can be tested against field occurrences? Many such diagrams do indeed lead to successful predictions, and serve as a convenient way to summarize large accumulations of data, but their correspondence with geologic observation is often far from perfect.

PROBLEMS

1. Complete and balance the following oxidation–reduction equations. (One method for balancing such equations is outlined in Appendix X.)

$$Fe^{2+} + UO_2^{2+} \rightleftharpoons Fe^{3+} + U^{4+} \qquad \text{(strongly acid solution)}$$

$$Cl^- + MnO_4^- \rightleftharpoons Cl_2 + MnO_2 \qquad \text{(weakly acid solution)}$$

$$V(OH)_3 + O_2 \rightleftharpoons VO_4^{-3} \qquad \text{(alkaline solution)}$$

$$CH_4 + SO_4^{2-} \rightleftharpoons CO_2 + S \qquad \text{(acid solution)}$$

2. From the data of Appendix IX, find $E°$ and calculate $\Delta G°$ for the following reactions:
   (a) $Mn + Cu^{2+} \rightleftharpoons Mn^{2+} + Cu.$
   (b) $MnO_2 + PbO + H_2O \rightleftharpoons Mn(OH)_2 + PbO_2.$
   (c) $3Cu + 2NO_3^- + 8H^+ \rightleftharpoons 3Cu^{2+} + 4H_2O + 2NO.$
   (d) $3HgS + 2NO_3^- + 8H^+ \rightleftharpoons 3Hg^2 + 4H_2O + 2NO + 3S.$
   (e) $2Ag + \frac{1}{2}O_2 + 2Cl^- + 2H^+ \rightleftharpoons 2AgCl + H_2O.$

3. On an Eh–pH diagram like Fig. 9-4, plot curves for the following half-reactions:
   (a) I^-–IO_3^-.
   (b) Cr^{3+}–$Cr_2O_7^{2-}$ and $Cr(OH)_3$–CrO_4^{2-}.
   (c) Sn–Sn^{2+} and Sn–$Sn(OH)_3^-$.
   (d) H_2S–S, HS^-–S, and S^{2-}–S.
   (e) H_2SO_3–SO_4^{2-} and SO_3^{2-}–SO_4^{2-}.
   Where it is necessary to assume an activity for some ion other than H^+ and OH^-, use 10^{-5}.

4. In lake water at a pH of 5 and a redox potential of +0.30 volt, what concentration of Cu^{2+} could exist in contact with metallic copper?

5. What redox potential must an environment possess in order for the activities of Fe^{2+} and Fe^{3+} to be equal?

6. The following questions refer to Fig. 9-5:
   (a) Describe the sequence of iron minerals that would form from a solution at a constant Eh of −0.2 volt, if its pH increases slowly from an initial value of 4.
   (b) Describe the sequence of oxidation products of siderite in contact with a solution whose pH is maintained at 9 while oxidation occurs.
   (c) Does the simple Fe(III) ion, Fe^{3+}, play a role in ordinary near-surface solutions? In what sort of geologic environment might its concentration become appreciable?

7. Arrange the following, insofar as possible, in order of (a) decreasing pH and (b) decreasing Eh. In places where the order is ambiguous, explain why.
   A. Seawater from near the surface in the tropics.
   B. Water from a temporary playa lake in Death Valley.
   C. Water from a lake in New York State.
   D. Water in a small stream draining an area of schist containing abundant pyrite.
   E. Water from a swamp in northern Canada.

8. Which of the following can be formed and continue to exist in near-surface environments? For those which cannot, give reasons to justify your answer.

$$\begin{array}{cccc} Ca(OH)_2 & MnO_4^- & CoCl_3 & Ag \\ UO_2^{2+} & K_2S & H_2CO_3 & SiO \\ SO_4^{2-} & Zn & Al^{3+} & Mn \end{array}$$

9. Construct Eh–pH and $\log f_{O_2}$–pH diagrams for solution–mineral phase relations in the system Fe–S–O–H at 25°C and 1 bar, for total dissolved sulfur of $0.1m$ and unit activity of water. Use data in Appendices VIII and IX to compute phase boundaries and lines of equal activities of aqueous species, and Figs. 9-5 and 9-8 as guides to constructing the diagrams.

REFERENCES AND SUGGESTIONS FOR FURTHER READING

Baas Becking, L. G. M., I. R. Kaplan, and D. Moore: "Limits of the natural environment in terms of pH and oxidation–reduction potentials," *Jour. Geology*, vol. 68, pp. 243–284, 1960. A tabulation of recorded measurements of Eh and pH in natural environments, and a discussion of the factors that limit these variables in different geologic situations.

Barnes, I., and W. Back: "Geochemistry of iron-rich groundwater of southern Maryland," *Jour. Geology*, vol. 72, pp. 435–447, 1964. A good example of careful measurements of Eh and pH in natural environments.

Garrels, R. M., and C. L. Christ: *Solutions, Minerals, and Equilibria*, Harper and Row, Publishers, New York, 1965. This is the standard reference on the measurement and use of oxidation potentials for geologic applications. The book has many Eh–pH diagrams, and gives detailed examples to show how the diagrams are constructed and interpreted.

Nordstrom, D. K., and J. L. Munoz: *Geochemical Thermodynamics*, The Benjamin/Cummings Publishing Co., Menlo Park, CA., 1985. Chapter 10 is a good review of oxidation–reduction equilibria using *pe* as a descriptive variable.

Sato, M: "Geochemical environments in terms of Eh and pH," *Econ. Geology*, vol. 55, pp. 928–961, 1960. Experimental evidence that the upper limit of oxidation potentials in nature is determined by a rate-controlling step involving hydrogen peroxide rather than by the oxidation of water.

Sillén, L. G.: "Stability constants of metal-ion complexes, Sec. I: Inorganic ligands," *Chem. Soc. London Spec. Pub.* 17, 1964, and Supplement 1, *Chem. Soc. London Spec. Pub.* 25, 1971. Contains a compilation of oxidation–reduction potentials.

Stumm, W., and J. J. Morgan: *Aquatic Chemistry, An Introduction Emphasizing Chemical Equilibria in Natural Waters*, 2d ed., Wiley, New York, 1981. Chapter 7 gives a more advanced treatment of oxidation and reduction reactions, including equilibrium relations and kinetics.

Thorstenson, D. C.: "The concept of electron activity and its relation to redox potentials in aqueous geochemical systems," U.S. Geological Survey Open-File Report 84-0072, p. 66, 1984.

ISOTOPE
GEOCHEMISTRY

Up to this point we have not questioned the familiar assumption of classical chemistry that atoms of any given element are all alike. All the atoms of sodium and chlorine in sodium chloride, for example, we have assumed to be identical, so that any sample of salt would show precisely the same solubility, melting point, heat conductivity, and so on as any other sample. About a century ago came the revelation that this is not quite true, and a great shock it was to chemists of that time. All atoms of an element are very nearly the same, but may show minor differences. Study of the differences has blossomed into a field of geochemistry that gives us a wealth of information on a variety of subjects: ages of rocks and minerals, temperatures of rock formation, places where rocks and ores and fluids have their origins. It is time that we look into this active kind of geochemical research.

10-1 KINDS OF ISOTOPES

The atoms of any given element all have the same number of electrons outside the nucleus and the same number of protons within the nucleus (this is the *atomic number*), but their nuclei may contain differing numbers of neutrons. Thus the atoms may differ in mass (*atomic mass* or *atomic weight* = number of protons plus number of neutrons), and hence differ slightly in properties. An element, in other words, is not necessarily a simple substance in the classical sense, but may consist of a mixture of substances with slightly different chemical and physical behavior. The separate varieties of an element are its *isotopes*. Some elements in nature consist of a single isotope (for example, F, Na, Co); most have at least two, and a few have eight or more (for example, Sn, Xe, Te). Artificially, by bombarding nuclei with fast-moving particles, additional isotopes of all elements except hydrogen can be prepared.

The isotopes of an element have different chemical and physical properties, but only very slightly different. Chemical properties depend largely on nuclear charge and arrangement of electrons, especially the outermost electrons; since atoms of isotopes differ only in nuclear mass and have the same electron distribution, their chemical properties are almost identical. Properties are so very similar that ordinary chemical reactions, either in the laboratory or in nature, lead to practically no separation of isotopes. Only a few of the lighter elements (for example, H, C, O, S) have isotopes with a sufficient percentage difference in atomic mass to cause appreciable separation in nature. Laboratory separation of isotopes of the heavier elements can be accomplished, but only with difficulty.

Some isotopes are *radioactive*, meaning that their atomic nuclei give out radiation spontaneously and thereby change to nuclei of other elements. Various kinds of radiation are observed, the commonest being *alpha particles* (helium nuclei, consisting of 2 protons and 2 neutrons) and *beta particles* (electrons); either of these may be accompanied by *gamma radiation* (high-energy x-rays). The radiation is sufficiently energetic to ionize atoms that it penetrates, hence is described as *ionizing radiation*. If the ionization occurs in living tissue, it may be damaging or even lethal to an organism. The new elements produced by radioactive decay are spoken of as *daughter products*, and in light of their origin are said to be *radiogenic* isotopes. Daughter products may be much more radioactive than their parents (for example, the dangerously active ^{226}Ra and ^{222}Rn are daughter products of feebly radioactive ^{238}U), and the quantity of such isotopes in nature is maintained by a balance between their rates of formation and decay. Some radioactive isotopes are continually being formed by reactions of stable nuclei with high-energy particles in the atmosphere (^{14}C, tritium, ^{10}Be). Isotopes produced artificially are for the most part radioactive, and a considerable quantity of such isotopes has been added to the surface of the Earth in recent decades by nuclear reactors and nuclear bombs. The great majority of naturally occurring isotopes are not radioactive, and are spoken of as *stable* isotopes.

The accumulation of radiogenic isotopes by the decay of radioactive parents provides a powerful means of establishing absolute ages of Earth materials. Study of radiogenic isotopes, particularly those of strontium, lead, and neodymium, has yielded also information on the differentiation of planetary substance, both in the early part of Earth history and in later episodes. Distribution of stable isotopes has proved to be a useful tool for measuring temperatures and drawing inferences about sources of rocks, ores, and fluids. In these and other ways, isotopes have provided new insights into the chemistry of the Earth and the processes that have shaped its surface and its materials.

10-2 RADIOGENIC ISOTOPES: AGE DATING

Because radioactive decay goes on at a constant rate that is unaffected by the temperatures, pressures, or chemical combinations found in geologic environments, it can serve as a reliable clock for measuring geologic time. Once a radioactive isotope is imprisoned in the structure of a growing crystal, its atoms decay to atoms of its daughter element or elements at a fixed rate. The ratio of daughter to parent thus steadily increases, and measurement of the ratio gives a number from which the time elapsed since the crystal was formed can be calculated. One must assume, of course, that the mineral has not been altered since its formation. If any amount of either parent or daughter has been added to or subtracted from the mineral, the calculated age would be false. Commonly freedom from alteration can be inferred from geologic evidence, or from the consistency of ages calculated using more than one parent–daughter pair. When alteration is evident or suspected, say as a result of metamorphism during the rock's history, the isotope ratio will not give a reliable age but may nevertheless provide useful information about the time and nature of the metamorphic event.

Equations of Decay

In any large number of atoms of a radioactive isotope, the decay follows a statistical rule: during any fixed time interval, a definite proportion of the atoms changes to the daughter product. Thus the rate of decay, or the number of atoms that decay, is simply proportional to the total number of parent atoms present:

$$-\frac{d\text{P}}{dt} = \frac{d\text{D}}{dt} = \lambda\text{P}, \tag{10-1}$$

where P is the number of parent atoms remaining at any time t, $d\text{D}/dt$ is the rate of formation of daughter atoms, and λ is a proportionality constant called the *decay constant*. Integration gives

$$-\ln \text{P} = \lambda t + \text{C}. \tag{10-2}$$

The integration constant C may be expressed in terms of the original number of parent atoms when $t = 0$ (time when the rock or mineral was formed), or $C = -\ln \text{P}_0$. Substitution of this value gives

$$\ln \text{P} - \ln \text{P}_0 = -\lambda t,$$

thus

$$\text{P} = \text{P}_0 e^{-\lambda t} \qquad \text{or} \qquad \text{P}_0 = \text{P}e^{+\lambda t}. \tag{10-3}$$

The number of daughter atoms produced by the time t (given the symbol D*) is the difference between P_0 and P:

$$\text{D*} = \text{P}_0 - \text{P} = \text{P}e^{\lambda t} - \text{P} = \text{P}(e^{\lambda t} - 1). \tag{10-4}$$

Now at time $t = 0$ some daughter atoms may have been present, and the total number existing at time t is therefore this original number plus those produced by radioactive decay. If the original number is D_0, the total will be

$$\underset{\text{total}}{D} = \underset{\text{initial}}{D_0} + \underset{\text{produced}}{D^*} = D_0 + P(e^{\lambda t} - 1), \tag{10-5}$$

and rearrangement gives an explicit formula for t:

$$t = \frac{1}{\lambda} \ln\left(\frac{D - D_0}{P} + 1\right). \tag{10-6}$$

This is the time during which an amount of the daughter represented by D has accumulated, leaving undecayed an amount of the parent represented by P. Values for D and P can be found by analyzing the rock or mineral in which the radioactive isotope occurs. If, then, we can also find values for λ and D_0, Eq. (10-6) will give us the age of the rock or mineral in years.

The decay constant λ is found by laboratory measurement of decay rate. For isotopes whose radioactivity is feeble, and whose decay is therefore very slow, precise determination of λ may be difficult. As methods of measurement improve, values of λ for the geologically important isotopes are refined, and calculated ages that depend on λ must be revised. Some of the discrepancies among ages published at different times can be ascribed to such changes in the accepted values for decay constants.

More easily visualized than λ is a related quantity called the *half-life*, which is the time required for half of any given amount of an isotope to decay. If P_0 has decreased to $1/2\ P_0$ in a time $t_{1/2}$, substitution in Eq. (10-3) gives:

$$\tfrac{1}{2}P_0 = P_0 e^{-\lambda t_{1/2}}, \qquad \text{or} \qquad t_{1/2} = \ln\frac{P_0}{\tfrac{1}{2}P_0} \cdot \frac{1}{\lambda} = \frac{\ln 2}{\lambda} = \frac{0.693}{\lambda}. \tag{10-7}$$

As an example, the measured value of λ for ^{238}U is 1.537×10^{-10} yr^{-1}, and the corresponding half-life is 4.51×10^9 years. For this isotope the half-life is enormously long, roughly equal to the age of the Earth.

The quantity D_0 in Eq. (10-5) (the amount of daughter isotope that may have been present initially when the rock or mineral was formed) can be estimated in various ways. If the parent isotope is known, on geochemical grounds, to be present in a mineral that ordinarily contains little or none of the daughter product, D_0 can be set equal to zero or given an arbitrary low value. Or if a nonradiogenic isotope of the daughter is present, its initial ratio to the radiogenic isotope can be estimated if the age of the rock or mineral is roughly known. Another procedure depends on the fact that Eq. (10-5) has the form $y = a + bx$: if D is plotted against P (y against x) for a number of rocks or minerals formed at the same time, the values should lie on a straight line whose intercept on the D axis is D_0 and whose slope is $e^{\lambda t} - 1$.

The straight-line plot of D versus P for samples of rock material formed at the same time provides a graphical method for finding t from analyses for the radioactive isotope and its daughter. The line is called an *isochron* (because it represents samples of the same age), and the steepness of its slope is a measure of the length of time during which decay has been going on. Examples of isochrons are shown in Fig. 10-1 for the Rb–Sr system discussed below.

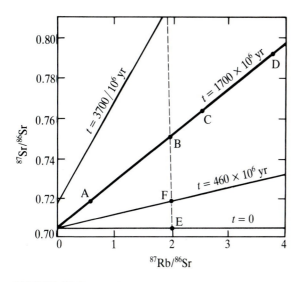

FIGURE 10-1

Four Rb–Sr isochrons ranging in age from early Archean to present. The heavy line is an isochron constructed from four hypothetical isotope analyses of a suite of igneous rocks represented by points A, B, C, D. Projection of the line to the $^{87}Sr/^{86}Sr$ axis gives an initial value for this ratio of 0.705, and the slope of the line ($e^{\lambda t} - 1$) gives an age of 1700 million years [Eq. (10-9)]. The lower light line is an isochron for a younger suite of rocks with the same initial Sr ratio. The upper light line is an isochron for rocks formed 3700 million years ago with an initial ratio of 0.715. The horizontal line is an isochron at $t = 0$ for rocks with an initial ratio of 0.705; in other words, it shows possible values of $^{87}Rb/^{86}Sr$ for such rocks at the time of their formation. For one such rock having an initial $^{87}Rb/^{86}Sr$ ratio of 2.0 (point E), the dashed line shows the changes in the two ratios over time, to point F after 460 million years and to point B after 1700 million years. (The data are plotted as ratios $^{87}Sr/^{86}Sr$ and $^{87}Rb/^{86}Sr$ rather than ^{87}Sr and ^{87}Rb alone, because the ratios are more easily measured.)

Use of the above formulas for calculating ages depends on two major assumptions: (1) The value of λ has not changed over geologic time. This seems reasonable since λ is a property of atomic nuclei, unaffected by extremes of temperature or pressure or by chemical combination in any ordinary geologic environment. (2) The rock or mineral has been a closed system since its formation; in other words, neither parent nor daughter has been added to or lost from the system. This assumption is often questionable, and many apparent anomalies in measured ages can be traced to its failure. Ages based on the assumption when its validity is uncertain are commonly called *model ages*, meaning that the ages refer only to a particular model of geologic history which may or may not be consistent with other geologic evidence.

Rubidium–Strontium Dates

The heavier of the two naturally occurring Rb isotopes, ^{87}Rb, decays by emission of an electron from its nucleus to stable ^{87}Sr. For this pair of isotopes Eq. (10-5) would be

$$\underset{\text{total}}{^{87}Sr} = \underset{\text{initial}}{^{87}Sr_0} + \underset{\text{produced}}{^{87}Rb(e^{\lambda t - 1})} . \qquad (10\text{-}8)$$

Because ratios of isotopes are more easily measured than numbers of atoms, it is customary to rewrite an equation of this sort in terms of ratios by dividing through by the amount of a stable and nonradiogenic isotope. For this system the isotope selected is ^{86}Sr, so that Eq. (10-8) becomes

$$\underbrace{\left(\frac{^{87}Sr}{^{86}Sr}\right)}_{\text{measured}} = \underbrace{\left(\frac{^{87}Sr_0}{^{86}Sr}\right)}_{\text{initial}} + \underbrace{\left(\frac{^{87}Rb}{^{86}Sr}\right)}_{\text{produced}}(e^{-\lambda t} - 1). \tag{10-9}$$

($^{86}Sr_0$ is the same as ^{86}Sr, because this isotope is nonradiogenic and does not change with time.) Eq. (10-6) for this system is then

$$t = \frac{1}{\lambda}\ln\left(\frac{\left(\dfrac{^{87}Sr}{^{86}Sr}\right) - \left(\dfrac{^{87}Sr_0}{^{86}Sr}\right)}{\left(\dfrac{^{87}Rb}{^{86}Sr}\right)} + 1\right). \tag{10-10}$$

The value of λ commonly used is 1.42×10^{-11} yr^{-1}, corresponding to a half-life of 4.88×10^{10} yr, but because the decay is so slow a precise value is difficult to obtain. For a mineral rich in ^{87}Sr (meaning a mineral of considerable antiquity that contained abundant Rb initially), the value of t is not very sensitive to the initial strontium ratio ($^{87}Sr_0/^{86}Sr$). A figure often used for this initial ratio is 0.704, an average of ratios found in recent volcanic rocks that have presumably come directly from the Earth's mantle. On the assumptions that this choice of an initial ratio is reasonable and that the mineral has remained a closed system with respect to Sr and Rb since its formation, Eq. (10-10) gives the mineral's age. Ages so calculated are especially useful for Rb-rich minerals (lepidolite, muscovite, biotite, K-feldspar) of Paleozoic age or older. Rb forms no minerals of its own, but substitutes extensively for potassium because of the close similarity in properties between Rb^+ and K^+.

Unfortunately the assumption of a closed system, i.e., that a particular mineral has neither gained nor lost Rb or Sr, is often not valid because some migration of one or both elements has been caused by later heating of the rock. It is often safer to assume that the entire rock has remained a closed system, and thus to use rock analyses rather than mineral analyses for determining ages. In this calculation, however, we can no longer choose an arbitrary low value for ($^{87}Sr_0/^{86}Sr$), because in the whole rock the concentration of Rb is small and the value of the present ratio ($^{87}Sr/^{86}Sr$) may not be very different from the initial ratio. We resort, then, to the plotting of an isochron by obtaining isotope analyses from several minerals in a rock specimen or from several rock specimens in a given suite. If, for example, we are examining a suite of igneous rocks which on geologic grounds we know to be of the same age and initial strontium ratio ($^{87}Sr_0/^{86}Sr$), different specimens will contain differing amounts of ^{87}Sr and ^{87}Rb, simply because the amounts of Rb originally present in the various minerals were different. Hence we can plot $^{87}Sr/^{86}Sr$ against $^{87}Rb/^{86}Sr$, and we should obtain a straight line because Eq. (10-9) has the form $y = ax + b$ when t is constant. The line can be extrapolated back to the $^{87}Sr/^{86}Sr$ axis to give the initial value of this ratio (Fig. 10-1). Furthermore the slope of the line is equal to $e^{\lambda t} - 1$, and from the slope the value of t can be calculated.

Uranium–Thorium–Lead Dates

Natural uranium consists chiefly of the isotope ^{238}U, with a small amount (about 0.72%) of ^{235}U. Present in addition is a tiny quantity of ^{234}U, formed as one of the steps in the decay of ^{238}U. Both ^{238}U and ^{235}U decay by series of short-lived intermediate radioactive isotopes to form stable isotopes of lead, ^{206}Pb from ^{238}U and ^{207}Pb from ^{235}U. Once U is incorporated in a mineral and decay is well under way, each intermediate substance decays as fast as it forms, so that for dating purposes (for times greater than about 1 million years) we may disregard the intermediates and write the overall decay schemes as

$$^{238}\text{U} \rightarrow {}^{206}\text{Pb} + 8\text{He} + 6\beta \tag{10-11}$$

and

$$^{235}\text{U} \rightarrow {}^{207}\text{Pb} + 7\text{He} + 7\beta. \tag{10-12}$$

The helium represents the alpha particles and β the beta particles (electrons) emitted in some of the decay steps. For example, the overall decay of one atom of ^{238}U [Eq. (10-11)] produces one atom of ^{206}Pb through emission of 8 alpha particles and 6 beta particles. The first reaction, as noted previously, is very slow (half-life of $^{238}\text{U} = 4.51 \times 10^9$ yr); the second reaction is faster (half-life of $^{235}\text{U} = 0.71 \times 10^9$ yr). Most of the Earth's original ^{235}U has vanished after 4.55 billion years of planetary history, but about half of the original ^{238}U remains.

Uranium-bearing minerals generally contain at least a little ^{232}Th, which also decays through a series of intermediates to form an isotope of Pb:

$$^{232}\text{Th} \rightarrow {}^{208}\text{Pb} + 6\text{He} + 4\beta. \tag{10-13}$$

Thus complete analysis of a U mineral can in principle provide data for three independent age determinations: ^{238}U–^{206}Pb, ^{235}U–^{207}Pb, and ^{232}Th–^{208}Pb. The helium produced in all three reactions would seemingly be the basis for still another age estimate, but the helium method has proved unreliable because the gas escapes from minerals too readily.

The amounts of Pb, U, and Th are customarily expressed as ratios of the isotopes to the nonradiogenic isotope of lead, ^{204}Pb. Thus Eq. (10-9) for the decay of ^{238}U to ^{206}Pb is written

$$\underbrace{\left(\frac{^{206}\text{Pb}}{^{204}\text{Pb}}\right)}_{\text{measured}} = \underbrace{\left(\frac{^{206}\text{Pb}_0}{^{204}\text{Pb}}\right)}_{\text{initial}} + \underbrace{\left(\frac{^{238}\text{U}}{^{204}\text{Pb}}\right)(e^{\lambda t} - 1)}_{\text{produced}} \tag{10-14}$$

and the expression corresponding to Eq. (10-10) is

$$t = \frac{1}{\lambda} \ln \left(\frac{\left(\dfrac{^{206}\text{Pb}}{^{204}\text{Pb}}\right) - \left(\dfrac{^{206}\text{Pb}_0}{^{204}\text{Pb}}\right)}{\left(\dfrac{^{238}\text{U}}{^{204}\text{Pb}}\right)} + 1 \right). \tag{10-15}$$

Similar equations can be set up for the other two pairs of isotopes, and from each of the three equations an age can be calculated—provided that a reasonable value for the amount and isotopic composition of any lead initially incorporated in the uranium mineral can be estimated.

To minimize possible inaccuracy due to initial lead content, it is desirable to use a mineral which contains substantial amounts of U and Th but is known to exclude most Pb during its crystallization. The mineral that best fits this description is zircon, and many U–Th–Pb dates are obtained using zircon separated from igneous rocks. Other minerals that give satisfactory ages include uraninite, titanite, apatite, and monazite.

The three independent ages calculated for a sample of zircon (or other mineral) should agree, provided that no U, Th, or Pb has been added to or subtracted from the mineral during its history. The ages often do not agree, suggesting that the system has not remained entirely closed. In this case a possible recourse is to calculate an age simply from the ratio of ^{207}Pb to ^{206}Pb. These two isotopes are produced at different rates from their parents, so that the ratio ^{207}Pb/^{206}Pb increases with time at a rate that can be calculated from the equations for the two U isotopes. This ratio is not sensitive to loss of Pb, because the two isotopes behave the same chemically and would be lost at the same rate. So if the discrepancy in calculated ages is due to loss of Pb, an age calculated from ^{207}Pb/^{206}Pb should be closer to the true age than any of the others.

Instead of the simple ratio of amounts of the two radiogenic isotopes, it is customary to use the ratios

$$^{206}\text{Pb}^*/^{238}\text{U} \quad \text{and} \quad ^{207}\text{Pb}^*/^{235}\text{U},$$

where ^{206}Pb* is the amount of radiogenic ^{206}Pb produced from ^{238}U (in other words, ^{206}Pb $-$ ^{206}Pb$_0$), and ^{207}Pb* is the amount of radiogenic ^{207}Pb produced from ^{235}U. These two quantities increase with time at different rates,

$$\left(\frac{^{206}\text{Pb}}{^{238}\text{U}}\right) = e^{\lambda_1 t} - 1 \qquad \left(\frac{^{207}\text{Pb}}{^{235}\text{U}}\right) = e^{\lambda_2 t} - 1$$

(where λ_1 is the decay constant of ^{238}U and λ_2 the decay constant of ^{235}U), and the increase may be displayed by plotting one ratio against the other (Fig. 10-2). For an ideal system in which no Pb is lost, the calculated ages would be concordant, and the curve representing the ages is called a *concordia* (Wetherill, 1956). If Pb has been lost at some time in the mineral's history, so that the calculated ages are discordant, a point for the two ratios would fall below the concordia. Lead loss would presumably not be the same for all samples of the mineral, so that several points can be plotted showing ratios for the different samples. If the loss of Pb has occurred in a single episode, the points should lie on a straight line (called *discordia*). If this line is extended until it intersects the concordia, the upper intersection gives the age of the mineral.

Still another method of handling U–Th–Pb data is to construct isochron diagrams for whole-rock samples, like the isochrons for the Rb–Sr system in Fig. 10-1. Such diagrams for the U–Pb pairs are generally not useful because U is too easily lost by oxidation and leaching from near-surface rock samples. Th and Pb are

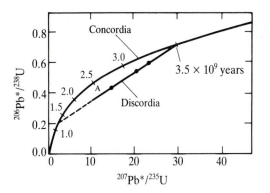

FIGURE 10.2

Concordia diagram for U–Pb. The concordia curve is the locus of points representing $^{206}Pb^*/^{238}U$ and $^{207}Pb^*/^{235}U$ for concordant U–Pb systems (systems that have remained closed to U and its daughters since their formation). The ages of such systems are indicated by the numbers along the curve (in billions of years); thus the two ratios for any concordant system 2.5×10^9 years old would be represented by point A. The discordia line is drawn through points plotted from (hypothetical) isotope analyses for samples of a mineral that have lost varying amounts of Pb. Extension of the line upward to its intersection with concordia gives the age of the mineral, in this case 3.5×10^9 years. The lower intersection is harder to interpret; if Pb loss occurred at only one brief period in the mineral's history, the point gives the time of that event, but the history is seldom so simple. (After Wetherill, 1956).

less subject to loss, so that isochrons for Th–Pb and $^{207}Pb/^{206}Pb$ have proved more satisfactory for age calculations.

Potassium–Argon Dates

The isotope ^{40}K, which makes up a minute fraction (about 0.01%) of naturally occurring potassium, undergoes several kinds of radioactive decay. The two principal kinds are decay to ^{40}Ca by electron emission (about 89% of the ^{40}K nuclei) and decay to ^{40}Ar by electron capture (about 11%). The decay to ^{40}Ca has little importance in dating because ^{40}Ca is the common isotope of calcium, and enough ordinary Ca is present in most K minerals that the tiny amount added by radioactive decay would be undetectable. The decay to ^{40}Ar, on the other hand, is the basis of one of the most common methods of age determination. The great advantage of the method is its applicability to very common and abundant minerals, since potassium is one of the major elements in the Earth's crust. Its major drawback is the fact that the radiogenic product is a gas, which may in part escape from the mineral in which it forms.

The total decay constant for ^{40}K may be written as the sum of the constants for the decay paths to Ca and Ar, and Eq. (10-5) for the decay to Ar alone is then

$$^{40}Ar = \, ^{40}Ar_0 + \frac{\lambda_a}{\lambda} \, ^{40}K(e^{\lambda t-1}), \qquad (10\text{-}16)$$

where λ is the total decay constant and λ_a the constant for decay to ^{40}Ar. Commonly used values are 5.305×10^{-10} yr^{-1} for λ and 0.585×10^{-10} yr^{-1} for λ_a, giving 0.110 for λ_a/λ. Since most potassium minerals have no original Ar, the value of ^{40}Ar$_0$ is generally zero, and the equation simplifies to

$$^{40}\text{Ar} = 0.110 \, ^{40}\text{K}(e^{\lambda t} - 1). \tag{10-17}$$

The equation giving the age explicitly [from Eq. (10-6)] is then

$$t = \frac{1}{\lambda} \ln\left(\left(\frac{^{40}\text{Ar}}{^{40}\text{K}}\right)\left(\frac{1}{0.110}\right) + 1\right), \tag{10-18}$$

and the necessary measurements are the amounts of ^{40}Ar and ^{40}K, generally determined on different samples of the potassium mineral. The age of a mineral calculated from Eq. (10-18) is the true age only if (1) no radiogenic Ar has escaped from the mineral since its formation, (2) no excess ^{40}Ar was present in the mineral when it was formed and none was introduced later, and (3) no K was added or removed. The first of these provisos is generally the most critical.

Escape of Ar is possible from any mineral, but experience has shown that minerals differ greatly in their ability to retain the gas depending on different rates of diffusion of Ar through the crystal lattice and on the rate of cooling. Among common minerals the one that generally gives the most reliable ages is hornblende—which at first sight seems strange, because the K content of hornblende is normally low. Biotite and muscovite retain Ar fairly well, but ages determined for micas are often somewhat lower than for hornblende in the same rock because Ar more readily diffuses through the layer silicate structure. Sanidine from volcanic rocks is highly retentive because of the rapid cooling rate, but K-feldspar from plutonic rocks, which cool very slowly, is not. Escape of Ar from any mineral is hastened by high temperature; the temperature at which rapid escape begins varies from one mineral to another, but for most minerals is of the order of a few hundred degrees or more. If an episode of metamorphism has temporarily maintained temperatures high enough for all Ar to escape from a rock, the ^{40}K geochronometer will be "reset," and age determinations will record the time of metamorphism rather than time of formation.

In addition to dating minerals, the K–Ar method can be used for dating some volcanic rocks, particularly basalts. The fine-grained aggregate of plagioclase and pyroxene in basalt contains only a little K, and the Ar produced by its decay is effectively retained by the two minerals. K–Ar ages for basalts have proved especially useful in dating times of reversal of the Earth's magnetic field. Lavas more felsic than basalt, especially if they contain devitrified glass, give less reliable ages because of Ar loss.

A useful variant of the K–Ar method depends on irradiation of a sample with neutrons, which convert some of the ^{39}K to ^{39}Ar, followed by measurement of the ^{40}Ar/^{39}Ar ratio with a mass spectrometer. The ^{39}Ar in effect is a measure of the potassium content, but no actual number for K need be found. One requires only a comparison of the ^{40}Ar/^{39}Ar ratio for the sample whose age is to be determined with the same ratio for samples of known ages which have been similarly irradiated, and the age is given by interpolation. This argon–argon procedure has advantages over the standard K–Ar method in that both isotopes are determined by a single measurement on the same sample, and the size of the sample can be much smaller. It is widely used for materials available in only small amounts. Application of the $^{40}Ar/^{39}Ar$ method of dating to argon partially released by stepwise heating of irradiated samples is useful in the analysis of heterogeneous minerals with complex metamorphic histories.

Samarium–Neodymium Dates

The decay of ^{147}Sm to ^{143}Nd provides a method of age determination often used in dating igneous rocks. Both elements belong to the rare-earth group, hence have very similar chemical properties and are not likely to be differentially removed by the accidents of weathering or hydrothermal alteration that the rock may undergo. Whereas a Rb/Sr date or a K/Ar date may be influenced by such later alteration, Sm–Nd analysis can still give a reliable figure for the age of original crystallization.

Equations may be set up for this system similar to those for Rb and Sr, using the nonradiogenic ^{144}Nd as the reference isotope:

$$\underbrace{\left(\frac{^{143}Nd}{^{144}Nd}\right)}_{\text{measured}} = \underbrace{\left(\frac{^{143}Nd_0}{^{144}Nd}\right)}_{\text{initial}} + \underbrace{\left(\frac{^{147}Sm}{^{144}Nd}\right)}_{\text{produced}}(e^{\lambda t} - 1) . \qquad (10\text{-}19)$$

The value of λ for ^{147}Sm is 6.54×10^{-12} yr^{-1} and the half-life 1.06×10^{11} yr. The initial ratio $(^{143}Nd/^{144}Nd)_0$ can be obtained from isotope analyses of several specimens and construction of an isochron using Eq. (10-19), in the same way that initial Sr ratios are found by constructing an isochron from Eq. (10-9). Dates obtained using Sm–Nd analyses have been especially useful for mafic rocks, in which zircon (for U–Pb dates) is scarce or absent, and in which Rb and Sr concentrations are so low that Rb–Sr dates are often unreliable.

Carbon-14 Dates

Atoms of ^{14}N in the atmosphere are converted to ^{14}C by reaction with neutrons from cosmic-ray collisions. Most of the ^{14}C is quickly oxidized to CO_2, which disperses through the atmosphere. The ^{14}C is radioactive, decaying with a half-life of 5730 years (to ^{14}N, by emission of an electron), and its amount in air is maintained at a constant value by the balance between rates of formation and decay. Carbon dioxide is consumed by plants in photosynthesis, and animals acquire ^{14}C both by eating plant material and by absorbing CO_2 from air and water. Thus organisms maintain a steady-state concentration of ^{14}C as long as life processes continue. When an organism dies the interaction with the atmosphere ceases, and the ^{14}C it contained at the time of death steadily decreases. The concentration of this isotope in the carbon of dead organic matter is therefore a measure of the time since death occurred. More broadly, the carbon in any compound formed in contact with air and then kept from further reaction with CO_2 can be dated by the proportion of ^{14}C in its makeup.

This method of age measurement differs from other isotope-based methods in that the radioactivity of a particular isotope is measured rather than the ratio of concentrations of a parent and daughter. If the radioactivity of C in organic matter (or another carbon compound) in equilibrium with atmospheric CO_2 is represented by A_0, and the activity of material that has not interacted with air for t years is represented by A, the equation of decay is

$$A = A_0 e^{-\lambda t}. \tag{10-20}$$

The A and A_0 are measured as numbers of disintegrations per minute per gram of carbon. Rearrangement of the equation and insertion of a numerical value for λ gives a simple expression for t in years:

$$t = 19,035 \log_{10}(A_0/A). \tag{10-21}$$

Because of the short half-life of ^{14}C, its activity in dead organic matter remains measurable for only about 70,000 years. For dates within this period, dates that are especially important for archeological and anthropological studies, the method has proved of great value.

Dates determined by ^{14}C are subject to minor corrections of several sorts. The dates can be valid only if the intensity of cosmic rays striking the upper atmosphere has remained constant (so that A_0 may be considered constant); since the cosmic-ray flux comes in large part from the Sun, whose activity is known to fluctuate, and since it is influenced by the Earth's magnetic field, which is by no means steady, the assumption of a constant rate of production of ^{14}C cannot be strictly correct. By comparing ^{14}C dates with dates determined in other ways it is possible to estimate the variations in cosmic-ray intensity over the last few thousand years, hence to correct ^{14}C dates for this period. Radiocarbon dates would also be affected by marked changes in the amount of organic activity on the Earth's surface, or by the introduction of unusual amounts of inert CO_2 into the atmosphere. Such changes from natural causes would presumably be small, but the marked increase of "dead" CO_2 in air over the past century due to the burning of fossil fuel has produced a detectable decrease in the amount of ^{14}C incorporated by organisms during the twentieth century. On the other hand, since about 1945 considerable amounts of ^{14}C have been added to the atmosphere by the explosion of nuclear devices. Thus radiocarbon dates are in principle simple to obtain, but making them precise requires a good deal of adjustment.

10-3   RADIOGENIC ISOTOPES: PETROGENESIS

Besides their use in dating rocks and minerals, radiogenic isotopes have found wide application to other geologic questions. An element like strontium or neodymium is a mixture of a radiogenic isotope and one or more nonradiogenic isotopes, and the amount of the radiogenic isotope in the world as a whole has increased over geologic time. The ratio of radiogenic to nonradiogenic isotopes in any given sample, however, will not have a fixed value, because it depends on the history of the sample—how much of the radioactive precursor was present in the sample originally, and how much of the radioactive element or Sr or Nd has been added to or removed from the sample at later times. Differences from the average ratio are small (as is indicated by the fact that atomic weights for the elements can be given in standard tables to two decimal places). But the differences are easily detectable, and isotope ratios for these elements have proved to be useful tools in studying the history of rocks and minerals that contain them.

Strontium

The isotope geology of Sr concerns the increases over time in the amount of ^{87}Sr produced by the radioactive decay of ^{87}Rb. The nonradiogenic isotope commonly used for comparison is ^{86}Sr, so that changes in composition are expressed as changes in the ratio ^{87}Sr/^{86}Sr.

To study Sr compositions, we need to know the initial value of ^{87}Sr/^{86}Sr in the material that made up the original Earth. This figure must be obtained indirectly, because the Earth's original substance is nowhere preserved. The best approximation we have to such material is meteorites—although even meteorites may have had a complex history, so that their validity as samples of the original solar nebula is not entirely without question. Some meteorites, however, have an extremely low content of Rb, so that their Sr cannot have changed greatly since their minerals crystallized from a melt. Analyses of many such meteorites agree on a figure of 0.699 as the most probable ^{87}Sr/^{86}Sr ratio for the Earth's original Sr. In effect, this number is the Sr isotope ratio for a hypothetical uniform reservoir (UR) of planetary material from which the Earth's mantle and all its derivative products were formed.

On the general assumption that the bulk of the Earth's crust differentiated from the mantle fairly early in geologic time, we can expect that the behavior of Rb and Sr during the differentiation would be markedly different. The ion Rb^+ is large and singly charged, hence would be concentrated, like its close relative K^+, in the relatively silica-rich differentiates that formed the crust. The smaller and doubly charged Sr^{2+} resembles Ca^{2+} in its geochemical behavior, and accordingly would distribute itself more evenly between crust and mantle. The greater concentration of Rb in the crust means that production of ^{87}Sr has been faster in crust material than in the mantle, and samples of rock that has been part of the crust for most of its history should in general have higher ^{87}Sr/^{86}Sr ratios than samples that come from long residence in the mantle. This difference in isotope ratios, then, gives us a means of distinguishing igneous rocks that have formed by partial melting of crustal rocks from those that have their origin in partial melting of mantle material.

The present ^{87}Sr/^{86}Sr ratio for mantle rock can be estimated by analyses of xenoliths or of recent basalts and gabbros from oceanic environments, where direct origin from the mantle can be assumed and contamination by continental material would be absent or nearly so. Most such analyses lie in the range 0.702 to 0.706, with an average of about 0.704; the fact that the values cover a considerable range is an indication that mantle material may not be entirely homogeneous. To produce a ratio of 0.704 from the original 0.699 during the 4.55×10^9 yr of Earth history would require an average Rb concentration in mantle rock of about 0.027%. Whether this amount has remained approximately constant through most of geologic time, or has decreased as additional crust was formed, is uncertain. For many purposes a simple linear interpolation between 0.699 and 0.704 gives a reasonable estimate of the Sr isotope ratio in the mantle at any time in the geologic past.

To draw inferences about the parentage of other igneous rocks, their isotope ratios are compared with these interpolated values for mantle material (Fig. 10-3). The ratios needed are the *initial* $^{87}Sr/^{86}Sr$ values, the values at the time the rocks crystallized, rather than present-day values; these values are found by constructing isochrons, as explained earlier (Fig. 10-1). If, for a suite of igneous rocks, the initial value is close to or below the corresponding mantle value (between 0.699 and 0.704), it is a reasonable inference that the magma was derived by differentiation or partial melting of mantle material (or of crustal material not long separated from the mantle). If it is much higher than mantle values, a conclusion seems safe that the magma originated by melting of crustal material, or was contaminated by assimilating crustal rocks after it was formed in the mantle.

Thus a study of Sr isotope ratios can shed light on the old geologic question as to the origin of granite. Initial values for various granites range from numbers identical with mantle values to numbers higher than 0.730. The high values clearly indicate derivation of the granite magma by partial melting of crustal material, and

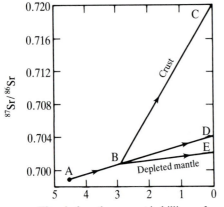

Time before the present in billions of years

FIGURE 10-3
Changes in strontium isotope ratios over time. Point A represents an $^{87}Sr/^{86}Sr$ ratio of 0.699 for a hypothetical uniform reservoir of meteoritic material from which the Earth was formed. Line AD shows approximate changes in this ratio in the part of the uniform reservoir that became the Earth's mantle. (The line is not strictly accurate, because the mantle's composition changed over time as crustal material was extracted from it.) If a batch of crustal material with an Rb/Sr ratio of 0.15 was separated from the mantle about 2.9×10^9 years ago, line BC shows the increase in its Sr isotope ratio—provided there was no later addition or subtraction of either element. Line BE shows the change of the ratio in the Rb-depleted part of the mantle from which the crustal material was derived. (After Faure, 1986.)

the low values suggest origin by fractional melting of the mantle or of material with only a short history in the crust. A great many values lie between the extremes, and for these a variety of interpretations is possible. Granitic rocks of the Sierra Nevada, for example, show initial values ranging from 0.703 to 0.709. The low values could represent material directly from the mantle, and the higher numbers suggest varying proportions of crustal material in the original mixture that melted to form the granites (Kistler and Peterman, 1973). Since initial ratios for different crustal rocks cover a wide range, the amount of crustal material involved in forming the magma cannot be estimated without additional information. We will return to this question of the origin of granites in a later section (Sec. 17-10).

Besides helping to unravel the history of igneous rocks, Sr isotope ratios are useful as tracers for identifying sources of other geologic materials. A good example is the metal-bearing hot brine in geothermal wells near the Salton Sea in southern California. Originally the surprisingly high metal content (more than 80 ppm of Pb and 500 ppm of Zn) was thought to be derived from rhyolitic magma at depth, related to small rhyolite domes in the vicinity. Isotope analyses for dissolved Sr and Pb in the brine, however, showed a much closer resemblance to ratios in the sediments through which the brine circulated than to those in the rhyolite. Very probably the brine derives its heat from the underlying magma, but at least a large part of the metal content comes from sediments rather than igneous material (Doe et al., 1966).

Neodymium

The change in amount of radiogenic neodymium over time is parallel to that of radiogenic strontium, and can be symbolized in much the same way. Here the radiogenic isotope is ^{143}Nd (formed from ^{147}Sm), the reference isotope is ^{144}Nd, and the ratio whose increase we follow is ^{143}Nd/^{144}Nd. Again we start with an estimate of the ratio at the Earth's beginning derived from analyses of meteorites, and again we assume that the Earth's original substance was a uniform reservoir with the composition of an average chondrite ("chondrite uniform reservoir," or CHUR), and ask how the ratio would have changed in such a reservoir during 4.55×10^9 yr. Just as for Sr, estimates of the Nd ratio in the original reservoir differ slightly because of variations in the meteorites selected by different analysts. A commonly used value for the ratio at the Earth's beginning is 0.5068, and the present-day value in the hypothetical reservoir would be 0.5126.

During the separation of the original meteoritic material into the Earth's mantle and crust, the Nd isotope ratio would have changed in a direction opposite to that of the Sr ratio. The Sr ratio grows larger in crustal material because the radioactive parent, Rb, tends to be more concentrated in silica-rich crustal material; for Nd, it is the radiogenic product rather than the parent that is so concentrated. In other words, Nd goes more readily than Sm into a felsic differentiate—an effect of the greater ionic radius of Nd, as we will see later (Secs. 17-11 and 20-3). Thus a diagram for Nd analogous to Fig. 10-3 for Sr shows that samples removed from the reservoir at a particular time undergo similar changes in rate of growth of the

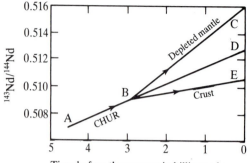

FIGURE 10-4

Changes in neodymium isotope ratios over time. Point A represents a $^{143}Nd/^{144}Nd$ ratio of 0.5068 for a hypothetical uniform reservoir of chondritic meteoritic composition (CHUR) at the Earth's beginning. Line AD shows changes in the average ratio for the reservoir over geologic time, to a present value of 0.5126. Line BE shows changes in the ratio for a sample of crustal material extracted from the mantle 2.9×10^9 years ago, and line BC the corresponding changes in the part of the mantle from which the Nd-enriched crust was derived. (After Faure, 1986).

isotope ratio, but in opposite directions—faster growth in mantle-like material, slower growth in crustal material (Fig. 10-4).

Evidently Nd isotope ratios can be used, like the Sr ratios, to draw inferences about the kind of material from which an igneous magma was formed, except that the conclusions are seemingly reversed. If the isotope ratio measured for a given igneous rock is high, the rock most probably was formed from a melt derived fairly directly from the mantle; if the ratio is low, a source in crustal material is indicated. Inferences based on high and low Sr ratios, of course, are just the opposite. Such conclusions from either element are at best only tentative, because the history of an igneous rock is often long and complicated and may involve other processes that can affect isotope ratios.

The relation of isotope ratios to igneous rock history can be conveniently displayed on a diagram showing changes in both Sr and Nd ratios (Fig. 10-5). Coordinates of such a diagram may be either the ratios $^{143}Nd/^{144}Nd$ and $^{87}Sr/^{86}Sr$ or the derivative quantities

$$\varepsilon_{Sr} = \left(\frac{\left(\frac{^{87}Sr}{^{86}Sr} \right)_m}{\left(\frac{^{87}Sr}{^{86}Sr} \right)_R} - 1 \right) \times 10^4 \qquad \varepsilon_{Nd} = \left(\frac{\left(\frac{^{143}Nd}{^{144}Nd} \right)_m}{\left(\frac{^{143}Nd}{^{144}Nd} \right)_R} - 1 \right) \times 10^4$$

where $(^{87}Sr/^{86}Sr)_m$ is the measured ratio in a rock and $(^{87}Sr/^{86}Sr)_R$ is the ratio that would exist in the uniform reservoir at present—and similarly for the Nd ratios. Thus zero on either epsilon scale represents samples whose isotope ratios are identical with ratios that would have been reached in a uniform reservoir of

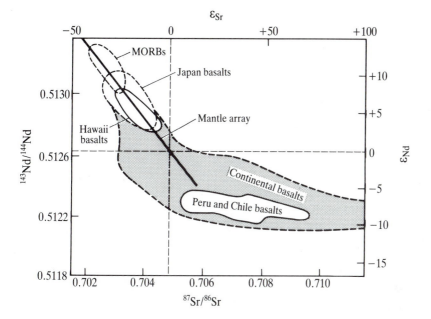

FIGURE 10-5

The "mantle array." For basalts with a presumed fairly direct origin in the mantle, an increase in [87]Sr should be accompanied by a decrease in [143]Nd, because in any liquid derived from the mantle the radioactive parent of one (Rb) should be higher, and of the other (Sm) lower, than in the original material. This relation is shown by the clustering of analyses of mantle-derived basalts (mid-ocean ridge basalts and oceanic island basalts) along a line descending from the upper left corner of the diagram. Basalts whose magma was derived from or mixed with crustal material, on the other hand, would show widely variable isotope compositions. Typical of these are continental basalts, whose analyses cover a broad swath including parts of both the upper left quadrant and the lower right quadrant. Outlines on the diagram show only a meager sampling of the many kinds of basalt; recorded analyses cover much of the diagram, but are scarce in the upper right and lower left quadrants. The position of an analysis on the diagram gives information, or at least a basis for speculation, about the origin of the magma. (On this kind of diagram, axes may be either isotope ratios or the derived ε values; here the former are shown on the bottom and left-hand sides, the latter on the top and right-hand sides.) (After Faure, 1986.)

meteoritic material after 4.55×10^9 yr. The epsilon scales have the advantage that positive values are a direct indication of enrichment in the radiogenic isotope, negative values a deficiency.

Rocks whose analyses plot in the upper left quadrant of Fig. 10-5, then, have high [87]Sr/[86]Sr ratios and low [143]Nd/[144]Nd ratios, characteristics of the mafic residual mantle material after separation of crustal constituents. Accordingly, the analyses that appear here are for basalts with a direct origin in the mantle, rocks from the mid-ocean ridges and some oceanic islands. The upper right quadrant includes rocks with high values of both [87]Sr/[86]Sr and [143]Nd/[144]Nd, a combination that would not be expected from normal igneous processes—and measured analyses in this quadrant are very few. Similarly for the lower left quadrant, where both radiogenic isotopes would be low, and rocks whose analyses appear here are also scarce. The lower right quadrant, with high $^{87}Sr/^{86}Sr$ and low $^{143}Nd/^{144}Nd$, has abundant analyses with a variety of isotopic compositions, as would be expected for igneous rocks formed from or contaminated by crustal material. The diagonal line in the upper left quadrant, around which basalts directly related to the mantle cluster, is often called the *mantle array*, and the position of basalt analyses elsewhere on the diagram relative to this line provides a basis for inferences about the origin and history of the corresponding magmas.

Lead

For lead also, it would seem that the increase of its radiogenic isotopes over time should make it possible to use the ratios $^{206}Pb/^{204}Pb$, $^{207}Pb/^{204}Pb$, and $^{208}Pb/^{204}Pb$ as an aid in identifying source materials for igneous rocks. The interpretation of these ratios is less straightforward, however, because of greater complications in the chemistry of lead and its radioactive parents during magmatic processes.

As for Sr and Nd isotopes, the initial isotopic composition of Pb incorporated into the primitive Earth can be estimated from meteorite analyses. The troilite (FeS) of iron meteorites contains Pb but practically no U or Th, so that its Pb presumably has the original composition of Earth and meteoritic material unmodified by later addition of radiogenic isotopes. Analyses of such Pb do indeed show lower isotope ratios than any terrestrial leads, and are commonly accepted as representing the primeval isotope distribution.

Stony meteorites, on the other hand, contain U and Th as well as Pb, and radioactive decay since the meteorites were formed has increased the amounts of the heavier Pb isotopes. Since the rates of decay are well-known, data from the two kinds of meteorites permit calculation of the time when the meteorites, and presumably also the Earth, were formed. This calculation is particularly important, because it leads to the generally accepted figure of $4.55 \pm 0.05 \times 10^9$ yr for the Earth's age. The calculation involves plotting an isochron (Fig. 10-6) for the $^{206}Pb/^{204}Pb$ and $^{207}Pb/^{204}Pb$ ratios of different meteorites, and finding the age from the slope of this line (Patterson, 1956). Support for the assumption of a common age for Earth and meteorites comes from the fact that similar ratios for an average sample of modern Pb (obtained from recent marine sediments, which should represent a well-mixed sample of Earth materials) locate a point that lies on the isochron (Fig. 10-6).

The $^{207}Pb/^{204}Pb$ and $^{206}Pb/^{204}Pb$ ratios can be used for determining ages not only of meteorites but also of ordinary lead in Earth materials. The Pb in a deposit of galena, for example, has presumably been out of contact with U or Th since the deposit was formed; before that time we may imagine that the Pb was disseminated in rock material, either in the mantle or in the lower crust, along with U and Th, and that the amounts of the three radiogenic isotopes increased steadily from radioactive decay. When the Pb was concentrated and segregated to form the ore deposit, its isotopic evolution ceased. The relation of the Th-derived ^{208}Pb to the other two

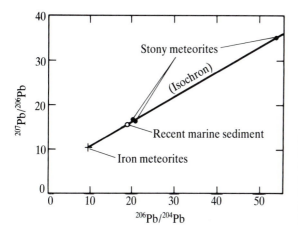

FIGURE 10-6

Lead isochron for analyses of meteorites and recent marine sediment. The slope of the line indicates an age of $4.55 \pm 0.07 \times 10^6$ years for the meteorites. (After Patterson, 1956.)

isotopes (hence ages from U/Pb and Th/Pb ratios) would depend on the relative amounts of U and Th in the source material; but the relation between ^{207}Pb and ^{206}Pb, since both come from isotopes of the same element, would depend *only* on the length of time decay had been going on (hence on the age of the deposit), and *not* on the amount of U present. In other words, the decay of ^{235}U and ^{238}U in the course of Earth history has produced a steadily increasing ratio of ^{207}Pb to ^{206}Pb, which at any one time should be the same for any U-bearing material. Thus we have here a possible method of dating an ore deposit simply from the isotopic composition of its Pb, without any reference to radioactive parents. We need only measure ^{207}Pb/^{206}Pb and ^{206}Pb/^{204}Pb in an ore sample, subtract from them the ratios in primeval Pb (which would make up part of the ore) to obtain the amounts due solely to radioactive decay, and then compare these ratios with the calculated growth curve to find the date when the ore was formed.

This procedure can give an accurate age only under rigidly prescribed conditions. In the source material from which the Pb came, U must have been quietly decaying since the Earth was formed, without any disturbance that would add to or decrease the amount of either U or Pb; then in a short time the Pb must have been separated cleanly to form the ore; and afterward the Pb must have remained isolated from any contact with U or additional Pb. Obviously such special conditions are seldom realized in nature. Because this method of dating depends on an idealized model, the numbers obtained are *model ages*, which may or may not have geologic significance.

Out of the many ore deposits for which Pb-isotope ages have been obtained, only a few give model ages that are probably close to real ages. Indications that conditions of the model were approximately fulfilled are constancy of isotope ratios in different samples from a deposit, agreement of ages calculated for different samples, and general agreement of the model age with ages calculated in other ways for the deposit and its surroundings. Pb from such deposits is often referred to as "ordinary lead," in contrast to "anomalous lead" in deposits for which the model ages are in apparent conflict with geologic evidence.

Anomalous leads may show model ages either greater or less than the true age. If the age is apparently too high, meaning a $^{207}Pb/^{206}Pb$ ratio that is too low, an obvious possibility is that the Pb has been remobilized from an earlier accumulation, so that the calculated age represents the time of this earlier event. Low ages are harder to explain, but suggest changes in amounts of U and Pb in the source material at some time, or more probably at several times, before the Pb was segregated into the ore.

Study of Pb isotope ratios is not limited to the lead of ore deposits, but can be extended to the small amounts of lead present in many ordinary rocks. Such studies have been especially provocative for the Pb in recent volcanic rocks from oceanic islands: the unexpected variation in isotope ratios both from one island to another and among the rocks of a single island is good evidence for considerable heterogeneity of upper mantle materials and for a multistage history of much of the lead.

10-4   STABLE ISOTOPES OF LOW ATOMIC NUMBER

Elements with isotopes suitable for dating, except for carbon, show practically no isotope separation as a result of ordinary geologic processes. We may confidently assume that U or Sr removed from a rock by weathering or hydrothermal alteration has the same proportion of isotopes as in the rock from which it came. Again, the reliability of the $^{207}Pb/^{206}Pb$ method of dating altered rocks depends on the fact that the ratio of these isotopes does not change in response to any geologic accidents that may befall the original rock material. For a few elements of low atomic number, however, whose stable isotopes have a large *proportional* difference in atomic mass, detectable changes in isotope ratios may result from very ordinary geologic processes (*isotopic fractionation*). By accurate measurement of isotope ratios, conclusions can be drawn about the conditions and reactions that produced the changes.

The difference in properties between two isotopes that may lead to slight separation is largely a result of the different *vibrational frequencies* of heavy and light atoms in a molecular or crystal structure. Atoms of a light isotope vibrate with higher frequencies, hence in general are less strongly bonded to other atoms, than atoms of a heavy isotope. The differences in bond strength are appreciable only for isotopes whose atoms have a large relative difference in mass: thus detectable separation might be expected for hydrogen and its compounds, since the stable hydrogen isotopes differ in atomic mass by a factor of 2 (^{1}H and ^{2}H), but would not be expected for tungsten, whose heaviest and lightest naturally occurring isotopes have a relative mass difference of only 1.03 (^{180}W and ^{186}W). This means that separation of stable isotopes is important in geologic environments only for elements of low atomic number. The most important of these elements are:

Hydrogen: Two stable isotopes, atomic masses 1 and 2, symbolized either ^{1}H and ^{2}H or H and D (the D standing for deuterium). A third isotope, ^{3}H or T (tritium), also exists in nature, but it is radioactive with a short half-life (12.26 yr).

Carbon: Two stable isotopes, ^{12}C and ^{13}C, plus the radioactive ^{14}C.

Oxygen: Three stable isotopes, with atomic masses 16, 17, and 18. The proportion of ^{17}O is very small, and ordinarily only variations in ^{16}O and ^{18}O are considered.

Sulfur: Four stable isotopes, with atomic masses 32, 33, 34, and 36. Commonly only variations in ^{32}S and ^{34}S are studied, because amounts of the others are small.

Differences in vibrational frequencies of particles of different masses become smaller at higher temperatures, and the separation of isotopes is correspondingly less pronounced. The ratio of isotopes of one of the light elements in a mineral, or the distribution of isotopes between two minerals that have formed at the same time, is therefore a possible measure of geologic temperatures. Some isotope distributions have been found to be characteristic of particular geologic environments or processes, so that measurements of isotope ratios can give clues about sources of minerals and fluids. For these reasons the study of stable isotopes has become a valuable adjunct to many geologic investigations.

Theoretical calculation of the equilibrium distribution of isotopes between different substances is possible for simple molecules, but conclusions about isotope separation are based largely on empirical observations rather than theory. Such experimental data have been obtained in large quantity, so that accurate predictions can be made about the degree of separation to be expected in many geologic situations.

Three mechanisms of isotope separation can be distinguished:

1. Mechanisms depending on physical properties, for example evaporation or precipitation. Evaporation of water, for example, leads to concentration of the light isotopes ^{1}H and ^{16}O in the vapor phase and the heavy isotopes in the liquid, because H_2O molecules containing the light isotopes move more rapidly and thus have a higher vapor pressure. Likewise, quartz precipitated from a solution will be enriched relative to the solution in the heavy isotope of oxygen.

2. Exchange reactions resulting in isotopic equilibrium between two or more substances. For example, if CO_2 containing only ^{16}O is mixed with water containing only ^{18}O, exchange will take place according to the reaction

$$\tfrac{1}{2}C^{16}O_2 + H_2{}^{18}O \rightleftharpoons \tfrac{1}{2}C^{18}O_2 + H_2{}^{16}O \qquad (10\text{-}22)$$

until equilibrium has been reached among the four species. At equilibrium the ratio ^{18}O/^{16}O will be nearly the same in CO_2 and H_2O, but not quite, because bond strengths in the two compounds are different. A geologic example is oxygen isotope exchange between low-^{18}O hydrothermal solutions and feldspar or quartz of a cooling granite pluton:

$$8H_2{}^{16}O + KAlSi_3{}^{18}O_8 \rightleftharpoons 8H_2{}^{18}O + KAlSi_3{}^{16}O_8 \qquad (10\text{-}23)$$

$\phantom{8H_2{}^{16}O + K}{}^{18}O\text{-feldspar}{}^{16}O\text{-feldspar}$

and

$$2H_2{}^{16}O + Si^{18}O_2 \rightleftharpoons 2H_2{}^{18}O + Si^{16}O_2 . \qquad (10\text{-}24)$$

$\phantom{2H_2{}^{16}O + }{}^{18}O\text{-quartz}{}^{16}O\text{-quartz}$

3. Separation depending on reaction rates. The rates of isotope reactions can vary greatly. For many exchange reactions at low temperatures, the rates are slow and equilibrium isotope fractionation is often not reached. The variable separation of isotopes depending on reaction rates is particularly noticeable in reactions catalyzed by bacterial activity. In the bacterial reduction of sulfate ion, for example, the production of sulfide (S^{2-}, HS^-, and H_2S) is faster for the light isotope of sulfur than for the heavy isotope, so that ^{32}S becomes concentrated to various degrees in the sulfide species and ^{34}S is enriched in the residual SO_4^{2-}.

Regardless of mechanism, the extent of isotope separation between two phases A and B can be expressed by a ratio called the *fractionation factor*:

$$\alpha_B^A = \frac{R_A}{R_B}, \qquad (10\text{-}25)$$

where R_A is the ratio of concentrations of heavy to light isotope in phase A ($^{18}O/^{16}O$ in liquid water, for example), and R_B is the same ratio in phase B ($^{18}O/^{16}O$ in water vapor). If equilibrium is established between water vapor and liquid at 25°C, the value of α is about 1.0092. Similar numbers, very slightly greater or very slightly less than 1, are obtained for other examples of isotope separation.

Because differences among such numbers are small and hard to visualize, another kind of symbolism is commonly used to describe isotope separation. The isotope ratio in a given sample is compared with the ratio in a standard by the expression

$$\delta_{heavy} = \frac{(R_{sample} - R_{standard})}{R_{standard}} \times 1000\,‰. \qquad (10\text{-}26)$$

The symbol ‰ stands for *per mil*, or parts per thousand. For water, as an example, the usual standard for both O and H is a sample of seawater, Standard Mean Ocean Water or SMOW; hence ^{18}O for any other water sample is

$$\delta^{18}O = \frac{(^{18}O/^{16}O)_{sample} - (^{18}O/^{16}O)_{SMOW}}{(^{18}O/^{16}O)_{SMOW}} \times 1000\,‰. \qquad (10\text{-}27)$$

A positive value of $\delta^{18}O$ indicates enrichment of a sample in ^{18}O relative to SMOW, and a negative value indicates depletion. Most samples of fresh water have negative values of $\delta^{18}O$ (because the light isotope is concentrated in vapor escaping from the sea), ranging down to about -60 per mil; oxygen in the air has a high positive value,

+23.5‰, and CO_2 in air a still higher value, +41‰. The relation between δ and α is given by the expression

$$\alpha_B^A = \frac{R_A}{R_B} = \frac{\delta^{18}O_A + 1000}{\delta^{18}O_B + 1000}, \tag{10-28}$$

where A and B refer to different samples of an oxygen-containing compound. For example, in the condensation of water vapor to liquid a typical pair of variables might be $\delta^{16}O = -5‰$ for the liquid (1) and $-14‰$ for the vapor (v); the corresponding fractionation factor would be

$$\alpha_v^l = (-5 + 1000)/(-14 + 1000) = 1.0092. \tag{10-29}$$

A few examples of geologic problems in which isotope studies have proved useful are described in the following paragraphs.

Oxygen and Hydrogen

When calcium carbonate precipitates from seawater under equilibrium conditions, exchange of oxygen isotopes between $CaCO_3$ and H_2O results in an equilibrium that can be expressed

$$\tfrac{1}{3}CaC^{16}O_3 + H_2{}^{18}O \rightleftharpoons \tfrac{1}{3}CaC^{18}O_3 + H_2{}^{16}O. \tag{10-30}$$

calcite   calcite

The equilibrium constant for this reaction is

$$K = \frac{\left(a_{CaC^{18}O_3}/a_{CaC^{16}O_3}\right)^{1/3}}{a_{H_2{}^{18}O}/a_{H_2{}^{16}O}}. \tag{10-31}$$

Concentrations, represented with brackets, are commonly used instead of activities in Eq. (10-31) because the ratios of activity coefficients for the substitution of oxygen isotopes in calcite and water are close to unity. The equilibrium constant now becomes

$$K \cong \frac{\left([CaC^{18}O_3]/[CaC^{16}O_3]\right)^{1/3}}{[H_2{}^{18}O]/[H_2{}^{16}O]} = \frac{R_{CaCO_3}}{R_{H_2O}} = \alpha, \tag{10-32}$$

where α is the fractionation factor for isotope distribution between $CaCO_3$ and H_2O [Eq. (10-25)].

The experimentally determined value of α for calcite–water equilibrium [Eq. (10-30)] at 25°C is 1.0286, meaning that calcite is enriched in ^{18}O relative to water when equilibrium has been established. For example, the $\delta^{18}O$ value of calcite in equilibrium with seawater, referred to the seawater standard, is +28.6‰ [Eq. (10-28)]. The fractionation factor depends on temperature [Eq. (8-61)]; since the $^{18}O/^{16}O$ ratio for seawater is approximately constant, this means that the isotope composition of the oxygen in $CaCO_3$ is also a simple function of temperature. Measurement of the ratio in calcareous sediments has given a detailed record of fluctuations in ocean temperature, and hence presumably of atmospheric temperature, over the last few hundred thousand years. The measurements depend on assumptions that (1) the $^{18}O/^{16}O$ ratio of seawater has been constant (which is not quite true, but probable variations can be at least roughly estimated), (2) $CaCO_3$ secreted as shell material by various organisms has oxygen of the same isotopic composition, and (3) the $CaCO_3$ has not changed in composition since it was developed. The third assumption would not be valid for $CaCO_3$ that has recrystallized since deposition, so that unaltered shells provide the best material for temperature measurement. Temperatures obtained for $CaCO_3$ older than Pleistocene become increasingly uncertain because of questions about the validity of these assumptions.

The isotopic compositions of other oxygen-containing minerals likewise are temperature dependent, and the amount of change with temperature varies from one mineral to another. If two minerals are formed at the same time, with the same access to a source of oxygen, the distribution of light and heavy oxygen between them should thus give a measure of the temperature of origin. The difference in fractionation of the oxygen isotopes between two minerals A and B may be expressed as

$$\delta^{18}O_A - \delta^{18}O_B = \Delta_B^A. \tag{10-33}$$

(Δ values for isotope fractionation may also be related to the equilibrium constant α [Eq. (10-32)] by the approximate equation $1000 \ln \alpha_B^A \cong \delta_A - \delta_B = \Delta_B^A$.) Numerous measurements of Δ values have shown that for many mineral pairs the relation with temperature is given to a good approximation by

$$\Delta = \frac{C}{T^2}, \tag{10-34}$$

where C is a constant. This approximate proportionality with the inverse square of temperature can be given theoretical justification by the methods of statistical mechanics. Temperatures indicated by the distribution of oxygen isotopes between representative mineral pairs are shown in Fig. 10-7.

This type of geothermometry is not limited to the isotopes of oxygen; it is often used, for example, with sulfur isotopes in estimating temperatures of ore formation (cf. Prob. 10-5 and Sec. 19-5). The method has an advantage over most others in that the volume changes of isotope exchange reactions must be very small [Eq. (8-71)], hence the results are not influenced by pressure. Success depends on the assumptions that isotopic equilibrium was maintained during formation of the minerals, and was not altered by isotopic exchange later. The assumptions do not always hold, but enough $^{18}O/^{16}O$ temperatures have been checked with temperatures obtained in other ways to indicate general reliability of the results, especially for mineral pairs in many types of igneous and metamorphic rocks.

The temperature dependence of stable isotope fractionation between mineral pairs, or minerals and solutions, is not without its complications. For example, the curves in Fig. 10-8 summarize experiments of D–H exchange between various

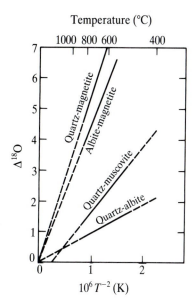

Temperature (°C)

FIGURE 10-7
Variations with temperature of ^{18}O distribution between mineral pairs. The vertical axis shows values of $\Delta^{18}O$, which are the differences in δ values for the various mineral pairs. Thus for the quartz–magnetite pair, a difference in $\delta^{18}O$ values of 5‰ indicates a temperature of about 800°C for isotopic equilibrium. (Plotted from data of Bottinga and Javoy, 1975.)

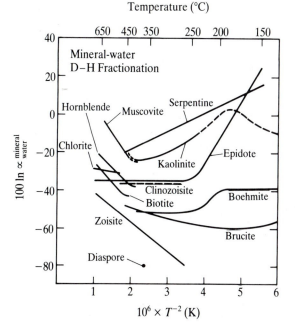

FIGURE 10-8
Experimentally determined fractionations of D and H between various minerals and water as a function of temperature. (Reprinted by permission from O'Neil, J. R.: "Theoretical and experimental aspects of isotopic fractionation," in Valley, J. W., H. P. Taylor Jr., and J. R. O'Neil, (eds): *Stable Isotopes in High-temperature Geological Processes, Reviews in Mineralogy,* Mineralogical Society of America, vol. 16, pp. 1–40, 1986.)

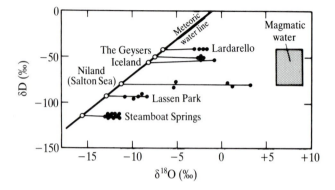

FIGURE 10-9

Values of δD and $\delta^{18}O$ for meteoric water, magmatic water, and several hot-spring waters. The diagonal line (*meteoric water line*) shows the relationship for many samples of water from rivers, lakes, and snow; points for individual samples are not shown. The open circles on this line are average values for the meteoric water in the different hot-spring areas. The solid dots represent analyses of hot-spring waters, and the light lines connecting them show the trend of isotope variation due to exchange of ^{18}O with adjacent rocks. The box for magmatic water encloses the area where analyses of biotite and hornblende from igneous rocks cluster. (After Craig et al., 1956; magmatic water from Taylor, 1974.)

minerals and water. The simple relationship of Eq. (10-34) obviously does not explain the minima and maxima in α at temperatures less than 500°C.

Oxygen isotope ratios in combination with hydrogen ratios have proved spectacularly successful as indicators of the source of hydrothermal fluids (Sec. 19-7). The long-standing argument among economic geologists as to a magmatic source versus a meteoric[1] source for the water in ore-forming solutions is in large degree settled: both kinds of water play a role, but in many deposits the role of meteoric water, or even of seawater, is dominant.

Evidence for this conclusion is based on differences in $\delta^{18}O$ and δD (expressed by Eq. (10-26) for D/H ratios in the sample relative to the seawater standard, SMOW) between hydrous minerals of igneous rocks and water in various solutions found near the Earth's surface. In minerals of unaltered igneous rocks such as biotite and hornblende the ratio has a fairly restricted range, as shown in Fig. 10-9. Meteoric water, on the other hand, shows a wide range of values. These represent the extent of isotope fractionation during successive episodes of evaporation and condensation of water originally evaporated from the sea. The D/H fractionation is proportional to the $^{18}O/^{16}O$ fractionation and about eight times larger, and both fractionations change proportionally as temperature changes. So the

---

[1] Meteoric water is surface water or groundwater that has fallen as rain or snow in fairly recent geologic time. It has nothing to do with meteorites. Its geochemistry is discussed in Sec. 12-2.

points for pairs of δ values distribute themselves for the most part along a straight line, which can be represented by the approximate equation

$$\delta D = 8\delta^{18}O + 10. \qquad (10\text{-}35)$$

In general the most negative δ values for both D and ^{18}O are obtained for samples from high latitudes, where the prevailing low temperatures lead to large fractionation factors between liquid and vapor.

Besides the line for meteoric water samples and the box for magmatic water, Fig. 10-8 shows values obtained for waters from hot-spring areas, in several of which deposition of sulfides is known to occur. Such measurements in a given area fall close to a horizontal line, indicating that the hot water contains an excess of ^{18}O over the meteoric water of the same region, but roughly the same deuterium content. This suggests that the water has reacted to varying degrees with the rocks through which it has traveled, exchanging some of its ^{16}O for ^{18}O from the minerals of the rock. A similar exchange of hydrogen isotopes is insignificant because most minerals contain only a little of this element. If the spring waters contained any considerable amount of magmatic water in addition to meteoric water, the mixing should be represented by values along lines connecting the meteoric-water points with the composition of magmatic water—which the diagram does not show. Thus the horizontal lines are good evidence for the derivation of most hot-spring waters chiefly from rainfall in the vicinity of the springs.

Further support for the meteoric origin of solutions responsible for many ore deposits has come from isotope analyses of minerals in alteration zones around ore accumulations (Taylor, 1974). The analyses commonly show lower $^{18}O/^{16}O$ ratios than those for unaltered rocks, and the difference often increases with the extent of alteration. This is not proof that the added water is all or mostly meteoric, because other sources of water are possible—connate water imprisoned in adjacent sediments, for example, or water released during low-grade metamorphism. But at least alteration entirely by magmatic water is ruled out. The extent to which meteoric water has taken part in alteration and ore deposition varies widely from one deposit to another. The significant result of isotope study is not to exclude magmatic water as an often important agent of ore formation, but simply to show that meteoric water (and perhaps other kinds of water) is far more important than was commonly thought by students of ore deposits a few decades ago.

Many shallow igneous intrusions that are not extensively altered and do not host ore deposits also have minerals with unusually low $^{18}O/^{16}O$ ratios, and distributions of oxygen isotopes between coexisting minerals that are far from the equilibrium fractionations indicative of magmatic conditions. Exposures of these intrusions commonly have the appearance of a pristine igneous rock, but upon close examination one finds extensive networks of thin veins and microscopic cracks. Apparently low-^{18}O meteoric solutions permeated the fractures as the intrusion cooled and exchanged oxygen with igneous minerals without causing extensive mineralogic alteration. Feldspars [Eq. (10-23)] seem to exchange oxygen with the solution much faster than do quartz [Eq. (10-24)] or pyroxene. This difference in the relative reaction rates produces the nonequilibrium isotope fractionations of feldspar–quartz and feldspar–pyroxene that are typical of these intrusions. The Idaho batholith and the Skaergaard intrusion in east Greenland are two examples where the low-^{18}O content of feldspars and nonequilibrium fractionations of ^{18}O in the mineral pairs feldspar–quartz and feldspar–pyroxene are the primary indicators of the vast amount of hydrothermal fluids that circulated through fractures as the intrusions cooled.

Sulfur

Sulfur isotope concentrations are expressed by δ^{34}S values, analogous to the expressions for δ^{18}O. The commonly used standard is the sulfur in the troilite (FeS) of meteorites, on the assumption that its composition should be close to that of the original sulfur in the material that formed the Earth. (The usual standard is troilite of the iron meteorite from Canyon Diablo whose ^{32}S/^{34}S ratio is 22.22.) Recorded values of δ^{34}S range from about $-30‰$ for some samples of coal and H_2S to values more positive than $+60‰$ for the sulfur in some sulfate minerals.

The sulfur of sulfide materials shows slight differences in isotope ratios, the amount of enrichment of ^{34}S in minerals precipitated from the same solution depending on relative bond strengths. For example, the metal–sulfur bond in sphalerite is stronger than that in galena, and the ratio ^{34}S/^{32}S is correspondingly greater for sphalerite (by about 3‰ at 200°C). The degree of enrichment of the heavy isotope changes with temperature, and the amount of change is different for different materials. Thus the isotopic composition of sulfur in coexisting sulfide minerals can serve as a measure of their formation temperature, just as oxygen isotopes provide temperature estimates for minerals that contain oxygen. Finding temperatures in this way depends on assumptions that (1) equilibrium was maintained during sulfide deposition and (2) the isotope composition was not affected by later exchange with sulfur from other sources. In practice, temperatures from isotopes have been found to agree reasonably well with those found by other methods, and discrepancies can be attributed to failure of these assumptions.

The reactions in which greatest separation of the sulfur isotopes occurs are oxidation–reduction reactions, especially those in which sulfides change to sulfates or vice versa. In such reactions the heavy isotope is enriched in compounds where sulfur has the highest oxidation number, the light isotope where the oxidation number is lowest. Of these reactions in nature, the major one at low temperatures is the reduction of SO_4^{2-} catalyzed by a very common kind of anaerobic bacteria ("sulfate-reducing bacteria"). The reaction produces H_2S or one of its ions, and in these substances ^{32}S is markedly concentrated; for example, dissolved H_2S in the organic-rich muds at the bottom of the Black Sea shows δ^{34}S values as low as $-31.9‰$. Because sulfur in near-surface environments undergoes frequent change between the oxidized and reduced forms, it can acquire a wide variety of ^{34}S/^{32}S ratios.

One might expect that isotope ratios could provide hints as to the source of sulfur in sulfide ore deposits. Sulfur in igneous rocks, especially those whose magma has come directly or fairly directly from the mantle, generally has δ^{34}S close to 0‰, consistent with the usual assumption that mantle and meteorites are similar in origin and composition. Ore deposits related to igneous rocks, then, might be expected to have sulfur isotope ratios close to this primitive value, while sulfide deposits associated with sedimentary rocks (the lead–zinc deposits of the Mississippi Valley, for example) could have a wide range of values because their sulfur has been exposed to surface processes. This inference, like many promising conclusions from isotope ratios, has not been borne out by observation. The difficulty is just the complexity of possible influences on isotope ratios: values of $^{34}S/^{32}S$, in both igneous and sedimentary environments, can change with so many variables (pH of solutions, oxygen fugacity, nature of other ions present, temperature, rates of reaction, to name a few) that a given measured ratio can have little genetic significance except in very simple situations.

The changes of sulfur isotope ratios in the sulfate of seawater over geologic time, as noted by analyses of the sulfur in evaporate deposits, have occasioned much interest. The $\delta^{34}S$ of seawater sulfate rose to a high of more than +30‰ at the start of the Cambrian, fell erratically to about 11‰ at the end of the Permian, and then climbed to its present value of about 20‰. This variation, as might be expected, has aroused a wealth of speculation, but remains without a generally agreed-upon explanation. One striking fact about the earlier history of marine sulfur, however, seems more readily explained. Until near the end of the Precambrian, $\delta^{34}S$ in the sea remained consistently within a modest range of about -10 to $+10$, as would be expected if the concentration of sulfate was low, and then rose to the very high figure mentioned for the Cambrian. The large increase suggests strongly that dissolved sulfate rather quickly reached an abundance like that of the present, perhaps because of the development of photosynthesis and the resulting great increase in concentration of O_2 in the atmosphere.

Carbon

Relative concentrations of the two stable isotopes of carbon, ^{12}C and ^{13}C, are expressed by $\delta^{13}C$ values analogous to the δ values for oxygen, hydrogen, and sulfur. The usual standard is the isotope ratio measured in the carbon of belemnites found in the Cretaceous Peedee formation of South Carolina, abbreviated PDB. Relative to this standard most $\delta^{13}C$ values for terrestrial materials are negative, ranging down to $-70‰$ for samples of methane, but high positive values are reported for some carbonates, especially the carbonates of meteorites.

Many reactions can lead to slight separation of the carbon isotopes in nature, the most effective ones being, as for sulfur, oxidation–reduction processes in which the heavier isotope is concentrated in the more oxidized forms. Particularly important is the reaction of photosynthesis in the leaves of green plants, in which the carbon of CO_2 is reduced and incorporated into organic compounds. Because the carbon in organic matter is thereby enriched to varying degrees in ^{12}C, an abundance of light carbon in bituminous material of doubtful origin in ancient rocks can serve as an indication that the source of the organic matter was living organisms rather than "juvenile" carbon gases from deep in the mantle. Thus the low values of $\delta^{13}C$ in the hydrocarbons of petroleum are one of the important bits of evidence for ascribing the origin of petroleum to the alteration of organic material rather than to condensation of primeval gases from the Earth's interior (Chap. 15).

At higher temperatures, equilibrium fractionation of isotopes may occur between CH_4 and CO_2 (or carbonate ions formed by dissolution of CO_2). Ohmoto (1972) has noted that this isotope separation, like the fractionation of sulfur isotopes between sulfate and sulfide species, is sensitive to changes in pH, oxygen fugacity, and ionic strength as well as to changes in temperature. Both carbon isotopes and sulfur isotopes may be useful as tools in assigning values to some of these variables, but relations are complicated enough to make extreme care necessary in attempting to use them.

SUMMARY

From this brief survey, it is evident that isotope ratios can be made to yield a variety of geologic information, from absolute ages to paleotemperatures to sources of rocks and fluids. Caution is needed in their use, because the necessary assumptions are not always obvious, many reactions are slow, and the number of variables on which values of isotope ratios depend may be very large. Calculation from theory is generally not helpful, and most of our knowledge about isotope ratios and their changes is empirical. But for many geologic purposes the study of isotope ratios has proved to be informative, and it is small wonder that isotopes have attracted increasing attention from geochemists of all stripes.

PROBLEMS

1. The following values of $^{87}Rb/^{86}Sr$ and $^{87}Sr/^{86}Sr$ were obtained by analysis of minerals in rocks of the Sudbury "nickel irruptive" (chiefly norite, or hypersthene gabbro) in Ontario. Use these values to construct an isochron diagram, and from the diagram find the age of the rock and its initial $^{87}Sr/^{86}Sr$ ratio. The decay constant for ^{87}Rb is 1.42×10^{-11} yr^{-1}. (Source: Hurst, R. W. and J. Farhat: "Geochronologic investigations of the Sudbury nickel irruptive," *Geochim. et Cosmochim. Acta*, vol. 41, pp. 1803–1816, 1977).

$^{87}Rb/^{86}Sr$	$^{87}Sr/^{86}Sr$
0.4009	0.7178
0.2983	0.7150
0.2074	0.7126
0.1080	0.7104
0.0458	0.7084

2. Another pair of elements that can be used for radiometric dating is lutetium–hafnium. The isotope ^{176}Lu decays by emission of a beta particle to form ^{176}Hf, with a decay constant

of 1.93×10^{-11} yr^{-1}. Set up an equation that could be used to find the age of a rock if analyses for these two isotopes were available. (Use ^{177}Hf as a reference stable isotope).

3. What method (or methods) of analysis would be most suitable for finding the radiometric ages of (a) a Precambrian granite, (b) an early Tertiary shale containing carbonized wood fragments, (c) an early Paleozoic evaporite deposit, (d) a Proterozoic basalt, (e) charcoal from a prehistoric Indian campfire, (f) garnet–pyroxene skarn from the metamorphic rock at a granite contact. Give reasons for your answers.

4. Clay minerals formed during weathering in humid environments have $\delta^{18}O$ and δD values that are largely determined by the isotopic composition of local meteoric water (the equation for the meteoric water line in Fig. 10-8 is $\delta D = 8\delta^{18}O + 10$). Lines showing the distribution of hydrogen and oxygen isotopes in different clay minerals, when plotted on a diagram like Fig. 10-8, are parallel to the line for meteoric water but have different intercepts. For kaolinite this line is $\delta D = 8\delta^{18}O - 220$ and the fractionation of oxygen between kaolinite and water is $\alpha = 1.027$. With this information calculate $\delta^{18}O_{kaolinite}$ and $\delta D_{kaolinite}$ formed by the weathering of feldspar in equatorial regions where $\delta^{18}O$ is -5 and in subarctic regions where $\delta^{18}O$ is -20. Construct a diagram similar to Fig. 10-8 showing the meteoric water line, the line for kaolinite, and isotopic compositions of the kaolinite and rain water computed above. On your diagram draw two tie-lines connecting the isotopic compositions of rain water and kaolinite formed in equatorial and subarctic regions. Using the results of your computations, what is the fractionation factor for the distribution of hydrogen isotopes between kaolinite and water?

5. The table below shows the isotopic composition of sulfur in three samples of galena and sphalerite from the Providencia mine in Mexico. Experimental work with the two minerals indicates that these values correspond with the temperatures of formation given in the last line of the table. Show by plotting a graph that these data are approximately consistent with the relation of Δ^{sl}_{gn} and T shown in the equation

$$\Delta^{sl}_{gn} = \delta^{34}S_{sl} - \delta^{34}S_{gn} \cong 1000 \ln \alpha^{sl}_{gn} = \frac{C}{T^2}$$

and evaluate the constant C. If analysis of another mineral pair from this deposit shows $\delta^{34}S_{sl} = +0.52$ and $\delta^{34}S_{gn} = -1.28$, what is the approximate temperature of its formation? (after Ohmoto and Rye, 1979).

	Sample 1	Sample 2	Sample 3
$\delta^{34}S_{sl}$	+0.62	+1.15	+1.56
$\delta^{34}S_{gn}$	−1.53	−1.62	−1.78
Temperature (°C)	300	230	185

6. The layered gabbro of the Skaergaard intrusion, East Greenland, intruded a thick pile of basalt lavas and upon cooling generated large-scale hydrothermal fluid circulation and alteration of basalts. Hydrothermal chlorite formed in the basalts has an average δD of $-136‰$ and $\delta^{18}O$ of $1.5‰$. The low δD value suggests that meteoric water was the source of the hydrothermal fluids. Assuming that the alteration occurred between 250 and 400°C, where H–D fractionation between chlorite and water is about 35‰ ($\Delta^{chlorite}_{water} = \delta D_{chlorite} - \delta D_{water} = -35‰$), compute the original isotopic properties (δD and $\delta^{18}O$) of the meteoric water.

REFERENCES AND SUGGESTIONS FOR FURTHER READING

Bottinga, Y., and M. Javoy: "Oxygen isotope partitioning among the minerals in igneous and metamorphic rocks," *Reviews of Geophysics and Space Physics*, vol. 13, pp. 401–418, 1975.

Bowen, R.: *Isotopes in the Earth Sciences*, Elsevier, 1988.

Craig, H., G. Boato, and D. E. White: "Isotopic geochemistry of thermal waters," *Proc. 2d Conf. Nuclear Processes in Geologic Settings, National Research Council, Nuclear Science Series Report* 19, pp. 29–38, 1956.

Dalrymple, G. B.: *The Age of the Earth*, Stanford University Press, California, 1991. The history of age determinations, description of modern radiometric methods, and a survey of age data for a variety of geologic and astronomic materials.

Faure, G.: *Principles of Isotope Geology*, 2d ed., Wiley, New York, 1986. An excellent treatment of all aspects of the study of isotopes as applied to geologic problems. This is the standard intermediate-level reference work on isotope geology.

Kistler, R. W., and Z. E. Peterman: "Variations in Sr, Rb, K, Na, and initial ^{87}Sr/^{86}Sr in Mesozoic granitic rocks and intruded wall rocks in central California," *Geol. Soc. America Bull.*, vol. 84, pp. 3489–3512, 1973. An interesting attempt to relate strontium isotope ratios in granitic rocks of the Sierra Nevada to the kinds of rocks which they intrude and from which they may have formed by partial melting.

Lipman, P. W., B. R. Doe, C. E. Hodge, and T. A. Steven: "Petrologic evolution of the San Juan volcanic field, southwestern Colorado: Pb and Sr isotopic evidence," *Geol. Soc. America Bull.*, vol. 89, pp. 59–82, 1978. A good example of applications of isotope study to petrologic problems in mineralized areas.

Ohmoto, H.: "Systematics of sulfur and carbon isotopes in hydrothermal ore deposits," *Econ. Geology*, vol. 67, pp. 551–578, 1972. A theoretical study of the effects on isotope distribution of the temperature, pH, oxygen fugacity, and ionic strength of hydrothermal solutions.

Ohmoto, H., and R. O. Rye: "Isotopes of sulfur and carbon," in *Geochemistry of Hydrothermal Ore Deposits*, H. L. Barnes (ed.), Wiley, pp. 509–567, 1979. A good summary of sulfur and carbon isotope geochemistry during ore-forming processes.

Patterson, C. C.: "Age of meteorites and the earth," *Geochim. et Cosmochim. Acta*, vol. 10, pp. 230–237, 1956. A classical paper in which the age of the Earth and the isotopic composition of primeval lead are derived from meteorite analyses.

Ralph, E. K., and H. N. Michael: "Twenty-five years of radiocarbon dating," *American Scientist*, vol. 62, pp. 553–560, 1974. Corrections of radiocarbon dates on the basis of tree-ring counts, archeological data, and fluctuations in the Earth's magnetic field.

Taylor, H. P., Jr: "The application of oxygen and hydrogen isotope studies to problems of hydrothermal alteration and ore deposition," *Econ. Geology*, vol. 69, pp. 843–883, 1974. One of Taylor's many papers on the use of isotopes in determining the source and history of hydrothermal fluids. This issue of *Economic Geology* (No. 6 of vol. 69) has many other papers on the application of isotopes to problems of metallic ores.

Taylor, H. P., Jr: "Water/rock interactions and the origin of H_2O in granitic batholiths," *Jour. Geol. Soc. London*, vol. 133, pp. 509–558, 1977. A good summary of stable isotope evidence for hydrothermal fluid circulation in batholiths.

Taylor, H. P., Jr., J. R. O'Neil, and I. R. Kaplan (eds): "Stable isotope geochemistry: a tribute to Samuel Epstein," *The Geochemical Society, Special Publication* No. 3, 1991. A collection of papers on experimental and theoretical isotopic fractionation, and the applications of isotope geochemistry to the hydrosphere, oceans, climates, glaciers, archaeology, petrology, and extraterrestrial bodies.

Valley, J. W., H. P. Taylor, Jr., and J. R. O'Neil (eds): "Stable isotopes in high-temperature geologic processes," *Reviews in Mineralogy*, vol. 16, p. 570, 1986. Papers on many aspects of the application of isotope study to geologic problems.

Wetherill, G. W.: "Discordant uranium–lead ages," *Trans. Amer. Geophys. Union*, vol. 37, pp. 320–326, 1956. A classical paper explaining the use of concordia diagrams in interpreting discordant uranium–lead ages.

CHAPTER 11

REACTION
RATES
AND
MASS
TRANSFER

During weathering, diagenesis, ore deposition, metamorphism, and the emplacement and crystallization of magmas, chemical components are redistributed among fluid and solid phases. How fast this redistribution occurs depends on rates of chemical reactions and on the rates of motion of the reacting materials as they are transported from one location to another. In this chapter we investigate the geochemistry of the redistribution process, looking first at the kinetics of chemical reactions and then studying the processes that move chemical components to and from the sites of reaction.

11-1 KINETICS AND EQUILIBRIUM

In preceding chapters we have emphasized equilibrium, the situation reached by a reversible process when its forward and reverse reactions are going at equal rates. By calculating equilibrium concentrations, we determine the state of a system when its free energy has been expended, when nothing more can happen unless a change in composition or conditions is introduced. Such calculations are powerful tools, as we have seen, for predicting the possible results of mixing any combination of materials at a given pressure and temperature. Unfortunately, however, the calculations often do not tell us how the world really works.

This is because a long time may be needed to attain equilibrium. Think of a piece of galena exposed to dry air in a museum display case. Equilibrium reasoning tells us that PbS in contact with O_2 in the atmosphere is unstable, and that $PbSO_4$ should form with a big loss of free energy—yet the galena remains shiny and unaltered for years. Or take a more striking example, the reaction of nitrogen and oxygen to form nitric acid. In the usual way we set up an equation:

$$H_2O + N_2 + {}^5/_2O_2 \rightarrow 2H^+ + 2NO_3^-, \tag{11-1}$$

for which $\Delta G°$ at 25°C and 1 bar is +19.7 kJ. The corresponding equilibrium constant for the reaction is

$$K = \frac{a_{H^+}^2 a_{NO_3^-}^2}{a_{H_2O} f_{N_2} f_{O_2}^{5/2}} = 10^{-3.4}. \tag{11-2}$$

For a crude calculation we use 0.8 bar as the fugacity ($\cong$ partial pressure) of N_2 in air and 0.2 bar as the fugacity of O_2, and let $x = a_{H^+} = a_{NO_3^-} (\cong m_{H^+})$:

$$x = [Ka_{H_2O} f_{N_2} f_{O_2}^{5/2}]^{1/4} = [(10^{-3.4})(1)(0.8)(0.2)^{5/2}]^{1/4} = 10^{-1.3}. \tag{11-3}$$

Thus the equilibrium concentration of H^+ and NO_3^- in water standing in contact with air is about $0.1m$. All water at the Earth's surface, therefore, according to equilibrium reasoning should be a tenth-normal solution of nitric acid. Yet somehow, even after more than 4 billion years of Earth history, equilibrium in Eq. (11-1) has—very fortunately—not yet been attained. Evidently the molecules O_2 and N_2 are so stable that the forward reaction of Eq. (11-1) is almost infinitely slow.

Equilibrium reasoning is useful in setting limits on possible processes in nature, but clearly reaction rates are often more important in determining what actually happens in reasonable times. We turn our attention now to this different aspect of geochemical processes.

11-2 REACTION RATES

The rate of a reaction, as noted long ago by Guldberg and Waage (Sec. 1-1), should depend on concentrations of the substances that react. If two compounds A and B combine to form a third,

$$A + B \rightarrow C, \tag{11-4}$$

the rate ideally would be

$$\frac{d[C]}{dt} = -\frac{d[A]}{dt} = -\frac{d[B]}{dt} = k[A][B], \tag{11-5}$$

where the brackets indicate concentrations and the k is a proportionality constant (the *rate constant*). Eq. (11-5) is accurate provided the reaction of Eq. (11-4) is ideally simple, meaning that reaction occurs whenever a certain fraction of the particles of A and B collide. One example is the precipitation rate of a simple mineral like barite

$$Ba^{2+} + SO_4^{2-} \rightarrow \underset{\text{barite}}{BaSO_4} \tag{11-6}$$

which is closely approximated by the rate law of Eq. (11-5) (Christy and Putnis, 1993).

Such equations seldom find much use in geochemistry, because they apply strictly only to ideal elementary processes. Many actual reactions take place in steps, and their overall rates are proportional to the squares or cubes or some other power of reactant concentrations. The sum of the exponents of the concentrations of substances that appear in the rate expression is referred to as the *order* of reaction. Thus the reaction of Eq. (11-4) is second order if the rate depends on the first power of the concentrations of A and B, as indicated by Eq. (11-5). It turns out, however, that the exponents may be fractional, and they need not correspond with the stoichiometric reaction coefficients.

Thus reaction of hydrogen and bromine to form hydrogen bromide,

$$H_2 + Br_2 \rightarrow 2HBr, \tag{11-7}$$

would appear to be a good illustration of Eq. (11-4). Yet when its rate is measured in the laboratory, the result (in the early part of the reaction) is

$$\text{rate} = -\frac{d[H_2]}{dt} = k[H_2][Br_2]^{1/2} \tag{11-8}$$

rather than the expected

$$\text{rate} = -\frac{d[H_2]}{dt} = k[H_2][Br_2]. \tag{11-9}$$

This is because the process takes place not by simple collisions of the diatomic molecules, but by a splitting of Br_2 followed by reactions of the atoms:

$$
\begin{aligned}
&Br_2 \rightarrow 2Br \\
&H_2 + Br \rightarrow HBr + H \\
&H + Br_2 \rightarrow HBr + Br \qquad\qquad (11\text{-}10)\\
&Br + Br \rightarrow Br_2 \\
&H + H \rightarrow H_2.
\end{aligned}
$$

Each of these step-reactions is an elementary process with a rate given by an expression like Eq. (11-5), and a combination of the different rates gives the overall rate [Eq. (11-8)]. Many apparently simple reactions, even when they take place under the carefully controlled conditions of a chemical laboratory, show similar complexity in their mechanisms. For most reactions that occur in nature we can expect the mechanisms to be still more complicated.

It follows that the expression for reaction rate given by Eq. (11-5) may be misleading if applied quantitatively to reactions of interest in geochemistry. This formulation is often cited in a qualitative sense, simply to express the fact that rates depend on concentrations, but the indicated first powers of the concentrations have meaning only for ideally simple reactions.

A simple process for which a quantitative treatment can be useful is a reaction in which a single substance breaks down into two or more others (a "first-order reaction"):

$$A \rightarrow B + C. \tag{11-11}$$

From Eq. (11-5),

$$\text{rate} = -\frac{d[A]}{dt} = k[A]. \tag{11-12}$$

Rearrangement gives

$$\frac{d[A]}{[A]} = -k\,dt, \tag{11-13}$$

which on integration becomes

$$\ln\,[A] = -kt + c, \tag{11-14}$$

where c is the integration constant. The constant may be evaluated by noting that when $t = 0$, c would be $\ln\,[A]^\circ$, the logarithm of the initial concentration of A. Thus

$$\ln\,[A] = \ln\,[A]^\circ - kt, \quad \text{or} \quad \ln\,\frac{[A]}{[A]^\circ} = -kt, \tag{11-15}$$

and

$$[A] = [A]^\circ e^{-kt}. \tag{11-16}$$

Reactions of this simple form—decomposition of a single substance into two or more others—are not common in nature. The anaerobic decay of organic materials can be suggested as one example, and another obvious one is radioactive breakdown. For radioactivity, in fact, we have already used expressions identical in form to Eqs. (11-14) and (11-16) [cf. Eqs. (10-2) and (10-3) in the last chapter]. In a radioactive process, the k in Eq. (11-16) simply becomes equal to λ, the radioactive decay constant, which we have used in calculating geologic ages.

One kind of fluid–solid reaction that would seem sufficiently simple for detailed study of rates is the dissolving of minerals in solutions like those responsible for weathering at the Earth's surface. Because such reactions play a role in the formation and alteration of agricultural soils, as well as in concentration and dispersal of metals during ore deposit weathering, study of their rates is important in practical applications of geochemical study. It is hardly surprising, then, that such reactions have been the subject of voluminous research. As an illustration of the problems involved and results obtained in experimental work on rates of these processes, we look briefly at the dissolution mechanism of the common mineral albite feldspar, $NaAlSi_3O_8$.

The problem seems simple enough: put the feldspar in various solutions, starting with pure water and then adding some of the solutes commonly found in weathering solutions in nature; let the solution flow slowly over the mineral, see how fast it dissolves, and measure the amounts of its constituents that go into solution. Unhappily the results prove to be anything but simple. The initial reaction is a rapid nonstoichiometric dissolving of a little feldspar, giving a solution containing a much higher Na/Si ratio and an Al/Si ratio slightly higher than those in the original mineral. Then the reaction slows, and after a number of days the dissolution becomes congruent.

An example of the results of such experiments for albite dissolution in water with a pH buffered at 5.6 is shown in Fig. (11-1). Measured amounts of Na, Al, and Si in solution are plotted as a function of time, so that slopes of curves drawn through the data are proportional to the rate of dissolution. Focusing on the slopes of the curves, we see that the initial dissolution rate is large (very steep slopes during the first 10 hours of dissolution), then decreases gradually. After 5 to 10 days the line through the data is nearly linear. This change is more clearly illustrated by the longer duration experiments shown in Fig. (11-2). The straight lines indicate that the dissolution rate is constant or at a *steady state*, and their slopes show that dissolution is congruent (for albite, $NaAlSi_3O_8$, the slopes for Na and Al release are equal and the slope for Si is three times that of Na and Al).

Large dissolution rates observed in the first days of the experiment are best explained by assuming formation of a surface layer of the feldspar that has been altered by partial loss of the alkali metals, presumably by exchange with H^+ or H_3O^+ ions in the solution, and by rapid dissolution of any very fine particles adhering to the surfaces of the reactant mineral. The reaction is not affected by stirring of the fluid, so dissolution rates must be controlled by reactions at the solution–mineral

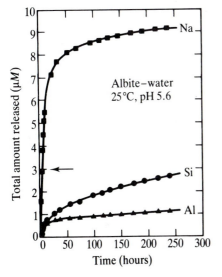

FIGURE 11-1

The total amount of Na, Al, and Si (in micromols per liter, $\mu M\ L^{-1}$) released to solution during albite dissolution in pure water at 25°C and pH = 5.6. Experiment is conducted in a fluidized bed reactor where water of a constant pH is continuously percolated through the crushed reactant mineral, and the discharge fluid is analyzed for the dissolved constituents. The arrow indicates an estimated amount of Na released due to exchange with H^+. (Reprinted by permission from Wollast and Chou 1985.)

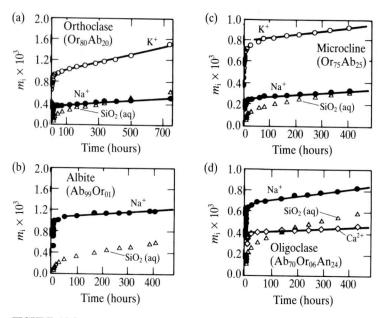

FIGURE 11-2

Changes in solution composition (m_i refers to the molality of total Na, K, Ca, and Si) as a function of time during dissolution of different kinds of feldspars at 25°C and pH = 5.1. The symbols represent experimental data reported by Busenberg and Clemency, 1976. The straight lines were generated from equations and data presented by Helgeson *et al.*, 1984. (Both sources by kind permission of Pergamon Press Ltd.)

interface and not by the rate of ion diffusion away from the mineral into the solution.

A rate constant (k) for steady-state dissolution of albite in water with pH = 5.6 can be estimated from the experimental data of Fig. (11-1). First we write a reaction representing the overall congruent dissolution, a reaction that describes material balance between dissolving albite and the aqueous species produced. At a pH of 5.6 nearly all of the Al in solution will be $Al(OH)_2^+$ (see prob. 4-2), so the reaction is

$$\underset{\text{albite}}{NaAlSi_3O_8} + 2H^+ + 6H_2O \rightarrow \underset{\text{solution}}{Na^+ + Al(OH)_2^+ + 3H_4SiO_4}. \qquad (11\text{-}17)$$

The absolute change in Na, Al, and Si concentrations during the experiment is proportional to the amount of albite present, and a rate equation for dissolution can be written as

$$\frac{d[Na^+]}{dt} = \frac{d[Al(OH)_2^+]}{dt} = \frac{1}{3}\frac{d[H_4SiO_4]}{dt} = k[\text{albite}], \qquad (11\text{-}18)$$

where k is the rate constant for the specified temperature, pressure, and pH of the experiment. As before, the brackets represent concentrations of reaction species, usually taken as molality for aqueous species (mols of solute per kg water) and for a mineral as its surface area (S_{albite}) per mass of water (M_{H_2O}), so Eq. (11-18) becomes

$$\frac{dm_{Na^+}}{dt} = \frac{dm_{Al(OH)_2^+}}{dt} = \frac{1}{3}\frac{dm_{H_4SiO_4}}{dt} = k\frac{S_{albite}}{M_{H_2O}}. \qquad (11\text{-}19)$$

This equation is used to compute a value for k from the slope of a line drawn through the linear portion of the experimental data (the slope of the line is equal to $k\,S_{albite}/M_{H_2O}$). In near neutral solutions the rate constant for albite dissolution is approximately $10^{-15.8}$ mol cm^{-2} sec^{-1} at 25°C. Similar analysis of other dissolution rate experiments in near-neutral dilute solutions at 25°C indicates that rate constants for the common rock-forming silicates have values that range over several orders of magnitude.

The rate of dissolving depends also on other ions in the solution, especially on the hydrogen ion. How dissolution rates of albite (expressed as mols albite dissolved per cm^2 surface area of albite per second) change with pH is shown by the experimental data of Chou and Wollast (1985) in Fig. (11-3). The rate has a minimum in near-neutral solutions and increases with either increasing or decreasing pH. To explain this behavior various mechanisms are possible. One can postulate effects on the dissolution resulting from properties of the feldspar surface: different degrees of coverage of the surface with an altered or leached layer, or different kinds of crystalline defects responsive to various concentrations of hydrogen ion. Or one can think of possible complexes of Al, Si, H$^+$, and OH$^-$ developed on the surface at different values of pH, leading to differences in the rate of breakdown of the feldspar structure. The mechanism is obviously complex. Chou and Wollast derive a mathematical expression for the change of rate with pH (shown by the line in Fig. (11-3)), using an assumption of transient complexes formed at the feldspar surface, but do not account for all details.

Such a relationship between pH and dissolution rate is not unexpected in light of the reactivity of H$^+$ and OH$^-$ ions with silicate mineral surfaces (Sec. 6-1) and the amphoteric nature of aqueous aluminum species (Sec. 4-2). Let us imagine, for illustrative purposes, that the elementary mechanism of albite dissolution is represented by the mass balance constraints of congruent dissolution (the rate

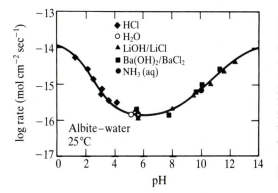

FIGURE 11-3
Logarithm of the rate of albite dissolution as a function of pH. The points are experimental values of the rate (expressed as mols of albite dissolved per cm^2 of surface per second) determined in various solutions. The symbols represent different kinds of solutions used to buffer pH. The line is based on an equation developed by assuming activated complexes on the feldspar surface. (Reprinted by permission from Chou and Wollast, 1985.)

of dissolution, R, is equal to the rate at which Na, Al, and 1/3 Si are released to solution per cm^2 of albite surface area). For acid solutions, where aluminum is present as Al^{3+} and silica as H_4SiO_4, the elementary reaction is

$$NaAlSi_3O_8 + 4H^+ + 4H_2O \rightarrow Na^+ + Al^{3+} + 3H_4SiO_4. \qquad (11\text{-}20)$$

albite   acid solution

In basic solutions at pH between about 10 and 12 aluminum is present as $Al(OH)_4^-$ [Eq. (4-17)] and silica as $H_3SiO_4^-$ [Eq. (4-3)]:

$$NaAlSi_3O_8 + 3OH^- + 5H_2O \rightarrow Na^+ + Al(OH)_4^- + 3H_3SiO_4^-. \qquad (11\text{-}21)$$

albite   basic solution

Application of the rate expression of Eq. (11-5) to these reactions gives

$$R_{\text{acid solution}} = k'[m_{H^+}]^4 \qquad \text{or} \qquad \log R = \log k' - 4\,pH, \qquad (11\text{-}22)$$

and

$$R_{\text{basic solution}} = k''[m_{OH^-}]^3 \qquad \text{or} \qquad \log R = \log k'' + 3\log K_W + 3\,pH,$$
$$(11\text{-}23)$$

where k' and k'' are rate constants for the model elementary reactions at a fixed ratio of S_{albite} to M_{H_2O}, and K_W is the equilibrium constant for the dissociation of water [Eq. (2-7)]. The two rate expressions indicate that dissolution rate will increase with decreasing pH in acid solutions [Eq. (11-22)] and with increasing pH in basic solutions [Eq. (11-23)], with a minimum rate in near neutral solutions.

Our model predicts the general trend of the pH dependence of dissolution rates, but it is not surprising that the model does not exactly conform to experimental results. For example, the data suggest that the dissolution rate for albite, and for many other silicate minerals, increases with increasing hydrogen ion concentration in acid solutions according to the relationship

$$R \propto (a_{H^+})^n, \qquad (11\text{-}24)$$

but the exponent n is not readily predictable. From Eq. (11-22) our model would suggest a value of 4 for albite [the stoichiometric reaction coefficient for H^+ in Eq. (11-20)], but experimental data show that the actual value is closer to 0.5. For K-feldspar, which has the same alkali : Al : Si stoichiometry as albite, n is about 1.0. The value of n appears to depend on the reaction mechanism and on the rate at which H^+ and OH^- react with different kinds of surface sites on the mineral. It is evidently not equal to the stoichiometric number of H^+ or OH^- ions in the overall reaction as suggested by Eqs. (11-22) and (11-23).

In addition to mineral surface area and pH, another determinant of reaction rates is the closeness of a reacting system to thermodynamic equilibrium. As a general rule, the rate of a reaction is faster under conditions far from equilibrium than close to equilibrium. For a simple illustration, we look briefly at an experimental study of the kinetics of the quartz–water system by Rimstidt and

Barnes (1980). The reaction is one we have considered before:

$$\underset{\text{quartz}}{SiO_2} + 2H_2O \underset{k_-}{\overset{k_+}{\rightleftharpoons}} \underset{\text{solution}}{H_4SiO_4} . \qquad (4\text{-}1)$$

Here k_+ is the rate constant for dissolution and k_- the constant for precipitation. The experiments demonstrated that the precipitation reaction closely follows a first-order rate law,

$$\left(\frac{dm_{H_4SiO_4}}{dt}\right)_{\text{precipitation}} = -k_- \frac{S_{\text{quartz}}}{M_{H_2O}} m_{H_4SiO_4} . \qquad (11\text{-}25)$$

The rate of dissolution can be expressed as proportional to the surface area of the dissolving mineral, as in Eq. (11-19):

$$\left(\frac{dm_{H_4SiO_4}}{dt}\right)_{\text{dissolution}} = k_+ \frac{S_{\text{quartz}}}{M_{H_2O}} . \qquad (11\text{-}26)$$

The total rate of change of H_4SiO_4 in solution due to quartz–water reaction, then, is the sum of Eqs. (11-25) and (11-26):

$$\left(\frac{dm_{H_4SiO_4}}{dt}\right)_{\text{total}} = \left(\frac{dm_{H_4SiO_4}}{dt}\right)_{\text{dissolution}} + \left(\frac{dm_{H_4SiO_4}}{dt}\right)_{\text{precipitation}} , \qquad (11\text{-}27)$$

$$\left(\frac{dm_{H_4SiO_4}}{dt}\right)_{\text{total}} = k_+ \frac{S_{\text{quartz}}}{M_{H_2O}} - k_- \frac{S_{\text{quartz}}}{M_{H_2O}} m_{H_4SiO_4} , \qquad (11\text{-}28)$$

and

$$\left(\frac{dm_{H_4SiO_4}}{dt}\right)_{\text{total}} = \frac{S_{\text{quartz}}}{M_{H_2O}} (k_+ - k_- m_{H_4SiO_4}) . \qquad (11\text{-}29)$$

At equilibrium, the concentration of H_4SiO_4 ($m_{H_4SiO_4}^{\text{equilibrium}}$) is equal to the equilibrium constant of Eq. (4-1), $K_{\text{quartz}} = m_{H_4SiO_4}^{\text{equilibrium}}$, and there is no change in this concentration with time. For these conditions the overall rate equation (11-29) becomes

$$0 = \frac{S_{\text{quartz}}}{M_{H_2O}} (k_+ - k_- m_{H_4SiO_4}^{\text{equilibrium}}) , \qquad (11\text{-}30)$$

from which

$$m_{H_4SiO_4}^{\text{equilibrium}} = \frac{k_+}{k_-} \cong K_{\text{quartz}} \qquad (11\text{-}31)$$

and

$$k_- = \frac{k_+}{K_{\text{quartz}}} = \frac{k_+}{m_{H_4SiO_4}^{\text{equilibrium}}} . \qquad (11\text{-}32)$$

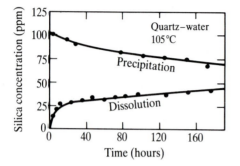

FIGURE 11-4
Measured concentrations of H_4SiO_4 as a function of time for two experiments of Rimstidt and Barnes (1980), one for dissolution and the other for precipitation of quartz in water at 105°C. For both reactions the concentration of aqueous silica approaches the equilibrium value of 58 ppm. Reaction rate is proportional to the slope of the curves, both becoming smaller and nearly linear as equilibrium is approached. (Reprinted by permission from Rimstidt and Barnes, 1980.)

The overall rate for the quartz–water reaction can then be written:

$$\left(\frac{dm_{H_4SiO_4}}{dt}\right)_{total} = k_+ \frac{S_{quartz}}{M_{H_2O}}\left(1 - \frac{m_{H_4SiO_4}}{m_{H_4SiO_4}^{equilibrium}}\right). \qquad (11\text{-}33)$$

The ratio in the right-hand parenthesis of Eq. (11-33) expresses how far the concentration of H_4SiO_4 differs from the concentration at equilibrium. If this ratio is small, meaning that the solution is very dilute, the rate $(dm_{H_4SiO_4}/dt)_{total}$ is large; as the ratio approaches 1, the rate steadily decreases and becomes zero at equilibrium. Then if the ratio becomes larger than 1, the right-hand side of the equation grows increasingly negative and the reaction rate becomes larger in the reverse direction (precipitation of quartz). These relations are illustrated by Fig. 11-4, showing the changes in dissolved silica concentration in two of Rimstidt and Barnes's experiments. In both runs the rate at which the concentration changes becomes more gradual as the concentration approaches the value at equilibrium (58 ppm). The very steep initial rise of the dissolution curve means that the reaction rate here is being affected by another factor besides the closeness to equilibrium: in this case, by the existence of a disturbed surface layer on the quartz which causes temporary rapid dissolution.

The effect of other variables on reaction rates can also be illustrated with the simple quartz–water system. For example, the influence of pH and of other ions in solution is nicely shown by the experiments of Dove and Elston (1992) as summarized in Fig. 11-5: the dissolution rate for quartz at 25°C is about

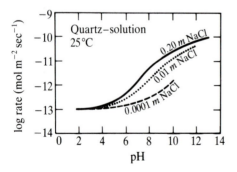

FIGURE 11-5
Dissolution rate of quartz at 25°C as a function of pH and NaCl concentration. Curves computed from experimental data using a surface reaction model discussed by Dove and Elston, 1992.

10^{-13} mol m^{-2} sec^{-1} in acid solutions, but in basic solutions it increases rapidly with both increasing pH and increasing concentrations of dissolved NaCl. Such complexities may be explained, as was suggested above for complexities in the dissolution rate of albite, by varying reactivity of the different kinds of surface complexes at the quartz–water interface. The Si–OH surface complex which is stable with acid solutions (see Sec. 6-1) appears to be resistant to hydrolysis, resulting in a minimum rate of dissolution in low-pH solutions. With increasing pH, protons are lost from the surface OH producing Si–O$^-$ complexes [Eq. (6-3)]. These surface complexes are more reactive and more readily hydrolyzed, thus increasing the dissolution rate. Addition of Na$^+$ apparently increases the dissolution rate by competing with H$^+$ for surface sites, leading to an increase in the abundance of the more reactive Si–Na$^+$ and Si–O$^-$ surface complexes.

It is evident from this work and many similar studies that rates of solid–fluid reactions in nature are a fruitful subject for research, but show complexities that make quantitative analysis and geochemical predictions difficult.

11-3 EFFECT OF TEMPERATURE ON RATES

Reaction rates depend on many variables, among which temperature is particularly important. We know, for example, that most reactions go faster at higher temperatures, and we learn as a rule of thumb that the rate for many processes, in a low-temperature range, approximately doubles with a $10°$ rise in temperature. Much of our knowledge about the effect of temperature on reaction rates is embodied in qualitative statements of this sort.

A brief look at reaction mechanics gives a basis for a somewhat more quantitative treatment. It seems obvious, to begin with, that any reaction like that in Eq. (11-4) can take place only when particles of A and B collide. Since particles move faster when they are heated, the qualitative effect of temperature on rates is readily explained. But details are less simple. If all or most particle collisions result in reaction, the overall rate should be very fast. Some reactions involving ions in solution are seemingly of this type; we recall the rapid precipitation that results when moderately concentrated solutions containing Ca^{2+} and CO$_3^{2-}$, or Ca^{2+} and SO$_4^{2-}$, are mixed. As a general rule, however, particle collisions do not all lead to immediate reaction. Only a fraction of the particles react, and presumably this is because only a fraction have the necessary energy. Thus we say that particles must somehow be *activated*, or supplied with energy, before they can react. The activation for reactions occurring in fluids normally comes simply from molecular motion. In the random motion of particles, some at any instant will acquire from collisions more energy than others, and only those with abnormally large amounts can take part in reaction. Because particles move faster at higher temperatures, addition of heat means that a larger fraction will acquire the necessary energy, and reaction rates are correspondingly greater.

Although the activation energy is most commonly supplied just by molecular motion, other means of activation are possible. Light energy is a common source, as is well illustrated by the formation of hydrogen chloride from its elements: although extremely slow at ordinary temperatures in the dark, the reaction of H_2 and Cl_2 goes with explosive rapidity when the gas mixture is illuminated. The photosynthetic reaction of CO_2 and H_2O in the green leaves of plants to form organic compounds is another reaction that requires light to make its rate appreciable. Activation energy can be supplied also by other forms of radiation, both fast-moving particles and electromagnetic radiation at frequencies different from those of visible light.

Activation energy with a source in molecular motion can be related to temperature by an equation suggested more than a century ago by Arrhenius:

$$k = Ae^{-E_a/RT}. \qquad (11\text{-}34)$$

Here k is the rate constant, A a constant specific for each reaction, E_a the activation energy, R the gas constant, and T the absolute temperature. In logarithmic form the relation is

$$\log k = \log A - \frac{E_a}{2.303R}\left(\frac{1}{T}\right), \qquad (11\text{-}35)$$

which suggests that the logarithm of a measured rate constant plotted against the reciprocal of absolute temperature should give a straight line with slope $E_a/2.303R$. From such a line values of E_a can be determined.

Eq. (11-35) is obviously similar in form to Eq. (8-61), the van't Hoff equation for the change of equilibrium constants with temperature. Just as the straight-line relationship for $\log K$ and $(1/T)$ depends on the assumed constancy of ΔH, so here the straight line for $\log k$ plotted against $(1/T)$ requires that E_a remain constant as temperature changes. This assumption is less often justified than it is for ΔH, and quantitative evaluation of E_a is less commonly attempted. One very simple example of an estimated activation energy is provided by experiments on the rate of dissolution of olivine sand grains (Grandstaff, 1986). Rates were measured by noting amounts of magnesium, iron, and silicon released to solution over a period of about 50 days at temperatures of 1, 26, and 49°C, and the three points obtained for each element are plotted in Fig. 11-6. The slopes of the lines are the same, within experimental error, and give an activation energy of 38.1 ± 1.7 kJ mol^{-1}. A few other estimated activation energies for dissolution reactions of geochemical interest are shown in Table 11-1. Lasaga (1984) notes that many mineral–solution reactions involved in rock alteration by metamorphic or diagenetic processes have activation energies in or near the fairly low range of 40–80 kJ, perhaps because reactions of this kind all show similar catalytic effects of sorption on mineral surfaces. Using the mean activation energy for mineral dissolution of about 60 kJ mol^{-1} (Table 11-1) and Eq. (11-35), Lasaga predicted that mineral dissolution rates will increase approximately an order of magnitude for a 30°C rise in temperature. Thus a geothermal gradient of 30°C km^{-1} in a sedimentary basin will cause reaction rates to increase 10-fold for every km of burial.

One way to visualize activation energy is to think of it as analogous to a hill that must be surmounted in a journey from a high level to a lower level (Fig. 11-7). The mixture of reactants acquires energy (from molecular motion or radiation) to

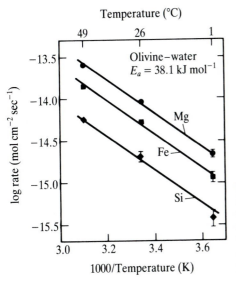

FIGURE 11-6
Logarithms of the rates of dissolution of Mg, Fe, and Si from olivine beach sand (in mol cm^{-2} sec^{-1}) plotted against the reciprocal of kelvin temperature for the determination of activation energy [Eq. (11-35), $E_a = 38.1$ kJ mol^{-1}]. (Reprinted by permission from Grandstaff, 1986.)

TABLE 11-1

Activation energies (E_a) of mineral dissolution reactions (kJ/mol)

Calcite	35.2 (5–50°C)
Quartz	67–75 (0–300°C)
Amorphous silica	61–65 (0–300°C)
Forsterite	38.1
Diopside	50–150
Enstatite	50
Augite	80

Reprinted by permission from Lasaga (1984).

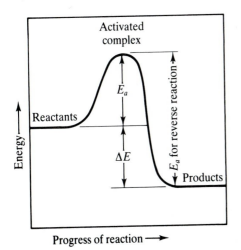

FIGURE 11-7
Diagram to show energy relations in a reaction requiring activation. The reaction shown is exothermic (the products have less internal energy than the reactants), but it can go forward only at the rate at which some of its particles acquire enough energy to get over the "energy hump" represented by the activation energy, E_a. For the reverse reaction (products → reactants), a larger total activation energy is required because it must include the difference in internal energy between reactants and products.

form an "activated complex" of its constituent particles, and until enough energy has built up to form this complex the reaction cannot get over the "energy hill"— even though the reaction as a whole (reactants → products) may be exothermic. If a reaction is reversible, the reverse process also requires activation energy, and for an exothermic process the amount required is greater than that for the forward reaction. For example, activation energies for the exothermic transformation of muscovite to kaolinite [Eq. (4-27)] were determined by Chermak and Rimstidt (1990) to be 101 kJ mol^{-1} for the forward rate and 155 kJ mol^{-1} for the reverse rate. The difference between the two activation energies is the difference in internal energy, ΔE, for the reaction, which in a reaction with no change in volume is equal to the enthalpy change, ΔH. [If there is a volume change but no pressure change, the two quantities are related by Eq. (7-23), $\Delta H = \Delta E + P\Delta V$.]

For the higher-temperature conditions of metamorphism only a few experimental and theoretical studies of reaction rates have been reported. One example is a study by Wood and Walther (1983). They estimated dissolution rate constants under near-equilibrium conditions at temperatures up to 750°C from results of published

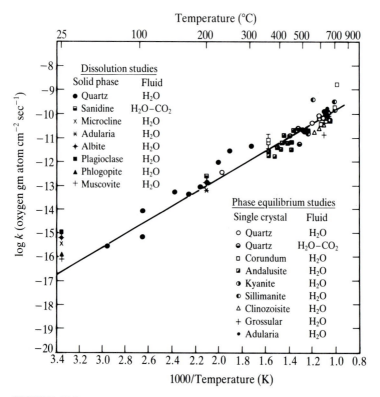

FIGURE 11-8

Logarithm of the dissolution rate constant (k, in oxygen gram atom cm^{-2} sec^{-1}) of silica and silicate minerals in near neutral solution as a function of the reciprocal of temperature in kelvin. The straight line corresponds to Eq. (11-36). (Reprinted by permission from Wood and Walther, 1983.)

phase equilibrium experiments where weight loss of a single crystal was used to monitor reaction progress. They found that most common rock-forming minerals dissolve at the same constant rate at high temperatures (Fig. 11-8), given by the equation

$$\log k = -\frac{2900}{T} - 6.85, \qquad (11\text{-}36)$$

where k is the dissolution rate constant expressed as gram atoms of oxygen per cm^2 per second. Results of the study demonstrate, not surprisingly, that many prograde metamorphic reactions are extremely fast, going to completion in times of 100's to 1000's of years, if all solutes are readily transported to and from the site of reaction.

11-4   OTHER INFLUENCES ON REACTION RATES

Besides the influence of temperature and various kinds of radiation on activation energies, reaction rates can be affected by many other agents and conditions of reaction. Especially important are the changes in rate that can be produced by water, by surface area, by organisms, and by catalysts.

The effect of water is familiar from everyday life. A knife blade remains shiny as long as it is kept dry, but rusts quickly if dampened; organic matter decomposes more rapidly when it is wet; rock surfaces weather faster in humid climates than in dry climates. In part the action of water may be explained simply by dissolution: if even a small amount of material is dissolved in the form of ions, reaction rates can be enormously increased. Or the water may participate directly in the reaction, as it does, for example, in some of the reactions of metamorphism.

Any reaction that takes place on surfaces is speeded up when surface area is increased, as shown, for example, by the acceleration in the rate of dissolution of salt or sugar if the solids are finely powdered. A more dramatic illustration comes from the disastrous "dust explosions" that sometimes result from the dispersion of large amounts of fine-grained inflammable material like powdered coal or flour in air. These substances burn slowly and quietly if ignited when in the form of compact piles, but react violently if their exposed surface is greatly increased by dispersion through a large mass of air.

Catalysts are more mysterious. A *catalyst* is a substance that merely by its presence can speed up a reaction, the amount of catalyst remaining the same after the reaction as before. If the catalyst enters the reaction at all, it is regenerated to the same extent as it is used. Two principal kinds of catalysts can be distinguished, *surface catalysts* and *chemical catalysts*. The former are substances that do not enter a reaction, but provide surfaces on which reactions take place with altered speeds; the latter are materials that take part in reactions and are regenerated as the reactions proceed.

Surface catalysts play an indispensable role in modern chemical industry. As a single example, the two gases N_2 and H_2 react hardly at all when placed together at moderate temperatures, but if in contact with metallic platinum will readily combine to form ammonia—the basis of the Haber process for producing ammonia, important in the manufacture of fertilizers and explosives. Many catalysts in nature operate similarly, probably by providing an electrically charged surface (Sec. 6-2) on which one or both reactants can be adsorbed in a state sufficiently activated for reaction to occur. In other words, sorption reactions at the surface lower the activation energy of the overall process. Clay minerals, with their enormous surface area, provide a good example in their influence on rates of diagenetic processes in sediments. Clays are also often cited as possible catalysts for the generation of petroleum hydrocarbons from buried organic matter (Sec. 15-4), and in speculative reconstructions of the reactions by which primitive organic molecules combined to form the first living organisms.

A chemical catalyst, by contrast, is part of a reaction mixture and controls rates while the mixture remains homogeneous. In some reactions the catalyst clearly acts by combining with a reactant to form an intermediate and is then regenerated, but in many processes the mechanism is not entirely clear. Chlorophyll, for example, is a catalyst for the process by which CO_2 and H_2O combine to form organic molecules and oxygen in the green leaves of plants, and large protein molecules called *enzymes* are known to catalyze many bodily processes in animals, but the pathways by which these complicated reactions occur are only partly understood. In geochemistry catalysts are often called upon to explain apparent anomalies in reaction rates, without complete understanding of the reaction mechanism.

Some reactions take place at appreciable rates only if aided by organic activity, for example the life processes of bacteria, algae, or yeast cells. An important example is the formation of nitrogen compounds from atmospheric nitrogen in the soil, a reaction that goes at significant rates only in the presence of a particular kind of bacteria that live on the roots of certain plants, particularly legumes. Bacteria also are notorious for speeding the oxidation of Fe^{2+} and its compounds, sometimes causing damage by clogging pipes with precipitated Fe(III) oxide, and creating unsightly scums of the precipitated oxide in streams draining ore deposits that contain pyrite or other iron sulfides. Still other bacteria provide an essential service by promoting decay of organic materials in the soil, in garbage, and in toxic waste dumps. In effect, the organisms act as a special kind of catalyst, by their life processes increasing rates of reactions that would be very sluggish without their aid.

In summary, reaction rates are clearly important in modifying many predictions based on the possible establishment of equilibria, but giving quantitative expression to this modifying influence in geochemical processes is difficult and generally possible only in fairly simple cases. Much of our information about kinetics is embodied in rough generalizations: rates increase with rising temperature, some rates are strongly influenced by light and other forms of radiation, water can dramatically speed up many processes, reactions involving ions are commonly rapid, many reactions that include the diatomic gas molecules in air are slow, heterogeneous reactions are generally speeded by an increase in surface area, catalysts and organisms control rates by mechanisms that remain in considerable part mysterious. The requirement that particles of reacting substances acquire activation energy before reaction can occur provides a convenient general explanation for much of the variety of observed reaction rates, but gives a basis for quantitative treatment only of relatively simple geochemical processes.

11-5 MASS TRANSFER

Chemical reaction rates are important in determining how fast events take place in geology, but they are only part of the story. Before a reaction can occur, the substances that react must be brought together, and the rate at which one reactant moves with respect to another is often the chief determiner of how fast the geologic event will proceed. The metamorphic change of calcite into wollastonite at high temperatures can be a rapid reaction in the laboratory, but how fast it will occur in a bed of limestone at an intrusive contact depends on how fast solutions from the intrusive can bring silica into the carbonate rock. We turn now to these processes of mass transfer, which are a necessary supplement to the rate chemistry we have been discussing.

Quite obviously this is a large and complex subject. For full treatment it would include not only the movement of solutions in rock, but currents in ocean, lakes, air, and magmas. Even if we restrict ourselves to interchange between solutions and rock, we would need to consider rapid movement of groundwater through open fissures, the slow penetration of water into interstices between mineral grains, and the even slower diffusion of ions and molecules through solutions and crystals— plus all gradations between these extremes. Much of this would be better called hydrology or fluid physics than geochemistry. We will avoid the complexities by limiting our consideration to a few examples where simple quantitative ideas can be applied.

Think first of the alteration of rock by reaction with constituents of a moving fluid. For a specific illustration we return to an example from the last chapter, the change of oxygen isotope ratios in the granite of a fractured and cooling pluton by low-^{18}O meteoric water moving through it (Sec. 10-4). The process can be visualized with the aid of Fig. 11-9: water moves into the pluton along fissures and mineral–grain boundaries, its contained oxygen exchanging isotopes with the rock minerals as it moves. The reaction can be symbolized

$$\mathbf{W}\delta^{18}O_{\text{fluid}}^{\text{initial}} + \mathbf{R}\delta^{18}O_{\text{rock}}^{\text{initial}} = \mathbf{W}\delta^{18}O_{\text{fluid}}^{\text{final}} + \mathbf{R}\delta^{18}O_{\text{rock}}^{\text{final}},$$

Total mass of ^{18}O in rock + fluid = Total mass of ^{18}O in rock + fluid
before reaction (nonequilibrium)   after reaction (equilibrium)

$$(11\text{-}37)$$

where $\mathbf{W}$ is the atom percent of oxygen in the system contributed by the meteoric water and $\mathbf{R}$ the atom percent of oxygen contributed by the rock. The superscript "initial" refers to the starting isotope properties of the rock and fluid, and the superscript "final" represents equilibrium values after reaction. Eq. (11-37) is a *material balance equation*, a simple statement that the number of mols of each element coming out of the reaction must equal the number going in—which is just another version of the familiar law of conservation of mass. The equation can be

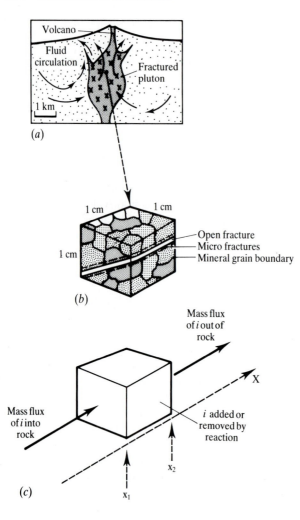

(a)

(b)

(c)

Open fracture
Micro fractures
Mineral grain boundary

Mass flux
of i out of
rock

Mass flux
of i into
rock

i added or
removed by
reaction

X

x_2

x_1

FIGURE 11-9
(a) Schematic cross section of a cooling and fractured subvolcanic pluton showing the pathways of meteoric hydrothermal fluid circulation. (b) Representative elementary volume of the pluton used to derive mass-balance constraints for the chemical component i. The cubic cm of rock is made of different types of minerals and pores consisting of throughgoing fractures, discontinuous microfractures, and grain-boundary pores. (c) Illustration of the mass-balance constraints for the flux of component i in the fluid as the fluid flows through the rock and reacts with its assemblage of minerals.

rearranged to give an expression for the atomic ratio of oxygen in water to oxygen in rock during the isotope exchange:

$$\frac{W}{R} = \frac{\delta^{18}O_{rock}^{final} - \delta^{18}O_{rock}^{initial}}{\delta^{18}O_{fluid}^{initial} - \delta^{18}O_{fluid}^{final}}. \tag{11-38}$$

Using Eq. (11-38), we can calculate the amount of liquid that is needed for the alteration of a given volume of rock. This requires measurement of the initial oxygen-isotope composition of both rock and water, the former accomplished by analyses of fresh unaltered rock samples and the latter by analysis of local meteoric water. A value for the final isotopic composition of the rock ($\delta^{18}O_{rock}^{final}$) comes from analysis of the altered material. Then to find a value for the final isotope ratio in the fluid, we assume that this is determined by isotopic equilibrium with the rock. The equilibrium is sensitive to temperature, and to make allowance for the variation with temperature we assume that it is similar to the variation for equilibrium between fluid and plagioclase feldspar—a variation that has been determined experimentally. These assumptions permit us to assign a value to $\delta^{18}O_{fluid}^{final}$ in Eq. (11-38) at any temperature that we know or assume for the alteration process. Thus values are measured or estimated for all terms on the right-hand side of Eq. (11-38), and the **W/R** ratio can be found as a function of temperature. From this kind of calculation, Taylor (1974) has found that the amount of fluid needed to accomplish the alteration seen in different hydrothermal areas varies widely, from very small up to 3 or 4 times the mass of rock. These numbers are only minimum values, because some fluid undoubtedly moves through the rock without reaching isotope equilibrium, especially if the flow is rapid. Thus the amounts of water needed to accomplish hydrothermal alteration are substantial.

To broaden the treatment of mass transfer, we now look more specifically at changes in the concentration of solutes in a fluid as it moves through rock. Again we think of the fluid as moving by roundabout paths through fractures, discontinuous microcracks, and spaces along grain boundaries [Fig. 11-9(b)]. The ratio of the volume of open space to the total volume of minerals plus openings is called *porosity*, a quantity that ranges from approximately 0.3 for uncompacted muds to values much less than 0.01 for some crystalline rocks. Porosity represents the maximum volume of fluids that can be participating in reactions with rock at any one time, and the connectivity of the openings determines the susceptibility of the rock to fluid flow. For simplicity, we consider mass transfer due to flow in only one direction, the x-direction in Fig. 11-9(c).

To express the change in concentration of a component i in the moving fluid, we use a new concentration unit, the *partial mass density*, ρ_i, defined as the grams of component i per cm^3 of fluid. The unit is preferable to the familiar g/kg or mols/kg, because we are concerned here initially with the volume of fluid flow.

The time-rate of change of the concentration of i in the fluid, then, is symbolized by the differential coefficient $(\partial \rho_i / \partial t)_{fluid}$. As illustrated in Fig. 11-9(c) for a specified volume of rock plus pores, this rate should equal the rate at which component i moves into the rock *minus* the rate at which it moves out, *plus* the amount added or removed by reaction within the rock. The mass of fluid moving through the rock is called the *mass flux*, symbolized by the vector **J**; it is measured as the grams of material passing through a unit area of the rock per second (g cm^{-2} sec^{-1}). For component i, the rate of change in concentration inside the rock of Fig. 11-9(c) due to mass transfer across faces x_1 and x_2 is equal to the change in its flux between the two faces, symbolized as the partial derivative $-(\partial \mathbf{J}_i / \partial x)_{flux}$. The negative sign tells us that the concentration of i decreases in the reaction volume if the flux at x_2 is greater than that at x_1.

The total flux of component i through the rock volume in Fig. 11-9(c) is made up of two parts: one, the flux due to bulk movement of fluid, called *advective flux*; and the other, flux due to molecular movement in the fluid, called *diffusion flux*. The concentration of i may also undergo change by reaction of the fluid with minerals in the rock, for which the rate is expressed as $(\partial \rho_i / \partial t)_{reaction}$ [an expression similar to

Eq. (11-27)]. Overall material balance, expressed in terms of rates, would then require that

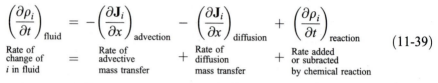

$$\left(\frac{\partial \rho_i}{\partial t}\right)_{\text{fluid}} = -\left(\frac{\partial \mathbf{J}_i}{\partial x}\right)_{\text{advection}} - \left(\frac{\partial \mathbf{J}_i}{\partial x}\right)_{\text{diffusion}} + \left(\frac{\partial \rho_i}{\partial t}\right)_{\text{reaction}}$$

(11-39)

| Rate of change of i in fluid | = | Rate of advective mass transfer | + | Rate of diffusion mass transfer | + | Rate added or subracted by chemical reaction |

This complex partial differential equation provides another means of predicting the nature and rates of element mass transfer during geochemical processes. Its application is difficult, requiring use of high-speed computers to calculate the time-rate of change of component concentrations for a specified set of physical and chemical conditions. Texts cited at the end of the chapter discuss in detail the use of the overall mass transfer equation. Here we limit discussion to a brief look at some qualitative aspects of the terms for advective flux and diffusion flux.

Advection

Mass transfer by advection is the most important process for redistributing chemical components in the Earth. On a global scale we think of mass movements in the core and mantle that give rise to the Earth's magnetic field and to plate tectonics. On a smaller scale we can list atmospheric movement of volcanic and industrial gases, ocean currents, stream and groundwater flow, hydrothermal fluid circulation, and the intrusion and extrusion of magma as processes where chemical components move by the action of fluids. Speeds of advective flow in nature range from very small up to the rates of explosive volcanic eruptions; rates of groundwater movement through rocks and sediments commonly range between 1 and 100 meters per year.

For conditions where fluid velocity is constant and there is no chemical reaction or transport by diffusion within our representative volume of rock, Eq. (11-39) reduces to

$$\left(\frac{\partial \rho_i}{\partial t}\right)_{\text{fluid}} = \left(\frac{\partial \mathbf{J}_i}{\partial x}\right)_{\text{advection}} = \mathbf{v}_{x,\text{fluid}}\left(\frac{\partial \rho_i}{\partial x}\right),$$

(11-40)

where $\mathbf{v}_{x,\text{fluid}}$ is the fluid velocity and $(\partial \rho_i / \partial x)$ is the gradient in fluid composition. For conditions specified above, the gradient in fluid composition must originate outside the representative volume of rock used to derive Eq. (11-40). Possible sources for the origin of the compositional gradient in the fluid moving through the rock might include mineral dissolution, or in the case of groundwater contamination a source for the contaminant such as a leaking chemical storage tank or a well used to inject toxic wastes into the ground.

Fluid velocity used in Eq. (11-40) reflects pressure and gravitational forces acting on the fluid. In the subsurface these forces result from hydrologic gradients produced by local topography and rock structure, from compaction during burial, and from density gradients caused by variations in salinity or temperature. The relationship between these forces and fluid flow through rocks and sediments is described by Darcy's law.

Noting that the fluid velocity (sometimes referred to as the Darcy velocity) is equal to the volume of fluid that passes through a unit area of rock [for example face x_1 in Fig. 11-9(c)] per second, we can express Darcy's law by the vector equation

$$\overrightarrow{v}_{\text{fluid}} = -\frac{k_{\text{rock}}}{\vartheta_{\text{fluid}}}(\overrightarrow{\nabla P} - \rho_{\text{fluid}}\mathbf{g}).\tag{11-41}$$

(This is a more generalized form of Darcy's law than the usual expression, $v_{\text{fluid}} = -K\,(\Delta h/\Delta 1)$, which is discussed in problem 5 at the end of the chapter.) Eq. (11-41) states that the direction and magnitude of the fluid velocity is proportional to the pressure and gravitational forces acting on the fluid. The expression $\overrightarrow{\nabla P}$ is a vector defining the direction and magnitude of the gradient in fluid pressure, and $\rho_{\text{fluid}}\,\mathbf{g}$ represents gravitational forces. For no fluid flow ($\overline{v}_{\text{fluid}} = 0$), Eq. (11-40) reduces to $\overrightarrow{\nabla P} = \rho_{\text{fluid}}\,\mathbf{g}$, an equation that defines hydrostatic equilibrium. The proportionality constant in Eq. (11-41), symbolized $k_{\text{rock}}/\vartheta_{\text{fluid}}$, is the ratio of rock permeability (k_{rock}, called the *intrinsic permeability* measured in units of cm^2) to fluid viscosity (ϑ_{fluid}, called the kinematic viscosity with units of $cm^2\ sec^{-1}$). The negative sign indicates that fluids flow in a direction opposite to the pressure gradient and in the direction of gravitational forces (i.e., fluids flow downhill).

Inspection of Eq. (11-41) shows that the magnitude of fluid velocity (1) decreases with increasing fluid viscosity (water will flow through a rock faster than petroleum when the same forces are acting on the fluids), and (2) increases with increasing rock permeability. Permeability is determined by the size, shape, and connectivity of the rock's porosity. It can be measured in laboratory and well-test experiments that produce a range in numbers from less than $10^{-16}\ cm^2$ for dense crystalline rocks and welded tuffs to more than $10^{-3}\ cm^2$ for unconsolidated, well sorted, clastic sediments. Rock permeability is a difficult property to quantify; it can vary by many orders of magnitude over short distances. Imagine, for example, the large increase in fracture permeability of igneous rocks within and near a major fault zone. For sediments the permeability is largely determined by the connectivity of intergranular pore spaces, and for crystalline igneous and metamorphic rocks by interconnected fractures. In outcrop and drillcore we see sediments whose porosity is filled with mineral cements, numerous generations of fractures in igneous and metamorphic rocks that are filled with vein minerals, and large conduits where magma once flowed (magma that is now frozen into dikes, sills, and plutons). Rock permeability must change dramatically over time in response to stresses that compact sediments or create rock fracture, and to chemical reactions that fill rock porosity with minerals.

Diffusion

Diffusion refers to the spontaneous dispersion of ions and molecules into a fluid or solid, as illustrated by the purple color that spreads slowly through a stationary beaker of water when a drop or two of potassium permanganate solution is added. Although not as efficient as advection in transporting components over long distances, it is nevertheless an important mechanism in the cementing of sediments, the growth of concretions during diagenesis, the formation of metasomatic mineral segregations in metamorphic and igneous rocks, and chemical zoning during crystal growth and exsolution.

The diffusive flux of a component i in a stationary fluid or solid is proportional to the concentration gradient,

$$\mathbf{J}_i = -D_i\left(\frac{\partial\rho_i}{\partial x}\right), \qquad (11\text{-}42)$$

a generalization known as Fick's first law. The law seems almost intuitively obvious: certainly there should be more movement of ions or molecules down a steep concentration gradient than down a more gradual one. The constant D_i is called the *diffusion coefficient*, and the minus sign indicates that movement is from regions of high concentration to those of low concentration. Measured values of D_i for common ions in dilute aqueous solutions under ordinary conditions range between 10^{-4} and 10^{-6} cm^2 sec^{-1}. Diffusion coefficients can take on a wide range of values, depending on temperature and on the nature and composition of the medium in which diffusion is occurring. In general, the diffusion coefficient increases as temperature increases.

To find how the concentration of component i at any point in a diffusing system changes with time, we combine Eqs. (11-39) and (11-42). On the assumption that there are no changes in advection or in reactions with the rock, Eq. (11-39) reduces to

$$\left(\frac{\partial\rho_i}{\partial t}\right)_{\text{fluid}} = -\left(\frac{\partial\mathbf{J}_i}{\partial x}\right)_{\text{diffusion}}. \qquad (11\text{-}43)$$

Substitution of $\mathbf{J}_i$ from Eq. (11-41) gives:

$$\left(\frac{\partial\rho_i}{\partial t}\right)_{\text{fluid}} = D_i\left(\frac{\partial}{\partial x}\left(\frac{\partial\rho_i}{\partial x}\right)\right) = D_i\left(\frac{\partial^2\rho_i}{\partial x^2}\right), \qquad (11\text{-}44)$$

provided that the diffusion coefficient is constant. Thus the time-rate of change of concentration of i in pore fluids is equal to the second derivative of the concentration gradient, a statement known as Fick's second law.

The relative importance of mass transfer by diffusion and advection in our example of a pluton undergoing alteration (Fig. 11-9) can be visualized with the aid of drawings of pores and fractures in Fig. (11-10). The chief mechanism of mass transfer in throughgoing fractures we can assume to be advection, by which fluid flow brings reactant elements into the rock volume and flushes out elements dissolved from the rock's mineral assemblage. In contrast, diffusion provides the primary means of elemental mass transfer on the grain boundaries that extend from the fracture into the rock matrix. Components brought into the rock by advection diffuse along the grain boundaries to reaction sites on mineral surfaces, and components dissolved from the minerals may diffuse along the same path in reverse, back to the fracture where they are advected out of this representative volume of rock. As the original minerals are replaced by alteration products, further reaction

Mineral grain
boundaries

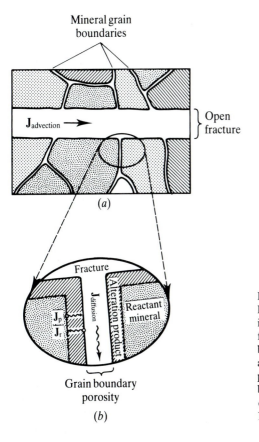

(a)

(b)

FIGURE 11-10
Details of mass flux in and near the fracture
illustrated in Fig. 11-9(b). Mass flux in the
fracture is by advection, and along the grain-
boundary pores by diffusion. The symbols J_r
and J_p represent diffusion of reactant and
product components between the grain-
boundary pore and the site of replacement
of the reactant mineral. (Modified from
Norton, 1979).

progress depends on diffusion of elements through the altered material to the still
fresh part of the minerals. This is a grossly simplified model to illustrate how
diffusion and advection work together in processes of rock alteration by circulating
fluids.

SUMMARY

Geochemical processes over the long term tend toward a state of equilibrium, but
the approach to equilibrium may be very slow. The rate of this approach is a major
determiner of geologic relations, and study of the rate has become an important part
of geochemistry. The chemistry of reaction rates, however, is less easily reduced to
simple quantitative rules than is the chemistry of equilibrium, and much of our
knowledge about it remains qualitative.

How fast a reaction will go in nature depends first on how rapidly constituents
of the reaction can move together—the rate of mass transfer—and secondly on
energy relations in the reaction itself—chemical kinetics. Mass transfer is accomplished in two ways, chiefly by the transportation of material in moving fluids

(advection) and to a lesser but often important degree by molecular diffusion. Both of these processes play obvious roles in such geologic processes as soil formation, diagenesis of sediments, rock alteration by hydrothermal fluids, and assimilation of xenoliths by magmas. Qualitatively their effects can often be described in minute detail, but quantitative expressions for their rates of activity are possible only in relatively simple situations.

Rates of simple chemical processes, once their components have been brought together, are generally describable in terms of concentrations and rate constants, but often so many steps are involved in a reaction mechanism that the resulting expression is too complex for easy use. Even for a reaction as simple as the dissolving of feldspar in acidic weathering solutions, the rate expressions are complicated and have occasioned much argument. Most reactions can take place only as their components gain sufficient energy for activation, and the amount of activation energy required is the chief determiner of rates. The energy for activation comes in part from random molecular motion, but an additional supply from external heat or various kinds of radiation is necessary for some reactions to take place at perceptible rates. The amount of activation required may be reduced, and the rate correspondingly increased, by catalysts or by the activity of organisms.

Even though strict quantitative treatment of the rates of natural processes is possible only in simple situations, study of the rates has proved to be a rewarding field of geochemical research.

PROBLEMS

1. It is often said that the rate of many reactions occurring at moderate temperatures approximately doubles with a $10°$ rise in temperature. How large are the activation energies for such reactions? To make a rough calculation, consider an elementary process, $A + B \rightarrow C$, for which the rate can be expressed as $dC/dt = km_A m_B$. The Arrhenius equation can be written

$$2.303 \log k = -E_a/RT + A.$$

Set up this equation for the two temperatures $25°$ and $35°C$ (298 and 308 K), and assume that both the activation energy E_a and the constant A are independent of temperature. If k is twice as great at $35°$ as at $25°C$, what is the value of E_a?

2. Nitrous oxide, one of the minor gases of the atmosphere, decomposes on heating: $2N_2O \rightarrow 2N_2 + O_2$. From the following data for this reaction, make a plot of $2.303 \log k$ against $1/T$ and find the activation energy.

$T(K)$	$k \ (\text{sec}^{-1})$
967	0.14
1000	0.38
1031	0.87
1052	1.67
1085	3.76
1125	11.60

3. Using the data given below, compute the mass of water necessary to dissolve one mol of quartz at 300°C and 0.5 kbar. If the final solution in equilibrium with quartz is a pore fluid in a rock with a porosity of 0.1, what is the volume of rock plus pores required to have one mol of $SiO_2(aq)$ in solution? Assume that $SiO_2(aq)$ has a negligible effect on the final volume of the solution, that is $V_{solution} = V_{H_2O}$.

$$SiO_2(\text{quartz}) \rightleftharpoons SiO_2(\text{aq})$$

$$\log K_{300°C, \, 0.5 \, kbar} = -1.92$$

$$V_{H_2O, \, 300°C, \, 0.5 \, kbar} = 23.17 \, \text{cm}^3 \, \text{mol}^{-1}.$$

4. In a cooling pluton, with an initial value of $\delta^{18}O_{rock}^{initial} = 6.5‰$, heated meteoric groundwater with $\delta^{18}O_{fluid}^{initial} = -15‰$ circulates through fractures and exchanges oxygen isotopes with the rock. Construct a graph of $\delta^{18}O_{rock}^{final}$ as a function of the ratio of the atom percent oxygen in water to rock [**W/R** in Eq. (11-38)]. On the graph show two curves for isotope exchange equilibrium, one for 200°C and another for 500°C, using the following temperature dependence of rock-to-water isotope fractionation

$$\Delta_{fluid}^{rock} = \frac{2.68 \times 10^6}{T^2} + 3.53,$$

where

$$\Delta_{fluid}^{rock} = \delta^{18}O_{rock} - \delta^{18}O_{fluid}.$$

5. One common form for expressing Darcy's law [Eq. (11-41)] is

$$v_{fluid} = -K\left(\frac{\Delta h}{\Delta l}\right),$$

where the ratio $\Delta h/\Delta l$ is the hydrologic gradient as illustrated in the cross-section below, and K is the hydrologic conductivity (measured in units of cm sec^{-1}; unlike intrinsic rock permeability, k_{rock} in Eq. (11-41), this proportionality constant includes both rock permeability and the transport properties of the fluid, such as density and viscosity). Using the above equation for Darcy's law, calculate the rate of fluid flow (cm sec^{-1}) in the sandstone aquifer of the cross-section, assuming that $\Delta h = 100$ m and $\Delta l = 10$ km for the hydrologic gradient and that the hydrologic conductivity of the aquifer is 10^{-1} cm sec^{-1}. How long will it take for rainwater falling at A to be transported to point B in the cross-section?

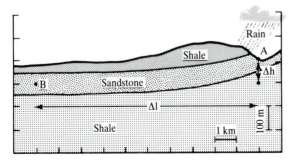

REFERENCES

Busenberg, E., and C. V. Clemency: "The dissolution kinetics of feldspars at 25°C and 1 atm CO_2 partial pressure," *Geochim. Cosmochim. Acta*, vol. 40, pp. 41–49, 1976.

Chermak, J. A. and J. D. Rimstidt: "The hydrothermal transformation rate of kaolinite to muscovite/illite," *Geochim. Cosmochim. Acta*, vol. 54, pp. 2979–2990, 1990.

Chou, L., and R. Wollast: "Steady-state kinetics and dissolution mechanisms of albite," *American Journal of Science*, vol. 285, pp. 963–993, 1985.

Christy, A. G. and A. Putnis: "The kinetics of barite dissolution and precipitation in water and sodium chloride brines at 44–85°C," *Geochim. Cosmochim. Acta*, vol. 57, pp. 2161–2168, 1993.

Dove, P. M., and S. F. Elston: "Dissolution kinetics of quartz in sodium chloride solutions: analysis of existing data and a rate model for 25°C," *Geochim. Cosmochim. Acta*, vol. 56, pp. 4147–4156, 1992.

Grandstaff, D. E.: "The dissolution rate of forsteritic olivine from Hawaiian beach sand," in Colman, S. M., and D. P. Dethier, (eds), *Rates of Chemical Weathering of Rocks and Minerals*, Academic Press, pp. 49–59, 1986.

Helgeson, H. C., W. M. Murphy, and P. Aagaard: "Thermodynamic and kinetic constraints on reaction rates among minerals and aqueous solutions. II. Rate constants, effective surface area, and the hydrolysis of feldspar," *Geochim. Cosmochim. Acta*, vol. 48, pp. 2405–2432, 1984.

Lasaga, A. C.: "Chemical kinetics of water–rock interactions," *Journal of Geophysical Research*, vol. 89, pp. 4009–4025, 1984. A review of reaction kinetics with application to bauxite formation by weathering of nepheline syenites.

Norton, D.: "Transport phenomena in hydrothermal systems: the redistribution of chemical components around cooling magmas," *Bull. Mineral.*, vol. 102, pp. 471–486, 1979. One of many papers by the author on heat and mass transfer in magma–hydrothermal systems.

Rimstidt, H. U., and H. L. Barnes: "The kinetics of silica–water reactions," *Geochim. Cosmochim. Acta*, vol. 44, pp. 1683–1700, 1980.

Taylor, H. P., Jr.: "The application of oxygen and hydrogen isotope studies to problems of hydrothermal alteration and ore deposition," *Economic Geology*, vol. 69, pp. 843–883, 1974. One of many papers by this author in which stable isotopes are used to evaluate the nature and amount of water–rock reactions and fluid flow in magma–hydrothermal systems.

Wollast, R., and L. Chou: "Kinetic study of dissolution of albite with a continuous flow-through fluidized bed reactor," in Drever, J. I., (ed.), *The Chemistry of Weathering*, Dordrecht/Boston/Lancaster, Reidel, NATO ASI ser. C149, pp. 75–96, 1985.

Wood, B. J., and J. V. Walther: "Rates of hydrothermal reactions," *Science*, vol. 222, pp. 413–415, 1983.

SUGGESTIONS FOR FURTHER READING

Bear, J.: *Dynamics of Fluids in Porous Media*, Elsevier, New York, p. 764, 1972. A standard textbook on fluid flow in porous media, with special application to problems in hydrology.

Colman, S. M., and D. P. Dethier, (eds): *Rates of Chemical Weathering of Rocks and Minerals*, Academic Press, p. 603, 1986. Contains many articles on the kinetics of chemical weathering.

Crank, J.: *The Mathematics of Diffusion*, 2d ed., Clarendon Press, Oxford, p. 410, 1983. A standard text for solving numerical problems of chemical diffusion.

Henderson, P.: *Inorganic Geochemistry*, Pergamon Press, p. 353, 1982. Chapter 8 on "Kinetic controls of element distribution" is a summary of the theory and application of diffusion to geochemical problems.

Hofman, A. W., B. J. Giletti, H. S. Yoder, Jr., and R. A. Yund, (eds): *Geochemical Transport and Kinetics*, Carnegie Institution of Washington, p. 353, 1974. Contains many papers on diffusion, reaction kinetics, and transport and reaction in rocks.

Lasaga, A. C., and R. J. Kirkpatrick, (eds): "Kinetics of Geochemical Processes," *Reviews in Mineralogy*, vol. 8, Mineralogical Society of America, 398 pp., 1981. Contains many papers on reaction kinetics.

Stumm, W., and J. J. Morgan: *Aquatic Chemistry*, 2d ed., Wiley, p. 780, 1981. Sections 2-14 through 2-16 are an excellent review of chemical kinetics, reaction rates, and heterogeneous processes.

Sverdrup, H. U.: *The Kinetics of Base Cation Release Due to Chemical Weathering*, Lund University Press, p. 246, 1990. A summary of experimental observations on weathering rates of the common rock-forming minerals.

Walther, J. V., and B. J. Wood: "Mineral–fluid reaction rates," in Walther, J. V., and B. J. Wood, (eds), *Fluid–Rock Interactions during Metamorphism*, Springer Verlag, New York, pp. 194–212, 1986. Provides several examples of the applications of reaction rates and mass transfer to metamorphic reactions.

CHAPTER

12

THE
FLUID
ENVELOPES

The Earth is fortunate among planets in having two fluid oceans, the ocean of water and the ocean of air. No other planet has a water ocean, and no other has a blanket of air with a composition anything like ours. It is the existence of these two fluid envelopes and their unique compositions that make possible the flourishing of life, another of our planet's peculiarities that, so far as we know, exists nowhere else in the universe. In this chapter we look briefly at the chemistry of these restlessly moving outer parts of the Earth.

12-1 AIR

Atmospheric chemistry seems remarkably simple (Table 12-1): just two diatomic molecules as major constituents ($N_2 + O_2 = 99\%$ by volume), a cluster of inert gases and carbon dioxide as minor ingredients, plus a highly variable content of water vapor. Only some of the gases that exist as minute traces present complications, and most of these we will ignore. Other constituents that locally and temporarily may play a role in the geochemistry of the atmosphere are aerosols—tiny suspended particles of dust, salt, and various liquids.

Measured proportions of the two major atmospheric gases under natural conditions show little change from place to place or from time to time. Some of the minor constituents vary a great deal, with water vapor as the most familiar example.

TABLE 12-1
Composition of air (near-surface), in volume percent

Nitrogen	Oxygen	Argon	CO_2	Neon	Helium	Krypton	Xenon
78.084	20.946	0.934	0.035	0.0018	0.0005	0.0001	0.00009

Reprinted by permission from Mason and Moore, 1982.

Carbon dioxide shows some change with the seasons and a disturbing long-time increase that will need our attention in a moment. Ozone, a modification of oxygen with three atoms per molecule rather than two (O_3), is another minor gas that shows large fluctuations. And the variability of rarer gases like SO_2 and HCl, depending on proximity to heavy industry or to volcanoes and hot spring areas, is well-known from common experience.

The rough constancy in concentration of O_2 and N_2 is easy to understand, just because of the enormous amount in the atmosphere and the rapidity with which it circulates. Many processes are at work producing and using up the two gases, but they are slow enough and on a scale small enough that overall amounts are not affected except very locally. Oxygen is produced chiefly by photosynthesis in the green leaves of plants, and is consumed by respiration in organisms and by a variety of oxidation processes that we have talked about before. Nitrogen is a less active gas, produced and consumed in much smaller amounts by a variety of complex organic and inorganic reactions, many of them catalyzed by micro-organisms.

The lesser activity of nitrogen means that the N_2 molecule is particularly stable, difficult to split into separate N atoms that can combine with other atoms, and that nitrogen compounds are unstable because N atoms readily combine to form N_2. Despite their instability, the formation of nitrogen compounds is all-important, because some of these compounds are essential constituents of living material. The compounds have their origin in the life processes of plants, and the plants acquire their nitrogen from simpler compounds dissolved in soil water. For life to persist requires that there be a continual renewal of these simple nitrogen compounds in the soil, because neither higher plants nor animals can use nitrogen directly from the air. The production and consumption of atmospheric N_2, then, is primarily a story of the maintenance of dissolved compounds available to plant roots.

The story can be summarized by a diagram like Fig. 12-1, in which the boxes represent the three reservoirs where nitrogen is stored near the Earth's surface (atmosphere, biosphere, and soils and water) and the arrows indicate processes by which nitrogen is transferred among the reservoirs. Two of the arrows show processes by which nitrogen is lost from the atmosphere and ultimately becomes part of living material (the biosphere). The more important of the two is *nitrogen fixation*, a series of complex reactions catalyzed by blue–green algae and by a kind of bacteria that live on the roots of leguminous plants, reactions by which N_2 molecules are broken and the atoms are fixed in organic compounds, chiefly as $-NH_2$ (amino) groups in the structure of proteins. The other arrow leading away

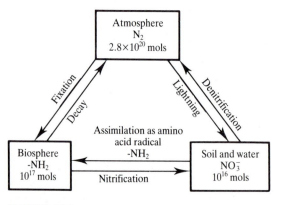

FIGURE 12-1
A simplified diagram of the biogeochemical cycle for nitrogen at the Earth's surface (referred to as the *exogenic cycle for nitrogen*). The boxes represent three natural reservoirs (atmosphere, biosphere, and hydrosphere including soils) where compounds of nitrogen reside for varying amounts of time. The most abundant nitrogen species and the total estimated mass of nitrogen are shown for each reservoir. Arrows indicate processes by which nitrogen moves from one reservoir to another. Net gains and losses of nitrogen from each reservoir are determined by the rates or overall flux of nitrogen due to each process, with the total mass of nitrogen in all the reservoirs remaining constant. The diagram does not show anthropogenic contributions to the flux of nitrogen, nor does it include nitrogen in sediments or sedimentary rocks (called the *endogenic cycle for nitrogen*). (Modified from Faure, 1991.)

from atmospheric N_2 symbolizes a reaction that occurs where air is locally heated, say in lightning discharges or in internal-combustion engines, a reaction that leads to combination of N_2 and O_2 to form nitrogen oxides. The oxides (chiefly a variable mixture of NO and NO_2, often symbolized NO_x) when dissolved in water may be further oxidized to nitrate ion, NO_3^-. This is one of the simple compounds that plants can take in through their roots and use in the manufacture of cell material. Thus there are two routes for atmospheric N_2 to nitrogen compounds in the biosphere: one direct, by the agency of blue–green algae or nitrogen-fixing bacteria, the other indirect (and relatively minor) via the formation of nitrogen oxides and nitrate ion.

The other principal simple compound that plants can take from soil water is ammonia (NH_3) in the form of ammonium ion, NH_4^+—not represented explicitly in Fig. 12-1, but properly placed in either the biosphere or soil and water box. Ammonia is generated when organic matter decays, in yet another reaction dependent on the activity of micro-organisms. Decay may follow other routes, for example recombination of N atoms and a return of N_2 to the atmosphere. Still another alternative for decay, shown by the arrow marked *nitrification*, is oxidation of some of the nitrogen to NO_3^-. Thus organic activity and decay partly balance each other, decay producing NH_4^+ and NO_3^- which are converted into living substance by organisms, and the living substance again generating these ions when it decays. The balance is not complete because some nitrogen along the way is lost to the air, both from decay of organic matter and directly from the solutes in soil water (*denitrification*). To make up for this loss, the natural organic world depends on the processes of nitrogen fixation by blue–green algae and bacteria and by the formation of NO_x in lightning discharges. Humans, of course, have learned to help in compensating for the loss of combined nitrogen by adding fertilizer to the soil. But the amounts of nitrogen involved in all these processes are so small in comparison with the great reservoir in the atmosphere that no long-term changes in the N_2 content of air are detectable.

With CO_2 the situation is different, because its concentration in air is small (about 0.035% by volume) and sensitive to changes in the rates of its formation and removal. The gas is produced both by the reaction of carbonates with acid and by the precipitation of calcium carbonate from Ca^{2+} and HCO_3^- [Eq. (3-1) and the reverse of reaction Eq. (3-4)], also by the respiration of organisms, by decay of organic matter, and by volcanic activity. It is removed from air during the photosynthetic production of organic compounds in plant leaves and by many of the reactions of weathering and diagenesis. These natural processes maintain a rough overall balance, although seasonal differences in the rate of photosynthesis are enough to cause measurable fluctuations in its concentration. Amounts of CO_2 are also markedly affected by human activities: great quantities are produced by the burning of fossil fuels, and the destruction of forests adds to the amount both by the burning of the wood and by the decrease in amounts of the gas that would normally be taken up by photosynthesis. Such human operations are almost certainly responsible for the observed increase in CO_2 over the past century (from about 0.029 to 0.035 volume percent).

This continuing increase is a cause for worry, because CO_2 is a major control of surface air temperatures. Heat energy from the sun, which reaches the earth chiefly in the form of short-wavelength radiation (visible and ultraviolet), is reflected by the Earth's surface back into space as long-wavelength radiation. Carbon dioxide helps to absorb this radiation, and thereby heats the atmosphere. The effect is similar to the trapping of heat in a greenhouse: the glass roof of the greenhouse permits light to enter, but the reflected longer-wavelength radiation cannot escape. Because of this analogy, the global warming of the atmosphere by increasing concentrations of CO_2 (and other radiation-absorbing gases, of which water vapor and methane are the principal ones) is often called the "greenhouse effect." The magnitude of the effect on Earth temperatures is difficult to predict, but a common estimate is an overall rise of 2–3°C by the middle of the 21st century if the present rate of increase in CO_2 concentration persists. Such a rise could have a marked effect on climate, with possible wide-scale melting of glaciers, a rise in sea level, and a drastic change in the distribution of arid and well-watered lands. The possibilities are sufficiently alarming to suggest a need for government action to attempt some control of CO_2 emissions.

This is by no means the only sort of long-term atmospheric change ascribable to human activity that may require political response. Another conspicuous example is the distribution of the minor gas ozone, O_3, normally produced from O_2 molecules in tiny amounts through a complex reaction aided by sunlight. Ozone has a Jekyll-and-Hyde personality. In the lower atmosphere, where we are in contact with it, ozone in more than trace amounts is a poison, and is also a major generator of smog. But in the high atmosphere ozone benefits us, by absorbing much of the ultraviolet part of the Sun's radiation and thereby screening the Earth's surface from radiation that is damaging to organisms and particularly to human skin. The reaction that produces ozone at lower levels is catalyzed by minor constituents of the atmosphere, especially by hydrocarbons and nitric oxide, two materials that are abundant in the exhaust gases of internal combustion engines. This means that ozone can reach dangerous levels in cities with heavy automobile traffic, unless cars are equipped with catalytic converters to reduce emission of NO. In the high atmosphere, on the other hand, ozone is an asset, and the maintenance of adequate concentrations for ultraviolet screening is a matter of great environmental importance. Unhappily, it has been found that some useful but highly volatile compounds produced by chemical industry can rise into the stratosphere and destroy ozone. Chief among these compounds are the chlorofluorocarbons (compounds of C, Cl, and F), whose disintegration products catalyze the conversion of O_3 back into O_2. The danger from decreasing amounts of high-level ozone is sufficient to require government action, and in recent years the production of chlorofluorocarbons has been sharply curtailed in most industrial countries. Thus ozone is a benefit to the environment in the high atmosphere and a detriment in the low atmosphere, and human activity is changing its concentration in the wrong direction at both levels.

By way of summary, the Earth's envelope of air consists chiefly of two gases, both of them important in life processes and both effectively balanced between production and removal. They are present in such huge amounts that their concentrations are roughly constant even over geologic times, and are not appreciably influenced by anything humans can do. Some minor gases in the atmosphere are more delicately balanced, and their concentrations can be measurably affected by human and geologic activities (see Prob. 12-2). Most notable of these is carbon dioxide, for which the rate of production is being increased and the rate of removal decreased sufficiently by human operations to make its growing concentration a serious concern for the future. Ozone is another of the problem gases, a threat to the environment both because of its human-generated decrease in the high atmosphere and its increase in the lower atmosphere.

12-2 THE WATER ENVELOPE: RAIN, STREAMS, AND LAKES

The Earth's envelope of water has many parts, of which by far the largest volume is occupied by the ocean (Table 12-2). The parts with which most people have more intimate contact are the waters that fall as rain and flow to the sea in streams or collect in lakes and swamps, and to the chemistry of these waters we give out first attention.

One can think of water at the Earth's surface as moving constantly in a huge cycle, powered by energy from the Sun (Fig. 12-2). Water is evaporated from the ocean, falls as rain or snow, collects in streams, in part finds its way underground, and ultimately returns again to the sea. Some of it may be held up temporarily in the form of glacial ice. En route, the moving water in rain and streams picks up a variety

TABLE 12-2
The water envelope

	Volume $\times 10^6$ km^3	Volume percent
Oceans	1370	97.3%
Ice caps and glaciers	29	2.1%
Groundwater	9.5	0.6%
Lakes	0.125	
Soil moisture	0.065	
Atmosphere (liquid equivalent of H$_2$O vapor)	0.013	<0.01%
Rivers	0.0017	<0.01%
Biosphere	0.0006	<0.01%

Reprinted by permission from Berner and Berner (1987, pp. 13, 63–69).

of soluble materials, first from the air, then from the rocks and soil of stream beds, and often a large amount from long contact with rock and mineral surfaces underground.

In the rainwater part of its cycle, water remains generally a very dilute solution, but surprisingly variable in composition from place to place (Table 12-3). Some of the variation, especially near the seacoast, is explained by the salt content of water droplets blown inland from breaking waves ("cyclic salt"). Other obvious sources of variation are volatile and particulate materials from industrial plants, automobile exhausts, and volcanic eruptions. The pH of rainfall is commonly in the

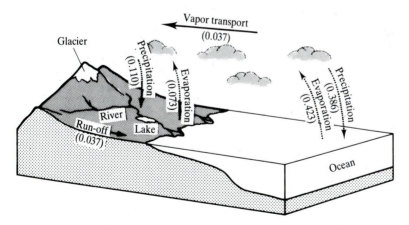

FIGURE 12-2
The water cycle. Arrows indicate processes of water transport, and the parenthetical numbers ($\times 10^6$ km^3 yr^{-1}) are the volume fluxes of H$_2$O due to evaporation, precipitation, atmospheric vapor transport, and runoff (includes surface and ground waters). (Figure modified from data summarized by Berner and Berner, 1987.)

TABLE 12-3
A sampling of rainwater analyses (parts per million)

	Minnesota	N.E.U.S.	New Mexico	Hawaii	France	Russia	Amazon
Na^+	0.20	0.27	0.07	5.46	0.92	2.4	0.23
K^+	0.17	0.16	0.12	0.37	0.16	0.7	0.012
Mg^{2+}	0.23	0.11	0.08	0.92	0.39	0.5	0.012
Ca^{2+}	0.69	0.60	0.70	0.47	0.68	2.0	0.008
Cl^-	0.15	0.45	0.33	9.63	2.13	1.8	0.39
SO_4^{2-}	1.53	4.3	3.29	1.92	2.8	5.7	0.28
NO_3^-	1.24	0.34	1.12	0.2	1.9	0.8	0.056
NH_4^+	0.60	0.22	—	0.1	0.29	0.6	—
pH	5.0	4.4	5.0	4.8	4.8	5.9	5.31

Reprinted by permission from Berner and Berner (1987, pp. 13, 63–69).

range 4.5–6; a figure of about 5.5 would be expected from the dissolving of atmospheric CO_2 in raindrops as they fall, but it often varies considerably from this because of other solutes, especially those generated by human activity.

Among the human-generated solutes, two kinds that have an especially pronounced effect on pH are the oxides of sulfur and nitrogen. The chief sulfur oxide is SO_2, produced in large amounts by the burning of fossil fuels and the smelting of sulfide ores. This compound is slowly oxidized in the atmosphere to SO_3, in a reaction catalyzed by minor constituents of air:

$$SO_2 + \tfrac{1}{2}O_2 \rightarrow SO_3. \qquad (12\text{-}1)$$

The trioxide reacts with the water of raindrops to make sulfuric acid:

$$SO_3 + H_2O \rightarrow 2H^+ + SO_4^{2-}. \qquad (12\text{-}2)$$

For nitrogen the reaction is slightly more complicated, because the oxide produced is a variable mixture of two gases, NO and NO_2. These compounds react with O_2, again with the aid of minor atmospheric catalysts, to form HNO_3:

$$2NO + \tfrac{3}{2}O_2 + H_2O \rightarrow 2H^+ + 2NO_3^- \qquad (12\text{-}3)$$

and

$$2NO_2 + \tfrac{1}{2}O_2 + H_2O \rightarrow 2H^+ + 2NO_3^-. \qquad (12\text{-}4)$$

In areas where industry is concentrated and automobile traffic is heavy, the amounts of these acids can be sufficient to lower the pH of rainwater by one or two units (to measured values below 4.0), making it potentially harmful to vegetation and to the aquatic life of lakes and streams. Such "acid rain," defined as rain with a pH lower than 5.5, has become notorious for the supposed killing of forests and the destruction of fisheries in eastern Canada and northern Europe, regions that are downwind from high concentrations of population and heavy industry. The sources of the acidity noted in surface waters and the amount of damage that can be attributed to it are still not entirely clear, but the evidence is sufficiently convincing to have instigated government action to reduce emissions of sulfur and nitrogen gases.

Rainwater that collects in streams retains whatever solutes it has picked up in falling through the atmosphere, and adds material it dissolves from rocks and soil. The total salt content of stream water is generally small, averaging something like 100 ppm and seldom exceeding 10,000 ppm (1%). Despite the low concentration, the amount of erosion accomplished by solution alone is far from negligible. An often quoted estimate of the average for dissolved material removed from the continents is about 30 tons/km^2 per year, which amounts to an average lowering of the land surface by 1 cm in every thousand years. Compositions of the dissolved material in a few rivers, and an estimate of the world average, are shown in Table 12-4.

The composition of stream water depends partly on climate and partly on the kinds of rock and soil in a drainage area. The latter is probably the most important, but tracing the origin of solutes in stream water is difficult and often a subject of debate. In the great majority of streams the ions with highest concentration are Ca^{2+} and HCO_3^-, probably because sedimentary rocks cover a large part (roughly 75%) of the Earth's land surface, and the $CaCO_3$ in these rocks is readily soluble in water

TABLE 12-4
Analyses of river waters, in parts per million

	A	B	C	D	E	F	World average
HCO_3^-	93	101	183	108	149.2	22.5	55.9
SO_4^{2-}	25	41	289	19	0.44	3.0	10.6
F^-	0.0	0.1	0.2	0.5	—	—	—
Cl^-	5.0	15	113	4.9	8	3.9	8.1
NO_3^-	1.2	1.9	1.0	0.3	0.44	0.2	0.84
Ca^{2+}	32	34	94	23	17.4	6.5	15.0
Mg^{2+}	4.9	7.6	30	6.2	5.2	1.0	3.9
Na^+	4.8	11	124	16	30.7	3.1	6.9
K^+	2.0	3.1	4.4	0.0	11.8	1.0	2.1
Fe	0.07	0.02	0.01	0.28	—	0.03	0.41
Al	0.30	1.0	0.01	0.24	—	0.04	—
SiO_2	4.9	5.9	14	13	25.6	11.2	13.1
Total	173	221	853	191	249	52.9	116.9

A. Hudson River, Green Island, NY
B. Mississippi River, Baton Rouge, LA
C. Colorado River, Yuma, AZ
D. Columbia River, 3 miles above The Dalles, WA
E. White Nile, near Khartoum, Sudan
F. Amazon River, Brazil, average
Sources: A–E, Livingston, D. A.: "Chemical composition of rivers and lakes," Chapter G in *Data of Geochemistry*, 6th ed., *U.S. Geol. Survey Prof. Paper* 440-G, 1963; F and World Average: Gibbs, R. J.: "Water chemistry of the Amazon River," *Geochim. Cosmochim. Acta*, vol. 36, pp. 1061–1066, 1972.

containing CO_2 [Eq. (3-4)]. More rarely the content of Na^+ exceeds that of Ca^{2+}, and either Cl^- or SO_4^{2-} becomes the dominant anion (e.g., Colorado River analysis, Table 12-4); such compositions are often explained by precipitation of $CaCO_3$ as a stream flows through an arid climate, but could equally well result from absence of $CaCO_3$ in the principal source material, or from contact of the stream with salt in adjacent beds or dissolved in groundwater. Magnesium is usually less abundant than calcium, presumably because common magnesium minerals are in general less soluble than those of calcium. Potassium is nearly always subordinate to Na^+, a fact for which at least four circumstances can be held accountable: (1) the much higher Na^+ in cyclic salt from the ocean, (2) the more rapid weathering of plagioclase feldspar than of K-feldspar, (3) the more extensive use of K^+ by vegetation, and (4) the greater sorption of K^+ on clays and organic materials. Thus explanations can be guessed for sources of and relations among the major solutes in stream water, but seldom with complete confidence.

Pollution from human activities may have a pronounced effect on stream water composition. Intense agriculture near a stream, with heavy use of fertilizers, may introduce large amounts of nitrogen, phosphorus, potassium, and organic matter. The mining of ores containing sulfide minerals, particularly pyrite, adjacent to a stream course is a notorious source of pollution (see Fig. 6-9). Oxidation of the ore or waste produces sulfuric acid and may lead to the formation of colloids or precipitates of Fe(III) hydroxides,

$$2FeS_2 + \tfrac{15}{2}O_2 + 7H_2O \rightleftharpoons 2Fe(OH)_3 + 4SO_4^{2-} + 8H^+, \qquad (12\text{-}5)$$

pyrite ferric hydroxide

a combination that ensures discoloration of the stream water and a drastic lowering of pH (see Sec. 6-2). If other sulfides are present in the ore, for example those of lead, cadmium, or mercury, these toxic metals may dissolve in amounts making the water unfit for use by humans or animals. Large amounts of pollutants, particularly organic matter and various materials yielding the ions Na^+, Cl^-, and SO_4^{2-}, are being added to streams in the form of industrial and human waste in densely populated areas. Contamination of these various kinds has become a major factor in the alteration of stream water compositions.

The composition of lake waters is also variable. Like stream water, the composition of water in a lake depends on complex interactions with the atmosphere, sediments, rocks, biological communities, and anthropogenic sources. In addition, physical processes in lakes related to seasonal changes in precipitation, surface evaporation, and deep circulation may also have marked effects on dissolved materials. In many lakes the composition and temperature of the water are stratified, with a layer of warmer low-density water floating on cold, higher-density water. There is little mixing between the two kinds of water, except in the fall and spring when temperature differences between the layers may be small or temporarily reversed. The cooler and more saline bottom waters have low concentrations of dissolved oxygen because of the decomposition of organic material that sinks to the floor of the lake; conditions become strongly reducing, and the water may have high concentrations of nitrite, ammonia, and Fe(II).

Although lakes make up only a small fraction of the water in the hydrosphere, they are important to humans as a source of water for domestic, industrial, and agricultural purposes, for recreation, and for sewage and refuse disposal. The study of lakes, called *limnology*, has become increasingly important in recent years, and requires an interesting combination of geochemistry, hydrology, and biology.

12-3   THE WATER ENVELOPE: OCEANS

Income and Outgo

Most of the water that falls as rain and collects in streams finds its way ultimately into the sea. The ocean is the Earth's great reservoir of water, covering nearly 70% of the surface and making up approximately 97% by volume of the hydrosphere (Table 12-2). Into the oceans go also the dissolved materials that rain and streams have collected from their contact with air, soil, and rocks. Some constituents of ocean salt come from other sources—windblown dust, the eruption clouds of volcanoes, reaction with sediments and basalt of the ocean floor—but far and away the major contributor is the water supplied by streams.

What happens to the dissolved material when it reaches the sea? Can we assume that it simply accumulates—that the present concentration of salts is just the result of constant addition from streams since the ocean was formed? This hardly seems likely, and a crude calculation shows that such a guess is clearly wrong. The total quantity of river water reaching the sea each year is estimated to be about 36,000 km^3, or roughly 36×10^{18} g, and we read from Table 12-4 that it contains an average of about 120 ppm of solute. The yearly addition of solute is then the product of these numbers, $36 \times 10^{18} \times 120 \times 10^{-6}$, or about 4×10^{15} g. The total quantity of seawater is 1370×10^6 km^3, or roughly 1.4×10^{24} g, and the amount of salt it contains is 3.5% of this, or about 49×10^{21} g. To accumulate this quantity of salt would then take $49 \times 10^{21}/4 \times 10^{15}$ or about 12×10^6 years. Because the ocean is considerably older than 12 million years, we infer that dissolved material is not simply accumulating, but must be continually removed from the ocean at a rate comparable to its rate of supply. And because the amounts of various solutes in stream water and seawater are so very different (Table 12-5), we know that some materials are being removed much more rapidly than others.

Evidently, then, an approximate balance has been established in the ocean between income and outgo of dissolved substances, a balance that can be quantitatively expressed by an explicit statement of the material balance of Eq. (11-39). Study of ancient salt deposits shows that this balance, or at least a rough state of near balance, has existed for a long time in the past (Sec. 14-6). How is this balance maintained? What are the processes by which substances brought in by streams are removed from the ocean? To attack this question requires a brief excursion into chemical oceanography, an enormously complex subject that would demand another book, or several books, for full treatment. Here we can do no more than touch on a few highlights.

TABLE 12-5

Principal dissolved substances in seawater, submarine hot springs, seawater–basalt experiments, and average river water (ppm)

Solute	Seawater	Submarine hot-spring	Seawater–basalt experiment	Average river water
Na^+	10800	11725	10468	5.15
K^+	407	1009	459	1.3
Ca^{2+}	413	805	1134	13.4
Mg^{2+}	1296	0	3	3.35
Cl^-	19010	20530	18500	5.75
SO_4^{2-}	2717	0	<1	8.25
HCO_3^-	137	0	—	52.0
SiO_2	0.5–10	325	828	10.4

Sources:
Seawater, Riley, K. P., and R. Chester: *Introduction to Marine Chemistry*, Academic Press, 1971.
Submarine hot spring, von Damm, K. L., J. M. Edmond, B. Grant, C. I. Measures, B. Walden, and R. F. Weiss: "Chemistry of submarine hydrothermal solutions at 21°N, East Pacific Rise," *Geochim. Cosmochim. Acta*, vol. 49, pp. 2197–2220, 1985. Data are for Sample NGS, T = 273°C.
Experimental seawater–basalt reaction, Seyfried, W. E., Jr., and J. L. Bischoff: "Experimental seawater–basalt interaction at 300°C, 500 bar, chemical exchange, secondary mineral formation and implication for the transport of heavy metals," *Geochim. Cosmochim. Acta*, vol. 45, pp. 135–147, 1981. Fluid composition reported for experiment with water-to-rock mass ratio of 10, at 300°C and 500 bar, after 1968 hours of reaction.
Average river water, Meybeck, M.: "Concentration des eaux fluviales en éléments majeurs et apports en solution aux océans" *Rev. de Géol. Dynam. et de Géogr. Phys*, vol. 21, pp. 215–246, 1979. Compare this world-wide average with that reported by Gibbs in Table 12-4.

Marine Geochemistry: General

The ocean as a chemical system consists of a complex solution with a total salt content, or salinity, of 3.5% (or, as usually stated, 35‰ or 35 parts per thousand), plus a variable amount of dissolved gases and organic matter. It is in contact with basalt (the basalt of the oceanic crust), detrital sediments, and a great number of precipitated materials on its floor. Its temperature is variable near the surface, but nearly constant and very cold (2–3°C) in deep water. The low bottom temperature is maintained by slow-moving currents that carry cold surface water from arctic and antarctic seas down along the sea floor toward the equator. The currents are part of a huge circulation system that brings dissolved oxygen and carbon dioxide to deep water and keeps the ocean stirred on a time scale of 1000–2000 years. Pressures in the sea depend chiefly on depth, reaching maximum values of about 400 bars.

Among the dissolved materials in seawater (Table 12-5), six ions stand out as the most abundant and the most constant in concentration in all parts of the ocean: Na^+, K^+, Ca^{2+}, Mg^{2+}, Cl^-, SO_4^{2-}. Because of their uniform concentrations and their only modest participation in organic processes, these ions are called *conservative*. Some of the less abundant substances (notably HCO_3^-, SiO_2, and the ions of nitrogen and phosphorus) are *nonconservative*, in the sense that they take a very active part in life processes and hence show widely different concentrations depending on the local abundance of living organisms and their debris.

Seawater pH is remarkably constant, seldom straying from the narrow range of 7.8 to 8.4. Reactions of carbonate species are the principal determiners of pH. In large part these are inorganic reactions involving equilibria in the system H_2O–CO_2–CaO, equilibria that we discussed at length in Chapter 3. The carbonate reactions maintain the ocean as a gigantic buffer system (Sec. 2-4), able to absorb large additions of acid or alkali without an appreciable change in pH. Helping with the buffer action in a minor way are compounds of boron:

$$H_3BO_3 + OH^- \rightleftharpoons H_2BO_3^- + H_2O, \tag{2-47}$$

and

$$H_2BO_3^- + H^+ \rightleftharpoons H_3BO_3. \tag{2-48}$$

Over long periods of time still another buffer system is probably important, a system involving silicates: one can suggest, for example, a reaction involving feldspar and clay minerals,

$$KAlSi_3O_8 + H^+ + \tfrac{1}{2}H_2O \rightleftharpoons K^+ + \tfrac{1}{2}Al_2Si_2O_5(OH)_4 + 2SiO_2(aq), \tag{8-4}$$
$$\text{K-feldspar} \qquad\qquad\qquad\qquad \text{kaolinite}$$

or alternatively a change from one clay mineral to another, which would mean consumption of hydrogen ions in one direction and release in the other. Such reactions would be very slow, and how active they are in controlling pH remains uncertain. Other reactions with some influence on pH involve organic materials—photosynthesis in the upper part of the sea, respiration of organisms, and decay of the bodies of dead plants and animals.

Living creatures play a major role in marine chemistry. In the uppermost layer of the oceans where sunlight can penetrate (down to depths of about 300 m), photosynthesis by phytoplankton produces the organic matter that serves as food for other creatures. The reaction is often expressed symbolically in the deceptively simple form

$$CO_2 + H_2O \rightarrow \underset{\text{organic matter}}{CH_2O} + O_2, \tag{12-6}$$

where CH_2O stands for a variety of complex compounds that become part of living tissue. This statement of the reaction shows the important bare essentials—the production of organic matter and free oxygen from carbon dioxide and water—but it is so simplified as to be misleading. It says nothing about the energy from sunlight that drives this endothermic process, or the catalysts without which it could not take place, or elements like nitrogen and phosphorus and a host of minor ones that are included in the organic matter as it forms. For example, one suggested reaction written for the stoichiometric composition of average marine plankton appears as

$$106CO_2 + 16NO_3^- + HPO_4^{2-} + 122H_2O + 18H^+ \rightarrow \underset{\text{average plankton}}{C_{106}H_{263}O_{110}N_{16}P} + 138O_2.$$

$$\tag{12-7}$$

The organic matter produced by marine photosynthesis undergoes many transformations: first a build-up of the many complex compounds needed for life processes, then when organisms die the breakdown of these compounds, partly by oxidation and partly by anaerobic decay, eventually to release their contained elements back to solution in seawater.

Nonconservative Elements

The elements that take part in organic reactions may be pictured as engaged in cycles that carry them through complex reactions and ultimately bring them back in large part to their original form. *Carbon*, for example, has its source in the CO_2 of air; it dissolves to form HCO_3^- and through photosynthesis becomes part of organic molecules; it is precipitated out of sea water both as calcium carbonate and as a constituent of dead organic matter; these partly accumulate on the sea floor and partly are decomposed (through decay and slight changes in pH), releasing the carbon once more as CO_2. The cycle can be extended for long times, so that ultimately, through geologic processes, even the carbon buried in limy sediments and dead organic matter will be returned to the atmosphere. *Nitrogen*, as N_2, also comes originally from air; blue–green algae among the near-surface phytoplankton and bacteria that grow on the roots of land plants (chiefly legumes) convert it to dissolved nitrate (NO_3^-) and ammonium ion (NH_4^+), which can be used by plants for the manufacture of protein; decay of the protein produces again one or both of the same two ions; both the organic matter and ions are unstable and readily decompose (usually aided by bacteria) to N_2 (Fig. 12-1). *Phosphorus* has its home in rocks, chiefly as a constituent of apatite (Sec. 14-4); it is dissolved and brought to the sea as one of the phosphate ions, chiefly HPO_4^{2-}; life processes incorporate it in protein, and decay of protein re-creates the ions; the phosphorus that is buried as a constituent of the organic matter of sediments becomes part of rock, and ultimately through diagenesis or metamorphism again is changed to apatite. *Silicon*, dissolved from rock, is used by some organisms (diatoms, radiolaria) to build shells, and the material of the shells after burial becomes part of the silica and silicate minerals in rocks.

Thus the residence of these elements in the sea is brief, only a part of recurrent cycles that make them for a time part of living matter and then return them to their inorganic forms as sediments or dissolved species. Because they enter the sea in only limited quantities and are used avidly by marine organisms, their concentrations fluctuate with the amount of organic activity. In places any one of them may be reduced to a concentration so low that growth of organisms is inhibited or temporarily stopped altogether. The element that most commonly serves in this growth-limiting role is phosphorus.

All of the nonconservative elements, it should be noted, may be taken out of their short-term cycles for geologically long periods by incorporation in sediments and rocks, then returned to the sea only by later uplift and erosion. This semipermanent removal is especially notable for two of the elements, carbon and silicon, simply because of the abundance of their sedimentary forms. For carbon the principal sediment is calcium carbonate, in the form of shell and skeletal material produced by many kinds of organisms, in largest amount by one-celled plants and animals. The shell material may be either calcite or aragonite or alternating layers of both, and commonly contains a few percent of Mg substituting for Ca. Familiar forms of limy sediments are coral reefs, accumulations of shells and shell fragments, and limy muds, the last being by far the most abundant. Limy muds cover large areas of the sea floor, down to depths of about 4 km; below this depth, called the *carbonate compensation depth*, the combination of high pressure and extreme cold makes $CaCO_3$ soluble (Fig. 3-4). Because most of the sea floor is above this depth, much of the limy sediment is stable and can accumulate in thick beds that ultimately may harden into limestone.

Silicon likewise is precipitated out of seawater by organisms, mostly tiny creatures (diatoms and radiolaria) that use opaline silica for their shells. Silica muds, like carbonate muds, accumulate in quantity on parts of the sea floor, and this time there is no depth restriction. Part of the precipitated silica may be in the form of a gel, and the hardening of these different kinds of silica produces ultimately the familiar sedimentary rocks called diatomite, radiolarite, and chert.

Conservative Elements

We now turn to the major elements, the conservative ones whose concentrations are so large that they are little affected by any organic processes in which the elements may take part. How are these concentrations preserved at such nearly constant values? As we have noted, this question refers chiefly to processes of removal. The major processes of supply are obvious, but what reaction or combination of reactions keeps concentrations from steadily increasing? Here we face a major problem in marine geochemistry: identifying the processes by which the elements Na, K, Ca, Mg, S, and Cl are taken out of seawater as fast as they are supplied to it by rivers and streams. It is a problem still not completely solved.

Some removal processes are obvious. We think, for example, of calcium precipitating as calcite or aragonite, generally in the form of shells built by organisms. Or we think of sulfur, converted to pyrite when sulfate is reduced by organic matter in bottom sediments. Magnesium, at least to a small extent, would form magnesian calcite or possibly dolomite. All these elements may be sorbed on mineral surfaces as discussed in Chapter 6, and all would be present in the interstitial liquid that is enclosed in the pores of sediments. All would also be removed in the salts formed when seawater evaporates in restricted basins—a process that is almost nonexistent today, but that certainly has been active at times in the geologic past (Sec. 14-5). To list these methods of removal is easy, but to evaluate their quantitative significance is a problem that has given much trouble and occasioned much argument.

Another means of removal has come to light in recent years, which at least for some of the elements promises a good answer to the problem: reaction with basalt of the oceanic crust. In part this may be a very slow general reaction as seawater drifts over the Earth's surface and penetrates into the pores and fractures of the cold basalt underlying the great ocean basins, but the reaction is more spectacular and probably more effective in submarine hydrothermal systems that develop along the mid-ocean ridges. These ridges, including the Mid-Atlantic Ridge, the East Pacific Rise, the Indian Ocean Ridge, and many connecting ridges, constitute the world's largest mountain range, extending over 55,000 km in length.

Beneath the mid-ocean ridges, basaltic magma wells up from the mantle as the oceanic plates on either side spread apart. Magma is injected into chambers deep beneath the ridges, and swarms of dikes above the chambers act as conduits for the flow of magma onto the ocean floor. Seawater penetrates into the solidified but cracked and permeable basalt, and is heated by the cooling magma; the buoyant fluid rises to form springs that erupt in the axial valleys along the mid-ocean ridges (Fig. 12-3). Many such springs have been observed, the water at times rising in great plumes carrying much dark particulate material (hence the common name "black smokers"). Measured temperatures are as high as 350°C. Convective circulation in deep fractures formed by the extensional tectonics ensures ample contact with hot rock, and analyses of the emerging water show that its composition is greatly altered. Much of its magnesium and sulfate are gone, and both calcium and silica have markedly increased (Table 12-5). Potassium is variably enriched relative to average seawater concentrations at the highest temperatures, but is to some extent diminished in water that has reacted with somewhat cooler basalt on the ridge flanks. In many cases the submarine spring waters are enriched in Fe, Mn, Cu, Zn, and Li relative to average seawaters. Measurements of pH indicate that the fluids are acid, with reported values commonly near 4.5.

The nature and extent of the seawater–basalt reaction at mid-ocean ridges are evident from rock samples obtained by deep drilling into the ocean floor. Another source of samples is a kind of rock sequence called *ophiolites*, consisting of

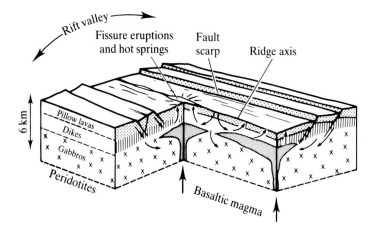

FIGURE 12-3
Schematic cross section of oceanic crust at a mid-ocean ridge spreading center, illustrating the major rock units (lavas, dikes, and gabbro plutons) and hydrologic features of submarine hydrothermal systems (represented by the small arrows).

gabbros, mafic dikes, and lavas together with deep-water marine sediments. Ophiolites are thought to represent exposed portions of ancient oceanic crust, altered by hydrothermal reactions like those we observe in progress near the modern springs. Extensive metasomatic replacement of igneous material of the lavas and dikes is evident, especially marked near pore spaces and fractures that acted as permeable structures for the flow of seawater through the rock. Common hydrothermal mineral assemblages include calcite, quartz, smectites, zeolites, prehnite, epidote, chlorite, alkali feldspars, and amphiboles in varying combinations, modal proportions, and parageneses that reflect the conditions of reaction. Even the deeper gabbro plutons have numerous veins filled with hydrothermal minerals, including high-temperature phases like hornblende and pyroxene, indicating that heated seawater penetrated deep into oceanic crust at unusually high temperatures. Similar types of basalt alteration are found in samples from drill holes into geothermal systems of the exposed mid-ocean ridge in Iceland, where subsurface temperatures in excess of 350°C have been recorded at depths less than 2 km.

Laboratory experiments (Bischoff and Dickson, 1975) provide additional evidence of the reactivity of seawater with basalt. Fresh basalt is placed in bombs with seawater and heated to temperatures simulating those measured along the ridges, fluids are extracted at intervals to measure the extent of reaction, and at the end of the experiment the basaltic material is analyzed to see what minerals have been produced or destroyed. The results of one such experiment are shown in Fig. 12-4. The changes in seawater composition during the short duration of the experiment seem surprisingly large, especially the almost complete removal of magnesium. Details of the experiments are difficult to determine and can still be questioned, but in general magnesium is precipitated as a hydrated sulfate salt or is incorporated into secondary silicates including smectite, chlorite, actinolite, or talc, sulfate is lost from solution chiefly by the formation of anhydrite, and calcium is added to solution during the hydrolysis of plagioclase feldspar in the basalt. Such experiments do not duplicate the many complex processes that occur in mid-ocean-ridge hydrothermal systems, but they do show the general trend and magnitude of

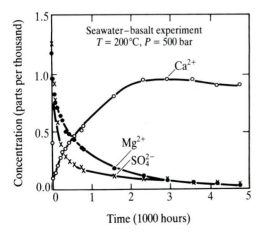

FIGURE 12-4

Concentrations of Ca^{2+}, Mg^{2+}, and SO_4^{2-} in seawater during reaction with crushed basalt at 200°C and 500 bars. Water-to-rock mass ratio in the experiment is 10 to 1. (Data from Bischoff and Dickson, 1975.)

reaction, specifically with regard to the major components of seawater as it is transformed from a slightly basic Na–Mg–Cl–SO$_4$ solution to an acid Na–Ca–Cl solution stripped of its magnesium and sulfate. (See also analysis 3 in Table 12-5.)

The number and size of the convection cells along ridges where such reactions take place are still uncertain, but it seems highly probable that this alteration of seawater by reaction with basalt has a major influence on the composition of the oceans, especially on the concentrations of magnesium and sulfur and perhaps potassium.

There remains the problem of the two most abundant elements, Na and Cl, which are not greatly affected by the basalt reaction. These two are in some measure removed by sorption and by the incorporation of seawater in sediments, but a part of the story almost surely is the crystallization of salts in evaporating basins. Such localized deposition is not evident in the present ocean except on a very small scale (shallow water exposed to desert sun along the Persian Gulf, for example), but certainly in the geologic past, even the fairly recent past, enormous quantities of very soluble materials have been removed from the sea by this mechanism. Can we infer that the concentrations of Na$^+$ and Cl$^-$ have not been strictly constant over long times, but have diminished somewhat in periods of extensive salt deposition and then increased again at times like the present when salt formation is minimal? A quantitative answer would be hard to get, because the fluctuations, if any, are slow and small.

We might broaden the question: the concentrations of *all* major ions in seawater are remarkably constant today, but have they remained constant over long reaches of geologic time? The probable answer is yes, at least approximately constant. From analyses of marine evaporites through the Phanerozoic it appears that concentrations of the major ions may have varied somewhat, but cannot have changed by more than factors of two or three. If deviations had been greater than this, for example, Mg^{2+} would at times have been high enough to make sepiolite a common evaporite mineral; or if sulfate had been much higher we would see evaporite beds of Na$_2$SO$_4$. In the absence of such anomalies, it seems a safe conclusion that seawater has had a composition much like the present one at least since the end of the Precambrian, but there may have been some variation timed with periods of extensive salt deposition, or periods of greater activity along midocean ridges. We will return to this question in Chapter 14.

A final question about seawater relates to the possible effect of human activities on its composition. We have noted the large and potentially harmful influence of man-made pollution on compositions of rain and stream waters; can similar comments be made about seawater? Offhand one would expect human influence to be slight, simply because of the ocean's enormous volume. It is hard to imagine that even the worst and largest of human enterprises can change appreciably the overall composition of so large a body of water. Locally, of course, contamination may be serious: the effects of oil spills and the desecration of beaches by garbage disposal make frequent newspaper headlines. For nitrogen and phosphorus, it is estimated that worldwide rates of addition to the sea from industry and from the greatly increased use of fertilizer have become at least equal to the rates of natural addition, so that some overall effect is expected eventually. One inevitable worldwide change in seawater composition has been suggested as a result of the buildup of CO_2 in the atmosphere that is responsible for the greenhouse effect. Certainly atmospheric CO_2 maintains an equilibrium with CO_2 dissolved in sea water, and certainly an increase in the former must lead to increase in the latter, with a resulting decrease in seawater pH. The decrease would be countered by the ocean's buffer system, and the resulting chain of reactions is so complex that there is no agreement as to just how great the effect would be. It will certainly be small, and probably the same can be said for any overall influence on seawater composition of anthropogenic pollution at present-day rates.

Summary

Seawater has a composition determined in large part by a balance between additions of dissolved material from stream water and various processes of removal. Among the latter, important ones are losses to accumulating sediments by precipitation, sorption, and organic activity, and reactions with basalt at midocean ridges. The balance maintains the major ions at nearly constant concentrations. Minor constituents that play a role in the life processes of organisms are a part of complex cycles, and vary widely in their concentrations from place to place and from time to time. The chemistry of the oceans has maintained a state of balance much like the present one, with only minor fluctuations, over most of the last half-billion years. Even the current high and increasing rate of human pollution is unlikely to affect the worldwide balance appreciably in the foreseeable future.

12-4   ENVIRONMENTAL GEOCHEMISTRY

The processes we have discussed in this chapter determine in large measure the chemical environment in which we live on the surface of our planet. They have been in operation for many millions of years, and evidently are part of a huge system that long ago reached a steady state. As a system, the fluid envelopes are remarkably stable toward natural events that might disturb it. Excess of any component sets in motion reactions that tend to cut down the excess, and a deficiency automatically leads to reactions that seek to restore the balance. The system, in other words, is an equilibrium on a grand scale, an equilibrium that has maintained conditions on the Earth's surface favorable to life processes for much of geologic history. We now stand off and look at this huge system as a whole, to see how the detailed processes we have been considering fit into the grand scheme that maintains the planet's surface environment. Is the equilibrium actually as stable as it appears, or are there natural processes that could result in permanent long-term change? To what extent has human activity perturbed the equilibrium? If population continues to increase and modern industry spreads to the less developed parts of the world, how serious will the effects of human disturbances become? We have looked at some specific examples of human influence on the environment earlier in this chapter, but now we try to see how these examples fit into an overall picture.

Natural Processes and Long-term Change

We consider first possible processes in nature that seem to go in one direction only, hence cannot be part of a large-scale equilibrium. The most obvious example is radioactive decay: uranium and thorium and other naturally radioactive elements have been steadily disappearing since the Earth was formed, and nowhere are they being produced except possibly in the interior of stars. Certainly one long-term change in Earth materials, then, is a decrease in the amounts of these elements in rocks and soils, and a decrease in the intensity of ionizing radiation to which creatures on the Earth's surface are exposed. Another effect of radioactive decay is a steady increase in the amount of argon in the atmosphere, produced from nuclei of ^{40}K in the crust and mantle.

We recall also that the two gases hydrogen and helium are too light to be retained by the Earth's gravitational field, hence have been steadily escaping from the high atmosphere all during geologic time. The loss is very small in comparison with the total quantity of these elements in the planet, but it does represent an uncompensated change in the Earth's overall composition.

More interesting from an environmental viewpoint are possible long-term changes in some of the major gases of the atmosphere. Oxygen, for example, is being lost in prodigious amounts through the processes of respiration, oxidation of organic matter, and reactions with reduced forms of iron and sulfur in rocks and soils. Is its rate of production by photosynthesis in the green leaves of plants sufficient to balance this steady depletion? Certainly the flourishing of complex life forms over many hundreds of millions of years has required at least an approximate balance in the latter part of Earth history, but there is good evidence for much lower oxygen concentrations at earlier periods (Sec. 21-5). Nitrogen maintains a seemingly precarious balance between the few processes by which it is snared from the atmosphere in forms usable by living things and the many reactions by which it is returned (Fig. 12-1), but there is no indication of any major long-term change.

Carbon dioxide is a more difficult problem. Many processes by which it forms in nature can be itemized and at least as many by which it is removed. At least one of the removal processes, the deposition of calcium carbonate in the sea, has certainly had long-term consequences, since the CO_2 so entombed has remained out of circulation for very long times, some from far back in the Precambrian. Whether enough is returned to air by the making over of crustal material during metamorphism and igneous activity to compensate for this steady drain is uncertain. So many processes contribute to the gain and loss of CO_2, and so many are influenced by changes in living forms and by changes in climate, that the history of this gas in the atmosphere is an intricate puzzle. There is abundant evidence in sedimentary rocks for much fluctuation in the concentration of CO_2 in air over geologic time, and for much larger amounts of this gas through a good deal of the Precambrian (see Sec. 21-5 and Prob. 12-2). As we have noted before (Sec. 12-1), CO_2 is one gas for which human activity may well have an important influence on its concentration, but to date the changes have not been large.

Rare gases, some of them known to be harmful to organisms, likewise have left no record of significant long-term changes in amount. Sulfur gases and halogen gases, for example, may on occasion be abundant and locally destructive in volcanic areas, but when dispersed in the air are quickly brought under control by reactions with other gases, aqueous solutions, and solid materials.

Much the same can be said of materials dissolved in water. The composition of salts in the sea has certainly changed somewhat over geologic time, but always within narrow limits. Stream water may be locally contaminated with solutes poisonous to organisms, but except in the immediate vicinity of industrial complexes the amounts have seldom grown large and are subject to control by natural processes of removal.

The salient fact about the chemistry of the natural environment in all of its parts is its constancy over long periods, a constancy preserved by a myriad of processes adding and subtracting individual substances. And perhaps most remarkable of all is the continued suitability of this nearly constant environment for the growth and flourishing of living things over an immense period, nearly four billion years of Earth history. Or should one say more simply that organisms have evolved in this favourable environment, hence are necessarily adapted to it and its slow changes, but could be harmed by any rapid major change?

Effects of Human Activity

With the appearance of homo sapiens on the Earth's surface a few million years ago there came a totally new sort of influence on the natural environment: creatures capable of radically altering some of the processes that control the long-established equilibrium. The effectiveness of these new creatures is all too evident in the huge cities they have built, the dense smog in the atmosphere of these cities, the toxic solutes in much groundwater and stream water, and the ever-increasing piles of waste materials on parts of the surface. For a long time in human history such changes in the environment seemed trivial, minor disturbances that would be self-corrected in a reasonable time by the natural processes that maintain the overall steady state. But with the recent burgeoning of human population and the spread of industrial civilization over ever larger areas, a possible major disturbance of the established equilibrium looms as a threat not only to humanity but to other parts of the biosphere. How serious is the threat, and are there counter-measures that could alleviate its consequences? This has become a much-argued sociopolitical question of the late 20th-century world. We look here at a few of its technical aspects.

Four major deleterious effects of human activity on the atmosphere have been mentioned in earlier discussions: the increase in acidity of rainfall by sulfur and nitrogen gases produced in automobile exhausts and by many industrial processes, the rise in CO_2 concentrations resulting from burning of fossil fuels and clearing of forests, the increase in smog-producing ozone at low levels by gases generated in the burning of fossil fuels, and the decrease of ultraviolet-screening ozone in the high atmosphere by volatile compounds produced by the organic chemical industry. Each of these atmospheric changes, of course, stimulates the counter reactions that in nature would reduce the harmful effects of the changes. How effective are such responses of the Earth's equilibrium system? Over the long term, will the effects of human activity build up to a dangerous degree, or will they be largely nullified by natural reaction? On the answers to such questions depend political careers and the expenditure of many millions of dollars, but convincing answers are hard to obtain.

The difficulty, as might be expected, lies in the complexity of environmental chemistry. That rainfall becomes acidified in heavily industrialized areas, for example, is a well established fact. But how much damage the acid rain actually does to soil and lakes and vegetation downwind from its source is much less easy to determine. One can point with horror at the areas of blighted vegetation that are often conspicuous near sulfide-ore smelters, where concentrations of sulfur oxides in the air are particularly high, but effects of the much lower concentrations produced by other industries are less clear. A recently reported decrease in the number of fish in lakes and streams of the northeastern United States and eastern Canada has been plausibly correlated with an increased acidity of the water due to acid emissions from industries that are concentrated south of the Great Lakes and in the eastern United States. Yet the correlation can be questioned, on the grounds that surface waters in humid parts of the north temperate zone are somewhat acidic even in their natural state. Soils of these regions commonly contain enough decaying vegetation to produce a soil atmosphere much richer in CO_2 than ordinary air; the resulting H_2CO_3, together with small amounts of organic acids, causes the pH of groundwater to drop to 4 or even lower. The soils and underlying rock also contain materials that should be effective in countering a great deal of increased acidity, particularly carbonate and silicate materials that readily dissolve in acid and adsorbed metal ions that can be replaced by H^+. Thus the pH values in nature are controlled by a complex of reactions, and they vary a good deal from place to place and from season to season. Perhaps the reported high acidity is no more than a minor fluctuation in the natural levels. So the magnitude and the causes of the low pH can be argued back and forth, the uncertainty stemming from the difficulty of obtaining a meaningful average pH over a large area and from the inadequacy of data on pH values of surface water and groundwater before the expansion of industry in modern times. How serious the problem is, how much is due to industry and how much to automobile exhausts, and how effective controls of sulfur and nitrogen emissions might be, remain matters of dispute.

Similar uncertainty beclouds the "greenhouse effect" of increasing atmospheric CO_2 (and other heat-absorbing gases). That the increase is occurring is beyond question: an abundance of measurements over the past century and a half has demonstrated a slow but steady climb in CO_2 levels. But predicting the environmental effects of the increase is much less straightforward. Offhand, the calculation would seem simple: we know the rise in temperature due to a given concentration of CO_2, the excess amount of CO_2 produced annually can be estimated, and multiplication of these two figures should give the amount by which the Earth's average temperature increases each year. Difficulties arise from uncertainties about the nature and long-time effectiveness of the back-reactions that in the Earth's near steady-state system are prompted by increasing CO_2. Some of the excess CO_2 would be used in additional photosynthesis of organic matter by plants; to what extent might this lead to changes in the amount and kind of vegetation? Probably more CO_2 would be removed by dissolution in seawater; how much would remain in water near the surface, how much would be carried to great depths? How much would be used by marine plants, how much by precipitation of $CaCO_3$? Answers to such questions remain unclear, and it is no surprise that attempted estimates of increased global temperatures fifty years hence cover a wide range, from small fractions of a degree to 5 or 6°C.

For ozone also, a question can be raised about the seriousness of the consequences of recent human-induced changes in concentration—the marked increases at low levels in the atmosphere, and the decrease at very high levels. Certainly dense smog can be unpleasant, and certainly oxidation by O_3 hastens the disintegration of some organic materials, but agreement has not been reached as to the magnitude of the danger to human health from current ozone concentrations near the Earth's surface. Nor is there agreement about effects on health of the recently observed "holes" in the ultraviolet screen furnished by ozone in the upper atmosphere. Long-term adverse consequences of both kinds of change are sufficiently probable to make protective measures desirable, but such counter-measures are expensive. Is the risk great enough to justify large expenditures? This, of course, is a political question rather than geochemical. The best a geochemist can do is to point out that processes exist in the lower atmosphere to keep ozone levels from rising too high, and in the upper atmosphere from falling too low—but just where the natural limits are we still do not know.

SUMMARY

All in all, this odd planet of ours seems remarkably well endowed with fluids that keep most of its surface a favorable place for life to flourish. Compositions of the fluids are well buffered against rapid changes, any alteration calling up immediately a process that helps to undo the change. This means that compositions of the major fluids—oceans and the atmosphere—change only slowly, and within narrow limits, over great reaches of geologic time. Possibly oxygen was lower in both oceans and atmosphere, and possibly carbon dioxide was higher, during much of the Pre-cambrian. Possibly some of the major-ion concentrations in seawater have fluctuated as hydrothermal activity waxed and waned along the midocean ridges. But for at least the last half billion years we have no record of major changes. In very recent geologic time some minor effects of human activity on the fluid envelopes have become detectable, notably the increase of carbon dioxide in air, the local development of acid rain, and changes in ozone concentrations at both high and low levels in the atmosphere. Over long times a continuation of such changes might affect environmental geochemistry adversely, but the changes are so slow that countermeasures or adaptation will minimize their effects. The fluid envelopes should be a stable and dependable part of our peculiar planet for a long time to come.

PROBLEMS

1. Draw a diagram of the *biogeochemical* cycle for carbon illustrating the distribution and transport of carbon species among the atmosphere, biosphere, hydrosphere, and the continental and oceanic lithosphere. As in the diagram for the nitrogen cycle in Fig. 12-1, use boxes to denote reservoirs for the various carbon species (e.g., CO_2), and use arrows to represent processes by which carbon is transferred from one reservoir to another. Be sure to represent processes by which human activity influences the carbon cycle, as well as the production of carbon compounds by volcanism, weathering, diagenesis, and metamorphic processes.

2. The figure below (from Berner, 1991) shows estimates of change in atmospheric CO_2 during Phanerozoic time. The variable, R_{CO_2}, is the ratio of estimated CO_2 to present-day CO_2. Write a short essay describing geochemical and geological processes that might explain the changes in CO_2 content of air over the past 570 Ma. Include in your essay suggestions as to when it is most likely that global-scale glaciation (Ice Ages) occurred and when the climates were warmest.

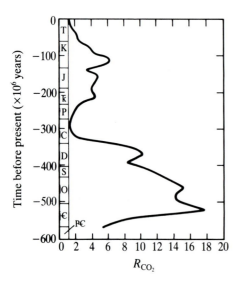

3. Suggest an explanation for the differences in composition of the water in the Nile, the Amazon, and the Colorado rivers (Table 12-4).

4. The annual input of SO_4^{2-} into the oceans by rivers is estimated to be 3×10^{12} mol/yr. If there were no removal of sulfate from seawater, how long would it take to increase the concentration of seawater sulfate by 5% (the total mass of seawater is approximately 1.4×10^{24} g)? List the natural processes for removing sulfate from seawater. How would you estimate the annual removal of sulfate for each of these processes?

5. In a hypothetical seawater–basalt experiment there is an initial decrease in pH and Mg^{2+}, after which pH and Ca^{2+} gradually increase as schematically shown below. Analysis of the altered basalt at the end of the experiment shows the presence of two new minerals, brucite $(Mg(OH)_2)$ and kaolinite $(Al_2Si_2O_5(OH)_4)$, and a decrease in the amount of plagioclase $(CaAl_2Si_2O_8)$ initially in the basalt. Which mineral is likely to form first in the

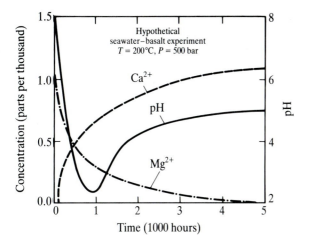

Time (1000 hours)

experiment, brucite or kaolinite? Write a balanced chemical reaction that represents the process causing the pH to decrease during the early stages of the experiment and a reaction describing the subsequent pH increase.

6. Why is the composition of rainwater in Hawaii so different from all the others listed in Table 12-3?

REFERENCES AND SUGGESTIONS FOR FURTHER READING

Berner, E. K., and Berner, R. A.: *The Global Water Cycle: Geochemistry and Environment*, Prentice-Hall, Englewood Cliffs, New Jersey, 1987. Excellent general textbook on the physical and chemical aspects of the water cycle.

Berner, R. A.: "A model for atmospheric CO_2 over Phanerozoic time," *American Jour. Sci.*, vol. 291, pp. 339–376, 1991. Changes in the concentration of CO_2 in air from the Cambrian period to the present.

Bischoff, J. L., and F. W. Dickson: "Seawater–basalt interaction at 200°C and 500 bar: Implications for the origin of sea-floor heavy-metal deposits and regulation of seawater chemistry," *Earth and Planetary Science Letters*, vol. 25, pp. 385–397, 1975. Experiments showing that Mg^{2+} and SO_4^{2-} are largely removed from seawater by reaction with hot basalt.

Faure, G.: *Inorganic Geochemistry*, Macmillan, New York, 626 pp., 1991.

Gregor, C. B., R. M. Garrels, F. T. Mackenzie, and J. B. Maynard, (eds): *Chemical Cycles in the Evolution of the Earth*, Wiley, 1988. A collection of five papers on cyclic processes in geology. Chapters 1 and 2 provide detailed information on geochemical cycles involving the ocean and the atmosphere.

Henderson, P.: *Inorganic Geochemistry*, Pergamon Press, 1982. Chapter 11 is a good review of chemical oceanography.

Holland, H. D.: *The Chemistry of the Atmosphere and Oceans*, Wiley, 351 pp., 1978. Changes in the composition of atmosphere and oceans over geologic time, as deduced from analyses of evaporites and ancient soils.

Mason, B., and C. B. Moore: *Principles of Geochemistry*, 4th ed., Wiley, 1982. Many analyses and detailed discussion of composition of air and water bodies.

O'Neill, P.: *Environmental Chemistry*, 2d ed., Chapman and Hall, London, 268 pp., 1993. A comprehensive text on environmental chemistry of major elements found in living matter and in the Earth's crust, and of environmental problems related to minor elements.

Spencer, R. J., and L. A. Hardie: "Control of seawater composition by mixing of river waters and mid-ocean ridge hydrothermal brines," in Spencer, R. J., and I-Ming Chou, (eds): *Fluid–Mineral Interactions: A Tribute to H. P. Eugster*, The Geochemical Society, Special Publ. No. 2, pp. 409–419, 1990.

Stern, P. C., O. R. Young, and D. Druckman, (eds.): *Global Environmental Change*, National Academy Press, Washington, D.C., 1992. Good discussion of the relation of human activities to such environmental changes as ozone depletion, deforestation, and increase in atmospheric CO_2.

Von Damm, K. L.: "Seafloor hydrothermal activity: black smoker chemistry and chimneys," *Annual Reviews in Earth and Planetary Sciences*, vol. 18, pp. 173–204, 1990.

Wetzel, R. G.: *Limnology*, Saunders, Philadelphia, 743 pp., 1975. A standard reference book on physical and chemical processes in lakes.

CHAPTER
13

WEATHERING
AND
SOILS

One geologic process in which we might expect to find chemical ideas particularly useful is the weathering of rocks at the Earth's surface. The reactions of weathering are in a sense the most familiar geochemical reactions of all, because they go on all around us, under conditions of normal temperature and pressure. The raw materials, the products, the agents of weathering are all accessible to study in the field and can all be brought easily into the laboratory.

Besides its scientific interest, the geochemistry of weathering has a practical side. Weathering produces the agricultural and forest soils that form the economic basis of many nations. It modifies the chemistry of groundwater and streams, and influences the nature of the landscape in which we live. An understanding of weathering rates allows estimates of ages of weathered materials, thus providing another tool for evaluating geologic hazards such as the frequency of volcanic eruptions, earthquakes, and landslides. And from our knowledge of conditions under which different kinds of weathering products are formed, we can draw inferences about the nature of paleoclimates based on the study of ancient soils and weathered surfaces.

13-1 GENERAL NATURE OF
WEATHERING REACTIONS

As often happens, the situation turns out to be less simple than it first appears. For one thing, weathering involves mechanical processes as well as chemical reactions—expansion of water on freezing, growth of roots, swelling of minerals due to hydration. We shall say little about these processes, simply because our focus here is on chemical details, but we must keep in mind that chemical weathering is only one aspect of the whole phenomenon. A second and more serious complication in applying chemical ideas is the extreme slowness of rock decay. Most weathering reactions are sluggish, incomplete, often irreversible. In the laboratory we can easily duplicate the materials and environments of weathering, but the times elude us. This is a common difficulty in geochemical experimentation: we need centuries, and we have only weeks or months at our disposal. The time limitation is especially troublesome when we deal with low-temperature processes, as we must in the study of weathering.

Estimating the time it takes for a rock to weather is difficult. Laboratory studies of mineral dissolution provide useful constraints, but apply strictly only to the carefully controlled conditions of the experiment (Sec. 11-2). Observations of the weathered rock used in buildings, statues, and tombstones of known age give some indication of weathering rates. Field studies of the rate of removal of elements from a watershed by percolating solutions or in the dissolved load of streams provide data on rates of weathering averaged over the area of a drainage basin. Results of such field studies, summarized in Table 13-1, permit estimates of the average amount of time required to weather 1 mm of rock to a kaolinitic saprolite. These estimates illustrate the approximate time scales involved in weathering, and highlight the importance of climate and rock type as determiners of weathering rates.

The chemical reactions of weathering are basically simple. The overall processes involve only reactions with which we are already familiar—ionic dissociation, addition of water and carbon dioxide, hydrolysis, oxidation, adsorption, and ion exchange. Some complications may arise because of the activity of living organisms, but the basic chemical relationships are straightforward.

From a different point of view, weathering means the approach to equilibrium of a system involving rocks, air, and water. An equilibrium assemblage in this

TABLE 13-1

Estimates of the time required to weather 1 mm of fresh rock to a kaolinitic saprolite

Rock Type	Climate	Time
Felsic	Tropical semi-arid	65 to 200 yrs.
	Tropical humid	20 to 70 yrs.
	Temperate humid	41 to 250 yrs.
	Cold humid	35 yrs.
Metamorphic	Temperate humid	33 yrs.
Mafic	Temperate humid	68 yrs.
	Tropical humid	40 yrs
Ultramafic	Tropical humid	21 to 35 yrs.

Reprinted by permission from Nahon (1991).

system is hard to define precisely, because the reactions are so very slow that we must deal often with processes only partly complete. For this reason our knowledge of weathering is largely qualitative rather than quantitative. We can decipher what happens chemically in the decay of a rock, but we cannot describe accurately the state of a rock at a particular time in the past or predict its state at a given time in the future.

We shall not attempt to define or delimit the term "chemical weathering" with any exactness. Reactions of rocks and minerals with the constituents of air and water at or near the Earth's surface—something of this sort is generally understood. But should weathering include the alteration of minerals in mines at depths of hundreds or even thousands of meters, evidently caused by groundwater that is largely of surface origin? Should it cover the alteration of rocks by seawater at great depths in the ocean, or alteration by the warm water of hot springs? Such semantic questions we shall sidestep, limiting our discussion to ordinary rocks within a few meters or tens of meters of the surface.

13-2   CHANGES IN ROCK COMPOSITION

An obvious approach to the chemistry of weathering is by a look at the overall changes in composition from fresh rock through its various stages of decay. A series of analyses showing such changes is given in Table 13-2, where concentrations of elements are expressed as weight percent of their oxides. Analyses in the table include as a last item "others," lumping together minor concentrations of TiO_2, MnO, BaO, P_2O_5, CO_2, and S. At the bottom of the table are approximate mineral (or "modal") compositions, in percent by volume.

From Table 13-2 the losses of some constituents are immediately evident: especially marked for Na, Mg, and Ca, somewhat less for K and Si. This order of loss is a common one for the weathering of many kinds of rock, but it is by no means universal. Just which constituents will disappear first in any particular case depends on a variety of factors—mineral composition, rock texture, climate, drainage, amount of exposure—so that useful gereralizations are hard to frame.

To determine *how much* of any one constituent has been lost is more difficult. Each analysis, of course, gives only the *relative* amounts of various elements present at a particular stage of weathering. If, for example, a weathered rock has lost most of its original Na and Ca but only a little of its Al and Fe, the analysis will show an apparent increase in the latter two constituents, as illustrated by analyses I and II in the table. To explain a pair of analyses showing a decrease in some constituents and an increase in others, we obviously can make a number of guesses. Possibly the increase and decrease are both real; in other words, weathering might involve loss of some elements and addition of others, the total mass of rock remaining approximately constant. Or, one constituent may be unaffected by weathering while everything else decreases; here the constant constituent and others whose loss is only slight will show an apparent increase. Or, finally, weathering may bring about a decrease in all elements, those which decrease least showing an apparent

TABLE 13-2

Analyses of quartz–feldspar–biotite gneiss and weathered material derived from it[1]

Column I gives the analysis of a sample of fresh rock, and columns II, III, and IV give analyses of weathered material. In general, the degree of weathering increases from II to IV, but there is no assurance that the original material was precisely the same or that IV represents a longer time of weathering than II or III.

Chemical composition (weight percent)				
	(I)	(II)	(III)	(IV)
SiO_2	71.54	68.09	70.30	55.07
Al_2O_3	14.62	17.31	18.34	26.14
Fe_2O_3	0.69	3.86	1.55	3.72
FeO	1.64	0.36	0.22	2.53
MgO	0.77	0.46	0.21	0.33
CaO	2.08	0.06	0.10	0.16
Na_2O	3.84	0.12	0.09	0.05
K_2O	3.92	3.48	2.47	0.14
H_2O	0.32	5.61	5.88	10.39
Others	0.65	0.56	0.54	0.58
Total	100.07	99.91	99.70	100.11

Approximate mineral composition (volume percent)				
Quartz	30	40	43	25
K-feldspar	19	18	13	1
Plagioclase	40	1	1	?
Biotite (+ chlorite)	7	Trace	Trace	0.2
Hornblende	1	None	None	Trace
Magnetite, ilmenite, secondary oxides	1.5	5	2	6
Kaolinite	None	36	40	66

[1] Reprinted by permission from Goldich (1938).

increase. Nothing in the analyses themselves enables us to choose between these alternatives.

If we had independent evidence as to how the mass or volume of rock has changed during weathering, our problem would be solved. Such evidence is usually not obtainable, so in most cases we can proceed only by making an arbitrary assumption that will allow computation of elemental gains and losses during weathering. One assumption commonly used is that alumina does not change appreciably during weathering—a guess that seems reasonable on the grounds that Al_2O_3 in analyses of weathered material generally shows the greatest *apparent* increase and that of all common rock constituents Al is least abundant in surface waters. Inasmuch as Al is not completely absent from stream and groundwaters, however, the assumption cannot be strictly accurate. In some weathering profiles there is good evidence that substantial Al has been removed from the rock, and for these it may be more appropriate to assume that a trace element like Ti or Zr is immobile during the weathering process for the purpose of calculating chemical gains or losses.

The calculation goes by the following steps, as illustrated in Table 13-3:

1. Recalculate analyses to 100 percent by distributing the analytical error (columns I and III).
2. Assume Al_2O_3 constant. During weathering, 100 g of fresh rock has decreased in weight so that Al_2O_3 has apparently increased from 14.61 to 18.40%. Hence the total weight has decreased in the ratio 14.61/18.40, or from 100 to 79.40 g. The amount of each constituent in the 79.40 g can be found by multiplying each number in column III by this same ratio. This gives the numbers in column A.
3. The decrease (or increase) in each constituent is found by subtracting the numbers in column A from those in column I, giving the numbers in column B.
4. The percentage decrease or increase of each constituent is computed by dividing the numbers in column B by those in column I, giving the numbers in column C.

This same method of calculation is often used with analyses showing other kinds of rock alteration, for example, hydrothermal alteration near veins and igneous intrusions. The assumption of constant aluminum is on shakier ground here,

TABLE 13-3
Calculation of gains and losses during weathering
Columns I and III, giving composition in weight percent, are repeated from Table 13-2, except that the analytical error in each has been distributed so that the totals are 100.00. Column A shows the calculated weight in grams of each oxide remaining from the weathering of 100 g of fresh rock, on the assumption of constant Al_2O_3. Column B shows the gains and losses of the different oxides in grams, and column C shows the same gains and losses in percentages of the original amounts.

	(I)	(III)	(A)	(B)	(C)
SiO_2	71.48	70.51	55.99	−15.49	−22
Al_2O_3	14.61	18.40	14.61	0	0
Fe_2O_3	0.69	1.55	1.23	+0.54	+78
FeO	1.64	0.22	0.17	−1.47	−90
MgO	0.77	0.21	0.17	−0.60	−78
CaO	2.08	0.10	0.08	−2.00	−96
Na_2O	3.84	0.09	0.07	−3.77	−98
K_2O	3.92	2.48	1.97	−1.95	−50
H_2O	0.32	5.90	4.68	+4.36	+1,360
Others	0.65	0.54	0.43	−0.22	−34
Total	100.00	100.00	79.40	−20.60	

but is partly justified by the lack of evidence that aluminum migrates extensively in the formation of many types of veins and contact metamorphic aureoles.

13-3   SEQUENCE OF SILICATE MINERAL ALTERATION

The usual rapid decrease of Na, Ca, and Mg, the slower loss of K and Si, and the still slower loss of Al and Fe indicated by the data in Tables 13-2 and 13-3 are, of course, reflections of the susceptibility of various minerals to weathering. Ordinarily mafic minerals decay more rapidly than felsic minerals (although by no means always), liberating Mg, Fe, and in lesser amounts Ca and the alkalies; Fe is in large part oxidized immediately to insoluble Fe(III) oxide, so remains with the weathered material instead of being carried off in solution. Among the feldspars plagioclase weathers faster than K-feldspar, and calcic plagioclase faster than sodic plagioclase, thus liberating Ca and Na more rapidly than K. When the cations are set free, the Al-Si-O frameworks of the original silicate minerals are in part decomposed, in part reconstituted into the structures of clay minerals, so that only a part of the Si and very little of the Al find their way into solution. Typical changes in the modal abundance of minerals during weathering are shown in the lower part of Table 13-2.

These conclusions about the weathering of various minerals are substantiated by examination of weathered material in hand specimen and in thin section, and by experimental work on artificial weathering of minerals and rocks. It is a common field observation, for example, that a weathered surface of granite shows its dark minerals largely converted to limonite while the feldspars remain comparatively fresh, and that plagioclase has a more chalky appearance than K-feldspar. The relatively fast decay of Na-Ca-feldspar is often conspicuous in thin section, where plagioclase crystals may be flecked with tiny grains of clay minerals and calcite while the K-feldspar is almost as clear and fresh as quartz. Experiments performed by letting powdered rocks and minerals stand in contact with water and dilute acids likewise show the faster decay of mafic minerals and calcic plagioclase, and the tendency of Na, Ca, and Mg to dissolve in larger amounts than the other cations. Still further evidence is provided by the common observation that the mafic mineral content of sandstones is generally smaller in older rocks, these minerals having disappeared by slow reaction with percolating groundwater.

From such observations and experiments, the common minerals of igneous rocks can be arranged in a series, or better in two parallel series, according to their decreasing rates of weathering:

Mafic minerals	Felsic minerals
Olivine	Ca-Na plagioclase
Pyroxene	Na-Ca plagiocase
Amphibole	K-feldspar
Biotite	Muscovite

These series have a striking resemblance to the order of crystallization of minerals from igneous melts (Sec. 17-6), the minerals that form at highest temperatures (olivine and calcic plagioclase) being those most susceptible to weathering processes. In a general way this seems reasonable, although it is not a necessary conclusion theoretically.

A word of caution is essential here. The above remarks on the relative rates of weathering are broad generalizations, applicable to a majority of rocks and a majority of weathering environments. But they must not be taken as universal rules, for exceptions are very common. Biotite may appear more decayed than hornblende, and the mafic minerals of a granite may look fresher than its feldspars. Rate of weathering is so dependent on such factors as grain size and amount of fracturing as well as on straight chemical susceptibility that generalizations to cover all cases are not possible.

13-4 AGENTS OF CHEMICAL WEATHERING

Dry air causes rock to decay only very slowly, as is attested by the marvelous preservation of carved inscriptions dating from three and four thousand years ago in the arid climate of Egypt. Moisture speeds up the process enormously, both because water itself is an active agent of weathering and because it holds in solution, and therefore in intimate contact with the rock surface, several substances that react with rock minerals. The more important of these substances are free oxygen, carbon dioxide, organic acids, and nitrogen acids.

Free oxygen is important in the decay of all rocks containing oxidizable substances, particularly iron and sulfur. At ordinary temperatures reactions involving free oxygen are slow (Sec. 11-4); any number of readily oxidizable materials such as wood, cloth, most metals, coal, and petroleum, can remain in contact with the air almost indefinitely if water is absent and the temperature remains low. Water speeds up oxidation, probably by dissolving minute quantities of minerals or other materials, for the reactions of oxygen are faster with dissolved substances, particularly with ions, than with solids. Water may enter the reaction itself, as in the formation of hydrates, but its role is largely that of a catalyst—simply to provide a favorable environment for otherwise extremely sluggish processes.

Carbon dioxide aids decay primarily by forming carbonic acid when it dissolves in water. All water exposed to air is a dilute solution of this acid (Sec. 2-1), and locally natural waters may contain other materials that increase their acidity. The decaying humus of soil, for example, adds substances to water that lower its pH very commonly to 4.5 or 5.0, sometimes to values under 4. The nature of these substances is not entirely clear. The acidity is due, in part, simply to abnormal amounts of CO_2 released by the decay; the concentration of CO_2 in soil water is commonly at least $0.1m$ and may be up to $1m$, in contrast to the $10^{-5}m$ at equilibrium in water exposed at the surface. The acidity may also be ascribed to minute amounts of well-defined simple organic acids (for example acetic acid, CH_3COOH). In geologic literature the role of *humic acids* is often emphasized, but the importance of these substances in this context is questionable. Strictly the term "humic acid" refers to one of an ill-defined group of high-molecular-weight compounds obtained by digesting wood in strong alkali and then neutralizing with acid; the humic acids appear as a gelatinous precipitate, which if carefully washed shows practically no acid properties at all. As commonly used in geology, the word has a wider and vaguer significance, referring to any indefinite, dark-colored, partly colloidal material derived from decaying organic matter and imparting an acidity to the solution. The acidity, however, is more likely due chiefly to carbonic acid and simple organic acids rather than to humic acids.

The nitrogen acids HNO_3 and HNO_2 must also play at least a minor role in most natural waters. These acids may be derived either from organic decay and bacterial action in soils, or by the dissolving in rainwater of nitrogen oxides formed during lightning discharges in the atmosphere (Fig. 12-1). More locally, particularly in volcanic regions and in the oxidized zones of sulfide ore deposits, the sulfur acids H_2SO_3 and H_2SO_4 become important, in some places lowering the pH even below 1. In recent decades both nitrogen and sulfur gases have become notorious as constituents of "acid rain," which forms where the gases are produced in large amounts by industrial processes and automobile exhaust and can be destructive because of their effect on organisms and the increase they cause in weathering rates (Sec. 12-4).

Sorting out the agents of weathering in this fashion is a useful introduction to weathering processes, but no inference should be drawn that the various agents function individually. In nature the processes of dissolution, hydration, acid attack, and oxidation take place simultaneously, and separating them into specific reactions is only a convenient way to pigeonhole them for discussion.

Weathering reactions are controlled, to a large extent, by the availability of water and the nature of gases and acid components dissolved in the water. Variations

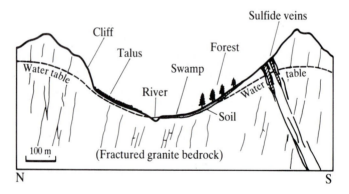

FIGURE 13-1

Schematic cross section of an alpine glacial valley illustrating the diverse environments of weathering that can be present on a local scale. Granite north of the river weathers primarily by mechanical processes (frost wedging), while granites near the river, under the swamp, and underlying the forest soils are weathered by chemical processes enhanced by organic and nitrogen acids produced by the decay of vegetation and by acids formed by weathering of the sulfide veins on the south slope of the valley.

in the average rainfall of the world's different climatic zones is an important factor in determining extent and type of weathering reactions. On a local scale topography and exposure to water can also cause extreme variations in weathering. This is illustrated in Fig. 13-1 with a schematic cross section of a glacial valley in fractured granite bedrock, in an area where the Sun angle at noon is approximately from the south. The granite forming the cliffs north of the river will be in contact with water only during storms; rapid drainage and evaporation on the south-facing cliffs will limit the time when meteoric water is in contact with the granite. Here mechanical weathering, such as frost wedging that forms the talus pile at the base of the cliff, will be the primary type of weathering. In contrast, the granite in the central portions of the valley near the river, under the swamp and underlying the forest soils, will be in contact with meteoric water for a longer period of time. In addition, organic and nitrogen acids produced within the swamp and forest soils will enhance the weathering here, as will acids produced by the weathering of the sulfide veins found on the southern ridge of the valley. Thus we can expect to find many varieties of conditions that control weathering, conditions that depend not only on climate but also on local landscape features.

13-5 DISSOLUTION OF CARBONATES

The chief reaction in the weathering of limestone is one that we have discussed at great length previously (Sec. 3-1):

$$CaCO_3 + H_2CO_3 \rightleftharpoons Ca^{2+} + 2HCO_3^-. \tag{3-4}$$
$$\text{calcite}$$

The equilibrium constant for this equation is 4.4×10^{-5}, which means that a solution in equilibrium with atmospheric CO_2 at 25°C and containing no HCO_3^- except that derived from the reaction itself can dissolve calcite up to a maximum Ca^{2+} concentration of $5.0 \times 10^{-4}m$. The amount would be greater at lower temperatures because more CO_2 can dissolve, giving higher concentrations of H_2CO_3; it would also be greater if the supply of CO_2 were unusually large, as it may be in spring water or in places where vegetation is rapidly decaying; it would be less at high temperatures and less also if the solution contained HCO_3^- from other sources. These are all familiar deductions from equilibrium reasoning.

In nature the dissolving of limestone in carbonic acid is evidenced by the fluted and pockmarked surfaces of limestone outcrops, by the widening of cracks in limestone, and by the high concentrations of Ca^{2+} (commonly 0.1 to $0.2m$) in groundwater in limestone regions. Despite this apparently rapid dissolution, limestone commonly forms prominent cliffs and ridges, especially in arid regions—probably because much limestone is a dense, relatively impermeable rock into which water cannot easily penetrate, so that weathering is confined to exposed surfaces and widely spaced cracks.

The dissolving of other simple carbonates in natural environments is the same kind of reaction, the extent of dissolution being roughly predictable from the solubility products given in Table 3-2. One can guess, for example, that dolomite should be less soluble than calcite, and this agrees with the common field observation that weathered limestone surfaces show deeper etching in calcite areas than in adjacent areas that have been dolomitized.

13-6 OXIDATION

Among the products of weathering, iron oxides are the most conspicuous because of their bright colors. All the oxides of iron formed in contact with air are Fe(III) oxides. They include two forms of the anhydrous compound Fe_2O_3 (the common mineral hematite and the less common magnetic oxide maghemite) and at least two hydrates, goethite ($HFeO_2$) and lepidocrocite ($FeOOH$). The material called limonite is chiefly fine-grained goethite, commonly mixed with more or less clay. The Fe(III) oxide precipitated in the laboratory by adding base to Fe^{3+} is conveniently written $Fe(OH)_3$, but a less definite formula like $Fe_2O_3 \cdot nH_2O$ would be more appropriate. The color of the simple oxide is characteristically red and of the hydrates yellow to brown, but the color is not a safe guide to composition because it depends at least as much on the state of subdivision and on minor impurities as on the degree of hydration. The precise conditions of formation of the different compounds and the conditions under which hydration or dehydration may take place are still imperfectly known. At the moment we are concerned with the change of Fe(II) compounds to Fe(III) oxide, and the particular form the oxide takes is of secondary importance.

An Fe(II) compound on prolonged exposure to the air is oxidized, according to reactions of the form

$$Fe_2SiO_4 + \tfrac{1}{2}O_2 + 2H_2O \rightarrow Fe_2O_3 + H_4SiO_4 \tag{13-1}$$
$$\text{fayalite} \qquad\qquad\qquad\qquad \text{hematite}$$

and

$$2FeCO_3 + \tfrac{1}{2}O_2 + 2H_2O \rightarrow Fe_2O_3 + 2H_2CO_3. \tag{13-2}$$
$$\text{siderite} \qquad\qquad\qquad\qquad \text{hematite}$$

Equations of this sort express only the overall result of the oxidation process: the tying up of iron in Fe(III) oxide, the setting free of silica as dissolved or colloidal SiO_2, the release of carbon dioxide to solution, and the ionizing of nonoxidizable metals like Ca and Mg that may be part of the primary Fe(II) silicate or carbonate mineral. The weathering of Fe(II) compounds probably takes place in steps, for example,

$$Fe_2SiO_4 + 4H_2CO_3 \rightarrow 2Fe^{2+} + 4HCO_3^- + H_4SiO_4 \tag{13-3}$$
$$\text{fayalite}$$

followed by oxidation of the Fe^{2+}

$$2Fe^{2+} + 4HCO_3^- + \tfrac{1}{2}O_2 + 2H_2O \rightarrow Fe_2O_3 + 4H_2CO_3. \tag{13-4}$$
$$\text{hematite}$$

The two steps may be widely separated, in case the original dissolving of iron takes place under reducing conditions (for example, through the agency of solutions containing or in contact with organic matter); the second step would follow only when the solution has moved into an oxidizing environment. Where the oxidation takes place on exposed surfaces, the two steps would not be distinguishable, for the Fe^{2+} would oxidize so rapidly that the amount present at any one time could not be detected.

Equilibrium in reactions of this sort [Eqs. (13-1), (13-2), and (13-4)], and the relation of concentrations to changes in Eh and pH, we have explored in Sec. 9-5. The fact that equilibrium is displaced far to the right is, of course, a reflection of the great stability and great insolubility of Fe(III) oxide. So stable is this substance that other Fe(III) compounds as products of weathering are relatively uncommon. Locally the slightly soluble Fe(III) phosphate (strengite), arsenate (scorodite), or basic sulfate (jarosite) may appear, but certainly the greatest part of the iron exposed to the atmosphere eventually goes into Fe(III) oxide in one of its various forms.

Closely paralleling the behavior of iron is the oxidation of manganese. The results of this oxidation are less conspicuous, because manganese is much less abundant than iron and because the products of oxidation are dark brown or black rather than bright red and yellow-brown. Nevertheless, films of manganese oxide on the walls of cracks and on rock surfaces in desert areas ("desert varnish") are very common. The complicated chemistry of manganese oxidation we will examine in a later section (Sec. 14-3).

A third common element that is oxidized during weathering is sulfur. In igneous rocks and in veins, this element occurs chiefly in sulfides—compounds with metals in which the sulfur has an oxidation number that is typically -2 or -1. Oxidation can change the number to any one of several higher values, but equilibrium in contact with air is reached only when the sulfur has reached its highest oxidation state, $+6$, in the form of sulfate. Equations for such reactions may be deceptively simple:

$$PbS + 2O_2 \rightarrow PbSO_4 \qquad (13\text{-}5)$$
$$\text{galena}$$

and

$$ZnS + 2O_2 \rightarrow Zn^{2+} + SO_4^{2-}. \qquad (13\text{-}6)$$
$$\text{sphalerite}$$

($PbSO_4$ is represented as undissociated because it is fairly insoluble and commonly appears as the mineral anglesite encrusting or embaying galena; $ZnSO_4$, on the other hand, is very soluble.) These reactions, like those for the oxidation of iron and manganese, take place with extreme slowness or not at all in the absence of water. The function of the water is probably to supply carbonic acid to dissolve minute amounts of the sulfides:

$$PbS + 2H_2CO_3 \rightarrow Pb^{2+} + H_2S + 2HCO_3^- \qquad (13\text{-}7)$$
$$\text{galena}$$

after which the hydrogen sulfide is oxidized:

$$H_2S + 2O_2 + Pb^{2+} + 2HCO_3^- \rightarrow PbSO_4 + 2H_2CO_3. \qquad (13\text{-}8)$$

This same relationship of dissolution and oxidation was suggested earlier in Chapter 6 [Eq. (6-13)].

Solutions resulting from oxidation of sulfides are usually acid because of hydrolysis of the dissolved metal ion. For example:

$$Zn^{2+} + H_2O \rightarrow ZnOH^+ + H^+. \tag{13-9}$$

The amount of acidity depends on the stability of the metal–hydroxy complex (Sec. 2-2). For metals that form very insoluble oxides and hydroxides, hydrolysis may lead to precipitation of the solid. This kind of reaction is particularly important, and the resulting acid particularly strong, in the oxidation of the common sulfide pyrite, FeS_2. The high acidity results from formation of the very insoluble Fe(III) oxide (or a hydrated oxide):

$$\underset{\text{pyrite}}{2FeS_2} + 15/2O_2 + 4H_2O \rightarrow \underset{\text{hematite}}{Fe_2O_3} + 4SO_4^{2-} + 8H^+. \tag{13-10}$$

Results of this reaction are evident at outcrops of pyritic veins and sulfide-rich mine tailings, where the rocks are heavily stained with yellow and brown Fe(III) oxides or hydroxides and where the groundwater has a sharply acid taste [see, for example, Fig. 6-10 and Eq. (6-12)].

Sulfur, iron, and manganese, plus the carbon in organic matter, are the only elements in ordinary rocks for which oxidation is an important part of weathering. Many of the less common elements—copper, arsenic, uranium, for example—are oxidized when their minerals are exposed to the atmosphere, but these reactions are best considered later on in connection with the chemistry of ore deposits (Chap. 19).

13-7 HYDROLYSIS OF SILICATES

The weathering of silicates is primarily a process of hydrolysis, a type of water–mineral reaction that produces an excess of H^+ or OH^- ions in solution (Sec. 2-2). As a simple example, the mineral forsterite (magnesium-rich olivine) hydrolyzes according to the equation

$$\underset{\text{forsterite}}{Mg_2SiO_4} + 4H_2O \rightarrow 2Mg^{2+} + 4OH^- + H_4SiO_4, \tag{13-11}$$

the hydrogen ion from water uniting with the silicate group to form the very weak silicic acid. More commonly in natural environments the proton is supplied by carbonic acid:

$$\underset{\text{forsterite}}{Mg_2SiO_4} + 4H_2CO_3 \rightarrow 2Mg^{2+} + 4HCO_3^- + H_4SiO_4. \tag{13-12}$$

This reaction is similar to those described earlier in Eqs. (3-4) and (13-3). Locally, if acids stronger than carbonic acid are present (for example, near a vein containing pyrite), the reaction is simply

$$\underset{\text{forsterite}}{Mg_2SiO_4} + 4H^+ \rightarrow 2Mg^{2+} + H_4SiO_4. \tag{13-13}$$

The abundance of hydrogen ions makes weathering in such places unusually deep and unusually complete.

The influence of solution pH on olivine hydrolysis is further illustrated by the experimental data in Fig. 13-2(*a*). These data come from experiments on rates of dissolution of olivine beach sands that we have discussed previously (Sec. 11-3). In the earlier discussion we were concerned with the effect of temperature on rates of dissolution of Mg, Fe, and Si at constant pH; here we note that the dissolution rates are markedly affected by pH, increasing by roughly an order of magnitude for a lowering of pH by one unit. It is worth noting that in these experiments, as in the ones described earlier, the dissolution is strictly *congruent* [Fig. 13-2(*b*)], meaning that the olivine releases its constituent elements into solution in proportions conforming to its stoichiometry (Sec. 3-10).

Aluminosilicates show more complicated hydrolysis behavior. As we have noted for the feldspars, dissolution begins as an incongruent reaction, becoming congruent in laboratory experiments only after a period of hours (Figs. 11-1 and 11-2); and the reaction rate increases with either an increase or a decrease in pH away from a minimum in approximately neutral solutions (Fig. 11-3). Furthermore, hydrolysis typically leads to formation of one or more clay minerals, compounds in

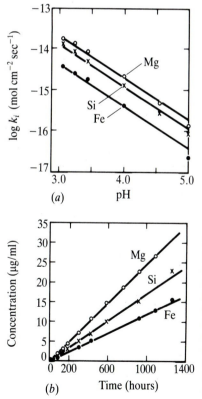

(*a*)

(*b*)

FIGURE 13-2
Experimental results on the dissolution of forsteritic olivine from Hawaiian beach sand in $0.1m$ KCl solutions at 26°C. (*a*) Dissolution rate, measured in mols of Mg, Si, and Fe released per cm^2 of surface area of reactant olivine per second, as a function of solution pH. (*b*) Changes in the concentrations of Mg, Si, and Fe during dissolution of 1 g of olivine in 500 ml of solution with a pH of 3.2 as a function of time. The different slopes of the curves in (*b*) are proportional to the stoichiometry of the reactant olivine. (Data from Grandstaff, 1986.)

which some of the aluminum and silicon of the primary mineral remain combined. For example, the weathering of K-feldspar may be symbolized by an equation we have studied exhaustively in an earlier chapter (Sec. 4-3):

$$2KAlSi_3O_8 + 11H_2O \rightarrow 2K^+ + 2OH^- + Al_2Si_2O_5(OH)_4 + 4H_4SiO_4, \quad (13\text{-}14)$$

K-feldspar   kaolinite

in which the clay mineral is represented by the formula of kaolinite. This equation describes, in a rough sort of way, the overall result of weathering in humid temperate environments—the slow ultimate conversion of feldspar to kaolinite, for which the obvious evidence can be found in the partially decayed feldspar crystals commonly seen in thin sections of weathered granite, and on a larger scale in the huge deposits of commercial *china clay* that in some places cap granite where weathering has been especially deep and long-continued (Sec. 4-2).

But the equation gives no hint of the complications involved in feldspar hydrolysis, particularly the possible formation of other clay minerals either as final products or as intermediaries en route to kaolinite. For a better understanding of aluminosilicate weathering, we pause now to look at these minerals which in aggregate make up the material called *clay*.

13-8 CLAY MINERALS

Geologists use the word "clay" in two senses: as a size term, referring to material of any composition whose average grain size is less than 0.004 mm (this figure varies somewhat with different authors), and as a mineralogic term, referring to a group of minerals with a specific range of composition and a particular kind of crystallographic structure. The two meanings often overlap or coincide, because the fine-grained part of a soil or sediment most commonly consists largely of clay minerals. The second meaning is the only one of concern here.

Clay minerals (with a few rare exceptions) are phyllosilicates (Sec. 5-6), silicates with continuous sheet structures like the micas. This could be anticipated from the flake-like form of clay particles. The resemblance to micas is apparent also in the hexagonal shapes which the flakes of some clays assume (Fig. 13-3). Both micas and clays have a characteristic structure, made up of two kinds of alternating layers or sheets. One sheet consists of the ions Al^{3+}, O^{2-}, and OH^-; in some clays the Al^{3+} is partly replaced by Mg^{2+} and other cations. The negative ions form octahedra around Al^{3+} (or Mg^{2+}), the relative numbers of O^{2-} and OH^- being adjusted to satisfy the valences of the entire structure. The O^{2-} and OH^- are shared between adjacent octahedra, so that the structure is continuous in two dimensions [Fig. 13-4(*a*)]. This pattern by itself, not as part of a clay structure, is identical with the structure of gibbsite [$Al(OH)_3$]; it is often referred to as the *gibbsite sheet*, the *octahedral sheet*, or the *aluminol layer* of the clay structure. The second kind of sheet is made up Si^{4+}, O^{2-}, and OH^-, ions; each Si^{4+} is in the center of a tetrahedron of oxygen ions; the tetrahedra all face the same direction, and the oxygens at their bases are linked so as to form hexagonal rings [Fig. 13-4(*b*)]. This sheet is called the *silica sheet*, the *tetrahedral sheet*, or the *siloxane layer* of the clay

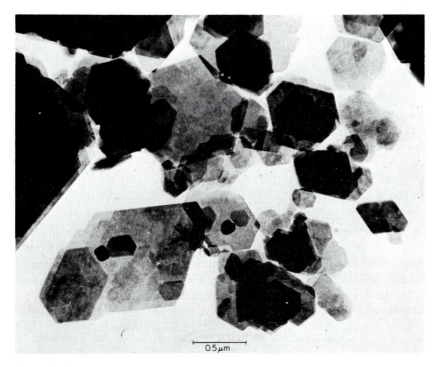

FIGURE 13-3
Transmission electron photomicrograph of kaolinite. Note the hexagonal shapes of many of the grains. The magnification is about 43,000 times. (Courtesy of Kenneth Towe, Smithsonian Institution, Washington, D.C.)

(*a*) Octahedral sheet

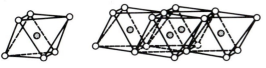

O and ◌ = Hydroxyls   ◎ Aluminums, magnesiums, etc.

(*b*) Tetrahedral sheet

O and ◌ = Oxygens   o and ● = Silicons

FIGURE 13-4
(*a*) Diagram of the octahedral sheet (also called the gibbsite sheet or aluminol layer) of phyllosilicate structures. The left-hand diagram shows a single octahedral unit. When Al^{3+} occupies the centers of the octrahedra, only two-thirds of the possible sites are filled; when Mg^{2+} occupies these positions, all sites are filled. (*b*) Diagram of the tetrahedral sheet (also called the silica sheet or siloxane layer). The left-hand diagram is a single tetrahedral unit. (Reprinted by permission from Grim, 1968.)

structure. The complete clay structure (or mica structure) consists of one of a few possible combinations of the octahedral and tetrahedral sheets.

The diagrams in Fig. 13-4 show the geometry correctly but give a false impression of ionic sizes. The big units of the structure are, of course, O^{2-} and OH^-, which have almost identical sizes (radius 1.40 Å) since the size of the added proton in OH^- is negligible. These large anions should be thought of as in contact or nearly so, clustered around the smaller cations. Only four can fit around the tiny Si^{4+} (0.26 Å), but six can be accommodated around the larger Al^{3+} (0.535 Å).

Kaolinite and Smectite

The simplest combination of octahedral and tetrahedral sheets is the structure of *kaolinite*. It has an octahedral sheet linked to a tetrahedral sheet by sharing some of the oxygen ions (Fig. 13-5).The double layer extends indefinitely in two dimensions, and the clay crystal consists of a succession of these layers one on top of another. Another possible arrangement of the octahedral and tetrahedral sheets is shown in Fig. 13-6. This pattern, with an octahedral sheet sandwiched between two tetrahedral sheets, characterizes a group of clay minerals called *smectites*, of which *montmorillonite* is the most familiar example. The difference between the kaolinite and smectite structures is easily seen by comparison of the schematic crystal structures shown in Figs. 13-5 and 13-6. Kaolinite is often said to have an unsymmetric or two-layer structure; it is commonly referred to as a 1:1 layer silicate to indicate the ratio of tetrahedral to octahedral layers (Table 13-4). A smectite has a symmetric or three-layer structure, referred to as a 2:1 layer silicate structure to emphasize the two tetrahedral sheets and one octahedral sheet (Table 13-4).

Off-hand one might expect that substitution would be common in the clay structures, since Si^{4+} and Al^{3+} should be replaceable by many other ions with coordination numbers 4 and 6. The expectation is abundantly fulfilled in the smectite clays but not in kaolinite. This difference in the amount of isomorphous

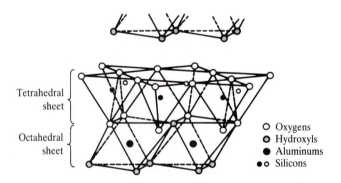

Tetrahedral sheet

Octahedral sheet

○ Oxygens
◉ Hydroxyls
● Aluminums
●○ Silicons

FIGURE 13-5
Diagram of the relationship of the tetrahedral and octahedral layers in the 1:1 layer silicate structure of kaolinite. (Reprinted by permission from Grim, 1968.)

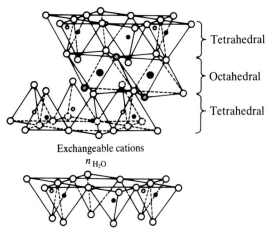

O Oxygens ⊙ Hydroxyls ● Aluminum, iron, magnesium
○ and ● silicon, occasionally aluminum

FIGURE 13-6
Diagram of the structure of montmorillonite, showing the 2:1 layer structure with two tetrahedral sheets and the central octahedral layer. (Reprinted by permission from Grim, 1968.)

substitution is another of the important distinctions between the two types of clay minerals: kaolinite *always* has a composition approximating the ideal formula, but smectite has a wide range in composition because of isomorphous substitution. The common substitution in the tetrahedral sheets is Al^{3+} for Si^{4+}, the amount of substitution being limited to about 15%. In the octahedral sheets a greater variety of substitution is possible, and the amount may be anywhere from very small to 100%. The most common substituents are Mg^{2+} and Fe^{3+}; rarer ones are Zn^{2+}, Ni^{2+}, Li^{+}, and Cr^{3+}. Thus a general formula for a smectite mineral can be written $(Al, Mg, Fe^{3+})_2((Si,Al)_4O_{10})(OH)_2$, and the various smectites differ in the nature of the

TABLE 13.4
Clay mineral groups

Group	Layer type	Layer charge	Typical chemical formula or range in composition
Kaolinite	1:1	<0.01	$Al_2Si_2O_5(OH)_4$
Illite	2:1	1.4–2.0	$K_{0.5-0.75}Al_2(Al_{0.5-0.75}Si_{3.25-3.5}O_{10})(OH)_2$
Smectite	2:1	0.5–1.2	$Na_{0.35}(Al_{1.65}Mg_{0.35})Si_4O_{10}(OH)_2 \cdot nH_2O$
Vermiculite	2:1	1.2–1.8	$(Mg,Ca)_{0.35}(Mg,Fe^{3+}, Al)_3((Al,Si)_4O_{10})(OH)_2 \cdot 4H_2O$
Chlorite	2:1 with hydroxide interlayer	variable	$(Mg,Al,Fe)_6((Al,Si)_4O_{10})(OH)_8$

The layer charge represents the typical range in mols of excess electron charge per chemical formula unit of the clay mineral produced by isomorphic substitutions. Modfied from Sposito (1989) and Deer (1971). Formula for smectite corresponds to an average montmorillonite. Compositions for illite, vermiculite, and chlorite illustrate typical ranges in compositions. Not all possible compositions are shown.

substitutions. In montmorillonite, for example, the main substitution is Mg for Al, in beidellite Al for Si, in nontronite both Al for Si and Fe^{3+} for Al.

In any clay mineral the structural layers are stacked one on top of another, and differences in the geometry of stacking give rise to clays with different properties. Thus the hydrothermal clay minerals dickite and nacrite have the same two-sheet structure as kaolinite but differ in the position of one layer with respect to the next. Whatever the arrangement, the distance from the O^{2-} and OH^- ions of one layer to the O^{2-} and OH^- of the next is greater than that between the anions within each layer, so that the mineral splits readily between the layers—which is reflected in the macroscopic (100) cleavage of the clay minerals and micas. Successive layers are more easily separated in the smectite structure than in kaolinite, since in kaolinite the O^{2-} ions of one layer face OH^- in the next while in smectite the O^{2-} of one layer faces O^{2-} of another (Fig. 13-6). The attraction of O^{2-} for OH^- (by a "hydrogen bond") is greater than that between two O^{2-} ions, so kaolinite layers are more firmly held together. Much of the difference in macroscopic properties between the two types of clay minerals can be ascribed to the greater ease with which the layers of the symmetric smectite structures separate from one another.

Some of the ionic substitutions that are common in the smectite clays leave the clay layers electrically charged, for example the substitution of Al^{3+} for Si^{4+}, or of Mg^{2+} for Al^{3+}. The deficiency of positive charge may be compensated in various ways: by the replacement of O^{2-} by OH^-, by introduction of excess cations into the octahedral layer (Al^{3+} in the ideal structure fills only two-thirds of the available positions), and by sorption of cations onto the particle surfaces or into the spaces between layers of the clay structure. Some of the compensation is always accomplished by sorption, and the adsorbed ions are particularly abundant on surfaces of the smectite clays—especially Ca^{2+}, Na^+, and K^+. The firmness with which the sorbed material is held varies with the kind of ion and its position on the clay particle surface, and the clay minerals have many different types of surface sites that can react with the adjacent solution. On the surfaces and edges of clay minerals some ions are easily exchanged, others strongly attached. Bonding is especially strong for ions that find their way into the interlayer spaces.

Because the layers of smectite, and a related clay called vermiculite (Table 13-4), are easily split apart, many ions can be accommodated in the interlayer spaces, even large ions like K^+ as illustrated in Fig. 13-7(*a*). Also water is readily admitted between layers, and the combination of water and cations can force the layers apart so that the clay readily expands [Fig. 13-7(*b*)]. Because of the adsorbed water the smectite clays are said to exhibit high "plasticity," a property that distinguishes them from kaolinite. In a mixture of clay minerals, even a small amount of smectite can greatly improve the plastic properties. Some smectites, notably the montmorillonite-rich material called *bentonite*, expand spectacularly on contact with water, to volumes more than fifty times that of the dry clay.

Illite

Clays with the smectite structure and an abundance of interlayer K^+ make up a third major group of clay minerals, the *illite* clays (Table 13-4). (The term *hydromica*,

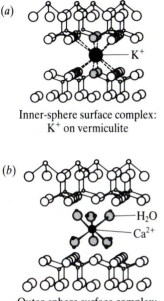

(a)

Inner-sphere surface complex:
K^+ on vermiculite

(b)

—H_2O
—Ca^{2+}

Outer-sphere surface complex:
$Ca(H_2O)_6^{2+}$ on montmorillonite

FIGURE 13-7
Diagram of interlayer surface complexes between metal cations and the tetrahedral sheets in 2:1 phyllosilicates. (Reprinted by permission from Sposito, 1989.)

often used in European literature, is practically a synonym.) In these clays the deficiency of positive charge results chiefly from substitution in the tetrahedral sheets, hence close to the surface of the three-part layers; this means that the interlayer K^+ is held especially tightly, and is only partly exchangeable for other ions. The illite structure is transitional to that of muscovite $[KAl_2(AlSi_3O_{10})(OH)_2]$, in which the amount of K^+ is greater and bears a constant relation to the amounts of Si and Al. The term illite is less a name for a definite mineral than a name for a group of substances with compositions intermediate between smectite and muscovite. Much so-called illite, in fact, is a mechanical mixture of fine-grained smectite and muscovite, or a clay with alternating layers having a smectite and a muscovite structure ("mixed-layer clay"). The name "illite" is convenient nevertheless as a general term to designate clay minerals with abundant potassium but less than that corresponding to the mica formula.

Other Clays and Related Minerals

Several other hydrous-silicate minerals produced by weathering, or at least near-surface processes, may be grouped with the clay minerals. *Vermiculite*, for example, is a three-layer clay with all octahedral positions occupied by Mg^{2+} and Fe^{2+} and with more substitution of Al for Si than is usual in smectite; it is much less expansible, but has a similar high ion-exchange capacity. *Glauconite* is a variety of illite, with considerable replacement of the Al of the octahedral sheet by Fe^{3+} and of the interlayer K^+ by Ca^{2+} and Na^+.

The *serpentine* minerals are related to clays in that they have the same structure as kaolinite but Mg^{2+} rather than Al^{3+} in the octahedral positions, as is shown by the formulas

$$Al_2Si_2O_5(OH)_4 \qquad \text{and} \qquad Mg_3Si_2O_5(OH)_4$$

kaolinite   serpentine

In kaolinite only two-thirds of the octahedral positions are filled by Al^{3+}, leaving the others vacant; in serpentine all are filled by Mg^{2+}. To mark this structural difference kaolinite and its relatives are often called *dioctahedral* clays and the serpentine minerals *trioctahedral*. Finally, the *chlorite* group of minerals have structures in which mica-like sheets with the general composition $(Mg,Fe)_3(Si,Al)_4O_{10}(OH)_2$ alternate with sheets having the structure of brucite [composition $(Mg,Al)_3(OH)_6$]. To what extent serpentine minerals and chlorite should be regarded as products of weathering is uncertain. Both are more abundantly formed by low-temperature hydrothermal alteration of mafic and ultramafic rocks, but it seems likely that they can also be produced in at least modest amount by long-continued weathering at Earth-surface temperatures.

The clay-like products of weathering include all these and several other more obscure varieties. In terms of mineral parageneses kaolinite and illite are formed during the weathering of felsic rocks, usually replacing K-rich micas and feldspars. Smectites are typically associated with the weathering of mafic rocks, and serpentine with the weathering of mafic and ultramafic rocks. The clay mineral vermiculite is formed as a weathering product of biotite, chlorite, and hornblende. Commonly the clay minerals are found not in pure form but as mixtures, the chief constituents of the mixture at any one point depending on the nature of the weathering rock, the prevailing conditions of temperature and moisture, and the length of time weathering has been in progress.

Interpreting the Clay Minerals

This long digression has introduced the more common clay minerals and their relatives, but it leaves many questions unanswered. Under what conditions do the different minerals form? Can they provide information about the nature of the rock from which they came, or about the conditions under which weathering took place? Once formed, are the clay minerals stable or can one variety change into another? Despite much field and laboratory work in recent decades, complete answers to such questions still elude us. Part of the trouble lies in the difficulty of working with such fine-grained material, part is in the complicated chemistry of the individual minerals and especially of mixtures, and part comes from the extreme slowness of the reactions by which the clay minerals are formed and change.

Seemingly a good way to start would be to establish conditions of equilibrium by looking up free energies for the clays and their precursors. For example, using the standard free energies in Appendix VIII we can compute the free energies and equilibrium constants for reactions involving K-feldspar and kaolinite [Eqs. (8-4) or

(13-14)]. These numbers will place constraints on the equilibrium values of pH and the activities of potassium and silica in the solution. They can be used in locating lines on a stability diagram showing log a_{K^+}/a_{H^+} plotted against log SiO_2(aq), for assumed equilibrium between K-feldspar and kaolinite in an aqueous solution (Fig. 4-13). How good is this assumption of equilibrium? It is certainly justified at temperatures of a few hundred degrees, where the reaction in both directions can be readily demonstrated experimentally (Fig. 4-10). At ordinary temperatures the existence of equilibrium is less assured: one cannot readily synthesize feldspar by leaving kaolinite in contact with a cold solution of appropriate pH, and the laboratory conversion of feldspar to kaolinite, although this can be accomplished, requires very long times. Nevertheless, it is not unreasonable to suppose that equilibrium is indeed approached at low temperatures, but impossible to demonstrate in times available for laboratory experiments because of the very slow reaction rates.

For reactions showing formation of the smectite clays the assumption of equilibrium is again reasonable, but harder to justify convincingly. For one thing, it is difficult to assign definite free energies to compounds whose chemistry is so variable. Experiments designed to simulate natural conditions give equivocal results, and extrapolation from higher temperatures is not meaningful because compositions of the clays do not remain constant. Reactions in which one clay mineral is converted to another are likewise slow, and equilibrium is hard to prove. An exception is the conversion of montmorillonite to illite, which can indeed be demonstrated in the laboratory at temperatures only slightly above normal and for which good evidence can be found in nature. One striking bit of evidence is the fact that clay mixtures brought to the Gulf of Mexico by the Mississippi River have a high proportion of smectite, while illite becomes increasingly dominant with depth in older sediments beneath the floor of the Gulf. A further strong hint of smectite-to-illite conversion is the simple observation that illite is the principal clay mineral in most pre-Pleistocene shales. Other inter-conversions among the clay minerals are more questionable: it seems reasonable to suppose that smectites and illite are unstable with respect to kaolinite under conditions of slightly acid groundwater and good drainage, and that illite should become dominant where temperatures are above normal and K^+ is available, but the persistence of all three types of clay in some older rocks suggests that the reactions do not occur readily.

These uncertainties mean that conclusions about conditions of formation and alteration of the clay minerals must be qualitative and hedged with provisos. Despite the lack of well demonstrated equilibrium, it is often useful to assume equilibrium and relate the clays to various combinations of cations and silica expected in weathering solutions. As a single example, Fig. 13-8 is an expanded plot similar to Fig. 4-13, showing fields of stability for clays and feldspar calculated (in part) from approximate free energies for a system in which the smectite mineral has K^+ as its principal exchangeable cation. The diagram summarizes relations that might be expected, kaolinite forming where acidity is high and dissolved silica is low, smectite appearing where silica is higher, and illite becoming dominant in an area of intermediate silica and high potassium. Boundaries of the stability fields, although

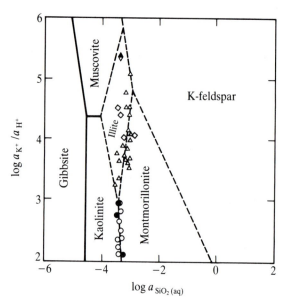

FIGURE 13-8

Activity–activity phase diagram showing roughly the compositions of solutions to be expected in equilibrium with minerals in the system $K_2O–Al_2O_3–SiO_2–H_2O$ at 25°C and 1 bar for unit activity of H_2O. Solid lines are located by calculation from free energies of gibbsite, kaolinite, muscovite, and K-feldspar. Dashed lines show roughly the positions of boundaries involving minerals of variable composition (illite and montmorillonite), for which precise values of free energies cannot be obtained. Symbols indicate measured compositions of solutions in contact with minerals and mineral combinations: open circles, kaolinite+montmorillonite; closed circles, kaolinite; diamonds, montmorillonite; triangles, montmorillonite + zeolite + kaolinite + K-feldspar. (Reprinted by permission from Garrels, 1984.)

not located with the precision we should like, are nonetheless reasonably consistent with the analyses of the many samples of soil groundwater shown on the diagram.

Summary

The weathering of an aluminosilicate mineral is a slow and complicated hydrolysis reaction in which the overall effect is removal of the cations and part of the silica in solution, while most of the aluminum and the remaining silica stay behind in a clay mineral. As a broad generalization, weathering may produce any one of the three principal kinds of clay minerals, or more commonly a mixture of two or three. Illite should be favored by high K^+ in the original rock or in the weathering solutions, kaolinite by acid solutions and good drainage, smectite by alkaline solutions containing abundant Ca^{2+}, Mg^{2+}, and Na^+. To what extent the clays may change their character after formation, in response to changes in their environment, is controversial. Some evidence suggests that the clay minerals adapt themselves fairly quickly to their surroundings, hence are sensitive indicators of temperature–

humidity conditions. Other work suggests the opposite, that clay minerals transported by wind or water into new environments retain their original character for long periods and so can serve as guides to their source.

13-9 FORMATION OF SOILS

Soils are the ultimate product of rock weathering. They have figured prominently in the preceding discussion, for the obvious reason that most soils contain clay minerals as major constituents. In this and the next section soils will be the focus of attention.

The partly decomposed debris of rock weathering is subject to continued attack by the atmosphere and by rainwater percolating through it (Fig. 13-9). Movement of water causes dissolved material and colloidal particles to be carried from one layer to another. Vegetation plays a role, particularly as it decays and so furnishes acids to the water; some of the organic matter may also be transported by the water and adsorbed by clays, thereby modifying their properties. Roots cause mechanical disruption, and upon their decay form open channels (root casts) for the downward transport of solutions and particulate matter. The activity of earthworms, ants, termites, gophers, moles, and ground squirrels causes continual mixing of the organic and mineral matter. Bacteria are a tremendous aid in decomposing the organic material, and very probably serve as catalysts for inorganic reactions also, particularly for the oxidation of iron and manganese. The result of this complex of reactions is the material called soil. Soil formation is not sharply distinguished from weathering, but is rather just the last stage of the weathering process.

The essential thing about soil formation is the transposition of material from one level in the weathered debris to another. This gives rise to layers of different composition called *horizons*, and the vertical sequence of horizons from the surface down to fresh rock is called the *soil profile* as illustrated by the drawing in Fig. 13-9. Horizons within the soil profile are designated by letters: the uppermost or A horizon is the soil layer from which downward-moving water has leached much of the more soluble and very fine-grained material, the B horizon is the intermediate layer in which some of the soluble and colloidal material is deposited, and the C horizon is the zone of fragmented but still largely unaltered debris that grades down into bedrock. An horizon designated O is often distinguished at the very top, representing an accumulation of decaying vegetation, and the letter E is sometimes used for a horizon at the base of A where extensive leaching is indicated by the light gray color and absence of organic material. Some soils have well developed profiles (meaning much transposition of material from the A to the B horizon), so that the horizons are clearly distinct as in the example of Fig. 13-9, others only a suggestion of profile formation.

The extent of profile development depends on a number of obvious variables. One is the kind of bedrock: clearly horizons will be more distinct in soils formed on granite or basalt than on a bed of kaolinitic clay or a pure quartz sandstone, because in the latter materials chemical weathering will be minimal and soil solutions will find little than can be moved from one horizon to another. A second variable is

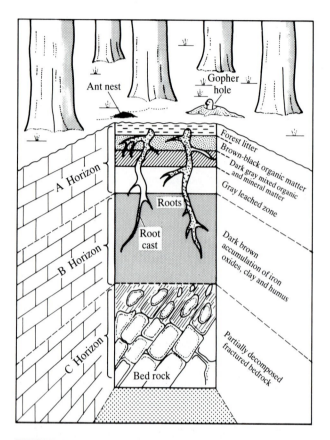

FIGURE 13-9

The best way to study soils is to dig a hole in the ground and square off the face to expose horizons of the soil profile. This is schematically illustrated here for a soil formed in humid temperate areas, a soil called podzol that displays sharp contrasts between the A, B, and C horizons. The uppermost horizon (A) consists of the forest litter, a zone of organic-rich debris in varying stages of decay, and a zone of extreme leaching of alkalies, alkaline-earth elements, and iron and aluminum. Iron, humus, and aluminum are deposited in the underlying dark brown middle horizon (B), which grades into the partially altered bedrock of the C horizon. Roots transect the profile, and upon their decay form open root casts that help in the downward transport of solutions and particulate matter. (Modified from Fig. 1-1 in FitzPatrick, 1980.)

climate, particularly amount of rainfall: the quantity of material moved, other things being equal, must be greater where rainfall is higher and the amount of soil water is larger. Topography as illustrated in Fig. 13-1 is a third factor, for on steep slopes soil is washed away before the A and B horizons can be clearly differentiated, and in swampy areas the movement of water is so slight that not much vertical transposition can take place. Other possible factors would be seasonal distribution of rainfall, length of the period of freezing, average temperature, and local sources of unusually acidic or basic solutions.

Observation indicates that soils developed under favorable conditions—on gentle slopes with adequate drainage, in a region where the climate has remained constant for a long time—acquire characteristics that depend chiefly on climate. The nature of the bedrock makes surprisingly little difference in the kind of soil ultimately formed (except for rocks of extreme composition, such as ultramafic rocks or salt beds). In the early stages of weathering, of course, soils developed on a granite, an andesite, and a gabbro will be very different, but as time goes on their characteristics converge to a soil type determined by the climate. Once the profile is fully formed and the soil is adjusted to the climate, it should be stable indefinitely. Slow erosion of the surface soil is compensated by progressive weathering of the bedrock beneath, and the soil profile moves gradually downward while maintaining the characteristics of its various horizons. Such a soil is described as a *mature* soil, meaning that it has passed through a sequence of developmental stages and is now no longer changing. In contrast, soils on steep slopes or soils in areas of recent tectonic movement, where the nature of the bedrock still markedly influences the character of the soil, are spoken of as *immature* soils. How long it takes in years to convert an immature soil to a mature soil is uncertain. Doubtless the time varies greatly from one kind of bedrock to another and from one climatic regime to another; it is certainly measured in at least hundreds of years, probably often in thousands of years.

13-10 CLASSIFICATION OF SOILS

Soils pose a particularly troublesome problem in classification. They differ in so many ways: grain size, composition, plasticity, "workability," mechanical strength, color, fertility, permeability, parent material, degree of maturity, nature of the profile. Which of these are the most important characteristics, the ones on which a usable classification can be based? Obviously there is room here for several classifications, one suited for engineering uses of soils, one for agricultural uses, one for geologists interested in soil genesis, and so on. The general problem of soil classification is not our concern, and would take us far afield from our purpose. We limit ourselves to a brief discussion of current thinking about classifications of geochemical significance.

Such classifications are very numerous. They have been attempted in many countries and at many times over the past century. Older classifications were based chiefly on ideas about responses of soils to different climates; more modern varieties put emphasis on observational details of soil profiles. Some words from the older groupings remain in use: for example, mature soils in regions of temperate climate with abundant rainfall show marked concentrations of iron oxides and aluminum silicates in the B horizon, hence are referred to as *pedalfer* soils (Greek *pedon* for soil, plus the Latin words for the two elements), while soils in areas of more modest rainfall show concentrations of calcium compounds (calcite or gypsum) in the soil profile and are given the name *pedocal*. The dividing line between the two is commonly set at an annual rainfall of 64 cm (25 inches), which in the United States coincides roughly with a north–south line along the Mississippi River. Another useful term from the older nomenclatures is *podzol*, referring to pedalfer soils of northern forests (like the soil depicted in Fig. 13-9), a common soil variety in northeastern Untied States and eastern Canada. Still another old term is *chernozem*, a kind of soil widely developed in the less arid parts of pedocal areas, where the fairly low rainfall does not remove the more soluble nutrient elements (e.g., Na, K, Ca, Mg) as effectively as in the podzols, and where the soils are consequently known for their fertility.

Such names, though useful, are inexact because soils of any climatic regime show much local variation, and soils of a given type may be found in different kinds of climate. More modern systems of classification seek to avoid these difficulties by naming soils according to characteristics of their profiles, without reference to climate or place of formation. The system currently used in the United States, proposed by the Department of Agriculture (USDA) in 1960 and revised in 1975, starts with ten "orders" of soils, differentiated chiefly by the degree of leaching of the upper layers, the amount and kind of substances added to the B horizon, and the amounts of organic material present (Table 13-5). Five of the orders (*entisol, inceptosol, alfisol, ultisol, oxisol*) can be grouped as a sequence of generally increasing amounts of weathering of the soil material, of leaching from the A horizon and additions to the B horizon. Entisols show only the beginning of soil development, while extreme weathering in ultisols and oxisols is indicated by strong leaching in the A horizon and much clay accumulating in the B horizon. Variants from this sequence are *spodosols*, with an accumulation of clay and iron oxides and hydroxides in the B horizon forming a conspicuous "hardpan"; *mollisols*, which

TABLE 13-5
Soil orders in the USDA classification

Entisol	Young soil with only minor leaching in A horizon and no B horizon, common in hilly and mountainous areas.
Inceptisol	Young soil with weakly developed A horizon and little clay enrichment in B horizon. Found in fairly humid upland areas.
Alfisol	Thin, strongly leached A horizon with organic matter; B horizon well developed with much clay. Often a light gray lower A (or E) horizon. One kind of pedalfer soil.
Ultisol	Strongly leached A horizon over weathered clay-rich B horizon. Typical of tropical and subtropical climates; common in piedmont and coastal plain of southeastern U.S.
Oxisol	Strongly leached A horizon, hydrated oxides of Al and Fe prominent in B horizon. Clay in B horizon partly decomposed and silica leached. Typical of the humid tropics.
Spodosol	Acid soil, clay and hydrated oxides accumulated as "hardpan" in B horizon. Much organic debris in A horizon. Typical of cool forest areas.
Mollisol	Thick dark A horizon with much organic matter. B horizon clay-rich with much Ca^{2+} and Mg^{2+}. May be calcium carbonate accumulation at depth. Typical grassland or steppe soil.
Aridosol	A and B horizons thin, with little organic material. Calcium carbonate accumulations (caliche) generally present. Dominant soils of deserts.
Vertisol	Old weathered soil with high content of clays that swell and shrink seasonally, causing wide vertical cracks.
Histosol	Thick O horizon of partly decayed organic debris. Typical of cool swamp environments.

Source: Soil Taxonomy, U.S. Dept. of Agriculture, Agricultural Handbook 436, 1975.

have a thick dark organic-rich A horizon and a clay-rich B horizon with deposits of calcium carbonate (caliche) in some layers; and *aridosols*, less weathered and showing more abundant carbonate deposits even in layers near the ground surface. The other two orders have more specialized compositions: *vertisols* are rich in swelling clay which leads to formation of wide vertical cracks, and *histosols* have a thick peat O horizon with much partly decomposed vegetation.

Some correspondence can be noted with earlier groupings based on climate. Entisols and inceptisols are immature soils, typical of hilly or mountainous areas with substantial rainfall, where weathered material is removed before much soil development has taken place. Alfisols and spodosols are pedalfer soils, widespread in the more humid parts of the northern temperate zones. In these soils abundant rainfall makes the soil water acid, pH values reaching 4.0 to 4.5 in the clayey part of the soil and as low as 3.5 in the humus. This acid water effectively leaches alkali and alkaline-earth ions from the A horizon, together with some of the iron and aluminum. These latter two, moving downward partly in the form of organic complexes and partly as colloidal material, are deposited in the B horizon as clay and hydrated iron oxides. Clay is more abundant in alfisols, iron oxide in spodosols. In part these are soils described by the older term podzol.

Ultisols and oxisols correspond to pedalfer soils of warmer regions, soils in which removal of alkali and alkaline-earth ions is extreme and concentration of iron oxides in the B horizon is conspicuous. In oxisols, which are abundant soils in the humid tropics, weathering in many places has progressed to the partial breakdown of clay minerals and removal of some of their silica in solution, leaving hydrated oxides of aluminum and iron as major constituents of the B horizon. Some of the layers rich in hydrated oxides have the property of hardening on exposure to air, hence can be used as construction material; such layers were long ago called *laterite* (from Latin for bricks), and this name is often applied to the entire mass of oxide-rich soil. Mineralogically, laterite consists of various mixtures of hematite, goethite, gibbsite, boehmite, more rarely diaspore—the first two predominating in *ferruginous laterite*, the latter three in *aluminous laterite* or *bauxite*. Generally oxisols contain kaolinite in addition to the oxides, and relative proportions vary widely in tropical soils; reasons for preservation of the clay in some areas and not in others are obscure. Ultisols, with greater amounts of clay and often described as *lateritic* soils, are common over large areas of subtropical climates, making up, for example, the famous red soils of the piedmont and coastal plain of Georgia and the Carolinas.

With increasing dryness of climate alfisols grade into mollisols and these into aridosols, two orders that would be described in the older terminology as pedocal soils. In contrast to soils of more humid areas, these are characterized by concentration of calcium salts somewhere in the profile. The reason is simply that water from rains is only rarely abundant enough to move completely through the soil and into drainage channels. Much of it, after rain has stopped, rises by capillarity back to the ground surface and evaporates. This means that only the most soluble soil constituents can be leached, and large parts even of these remain behind as adsorbed material on mineral surfaces. Calcium in particular becomes concentrated in the residual solution as capillary water evaporates, and may be precipitated as one of its insoluble salts—most commonly as calcite, in places as gypsum. In mollisols, which are grassland soils formed in areas where the rainfall is just under that needed to produce alfisols, the deposition of calcium salts takes place far down in the B horizon, at depths of a few meters or tens of meters. In regions of greater aridity, where aridosols are the common soils, calcite deposits are closer to the surface and often in well defined layers; the "caliche" crusts so common at or near the surface in soils of the western states are a good example.

Vegetation growing on mollisols and aridosols is neither so abundant nor so rapidly decayed as in more humid regions. The slowness of decay permits much organic matter to accumulate in the A horizon of mollisols, but in aridosols the scarcity of vegetation means that little organic debris appears anywhere in the profile. The meager vegetation on soils of both orders provides a scant source of acidity, so that soil water tends to retain the slight alkalinity resulting from hydrolysis of silicate and carbonate minerals. Alkalinity is especially pronounced in some aridosols. The common clay mineral of mollisols and aridosols, as might be expected, is montmorillonite or another smectite, or illite in soils where potassium is abundant. Mollisols are roughly equivalent to the older term *chernozem*, or "black earth," soils of the grasslands and steppes, prized for agricultural use because the scant rainwater removes less of the principal nutrient elements than does the heavier rainfall of more humid regions. Such soils form a wide belt in the United States just west of the Mississippi River, and in Eurasia a still larger area that includes the enormously productive soils of Ukraine and northern China.

The two remaining soil orders, as noted above, are specialized types that need only brief mention: *vertisols* are characterized by an abundance of clays that expand and shrink with seasonal change in moisture, leading to growth of vertical cracks, and *histosols* are cool-climate marshland soils with thick O horizons rich in partly decayed organic matter.

This is a very brief survey of an exceedingly complex subject. No mention has been made of the intricate divisions of soil orders in the USDA classification—suborders, great groups, subgroups, families and series. Nor has any hint been given of the equally complex systems devised by soil scientists in other countries. But for a general understanding of the geochemistry of soil formation and soil relationships, an expansion into such details seems hardly warranted.

PROBLEMS

1. Attempts have been made to devise a *weathering index* by which the concept of the *degree of weathering* can be made quantitative. For example, one suggested index is the ratio

$$\frac{\text{mols of } CaO + MgO + Na_2O + K_2O - H_2O}{\text{mols of } SiO_2 + Al_2O_3 + Fe_2O_3 + CaO + MgO + Na_2O + K_2O - H_2O}$$

As the weathered material loses Ca, Mg, Na, and K and appears to gain Al and Fe, this fraction will obviously decrease. By calculating such ratios for a series of rocks, we can compare their degrees of decay more accurately than is possible by the usual qualitative terms *lightly weathered, extensively weathered*, and so on. Do you think that this is a good idea? What advantages and what difficulties in the calculation of weathering indices can you point out?

2. For the analyses of fresh and weathered diabase in the table below, calculate the loss or gain in each constituent on the assumption that Al_2O_3 remains constant. Make the calculation in terms both of grams per 100 g and of percent of original amounts. Note that the behavior of Na and K in this example is different from that in the weathered gneiss of Sec. 13-2. (Analyses given in weight percent.)

	Fresh	Weathered
SiO_2	47.28	44.44
Al_2O_3	20.22	23.19
Fe_2O_3	3.66	12.70
FeO	8.89	—
MgO	3.17	2.82
CaO	7.09	6.03
Na_2O	3.94	3.93
K_2O	2.16	1.75
H_2O	2.73	2.73
Others	1.45	1.22
Total	100.59	98.81

3. You are comparing the heavy-mineral content of two samples of sand, one taken from near the top of a terrace and one from near the bottom. The formation of the terrace required at least several hundred thousand years, and as nearly as you can tell, the source of the terrace material did not change during that time. Of the following minerals, which would you expect to be markedly depleted in the lower sample: zircon, hornblende, olivine, biotite, labradorite, augite, tourmaline, magnetite, apatite?

4. Write balanced equations for the following reactions. Be sure that your equations are geologically reasonable, in the sense that the products are stable and can coexist in natural environments in the pH range in which the reactions would occur. Assume that the reactions take place at ordinary temperatures and pressures. (*a*) Calcite dissolves in carbonic acid. (*b*) Grossularite ($Ca_3Al_2Si_3O_{12}$) reacts with carbonic acid. (*c*) Sphalerite dissolves in carbonic acid. (*d*) Nepheline hydrolyzes. (Assume that kaolinite is one product and that nepheline can be represented by the simple formula $NaAlSiO_4$.)

5. List the chief products of chemical weathering of (*a*) basalt and (*b*) dolomite.

6. The hydrolysis of silicate minerals during weathering leads to the formation of OH^- [Eq. (13-14)]. Explain why this process generally does not give solutions with a pH higher than 9 during the weathering of ordinary silicate rocks.

7. The mean composition of dissolved material in 56 perennial springs in granitic rocks of the Sierra Nevada is reported below. The climate is temperate and fairly humid. The rock is chiefly granodiorite and quartz monzonite, with quartz, orthoclase, andesine, biotite, and hornblende as principal minerals. No average analysis of the rock is available, but the single analysis given below is probably representative. Suggest a source for each of the dissolved substances in the spring waters, and suggest reasons why the relative amounts are so different from the relative amounts of the corresponding oxides in the rock.

Composition of rock (weight percent)		Dissolved material in springs (ppm)	
SiO_2	67.4	SiO_2	24.6
Al_2O_3	15.8	Al	0.02
Fe_2O_3	1.7	Fe_{total}	0.03
FeO	2.2	Mg	1.7
MgO	1.3	Ca	10.4
CaO	3.1	Na	6.0
Na_2O	3.5	K	1.6
K_2O	4.2	HCO_3^-	54.6
		SO_4^{2-}	2.4
		Cl	1.1
		pH	6.8

Source: Feth, J. H., C. H. Roberson, and W. L. Polzer: "Sources of mineral constituents in water from granitic rocks," Sierra Nevada, U. S. Geological Survey Water-Supply paper 1535-I, 1964.

REFERENCES

Deer, W. A., R. A. Howie, and J. Zussman: *Rock-forming Minerals*, vol. 3, *Sheet Silicates*, Longman, London and New York, 1971.

FitzPatrick, E. A.: *Soils, Their Formation, Classification, and Distribution*, Longman, London and New York, 1980.

Garrels, R. M.: "Montmorillonite/illite stability diagrams," *Clays and Clay Minerals*, vol. 32, pp. 161–166, 1984.

Goldich, S. S.: "A study in rock weathering," *Jour. Geology*, vol. 46, pp. 17–58, 1938.

Grandstaff, D. E.: "The dissolution rate of forsteritic olivine from Hawaiian beach sand," in Colman, S. M., and D. P. Dethier, (eds): *Rates of Chemical Weathering of Rocks and Minerals*, Academic Press, New York, 1986.

Grim, R. E.: *Clay Mineralogy*, 2d ed., McGraw-Hill, New York, 1968.

Nahon, D. B.: *Introduction to the Petrology of Soils and Chemical Weathering*, translated by A. V. Carozzi, Wiley, New York, 1991.

Sposito, G.: *The Surface Chemistry of Soils*, Oxford University Press, 1989.

SUGGESTIONS FOR FURTHER READING

Chesworth, W., and F. Macias-Vasquez: "pe, pH and podzolization," *American Jour. Sci.*, vol. 285, pp. 128–146, 1985. Separation of Si from Al and Fe during formation of podzol soils as influenced by organic acids and complexants.

Colman, S. M., and D. P. Dethier, (eds): *Rates of Chemical Weathering of Rocks and Minerals*, Academic Press, New York, 1986. Papers on experimental and theoretical studies of weathering rates.

Drever, J. I., (ed.): *The Chemistry of Weathering*, NATO ASI Series C149 (Advanced Study Institute on the Chemistry of Weathering), Kluwer Academic, Dordrecht, 1985. Theoretical and experimental papers on weathering chemistry.

FitzPatrick, E. A.: *Soils, Their Formation, Classification and Distribution*, Longman, London and New York, 353 pp, 1980. An excellent elementary text on the physical and chemical processes of soil formation.

Harriss, R. C., and J. A. S. Adams: "Geochemical and mineralogical studies on the weathering of granitic rocks," *American Jour. Sci.*, vol. 264, pp. 146–173, 1966. A detailed study of soil profiles developed on five kinds of granitic rock.

Law, K. R., H. W. Nesbitt, and F. J. Longstaff: "Weathering of granite tills and the genesis of podzol," *American Jour. Sci.*, vol. 291, pp. 940–976, 1991. Chemical details in the development of a podzol from granitic material.

Lerman, A. and M. Meybeck, (eds): *Physical and Chemical Weathering and Geochemical Cycles*, NATO ASI Series C149 (Advanced Study Institute on the Chemistry of Weathering), Kluwer Academic, Dordrecht, 1988. Papers on details of weathering as related to many kinds of geochemical cycles.

Nesbitt, H. W., and G. M. Young: "Prediction of some weathering trends of plutonic and volcanic rocks based on thermodynamic and kinetic considerations," *Geochem. et Cosmochim. Acta*, vol. 48, pp. 1523–1534, 1984. An attempt to make quantitative predictions about weathering reactions.

Sposito, G.: *The Chemistry of Soils*, Oxford University Press, New York, 277 pp, 1989. An advanced text on the chemical processes of soils and their interactions with solutions.

CHAPTER
14

SEDIMENTATION AND DIAGENESIS: INORGANIC GEOCHEMISTRY

The debris of weathering, the heterogeneous assortment of materials resulting from the action of water and the atmosphere on pre-existing rock, makes up the inorganic part of sediments and sedimentary rocks. Mixed with the inorganic debris is usually some organic material, from the life processes of plants and animals. The organic fraction of sediments and sedimentary rocks ranges from zero for pure quartz sand to nearly 100% for a bed of coal; in sediments as a whole, the fraction is estimated to be about two percent. To the organic part we devote the next chapter, but for the moment we ignore organic compounds except as they may influence the deposition or alteration of inorganic material.

Emphasis here will be on chemical processes that form sediments, not the mechanical processes leading to separation of sands, gravels, and clays. We consider first six kinds of inorganic chemical sediments, sediments formed primarily by deposition of compounds out of solution. These include carbonate sediments, silica sediments, sediments of iron and manganese, phosphate, and evaporites. In the last part of the chapter we look at the chemistry of diagenesis, the reactions by which chemical and detrital sediments are altered after deposition and ultimately converted into sedimentary rock.

14-1 CARBONATE SEDIMENTS

For details of the chemistry of carbonate sediments we need little more than a reference back to the discussion in Chapter 3. That chapter, we recall, is largely concerned with the carbonate of calcium, which is by far the most abundant compound in most carbonate sediments.

Calcium carbonate has two principal crystalline forms, calcite and aragonite. The former is the more stable polymorph under usual surface conditions of temperature and pressure, but both are common in recent sediments. Aragonite changes ultimately to the more stable form, so that older sediments and limestones are almost exclusively calcite. Calcium carbonate in one or the other of these forms precipitates when a solution containing Ca^{2+} and HCO_3^- is evaporated or made alkaline, or when the solution is heated:

$$Ca^{2+} + 2HCO_3^- \rightarrow CaCO_3 + H_2O + CO_2, \qquad (14\text{-}1)$$
$$\Delta G^\circ = -33.1 \text{ kJ}, \qquad \Delta H^\circ = +40.6 \text{ kJ}.$$

The reaction in many natural environments is slow, and solutions super-saturated with $CaCO_3$ are often encountered. Surface seawater, for example, may have an ion activity product, $a_{Ca^{2+}} \cdot a_{CO_3^{2-}}$, as much as 5 times the equilibrium value. Nevertheless, precipitation is a common occurrence in lakes, soils, caves, and at the mouths of hot and cold springs. But the place where $CaCO_3$ sediments accumulate most abundantly is on the floor of the ocean. Here the sediment consists in large part of the remains of organisms that used this compound for constructing their shells. Much of it is fine-grained, both because many of the original organisms are tiny and because larger shells have been pulverized by the action of waves and currents. Some shell-forming creatures favor aragonite as building material, others prefer calcite, and a few use both in successive layers. The calcite generated by some of the organisms has a large admixture (up to 20%) of magnesium replacing some of the calcium. This high-Mg calcite is unstable, gradually losing its magnesium back to the surrounding water. A fresh carbonate sediment on the seafloor thus may be a mixture of calcite, aragonite, and magnesian calcite, but with time the latter two minerals react and disappear. Limestone eventually formed from such a sediment would consist almost wholly of calcite with no more than one or two percent of magnesium.

A recently deposited limy sediment contains much interstitial water, and may be porous enough to permit this and additional water to circulate freely. The water may change its temperature and CO_2 content from time to time and from place to place, so that the $CaCO_3$ grains among which it moves are in contact alternately with unsaturated and supersaturated solutions. Thus ions of the compound can move from one grain to another, permitting large crystals to grow at the expense of the more soluble small ones and gradually filling the inter-grain spaces. By this mechanism limy sediments may be converted slowly to limestone, and it is easy to see why many older limestones are coarse-grained and relatively impermeable.

Water moving through a limy sediment may carry additional dissolved materials capable of reacting with the $CaCO_3$ particles. One such solute is

Mg^{2+}, which under some conditions can react with calcite to form dolomite. Just what the necessary conditions are for this reaction to take place remains a major unknown of carbonate geochemistry, but certainly much older limestone has been replaced partially or wholly by the double carbonate (Sec. 3-10). (Dolomite, it should be recalled, is distinct from the high-Mg calcite mentioned above that is used as shell material by some organisms: in the latter the Mg^{2+} ions randomly replace some Ca^{2+} without altering the calcite structure, while dolomite has a crystal structure of its own with regular alternation of the two cations. Unlike dolomite, magnesian calcite is unstable and does not persist in a limy sediment.) Few pronouncements about dolomite can be made with complete assurance: the replacement reaction just described is a plausible explanation for the origin of many older dolomites, perhaps the majority, but a good case can be made for thinking that some dolomites were formed by direct precipitation of a dolomitic sediment.

The only other carbonate mineral that is found in extensive sedimentary beds is siderite, $FeCO_3$. Because Fe(II) carbonate is so readily oxidized to hematite or limonite (Sec. 9-5), siderite cannot be a common sediment in most surface environments. It could not precipitate from present-day seawater, for example, because in most places the dissolved oxygen content is too high; and where the water does become reducing, its high sulfur content would lead to precipitation of sulfide rather than carbonate. Conditions required for iron carbonate to precipitate in large quantities can exist only where organic matter is abundant and oxygen is largely excluded, say an extensive freshwater swamp. This is the kind of environment generally pictured to explain the common association of siderite strata ("clay ironstones") or "ironstone concretions" with black shales and coals.

14-2 SILICA SEDIMENTS

More difficult to understand than the carbonates are sediments that consist entirely, or almost entirely, of SiO_2. These can be observed forming today in several kinds of natural environments, for example hot-spring areas and lakes of both fresh and salt-water, but like the carbonates most older deposits of silica have the clear stamp of marine origin. The problem in accounting for their origin and their abundance is the very low content of silica dissolved in seawater, generally no more than one or two parts per million. As noted in Chapter 4, different forms of silica have widely different solubilities at usual conditions of temperature and pressure, from about 6 ppm for quartz to more than 100 ppm for amorphous silica, but the amount in seawater is less than any of these. How, then, can a precipitate of silica form in the sea?

The answer, as for the precipitation of much calcium carbonate, depends on organisms, on the life processes of creatures that use silica for construction of shells or skeletal parts. Such organisms—radiolaria (single-celled animals), diatoms (single-celled plants), siliceous sponges—can somehow precipitate SiO_2 from an unsaturated solution and protect it with an organic membrane while they are alive. After death the diatom and radiolaria shells and sponge spicules dissolve so slowly, and the organisms have flourished in such numbers, that thick deposits of the siliceous debris can accumulate. Broad areas of the present ocean floor are covered with such material, and beds of diatomite interlayered with older sedimentary strata are evidence of similar deposits in the past.

More problematic are the numerous occurrences of apparently amorphous silica in the form that goes under the general name of *chert* and includes specific varieties such as agate, jasper, and onyx. Chert may appear in beds interstratified with other sedimentary rocks, or as nodules with rounded and irregular shapes. Its amorphous or cryptocrystalline appearance, the rounded nodular forms, the intricate crumpling of much bedded chert, and the frequent appearance of colored bands resembling Liesegang rings (Sec. 6-4), are good evidence for colloidal phenomena as somehow playing a part in its origin, but how and where the colloidal silica accumulates remain only partially answered questions.

Silica, as noted in Chapter 6, readily exhibits colloidal behavior. If a supersaturated solution of silica is prepared in the laboratory (either by adding acid to a solution of sodium silicate or by cooling a solution of silica gel prepared at high temperature) and allowed to stand for a period of days or weeks, silica does not crystallize but slowly forms a sol. A sol with a few hundred ppm of excess silica is stable indefinitely, but a more concentrated sol eventually forms a gelatinous precipitate (in alkaline solution) or sets to a uniform gel (in acid solution). As with other colloids, the coagulation is aided by electrolytes. Sols with much less than 100 ppm of SiO_2 can also be prepared, and they too can exist for periods of days or weeks. The laboratory formation of silica gels depends on changes in temperature, silica concentration, and amount of electrolytes, changes that can be readily imagined in natural environments.

Stream water and groundwater contain silica in true solution derived from the dissolution of silica and silicate minerals in rocks and soils. Concentrations are generally between 10 and 60 ppm of SiO_2, well below the equilibrium solubility of amorphous silica (and far above the solubility of quartz), in the range where silica can form a gel or be dissolved in response to various changes of conditions. Thus the chemistry of silica sediments is peculiar in that it is largely concerned with the behavior of an unstable form (a gel, or the amorphous solid silica formed from the gel) rather than the stable form (quartz).

The apparently amorphous material that makes up the various forms of chert in nature is partly amorphous (opal) and partly cryptocrystalline quartz (often called *chalcedony*). From their structure, both materials are assumed to represent hardened forms of silica gel, the opal being generated by loss of water from the gel, and the chalcedony by additional very slow crystallization of the gel to form a network of tiny quartz crystals. How can such gels originate in nature? One obvious mechanism is the cooling of hot solutions associated with volcanic activity, either solutions directly accompanying eruptions of lava or the later solutions that feed hot springs and geysers. This is an easy explanation for the abundant chert often found in sedimentary beds associated with volcanic rocks. But much chert has no obvious connection with volcanic activity, and its origin is far less clear. Perhaps some is formed in small amounts by flocculation of colloidal silica from lower-temperature solutions. For generating larger quantities, a likely possibility is the alteration of deposits of radiolaria, diatoms, and sponge spicules: groundwater moving through such deposits would presumably dissolve some of the silica, particularly that on the exposed corners and edges of the shell fragments, and deposit it on the larger fragments—essentially the same process as the growth of large crystals at the expense of small ones in deposits of calcium carbonate. In this way a uniform mass of amorphous or nearly amorphous silica could be formed from original fragmental material, and later crystallization could produce the cryptocrystalline quartz of chalcedony. Some steps in this sequence have been demonstrated in the diatomite beds of the Monterey formation of California—change of the original opal of the shell fragments to more massive opal, then change to a cryptocrystalline variety of cristobalite, which would require little further alteration to become chalcedony. Such a sequence is certainly plausible for other cherts containing recognizable shell fragments, and may well be applicable even if the shapes of fragments have been obliterated.

One indication that solutions containing abundant silica are not uncommon in nature is the fact that chert extensively replaces other rock and organic materials, often preserving original structures and textures with marvelous fidelity. Petrified wood is an outstanding example. Much of the chert in sedimentary rocks can probably be accounted for by this sort of replacement, although details of the replacement raise troublesome questions. Why, for example, do silica solutions in some rocks replace earlier minerals with opal or chalcedony, and in other rocks cause overgrowths of quartz around original sand grains? Perhaps the grains serve as nuclei for crystallization of additional quartz, or perhaps slightly higher temperatures lead to crystallization of the most stable polymorph. Again, why do solutions in some places replace calcite by silica, and in other places silica by calcite? We can speculate that slight changes in pH and silica concentration may be responsible: an acid solution supersaturated with silica could cause calcite to dissolve and silica to be deposited, while a slightly basic solution undersaturated with silica could be responsible for the opposite change. But such statements are little more than plausible guesses.

Thus the geochemistry of siliceous sediments is not as well understood in its quantitative details as the geochemistry of the carbonates, primarily because reactions are very slow and involve the formation of sols and gels. The broad outlines are clear, but colloidal processes introduce a considerable degree of uncertainty.

14-3   IRON SEDIMENTS

Sedimentary rocks consisting chiefly of iron minerals are well-known. One thinks, for example, of the widespread Precambrian *iron formation* of the Lake Superior district, with its delicately banded sequences of dark iron-rich layers alternating with layers rich in chert, or of the oölitic hematite ore of the Clinton formation (Silurian) in the southeastern states. Iron minerals in great variety are also minor constituents of other sedimentary rocks. The essential chemistry that governs iron deposition in sedimentary environments can be read from the Eh–pH diagrams of Chapter 9. Here we review briefly the interpretation of these diagrams, then consider other aspects of the behavior of iron which are not so easily expressed in quantitative terms.

The iron in minerals of igneous and metamorphic rocks is partly oxidized during weathering, giving the familiar brown, yellow, and red colors of exposed rock surfaces, and partly dissolved in the form of Fe^{2+}. This ion can be transported long distances if the solution stays reducing and slightly acidic, and if it is not admixed with other ions that form insoluble compounds. A great variety of ions may cause precipitation, the most common being carbonate, sulfide, and silicate. The precipitation reaction for any of these is aided by an increase in alkalinity of the solution. Oxidation followed by precipitation of hematite or goethite may take place at any time during the transportation of Fe^{2+}, in solutions of any acidity within the normal range, but the reaction occurs most readily under alkaline conditions. These are the geologically useful conclusions that can be read from Fig. 9-5 and similar diagrams.

The geologic conditions in which iron deposition occurs are well-known: by oxidation at the site of the original iron mineral, before any transportation has taken place; by oxidation in soil derived from the original rock, after only minor movement within the soil layer; by oxidation in streams, lakes, or swamps, when the water is aerated and loses its contained organic matter; in seawater, as oxide if the water is aerated, as hydrous silicate if the water is mildly reducing, or as sulfide if the redox potential is low and sulfur is abundant. Fe(II) carbonate precipitates in reducing freshwater sites, but probably not in present-day marine environments because seawater contains abundant dissolved sulfur and the sulfide forms rather than the carbonate. Iron sulfide appears initially as one of several forms of FeS, which changes on standing to pyrite, FeS_2. Especially favorable conditions for these reactions exist in shallow-water marine sediments at depths of a few centimeters below the sediment surface, where organic matter keeps the environment reducing and the sulfate of interstitial seawater is reduced by bacterial action to form H_2S. As a minor constituent formed in this manner, pyrite is very common in organic-rich marine sedimentary rocks.

The characteristics of modern sites of iron deposition make possible a plausible reconstruction of the environments where iron-bearing sediments of the past have accumulated. For example, facies of the Precambrian "iron formation" near Lake Superior have been identified with sediments formed at different depths in an ancient sea: hematite-bearing rocks deposited in an oxidizing environment near shore, magnetite and greenalite ($Fe_3Si_2O_5(OH)_4$) in deeper water under mildly reducing conditions, and pyrite in still deeper water where organic matter was abundant. Siderite-bearing rocks of the iron formation have been interpreted in two ways, either as swamp deposits from freshwater basins near the old shoreline, or as marine deposits formed at moderate depths in an ocean with less dissolved sulfur and oxygen and more dissolved carbon dioxide than the ocean of modern times.

Although phase diagrams allow us to evaluate possible chemical conditions for iron deposition in different sedimentary environments, numerous questions remain to be clarified. One concerns the relation of the anhydrous oxide hematite to the hydrated oxide goethite, which is the chief constituent of limonite. The free energy change for the reaction

$$Fe_2O_3 + H_2O \rightleftharpoons 2FeOOH \qquad (14\text{-}2)$$

$$\underset{\text{hematite}}{} \underset{\text{goethite}}{}$$

is small; reported values are only a few hundred joules plus or minus, the exact number and the sign depending on whether one mineral or the other is very finely divided (and hence less stable). Hematite is normally more coarsely crystalline than limonite, and this fact together with the geologic observation that limonite is scarce in older rocks suggests that the equilibrium in most natural environments is displaced to the left. Both forward and reverse reactions are slow, however, so that limonite can persist for geologically long times.

A second item that needs attention is the role of colloids. When conditions are right for Fe(III) oxides or hydroxides to precipitate, they often form instead a stable sol, and as a sol may be transported long distances. This is, in fact, the chief method of transportation of iron in surface waters. As discussed in Chapter 6, the sol is flocculated by electrolytes, for example where streams carrying iron enter the sea. Probably most of the iron that reaches the sea is precipitated first as hydrous Fe(III) oxide by flocculation of the colloid in oxygen-rich near-shore waters, and is only later reduced to Fe^{2+} by contact with organic matter in or near the bottom sediments.

A third gap in the previous discussion is a consideration of bacteria as possible precipitating agents. Certain species of bacteria have the ability to use the slow oxidation of iron compounds as an energy-producing reaction for their life processes. They probably do not cause any reaction to occur that would not happen anyway if given time enough, but they serve as efficient catalysts in speeding up a process like

$$4Fe^{2+} + O_2 + 4H_2O \rightarrow 2Fe_2O_3 + 8H^+. \qquad (14\text{-}3)$$

$$\underset{\text{hematite}}{\phantom{4Fe^{2+} + O_2 + 4H_2O \rightarrow 2}}$$

Bacteria are active in forming "bog iron ores," deposits of limonite that develop in swamps and lakes, and also in causing precipitation of Fe(III) oxide around springs. Bacterial deposition of Fe(III) oxide can be troublesome when quantities of the compound appear on the walls of pipes or on turbine blades. Other species of bacteria aid in the formation of pyrite in reducing environments.

Precipitation of iron may be hindered by the formation of complexes (of both Fe^{2+} and Fe^{3+}) with certain organic compounds, particularly in fairly acid water containing abundant organic material. Such complexes help to explain the apparently anomalous concentrations of dissolved iron sometimes reported in lakes and swamps of north temperate and subarctic areas, where precipitation would be expected.

The most serious remaining question about iron is the mechanism by which iron sediments become segregated from other sedimentary materials. The various processes we have considered explain clearly how iron has been transported and precipitated in nature, and we may reasonably suppose that such processes operate efficiently in surface waters of the present day. But most surface waters carry little iron, and that which does precipitate forms only a minor part of the accumulating sediments. At some times in the past, however, iron-rich sediments have been deposited in large amounts and over long periods, without admixture of much other material. The Clinton ores of Silurian age in the southeastern United States and the Precambrian iron formation of the Lake Superior district are good examples. How was it possible for so much iron to be concentrated in one place?

A convincing answer is hard to give, because we find no example in the modern world of such large-scale segregation and precipitation of iron compounds. Explanations have been attempted in three ways. According to the first, a source of iron in greater abundance than usual is necessary. The source commonly suggested is volcanoes—submarine lava flows, submarine springs, or perhaps just the rapid weathering of mafic lavas and tuffs. As corroboratory evidence, the common occurrence of iron-rich springs near modern volcanoes can be cited. The second kind of explanation supposes that ordinary weathering, given a low-lying land mass and a long-continued absence of orogenic movement, might serve to concentrate iron—provided that the climate was warm and humid and the land supported lush vegetation, so that the Eh and pH of stream water and groundwater would be low. The third explanation, applicable to the Precambrian ores, invokes a possible difference in atmospheric composition that would make dissolved Fe(II) more stable: if oxygen were less abundant and carbon dioxide more abundant, the dissolving of iron would be more rapid and its oxidation inhibited, so that iron in the Fe(II) form could be transported and deposited in large amounts. For such a composition of the Precambrian atmosphere there is much other evidence (Sec. 21-5), and the third explanation is commonly accepted for the concentration of iron in the enormous deposits of this early time.

14-4 MANGANESE SEDIMENTS

Manganese minerals in minor amounts are common constituents of sedimentary rocks, but rocks consisting chiefly of these minerals are rare. The most famous of such manganese-rich sedimentary beds is an early Tertiary formation in southern Ukraine, which for a long time has been a major supplier of manganese for special steels to the countries of Europe. One part of the iron formation in northern Minnesota (Cuyuna range) is exceptionally rich in manganese, but otherwise sedimentary manganese ore deposits in North America are few and small.

The chemistry of manganese resembles that of iron very closely. In solutions with low redox potential and low pH, both elements are stable as the divalent ions. Both form carbonates, sulfides, and silicates that are fairly insoluble in neutral or basic solutions. Both are readily oxidized under surface conditions to give very insoluble oxides. Details of the chemistry are different: manganese in nature shows two higher oxidation states, +3 and +4, whereas iron has only +3; oxidation of Mn(II) compounds requires higher oxidation potentials than does oxidation of Fe(II) compounds; the simple sulfide of manganese, MnS, is more soluble than FeS, and the double sulfide MnS_2 is far less stable than FeS_2 (pyrite or marcasite). In the laboratory, where redox potentials higher than those found in nature are easily attained, manganese also differs from iron in its ability to form stable compounds in which it has oxidation numbers of +6 (manganates, for example Na_2MnO_4) and +7 (permanganates, for example $KMnO_4$).

The simplified Eh–pH diagram for manganese in Fig. 14-1 is drawn with the same fixed concentrations of total dissolved carbonate ($1m$) and sulfur ($10^{-6}m$) that were used in Fig. 9-5 for iron, to facilitate comparison of the two elements. The diagram shows that MnO_2 is the stable compound of manganese at high redox potentials, regardless of pH; that other oxides can form at lower potentials in basic solutions; that $MnCO_3$ (rhodochrosite) is stable over a wide range of Eh and pH if dissolved carbonate is high; and that MnS does not form unless sulfide concentrations are high and carbonate low (in fact, not unless total dissolved sulfur exceeds total carbonate by a factor of at least 100). Manganese silicates (for example, rhodonite, $MnSiO_3$) would occupy a prominent place on the right-hand side of the diagram for solutions with high silica concentrations and low carbonate.

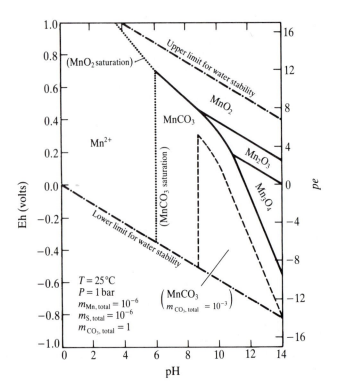

FIGURE 14-1
Theoretical Eh–pH phase diagram for the system MnO_2–CO_2–H_2O–S_2–O_2, showing stability fields of the common manganese minerals. The thick lines are phase boundaries for conditions where total carbonate is $1m$ and total sulfur is $10^{-6}m$ (the same conditions as in Fig. 9-5), and the dotted lines represent mineral saturation for a solution containing $10^{-6}m$ total Mn. The thin dashed lines show the stability of $MnCO_3$ for a total carbonate concentration of $10^{-3}m$.

To bring out the relationship between manganese and iron minerals, think of superimposing Fig. 14-1 over Fig. 9-5. The most immediately obvious difference is the enormously greater field of the carbonate in the manganese diagram and of the highest oxide (hematite) in the iron diagram—a pictorial representation of the fact that iron compounds oxidize more readily in natural environments. The large size of the $MnCO_3$ field, of course, is a consequence of the high assumed total carbonate; to indicate how the field would shrink at lower carbonate concentrations, the light dashed lines in the $MnCO_3$ field show its outline for total carbonate equal to $0.001m$. At this low value for carbonate the oxides would have larger fields, and over much of the normal Eh–pH range no manganese minerals could form at all unless the concentration of Mn^{2+} is abnormally high. Note that hematite is stable in the presence of Mn(II) compounds over a considerable Eh–pH range and that, in a large area of acid and moderately oxidizing conditions, hematite could precipitate while manganese remains in solution as Mn^{2+}.

The sedimentary behavior of manganese, as reconstructed from Fig. 14-1, is much like that of iron. The metal is dissolved from its compounds by weathering solutions as Mn^{2+}, and remains in this form as long as the solution is slightly acid and not too oxidizing. Precipitation will occur when pH increases, provided carbonate or silicate is present in sufficient concentration. If the solution becomes more oxidizing, say by prolonged exposure to air so that any organic material is oxidized, the manganese precipitates as one of the oxide minerals. With sufficient exposure to atmospheric oxygen, pyrolusite (MnO_2) should be ultimately the most stable mineral. This general sequence of events can be observed wherever manganese deposits are forming at the present time, and can be reasonably assumed for sedimentary manganese deposits of the past. The readiness of the dioxide to precipitate (as pyrolusite or a more complex mineral) on exposure of manganese solutions to air is evident in the black films, often showing intricate dendritic forms, deposited on the surfaces and in cracks of all kinds of rocks exposed to the atmosphere.

Like the oxides of iron, the higher manganese oxides commonly form first as colloids and may be transported long distances as sols. These are unusual among oxide sols in that their particles carry a negative charge over a large pH range, and therefore preferentially adsorb cations out of the solution. The particular assortment of cations commonly found with manganese oxide sediments is an odd one—K^+, Ni^{2+}, Co^{2+}, Pb^{2+}, Ba^{2+}, Cu^{2+} are especially abundant—and probably indicates that sorption is supplemented by formation of definite compounds.

Manganese resembles iron in the ability of its divalent ion to form complexes with organic compounds, and also in the speeding up of its oxidation by bacteria. Some of the same bacteria that help to precipitate iron [for example, Eq. (14-3)] can also derive energy from the oxidation of manganese, and at least two species apparently prefer manganese. The bacterially catalyzed precipitation of manganese oxide has proved a nuisance in pipes used for industrial and municipal water supplies, and in nature is probably in large part responsible for the formation of manganese ores in bogs and lakes.

The great puzzle in the sedimentary geochemistry of manganese, just as for iron, is not the chemical processes as such—these are understood in all but minor details—but the sequence of geologic processes by which compounds of manganese are separated from other sediments, especially from iron-rich sediments. In solutions resulting directly from the weathering of igneous rocks, or from the action of hot acids on rocks in the laboratory, the two elements are normally present in a ratio of about 50:1, which is the average ratio of iron to manganese in the Earth's crust. Exposure of such solutions to air and a rapid increase in pH cause both metals to precipitate, resulting in the formation of iron minerals with a small admixture of manganese—corresponding to the compositions of many sedimentary iron ores. If, on the other hand, the pH increases very slowly, iron compounds reach their limit of solubility before manganese compounds and so can precipitate while manganese is left in solution. This process of isolating manganese in solution can be demonstrated in the laboratory and is also a commonplace in nature: for example, spring deposits are often reported where an iron-rich precipitate accumulates near the spring and a manganese-rich precipitate farther away.

This is demonstrably an effective separation process, but can it be postulated as a general method of isolating manganese in near-surface solutions? It requires, of course, that formation of sedimentary manganese deposits should be accompanied by precipitation of much greater quantities of iron minerals somewhere in the vicinity, either on the surface or underground. Field evidence on this point is equivocal: in some places abundant iron minerals do appear in the vicinity of sedimentary manganese deposits, but by no means always. The incomplete evidence leaves this mechanism for large-scale isolation of manganese as an attractive but not entirely convincing hypothesis.

Other suggested mechanisms depend on differences in colloidal behavior of the two metals, differences in stability of their organic complexes, and differences in their response to bacterial action. Any or all of these may be effective locally, but as general explanations for the origin of relatively pure iron deposits and relatively pure manganese deposits they seem less probable than separation by differential oxidation and differential solubility.

The differences in solubility of compounds and in oxidation potential likewise provide at least a partial explanation for differences in behavior of iron and manganese in the sea. Some of the iron carried to the sea is precipitated in shallow water, as Fe(III) oxide where the water is oxidizing and as pyrite in the more reducing water near or within the bottom sediments. Manganese cannot follow these precipitation processes in any quantity because its oxide and sulfide are too soluble under prevailing shallow-water conditions. Rather, most manganese brought into the sea should be widely dispersed as Mn^{2+} (or a complex of this ion), and perhaps is ultimately precipitated as oxide in deep water where dissolved oxygen is high and organic matter is scarce. Some of this precipitated oxide may appear in manganese nodules, which are rounded masses up to a meter or so in diameter that form conspicuous parts of the present-day deep-water sediment over large areas. So abundant are the nodules in places that commercial exploitation seems a possibility. The nodules always contain iron oxide as well as manganese, in amounts up to 50%, and commonly also have a percent or more of rare metals like Cu, Co, and Ni. The source of manganese in the nodules is still a matter of dispute, but the movement of Mn^{2+} from shallow to deep water, as just suggested, is a likely hypothesis. The common association of abundant nodules with mafic volcanic rocks on the sea floor strongly suggests that other sources of the metal may be submarine weathering and hydrothermal alteration of the volcanic material.

14-5 PHOSPHATE SEDIMENTS

Phosphorus is an element essential to all forms of life, one whose scarcity often sets a limit on the ability of organisms to multiply. It is found in some kinds of protein, one of the major constituents of living substance, and is also important to some organisms in the construction of solid materials such as the bones and teeth of vertebrates and the shells and chitinous exoskeletons of some invertebrates. The phosphorus needed by plants is obtained from solution in soil water or the water of lakes and oceans, where the element exists chiefly as one or more of the phosphate ions (PO_4^{3-}, HPO_4^{2-}, $H_2PO_4^-$). When organic material decays, most of its phosphorus dissolves in this form. The inorganic chemistry of phosphorus is largely concerned with the behavior of these ions and the compounds they may form.

Sedimentary beds consisting chiefly of phosphate minerals are not common, but in a few places are thick and widespread. Such deposits have long been known, and have been exploited for the production of fertilizer, in the southeastern United States and in an area of the west including southern Idaho and adjacent states. More commonly phosphate minerals appear as isolated nodules or minor lenses in other sedimentary rocks, especially limestones.

The chemistry of phosphates in many ways resembles that of the carbonates. Like carbonates, most phosphates except those of the alkali metals are nearly insoluble in neutral and alkaline solutions. In acid solutions they dissolve, because phosphoric acid, like carbonic acid, is a fairly weak acid. The reactions are a bit more complicated, because phosphoric acid has three hydrogens per formula unit instead of two:

$$H_3PO_4(aq) \rightleftharpoons H^+ + H_2PO_4^- \qquad K_1 = 10^{-2.1}, \qquad (14\text{-}4)$$

$$H_2PO_4^- \rightleftharpoons H^+ + HPO_4^{2-} \qquad K_2 = 10^{-7.2}, \qquad (14\text{-}5)$$

and

$$HPO_4^{2-} \rightleftharpoons H^+ + PO_4^{3-} \qquad K_3 = 10^{-12.4}. \qquad (14\text{-}6)$$

From these activity constants we can infer that the acid itself is somewhat stronger than carbonic acid, and that the principal dissolved form of phosphate in geologic environments will be $H_2PO_4^-$ in acid solutions and HPO_4^{2-} in alkaline solutions. The unhydrolyzed ion PO_4^{3-} becomes dominant only at pH's over 12.4, hence in geologic situations is always minor; but the precise amount of this ion at different pH's is of critical importance in discussions of phosphate precipitation, because it is chiefly PO_4^{3-} that reacts with cations to form the common sedimentary phosphates.

The dissolving of a simple phosphate in acid may be represented by equations of the form

$$Ca_3(PO_4)_2(s) + 2H^+ \rightarrow 3Ca^{2+} + 2HPO_4^{2-} \tag{14-7}$$

and

$$Ca_3(PO_4)_2(s) + 4H^+ \rightarrow 3Ca^{2+} + 2H_2PO_4^-, \tag{14-8}$$

which are analogous to Eqs. (3-2) and (3-1), respectively, for the action of acids on calcium carbonate.

When Ca^{2+} is added to a phosphate solution, the immediate precipitate is $Ca_3(PO_4)_2$ or $CaHPO_4$, depending on pH. Both of these compounds are known as minerals, but are very rare. The chemistry of calcium phosphate is complicated by its ability to react with other materials in solution to form a still more insoluble compound called *apatite*, which is the most abundant of the phosphate minerals. The formula of the most common variety of apatite is $Ca_5(PO_4)_3F$, but much substitution is possible: Cl^- and OH^- for F^-, CO_3^{2-} and SO_4^{2-} for PO_4^{3-}, and cations like Sr^{2+}, Y^{3+}, and Mn^{2+} for Ca^{2+}. The sedimentary apatites of various compositions are generally not distinguishable under the microscope, even with high magnification; they often appear completely amorphous, and their crystalline structure is revealed only by x-rays.

The great insolubility of compounds with the apatite structure is indicated by the measured solubility constants of fluorapatite and hydroxyapatite:

$$Ca_5(PO_4)_3F \rightleftharpoons 5Ca^{2+} + 3PO_4^{3-} + F^- \qquad K = 10^{-60.4} \tag{14-9}$$
fluorapatite

and

$$Ca_5(PO_4)_3OH \rightleftharpoons 5Ca^{2+} + 3PO_4^{3-} + OH^- \qquad K = 10^{-57.8}. \tag{14-10}$$
hydroxyapatite

Experimentally, apatite has been formed in a variety of ways: by slow direct precipitation, with control of pH and ion concentrations; by slow reaction of solutions with freshly precipitated calcium phosphate; and by replacement of calcium carbonate in a reaction with dissolved phosphate. Furthermore, one kind of apatite can be changed to another by slow reaction with appropriate solutions. All these processes probably go on in nature. The replacement of calcium carbonate is strikingly illustrated by phosphate beds consisting largely of shell fragments, the original calcareous shell material being now entirely converted to apatite. Direct precipitation is suggested by the occurrence of phosphate in thin, uniform, largely nonfossiliferous strata interbedded with such common sedimentary materials as shale and shaly limestone. The slow conversion of hydroxylapatite to fluorapatite is illustrated by the fluorine content of bones: fresh bone material has little fluorine, but buried bones exposed to groundwater show increasing amounts with age, the increase being so regular that in favorable locations the analysis of bones for fluorine can give a rough estimate of their antiquity.

The existence of thick beds of nearly pure sedimentary apatite ("phosphorite" or "phosphate rock") has long been a geologic puzzle. The difficult thing to understand is not so much the mechanism of precipitation, but rather the sequence of processes that lead to accumulation of such large quantities of phosphorus over long periods of time in particular areas.

The explanation probably involves the concentration of phosphorus by organisms. Living things flourish most luxuriantly in the upper few tens of meters of the sea, where sunlight can penetrate. As dead organisms and organic debris sink below this zone and decay, phosphorus from the organic matter changes rapidly to dissolved phosphate. This means that parts of the sea below the illuminated zone are relatively rich in phosphate, the concentration remaining fairly constant to great depths. So in areas of the sea where upwelling takes place— where water from depth rises toward the surface as illustrated in Fig. 14-2—the phosphate-rich water should promote an unusual abundance of organic activity. This prediction is fulfilled in the present oceans, for areas of upwelling, chiefly on the west side of continents in subtropical latitudes, are noted for their organic "productivity." These are also parts of the ocean where phosphate crusts and nodules are found in recent sediments on the continental shelf.

One can suppose that the accumulation of phosphate on the shelf is a direct response to organic activity: organic matter from both pelagic and bottom-dwelling organisms piles up so rapidly that its phosphorus content is partly converted to apatite before it is entirely consumed by scavengers. An alternative hypothesis makes the role of organisms in phosphate deposition largely indirect. Conditions in upwelling water moving onto the continental shelf are favorable also for the inorganic precipitation of apatite: cold water moving from depth toward the surface is heated and tends to lose carbon dioxide, both because pressure decreases and because plant activity in the zone of photosynthesis consumes it. Thus water unusually rich in phosphorus moves into a region of increasing pH, which should favor deposition of calcium phosphate even without organisms acting as intermediaries.

Whether the influence of organisms on phosphate precipitation is largely direct or indirect, and whether the formation of apatite is chiefly by direct

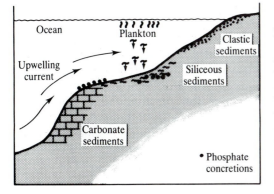

FIGURE 14-2

Schematic cross section illustrating conditions favorable for phosphorite formation in areas of upwelling currents near a continental shelf. The phosphate-rich upwelling currents bring phosphate to the shelf environment where it may form concretions near the shelf margin or be consumed by organisms and later deposited with the biogenic detritus where it is concentrated by diagenetic processes. (Diagram modified from Baturin, 1982.)

precipitation or by diagenetic replacement of first-formed calcium carbonate or tricalcium phosphate, are questions not yet settled. Perhaps, as in so many geologic arguments, all the postulated processes of origin play a role, and local circumstances determine the one that predominates. In any event, all current hypotheses suggest for the most likely place of phosphate accumulation the outer part of the continental shelf, in places where detrital material is scarce and in latitudes where upwelling water brings abundant phosphate from depth. These conditions are sufficiently unusual to account for the limitation of thick phosphate beds to a few geographic areas and a few periods of geologic history.

14-6   EVAPORITES

The sediments we have considered up to this point—carbonates, silica, iron, manganese, and phosphate sediments—have this in common: their formation requires precipitation of a relatively insoluble substance from aqueous solution. To complete our survey of common chemical sediments, we now look briefly at those made up of the more soluble constituents of natural solutions, those that do not precipitate until long evaporation has removed most of the water. Such sediments we see forming today in desert basins and locally on shallow margins of the sea in hot climates, and their older equivalents we find occasionally as beds or lenses of evaporites interstratified with other sedimentary rocks.

The fact that we can easily observe and study evaporite formation in the present world helps greatly in the interpretation of older deposits. For the most part the conclusions to be drawn from places where evaporite minerals are now being deposited are pretty obvious: streams carry dissolved material into a basin, and the salts formed are a direct reflection of the composition of the water. Stream water, of course, has a variety of compositions, depending on climate and on the nature of underlying rock; a sampling of streams from different areas is shown in Table 12-4. Evaporation of such waters would be expected to produce $CaCO_3$ as a first precipitate, generally followed by $CaSO_4$ and $NaCl$, simply on the basis of solubilities; and the expectation is fulfilled by the common finding of calcite, gypsum or anhydrite, and halite as major evaporite minerals. Locally there will be variation, depending on the immediate surroundings: for example, sodium sulfate may be the principal type of salt in a small basin among hills composed chiefly of granodiorite containing pyrite, and magnesium sulfate may be prominent in a similar basin surrounded by pyritic greenstones. In regions of volcanic activity and hot springs, desert basins may have salts of unusual composition, like the famous lithium salts in Searles Lake in California and the borate deposits of Death Valley. Generally the more soluble salts of magnesium, potassium, and bromine appear only where evaporation has gone to extremes. Thus the origin of the salts in most basins of arid regions can be worked out from knowledge of the geology and stream compositions. Surprises there may be in details of a salt-mineral sequence, but the overall nature of the deposit is seldom a mystery. And generalities derived from studies of such recent deposits can be readily applied to the interpretation of older small deposits of this sort.

Much more complicated is the study of larger deposits in the geologic record that are commonly assumed to result from evaporation of seawater. These are of particular interest because seemingly they should provide information about the composition of seawater in the past. Here we do not have the great advantage of a modern example to study, of large-scale salt deposition from the sea. In a few places, of which the shores of the Persian Gulf and South Australia are the best illustrations, broad shallow stretches of seawater undergo intense evaporation under a desert sun (the "sabkha" environment), and some salts are being deposited; but this is hardly a duplicate of the conditions under which the thick and widespread evaporite beds of the past have accumulated. Think, for example, of the rock salt and anhydrite of the Silurian Salina formation that is found over much of New York and Michigan, or the even more extensive Permian salt beds of Kansas, Texas, and New Mexico. To form such beds, enormous amounts of seawater must have evaporated over long periods of time, a possibility only if a large basin of the sea remained for long intervals separated from the rest of the ocean by a shallow bar, over which water could move into the basin as evaporation was taking place (as illustrated in Fig. 14-3). This situation has no counterpart in the present world.

In the absence of such a contemporary example, the most promising method of study would seem to be a comparison of salts obtained by evaporating present-day seawater in the laboratory with the composition of the ancient deposits. This has been attempted at many times and in many places over the past century and a half. The most elaborate of these attempts is a study based on the famous salt beds of the Zechstein formation in northern Germany—beds of Permian age that extend widely over northern Europe from Poland to Holland and the British Isles. These beds are ideally suited for such a study, since they include not only anhydrite and halite, the most abundant of marine evaporite minerals, but also large amounts of the more soluble salts of magnesium and potassium. The sequence of minerals as exposed near the city of Stassfurt is shown in Table 14-1. In very general terms it is the sequence that might be expected from long-continued evaporation of seawater: first rock salt and anhydrite, then a sulfate of Ca, Mg, and K, then chiefly magnesium sulfate minerals, and finally chlorides of Mg and K. Thus a full panorama of salts from earliest to latest stages is available, for comparison with salts that could be

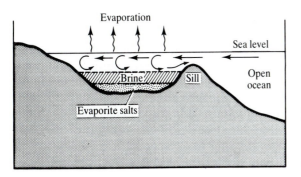

FIGURE 14-3
Schematic cross section illustrating a possible depositional environment of marine evaporites in a basin separated from the main ocean by a shallow bar over which seawater moves into the basin as evaporation goes on.

TABLE 14-1

Generalized section of salt deposits of the Stassfurt series (Zechstein 2) in the interior of the Zechstein basin near Stassfurt, Germany.

There is much local variation in thickness of the different units. Toward the margin of the basin the minerals containing $MgSO_4$ become less abundant and the chloride minerals more abundant. An important constituent of all the units, whether mentioned explicitly or not, is halite.

Mineralogy	Thickness
Gray salt clay.	
Discontinuous zone of "Hartsalz" (mixture of sylvite, kieserite, halite). Locally also schoenite, langbeinite, kainite.	0–20 m
Carnallite zone. Chiefly carnallite, some kieserite.	15–40 m
Kieserite zone. Some carnallite and anhydrite.	30–40 m
Polyhalite zone. Some anhydrite.	5–10 m
Older rock salt and anhydrite.	100–1000 m

Mineral formulas			
anhydrite	$CaSO_4$	kieserite	$MgSO_4 \cdot H_2O$
carnallite	$KMgCl_3 \cdot 6H_2O$	langbeinite	$K_2Mg_2(SO_4)_3$
gypsum	$CaSO_4 \cdot 2H_2O$	polyhalite	$K_2MgCa_2(SO_4)_4 \cdot 2H_2O$
halite (rock salt)	$NaCl$	schoenite	$K_2Mg(SO_4)_2 \cdot 6H_2O$
kainite	$KMgClSO_4 \cdot {}^{11}/_4H_2O$	sylvite	KCl

Source: Stewart (1963). This paper is an excellent summary of the research by van't Hoff and his followers on the geochemistry of marine evaporites, especially the Zechstein evaporites at Stassfurt.

obtained by evaporating modern seawater or that could be predicted from chemical study of the precipitation process.

Serious study of the Zechstein salts at Stassfurt was begun by the Dutch chemist Jacobus H. van't Hoff in the early 1900's. Van't Hoff's goal was an ambitious one: starting with simple salt solutions, to learn the physical chemistry of crystallization of each salt individually, then salts in combination, finally working up to the complex assortment of salts in present-day seawater, and then to compare his results with the sequence he observed at Stassfurt. It was an enterprise fraught with difficulty because of slow reactions, metastability of many salt minerals, and effects of minor temperature variation. The work was only partly successful, and arguments about the validity of some of van't Hoff's conclusions persist to the present. But there is general agreement that he did show an impressive general correspondence of the experimental results with the Zechstein salts. It seems clear that the Permian sea, or at least the part of the sea that evaporated to form the salts at Stassfurt, had a composition much like that of the present ocean.

But this is only the beginning of speculation about possible changes in the composition of seawater. It turns out that many evaporite deposits of presumed marine origin have salt assemblages different from those at Stassfurt; in fact, the Stassfurt deposit itself shows differences in the sequence of salt minerals from place to place. Particularly troublesome is the variation in proportions of magnesium sulfate minerals [kieserite, $MgSO_4 \cdot H_2O$); schoenite, $K_2Mg(SO_4) \cdot 6H_2O$; langbeinite, $K_2Mg_2(SO_4)_3$]. These minerals are abundant near the city of Stassfurt, roughly in the center of the Zechstein basin, but toward its margins give way to choride minerals like carnallite ($KMgCl_3 \cdot 6H_2O$) and sylvite (KCl). Amounts of magnesium sulfate show variation not only locally within a given deposit, but also on a larger scale from one deposit to another. Some evaporite deposits commonly thought to be of marine origin have only a very small overall content of $MgSO_4$ minerals; Pennsylvanian evaporites of the Paradox basin of Utah and the Devonian evaporites of western Canada are examples. Does this mean that the concentrations of Mg^{2+} and SO_4^{2-} in seawater have changed over geologic time?

To approach an answer, we recall from Chapter 12 the discussion of probable sources of ions in seawater. Certainly a major source is stream water, and much about the composition of the ocean can be understood simply by regarding it as average stream imput (Table 12-4) modified by precipitation of $CaCO_3$ and adsorption of ions on bottom sediments. But another important source is hydrothermal solutions, solutions consisting of seawater that has reacted with hot basalt in submarine magma-hydrothermal systems associated with mid-ocean ridges (Fig. 12-3). This reaction, as inferred from analyses of the hydrothermal solutions and from experiments with seawater and hot basalt, removes the two ions of our particular concern, Mg^{2+} and SO_4^{2-}, and adds to the seawater Ca^{2+} and Na^+. Thus the composition of seawater at any time should depend on a balance between additions from stream water and exchanges of material with deeply circulating hydrothermal solutions along ridges. Because the intensity of activity along the ridges varies over geologic time with motions of the lithospheric plates and deeper motions in the mantle, the composition of seawater and of evaporites precipitated from it should change also: a greater concentration of Mg^{2+} and SO_4^{2-} when activity along mid-ocean ridges is weak, a smaller concentration when, as at present, ridge activity is relatively strong. This does not mean that concentrations have changed radically; as noted in Chapter 12, the overall composition of evaporites down the ages tells us that major ions cannot have changed by more than factors of two or three. But very probably the changes have been sufficient to account for the observed variation in amounts of magnesium sulfate salts.

Thus a study of marine evaporites in its details can provide much information about the physical chemistry of salt-bed formation and diagenesis, and in its broader aspects also an overview of changes in seawater composition and in activity along mid-ocean ridges through geologic history.

14-7 DIAGENESIS

Much sediment after deposition becomes ultimately a sedimentary rock, and the processes involved in this change go by the name of *diagenesis.* As commonly used, the term includes not only the immediate consolidation of loose material into a coherent solid, but also the longer-term changes that may affect a rock after its formation and burial beneath other sediments, as well as alteration following uplift.

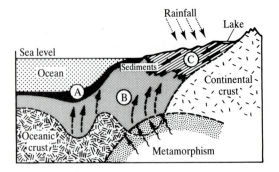

FIGURE 14-4
Cartoon profile of a continental margin illustrating three kinds of diagenetic environments: A, depositional and early burial; B, deep burial; and C, emergent, pre-erosion phase. The arrows schematically represent the movement of water. The upward pointing dashed arrows denote the flow of pore fluids (connate waters) due to compaction and porosity reduction in the deeper sediments. In the lowest portions of the sedimentary basin these fluids may mix with metamorphic waters (wavy arrows) derived from dehydration and decarbonation reactions deeper in the crust. Downward pointing arrows represent the flow of meteoric groundwaters. (Figure modified from Fig. 2-5, p. 34 of Fairbridge, 1983.)

These different geologic environments are illustrated in Fig. 14-4. The changes are restricted to those occurring under conditions of temperature and pressure found in upper-crustal environments, exclusive of changes due to weathering. Pressure and temperature limits commonly specified for diagenetic processes are suggested in Fig. 14-5. Changes at higher P and T are considered metamorphic rather than diagenetic, but fixing the boundary is difficult. Reactions involved in diagenesis are slow and are dependent on factors like grain size, the presence or absence of fluids, fluid composition, organic matter, temperature, and pressure, so arguments often arise as to whether a given mineral assemblage is best called diagenetic or metamorphic.

Diagenetic changes are a complex of physical, chemical, and biologic

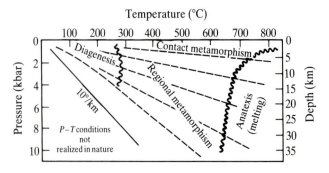

FIGURE 14-5
The approximate range in pressure and temperature for diagenetic processes. (After Winkler, 1967.)

processes. Compaction of clay-rich sediments and recrystallization of fine-grained evaporites are examples of physical changes; bioturbation and bacterial alteration of organic matter are signs that organisms have been active. Chemical changes come in great variety and complexity: alteration of feldspars to clays, reduction of iron and its precipitation as sulfide, conversion of kerogen to methane and other hydrocarbons, cementation of mineral grains by calcite or silica—the list could go on and on. We will limit the discussion here to a few examples.

Most of the chemical changes in diagenesis are very slow, and their rates depend on many factors. In a typical sediment accumulating in shallow water as a mixture of coarse and fine mineral grains plus a small amount of organic material, the chemical environment will be different at different levels. Near the depositional surface the freshly deposited material is porous and steadily exposed to water and to active dissolved gases like oxygen and carbon dioxide; below a depth of a few centimeters the porosity is reduced by compaction, and contact with the atmosphere is lost; at still deeper levels the older sediment is affected by reactions induced by increasing pressure and temperature and by infiltration of reactive pore fluids (Fig. 14-4). The kind of reactions going on in this accumulating pile will obviously change with depth: oxidation in the fresh material near the sediment–water interface, reduction at lower levels if organic matter is present, perhaps the slow formation of pyrite. Then in the oldest part of the pile new reactions are promoted by the higher temperature. Here silica, as quartz or chalcedony, and carbonates including calcite, dolomite, siderite, and ankerite form in pores cementing the sediment into rock. Kaolinite, smectite, illite, chlorite, and possibly at higher temperatures albite and K-feldspar, are produced by reactions between sediment and pore fluids. The reactions change both with depth and over the course of time. Diagenesis, in effect, is a slow approach of deposited material toward a new equilibrium, or a series of new equilbria, as conditions of temperature, pressure, and access of reactive pore fluids change during burial and subsequent uplift.

In the upper oxidizing layer of a sediment the most prominent diagenetic reactions are the oxidation of iron and the partial destruction of organic matter. These, of course, are similar to reactions of weathering, and the line between diagenesis and weathering is a hazy one. We note only that such reactions may occur under water as well as in a sediment accumulating on land, and in a porous sediment may persist to depths below what is normally thought of as the zone of weathering. The oxidation of iron may be reversed as sediment accumulates and oxygen is excluded, especially if the sediment contains organic matter, so that the oxidized layer changes to a reduced layer at a depth generally no more than a few tens of centimeters.

Reactions in the sedimentary pile beneath the oxidizing layer are more complex. In many of them organic matter plays an important role, and various species of bacteria serve as catalysts. For example, bacteria can derive energy for their life processes from the oxidation of organic matter by Fe(III) oxide

$$CH_2O + 2Fe_2O_3 + 3H_2O \rightarrow 4Fe^{2+} + HCO_3^- + 7OH^-, \qquad (14\text{-}11)$$

organic   hematite
matter

or by sulfate ion

$$2CH_2O + SO_4^{2-} \rightarrow 2HCO_3^- + H_2S. \tag{14-12}$$
$$\text{organic}$$
$$\text{matter}$$

(The formula CH_2O in these equations is not a specific compound, but represents the average proportions of elements in much organic material found in sedimentary rocks.) The Fe^{2+} and H_2S produced in these reactions may then combine in a series of reactions whose ultimate product is pyrite. The overall process may be represented by the generalized reaction

$$15CH_2O + 2Fe_2O_3 + 8SO_4^{2-} \rightarrow 4FeS_2 + 7H_2O + 15HCO_3^- + OH^-. \tag{14-13}$$
$$\text{organic} \quad \text{hematite} \qquad\qquad \text{pyrite}$$
$$\text{matter}$$

Pyrite produced in this way is a common minor constituent of black shales, especially of shales formed from organic-rich sediments in a marine or brackish water environment, where SO_4^{2-} is a major ion of the interstitial water.

Organic matter, again with the aid of bacteria, in a restricted temperature range (roughly 70–130°C) may simply break down with the formation of hydrocarbons

$$2CH_2O \rightarrow CH_4 + HCO_3^- + H^+. \tag{14-14}$$
$$\text{organic}$$
$$\text{matter}$$

This is a crude symbolization of the important diagenetic reaction by which the components of natural gas and liquid petroleum are generated from the kerogen of organic-rich sedimentary rocks (Sec. 15-4). During any of these reactions involving organic compounds, the less abundant constituents may be set free in solution, notably nitrogen in the form of ammonia or ammonium ion and phosphorus in one or more of the phosphate ions.

The silicate minerals of accumulating sediments react only slowly during diagenesis, most commonly to form one or another of the clay minerals. The reaction, symbolized for K-feldspar in Eqs. (4-26) and (8-4), is one that we have discussed at great length earlier. The clay minerals, whether formed in place (authigenic) or deposited as a constituent of the sediment (detrital), obviously must be heavily involved in diagenetic change. The three major groups of clay minerals— kaolinite, smectite, illite—are somewhat reactive with ions in solution at low temperatures, but just how reactive is a matter of considerable uncertainty and much argument (Secs. 6-3 and 13-8). There is good evidence, from both experiment and observation, that smectite clays in sediments change slowly over the temperature range of 70 to 150°C to mixed-layer smectite–illites and ultimately to illite, if the pH is right and the necessary K^+ is available. Conditions under which kaolinite will undergo similar changes are less clear. On purely thermodynamic grounds, free energies indicate that the three kinds of clay have definite fields of stability defined by concentrations of silica, H^+, and K^+, as illustrated by Fig. 13-8, but the reactions are slow enough and the free energy data uncertain enough to make strict application of this figure to clay mineral diagenesis somewhat uncertain. The fact that each of the three clay varieties is dominant over a large area of present-day marine sediments suggests that the clays do not change rapidly at low temperatures. But observations that illite is the principal clay in older shales, and that in Cenozoic sediments and sedimentary rocks recovered from boreholes illite commonly becomes more abundant with depth, provide evidence that in the warmer parts of diagenetic environments other clay minerals alter slowly to illite.

A mineral related to the clays that forms during diagenesis is *chlorite*, a hydrous silicate with a formula like that of kaolinite except that much of the Al is replaced by Mg and Fe: $(Mg,Al,Fe)_3[(Si,Al)_2O_5](OH)_4$. Chlorite is an alternative product of the alteration of smectite, if the smectite contains abundant Mg and Fe relative to K. Or, more commonly, it forms from reactions of other clay minerals with carbonates such as dolomite, ankerite, or siderite. A simple form of this reaction is the breakdown of kaolinite and dolomite,

$$Al_2Si_2O_5(OH)_4 + 5CaMg(CO_3)_2 + SiO_2 + 2H_2O \rightleftharpoons$$

kaolinite   dolomite

$$Mg_5Al(AlSi_3O_{10})(OH)_8 + 5CaCO_3 + 5CO_2. \qquad (14\text{-}15)$$

chlorite   calcite

These types of clay–carbonate reactions generate CO_2 and are sometimes referred to as decarbonation reactions to emphasize the release of CO_2 from the rock.

In some deeply buried sandstones, such as those along the Gulf Coast of the United States and in the North Sea near Norway, there is a close correlation between measured cation to hydrogen ion activity ratios in pore fluids and similar ratios computed from hydrolysis reactions among authigenic and detrital silicates. An example of this correlation is shown in Fig. 14-6 for the North Sea sediments. Here diagenetic minerals consist of varying combinations of kaolinite, illite, albite, quartz, chlorite, and calcite; and K-feldspar is a common detrital mineral. Because the enthalpy of reactions shown in the figure are negative, the activity ratio of Na^+ to H^+ decreases with increasing temperature [see, for example, Eq. (8-61)]. This is the same trend as measured for the pore fluids, suggesting that the cation activity ratio is buffered by hydrolysis reactions among silicates.

Silicate hydrolysis reactions may have an important influence on solution pH and subsequent reactivity of pore fluids with the sediments. For example, consider again the relationships given in Fig. 14-6. If the activity of Na^+ in the fluid is large, and thus does not change appreciably during diagenetic reactions over a specified temperature interval, say 100 to 150°C, the change in solution pH with temperature will be determined by the enthalpy of silicate hydrolysis. We illustrate this with the reaction of kaolinite and K-feldspar to form illite and albite,

$$\tfrac{3}{2} Al_2Si_2O_5(OH)_4 + KAlSi_3O_8 + Na^+ \rightleftharpoons$$

kaolinite   K-feldspar

$$KAl_2(AlSi_3O_{10})(OH)_2 + NaAlSi_3O_8 + \tfrac{3}{2}H_2O + H^+ \qquad (14\text{-}16)$$

illite   albite

[this reaction is the reverse of the one used to compute curve (*b*) in Fig. 14-6]. The equilibrium constant for Eq. (14-16) is $\log K = -\log(a_{Na^+}/a_{H^+})$, so for constant

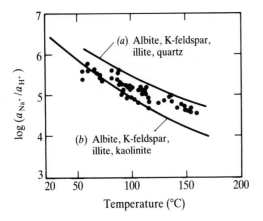

FIGURE 14-6
Temperature dependence of the activity ratio of Na^+ to H^+ in pore fluids from North Sea sediments (symbols) and computed equilibria for the assemblages: (*a*) albite, illite, K-feldspar, quartz, and (*b*) albite, illite, kaolinite, K-feldspar. Reaction curves computed for muscovite with an activity of 0.5 to approximate the mineral illite. The reactions used to represent silicate mineral equilibria are: curve (*a*),

$$\underset{\text{albite}}{3NaAlSi_3O_8} + K^+ + 2H^+ \rightleftharpoons \underset{\text{illite}}{KAl_2(AlSi_3O_{10})(OH)_2} + \underset{\text{quartz}}{6SiO_2} + 3Na^+,$$

where the ratio of a_{K^+} to a_{H^+} employed in the calculation was determined by least squares regression of this cation-to-hydrogen-ion ratio in the pore fluids; curve (*b*),

$$\underset{\text{illite}}{KAl_2(AlSi_3O_{10})(OH)_2} + \underset{\text{abite}}{NaAlSi_3O_8} + 3/2H_2O + H^+ \rightleftharpoons \underset{\text{kaolinite}}{3/2Al_2Si_2O_5(OH)_4} + \underset{\text{K-feldspar}}{KAlSi_3O_8} + Na^+.$$

This reaction is metastable with respect to quartz, as values of $SiO_2(aq)$ required for silicate equilibrium exceed quartz saturation. As written above, both of these reactions have negative enthalpies of reaction, so that the activity ratio of Na^+ to H^+ decreases with increasing temperature [Eq. (8-61)]. (Reprinted by permission from Aagaard et al., 1992.)

values of a_{Na^+} and activities of the mineral components, Eq. (8-61) becomes,

$$\left(\frac{\partial \text{pH}}{\partial T}\right)_P = \frac{-\Delta H_r}{2.303RT^2}. \tag{14-17}$$

As the reaction is written in Eq. (14-16), its enthalpy change is positive. Thus we predict that pH will decrease with increasing temperature, a prediction borne out by many observations. What is important here is not that a single reaction exactly fits the observed trends in fluid composition, but rather that the observed changes in cation to hydrogen ion activity ratios in the pore fluids parallel theoretical reaction boundaries for hydrolysis reactions among the authigenic and detrital phases.

A common effect of diagenesis is the cementation of a rock by the filling of interstices and the enlargement or replacement of crystals by material deposited from solution. The carbonate minerals, especially calcium carbonate, are the commonest cementing agents, as might be expected from the abundance of bicarbonate ions generated in reactions like Eqs. (14-11) to (14-14). The amounts of cement are generally far too large to assume that the required Ca^{2+} (or Mg^{2+} or Fe^{2+}) is generated locally near the grains that the cement binds together. The existence of abundant well cemented rocks means, then, that water carrying the cement constituents must have traveled in large amount through interstices of the sediment during its diagenesis.

Conditions that would lead to precipitation of one or another of the common carbonates can readily be suggested from our earlier discussion of carbonate behavior in Chapter 3 and reexamination of the reaction,

$$Ca^{2+} + CO_2 + H_2O \rightarrow CaCO_3 + 2H^+. \qquad (14\text{-}18)$$
calcite

Abundant calcite cement would be expected to form, for example, in the warmer parts of diagenetic environments (because calcite solubility decreases as temperature rises), or in places where the pH is higher than usual because ammonia is being liberated from organic decay. The necessary calcium comes chiefly from irreversible hydrolysis of detrital plagioclase. Carbon dioxide is a variable that may affect calcite stability in a variety of ways. If CO_2 increases, the solution pH may decrease because of the reaction $CO_2 + H_2O \rightarrow HCO_3^- + H^+$, thus destabilizing calcite [the reverse of Eq. (14-1)]. However, if pH is buffered by other equilibria, say silicate hydrolysis reactions or reactions involving organic acids, increasing CO_2 may lead to calcite precipitation [Eq. (14-18)].

In some sedimentary basins the CO_2 concentration changes systematically with increasing depth and temperature. For example, in the Gulf Coast and North Sea basins discussed above, CO_2 content of the pore fluids increases with increasing temperature as illustrated in Fig. 14-7. Also shown in the figure is a calculated curve for equilibrium of kaolinite, illite, and calcite, a reaction that closely parallels the higher temperature portion of the fluid measurements. It thus appears that reactions involving clay and carbonate minerals may influence the distribution of CO_2 in

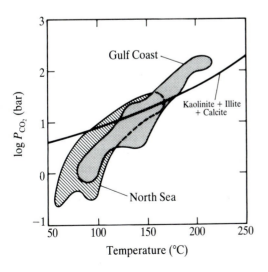

FIGURE 14-7
Partial pressures of CO_2 in pore fluids from deep sediments in the Gulf Coast and North Sea basins. The curve represents computed values of the partial pressure of CO_2 required for equilibrium of calcite, kaolinite, illite (approximated by the thermodynamic properties of muscovite), and the deep pore fluids of the Gulf Coast sediments. Details of this calculation are given in Hutcheon (1992). Partial pressures of CO_2 in the Gulf Coast and North Sea sediments are from Smith and Ehrenberg (1989).

some deep sedimentary basins. We also note from Eq. (14-18) that to maintain carbonate–solution equilibrium under the conditions of increasing CO_2 requires a decrease in one or both of Ca^{2+} and pH.

Another common cementing material in sediments is silica, usually in the form of quartz but sometimes as opal or chalcedony, and again our previous treatment of the solubility of this substance makes possible good guesses about the conditions of temperature, pressure, and concentration of solutions that would lead to its deposition as a cement. It is not unusual for silica concentrations to be in excess of those required for quartz saturation during early stages of diagenesis. This is evident from fluid compositions in modern marine shales shown in Fig. 14-8. High silica concentrations at these low temperatures are probably due to the abundance of biogenic silica and to reactions involving silicates [for example, the irreversible replacement of K-feldspar by kaolinite, Eq. (8-4)]. As temperatures increase in deep sedimentary basins, the measured silica concentrations commonly approach those for equilibrium with quartz (Fig. 14-8).

A great variety of other substances (gypsum, phosphate, iron oxide, zeolites, to name a few) have been reported as less common cementing materials in sedimentary rocks. These presumably can form only under diagenetic conditions rather different from the ordinary.

One product of diagenesis that involves cementation is the structure called *concretions*. These are hard, rounded masses that differ from the enclosing rock material only in their much better cementation. They are often conspicuous in

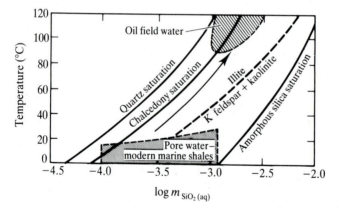

$$\log m_{SiO_2\,(aq)}$$

FIGURE 14-8

Concentrations of dissolved silica in pore fluids from modern marine shales and from deeper sediments associated with oil fields as a function of temperature. Theoretical curves for the solubility of silica polymorphs are shown by solid lines, and the dashed line represents computed equilibrium among K-feldspar, kaolinite, illite (muscovite with an activity of 0.3), and an aqueous solution represented by the reaction,

$$KAlSi_3O_8 + Al_2Si_2O_5(OH)_4 \rightleftharpoons KAl_2(AlSi_3O_{10})(OH)_2 + 2SiO_2(aq) + H_2O.$$

K–feldspar   kaolinite   illite

The arrow denotes a possible change in pore fluid composition with increasing temperature and depth in sedimentary basins. (Reprinted by permission from Bloch and Hutcheon, 1992.)

outcrops because of their greater resistance to weathering, and are prized by rock collectors because of the odd shapes they sometimes assume and the well preserved fossils they may contain. Most commonly they are roughly spheroidal or ellipsoidal, and may be of any size up to a few meters in diameter. They are obviously formed by the slow progressive precipitation of a cementing agent, most commonly calcite, around some sort of nucleus in the sediment, but the nature of the nucleus is often a puzzle. Not infrequently a foreign object can be found at the center of a concretion—a fragment of shell or bone or plant material, sometimes just a trace of organic matter—which suggests that a possible explanation may lie in the HCO_3^- [Eq. (14-14)] and ammonia generated by bacterial activity: a small amount of abnormally alkaline water could have formed around the organic fragment, leading to precipitation of a little $CaCO_3$ from slowly moving groundwater, the crystals then serving as a nucleus for further deposition of carbonate. This seems reasonable enough for concretions that have visible nuclei, but is less certain for those without nuclei and those that are cemented by compounds other than carbonates.

The movement of solutions that cause cementation and dissolution of minerals during diagenesis may have a pronounced effect on the porosity and permeability of the sediment or sedimentary rock. Solutions containing dissolved CO_2, for example, may be undersaturated with respect to calcium carbonate and may widen cracks or intergranular spaces in parts of a limestone through which they move; the same solutions, once they have dissolved some of the carbonate, may deposit it in other parts where conditions are different, and thus decrease porosity by filling some of the openings. This process is also effective in removing previously formed carbonate cement in sandstones, locally increasing the sediment's porosity. It is small wonder that diagenesis is important to the petroleum industry, for the effects of this dissolving and reprecipitating are major controls on the migration of oil and the size of reservoirs available for its accumulation.

By way of summary, diagenesis is a complex part of low-temperature geochemistry, interesting for what it can tell us about the history of sedimentary rocks, and important to understanding groundwater quality and to the petroleum industry for the insights that its study can give about the environments in which petroleum and natural gas are formed and accumulate. It is a difficult subject, both because reactions are slow and because the materials involved are commonly very fine-grained. Difficulties are compounded by uncertainties in thermodynamic data for some of the minerals that take part in diagenetic reactions, because of their variable compositions and because of sluggish reaction rates. Despite these obstacles, the chemistry of diagenesis continues to be an active field of research, aided spectacularly in recent years by the development of powerful new techniques for the analysis of fine-grained materials.

PROBLEMS

1. Construct a diagram similar to Fig. 2-3(a) showing concentrations of H_3PO_4, $H_2PO_4^-$, HPO_4^{2-}, and PO_4^{3-} as a function of pH in solutions with a total phosphate concentration of $0.0001m$ at 25°C and 1 bar. If you wanted to make the diagram for a higher temperature, what additional information would you need?

2. Show that hydroxylapatite is more stable in contact with seawater than either of the two simple calcium phosphates, $Ca_3(PO_4)_2$ and $CaHPO_4$, given the approximate solubility products

$$Ca_3(PO_4)_2 \rightleftharpoons 3Ca^{2+} + 2PO_4^{3-}, \qquad K = 10^{-26}$$

and

$$CaHPO_4 \rightleftharpoons Ca^{2+} + HPO_4^{2-} \qquad K = 10^{-7}.$$

Using these constants and the approximate constant for hydroxylapatite in Eq. (14-10), find constants for the reactions

$$Ca_5(PO_4)_3OH + 4H^+ \rightleftharpoons 3CaHPO_4 + 2Ca^{2+} + H_2O$$

and

$$2Ca_5(PO_4)_3OH + 2H^+ \rightleftharpoons 3Ca_3(PO_4)_2 + Ca^{2+} + 2H_2O,$$

and compare the equilibrium values of $a_{Ca^{2+}}/a_{H^+}^2$ that you obtain with the value of this quotient in seawater. In which direction must the two reactions go in order to reach these equilibrium values? Which of the three compounds, then, is most stable?

3. What concentration of Fe^{2+} can exist in the presence of $Fe(OH)_3$ at a pH of 6.5 and an Eh of 0.30 volt? At a pH of 8.4 and an Eh of -0.30 volt?

4. Show that MnO_2 is a stronger oxidizing agent than $Fe(OH)_3$ at both pH 8 and pH 4.

5. Which of the following pairs of minerals could exist together at equilibrium at 25°C and 1 bar? (Use Figs. 9-5 and 14-1.)

Pyrite and pyrolusite (MnO_2).	Hausmannite (Mn_3O_4) and siderite ($FeCO_3$).
Hematite and pyrolusite.	Rhodochrosite ($MnCO_3$) and pyrite.
Magnetite and braunite (Mn_2O_3).	Rhodochrosite and hematite.

6. Suppose that water emerging from a spring with a pH of 4 and an Fe^{2+} concentration of 1000 ppm flows downhill over a limestone surface. Describe what you would see, and explain what is happening.

7. In well aerated seawater the Eh is generally about $+0.40$ volt and the pH about 8.2. Under such conditions, would manganese precipitate as MnO_2 if an amount of manganese is added sufficient to give a concentration of 1000 ppm Mn^{2+}? 1 ppm? 0.001 ppm?

8. In a sedimentary basin filled with arkosic sands, early formed kaolinite and detrital K-feldspar are found to coexist with authigenic illite and albite over a temperature interval of 100° to 150°C. Using the reaction and thermodynamic data given below, compute the pH in the pore fluids at 100°C and at 150°C, assuming that the activities of the sodium ion and all mineral components are equal to unity throughout this temperature range:

$$\tfrac{3}{2}Al_2Si_2O_5(OH)_4 + KAlSi_3O_8 + Na^+ \rightleftharpoons$$
$$\text{kaolinite} \qquad \text{K–feldspar}$$

$$KAl_2(AlSi_3O_{10})(OH)_2 + NaAlSi_3O_8 + \tfrac{3}{2}H_2O + H^+$$
$$\text{illite} \qquad \text{albite}$$

$$\log K_{100°C} = -4.84 \qquad \Delta H^\circ_{100°C} = 45 \text{ kJ.}$$

If calcite is in equilibrium with this mineral assemblage, what is the fugacity of CO_2 in the pore fluids at 100°C and at 150°C? In your calculations assume that $a_{Ca^{2+}}$ is equal to 10^{-1}

and that the activities of calcite and of water are both unity at these two temperatures. Use the following equilibrium constants:

$$CaCO_3 + 2H^+ \rightleftharpoons Ca^{2+} + CO_2 + H_2O.$$
calcite

$$\log K_{100°C} = 9.13 \qquad \log K_{150°C} = 8.87$$

REFERENCES

Aagaard, P. A., J. S. Jahren, and P. Kr. Egeberg: "North Sea clastic diagenesis and formation water constraints," in Kharaka, Y. K., and A. S. Maest, (eds): *Proceedings of the 7th International Symposium on Water–Rock Interaction—WRI-7, Park City, Utah, USA,* vol. 2, A.A. Balkema, Rotterdam, Brookfield, pp. 1147–1160, 1992.

Baturin, G. N.: *Phosphorites on the Sea Floor, Origin, Composition and Distribution: Developments in Sedimentology 33,* Elsevier, 1982.

Bloch, J., and I. E. Hutcheon: "Silica mass transport during shale diagenesis: are shales a source or sink for silica?" in Kharaka, Y. K., and A. S. Maest, (eds): *Proceedings of the 7th International Symposium on Water–Rock Interaction—WRI-7, Park City, Utah, USA,* vol. 2, A.A. Balkema, Rotterdam, Brookfield, p. 1158, 1992.

Fairbridge, R. W.: "Syndiagenesis–anadiagenesis–epidiagenesis phases in lithogenesis, in Larsen, G., and G. V. Chilingar, (eds): *Diagenesis in Sediments and Sedimentary Rocks,* Elsevier, New York, pp. 20–90, 1983.

Hutcheon, I.: "Clay-carbonate reactions in the Venture area, Scotian Shelf, Scotia, Canada," in Spencer, R. J., and I-Ming Chou, (eds): *Fluid–Mineral Interactions: a Tribute to H. P. Eugster,* The Geochemical Society, Special Publication No. 2, pp. 199–212, 1992.

Smith, J. T., and S. N. Ehrenberg: "Correlation of carbon dioxide abundance with temperature in clastic hydrocarbon reservoirs: relationship to inorganic chemical equilibrium," Marine and Petroleum Geology, vol. 6, pp. 129–135, 1989.

Stewart, F. H.: "Marine Evaporites," Chapter Y in Data of Geochemistry, 6th ed., U.S. Geological Survey Prof. Paper 440-Y, 1963.

Winkler, H. G. F.: Petrogenesis of Metamorphic Rocks, 2d ed., Springer Verlag, New York, Fig. 12, p. 4, 1967.

SUGGESTIONS FOR FURTHER READING

Berner, R. A.: *Early Diagenesis,* Princeton University Press, 1980. A theoretical study of reactions in the early stages of diagenesis.

Berner, R. A.: "Sedimentary pyrite formation: an update," *Geochim. Cosmochim. Acta,* vol. 48, pp. 603–615, 1984. Conditions for formation of pyrite during early diagenesis of marine and nonmarine sediments, and effect of pyrite formation on the concentration of oxygen in the atmosphere and of sulfate in seawater.

Boggs, S., Jr.: *Petrology of Sedimentary Rocks,* Macmillan, New York, 1992. A standard text. Diagenesis is discussed fully, for sandstones and shales in Chapter 9 and for limestones in Chapter 12.

Boyle, E. A., J. M. Edwards, and E. R. Sholkovitz: "Mechanism of iron removal in estuaries," *Geochim. Cosmochim. Acta,* vol. 41, pp. 1313–1324, 1977. A field and experimental study of the precipitation of $Fe(III)$ oxide colloids on contact with seawater.

Drever, J. I.: "Geochemical model for the origin of Precambrian banded iron formation," *Geol. Soc. America Bull.,* vol. 85, pp. 1099–1106, 1974. Chemical factors in the deposition of iron minerals at different depths and different distances from shore in a Precambrian sea, with postulated lower O_2 and higher CO_2 than the present ocean.

Force, E. R., and W. F. Cannon: "Depositional model for shallow-water manganese deposits around black-shale basins," *Econ. Geology,* vol. 83, pp. 93–117, 1988. A novel hypothesis for the origin of large sedimentary manganese deposits, with applications to many known manganese occurrences.

Hem, J. D., and C. J. Lind: "Nonequilibrium models for predicting forms of precipitated manganese oxides," *Geochim. Cosmochim. Acta,* vol. 47, pp. 2037–2046, 1983. Experimental precipitation of manganese oxides at different values of pH and temperature, and effects of aging on the precipitates.

James, H. L.: "Chemistry of the iron-rich sedimentary rocks," Chapter W in *Data of Geochemistry,* 6th ed., *U.S. Geol. Survey Prof. Paper* 440-W, 1966. A comprehensive review of the long controversy over the origin of sedimentary iron ores, including many analyses, a theoretical discussion of facies based on Eh–pH diagrams, and a hypothesis of origin based on much field work and a critical evaluation of the literature.

Langmuir, D.: "Particle-size effect on the reaction goethite = hematite + water," *American Jour. Science,* vol. 271, pp. 147–156, 1971. Experimental study of goethite–hematite relations, with emphasis on the effect of particle size on stabilities.

Sass, B. M., P. E. Rosenberg, and J. A. Kittrick: "The stability of illite/smectite during diagenesis: an experimental study," *Geochim. Cosmochim. Acta,* vol. 51, pp. 2103–2115, 1987.

Spencer, R. J., and Hardie, L. A.: "Control of seawater composition by mixing of river waters and mid-ocean ridge hydrothermal brines," in Spencer, R. J., and I-Ming Chou, (eds): *Fluid Mineral Interactions: a Tribute to H. P. Eugster,* Geochem. Soc. Special Publication No. 2, pp. 409–419, 1990. Effects on seawater composition of variations in the proportions of stream water and hydrothermal brine.

Warren, J. K.: *Evaporite Sedimentology,* Prentice-Hall, Englewood Cliffs, New Jersey, p. 285, 1989. Excellent textbook on the nature and formation of evaporite deposits.

CHAPTER
15

SEDIMENTATION AND DIAGENESIS: ORGANIC GEOCHEMISTRY

Organic materials were mentioned frequently in the last chapter, chiefly in their role as modifiers of conditions under which inorganic compounds form or dissolve. But organic substances are geochemically interesting for many other reasons, and here we set out to look in some detail at their behavior in sedimentary processes. Diagenesis of organic matter in sediments and sedimentary rocks is important in a practical sense, because it is the source of such fuels and raw materials as coal, petroleum, and natural gas.

The carbon compounds manufactured by organisms are in general not very stable. Exposed to oxygen, all but the most resistant decay in times brief by geologic standards, their carbon returning to the great reservoir of carbon dioxide in the atmosphere. Kept away from oxygen in stagnant water or by burial under accumulating sediments, organic compounds cannot decay completely, but decompose into other compounds generally more stable than those in the original organisms. Such changes will be our focus in this chapter, but first we need a brief look at the properties of compounds in which carbon is a major constituent.

15-1   THE CHEMISTRY OF CARBON COMPOUNDS

Carbon is unique among the elements in the number and complexity of its compounds. Of known compounds, those containing carbon are at least ten times as numerous as compounds of all other elements combined. The study of these compounds is a special branch of chemistry called *organic chemistry*, and the carbon compounds (except for a few simple ones like the oxides and carbonates) are generally referred to as *organic compounds*. Organic chemistry is a complex subject, but fortunately we need here only the bare rudiments for an understanding of the general geologic behavior and sources of carbon compounds in sediments.

The ability of carbon to form so many compounds may be correlated with two properties of the carbon atom: its small size and its possession of four valence electrons. With this number of electrons its bonds with other atoms are covalent rather than ionic, and the small size of the atom makes these bonds very strong. How strong they can be is attested by the hardness of diamond, which consists of carbon atoms linked together, each one to four others, by covalent bonds. Unlike the atoms of most elements, carbon atoms can join with each other in compounds as well as in the structure of the element itself, and this ability makes possible the formation of an almost indefinite number of molecular structures.

The existence of strong covalent bonds in carbon compounds not only helps to account for their enormous number, but explains also some of their general properties. Most carbon compounds, for example, are not very soluble in water, and those that do dissolve are not dissociated into ions, or dissociate only slightly. We shall meet a few exceptions in which special structures make possible the formation of partly ionic bonds. The strong bonds and lack of dissociation mean also that reactions of carbon compounds are generally slow at ordinary temperatures.

Hydrocarbons

Simplest of organic substances are the *hydrocarbons*, compounds that contain only the elements carbon and hydrogen. A familiar example is methane, CH_4, the chief constituent of natural gas and a product of partial decay of organic material in stagnant water ("marsh gas"). Chemically similar to methane are other hydrocarbons found in natural gas and petroleum (Table 15-1). Mixtures of these hydrocarbons are well-known substances: for example, the first four make up most natural gas, those from hexane through decane are the chief constituents of gasoline, those from $C_{14}H_{30}$ to $C_{18}H_{38}$ make up diesel fuel, and those from $C_{26}H_{54}$ to $C_{40}H_{82}$ are constituents of lubricating oil.

Note that the melting point, boiling point, and density of these substances increase with their molecular weight. Many other related organic compounds can be arranged in series of this sort, and as a general rule physical properties for such series change regularly as the molecular weight increases.

The simplest hydrocarbons, called *normal alkanes*, have their carbon atoms

TABLE 15-1
The normal alkane (or paraffin) series of hydrocarbons

Formula	Name	Freezing point (°C)	Boiling point (°C)	Density (g/cm³)	Type
CH_4	Methane	−182	−161		Fuel gases
C_2H_6	Ethane	−183	−89		
C_3H_8	Propane	−190	−45		
C_4H_{10}	Butane	−138	−1		
C_5H_{12}	Pentane	−130	36	0.626	Petroleum ether (naphtha)
C_6H_{14}	Hexane	−95	68	0.659	
C_7H_{16}	Heptane	−90	98	0.684	
C_8H_{18}	Octane	−57	125	0.703	Gasoline
C_9H_{20}	Nonane	−51	151	0.718	
$C_{10}H_{22}$	Decane	−30	174	0.747	
					Kerosene
$C_{16}H_{34}$	Hexadecane	+18	287	0.773	

arranged in a linear chain. An example is butane, C_4H_{10}, for which a diagrammatic representation is

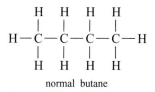

normal butane

Here each dash indicates a covalent bond, in other words a pair of electrons shared between two atoms. Carbon can form a total of four such bonds, so that each carbon atom must be the center of four dashes; hydrogen, with a single electron per atom, can form only one bond, so that every H is drawn with a single dash. A little juggling of atoms and dashes shows that these requirements can be met also with a different sort of diagram:

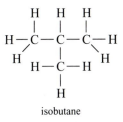

isobutane

Corresponding to these two representations, we find that two gases exist with the formula C_4H_{10}, and the different molecular structures give the two compounds slightly different chemical and physical properties. To distinguish them, we call the first gas *normal butane* and the second one *isobutane*.

Obviously, the possibilities of forming molecules with different structures that satisfy the rules of bonding will increase rapidly as the number of carbon atoms per molecule becomes larger. The existence of more than one compound with the same simple molecular formula but different molecular structures is extremely common in organic chemistry, and accounts in large part for the enormous number of organic compounds. The phenomenon is called *isomerism*, the different substances with the same composition but different molecular architecture are called *isomers*, and expanded formulas like those just used for the two butanes are called *structural formulas*. Such architectural formulas are the organic chemist's shorthand for describing some of the properties of organic compounds. Their usefulness in distinguishing isomers is only one example of the manifold applications of these formulas.

To summarize, the hydrocarbons so far considered are characterized by chains of carbon atoms in their molecules. The chains are straight in the normal compounds, branched in the isomers. These compounds are also characterized by the fact that the link between each pair of atoms is a single electron-pair bond. Hydrocarbons with these characteristics are called *alkane* or *paraffin* hydrocarbons. They are prominent constituents of petroleum and natural gas, and are likewise obtainable in large amounts by distillation of coal.

Carbon atoms can be linked into rings as well as open chains. The two simplest common compounds with ring structures are cyclopentane, C_5H_{10}, and cyclohexane, C_6H_{12}:

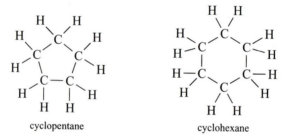

cyclopentane   cyclohexane

These two are prominent ingredients of some petroleums. Collectively, compounds with this kind of ring are called *naphthenes* or *cycloparaffins*. They resemble the alkane hydrocarbons in that all bonds are formed by sharing single electron pairs, a characteristic we describe by saying that all compounds discussed so far are *saturated* compounds.

In contrast to saturated hydrocarbons are those in which one or more of the carbon–carbon bonds has two shared electron pairs. The simplest example is the gas ethylene (or ethene), C_2H_4:

ethylene

The double bond between the C atoms in this formula symbolizes the two shared pairs of electrons, and with these shared pairs each C atom has its normal total of four bonds. Compounds with double bonds are called *unsaturated*, and compared with the saturated variety they are in general much more reactive. Ethylene, for example, reacts with acids,

$$C_2H_4 + HCl \rightleftharpoons C_2H_5Cl \tag{15-1}$$

while the saturated compound ethane does not. Unsaturated compounds with straight chains and branching chains of carbon atoms, called *alkene* or *olefin* hydrocarbons, form a series (C_2H_4, C_3H_6, C_4H_8, . . .) analogous to the alkane series, and still other series are possible with compounds containing two or more double bonds per molecule. Compounds also exist with a higher degree of unsaturation, like acetylene, C_2H_2,

$$H-C\equiv C-H$$
acetylene

in which C atoms are linked by a triple bond. Olefin hydrocarbons are produced by the life processes of some plants and animals, but are generally absent in older organic material because they are too reactive to persist in natural environments. They are formed from other organic compounds during the destructive distillation of coal and in petroleum refining.

Certain unsaturated hydrocarbons with rings of carbon atoms are more stable than the olefins, sufficiently stable that they are important constituents of some petroleums. These are the *aromatic* hydrocarbons, of which the simplest example is benzene, C_6H_6. The structure of benzene may be shown in two ways:

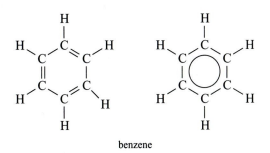

benzene

The second (and preferable) formula shows that the double bonds are not actually positioned between specific pairs of carbon atoms, as the first formula suggests; rather, the six extra electrons of the carbon atoms form a generalized bond (a so-called π-bond) that stabilizes the ring structure. Other aromatic hydrocarbons consist of a benzene nucleus with "side chains" of carbon atoms, and still others of two or more benzene rings joined together:

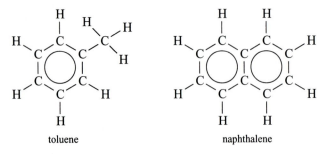

toluene

naphthalene

The name "aromatic" refers to the fact that many of these compounds with molecules containing the benzene ring have pleasant odors.

This by no means exhausts the catalog of known hydrocarbons, but for geologic purposes the four groups mentioned above—the alkane, naphthene, alkene, and aromatic hydrocarbons—are the most important.

Alcohols

Structural formulas of alcohols are similar to those of hydrocarbons, but one or more of the H atoms is replaced by an OH group. These compounds are named with the suffix -ol. The two simplest and most familiar alcohols are methanol and ethanol:

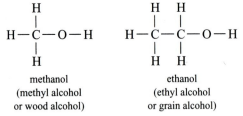

methanol
(methyl alcohol
or wood alcohol)

ethanol
(ethyl alcohol
or grain alcohol)

The simple alcohols appear in nature only ephemerally, as a result of some kinds of organic decay. Many other compounds with OH groups attached to hydrocarbon structures, however, are stable in themselves or as parts of larger molecular structures. Two examples are phenol,

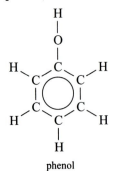

phenol

which is a structural unit in lignin and chitin, and the trihydric alcohol glycerol or glycerin,

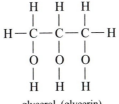

glycerol (glycerin)

a structural unit in fats and oils. *Sterols* are particularly stable alcohols with large and complex molecules, found in organic materials as structural units in waxes.

The OH in alcohols is referred to as a *functional group* because it is the part of the molecule that is reactive and takes part in chemical reactions with ions in solution or with functional groups on other organic compounds. We have encountered this concept previously in Sec. 6-1, where surface hydroxyl ions attached to metal cations on a mineral surface are sites for reactions with ions in the fluid. Thus the Si-OH and A1-OH sites in Eqs. (6-1) through (6-4) are two types of functional groups on a mineral surface.

Organic Acids

Most organic compounds that behave as acids contain in their molecules the functional group

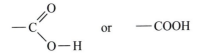

called the *carboxyl* group. Two simple examples of carboxyl acids are

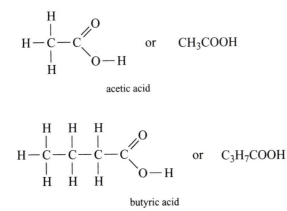

The first is the acid of vinegar, the second the acid of rancid butter. The H-atom of the carboxyl group is capable of dissociating slightly, the degree of dissociation (and thus the strength of the acid) depending on the length of the carbon chain and the kinds of atoms attached to it. Organic acids are found in soils and in decaying organic matter; at least locally, they can be effective agents of weathering. They are also important as intermediates in the synthesis of some of the complex molecules of living creatures.

Humic acids are an ill-defined and structurally diverse group of organic acids that may be extracted from sediments with solvents or formed in the laboratory by treating wood with strong alkali and neutralizing with a mineral acid. The humic acids themselves are only feebly acidic; probably most of the materials called "humic acids" in geologic literature contain also molecules of smaller size and greater acidity (Sec. 13-4).

The group called *fatty acids* includes those whose molecules have a carboxyl group attached to a paraffin hydrocarbon. Acetic acid and butyric acid are simple members of this group. The designation "fatty" is appropriate because some members of the group serve as structural units in the complex molecules of fats.

Fats, Oils, and Lipids

The reaction between the functional groups in an alcohol and an acid (the acid may be either organic or inorganic) is called *esterification*, and the products are water and a compound called an *ester*. For example,

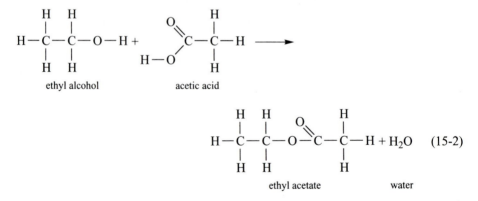

Superficially the reaction resembles neutralization, in that OH from one molecule reacts with H from another to form water. The product of esterification, however, is commonly an undissociated, sweet-smelling liquid rather than a crystalline salt. As carried out in the laboratory esterification is generally a slow process, but in the life processes of organisms some forms of esterification are catalyzed by enzymes and may proceed rapidly (Sec. 11-4).

Geologically the most important esters are fats and oils, compounds whose molecules consist of glycerin combined with one or more long-chain fatty acids. A typical one is glyceryl palmitate:

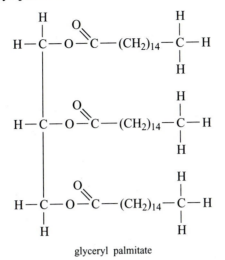

glyceryl palmitate

As a general rule the carbon chains of solid fats are saturated and those of liquid oils unsaturated, but there are many exceptions. The chains are for the most part unbranched, and the more abundant ones contain even numbers of carbon atoms in the range 12 to 18. Fats and oils unite slowly with water to set free glycerin and the fatty acids, a reaction that is the reverse of esterification [Eq. (15-2)]. This reaction is called *hydrolysis*. It resembles inorganic hydrolysis (Sec. 2-2) in that H_2O is broken up by the formation of weakly dissociated compounds, but little H^+ or OH^- is set free because the acid and "base" (alcohol) formed are both weak. Organic hydrolysis reactions may be speeded up by heating, by bacteria, or by the catalytic action of enzymes.

Some other complex compounds are included with fats and oils under the more general term *lipids*, which applies to all compounds soluble in such organic solvents as ether, chloroform, and benzene. Among these non-fat lipids an important group is the *waxes*, which are in large part esters of alcohols more complex than glycerol (sterols) but include also hydrocarbons of high molecular weight. *Phospholipids* are a resistant variety containing phosphorus and nitrogen in addition to glycerol and fatty acids.

Carbohydrates

Another major class of compounds produced by organisms is given the name *carbohydrate*, because their formulas look as if they consist of carbon united with water; in other words, they consist of carbon, hydrogen, and oxygen with the number of H atoms always twice the number of O atoms. Simplest of the carbohydrates are sugars like glucose, $C_6H_{12}O_6$. More complex carbohydrates may be regarded as chains of these 6-carbon units, each link in the chain after the first having lost a molecule of water: $C_6H_{12}O_6 \cdot C_6H_{10}O_5 \cdot C_6H_{10}O_5 \cdots$ For example, the formation of sucrose (ordinary sugar) from glucose, a reaction that is common in some plants, would be represented by:

$$2C_6H_{12}O_6 \rightarrow C_{12}H_{22}O_{11} + H_2O. \tag{15-3}$$

glucose   sucrose

Carbohydrates with only a few 6-carbon units per molecule (generally three or less) are called *sugars*; those with longer chains, represented by the general formula $(C_6H_{10}O_5)_x$, are *starches*; and those in which the subscript x attains values of the order of a few thousand are *celluloses*.

Occurring with cellulose in woody plants are more rigid compounds called *lignins*, made up of the same three elements but with higher ratios of H and O to C and smaller ratios of O to H. Their structure is complex, including chains of benzene rings with functional groups of OH and OCH_3 attached. In general the lignins are more resistant than carbohydrates to decay.

Proteins

Among the most complex of organic compounds are the *proteins*. These always contain nitrogen in addition to carbon, hydrogen, and oxygen, and some contain sulfur as well. Their molecular structure is built up of units called *amino acids*, of which two examples are glycine and alanine:

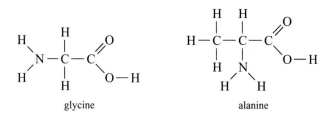

glycine   alanine

The functional group —NH_2 ("amino group") in one molecule can react with the carboxyl group of another molecule, giving a mechanism by which plants and animals can link together long chains of amino acids into the enormous molecules of proteins. A simple example of this linking mechanism is the reaction between two glycine molecules:

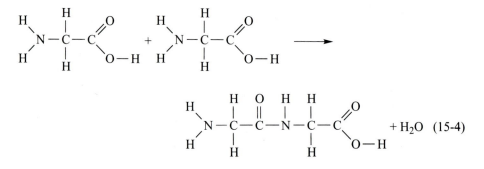

Chains of amino acids built up in this manner by living organisms contain thousands or tens of thousands of atoms. The process can be duplicated in the laboratory to the extent of forming chains of moderate length, but not the huge molecules of natural proteins.

Enzymes, the organic catalysts we have mentioned several times, are proteins with very large molecules. The substance called *chitin*, often described as a very resistant protein, is more properly regarded as a derivative of cellulose with an $NHCOCH_3$ group substituted for an H-atom in each $C_6H_{10}O_5$ unit. Chitin serves as a structural material in many organisms, and is so resistant to decay that some has been found in the remains of invertebrates from rocks as old as Cambrian.

Optically Active Compounds

The amino acid alanine, whose formula is given above, provides an example of the many carbon compounds in which *four different* groups in the molecule are attached to the same carbon atom. Such compounds, as was discovered long ago by Pasteur, have the peculiar property of rotating the plane of polarization of a beam of polarized light that passes through them. This property, called *optical activity*, depends on the two different possible arrangements of the four groups around the carbon atom as illustrated in Fig. 15-1. Note that one of these arrangements is the mirror image of the other, so that the two cannot be superimposed.

Two isomers of this sort, called *optical isomers*, rotate the plane of polarized light by equal amounts in opposite directions. When such compounds are prepared in the laboratory, the result is a mixture of equal amounts of the two isomers (a *racemic* mixture), so that no optical activity is apparent. Living organisms, however, have the strange ability to manufacture one isomer to the exclusion of the other. Hence if a natural material containing carbon compounds can be shown to possess optical activity, the conclusion seems inescapable that living organisms played a role in its genesis. The fact that petroleum contains optically active compounds is one of the most convincing pieces of evidence that it was formed by organic rather than inorganic processes.

Metals in Organic Compounds

Organic compounds containing metals have much geochemical interest because some of the rarer metals, particularly vanadium, molybdenum, and nickel, show remarkable enrichment in the organic matter of certain sediments. Three possible

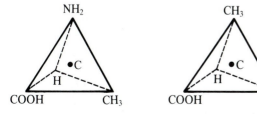

FIGURE 15-1
Optical isomers of the amino acid alanine.

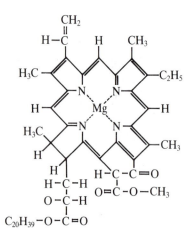

FIGURE 15-2
The structure of chlorophyll, a typical chelate compound. Each unlabeled corner of the complex ring structure is occupied by a carbon atom. The porphyrins of petroleum have similar basic structures, but (1) some of the side chains are modified or missing, and (2) the magnesium in the chelate ring is either absent or replaced by nickel or vanadium.

positions can be occupied by a metal atom in an organic molecule: (1) It may substitute for the hydrogen of a carboxyl group, forming a salt like sodium acetate. Many such salts are soluble in water, dissociating into the metal cation and an organic anion. (2) It may be united directly to the carbon atom of an organic radical, forming a so-called *metallo-organic* compound; a familiar example is lead tetraethyl, $Pb(C_2H_5)_4$, the antiknock ingredient in ethyl gasoline. (3) It may form the center of a complex ring structure, being attached to the carbon atoms of the ring through intermediate N, O, or S atoms. These are the *chelate* compounds, of which the chlorophyll of plants (Fig. 15-2) and the hemin in the blood of animals are familiar examples. The group of chelate compounds called *porphyrins* are of special interest, because representatives of this group that are found in petroleum furnish a second important piece of chemical evidence that petroleum must be chiefly of organic origin.

15-2 ORGANIC REACTIONS

As should be evident from this long review of carbon compounds, organic processes in nature differ in several important respects from the inorganic reactions we have discussed hitherto. Starting materials are generally complex mixtures rather than single compounds, and precise formulas for at least some constituents of the mixtures are uncertain. Reactions are slow and often incomplete. The general way in which components of a mixture react can be described, but details of the reaction for each substance are often obscure. Furthermore, a complex organic molecule can generally react in different ways, the different reactions being roughly equivalent energetically and one or another being favored by organic catalysts or by the activity of bacteria. These complications make it difficult to frame useful generalizations about organic reactions, but a few major types are worth noting because they recur again and again in living organisms and in geologic environments.

Photosynthesis is a general term referring to any chemical combination aided by light, but if used without qualification it means the reaction by which green plants manufacture carbohydrates from CO_2 and H_2O, using sunlight to supply the necessary energy for this endothermic reaction, and using the green pigment chlorophyll as a catalyst (Sec. 12-3). The overall process can be symbolized,.

$$6CO_2 + 6H_2O \rightarrow C_6H_{12}O_6 + 6O_2, \tag{15-5}$$

but it is actually much more complicated, involving a long series of steps. This reaction is the ultimate source of nearly all the organic material in plants and animals. Not only carbohydrates, but fats and proteins as well, are manufactured from carbon fixed in this way. The reaction is essential also in continually renewing the supply of free oxygen in the atmosphere.

Aerobic decay, in the presence of ample free oxygen, can be symbolized as the reverse of photosynthesis. Like the forward reaction, it may take place in many complicated steps, but the end products are chiefly carbon dioxide and water. The same reaction proceeds more rapidly when organic matter is burned, and it is likewise used by animals to furnish energy in the process of respiration. The three processes of decay, combustion, and respiration, together with volcanic activity, are largely responsible for maintaining the supply of carbon dioxide in the atmosphere.

Anaerobic decay, decay in the absence or near absence of oxygen, is a more complicated process in which organic compounds undergo a sort of partial internal oxidation. Much of the oxygen present in the original compounds combines with carbon to form carbon monoxide; the remainder of the carbon may go into hydrocarbons, or it may unite with hydrogen and a little of the oxygen to form new compounds of complex structure. The direction taken by anaerobic decay depends on many factors, for example temperature, the nature of the original material, and the degree of exclusion of oxygen. The number and kinds of bacteria present are especially important. Bacterial action may involve not only organic material itself, but its reactions with other substances, for example the reduction of sulfur from SO_4^{2-} to form pyrite.

Reduction, in reference to organic reactions, means any process in which the amount of oxygen in a compound is decreased or the amount of hydrogen increased. If an organic compound containing oxygen is converted into a hydrocarbon, no matter by what method, the reaction involves reduction. If hydrogen is added to an unsaturated compound so as to form a saturated compound, the process is also reduction. Reduction reactions in which hydrogen is added are often spoken of as *hydrogenation*.

Polymerization is a general name for any process in which large molecules are formed by the joining together of small ones. The reactions in plants by which 6-carbon sugar units are joined to form starches and cellulose, and those by which amino acids are linked into long chains of proteins, are typical polymerizations. The opposite sort of reaction, the splitting of big molecules into small units, is called *depolymerization* (or *cracking*). Both polymerization and depolymerization reactions are essential in the life processes of organisms. The sequence generally begins with depolymerization of complex compounds into small units, then the buildup of new kinds of large molecules (*biopolymers*) by the joining together of these separated units. The digestion of food by animals, followed by buildup of bodily structures, is an example of this pattern.

The compounds produced by these organic reactions—what can be said about where they form in organisms, and how durable they are as constituents of sediments undergoing diagenesis? This is obviously a complicated subject, but a few generalizations can be attempted. The organic materials of plants are characterized by a predominance of carbohydrates and lignins, those in animals by lipids and proteins. Marine organisms, both plants and animals, differ from terrestrial organisms in the greater proportion of lipids and proteins in their makeup. The protoplasm of cells is chiefly protein, the material of cell membranes a layered mixture of protein and lipids. Some plant cells have also outer walls consisting chiefly of cellulose and lignin. The woody part of plants is mostly lignin, their seeds and spores largely protein and lipids. As a rule the lignins and the lipids are the longest-lasting of the major organic materials, the carbohydrates and proteins the ones least likely to survive in geologic environments. But to all such general statements there are many exceptions: some proteins of cartilage and connective tissue, for example may be very long-lived. The organic matter found in sediments, then, will in general be a mixture of all the major kinds of organic material (carbohydrates, lignins, lipids, proteins), and differences in the proportions present in a particular sample of organic matter often make possible good guesses about its source.

Summary

This very sketchy review of elementary organic chemistry has been limited strictly to the compounds and processes of interest in the study of geochemical phenomena. Hydrocarbons have occupied a prominent place, because they are the chief constituents of oil and natural gas and are likewise among the important substances given off when coal is heated. Of the more complex compounds, only those of possible interest as source materials of oil and coal have been emphasized: carbohydrates, lipids, proteins, and a few less abundant but more resistant materials. The general processes of photosynthesis, decay, polymerization, and depolymerization are those that play a major role in the origin and transformations of geologically important substances.

In the study of organic chemistry the unifying, fruitful ideas that have geologic applications are concerned with the architecture of molecules: the way various carbon groups fit together to form molecular structures, and the properties that these groups impart to the molecules in which they occur. The ability to predict the behavior of organic materials in geologic environments depends chiefly on an understanding of how broad classes of molecules with similar structures behave when conditions of temperature, pressure, and compositions of solutions change. For organic reactions in the laboratory, where individual substances can be isolated and variables can be controlled, the concepts of reaction rate, equilibrium, free energy, and oxidation potentials, on which we have relied heavily in previous discussions, are still applicable. But for the complex reactions in nature these quantitative concepts are difficult to apply and usually give us little new information that we could not guess from qualitative observations.

15-3 CARBON IN ROCKS

Igneous and Metamorphic Rocks

Elemental carbon occurs in igneous rocks as the two crystal forms graphite and diamond. Under ordinary pressures graphite is the stable form at all temperatures; diamond is truly stable only at high pressures. Appropriately, the only igneous rocks with which diamond is found are ultramafic bodies whose pipelike form and brecciated structure suggest that they have moved rapidly upward from great depths. At the level where the diamond-bearing rocks presumably form, the pressure due simply to the weight of overlying rock would be sufficient to make diamond stable.

The source of the carbon in igneous rocks remains an open question. In rocks derived by differentiation from deep parts of the crust or upper mantle the carbon may well be an original constituent of the magma ("juvenile carbon"). In rocks formed from the partial melting of preexisting crustal rocks an obvious alternative source is the carbon of the original sediments, contained either in organic matter or in carbonate minerals. A decision about source can sometimes be based on measurements of ratios of the stable carbon isotopes, ^{12}C and ^{13}C, since the lighter isotope shows a slight preferential concentration in organic materials (Sec. 10-4).

In the form of carbonate, carbon occurs sparingly in igneous rocks as CO_3 groups in the structure of apatite and scapolite crystals. Calcite is not unknown as a minor constituent of ordinary igneous rocks, but a question generally arises as to whether it is a primary constituent, a remnant of partly assimilated limestone xenoliths, or a mineral introduced later during alteration of the rock by solutions. The peculiar rocks called *carbonatites*—masses up to several square kilometers in cross section consisting largely of simple carbonates and showing intrusive relations with their surroundings—almost surely represent igneous carbonate on a large scale, in the sense that they solidified from a magma consisting chiefly of carbonate material. This conclusion was for a long time in doubt: carbonate magmas seemed impossible, because carbonate minerals at magmatic temperatures are known to decompose into CO_2 and refractory oxides. Recrystallization of big limestone inclusions, or later introduction of carbonate by hydrothermal solutions, were suggested as more likely hypotheses. But experimental work showing that water-rich carbonate liquids can indeed exist under geologically reasonable conditions at high pressures, together with observations of carbonate-rich lavas extruded from East African volcanoes, has amply justified the assumption that most carbonatites have formed from bodies of magma.

Carbon in metamorphic rocks is chiefly in the form of graphite and carbonate minerals, both generally of ultimate sedimentary origin. Increasing metamorphic grade is often reflected in the degree of perfection of the crystal structure of graphite. *Anthracite* is a product of mild thermal or regional metamorphism of coal; increasing temperature during metamorphism converts anthracite into graphite. The hard, black, carbon-rich materials called *anthraxolite* and *impsonite* are probably metamorphic equivalents of bitumen.

Sedimentary Rocks

Elemental carbon in sedimentary rocks occurs only as detrital grains of diamond or graphite. The major part of sedimentary carbon is in the form of carbonates, but a large amount occurs also in various mixtures of organic compounds. It is this organically derived material that needs attention here. Amounts of the organic carbon range from practically none to nearly 100% in coal, with a worldwide average in sedimentary rocks of about 2%. As a general rule, organic matter is more abundant in fine-grained clastic rocks than in coarse-grained rocks or limestones.

The nomenclature of sedimentary organic materials is ambiguous and redundant. This is almost inevitable in view of the wide range in characteristics of the different organic mixtures, the extreme local variability, and the general difficulty of finding clearly discriminating properties of either a petrographic or a chemical nature. The two words *bitumen* and *kerogen* are often used loosely for practically any organic material in rocks; more strictly the former refers to organic material that is largely soluble in organic solvents, and the latter to material that is insoluble. Thus *petroleum* is a liquid bitumen. *Asphalt* is a solid or semisolid bitumen that liquefies on heating; the term is sometimes restricted to solid material associated with or derived from oils with a high content of cycloparaffin and aromatic hydrocarbons ("asphalt-base crudes"). The corresponding solid or semisolid bitumen associated with petroleum rich in alkane hydrocarbons ("paraffin-base crudes") is called *ozokerite*. Kerogen is made up of two kinds of material, *sapropelic* and *humic*; the former is derived from organic matter consisting largely of lipids (fats and waxes) and proteins, the latter from cellulose and lignin. Sapropelic kerogen is the source material from which most petroleum is formed, and humic kerogen is the major constituent of most coals.

The production of petroleum from one kind of kerogen, and of coal from another kind, are diagenetic processes that take place in sedimentary rocks under conditions of temperature and pressure similar to those that promote such inorganic reactions as the formation of illite or mixed-layer clays from smectites. We now look in more detail at these important diagenetic reactions that transform the organic materials of sediments.

15-4   ORIGIN OF PETROLEUM

Petroleum is a complex and variable mixture of hydrocarbons. Some idea of the extent of variation in composition is conveyed by Table 15-2, which shows the compositions of "straight-run" gasolines from different petroleums at temperatures between 40 and 102°C. Besides the hydrocarbons, most petroleums contain small amounts of other substances, notably porphyrins and optically active compounds. Olefin hydrocarbons, free hydrogen, and carbon monoxide are nearly or completely lacking. This particular kind of hydrocarbon mixture has not been observed actually forming in nature, nor can all stages in its formation be reproduced in the laboratory under conditions similar to those of natural environments. Nevertheless, many years of chemical and geologic research on the occurrence and details of composition

TABLE 15-2

Percentage composition of straight-run gasolines obtained by fractional distillation of petroleum from various sources. The compositions refer to the fractions obtained between 40 and 102°C

Kind of hydrocarbon	Ponca City, Okla.	Winkler, Tex.	Conroe, Tex.	Midway, Calif.	Greendale, Mich.	Bradford, Pa.	East Texas
Normal paraffin	35.3	9.5	17.6	9.9	62.2	34.1	24.6
Branched paraffin	20.2	61.4	19.6	21.4	13.0	32.0	27.1
Alkyl cycloparaffin	43.3	28.6	59.2	68.1	23.3	33.4	47.8
Aromatic	1.1	0.5	3.6	0.6	1.5	0.5	0.5

Source: *American Chemical Society Monograph* 121, 1953. The term *alkyl* refers to a radical obtained by removing one H from an alkane hydrocarbon, thus methyl ($—CH_3$), ethyl ($—C_2H_5$). An alkyl cycloparaffin is a hydrocarbon whose molecule consists of a cycloparaffin ring (or rings) with one or more of the hydrogens replaced by alkyl groups. Courtesy of American Chemical Society.

have given us a clear, consistent, and generally accepted idea as to how this valuable material has been generated.

It was not always so. Early in this century many workers favored an inorganic hypothesis for the origin of petroleum. Noting that methane is found as a trace constituent of volcanic emanations and of fluid inclusions in igneous rocks, advocates of an inorganic origin assumed that this gas was sweated out of the Earth's interior during all of geologic time, rose into the crust, polymerized into heavier hydrocarbons, and collected into the accumulations of oil and gas we now find in crustal rocks. Opponents of this view could raise weighty objections. On the chemical side, the presence of optically active compounds and of porphyrins in petroleum is hard to reconcile with an inorganic origin, and a plausible mechanism for polymerization of methane under natural conditions is not known. From a geologic standpoint the most convincing evidence is the simple fact that petroleum occurs almost always in sedimentary rocks; in the few places where oil is pumped from fractured igneous and metamorphic rocks, its source in nearby sediments is clear. Proponents of an inorganic origin are still heard, but only faintly. The great preponderance of evidence indicates that petroleum as we find it today formed chiefly by anaerobic diagenesis of compounds produced by living organisms.

Some of the petroleum hydrocarbons are generated by plants and animals as part of their life processes, and collect in sediments when the organisms are buried. Such direct accumulation cannot be a major source of petroleum constituents, however, because the amount is too small and the proportions of different hydrocarbons are unlike those in petroleum. Most of the compounds in petroleum must come from the diagenetic alteration of other kinds of organic matter after deposition, presumably from the kerogen that forms from plant and animal remains after their burial in sediments. Many details of the slow alteration of kerogen have been worked out by laboratory research, using kerogen both from recent sediments and from older sedimentary rocks associated with petroleum occurrences (Hunt, 1979). The kerogen changes in composition with time, at a rate depending on temperature and therefore on depth of burial of the sediment; the chief overall change is progressive loss of hydrogen and oxygen relative to carbon, a loss accompanied by changes in color and reflectance of the material, from which the extent of change, or *maturity*, of the kerogen can be estimated. Hydrocarbons may be generated at any stage of the alteration, but most abundantly in certain temperature ranges—the hydrocarbons of liquid oil chiefly between 60 and 130°C, and the smaller molecules of natural gas between 100 and 200°C. Thus a study of the thermal history of a sedimentary rock, particularly of the "maturity" of its organic content, can give an indication as to whether the kerogen has served, or is capable of serving, to generate oil—in other words, whether the rock has been a "source rock" for existing petroleum reservoirs, or might become a source rock in the future.

Specific hydrocarbons obtained from kerogen in laboratory experiments, or found in petroleum samples, can often provide information about the kind of organic material from which they came. For example, alkane hydrocarbons of odd carbon-number ($C_{15}H_{32}$, $C_{17}H_{36}$, etc.) in the range C_{15} to C_{21} are known to be synthesized by marine plants, and those in the range C_{27} to C_{35} by land plants. These groups of odd-carbon-number compounds are thus *biogeochemical markers*, or *biomarkers*, like the optically active compounds and porphyrins mentioned above. They not only add further proof that petroleum must have an organic source, but give more specific evidence about the kind of environment that produced them. The study of these and many other biomarkers in trace amounts, made possible by recent advances in analytical capability, has become a lively part of research in organic geochemistry.

If petroleum is assumed to form by the slow alteration of complex organic materials, what kinds of compounds would be the most likely sources for its constituents? Because petroleum occurs with sediments formed in widely different environments, it seems probable that almost any kind of organic matter can serve as a source. On the other hand, the greater abundance of petroleum in sedimentary rocks of marine rather than freshwater origin suggests that marine organisms play a greater role than terrestrial organisms as source materials. Since plants and animals of the sea contain a higher proportion of lipids and proteins than those that live on land, these compounds are perhaps more important than carbohydrates and lignin as precursors of petroleum. In other words, petroleum is formed in larger amount from the sapropelic part of kerogen than from the humic part—a conclusion amply borne out by experimental work on the degradation products of kerogen.

Among the major chemical constituents of kerogen, the fats and waxes that make up lipids seem particularly well suited to serve as a major source of petroleum hydrocarbons. They have a molecular structure from which hydrocarbons can be derived by fairly simple chemical changes (Sec. 15-1). And they are stable but only moderately so, hence should be able to survive initial decay and burial but would alter slowly in the anaerobic environment of a buried sediment. Carbohydrates, at least the simpler ones, can also make substantial contributions to petroleum formation, but are less stable than the lipids and require more extensive chemical change for conversion into hydrocarbons. The more resistant carbohydrates and the lignins go eventually into coaly material rather than fluid hydrocarbons. Proteins, like fats, have structures from which hydrocarbons are derivable by simple degradational processes, but in general are less stable and many do not survive after burial. Thus petroleum hydrocarbons may be produced from all the major constituents of sapropelic kerogen, the relative contributions of the different constituents in any particular petroleum depending on the nature of the original organic matter and the conditions under which diagenesis took place.

The conditions needed for generation of the hydrocarbons that make up petroleum have been the subject of extensive research. Temperatures required can be estimated in various ways, one of the best depending on analyses of hydrocarbons in fine-grained, organic-rich sediments at different depths in a subsiding sedimentary basin. Such sediments are assumed to be the source rock in which hydrocarbons are generated before they move into petroleum reservoirs. The source beds have experienced slowly rising temperatures as they were buried more and more deeply, and the depth to a given sample thus gives a measure of the temperature to which its contained kerogen has been subjected. Amounts of hydrocarbons in the samples increase with depth up to a maximum and then decrease, the maximum representing the depth, and therefore the temperature, at which generation of hydrocarbons was most intense. Such analyses indicate that formation of hydrocarbons begins at temperatures between 50 and 130°C, and is substantially complete below 200°C. Values at particular places, of course, depend on the nature of the kerogen, the chemical environment of the sediment, and the rate at which the sediment accumulated, but the general range of 50–200°C is established as the temperature of formation of most petroleums.

The low temperatures pose a problem, because the breakdown, or "cracking," of kerogen to form hydrocarbons in the laboratory requires heating to 300–400°C. How can the difference be explained? One factor is probably time: the thermal decomposition that goes only at high temperatures in short-term laboratory experiments may proceed at much lower temperatures over geologic times. Or the natural environment may provide other means for speeding up the low-temperature conversion. Bacteria, for example, are known to produce methane from organic acids at ordinary temperatures, and may be responsible for other hydrocarbons; but the generation by bacterial activity alone of any hydrocarbon except methane has not been conclusively demonstrated. A better possibility is the catalytic action of clay minerals: aluminosilicates are widely used as catalysts commercially in the high-temperature cracking of large hydrocarbon molecules, and the abundant clays in fine-grained sediments may promote similar reactions at lower temperatures. Although not all details in the formation of petroleum hydrocarbons in natural environments have been worked out, the general nature of the process seems clear.

In summary, petroleum and natural gas are in large part products of low-temperature anaerobic decay of organic matter buried with sediments. As the principal substances from which petroleum hydrocarbons arise, lipids and some proteins seem likely choices, but other constituents of kerogen contribute substantially. The reactions that produce hydrocarbons in source beds take place slowly at relatively low temperatures, probably aided by catalysis and possibly by bacterial activity.

15-5 ORIGIN OF COAL

Coal is a mixture of compounds of high molecular weight and complex structure, containing a large percentage of carbon and small amounts of hydrogen, oxygen, sulfur, and nitrogen; most coals contain also phosphorus and traces of many other elements. Most of the specific compounds in coal have not been identified. Free carbon 'is not present except in coals that have been metamorphosed. The various hydrocarbons and other simple organic compounds obtained from coal by heating are probably not present as such in the coal, but form at high temperatures by decomposition of complex original substances. In addition to organic material, most coal contains inorganic substances like quartz, feldspar, and clay minerals.

In contrast to petroleum, coal betrays its origin by fossils, particularly by plant fragments (*macerals*) discernible under a petrographic microscope. There seems no possible doubt that coal is a product of partial decomposition, under anaerobic conditions, of buried vegetation, largely terrestrial vegetation from a swamp environment. The chief geochemical questions about this kind of diagenesis are matters of detail: how the transformation of plant material takes place, how the composition of coal is influenced by the nature of the vegetation and by the specific conditions of decay, what are the recognizable stages of the alteration, and so on. A further important question, but one that has no direct bearing on the formation of coal, is the reason for abnormal concentrations of rare metals like uranium and germanium in some coal beds.

Despite the difficulty of identifying particular compounds in coal, gross chemical composition is often useful in characterizing different coals and different megascopic constituents of coal. The ratio of carbon to the volatile constituents (H, O, and N) permits coals to be arranged in a general order of increasing carbon content, an order that roughly parallels an arrangement according to disappearance of recognizable plant remains, or according to increase in hardness, or according to increase in ash content. Steps in this progression are lignite, subbituminous, bituminous, and anthracite coals. The suggestion is obvious that these products represent stages in the progressive alteration of plant material, bituminous coals having been subjected to alteration over longer times and at higher temperatures and pressures than lignite. This assumption of a continuous sequence of changes is implicit in the term *rank*, lignite being a low-rank coal and bituminous coal relatively high-rank. It is by no means certain that *all* lignite would ultimately become bituminous coal, or that *all* bituminous coal has passed through a stage like present-day lignite, but the concept of progressive change is probably valid for most coals.

Many coals are conspicuously banded, the bands differing in such obvious characteristics as luster, hardness, and kind of fracture. On the basis of chemical composition and the nature of enclosed fossil remains, the bands can be grouped into two major kinds of material: the hard, shiny, obsidian-like coal derived principally from lignin and cellulose, and the dull, dirty coal derived in large part from fats, waxes, resins, and resistant proteins. The former, in the language of a previous paragraph, is humic kerogen and the latter is sapropelic. The humic variety is generally more abundant, as would be expected since coal is formed chiefly by the alteration of material from land plants. Sapropelic coals are derived from organic matter similar to that which forms bitumens and, in fact, grade into bitumens through hydrocarbon-rich varieties of coal called cannel coal and boghead coal.

The relative importance of lignin and cellulose in the origin of humic coal has occasioned much argument. Most cellulose decays so rapidly when a plant dies that its preservation by burial seems unlikely, but on the other hand plant structures are often found in coal that almost certainly were cellulose initially. Very likely both materials play a part in the formation of coal, and there seems no need to pursue the argument here.

Thus anaerobic decay of organic matter during diagenesis accounts for the origin of both coal and petroleum. What determines whether one or the other will form in a particular situation? The most important factor is surely the kind of organic matter, since coal is derived from land vegetation consisting largely of lignin and cellulose, whereas oil comes chiefly from the remains of marine organisms which contain a greater proportion of lipids and proteins. The conditions of alteration—pressure, temperature, presence of solutions—probably also play a role, but how much of a role is not clear.

15-6 ORGANIC MATTER IN BLACK SHALES

In black shales the amount of organic matter is commonly a few percent, but may be much higher. With increasing content the shales grade into pure organic matter, into coal on the one hand and bitumen on the other. A particularly important variety of black shale (which is not necessarily black, but often brown) is the kind called *oil shale*, a rock containing large amounts of bituminous material from which hydrocarbons like those of petroleum can be distilled. The origin of the organic materials in black shales poses no problems we have not already considered in talking about coal and petroleum. Again we are dealing with partial anaerobic decay of buried organic matter, the only difference being that now an abundance of inorganic material has been deposited together with dead organisms.

In contrast to coal and oil, black shales can be observed in process of formation today. More accurately, places are known where organic matter in large amount is being deposited with fine-grained sediment that seems to have all the qualifications needed as the starting material for black shale. Such places are stagnant marine basins: deep spots in Norwegian fjords, the bottom of the Black Sea, the Cariaco trench off the coast of Venezuela, or some shallow marginal basins in particularly quiet lagoons and estuaries. The necessary conditions are water that does not circulate, or that circulates only rarely, and a very slow deposition of clastic debris. The restricted circulation means that oxygen in the bottom water is used up in oxidizing organic matter and is not replenished, so that dead organisms falling to the bottom cannot decay completely to CO_2 and H_2O. Instead, the carbon compounds of their bodies only partially decompose, through reactions aided by anaerobic bacteria, and the products of these reactions accumulate as black mud. The water column in such environments shows a decrease in dissolved oxygen from the aerated surface waters downward, then an increase in hydrogen sulfide towards the bottom. As an example, Table 15-3 lists analyses of water samples taken at increasing depths in a Norwegian fjord.

A few excellent detailed studies have been made of present-day black muds, but our chemical information is still limited largely to overall characteristics rather than specific reactions. The pH of water in contact with black mud is generally between 5.5 and 7, somewhat more acidic than normal seawater. The redox potential varies, depending on whether oxygen is wholly or only partly excluded; it may reach extreme reducing values in the neighborhood of -0.5 volts. Below the surface of the mud, where decomposition of organic material has gone on longer, the pH tends to rise and the Eh to become more reducing, but these changes are reversed in some layers. The particular organic compounds produced in the black mud are largely unknown. Methane is generally one recognizable product, but heavier hydrocarbons, if formed at all, appear only as traces. Sulfur from original protein and from SO_4^{2-} is largely converted to hydrogen sulfide, which gives the mud its characteristic odor—hence the common expression for such areas, "foul bottoms." Some of the sulfur combines with iron to form iron sulfide, perhaps first as fine-grained hydrotroilite ($FeS \cdot nH_2O$), but this compound changes quickly to the more stable pyrite. Nitrogen appears chiefly in the form of ammonium ion, and phosphorus as one of the phosphate ions.

All gradations can be found, of course, between such extreme reducing environments and the oxidizing environment of shallow seawater on an open shelf. In the latter environment most organic material decays to carbon dioxide and water before burial can take place, regardless of how abundant organisms may be. A little of the more resistant material generally persists long enough to be buried, and local accidents of sedimentation may incorporate lenses and pockets of undecayed

TABLE 15-3
Chemical characteristics of water in Bolstadsfjord, western Norway
Measurements made 17 June 1932, at a spot where the depth of the fjord is 138 meters.

Depth (meters)	Temperature (°C)	Salinity (g/1000 g)	Oxygen (cm³/liter)	Oxygen (% of saturation)	H_2S (cm³/liter)	pH
1	12.90	0.47	7.99	106.2	0	6.80
10	8.43	0.68	8.40	101.5	0	6.85
15	7.18	1.34	8.61	101.4	0	7.05
20	5.06	14.64	7.89	96.0	0	7.85
40	4.33	18.08	6.20	75.8	0	7.70
80	4.03	20.79	0	0	0.15	7.00
130	4.09	21.09	0	0	1.05	6.90

Source: Strøm (1936).

organic material in the accumulating muds and sands. Once buried to depths of a few centimeters, the carbon compounds are largely protected from oxidation and undergo the same sort of anaerobic decay as in the black-mud environment.

The various kinds of organic matter found in sediments can often be correlated, at least tentatively, with the nature of the original site of deposition. A marine black-mud environment is suggested by an abundance of bitumen containing scattered pyrite crystals; such material, by analogy with chemically similar material in coal, is often called *sapropel*, and the inferred environment is spoken of as sapropelic. A dark shale with less abundant organic matter and without pyrite may have formed under less strongly reducing conditions, perhaps where oxygen was nearly but not quite removed from bottom waters; a sediment and an environment of this sort are described by the term *gyttja*. Shiny coal-like material in terrestrial sediments is doubtless an alteration of wood fragments or other debris from land plants.

Such general correlations and inferences are pretty obvious, but many knotty chemical problems regarding organic materials in sediments remain to be unraveled. We need, for example, more information about pH and Eh in the various kinds of environment where organic matter accumulates and about the relation of these quantities to the kinds of organic compounds formed. Useful also would be pH and Eh measurements in water standing in contact with organic materials of ancient sediments. Perhaps with enough data of this sort the "reducing action of organic matter," so often referred to in geologic discussions, could be given a more precise meaning. Bacteria obviously play an important part in both aerobic and anaerobic decay. To what extent do the nature and number of bacteria in a particular environment determine the chemical changes that take place? Are the same bacteria always present in a given environment, so that they can be considered a constant factor, or is it possible for two environments, otherwise identical, to have different kinds of bacteria and therefore to produce radically different kinds of organic material? The problem of experimentally distinguishing different kinds of organic matter in sediments has been attacked with vigor in recent years, but still remains troublesome. And finally there is the cluster of problems mentioned in preceding sections regarding the formation of coal and oil: what are the specific reactions that take place, why does coaly material form from one kind of organic debris (or in one kind of environment) and bituminous material from another, what determines the extreme variations in composition of different coals and oils, and so on and on.

15-7 CARBON COMPOUNDS AS REDUCING AGENTS

We have referred frequently to the reducing nature of the environments in which carbon compounds accumulate. To some degree this manner of speaking puts the cart before the horse: carbon compounds do not accumulate *because* an environment is reducing, but more correctly an environment is reducing *because* carbon compounds have accumulated there. Other reducing agents that may be present—methane, hydrogen sulfide, Fe(II) ion—are by-products of the diagenetic alteration of the organic material which establishes the redox potential of the environment. Both deposits of plant and animal debris in the modern world and ancient accumulations of organic compounds such as coal beds, oil pools, and the flakes and lenses of organic matter in ordinary sedimentary rocks represent some of the most reducing parts of the sedimentary record. It would simplify sedimentary geochemistry greatly if redox potentials of such environments could be specified numerically, but to do this in any meaningful way is difficult because of the number and complexity of the compounds present and the slowness with which most of them react.

The few direct measurements of Eh that have been reported from present environments of organic deposition give potentials of the order of -0.1 to -0.5 volt. Such figures are questionable, because for slow reactions of compounds held together largely by covalent bonds it is never certain to what extent they represent equilibrium potentials and to what extent they are merely effective potentials, changing slowly with time, determined by rates of particular reactions. The numbers do correlate fairly well with the kinds of metal ions and metal compounds found associated with the organic matter, and so give at least a rough measure of reducing conditions in these environments. One example of such compounds is pyrite, which can form only at Eh values below about $+0.2$ volt in acid solutions and -0.2 in basic solutions (Fig. 9-5), so that its presence in an organic-rich sediment means immediately that the redox potential is below these maxima. Pyrite, as mentioned in the last section, is the principal distinguishing feature of the "sapropelic" environment as opposed to the less reducing "gyttja" environment. The occasional presence of metallic copper in swamp muds gives a similar limiting value to the Eh of the environment.

The pH of an organic-rich environment is an important determiner of the redox potential, and it in turn depends in large measure on the substances produced as the organic matter decays. Here again the complexity of the possible reactions prevents a satisfactory numerical analysis. Under aerobic conditions CO_2 would be produced in abundance, leading to moderate acidity in adjacent solutions; the limiting amount of CO_2 would be determined by atmospheric pressure, and the limiting pH would be that of a carbonic acid solution in equilibrium with CO_2 at 1 bar, or about 4.0. Locally the pH could go lower if organic acids are produced as intermediates. In anaerobic decay several possible reactions may influence pH: a little CO_2 may appear if oxygen is not altogether excluded, organic acids may form, the sulfur of proteins may appear as H_2S, and nitrogen may counteract the effect of other elements by increasing pH with the formation of ammonia. The net effect of these reactions is not readily predictable, but measurements of pH in swamps and stagnant marine basins give figures most commonly just under neutral, in the range 6 to 7.

Thus places where organic matter accumulates, or has accumulated in the past, are generally the most reducing of sedimentary environments, but their reducing ability is hard to express in quantitative terms. Meaningful redox potentials cannot be calculated from thermodynamic data because formulas of all substances present are usually not known, and experimentally determined redox potentials are open to question because of the slowness of organic reactions. Potentials can be roughly estimated from the kinds of inorganic substances associated with the organic material, and these estimates corroborate the conclusion from direct measurement that Eh's are in the range -0.1 to -0.5 volt. The pH of solutions in contact with decaying organic matter is generally less than 7, reaching a minimum of about 4 where decay takes place under aerobic conditions.

PROBLEMS

1. In the inorganic hypothesis for the origin of petroleum, methane molecules are supposed to polymerize into heavier hydrocarbons. The organic hypothesis, on the other hand, pictures a breakdown of more complex compounds into relatively simple ones. A reaction like the following symbolizes the two points of view:

$$2C_2H_6 \rightleftharpoons C_4H_{10} + H_2,$$

the inorganic hypothesis favoring the forward reaction and the organic hypothesis the reverse. Using free energies given in Appendix VIII, determine which reaction is more probable from the standpoint of energy alone.

2. As another alternative, the buildup of heavy hydrocarbons may be thought to involve unsaturated compounds, for example:

$$C_2H_4 + C_2H_6 \rightarrow C_4H_{10}.$$

Would it be necessary to supply energy to make the forward reaction go?

3. Write structural formulas for three isomers of pentane.

4. Write structural formulas for normal hexane, cyclohexane, and benzene.

5. The sulfur associated with salt domes is often assumed to be a product of reduction of $CaSO_4$ (in the form of gypsum or anhydrite) by petroleum hydrocarbons, with bacteria performing the function of a catalyst. Write an equation for this reaction, using $CaSO_4$ as the formula of the sulfate, letting CH_4 represent the hydrocarbons, and assuming that accompanying solutions are acid.

6. Barium ion, Ba^{2+}, is often noted as a constituent of groundwater in and near oil fields, but is practically never found in appreciable amounts in groundwater elsewhere. Remembering that $BaSO_4$ is an extremely insoluble compound, suggest a reason for this peculiarity in the composition of groundwater.

7. To which group of hydrocarbons does each of the following belong: (a) C_6H_{14}, (b) C_4H_8, (c) C_7H_8? Of these three compounds, which might you expect to find among the products of distillation of petroleum? Of destructive distillation of coal tar? Among the products of either?

8. What unsaturated hydrocarbon has the same molecular formula as cyclohexane?

9. To what group of organic compounds does each of the following belong? (a) $C_{12}H_{22}O_{11}$, (b) C_4H_9OH, (c) $C_6H_5CH_2COOH$, (d) CH_3CHNH_2COOH, (e) C_6H_5OH, (f) $CH_3COOC_2H_5$, (g) $C_3H_5(OOCC_{17}H_{35})_3$?

10. What hydrocarbon would be produced by the decomposition (driving off CO_2) of butyric acid, C_3H_7COOH?

11. During the aerobic decay of an animal body, what happens to the following elements originally contained in the organic compounds: C, H, N, S, P, Mg, Fe? Answer the same question for anaerobic decay.

12. Name three common inorganic compounds of carbon that occur in nature. Does carbon ever occur naturally as a constituent of compounds containing silicon?

13. Why are lipids commonly regarded as an important source material for petroleum?

14. Explain and compare the various uses of the term sapropel.

15. From thermodynamic data (Appendix VIII), show that the concentration of CO in volcanic gases is always low relative to CO_2 and CH_4.

REFERENCES AND SUGGESTIONS FOR FURTHER READING

Beaumont, E. A., and N. H. Foster, (eds): *Geochemistry,* Treatise of Petroleum Geology, Reprint Series, No. 8, Am. Assoc. Petroleum Geologists, Tulsa, Oklahoma, 1988. Many papers on the geochemistry of petroleum formation, the use of geochemistry in prospecting, and the geochemistry of petroleum migration.

Engel, M. H., and S. A. Macko, (eds): *Organic Geochemistry,* Plenum, New York, 1993. Many articles covering all aspects of organic geochemistry.

Hedberg, H. D.: "Geological aspects of the origin of petroleum," *Am. Assoc. Petroleum Geologists Bull.,* vol. 48, pp. 1755–1803, 1964. A classical paper reviewing the development of ideas about the origin of petroleum.

Hoering, T. C., and P. H. Abelson: "Hydrocarbons from kerogen," *Geophys. Lab. Ann, Rept.* 1962–1963, pp. 229–234, 1963.

Hunt, J. M.: *Petroleum Geochemistry and Geology,* W. H. Freeman, New York, 1979. A standard reference book on all aspects of petroleum geochemistry.

Ishiwatari, R., M. Ishiwatari, B. G. Rohrback, and I. R. Kaplan: "Thermal alteration experiments on organic matter from recent marine sediments in relation to petroleum genesis," *Geochim. et Cosmochim. Acta,* vol. 41, pp. 815–828, 1977.

Peters, K. E., and J. M. Moldowan: *The Biomarker Guide,* Prentice-Hall, Englewood Cliffs, New Jersey, 1993. Details of the nature, origin, and interpretation of biomarkers.

Philippi, G. T.: "The influence of marine and terrestrial source material on the composition of petroleum," *Geochim. et Cosmochim. Acta,* vol. 38, pp. 947–966, 1974.

Sato, M.: "Thermochemistry of formation of fossil fuels," in The Geochemical Society Special Publication No. 2, R. J. Spencer, and I-Ming Chou, (eds): *Fluid–Mineral Interactions: A Tribute to H. P. Eugster,* pp. 271–283, 1990. An example of the applications of chemical thermodynamics to reactions leading to formation of petroleum.

Strøm, K. M.: "Land-locked waters," *Norske Videnskaps Akad., Mat.-Naturv. Klasse,* No. 7, 1936. An old but often-quoted paper describing a detailed study of recent organic-rich sediments formed in the stagnant water of a Norwegian fjord.

Tissot, B. P., and D. H. Welte.: *Petroleum Formation and Occurrence,* 2d ed., Springer Verlag, p. 699, 1984. A standard text on the production and accumulation of organic matter in sediments.

Yen, T. F., and J. M. Moldowan, (eds).: *Geochemical Biomarkers,* Harwood Academic, 1988. Many papers describing the use of biomarkers in tracing the origin and history of various petroleums.

Ziegler, D. L., and J. H. Spotts.: "Reservoir and source-bed history of the Great Valley, California," *Am. Assoc. Petroleum Geologists Bull.,* vol. 62, 813–826, 1978. Estimates of temperatures, pressures, and kinds of source material for generation of oil and gas in Cretaceous and Tertiary rocks of California.

CHAPTER 16

METAMORPHISM

Near intrusive granite contacts, rocks commonly show pronounced changes in color and texture. Shale and sandstone become hard, brittle rocks, and limestone is converted into coarsely crystalline marble. New minerals may appear in the altered rocks, evidently resulting from the high temperatures produced by the intrusive. Elsewhere, over broad areas not visibly associated with igneous activity, rocks may show a different kind of alteration: shales transformed into slates and mica schists, lavas into chlorite and amphibole schists, limestone again into marbles. All such changes, which may be reasonably ascribed to the action of heat and pressure beneath the Earth's surface, go by the name of metamorphism.

To frame a more precise definition is difficult. On the one hand we need a distinction between metamorphism and sedimentary processes, and the problems in making this distinction we have already noted in trying to define diagenesis (Chap. 14). At the opposite temperature extreme we need a criterion to differentiate metamorphic processes from igneous processes. This also is troublesome, because rocks on sufficient heating will melt to form magmas, and the melting generally takes place over a considerable range of temperature and pressure. Recognizing that the definition must be somewhat arbitrary, we set out in this chapter to study the chemistry of metamorphic processes—processes that grade imperceptibly into diagenetic reactions at one end of the temperature scale and into igneous reactions at the other.

16-1 CONDITIONS OF METAMORPHISM

Metamorphic reactions differ from igneous processes in that they involve chiefly changes in solid rock rather than a silicate melt. Fluids are often present, to be sure, but there is no actual melting of rock material except at the very highest temperatures. The fluids may be the small amount of aqueous solution in fractures, pores, and interstices between mineral grains, or a CO_2–H_2O mixture from the breakdown of mineral hydrates and carbonates, or hydrocarbons from reactions of organic material, or moving groundwater that has come from a distance. Such fluids have an important function in speeding up the processes of metamorphism, which are slow when they involve solid minerals alone (Sec. 11-4). Fluids are important also because in their motion through fractures and pore networks they transport chemical components to and from the site of metamorphic reactions (Sec. 11-5). But the bulk of the rock remains solid during metamorphic changes.

What are the temperature–pressure conditions during these changes? As to temperatures the answer is easy but not very precise: the range must lie between the upper limit of diagenetic temperatures (150–200°C) and the temperatures at which rocks begin to melt–which for most ordinary kinds of rock, under reasonable pressures of water vapor, are between 650 and 1000°C. Regarding pressures, we could set an upper limit by simply noting that the most common metamorphic rocks are formed within the crust, hence at depths generally not exceeding 40 km, a level at which pressure from the weight of overlying material is about 12 kbar. Locally and temporarily pressures resulting from orogenic movement may exceed this figure, and abnormally high pressures are indicated also for odd metamorphic rocks that come from sources deeper than usual crustal levels. The lower limit of pressure, if we count hydrothermal alteration around fumaroles as a form of metamorphism, is just the ordinary pressure of the atmosphere. So the reactions we are to consider here take place in solid rock containing variable amounts of fluid, over a wide range of temperatures and pressures.

Pressure–temperature relations are conveniently displayed by plotting one variable against the other, as in Fig. 16-1. The diagram has more detail than we need at the moment, but it shows the general relation of metamorphism to diagenesis in the lower left corner and to the broad area of rock melting (defined by the two dashed lines) on the right side. The solid lines in Fig. 16-1 represent typical geothermal gradients in the Earth's crust. Names on the diagram are different kinds of metamorphic rock produced at various pressure–temperature combinations: near the bottom margin those formed by heating at the contacts of igneous rocks intruded into shallow levels of the crust (contact metamorphism), near the left margin some of the rare rocks that have been deep within the crust but were never strongly heated, and in the middle of the diagram the abundant rocks that have undergone both strong heating and strong pressures (regional metamorphism). We will return to the names on this diagram in a later section.

Pressure and temperature change with time during metamorphism. The rate of change and the maximum values attained in a particular rock are determined by the geologic setting and geologic history. Examples of two different geologic settings are shown in Fig. 16-2, which illustrates pressure–temperature paths during a long-term history of burial and uplift [path (*a*)], and a relatively shorter period of heating and cooling near an intrusion of magma [path (*b*)]. Path (*a*) shows the conditions for

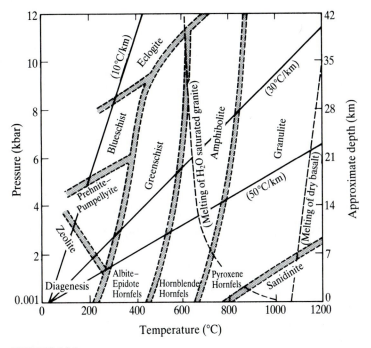

FIGURE 16-1
Approximate pressure–temperature fields of the principal metamorphic facies (the "petrogenetic grid"). The three solid lines show possible values of the geothermal gradient: the mean of measured values is about 30°C/km, the maximum about 50°C, the minimum about 10°C. Geothermal gradients greater than 50°C/km may occur near the contacts of igneous intrusions. The dashed lines show temperatures of incipient melting of a water-saturated granite (water-vapor pressure = total pressure) and of a basalt that does not contain water. Between these lines are the temperature–pressure conditions where differential melting of high-grade metamorphic rocks may occur.

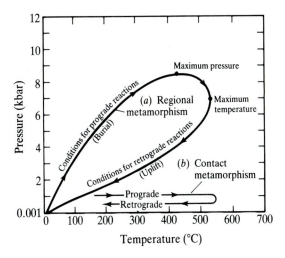

FIGURE 16-2
Possible pressure–temperature paths for a metasediment undergoing regional metamorphism during burial and later uplift [path (a)], and for a rock heated by a nearby igneous intrusion and allowed to cool [contact metamorphism, represented by path (b)]. The prograde portion of each path marks conditions for metamorphism during an increase in temperature and/or pressure, and the retrograde segment shows conditions for possible reactions during a decrease in temperature and/or pressure.

regional metamorphism, say of a metasediment, as it is carried deep within the crust by orogenic movement and then later brought back near the surface. The time required to follow this path is variable, of course, but may be as long as 10's to 100's of millions of years. Pressure increases during burial, as does temperature because of the influx of heat related to the geothermal gradient, to intrusions of magma, or to decay of radioactive elements in the sediment. During the burial stage the rock undergoes a series of metamorphic changes collectively referred to as *prograde* reactions, signifying reactions that result from increasing pressure and temperature. After peak metamorphism, as the metasediment is uplifted and its pressure and temperature decrease, the high-pressure–temperature metamorphic minerals may adjust to these new conditions by *retrograde* reactions, particularly with fluids near grain boundaries, fractures, and faults. Path (*b*) in the figure illustrates a similar series of prograde and retrograde reactions associated with thermal metamorphism at relatively low pressures near an intrusive igneous body. Here the time-scale related to the pressure–temperature path may be as short as 10's to 100's of years for small dikes and sills.

The pressure variable has an added complication, in that two kinds can be distinguished: uniform pressure and directed pressure. The former is the pressure resulting simply from the weight of overlying material, the only kind of pressure that can be sustained in a fluid medium. In water this uniform or nondirected pressure is called *hydrostatic*, and the same term is often used to refer to nondirected pressure in other media, even to pressure in rocks. A more specific term like *lithostatic* is preferable in describing uniform pressure from overlying rocks, to avoid ambiguity. Fluid confined in closed pores of a metamorphic rock will usually be under lithostatic pressure; but if the pore-networks are open to the surface the fluid pressure will be less, approximately hydrostatic. Fluids in confined pores, on heating during prograde metamorphism, may reach pressures greater than lithostatic because the fluids expand at a greater rate than surrounding minerals, thus leading to instability and possible fracturing.

In contrast to lithostatic pressure, directed pressure in rocks is pressure stronger in one direction than another. Evidence that directed pressure has been active in the history of a metamorphic rock lies in large-scale folding and in such mechanical effects as granulation, fracturing and bending of grains, streaking out of fine material, and rotation of large crystals, but most notably in the development of *foliation*—the arrangement of new-formed platy crystals like mica in planes, and of elongated crystals like amphibole either in planes or in linear patterns. Foliation results from intimate movement and granulation in a rock undergoing deformation, the new minerals as they grow being pushed or rolled into orientations that offer the least resistance to movement. The orientations are guided by any planes of weakness, either original or developed by the movement, that the rock or its minerals may possess. In other words, directed pressure acts as a catalyst, bringing mineral grains in contact and exposing them to interstitial solutions, thus speeding up some reactions and inhibiting others, and favoring growth of platy and needle-shaped crystals.

Can directional pressure have other effects, say by promoting the growth of specific minerals that could not form without it? This is an old question, much argued in the early development of metamorphic petrology. Recent work, both experimental and observational, has answered it with a definite negative: directed pressure is an effective catalyst, but is in no sense an additional thermodynamic variable. Minerals do not exist that can form only when directed pressure is active. In diagrams like Figs. 16-1 and 16-2 the indicated pressures may be either lithostatic or directed or both, and no distinction is made between them.

To distinguish in the field between the effects of regional metamorphism and contact metamorphism, one commonly depends chiefly on the presence or absence of foliation in the rocks of a metamorphic assemblage, hence on the presence or absence of directed pressure during their formation. Both kinds of rock have been subjected to high temperature and at least moderate pressure, but most regionally metamorphosed rocks have additionally undergone intimate movement within the rock fabric because of strong directed pressure. This is not universally true, because rocks locally have been metamorphosed on a large scale simply by the effects of deep burial, without appreciable movement (*burial metamorphism*). Also the distinction based on foliation is not clearcut, because some rocks (limestone and pure quartz sandstone, for example) do not have the ingredients from which platy or needle-shaped crystals can grow. Despite these ambiguities, the distinction between contact metamorphism and regional metamorphism is a useful one, and foliation is the commonest easily recognizable sign of the latter.

16-2 EQUILIBRIUM RELATIONS

Metamorphism we can think of as a progressive adjustment of rocks to changing conditions. Ideally, from a rock of given initial composition, metamorphic reactions should give a sequence of mineral combinations, each representing the stable assemblage of minerals at equilibrium in a certain range of temperature and pressure. Slow reactions or rapid changes in temperature or pressure may prevent attainment of some equilibria, and permeation of the rock by fluids may change its bulk composition and so alter the equilibrium assemblages. Despite such complications, it is customary to consider metamorphic processes in terms of changes in idealized equilibria. Below, we use the principles of equilibrium thermodynamics developed in Chapters 7 and 8 to help in estimating the conditions under which various metamorphic rocks have formed.

The Phase Rule

How complex can a metamorphic rock be? The phase rule (Sec. 8-6) should help in fixing the maximum number of minerals that can exist at equilibrium for a given composition. Theoretically the number of phases (p) would be a maximum when the number of degrees of freedom (f) is zero, and should be two more than the number of components [cf. Eq. (8-81)]:

$$p = c + 2 - f, \tag{16-1}$$

whence

$$p_{max} = c + 2. \qquad (16\text{-}2)$$

This maximum could be attained only at a fixed point, a point where both temperature and pressure have unique values. One example is shown in Fig. 5-6, for equilibrium among andalusite, kyanite, and sillimanite in the system Al_2SiO_5. To find such a special assemblage preserved in a metamorphic rock would be possible, but extremely unlikely. In general, an actual rock has a mineral assemblage that is stable over a considerable range of both temperature and pressure, in other words with a minimum of two degrees of freedom (f = 2). This means that the maximum number of solid phases must be equal to the number of components, a generalization that was first suggested by Goldschmidt and is often given his name. It holds for igneous rocks as well as metamorphic rocks, hence is commonly called the *mineralogic phase rule*.

In ordinary shale or arkose or lava the number of major components may be six or seven, so the rule tells us that many phases should be possible. But application of Goldschmidt's rule to determine the maximum number commonly serves little purpose, because metamorphic rocks generally have only a small number of minerals anyway. A striking characteristic of most metamorphic rocks, in fact, is the relative simplicity of their mineral assemblages. Such very abundant rocks as quartz–mica schists and amphibolites, for example, have only two or three minerals of any importance despite the presence of seven or eight components.

Volatiles pose another stumbling block in applying the phase rule to metamorphic processes. A volatile substance would count as a component in the phase rule if it is present in limited amount and if it is confined within the system (a "closed system"), but not if it is free to move in and out of the system (an "open system"). Volatile components in open systems are commonly referred to as *mobile* components. Now in trying to work out the history of a metamorphic rock collected in the field, we seldom know how many mobile components were present or how nearly conditions approached the ideal closed or open system during metamorphism; hence the number of phases to be predicted from the phase rule is always to some extent indeterminate.

Nevertheless, the phase rule does provide a useful starting point for the study of metamorphic reactions. To illustrate what can be learned from the rule, we consider a simple example of thermal (contact) metamorphism of a sandy limestone. This rock, consisting of calcite with a minor amount of quartz, we will assume has pores filled with a fluid mixture of H_2O and CO_2. Compositions of these phases are described by the components $CaO–SiO_2–H_2O–CO_2$, so equilibrium among calcite, quartz, and fluid requires that we specify three intensive thermodynamic variables (f = c + 2 − p = 4 + 2 − 3 = 3). These could be temperature, pressure, and chemical potential of CO_2 in the fluid. The latter we represent by the mol fraction of CO_2 (X_{CO_2}, noting that $1 = X_{CO_2} + X_{H_2O}$), which is related to chemical potential by Eqs. (8-31) and (8-32).

Now imagine that a large pluton of granodiorite intrudes the limestone and thermal energy from the cooling magma causes its temperature to rise. Pressure

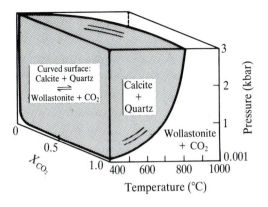

FIGURE 16-3

Three-dimensional diagram for equilibrium of calcite, quartz, wollastonite, and fluid in the system $CaO–SiO_2–H_2O–CO_2$ as a function of pressure, temperature, and fluid composition. The surface marks conditions for $\Delta G = 0$ (equilibrium) for Eq. (16-3).

during the thermal metamorphism is essentially constant (Fig. 16-2). A reaction may occur between calcite and quartz,

$$\underset{\text{calcite}}{CaCO_3} + \underset{\text{quartz}}{SiO_2} \rightleftharpoons \underset{\text{wollastonite}}{CaSiO_3} + CO_2, \tag{16-3}$$

forming the metamorphic mineral wollastonite and releasing CO_2 to the fluid. The reaction is endothermic, so wollastonite should be favored by high temperatures. Of the four substances that appear in the reaction, three are solids and CO_2 is a gas, so that the formation of wollastonite should be inhibited by high partial pressures of CO_2. The effect of these variables is shown in Fig. 16-3, which graphically represents the equilibrium as a surface in a three-dimensional diagram with axes of pressure, temperature, and fluid composition. Note that the three solids can exist together at equilibrium over a considerable range of the three variables.

The equilibrium surface in Fig. 16-3 corresponds to conditions where $\Delta G = 0$, as determined by experimental measurements of the equilibrium constant for Eq. (16-3):

$$\log K = \log f_{CO_2}. \tag{16-4}$$

(We assume that the minerals are pure so their activities are unity.) We recall that K is a function of temperature and pressure [Eq. (8-55)], and that f_{CO_2} is a function of fluid composition [Eq. [8-31)]. The reaction is divariant because we must specify two degrees of freedom to locate a point on the equilibrium surface ($f = c + 2 - p = 4 + 2 - 4 = 2$). If one degree of freedom is fixed during metamorphism, say by maintaining constant pressure, the reaction (Eq. 16-3) can be represented by a single curve in a temperature–fluid composition diagram like Fig. 16-4. Only one more degree of freedom, represented by T or X_{CO_2}, is needed to define equilibrium.

Phase diagrams like Fig. 16-4 permit us to graphically illustrate various possible metamorphic reaction paths. We consider two examples, one for a system closed to the movement of fluids, another for a system where fluid can move in and out, in other words where H_2O and CO_2 are mobile components. To define the initial conditions of equilibrium before intrusion of the granodiorite pluton, we assume that the sandy limestone is at a pressure of 1 kbar and a temperature of

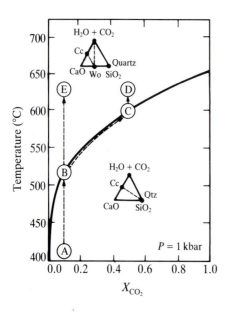

FIGURE 16-4

Temperature–fluid composition diagram for equilibrium among calcite, quartz, wollastonite, and fluid [Eq. (16-3)] at a total fluid pressure fixed at 1 kbar. The curve is computed from experimental data data reported by Greenwood (1967). Arrows mark various reaction paths for the heating of a calcite and quartz metasediment at point A to a temperature of 630°C. The path A → B → C → D represents the reaction for a closed system, in which fluids cannot enter or leave the rock, and path A → B → E represents an open system where a fluid composition of $X_{CO_2} = 0.1$ is externally buffered.

400°C, and that the pore fluids have a composition of $X_{CO_2} = 0.1$ (point A in Fig. 16-4). Heating of the rock and fluid by the intrusion is represented in the diagram by arrow A–B. At point B calcite and quartz begin to react, forming wollastonite and CO_2. As the reaction proceeds with rising temperature, the concentration of CO_2 will increase if the fluids are confined to the rock (closed system). The reaction follows the equilibrium line until all of the quartz (which is the minor phase in the limestone) is consumed, a reaction path schematically represented by the line B–C. During reaction the fluid is said to be *buffered* by the divariant assemblage calcite, quartz, and wollastonite. On further heating, after all the reactant quartz is gone, there is no change in the final assemblage of wollastonite, calcite, and fluid (arrow C–D).

Now consider the same reaction if the system is open. Imagine that the pores are connected to the exterior, allowing CO_2 produced by reaction to escape from the system. This would occur, for example, if the rock is continually flushed by a fluid of constant composition, a fluid referred to as an external buffer. The reaction path followed for a fluid externally buffered at $X_{CO_2} = 0.1$ is shown by arrow A–E in Fig. 16-4. The reaction will begin and end at point B, provided the rate of temperature increase is small relative to the rate of reaction.

The occurrence of reaction (16-3) in nature is shown by the bands of wollastonite rock often found at contacts between granite and limestone. The silica in the wollastonite comes either from quartz grains in the original limestone or from silica-rich solutions permeating the contact rock from the molten granite. From our application of the phase rule we see that the presence of wollastonite at an igneous contact provides little information about the maximum temperature of metamorphism unless, of course, we have information from other sources about pressure and the composition and mobility of the fluid phase. In nature the system would probably be neither open nor completely closed, so that equilibrium would be reached somewhere between the points B and C in Fig. 16-4. This, of course, introduces further uncertainty into any attempt to use the presence or absence of wollastonite as an indicator of temperature at an intrusive contact.

This simplified example illustrates the importance of temperature, pressure, and composition in determining metamorphic mineral assemblages. Below we take a closer look at these variables, focusing first on pressure and temperature.

Pressure and Temperature

In relating metamorphism to tectonic processes in the Earth, one important question is how metamorphic equilibria change with temperature and pressure. For a reaction mixture we know from Sec. 8-1 that the mineral assemblage stable at high temperatures will be the one with the largest entropy, and the assemblage stable at high pressures the one with the smallest molal volume (or the highest density). We will now use these variables, entropy and volume, to predict the relationship between pressure and temperature for a metamorphic reaction.

Consider the reaction of kyanite to form sillimanite in the system Al_2SiO_5, a reaction we have discussed before in Sec. 5-7. Equilibrium at a fixed temperature and pressure requires the free energy of reaction to be zero (Sec. 7-4) and the chemical potential of Al_2SiO_5 to be the same in each phase (Sec. 8-1). If temperature and pressure are changed by a small amount and equilibrium is maintained, the corresponding change in chemical potential must be the same in both phases, that is

$$d\mu_{Al_2SiO_5}^{kyanite} = d\mu_{Al_2SiO_5}^{sillimanite}. \tag{16-5}$$

Because the chemical potential is a function of temperature and pressure $[\mu_i = F(T,P)]$, we can substitute expressions of its total derivative (see Fig. 7-2) into Eq. (16-5) to obtain the following equality:

$$\left(\frac{\partial \mu_{Al_2SiO_5}^{kyanite}}{\partial T}\right)_P dT + \left(\frac{\partial \mu_{Al_2SiO_5}^{kyanite}}{\partial P}\right)_T dP = \left(\frac{\partial \mu_{Al_2SiO_5}^{sillimanite}}{\partial T}\right)_P dT + \left(\frac{\partial \mu_{Al_2SiO_5}^{sillimanite}}{\partial P}\right)_T dP.$$

$$\tag{16-6}$$

We recognize the partial derivatives in this equation as partial molal analogs of the identities given by Eqs. (8-17a, b); that is, the change in chemical potential with temperature at constant pressure is the partial molal entropy $(\overline{S})$ and the change with pressure at constant temperature is the partial molal volume $(\overline{V})$. Note that these new identities are obtained by differentiating Eqs. (8-17a, b) with respect to composition. We now rewrite Eq. (16-6) as

$$-\overline{S}_{kyanite}dT + \overline{V}_{kyanite}dP = -\overline{S}_{sillimanite}dT + \overline{V}_{sillimanite}dP \tag{16-7}$$

and solve for the slope of the equilibrium curve on a pressure–temperature diagram

$$\frac{dP}{dT} = \frac{\overline{S}_{\text{sillimanite}} - \overline{S}_{\text{kyanite}}}{\overline{V}_{\text{sillimanite}} - \overline{V}_{\text{kyanite}}} = \frac{\Delta S}{\Delta V}, \tag{16-8}$$

where ΔS and ΔV are the entropy and volume change of reaction. This equation, referred to as the Clapeyron equation, tells us that the change of pressure with respect to temperature for equilibrium between kyanite and sillimanite is equal to the ratio of the reaction entropy to the reaction volume. The relation can also be cast in terms of reaction enthalpy by noting that $\Delta G = \Delta H - T\Delta S = 0$,

$$\frac{dP}{dT} = \frac{\Delta H}{T\Delta V}. \tag{16-9}$$

The Clapeyron equation is useful because it allows us to predict the slope of univariant equilibrium curves on pressure–temperature diagrams from calorimetric and volumetric data alone. For reactions that involve only minerals, such as polymorphic transformations in the systems SiO_2 (Fig. 4-1) or Al_2SiO_5 (Fig. 5-6), or more complex reactions like albite–jadeite–quartz equilibrium [Eq. (8-75)], the entropies and volumes of reaction are usually small numbers with the same algebraic sign and they change in value only slightly over broad ranges of temperature and pressure. As a consequence, equilibrium curves for most solid–solid reactions appear as straight lines with positive slopes in pressure–temperature plots (see for example Figs. 4-1 and 16-5). One notable exception is the andalusite–sillimanite equilibrium shown in Fig. 5-6.

Because ΔS and ΔV for most solid–solid reactions change only slightly with increasing pressure and temperature, a rough estimate of the slope of an equilibrium curve can be made using volumes and entropies of minerals tabulated at 25°C and 1 bar. Consider as an example the equilibrium

$$\underset{\text{jadeite}}{NaAlSi_2O_6} + \underset{\text{quartz}}{SiO_2} \rightleftharpoons \underset{\text{albite}}{NaAlSi_3O_8} \tag{8-75}$$

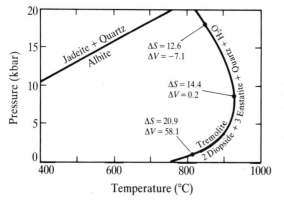

FIGURE 16-5
Pressure–temperature diagram showing the typical form of univariant equilibrium curves for solid–solid reactions and for dehydration reactions. Calculated values of ΔS ($J \text{ mol}^{-1} K^{-1}$) and ΔV ($cm^3 \text{ mol}^{-1}$) are shown at 1 kbar, 8 kbar, and 18 kbar for the tremolite dehydration reaction. Reactions are computed from equations and data reported by Helgeson et al. (1978).

shown in Fig. 16-5. The entropy and volume change of reaction at 25°C and 1 bar are (using the entropy and volume for the ordered form of $NaAlSi_3O_8$ called low albite, see Sec. 5-8)

$$\Delta S_{25°C,\,1\,bar} = \overline{S}_{albite} - \overline{S}_{jadeite} - \overline{S}_{quartz}$$

$$\Delta S_{25°C,\,1\,bar} = 207.2 \text{ J mol}^{-1} \text{ K}^{-1} - 133.5 \text{ J mol}^{-1} \text{ K}^{-1} - 41.3 \text{ J mol}^{-1} \text{ K}^{-1}$$

$$= 32.4 \text{ J mol}^{-1} \text{ K}^{-1} \tag{16-10}$$

and

$$\Delta V_{25°C,\,1\,bar} = \overline{V}_{albite} - \overline{V}_{jadeite} - \overline{V}_{quartz}$$

$$\Delta V_{25°C,\,1\,bar} = 100.07 \text{ cm}^3 \text{ mol}^{-1} - 60.4 \text{ cm}^3 \text{ mol}^{-1} - 22.69 \text{ cm}^3 \text{ mol}^{-1}$$

$$= 16.98 \text{ cm}^3 \text{ mol}^{-1}. \tag{16-11}$$

Substituting these values into Eq. (16-8) and noting that 1 J = 10 cm³ bar gives the predicted slope of the equilibrium curve,

$$\frac{dP}{dT} = \frac{32.4 \text{ J mol}^{-1} \text{ K}^{-1} \times 10 \text{ cm}^3 \text{ bar J}^{-1}}{16.98 \text{ cm}^3 \text{ mol}^{-1}} = 19 \text{ bar K}^{-1}$$

If the disordered form of $NaAlSi_3O_8$ is used in the calculation the slope is 25 bar K^{-1}. These numbers are comparable to the range in slopes measured from experimental data, 21 to 29 bar K^{-1}. For many geologic purposes such a simplified calculation is useful for graphically representing metamorphic phase relations.

In contrast to solid–solid reactions, the equilibrium curve for a reaction involving volatile components like H_2O or CO_2 may exhibit a wide range in slopes because the volumetric and calorimetric properties of volatiles vary greatly with changing temperature and pressure (Fig. 16-6). The curvature typical of these reactions at low pressures is illustrated in Fig. 16-6(*a*) by the experimental data and computed equilibrium curve for the dehydration of brucite to form periclase:

$$\underset{\text{brucite}}{Mg(OH)_2} \rightleftharpoons \underset{\text{periclase}}{MgO} + H_2O. \tag{16-12}$$

The steepening of slope with increasing pressure and temperature means that the ratio $\Delta S/\Delta V$ is increasing also [Eq. (16-8)]. Changes in this ratio are due largely to changes in the molal entropy and molal volume of H_2O, because the corresponding changes for the two solids at low pressures are relatively small. For water, both variables decrease as pressure increases, and the decrease in volume is faster than the decrease in entropy. Thus the ratio $\Delta S/\Delta V$ becomes steadily larger as pressure rises, and the reaction curve grows correspondingly steeper. The slope becomes infinite when $\overline{V}_{MgO} + \overline{V}_{H_2O} = \overline{V}$ $Mg(OH)_2$, and at still higher pressures is negative.

These relations are further illustrated by the reaction

$$\underset{\text{tremolite}}{CaMg_5Si_8O_{22}(OH)_2} \rightleftharpoons \underset{\text{diopside}}{2CaMgSi_2O_6} + \underset{\text{enstatite}}{3MgSiO_3} + \underset{\text{quartz}}{SiO_2} + H_2O \tag{16-13}$$

shown in Fig. 16-5. This is essentially a dehydration reaction, and again the changes in entropy and volume for water are greater than those for any of the solid minerals.

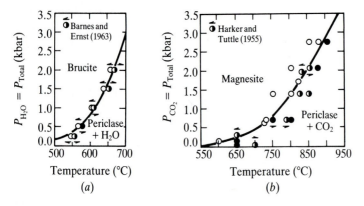

FIGURE 16-6

Univariant equilibrium curves and experimental observations (symbols) for the dehydration of brucite to form periclase and water and the decarbonation of magnesite to form periclase and carbon dioxide as a function of pressure and temperature. The symbols show the results of the experimental studies: open symbols represent stability of the reactant phase, brucite in (a) and magnesite in (b), and the solid symbols indicate the stability of the products, such as periclase and water in (a). For experiments where the reaction did not go to completion, the proportions of reactant and product phases present at the end of the experiment are represented by the partial shading of the symbols, and the small arrow indicates the direction of reaction. (Reprinted by permission from Helgeson *et al.*, 1978, Fig. 20, p. 87. Complete citations for the experimental studies are given in this paper.)

The equilibrium curve rises steeply as pressure and temperature increase, reversing slope at the point where $\Delta V = 0$; this point is the maximum temperature for the stability of the mineral or mineral assemblage with the smaller entropy (the phase stable at lower temperature, in this case tremolite). Similar features are characteristic of many other dehydration reactions involving common metamorphic minerals, the maximum temperatures of equilibrium and changes in the slopes of reaction curves occurring at pressures between 5 and 25 kbar. One consequence of the reversal in reaction slope at high pressures is that hydrous minerals formed by mid-ocean ridge hydrothermal systems will dehydrate during subduction of oceanic crust, thus liberating water to the upper mantle beneath convergent plate margins.

From the variations in slope of metamorphic reactions on pressure–temperature plots we infer that reaction assemblages with shallow slopes provide useful indicators of metamorphic pressure, while reaction assemblages with steep slopes are better indicators of metamorphic temperature. In a later section we explore in greater detail the use of mineral assemblages to predict pressures and temperatures of metamorphism.

Composition

We now look at composition as a variable for determining metamorphic mineral assemblages. In our example of quartz–calcite–wollastonite–fluid equilibrium, the effect of changes in composition can be represented by the simple diagrams of Fig.

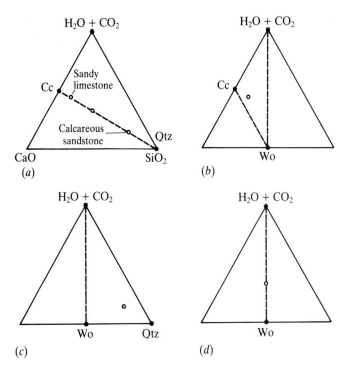

FIGURE 16-7

Ternary composition diagrams showing metamorphism of a metasedimentary rock consisting of varying proportions of calcite and quartz in the system $CaO–SiO_2–H_2O–CO_2$. The variables at the apices of the triangle correspond to molal proportions of the components $H_2O + CO_2$, CaO, and SiO_2. Solid symbols represent the compositions of phases and open symbols are bulk-rock compositions. Diagram (a) shows rock compositions with varying amounts of quartz (Qtz) and calcite (Cc). Diagram (b) illustrates the metamorphic mineral assemblage for a sandy limestone, and (c) shows the assemblage for a calcareous sandstone. In (d) the sediment is assumed to have equal molal proportions of calcite and quartz, so that the final metamorphic assemblage is only wollastonite (Wo) plus fluid.

16-7. Triangle (a) shows the initial compositions of minerals and sediments, triangle (b) the metamorphic assemblage (calcite–wollastonite–fluid) obtained from a sandy limestone consisting of calcite plus minor quartz, triangle (c) the assemblage of quartz–wollastonite–fluid obtained from a calcareous sandstone with quartz in excess of calcite, and triangle (d) the combination of wollastonite with fluid when original quantities of quartz and calcite are equimolal.

For rocks of more complex composition, the problem of diagramming possible metamorphic assemblages obtainable from different proportions of starting materials is not as easily solved. The problem is one we have met before: how to represent the effect of variations in seven or eight components on a two-dimensional diagram. A convenient solution, here as elsewhere, is to reduce the number of significant variables by combining the components in groups and selecting the more important ones—as in the previous example we combined the fluid components H_2O and CO_2 in a single variable (Fig. 16-7). Various ways of combining components or projecting graphs through various axes have been tried, of which the one that has proved most generally useful is the *ACF* diagram devised by Eskola (1939, 1960).

Constructing an *ACF* diagram from a chemical analysis involves the following steps:

1. Recalculate weight percentages of oxides to molecular percentages.
2. Disregard constituents of accessory minerals and Al_2O_3 in alkali feldspars (in other words, subtract from the Al_2O_3 percentage an amount equivalent to that combined with Na_2O and K_2O in $Na_2O \cdot Al_2O_3 \cdot 6SiO_2$ and $K_2O \cdot Al_2O_3 \cdot 6SiO_2$).
3. Add the remaining Al_2O_3 to Fe_2O_3 to get a quantity called *A*.
4. Let CaO be a quantity *C*.
5. Let *F* be the sum MgO + FeO + MnO.
6. Recalculate $A + C + F$ to 100%, and plot on a triangular diagram.

Additional rules are needed for more exact computation to take account of the accessory minerals, and for rocks of unusual composition. The calculation assumes that silica and alkali feldspar are present in excess. This assumption is justified by the observation that these substances are present in a majority of metamorphic rocks, hence are not significant as phases differentiating one mineral assemblage from another. For rocks that do not have excess silica and alkali feldspar, a different choice of variables can be made.

A classic example of the use of *ACF* diagrams is provided by the lime–silicate hornfelses of the Oslo region in southern Norway. Here quartz and K-feldspar are in excess, so that possible equilibrium assemblages may be adequately represented by variations in *A*, *C*, and *F* (Fig. 16-8). The ten varieties of hornfels that can be distinguished are shown by number on the diagram and in the figure caption (quartz

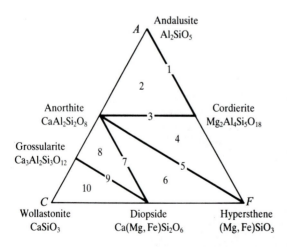

FIGURE 16-8

ACF diagram for hornfelses of the Oslo region, Norway (pyroxene hornfels facies, after Eskola, 1960). Numbers indicate the following metamorphic mineral assemblages: (1) andalusite–cordierite, (2) andalusite–cordierite–plagioclase, (3) cordierite–plagioclase, (4) cordierite–plagioclase–hypersthene, (5) plagioclase–hypersthene, (6) plagioclase–hypersthene–diopside, (7) plagioclase–diopside, (8) plagioclase–diopside–grossularite, (9) diopside–grossularite, (10) diopside–grossularite–wollastonite.

and K-feldspar are also present in each variety). From the assemblages found in the hornfelses we can guess that the original rocks undergoing metamorphism were shales, ranging in composition from lime-free (high *A*) to lime-rich (high *C*).

Note carefully what a diagram like Fig. 16-8 signifies. It shows the possible equilibrium assemblages for different initial compositions in a certain range of temperature and pressure—no more than this. It cannot be interpreted as showing possible paths of crystallization. Nor can it serve as a basis for predictions except in a negative sense; one could venture a guess on the basis of Fig. 16-8, for example, that the combinations cordierite–diopside and hypersthene–grossularite would not occur in the Oslo hornfelses. (If such combinations were found, they would be interpreted as equilibrium assemblages in a different pressure–temperature range, for which a new *ACF* diagram would have to be constructed.) In a word, these diagrams are simply a means of summarizing concisely a mass of analytical data. It should be noted that *ACF* diagrams, or similar triangular diagrams with other variables, are useful only for groups of metamorphic rocks whose compositional relations can be adequately expressed by variations in three components. A great many metamorphic rocks meet this requirement, but by no means all.

Recognition of Equilibrium Assemblages

We have been assuming that the minerals of metamorphic rocks form equilibrium assemblages, but the assumption may well be questioned on the basis that metamorphic reactions are generally slow. How is the assumption justified? And how can we distinguish observationally between equilibrium and nonequilibrium assemblages?

Such questions can be approached by experimental duplication of metamorphic mineral combinations, by constructing theoretical phase diagrams of equilibrium assemblages, and by study of petrographic data. One petrographic indication that equilibrium is attained in most metamorphic rocks is the fact that the number of minerals does not exceed the number permitted by the phase rule—but because the number is often well below the maximum, this is hardly a rigorous test. Another observation suggesting attainment of equilibrium in metamorphic mineral assemblages is the usual lack of composition zoning in minerals such as plagioclase and pyroxene. In igneous rocks these minerals often betray failure of the melt to maintain equilibrium on cooling by showing a different composition between the centers and margins of crystals, but in metamorphic rocks their generally uniform composition must mean that the crystals have adjusted themselves completely to the temperature and pressure of their environment. Perhaps the most convincing evidence for equilibrium is the presence of the same mineral assemblage in different rocks of similar overall composition, regardless of the ages of the rocks or the localities from which they come. Surely, if nonequilibrium were common, one would expect rocks of similar composition to show a variety of minerals depending on how closely equilibrium was approached.

Rather surprisingly, then, we find that equilibrium assemblages are pretty much the rule among metamorphic rocks. Exceptions are by no means lacking, especially exceptions in which relics of original minerals can still be recognized, or where partial replacement of a rock occurs near hydrothermal veins. In general, such cases are rare in comparison with rocks in which the metamorphic transformation has gone to completion.

Preservation of High-Temperature Equilibrium Assemblages

If equilibrium is so much the general rule, a troublesome question presents itself. After a rock has been metamorphosed at high temperature and pressure, it must undergo a gradually decreasing temperature and pressure in order to appear finally in surface outcrops (such as the pressure–temperature–time path for regional metamorphism in Fig. 16-2). Why does its mineralogy not readjust itself so as to be in equilibrium with the lower temperature–pressure conditions? How are the high-temperature mineral assemblages preserved? Why, to put it baldly, do we ever find metamorphic rocks at all?

Metamorphic reactions are known to be sluggish, and are known to increase in rate as the temperature rises. If a rock is heated slowly, say to 500°C, and then cooled rapidly, an assemblage characteristic of 500°C might form which would not have time to readjust itself to low-temperature conditions. But this seems like special pleading; nothing in current knowledge of metamorphism would suggest that temperatures always or even commonly rise more slowly than they fall. Why do we not find at least incipient adjustments to lower temperatures? Partial conversion of high-temperature minerals to low-temperature minerals (retrograde metamorphism) is indeed sometimes noted, but it is rare rather than commonplace unless fluids are involved. Differing rates of temperature rise and fall thus seem a poor explanation for preservation of high-temperature assemblages.

For thermally metamorphosed rocks a possible answer lies in the observation that metamorphic zones at intrusive contacts are widest where veins and abundant new minerals show addition of fluids during alteration. Perhaps prograde metamorphism always requires the presence of at least minor amounts of fluid, and the metamorphic reactions stop abruptly when the supply of fluid is cut off. For regional metamorphism an additional explanation is suggested by the fact that this kind of metamorphism is generally an accompaniment to orogenic movement. Reactions may occur during the movement itself, in response to intimate crushing and granulation of the rock; perhaps reaction ceases when the movement ceases, preserving the mineral assemblage formed during the orogeny. One or both of these explanations—the speeding up of prograde reactions by the presence of fluids, or by granulation due to orogenic movement—probably accounts for the formation and subsequent preservation of metamorphic rocks.

16-3 METAMORPHIC FACIES

So far we have been discussing equilibrium relations in metamorphic processes from a very general standpoint. Implicit in the discussion was the idea that the mineral makeup of a metamorphic rock changes progressively as temperature and pressure rise, but we have not examined such changes in detail. We turn our attention now to specific examples of metamorphic reactions, and in particular to the sequences of rocks that have been ascribed to increasing temperature and pressure.

A striking feature of the thermally metamorphosed rocks near an intrusive contact is the change in their appearance and their mineral composition as the contact is approached. Shales, for example, first become harder, more brittle, less fissile; closer to the contact they may develop small rounded spots as the first sign of growth of new minerals; then flakes of sericite and biotite become prominent, and still closer to the contact andalusite and feldspar may appear. Through all these changes the overall chemical composition may have remained practically constant, the different minerals simply representing alternative combinations of the same chemical components. It seems clear that we are observing here the effect of higher and higher temperatures as we go from unaltered shale to the andalusite rocks near the contact.

A zoning of rock types is often equally prominent in regional metamorphism. If the original rock is andesite or andesite tuff, for example, the first sign of change may be a greenish color due to development of chlorite, and chlorite schists may form a band that can be followed for many kilometers. Going across the band, one may find adjacent to it a zone of actinolite schists, and beyond that an area of amphibolite. Again, the overall composition of the rock need not have changed perceptibly through this sequence. The progression from unaltered tuff to amphibolite will also generally be in the direction of increasing deformation, of increasing grain size, and of increasing frequency of quartz veins and small intrusive bodies. The field relations seem most plausibly explained by assuming that the different mineral assemblages have developed in response to an increase in temperature and pressure toward the amphibolite.

Metamorphic sequences in nature are seldom as simple and regular as those just described. Outcrops all too often are inadequate to show the entire sequence; locally the sequence may be reversed or one of its members may be lacking; initial rock composition may be far from uniform; and complications arise where fluids have made notable alterations in composition. Nevertheless, sequences of this sort are sufficiently well defined and sufficiently similar from one area to another that they must be regarded as a normal phenomenon of metamorphism. They are universally recognized as due to progressive increase in one of the factors of metamorphism, but just which factor or which group of factors is most important may be not entirely certain. Often a sequence is merely ascribed to a change in "intensity" of metamorphism—this being a conveniently vague term that has connotations of temperature, pressure, depth of burial, action of fluids, and degree of deformation, without specifying any one of them.

The regularity of sequences led Eskola long ago (1939) to a now generally accepted classification of metamorphic rocks according to *metamorphic facies*. A facies includes all rocks with equilibrium mineral assemblages in a certain range of temperature and pressure, temperature being the more critical variable. Successive layers of shale and tuff, for example, may be altered to quartz–sericite schist and chlorite–albite–epidote schist, respectively; these two mineral combinations would belong to one facies, because they represent adjustments of two different rocks to the same conditions of temperature and pressure. Interbedded limestone might be converted to fine-grained marble, and sandstone to quartzite; these would also belong to the same facies. A biotite–garnet schist, on the other hand, would belong to a different facies representing a higher temperature. Facies are named from characteristic minerals or rocks that are sensitive to changes in temperature.

Metamorphic grade is a term used, often rather loosely, to describe the relation of one facies to another. Low-grade metamorphic rocks are rocks belonging to facies formed at low temperatures and pressures, high-grade rocks those formed at higher temperatures and pressures. In approaching an igneous contract, we often say that metamorphism increases in intensity and that the rocks represent assemblages of increasing metamorphic grade.

One suggested grouping of metamorphic facies is given in Table 16-1. The primary breakdown is into three temperature ranges, but the ranges overlap extensively because limiting temperatures depend somewhat on total pressure

TABLE 16-1
Some common metamorphic facies, with examples of typical mineral assemblages

Temperature (°C)	Low pressure Lithostatic pressure generally less than 3 kbar (depth <10 km). Water pressure variable. Conditions typical of contact metamorphic zones.	Moderate to high pressure Lithostatic pressure and water pressure approximately equal, generally 3–10 kbar (depth 10–35 km). Conditions typical of regional metamorphsim.
350–550	*Albite–epidote hornfels facies* Sh: quartz–albite–muscovite–biotite Ma: albite–epidote–actinolite–chlorite	*Greenschist facies* Sh: quartz–albite–muscovite–biotite–chlorite Ma: albite–epidote–actinolite–chlorite
500–700	*Hornblende hornfels facies* Sh: quartz–plagioclase–K-feldspar–muscovite–biotite Ma: plagioclase–hornblende	*Amphibolite facies* Sh: quartz–plagioclase–muscovite–biotite–almandine garnet Ma: plagioclase–hornblende
Over 700	*Pyroxene hornfels facies* Sh: quartz–plagioclase–K-feldspar–cordierite–andalusite Ma: plagioclase–diopside–hypersthene	*Granulite facies* Sh: quartz–plagioclase–K-feldspar–garnet–sillimanite Ma: plagioclase–diopside–hypersthene–garnet

After Turner (1968).
Limits of temperature and pressure ranges are rough approximations only. Sh means metamorphic assemblages derived from shale, and Ma means assemblages derived from mafic igneous rocks. See also Fig. 16-1.

and water-vapor pressure. The facies are subdivided further according to pressure, the lower pressures being those characteristic of contact metamorphic zones and the higher pressures those typical of regional metamorphism. Differences between typical high-pressure and low-pressure assemblages are not very great, except that at high pressures one finds relatively dense minerals like garnet and sillimanite, at low pressures lighter minerals like cordierite and andalusite. The principal changes from the first temperature range to the next are the formation of Ca-plagioclase from albite and epidote, the change of actinolite to hornblende, the disappearance of chlorite and formation of garnet; and from the second range to the third, the disappearance of mica in favor of orthoclase, cordierite, andalusite, and sillimanite, and the conversion of hornblende to pyroxene. As might be expected, hydrous minerals become progressively less prominent in the higher temperature facies.

An additional facies is often added at the beginning of such a table, the zeolite facies, to include mineral assemblages formed at temperatures only slightly above those near the Earth's surface. Shales in this temperature range might have illite formed from kaolinite or montmorillonite, plus quartz and chlorite; mafic rocks would have chlorite and one or more zeolite minerals (laumontite, heulandite, analcite, rarely others). All rocks of the zeolite facies represent a low-to-moderate-pressure environment, so there would be no distinction between contact and regional metamorphism.

Relations between facies can be represented conveniently by means of *ACF* diagrams, or similar diagrams with other variables. Two typical plots for rocks of the greenschist and amphibolite facies are shown in Fig. 16-9. Point X in Fig. 16-9 might represent the composition of a shale, with quartz and clay minerals as the chief original constituents, and with a low content of calcium, iron, and magnesium. In the greenschist facies such a rock would consist chiefly of muscovite and quartz, with a little albite, chlorite, and epidote or clinozoisite. In the amphibolite facies the mineral composition would have changed (point X') to sillimanite, quartz, and microcline, with a little plagioclase and almandine. A metamorphosed tuff might be represented by points Y and Y': chiefly epidote, actinolite, chlorite, and albite in the greenschist facies, and plagioclase, hornblende, and diopside in the amphibolite facies. Formally the changes from one facies to the other can be represented by equations like

$$\underset{\text{muscovite}}{KAl_3Si_3O_{10}(OH)_2} + \underset{\text{quartz}}{SiO_2} \rightleftharpoons \underset{\text{sillimanite}}{Al_2SiO_5} + \underset{\text{microcline}}{KAlSi_3O_8} + H_2O \qquad (16\text{-}14)$$

and

$$\underset{\text{epidote}}{6Ca_2Al_3(SiO_4)_3OH} + \underset{\text{chlorite}}{Mg_5Al_2Si_3O_{18}(OH)_8} \rightarrow$$

$$\underset{\text{hornblende}}{Ca_2Mg_5Si_8O_{22}(OH)_2} + \underset{\text{anorthite}}{10CaAl_2Si_2O_8}, \qquad (16\text{-}15)$$

but such formalism gives little information not already conveyed by the diagrams.

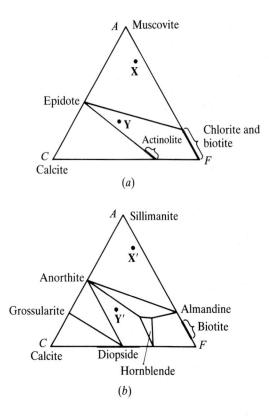

FIGURE 16-9
Some possible mineral assemblages in the greenschist facies (*a*) and amphibolite facies (*b*). Quartz and alkali feldspar are possible additional phases. Points **X** and **X'** represent the bulk composition of shale, and points **Y** and **Y'** the composition of a mafic volcanic tuff. (After Turner, 1968.)

Not all geologists, by any means, would separate facies in precisely the manner shown by Table 16-1 and Fig. 16-9. The general principle of facies classification is widely accepted, but much divergence of opinion persists about the specific minerals or mineral combinations best suited to establish boundaries between facies. The rocks commonly used to define boundaries are shales and mafic lavas or tuffs, because these are both widespread and sensitive to changes in temperature and pressure; but it is not always certain that boundaries based on a metamorphic sequence derived from shale will coincide with boundaries based on metamorphosed basalt. Such difficulties are inevitable in a classification based entirely on observation of complex natural materials.

16-4   EXAMPLES OF EXPERIMENTAL STUDIES ON METAMORPHIC REACTIONS

The facies classification could be put on a sounder basis if it were possible to establish experimentally the phase relations in metamorphic systems, in other words to determine the conditions of temperature, pressure, and composition under which various mineral assemblages are stable. Given this information, we could define facies unequivocally in terms of pressure and temperature limits, and designate specific equations as showing the transition from one facies to another. Such experimental study of metamorphic systems has lagged behind similar work on igneous rocks, for the understandable reason that common metamorphic reactions involve solids at relatively low temperatures, hence are slow and sensitive to small changes in the amount and character of associated fluids. Nevertheless progress in recent years has been rapid, and the chemistry of many metamorphic phase changes is known in satisfactory detail (see for example Figs. 4-6, 4-10, 5-6 through 5-9, and Fig. 8-5). In the following paragraphs we review briefly two such studies we have already considered, the calcite–wollastonite reaction in Sec. 16-2 and the kaolinite–mica–feldspar reaction in Sec. 4-3, then look at a few other experimental studies pertinent to facies relationships.

Calcite–Quartz–Wollastonite

For this very simple reaction [Eq. (16-3)], the effects of changing temperature, pressure, and fluid composition on the formation of wollastonite from calcite and quartz are well displayed by Fig. 16-3. Equilibrium among the three solids, shown by the experimentally determined curved surface in the figure, can exist over a wide range of the three variables. With pressures near atmospheric, the equilibrium can be shifted toward wollastonite even at temperatures below 400°C, but at a CO_2 pressure of 3 kbar (corresponding to a depth of about 10 km) the combination calcite–quartz is stable up to well over 800°C. The presence of wollastonite at an igneous contact, then, tells us only that the maximum temperature during metamorphism was somewhere above 400°C—unless, of course, we have other information about the partial pressure of CO_2, which would enable us to fix the temperature limit more precisely.

Kaolinite–Mica–Feldspar

This combination we have considered before, in connection with the stability of kaolinite and feldspar in aqueous solutions (Sec. 4-3). Here we examine it in more detail with reference to the progressive metamorphism of shale, a series of reactions in which the clay minerals play an important role. The basic data, in this and the previous discussion, come from experimental work by Hemley and his colleagues, reported in Hemley and Jones (1964), Montoya and Hemley (1975), Hemley, *et al.* (1980), and Sverjensky, Hemley, and D'Angelo (1991).

Kaolinite, the simplest of the clay minerals, is known to break down on heating into pyrophyllite, alumina, and water:

$$2Al_2Si_2O_5(OH)_4 \rightarrow Al_2Si_4O_{10}(OH)_2 + 2AlO(OH) + 2H_2O, \qquad (16-16)$$

$$\underset{\text{kaolinite}}{} \qquad \underset{\text{pyrophyllite}}{\phantom{Al_2Si_4O_{10}(OH)_2}} \qquad \underset{\text{diaspore}}{}$$

the alumina forming one of the hydrates diaspore or boehmite. The appearance of a single volatile substance on the right-hand side of the equation means that the equilibrium temperature for the decomposition of kaolinite would rise with increasing pressure of H_2O [Eqs. (8-56), (8-61), and (8-71)], just as the temperature of wollastonite formation rises with increasing CO_2 pressure.

Under natural conditions the reaction would ordinarily not be this simple. If silica is present with kaolinite, no free alumina is produced:

$$Al_2Si_2O_5(OH)_4 + 2SiO_2 \rightarrow Al_2Si_4O_{10}(OH)_2 + H_2O. \tag{16-17}$$

kaolinite   quartz   pyrophyllite

Or if the clay contains adsorbed potassium ions, as most clays do, muscovite would form instead of pyrophyllite:

$$3Al_2Si_2O_5(OH)_4 + 2K^+ \rightarrow 2KAl_2(AlSi_3O_{10})(OH)_2 + 2H^+ + 3H_2O. \tag{16-18}$$

kaolinite   muscovite

Equilibrium in this reaction depends not only on temperature, total pressure, and partial pressure of H_2O, but on the concentrations of H^+ and K^+ as well. Hemley's data for the reaction are summarized in Fig. 4-10. This figure shows also similar experimental results for the further change of muscovite to K-feldspar:

$$KAl_2(AlSi_3O_{10})(OH)_2 + 2K^+ + 6SiO_2 \rightarrow 3KAlSi_3O_8 + 2H^+. \tag{16-19}$$

muscovite   quartz   K-feldspar

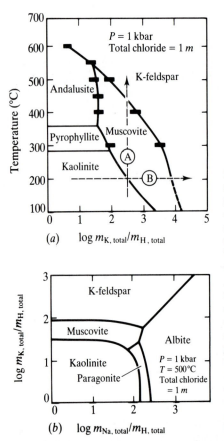

(a) $\log m_{K, total}/m_{H, total}$

(b) $\log m_{Na, total}/m_{H, total}$

FIGURE 16-10

(a) Equilibrium curves for the system K_2O–Al_2O_3–SiO_2–H_2O at a fluid pressure of 1 kbar with total chloride equal to $1m$, as a function of temperature and the total concentration of potassium and hydrogen in the solution. Quartz is present in the reaction mixture. The solid rectangles represent experimental data and the curves are computed from thermodynamic data. Arrow A shows metamorphic changes kaolinite → mica → feldspar caused by increasing temperature at constant $m_{K,total}/m_{H,total}$ ratios and arrow B shows the same progression caused by change in fluid composition alone. (Reprinted by permission from Sverjensky *et al.*, 1991.) (b) Experimental phase diagram for the system Na_2O–K_2O–Al_2O_3–SiO_2–H_2O at 500°C and 1 kbar in the presence of quartz. The curvature of the phase boundaries is due to minor substitution of Na for K in K-feldspar and muscovite and of K for Na in albite and paragonite. (Reprinted by permission from Hemley and Jones, 1964, p. 549.)

Still another variable appears in this reaction, the activity of silica, which was kept constant in the experimental work by including quartz in the reaction mixtures. Experiments conducted at temperatures greater than those in Fig. 4-10 are illustrated in Fig. 16-10(a), which shows the decreasing size of the stability field of muscovite as temperature increases above about 400°C. At 540°C and 1 kbar muscovite reacts in the presence of quartz to form K-feldspar and andalusite (Fig. 16-10a).

Hemley and his colleagues extended their work to systems including compounds of calcium and sodium instead of potassium, and obtained strikingly similar results. The diagram for the sodium system at 500°C and 1 kbar is shown in Fig. 16-10(b) together with phase boundaries for the potassium system. Note that the stability field of the Na-mica paragonite is much smaller than that of K-mica, as would be expected from the fact that muscovite is a far commoner mineral.

In these experiments Hemley and his colleagues have, in effect, isolated some of the essential reactions in the facies changes from unaltered shale to greenschist facies (kaolinite → muscovite), and from greenschist to amphibolite facies (muscovite → K-feldspar). Particularly striking is the wide temperature range over which these facies changes can occur: both of the alkali feldspars are stable down to at least 200°C if the molal ratios K/H and Na/H are high, and both micas are stable to well over 500°C if the ratios are small. Thus facies changes represented by the progression kaolinite → mica → feldspar could be brought about either by a change in temperature alone [arrow A on Fig. 16-10(a)] or by a change in composition alone (arrow B), or by a combination of these factors. Changes in other variables also—total pressure, H_2O pressure, SiO_2 activity, mineral composition—would cause the field boundaries to shift, and hence could bring about changes of facies. Temperature may well be the most important determiner of facies in nature, as it is commonly thought to be, but other variables can also have a strong influence.

Experimental Metamorphism of Shale

Hemley's work is nicely supplemented by older experiments on natural materials described by Winkler (1957, 1958). Instead of pure minerals, Winkler started with natural clays—fairly simple clays, to be sure, but much closer than Hemley's materials to the usual makeup of shales. The reactants were various mixtures of quartz, illite, and kaolinite, with traces of several other minerals; to keep complications within bounds, Winkler was careful to select material low in calcium. Experimental temperatures were in the range 400 to 750°C, and water-vapor pressure was maintained at 2 kbar. Potassium was present as adsorbed ions and as an essential constituent of the illite, but no attempt was made to control or determine the cation to hydrogen ion ratio. When a clay containing both illite and kaolinite was heated, the former recrystallized to muscovite and set free some of its silica; the kaolinite reacted with this silica and with original quartz to form prophyllite, the reaction completing itself below 420°C. No K-feldspar appeared until the temperature reached 665°C. Starting with a sample whose clay was almost entirely illite, Winkler obtained evidence for other reactions: as the illite changed to muscovite, iron and magnesium were set free in addition to silica, and these formed chlorite; at about 550°C the chlorite and some of the muscovite reacted to form biotite and cordierite. The temperature of appearance of K-feldspar was a trifle lower than in the first clay sample, about 620°C, and the temperature of final disappearance of muscovite about 665°C. On the basis of these data, for the particular conditions of his experiments, Winkler sets the upper boundary of the greenschist facies (marked by disappearance of chlorite) at 550°C, and the upper boundary of the amphibolite facies (disappearance of muscovite) at 665°C.

In a further series of experiments, Winkler came still closer to natural conditions by adding a few percent of NaCl to the clay samples, on the grounds that pore spaces of deeply buried shales are commonly filled with saline solution. This made possible the formation of albite from kaolinite and quartz at temperatures lower than 400°C; the albite also took up whatever small amount of calcium the clay contained to form plagioclase. K-feldspar appeared at a slightly lower temperature than in the absence of NaCl, about 600°C, and included a good deal of albite in solid solution. Winkler's experiments lacked the close control of variables that Hemley's had, but by working with more complex systems he came closer to duplicating natural metamorphic assemblages.

Experimental Study of Mafic Rocks

Similar laboratory studies on natural materials of mafic composition have been carried out by several investigators. Particularly informative is work by Liou, Kuniyoshi, and Ito (1974) on the transition between the greenschist and amphibolite facies for rocks with the original composition of basalt. The materials used were samples of greenschist (albite–epidote–chlorite–actinolite) and amphibolite (plagioclase–hornblende) from an area of progressive metamorphism on Vancouver Island. Separate samples and mixtures of the two rocks were sealed with excess water in capsules that could be held at pressures of 2 kbar and 5 kbar and at various temperatures between 450 and 800°C for periods up to three months. Changes of minerals from those typical of one facies to those of the other could be observed in both directions. In the direction of increasing temperature, albite and epidote reacted to form intermediate plagioclase (An_{40-52} in the amphibolite), and hornblende was formed by reactions among actinolite, chlorite, and epidote. At 2 kbar, chlorite decreased sharply at 475°C, but did not disappear completely until 550°C, in general agreement with Winkler's experiments.

The change from one facies to the other, according to this study, is complex. A temperature of 475°C can be taken as a rough upper limit for the greenschist facies at 2 kbar, and 550°C as a lower limit for the amphibolite facies; between the two is a transitional assemblage plagioclase–actinolite–chlorite. At higher pressures the transition zone probably has the composition albite–epidote–hornblende, characteristic of an "epidote amphibolite facies" which is often distinguished in natural occurrences between greenschist and amphibolite. Liou *et al.* point out the many factors on which the width and the temperature boundaries of the transition zone depend: the overall pressure, the fraction of the pressure due to water vapor, the extent of isomorphic substitution in the minerals, and the fugacity of oxygen (controlled in their experiments by a quartz–fayalite–magnetite buffer). They suggest that an increase in oxygen fugacity could lead to narrowing and disappearance of the transition zone, hence a sharp break between the greenschist and amphibolite facies, as is sometimes observed in nature.

Serpentine

A further example of the application of experimental work to metamorphic equilibria is provided by the $MgO–SiO_2–H_2O$ system, which includes the principal constituents of ultramafic igneous rocks and the metamorphic rock serpentine. Results of one experimental study (Hemley et al., 1977) are shown in Fig. 16-11(a), where the stability of the serpentine mineral chrysotile $[Mg_3Si_2O_5(OH)_4]$ and other minerals in this system is plotted as a function of temperature and the concentration

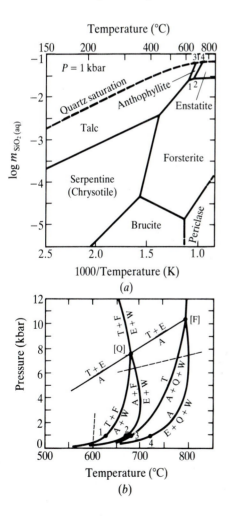

FIGURE 16-11
Equilibrium curves for reactions in the system $MgO–SiO_2–H_2O$. (a) Phase relations at 1 kbar fluid pressure as a function of temperature and aqueous silica concentration. (Reprinted by permission from Hemley et al., 1977.) (b) Univariant reactions illustrating mineral stabilities at temperatures greater than 500°C. Each univariant curve shows equilibrium for the reaction specified. Abbreviations: Q = quartz, P = periclase, B = brucite, W = vapor (chiefly H_2O), E = enstatite (or pyroxene), F = forsterite (or olivine), A = anthophyllite, T = talc, [Q] invariant point for assemblage T + E + A + F + W, [F] = invariant point for assemblage T + E + A + Q + W. (Reprinted by permission from Chernosky et al., 1985.) The two dashed lines show the position of the solid–solid reaction of anthophyllite to form talc and enstatite proposed by Hemley et al., 1977 (lower dashed line) and Helgeson et al., 1978 (upper dashed line). Points 1 through 4 are the same in both figures.

of aqueous silica at 1 kbar fluid pressure. Apparently serpentine is stable only in solutions undersaturated with respect to quartz and at temperatures less than 500°C.

The lines in the figure can be taken as a model for the conversion of an ultramafic rock (pyroxene plus olivine) to serpentine on cooling and reaction with water, or for the opposite metamorphic change in serpentine when it is heated at an igneous contact. Particularly significant is the demonstration that pyroxene (enstatite, $MgSiO_3$) and olivine (forsterite, Mg_2SiO_4) are stable in the presence of water vapor over a wide temperature range. Even at temperatures well over 1000°C the two solid minerals remain unaffected by high water-vapor pressures, in sharp contrast to the behavior of the quartz–feldspar mixtures of granitic rocks. This means that an ultramafic magma cannot exist at temperatures normally encountered in the Earth's crust, no matter what the amount of water. The familiar conversion of ultramafic rocks to serpentine can occur only at temperatures below about 500°C; serpentine cannot exist, even at high pressures, if the temperature is greater than this, and a "serpentine magma" is impossible.

The numerous experimental and theoretical studies of mineral stabilities in this system have shown little agreement concerning the relative stability of the different serpentine minerals or regarding the stability of the Mg-mica, talc $[Mg_3Si_4O_{10}(OH)_2]$, and the Mg-amphibole, anthophyllite $[Mg_7Si_8O_{22}(OH)_2]$, relative to that of olivine and pyroxene (see summary in Berman *et al.*, 1986). As an example, results of one such study (Chernosky, Day, and Caruso, 1985) are represented by the pressure–temperature plot in Fig. 16-11(*b*), the lines showing equilibrium conditions of specific reactions involving talc and anthophyllite [numbered 1 through 4 in Figs. 16-11(*a*) and (*b*)]. With the exception of the solid–solid reaction of anthophyllite to form talc and enstatite, all the reactions in Fig. 16-11(*b*) are dehydration processes, with a single gas (water vapor) as one of the products. They show equilibrium temperatures increasing with pressure, the rate of increase falling off at higher pressures, then at even higher pressures the equilibrium temperature decreases with increasing pressure (see Fig. 16-5). The position of the solid–solid reaction, anthophyllite to form talc and enstatite, is not well constrained because of the sluggish rate of the reaction mixture and the small entropy and volume of reaction (Sec. 16-2). In addition, curves for the dehydration reactions are so closely spaced and intersect at such shallow angles that precise coordinates of the invariant points marked [Q] and [F] in the figure are difficult to locate and differ considerably from one experimental study to another, as indicated by the dashed lines in the figure which mark the position of the anthophyllite-talc-enstatite reaction proposed by two other studies. Use of these experiments is further complicated by isomorphous substitution of Fe^{2+} for Mg^{2+} in metamorphic minerals.

16-5 HIGH PRESSURE FACIES

It is commonly said, as we have noted, that temperature is a more important variable than pressure in determining what metamorphic assemblage will form from a given rock composition. This is a very loose generalization that needs refinement.

The most often observed effects of pressure are embodied in the two columns of Table 16-1: mineral assemblages formed by thermal metamorphism at shallow intrusive contacts represent conditions of lower pressure than the foliated rocks that have undergone regional metamorphism at deeper levels. As the table shows, common assemblages from the two environments are not very different, except for low-density minerals like cordierite and andalusite in the former and higher density minerals like garnet, kyanite, and sillimanite in the latter. The general similarity, of course, is the basis for the generalization about the greater importance of temperature than other variables in determining metamorphic facies.

In the usual sequence of regionally metamorphosed rocks (zeolite, green-schist, amphibolite, granulite facies) we think of pressure and temperature as increasing together, presumably in rough correspondence with the normal geothermal gradient of about 30°C per kilometer of depth. In many places the geothermal gradient strays far from the average, and facies different from the "normal" ones can develop. Where the gradient during metamorphism was markedly greater than 30°C/km, transitional facies may appear between those typical of contact zones and those of normal regional metamorphism.

More interesting are the facies produced where the gradient was exceptionally low, where conditions of high pressure were attained at relatively low temperatures. Under these conditions a very different set of metamorphic minerals can develop: prehnite and pumpellyite at low temperatures, glaucophane, lawsonite, and jadeite at higher temperatures. Correspondingly, new facies can be distinguished, of which the most striking is the *blueschist facies*, named from the color of its characteristic mineral glaucophane.

To show the relation of the blueschist facies to more common ones, we return to the pressure–temperature plot in Fig. 16-1—a kind of plot often called a "petrogenetic grid." As noted earlier, the various low-pressure facies (different varieties of hornfels) appear at the bottom of this diagram, and the "normal" series of regional metamorphism occupies a broad zone in its central part. The high-pressure, low-temperature facies, from prehnite–pumpellyite to blueschist to eclogite, form a separate sequence near the left margin. Facies representing various intermediate pressure–temperature combinations have been distinguished in recent studies of metamorphic terrains, so that a complete petrogenetic grid becomes a complicated diagram with very numerous facies and subfacies.

Rocks of the high-pressure blueschist facies are not common. They are restricted in time to the Phanerozoic and late Proterozoic, and appear mostly in rocks of the Mesozoic and Cenozoic eras. They are restricted in occurrence to orogenic zones, and often show strong folding and fragmentation. These facts, plus the observation that such rocks commonly occupy a belt parallel to an adjacent belt of more normal metamorphic rocks, suggest that their origin is related to plate tectonics, particularly to the subduction of crustal plates at convergent plate boundaries (Fig. 16-12).

We can imagine a lithospheric plate moving downward along an inclined Benioff zone, carrying cool rock and sediments down to depths below the bottom of the normal crust. The moving slab heats up only gradually, and if its descent is fast

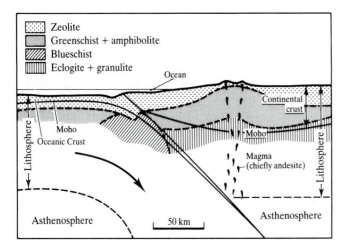

FIGURE 16-12
Schematic cross section of the upper part of a subduction zone, showing the relation of high-P, low-T metamorphism (blueschist facies) to high-T, low-P metamorphism (zeolite to amphibolite facies). Subducting crust and lithosphere (left side) carry relatively cool material to depths >100 km. Magma generated at high temperature above the subducting plate helps to keep temperatures high along its path to the surface. (After Ernst, 1990.)

enough the cool material becomes subject to enormous pressure before its temperature has risen markedly. Because the top of the plate is being sheared and crushed as it moves, conditions are right for high-pressure, low-temperature metamorphism. By contrast, in the mantle wedge and crust above the moving plate, where the geothermal gradient is normal (or greater than normal because of bodies of magma moving upward), the usual facies of medium-pressure, medium-temperature conditions can develop. These would constitute the underpinnings of an island arc, or a volcanic chain on a convergent continental margin. Then many millions of years later, when faulting and erosion have exposed all these rocks at the surface, a belt of deformed high-pressure, low-temperature metamorphic rocks will be found paralleling a strip of more normal rocks. A good example is the pair of metamorphic belts in California, with high-pressure assemblages in the Coast Range and more ordinary metamorphic types in the Sierra Nevada, probably produced by motion of an oceanic plate eastward under the continental margin in the late Mesozoic.

These inferences from plate tectonics can be carried a step further. The apparent restriction of most blueschist-facies rocks to the Mesozoic and Cenozoic suggests that the character of lithosphere plates and the nature of their movement may have changed during geologic time. The typical high-pressure, low-temperature assemblages can form only if the moving plate is thick enough, and its motion is fast enough, that its temperature does not increase rapidly. Perhaps, then, thick lithosphere plates and rapid plate movement have become common only in post-Paleozoic time (Ernst, 1973). Such a conclusion is highly speculative, but is consistent with other inferences about the Earth's past.

16-6 ULTRAMETAMORPHISM

As temperature rises in the normal course of metamorphism, many sedimentary rocks are converted into mineral assemblages that resemble more and more closely the assemblages in igneous rocks. With igneous rocks as starting material, metamorphism produces at first aggregates of hydrous, "low-temperature" minerals, then at higher grades assemblages that are increasingly close to the original mineral composition. The distinction between igneous rocks and these products of high-grade metamorphism is based more on texture than on mineral composition (although significant differences in composition may exist, particularly if the initial rock was at all out of the ordinary chemically). In texture a metamorphic rock consists typically of rounded, intergrowing grains, no mineral showing much tendency to develop euhedral crystal outlines; igneous textures, on the other hand, are characterized by good or fairly good crystal outlines for some minerals, notably mafic minerals and plagioclase, and complete lack of euhedral shape for others, particularly quartz and K-feldspar. There are obvious and numerous exceptions to these generalizations, but not enough to destroy their validity for typical specimens of the two kinds of rock. The generalizations, furthermore, seem to have genetic significance: the differences in crystal shape found in igneous rocks are what one might expect during the freezing of a melt, where some constituents crystallize before others, and the lack of crystal form in metamorphic textures could well result from reactions in a solid where all constituents crystallize simultaneously and interfere with each other's growth.

The typical "metamorphic" texture, however, may not be the end result of metamorphism. The highest temperatures of metamorphism overlap the lowest temperatures of igneous activity, and sufficient heating would certainly produce an interstitial melt with a quartz-plus-alkali-feldspar composition. Only a small percentage of the rock needs to liquefy to make it lose its coherence and become a mush of crystals, and this material could ultimately freeze to give a typical "igneous" texture. Thus the distinction between metamorphic and igneous rocks becomes hazy.

Rocks showing an apparent transition between metamorphic and igneous varieties are common in areas of medium- and high-grade metamorphism. The descriptive literature about the processes and products in such areas is cluttered with long words of Greek derivation, as is usual in parts of geology where basic ideas are hard to define and decisions between rival hypotheses are difficult. *Ultrametamorphism* refers to any alteration that takes place at temperatures higher than those in the metamorphic range, hence alteration that involves the production of igneous-appearing textures and partial or complete melting. *Anatexis* and *palingenesis* emphasize the production of new magma by the partial or complete melting of preexisting rocks; in recent literature anatexis is the commoner term. *Migmatitiza-tion* signifies the formation of a migmatite, or "mixed rock," consisting partly of igneous material and partly of high-grade metamorphic remnants; the process may involve either anatexis in place, or the injection of fluid material along layers or cracks in an adjacent metamorphic rock. *Mobilization* is a marvelously indefinite word that refers to any process by which either ions or the products of partial fusion are put in a condition that enables them to move from one place to another. These are the commonest terms in general use, but the list does not exhaust the number of long words invented to describe vaguely defined aspects of extreme metamorphism.

The relation of anatexis to high-grade metamorphism is shown on the right-hand side of Fig. 16-1, where melting curves for granite and basalt are drawn through the high-grade metamorphic assemblages. The left-hand curve shows temperatures of initial melting of granite in contact with water vapor at a pressure equal to the total pressure; under these conditions the melting temperature falls as pressure rises. The right-hand curve shows the temperature of incipient melting for dry basalt, in which the change of melting temperature with pressure depends chiefly on the difference in density between solid and liquid rock; since the liquid has the lower density, the melting point rises slightly as pressure increases. In nature the melting of most rock material would follow a path between these extremes, and the pressure–temperature conditions of anatexis would thus be represented by the area between the curves. Note that areas of the granulite, amphibolite, and pyroxene hornfels facies straddle parts of the melting-point curve for granite under hydrous conditions. Rocks of these facies can be strictly metamorphic if conditions are relatively dry, or they can be constituents of a migmatite if much water is present and the rock has been partly melted. This means that migmatites may be formed with a great variety of metamorphic rocks.

Thus the possible products of ultrametamorphism can be easily imagined from a theoretical standpoint, but recognizing such products in the field and sorting the processes that have operated in a particular area is often a real challenge.

PROBLEMS

1. What kinds of rocks would you expect to form from the progressive metamorphism of (a) a rhyolite consisting chiefly of quartz, potassium feldspar, silica-rich glass, and a little biotite, (b) a shale consisting chiefly of montmorillonite and quartz?

2. Equilibrium pressure–temperature relations for the reaction

$$\underset{\text{calcite}}{CaCO_3} + \underset{\text{quartz}}{SiO_2} \rightleftharpoons \underset{\text{wollastonite}}{CaSiO_3} + CO_2, \tag{16-3}$$

are shown in Fig. 16-3. Consider mixtures in which CO_2 pressures are equal to total pressure $(P_{CO_2} = 1)$.

(a) What combinations of minerals (calcite–quartz, calcite–wollastonite, quartz–wollastonite) are stable at 100°C and 1 kbar? At 700°C and 2 kbar? At 900°C and 3 kbar?

(b) What is the sign of the free energy change, for the reaction as written above, at 600°C and 3 kbar?

(c) Show that the shape of the curve on the front face of Fig. 16-3 is qualitatively what you might expect from the equilibrium constant.

(d) If a dike intrudes limestone and if geologic evidence permits you to guess that the intrusion took place at a depth of about 4 km, what conclusions, if any, could you draw from (1) the presence of wollastonite, (2) the absence of wollastonite, at the contact?

3. In general, how would the chemical composition of a typical shale differ from that of a diorite or quartz diorite? How might you distinguish a gneiss formed by the metamorphism of shale from a gneiss formed by flow in a crystallizing quartz–diorite magma? Point out possible criteria of a chemical, mineralogical, and textural nature.

4. What substances must be added to (a) limestone to convert it into grossularite–diopside hornfels; (b) pyroxenite to convert it into serpentine; (c) shale to convert it into tourmaline–lepidolite rock; (d) gabbro to convert it into scapolite–epidote rock?

5. Amphibolites consisting chiefly of hornblende and plagioclase (andesine), often with minor epidote and titanite, are a very common type of metamorphic rock. From what kinds of sedimentary or igneous rocks might such metamorphic rocks originate? What criteria could you use to decide in a particular case what kind of original rock is most probable?

6. The green color so common in low-grade metamorphic rocks derived from lavas and tuffs is generally due to one of the four minerals chlorite, epidote, actinolite, or antigorite. From what original minerals are these derived? Express the formation of these minerals by symbolic equations, using oxides to represent constituents added from or removed by interstitial fluids.

7. Would the following be possible equilibrium mineral assemblages in the metamorphism of a shaly limestone consisting initially of kaolinite and calcite?

(a) Calcite–grossularite.

(b) Wollastonite–anorthite–quartz–andalusite.

(c) Anorthite–grossularite–calcite.

(d) Andalusite–grossularite–quartz.

(e) Corundum–wollastonite–grossularite.

Represent the possible mineral combinations by means of a triangular diagram with CaO (or calcite) as one corner, Al_2O_3 as a second, and SiO_2 as the third.

8. Shown below is the univariant curve for equilibrium of reaction (1–48),

$$KAl_2(AlSi_3O_{10})(OH)_2 + SiO_2 \rightleftharpoons Al_2SiO_5 + KAlSi_3O_8 + H_2O. \qquad (16\text{-}14)$$

muscovite   quartz   sillimanite   microcline

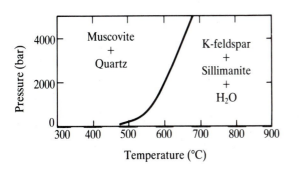

If the change in the standard molal volume for this reaction at 3 kbar and 630°C is $21.5 \text{ cm}^3 \text{ mol}^{-1}$, what is the entropy change ($\Delta S°$) of the reaction at this pressure and temperature? Qualitatively, what would be the effect on the equilibrium of muscovite + quartz + sillimanite + K-feldspar + water (*a*) if paragonite [$NaAl_2(AlSi_3O_{10})(OH)_2$] substitutes into the muscovite structure to form a muscovite solid solution, and (*b*) if the fluid was a mixture of H_2O and CO_2?

REFERENCES

Berman, R. G., M. Engi, H. J. Greenwood, and T. H. Brown: "Derivation of internally-consistent thermodynamic data by the technique of mathematical programming: a review with application to the system MgO–SiO$_2$–H$_2$O," *Journal of Petrology*, vol. 27, pp. 1331–1364, 1986.

Chernosky, J. V., Jr., H. W. Day, and L. J. Caruso: "Equilibria in the system MgO–SiO$_2$–H$_2$O: experimental determination of the stability of Mg-anthophyllite," *American Mineralogist,* vol. 70, pp. 223–236, 1985.

Ernst, W. G.: "Blueschist metamorphism and pressure–temperature regimes in active subduction zones," *Tectonophysics*, vol. 17, pp. 255–272, 1973. Experimental evidence regarding the physical and chemical conditions of blueschist formation, and speculation about the relation of blueschist to subduction at convergent plate junctions.

Ernst, W. G.: *The Dynamic Planet*, Columbia University Press, New York, 1990. An elementary textbook, with clear descriptions in Chapter 4 of igneous and metamorphic phenomena as related to plate tectonics.

Eskola, P.: in *Die Entstehung der Gesteine* (Barth, Correns, Eskola), Springer Verlag OHG, Berlin, 1939; reprinted 1960. A summary of Eskola's many years of work on metamorphic rocks. Eskola's concept of facies is described on pages 336–368, and the construction of *ACF* diagrams on page 347.

Greenwood, H. J.: "Wollastonite: stability in H$_2$O–CO$_2$ mixtures and occurrences in a contact-metamorphic aureole near Almo, British Columbia," *American Mineralogist,* vol. 52, pp. 1669–1680, 1967.

Helgeson, H. C., J. M. Delany, H. W. Nesbitt, and D. K. Bird: "Summary and critique of the thermodynamic properties of rock-forming minerals." *American Jour. Sci.,* vol. 278A, 229 pp., 1978.

Hemley, J. J., and W. R. Jones: "Chemical aspects of hydrothermal alteration with emphasis on hydrogen metasomatism," *Econ. Geol.,* vol. 59, pp. 538–569, 1964. A detailed experimental study of clay–mica–feldspar transitions.

Hemley, J. J., J. W. Montoya, D. R. Shaw, and R. W. Luce: "Mineral equilibria in the MgO–SiO$_2$–H$_2$O system: II. Talc–antigorite–forsterite–anthophyllite–enstatite stability relations and some geologic implications in the system." *American Jour. Sci.* vol. 277, pp. 353–383, 1977.

Hemley, J. J., J. W. Montoya, J. W. Marinenko, and R. W. Luce: "Equilibrium in the system Al$_2$O$_3$–SiO$_2$–H$_2$O and some general implications for alteration/mineralization processes," *Econ. Geol.,* vol. 75, pp. 210–228, 1980.

Liou, J. G., S. Kuniyoshi, and K. Ito: "Experimental studies of the phase relations between greenschist and amphibolite," *American Jour. Sci.,* vol. 274, pp. 613–632, 1974.

Montoya, J. W., and J. J. Hemley: "Activity relations and stabilities in alkali feldspar and mica alteration reactions," *Econ. Geol.,* vol. 70, pp. 577–594, 1975.

Sverjensky, D. A., J. J. Hemley, and W. M. D'Angelo: "Thermodynamic assessment of hydrothermal alkali feldspar–mica–aluminosilicate equilibria," *Geochim. Cosmochim. Acta,* vol. 55, pp. 989–1004, 1991.

Turner, F. J.: *Metamorphic Petrology*, McGraw-Hill, New York, 1968. A standard textbook, especially good in relating theoretical concepts to field observations.

Winkler, H. G. F.: "Experimentelle Gesteinsmetamorphose," *Geochim. Cosmochim. Acta,* vol. 13, pp. 42–69, 1957 and vol. 15, pp. 91–112, 1958.

SUGGESTIONS FOR FURTHER READING

Ferry, J. M.: "Regional metamorphism of the Vassalboro Formation, south-central Maine: a case study of the role of fluid in metamorphic petrogenesis," *Jour. Geol. Society, London,* vol. 140, pp. 551–576, 1983.

Miyashiro, A.: "Metamorphism and related magmatism in plate tectonics." *American Jour. Sci.,* vol. 272, pp. 629–656, 1972. The relation of plate tectonics to paired metamorphic belts and to associated igneous rocks.

Morgan, B. A.: "Mineralogy and origin of skarns in the Mount Morrison pendant, Sierra Nevada, California," *American Jour. Sci.,* vol. 275 pp. 119–142, 1975. High-temperature contact-metamorphic rocks at a granodiorite–limestone contact with emphasis on the effects of oxidation state and retrograde metamorphic reactions.

Philpotts, A. R.: *Principles of Igneous and Metamorphic Petrology,* Prentice-Hall, New Jersey, 1990. A standard textbook, with a particularly clear treatment of the thermodynamics of metamorphic processes.

Vance, J. A., and M. A. Duncan: "Formation of peridotites by deserpentinization in the Cascade Mountains, Washington," *Geol. Soc. America Bull.,* vol. 88. p. 1497–1508, 1977. A field and laboratory study of an unusual kind of metamorphism in the amphibolite facies.

Winkler, H. G. F.: *Petrogenesis of Metamorphic Rocks,* 3d ed., Springer Verlag, New York, 1974. A standard textbook. Chapter 18 is an excellent discussion of anatexis.

CHAPTER
17

FORMATION AND CRYSTALLIZATION OF MAGMAS

The rocks we call *igneous*—rocks that form by the cooling and crystallization of molten material—appear in bewildering variety. One need think only of the many kinds of common lavas, ranging from basalt to rhyolite, and of intrusive rocks ranging from peridotite to granite, to appreciate the multiplicity of mineral and chemical compositions. Add the more exotic kinds, like phonolite and carbonatite, and the number becomes discouragingly large. Early petrographers delighted in distinguishing each minor variant, and the resulting long catalog of names, mostly of Greek derivation, continues to appall beginners in petrology.

Chemical questions pop into mind at once. Why should there be such seemingly infinite variety? Does each minor variant correspond to a particular kind of molten material (*magma*)? Or can one kind of magma produce rocks of different compositions? At what depth in the Earth is the molten material generated? In the last chapter we noted one way to form a melt—simply by increasing the temperature of a metamorphic rock beyond the melting point of one of its minerals or mineral combinations. Are all magmas, or most magmas, formed in this way, or do reservoirs of silicate liquid exist somewhere within the Earth? If a magma is formed by melting of pre-existing rock, does the composition of the magma depend on the rate or extent of melting? To such questions we turn our attention in this chapter.

17-1 THE PHASE RULE APPLIED TO MAGMAS

We start with details of the process of crystallization by which liquid magma freezes into solid rock. As freezing goes on, the originally homogeneous magma becomes a complex mixture of crystals plus remaining liquid, in other words a constantly changing system of several phases in a material made up of many components. The phase rule obviously should serve as a guide in studying the sequence of events, particularly the changes in composition of liquid and crystals in response to changes in temperature and pressure.

Consider, for example, the basaltic magma of a moving lava flow. As it solidifies the pressure remains approximately constant, so that the phase rule can be stated $f = c - p + 1$. The number of major components (as suggested, say, by the number of oxides in a chemical analysis) is 9 or 10, and this is likewise the maximum number of degrees of freedom—since there must be at least one phase present, so that $f = 9 - 1 + 1$. This means that to study the system directly would mean to follow the effects on crystallization of at least 9 variables acting together, which is well-nigh impossible. So, as we have done with other complex problems, we abandon the direct approach and look for simpler systems that can be worked out in detail and then combined to give an approximate picture of the overall system. Fortunately, it turns out that the actual crystallization of a magma takes place in large measure as a series of steps, or in two or three independent series of steps, each one involving only a few of the many components. Thus an approach through simpler systems can in this case give a satisfyingly realistic picture of the overall process.

So we begin our inquiry into magmatic geochemistry by looking at the behavior of very simple artificial silicate liquids as temperature falls and one or more kinds of crystals settle out. Compositions of both liquid and crystals may change during the freezing process, and such changes will be especially important to follow. As usual, in these experiments equilibrium conditions will be assumed at first, so that the phase rule can be readily applied, and the possible effects of deviations from equilibrium will be noted later. The number of components and phases will be kept low enough to permit representation by two-dimensional diagrams. Pressure will be assumed constant and roughly equal to atmospheric, so that conclusions will be applicable strictly only to magmas in surface or near-surface environments. Later on, pressure will become an important variable when we consider processes at substantial depths.

17-2 EUTECTICS AND SOLID SOLUTIONS

Artificial melts containing only two components show two extreme types of behavior. A melt may freeze to give a mixture of separate crystals of the two component substances, or it may form a single kind of crystal representing a solid solution of the two components. Most melts show behavior between these extremes, but intermediate cases are best understood as modifications of the extremes. For examples of the two types, we consider systems with a direct bearing on crystallization processes in magmas.

Eutectics

When a mixture of anorthite ($CaAl_2Si_2O_8$) and diopside ($CaMgSi_2O_6$) is melted in a platinum crucible and allowed to cool slowly, the solid produced is a mixture of separate crystals of the two minerals. From melts rich in diopside the first crystals to form are diopside; if more and more anorthite is included in the melt, the temperature of first crystallization drops steadily. Melts rich in anorthite give anorthite as the earliest-formed crystals, and the melting point becomes progressively lower as diopside is added. These experimental data are conveniently represented on a temperature–composition plot (Fig. 17-1). Here the line DE shows temperatures of first crystallization of diopside-rich mixtures, the line AE those of anorthite-rich mixtures. The curves intersect at E, which is the lowest temperature at which a liquid phase can exist in this system.

The curves show not only temperatures of first appearance of crystals, but also the resulting changes in composition of the liquid. A melt with 20% anorthite, for example, cools to give the first diopside crystals at 1350°C (line MN). As diopside continues to crystallize, the composition of the liquid changes and the temperature decreases, as shown by points on the line NE. When the liquid composition has reached 58% diopside and 42% anorthite (corresponding to E), anorthite begins to crystallize together with the remainder of the diopside, and the temperature remains constant until crystallization is complete. The lowest melting temperature E is called the *eutectic* temperature, and the composition corresponding to E is a eutectic mixture.

The melting of a solid anorthite–diopside mixture on heating follows the same course in reverse. Liquid appears first at the eutectic temperature, and the first liquid has the eutectic composition. The temperature and liquid composition remain constant until all of one mineral has melted. Then the temperature rises as the

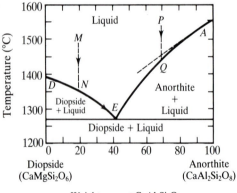

FIGURE 17-1
Temperature–composition diagram for the system $CaMgSi_2O_6$–$CaAl_2Si_2O_8$ at 1 bar. (Reprinted by permission from Bowen, 1928, p. 26.)

residue of the other mineral melts, the melting point changing along *ED* or *EA* as the composition of the liquid changes.

Solid Solution

A mixture of albite ($NaAlSi_3O_8$) and anorthite melted and allowed to cool very slowly gives crystals of only a single kind, a "solid solution" of the two minerals called plagioclase (see Sec. 5-8). Details of the crystallization, somewhat more complicated than for systems with a eutectic, are shown in Fig. 17-2. Pure albite melts at 1118°C, pure anorthite at 1557°C, and mixtures have intermediate melting ranges. A melt with 40% albite, for example, cools to about 1470°C before crystals appear (point *X*). The composition of the first-formed crystals is not the composition of the melt nor the composition of pure anorthite, but somewhere between; the experimentally determined composition is shown by point *Y*. As crystals of composition *Y* are formed, the melt is impoverished in anorthite and enriched in albite, so that its melting point must steadily decrease (line *XK*). If the first formed crystals remain suspended in the melt, they are out of equilibrium as the temperature falls and must react with the melt to form solid solutions of higher albite content. Thus at any temperature along *XK* the remaining melt is in equilibrium with a solid rich in anorthite, and the compositions of both liquid and solid change progressively. Crystallization ends when the last remaining melt reaches the composition *K* and the crystals have composition *L*—identical, of course, with the composition of the original melt.

When plagioclase crystals are heated, the first liquid to form has a composition more soda-rich than the crystals. For example, if the crystals contain 50% albite, the first melt contains 87% albite and 13% anorthite (points *F* and *G*, respectively). As the temperature rises, both liquid and solid change their

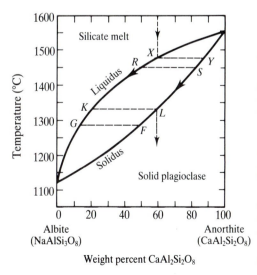

FIGURE 17-2
Temperature–composition diagram for the plagioclase feldspars at 1 bar. The upper curve, the *liquidus*, shows compositions of the liquid for temperatures at which the first solid crystals would appear as a melt is cooled; the lower curve, the *solidus*, shows compositions of the solid for temperatures at which melting would start as the solid is heated. No solid would exist at temperatures above the liquidus, and no liquid at temperatures below the solidus. Note that in Fig. 17-1 the liquidus is the pair of lines that meet at the eutectic, and the solidus consists of the two vertical lines representing pure diopside and pure anorthite. (After Bowen, 1928, p. 34.)

composition (along *FLS* and *GKR*, respectively), until the last crystals melt to give a liquid with 50% albite, the same composition as the original solid.

Fractional Crystallization

The preceding discussion has been based on the assumption that equilibrium is maintained during cooling. This means that crystals must remain in contact with the melt until crystallization is complete, a condition not necessarily fulfilled in geologic environments. What effect would lack of equilibrium have?

Consider first the albite–anorthite system under conditions in which crystals are removed as fast as they are formed, say by gravitative settling of crystals out of the liquid. In this case the enrichment of albite in the liquid is greater than before, since the anorthite-rich crystals cannot react with it. For an original composition of 40% albite, the last liquid to crystallize may have a composition far more albite-rich than point *K*; in fact, if removal of crystals as they form is complete and continuous, the last liquid may be practically pure albite. The net result of such failure of equilibrium would be a pile of plagioclase crystals of nonuniform composition, rich in anorthite toward the bottom and in albite toward the top. If removal of crystals is less rapid, they may have an opportunity for partial reaction with the liquid, giving rise to the zoned crystals which are so prominent in many natural plagioclases.

For mineral pairs that form eutectics, the early removal of crystals would have less influence on the course of crystallization. Cooling progresses to the eutectic temperature as before, and the last liquid to crystallize would have the eutectic composition. In the resulting solid aggregate, the mineral whose crystals form first would be more abundant toward the bottom; but the bulk composition near the top would show no enrichment (beyond the eutectic) toward the other mineral, and the composition of the individual minerals would of course show no change.

17-3   THEORETICAL CRYSTALLIZATION CURVES

Phase diagrams like Figs. 17-1 and 17-2 are drawn on a purely empirical basis. The technique is tedious but simple: various mixtures of the end members are heated until they form homogeneous melts, then as the melts cool the sequence of solid phases crystallizing out is recorded. In actual practice this seemingly straightforward procedure is often beset with difficulties, arising from trouble in identifying the solid phases and in preserving them down to low temperatures where they can be examined carefully, and from sluggish crystallization or crystallization of metastable phases rather than stable ones. But basically the operation is not complicated. Lines on the diagrams merely summarize the results of large numbers of such experiments. It is reasonable to ask if the procedure might be shortened, or the results systematized, by handling the problem theoretically.

At first glance a theoretical treatment seems easy. Take, for example, the right-hand side of the anorthite–diopside diagram (Fig. 17-1): the line is drawn through points showing compositions of a liquid in equilibrium with pure anorthite over a range of temperature. The curve represents the conditions of temperature, pressure, and composition where the chemical potential of $CaAl_2Si_2O_8$ in the melt and in the solid are equal ($\mu^{\text{anorthite}}_{CaAl_2Si_2O_8} = \mu^{\text{melt}}_{CaAl_2Si_2O_8}$). To express the equilibrium more formally,

$$CaAl_2Si_2O_8 \rightleftharpoons CaAl_2Si_2O_8 \qquad (17\text{-}1)$$

solid anorthite $\qquad$ anorthite in melt

for which the equilibrium constant is the quotient of the two activities. The activity of the pure solid, by convention, is taken as 1, so the expression for the constant is merely

$$K_a = a^{\text{melt}}_{CaAl_2Si_2O_8}. \qquad (17\text{-}2)$$

Variation of the constant with temperature is given by the van't Hoff equation [Eq. (8-61)]

$$\left(\frac{\partial \log K}{\partial T}\right)_P = \left(\frac{\partial \log a^{\text{melt}}_{CaAl_2Si_2O_8}}{\partial T}\right)_P = \frac{\Delta H}{RT^2}, \qquad (17\text{-}3)$$

which integrates to gives,

$$\log K = -\frac{\Delta H}{2.303R}\frac{1}{T} + C. \qquad (17\text{-}4)$$

The integration constant C can be evaluated by noting that $K = 1$ and $\log K = 0$ at the melting point of pure anorthite, T_m. Solving for C in this manner gives

$$\log K = \frac{\Delta H}{2.303R}\left(\frac{1}{T_m} - \frac{1}{T}\right). \qquad (17\text{-}5)$$

This equation has the form

$$\log K = \log a^{\text{melt}}_{CaAl_2Si_2O_8} = A - B\frac{1}{T}, \qquad (17\text{-}6)$$

where A and B are constants if ΔH is constant. Thus $\log a^{\text{melt}}_{CaAl_2Si_2O_8}$ plotted against $^1/_T$ should give a straight line, and a plot of $a^{\text{melt}}_{CaAl_2Si_2O_8}$ against T should be a curve. To see how closely the curve would fit the experimental curve, a value of ΔH can be assumed (or the heat of fusion of anorthite may be obtained from tables) which will make the curve fit the upper part of AE in Fig. 17-1. Then if $a^{\text{melt}}_{CaAl_2Si_2O_8}$ is equated to the concentration of anorthite in the melt ($a^{\text{melt}}_{CaAl_2Si_2O_8} = X^{\text{melt}}_{CaAl_2Si_2O_8}$), a curve is obtained shown by the light dashed line on the drawing. The fit of the two curves is reasonably good only for mixtures with more than 80% anorthite.

Why is the agreement so poor for all but anorthite-rich liquids? Note that three assumptions are hidden in the last paragraph: (1) ΔH, the heat of melting of anorthite, is assumed constant over a temperature range of several hundred degrees; (2) ΔH is assumed not to be affected by the presence of diopside in the melt; (3) the activity of anorthite in the melt is assumed to be measured by its concentration. The first assumption is probably not a serious source of error. The other two are equivalent to assuming that diopside–anorthite mixtures behave as ideal solutions, in other words that each component is completely independent of the other and is merely diluted by it. The assumption of ideal behavior is generally far from true for silicate mixtures: ΔH is modified by heats of solution, and activity deviates markedly from concentration except in mixtures near the pure components. Enough is known about some mixtures so that corrections can be applied to ΔH and a, and the fit of theoretical and experimental curves can be greatly improved.

The slope of melt–mineral equilibrium curves in diagrams like Fig. 17-1 is largely determined by the enthalpy and entropy of melting (note that $\Delta H = T\Delta S$ because ΔG equals zero along the melting curve). The curve for minerals with large enthalpies and entropies of melting will have a shallow slope on a temperature–composition diagram like Fig. 17-1; thus the melt concentration of the mineral component will decrease rapidly with decreasing temperature. On the other hand, if the enthalpy and entropy of melting are small the concentration of the mineral component will decrease only slightly in the melt with decreasing temperature. This is why silica, which has a small enthalpy and entropy of melting relative to other igneous minerals, is concentrated in the last liquids to crystallize, a prediction borne out by experimental and geologic observations presented below.

One other important simple generalization can be drawn from Eqs. (17-5) and (17-6). Remember that the temperature T represents equilibrium between solid anorthite and anorthite in the melt. As T decreases, the fraction $B(1/T)$ becomes larger and $A - B(1/T)$ becomes smaller. Thus the activity of anorthite in the melt decreases as T decreases, meaning that the left-hand curve in Fig 17-1 slopes downward rather than upward. Or in general, the melting temperature of a substance is lowered by mixing it with another.

Relations similar to those summarized above for temperature and composition can be derived for pressure and composition from Eq. (8-71),

$$\left(\frac{\partial \log K}{\partial P}\right)_T = \left(\frac{\partial \log a^{\text{melt}}_{\text{CaAl}_2\text{Si}_2\text{O}_8}}{\partial P}\right)_T = -\frac{\Delta V}{2.303 RT}, \tag{17-7}$$

which shows that the pressure dependence of the melt equilibrium is determined by the volume of reaction. The effect of pressure on melt crystallization is discussed in a later section.

17-4 COMPLEX BINARY SYSTEMS

Double Eutectic

In the system silica–nepheline, solids of three different compositions may form: SiO_2 (as tridymite in laboratory experiments with dry melts, as quartz in nature); nepheline, $NaAlSiO_4$ (nepheline in nature has a somewhat more complex formula); and albite, $NaAlSi_3O_8$. In phase-rule language, however, only two components are present, since one of these substances is derivable from the other two:

$$\underset{\text{tridymite}}{2SiO_2} + \underset{\text{nepheline}}{NaAlSiO_4} \rightleftharpoons \underset{\text{albite}}{NaAlSi_3O_8}. \tag{17-8}$$

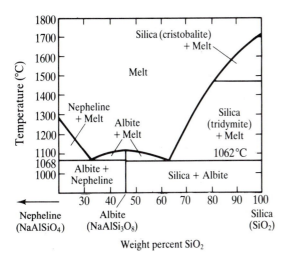

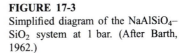

FIGURE 17-3
Simplified diagram of the NaAlSiO₄–
SiO₂ system at 1 bar. (After Barth,
1962.)

According to the phase rule the three solids cannot exist together, because there would then be a total of four phases (3 solids + 1 liquid), leading to a negative value for f (f = $c - p + 1 = 2 - 4 + 1$). Either of the pairs silica–albite or nepheline–albite may exist together, and both pairs form simple eutectics, as shown in Fig. 17-3. Interpretation of this diagram is no different from that for Fig. 17-1: any melt on cooling approaches one of the eutectic mixtures, the choice depending on whether the original composition is more rich or less rich in silica than pure albite. Removal of early-formed crystals cannot affect the course of crystallization and can at best lead only to a solid having a eutectic mixture in one part and a concentration of one of the minerals in another. The diagram illustrates, as of course it must, the impossibility of the three possible solids existing together, or of silica and nepheline crystallizing together from the same melt. (We ignore here some minor complications resulting from solid solutions at the nepheline end of the diagram.)

Eutectic with Incongruent Melting

The silica–leucite system is another example of a two-component system with three possible solids (silica, K-feldspar, leucite) but the relation between the solids is different. Potassium feldspar is unstable at temperatures above 1150°C, decomposing into leucite and a silica-rich liquid:

$$KAlSi_3O_8 \rightleftharpoons KAlSi_2O_6 + SiO_2 .\qquad(17\text{-}9)$$

K-feldspar   leucite   in liquid

The K-feldspar is said to *melt incongruently*. Liquids with compositions near that of K-feldspar, then, cannot yield this mineral as a direct product of crystallization from a melt; leucite must crystallize first and then change to K-feldspar on cooling by reaction with the liquid. This behavior is diagramed in Fig. 17-4.

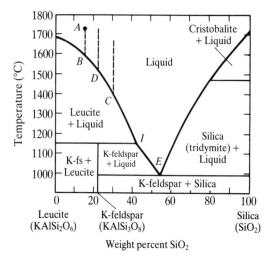

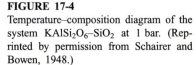

FIGURE 17-4

Temperature–composition diagram of the system $KAlSi_2O_6$–SiO_2 at 1 bar. (Reprinted by permission from Schairer and Bowen, 1948.)

The right-hand part of the diagram (compositions 0 to 58% $KAlSi_2O_6$) is obviously a simple eutectic, either silica or K-feldspar precipitating first and the liquid thereafter changing in composition toward the eutectic point. Melts with more than 58% $KAlSi_2O_6$, however, have a more complicated cooling history. Consider first a liquid with only 15% SiO_2 (point A). As it cools, leucite begins to crystallize at about 1590°C (point B). Removal of leucite means that the liquid becomes richer in silica, its composition changing along the line BI. When the temperature falls to 1150°C, leucite is no longer stable in the presence of the melt; it reacts with the melt to form K-feldspar, and more K-feldspar precipitates directly, so that the composition stays at point I until the liquid is used up. The final solid, then, consists of K-feldspar plus the remaining leucite which did not react. If the original melt had 30% SiO_2 instead of 15% (point C) the cooling history would be similar except that now all the leucite would react at I and some liquid would be left over. As cooling continued, K-feldspar would crystallize from this liquid, and the composition of the liquid would change along the line IE until the eutectic composition was reached. The final solid would consist of K-feldspar plus silica, all trace of the early-formed leucite having disappeared. If, finally, the original melt has precisely the composition of K-feldspar (point D), leucite again crystallizes first; at point I it all reacts with the liquid, and this time no liquid is left over. The resulting solid would be, of course, pure feldspar. Thus the net product of crystallization in this system is similar to that for the silica–nepheline system: silica and feldspar if the original composition is more silica-rich than feldspar, leucite and feldspar if the original composition is less silica-rich.

The important difference between the two kinds of crystallization lies in the greater possibility of separation of minerals in the leucite system *when equilibrium is not maintained*. The discussion above assumes equilibrium; early-formed leucite is available to react with melt, and cooling is so slow that all possible reactions can go to completion. But now suppose that, in a melt containing initially 15% SiO_2, leucite crystals are removed as fast as they form. When the temperature drops to 1150°C (point *I*), leucite is not available; K-feldspar crystallizes from the melt, and the composition changes toward the eutectic. Ultimately point *E* is reached, and the last liquid to crystallize is a eutectic mixture. The resulting solid would have leucite concentrated in one part, and a silica–feldspar mixture in another. In other words, two "rocks" of very different composition may be formed from the same melt. Silica may crystallize from a liquid initially rich in leucite, and leucite may be preserved from initial crystallization in a melt more silica-rich than K-feldspar—possibilities that have no counterpart in the silica–nepheline system.

A similar, and petrologically more important, system is the combination silica–forsterite:

$$Mg_2SiO_4 + SiO_2 \rightleftharpoons 2MgSiO_3. \qquad (17\text{-}10)$$

forsterite   cristobalite   enstatite

(Recall that forsterite is an end member of the olivine series and that enstatite is a pyroxene. Hence this system is a simple analog of reactions that are important in the crystallization of mafic igneous rocks like gabbro and basalt.) The artificial system (Fig. 17-5) differs from natural rocks in that clinoenstatite rather than enstatite is the intermediate substance formed and silica appears as cristobalite (because of the higher temperatures required in a dry melt), but the phase relations are similar to those observed in nature. The clinoenstatite melts incongruently at 1557°C, so that forsterite is the first substance to crystallize from melts containing initially less than 51 mol percent SiO_2. If equilibrium is maintained, the early-formed forsterite would react with the liquid at 1557°C to form clinoenstatite; whether or not the forsterite is all changed to clinoenstatite at this point depends on the original composition of the melt. But if equilibrium is not maintained, early-crystallized forsterite may settle out

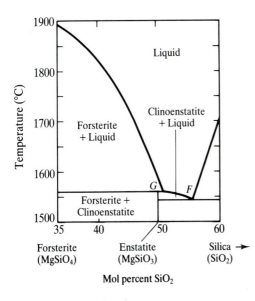

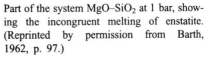

FIGURE 17-5

Part of the system MgO–SiO_2 at 1 bar, showing the incongruent melting of enstatite. (Reprinted by permission from Barth, 1962, p. 97.)

and the remaining liquid may reach the clinoenstatite-silica eutectic, even if the initial composition is low in silica. The analogy to crystallization in mafic rocks is obvious: early-formed olivine may fail to react completely with the magma to form pyroxene, so that olivine crystals may be preserved in a rock apparently saturated with silica, and quartz may form from a magma initially deficient in silica. Thus thick basalt flows and dolerite sills may have accumulations of olivine near their base and interstitial quartz near their top; and olivine grains in a basalt may be surrounded by tiny pyroxene crystals ("reaction rims"), as if they were caught in the process of reacting with the liquid.

The phenomenon of incongruent melting is obviously analogous to incongruent dissolution discussed in Chapter 13. One refers to the breakup of a complex compound by partial melting, the other to breakup by partial dissolving in water.

Solid Solution with a Minimum Melting Point

Albite and K-feldspar form a continuous series of solid solutions, just as do albite and anorthite, but mixtures of intermediate composition have lower melting points than either of the end members (Fig. 17-6).The diagram for this system, unlike the preceding ones, is drawn for a water-vapor pressure of 2000 kg/cm^2 rather than for a dry melt, because in the dry-melt diagram the minimum melting-point relation is obscured by complications due to formation of leucite in K-feldspar-rich mixtures. The lower curve in the diagram (*solvus*) shows "unmixing" of the two feldspars in the solid state (Fig. 5-7): the solid solution is stable only at high temperatures, breaking up on cooling into separate crystals which approach the compositions of the end members more and more closely as the temperature falls. Unmixing may lead to intimate, very fine-grained intergrowths like those that make up anorthoclase, or to the coarser intergrowths called perthite and antiperthite, or possibly even to large separate crystals. An alkali feldspar that has cooled very rapidly may remain in a metastable solid solution; sanidine crystals in lavas are a good example.

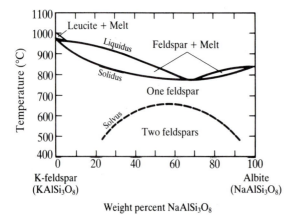

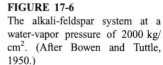

FIGURE 17-6
The alkali-feldspar system at a water-vapor pressure of 2000 kg/cm². (After Bowen and Tuttle, 1950.)

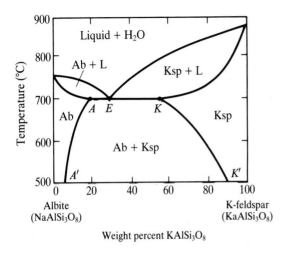

FIGURE 17-7
Water-saturated phase diagram for the system $NaAlSi_3O_8-KAlSi_3O_8-H_2O$ at a pressure of 5 kbar. Na-rich feldspar is indicated by the symbol Ab, and K-rich feldspar by the symbol Ksp. (After Morse, 1970.)

Combinations of Eutectics and Solid Solutions

The simplest combination of a eutectic and solid-solution diagram is shown in Fig. 17-7. This diagram is similar to Fig. 17-6 except that water pressure has been increased. The excess water lowers the temperature of the liquidus and solidus curves and they intersect the solvus. As in Fig. 17-6, each substance lowers the melting point of the other, and the solids that crystallize out are not the pure end members. Crystallization of a melt of any composition would end at the eutectic point E, and the last crystals to form from this liquid would be a mixture of composition A and composition K. The lines AA' and KK' are portions of the solvus curve (Fig. 5-7) and represent changes in the compositions of the solids at temperatures below the melting point.

In summary, the freezing of a binary, or two-component system may be represented by a temperature–composition diagram on which combinations with one degree of freedom are represented by lines, and combinations with zero degrees of freedom by points. These diagrams show either eutectic mixtures, or solid solutions, or combinations of the two, and may be complicated by incongruent melting and by solid solutions with minimum melting points. Although binary systems are far simpler than natural lavas, they help to explain some of the petrographic relations found in igneous rocks.

17-5 TERNARY SYSTEMS

A closer approach to natural silicate liquids can be obtained with artificial melts having three components. Here we face a greater difficulty in representation, for ternary systems have a maximum of three degrees of freedom (temperature and two ratios expressing composition) and hence require three-dimensional diagrams for complete display. For many purposes a two-dimensional plot showing only compositions is sufficient; temperature may be added in the form of contour lines if needed. The representation most commonly used is a triangular diagram like Fig. 17-8(*b*) in which each corner indicates 100% of the named component and the opposite side 0%.

Figure 17-8 is a hypothetical diagram for the simplest kind of ternary system, one in which the three components crystallize from a melt without forming solid solutions. Each side of the triangular prism in Fig. 17-8(*a*) shows a binary eutectic; in Fig. 17-8(*b*) we look down vertically along the temperature axis so that temperature is represented by contour lines (isotherms) on the liquidus surfaces. From a melt containing all three components, one of the three, in general, begins to crystallize first. As cooling proceeds, the crystallizing component is joined by a second, and in the last stage all three crystallize together. This sequence is illustrated by the mixture *X* (about 65% *B*, 20% *C*, 15% *A*). First crystals to form are *B*; as these are removed, the composition of the liquid changes directly

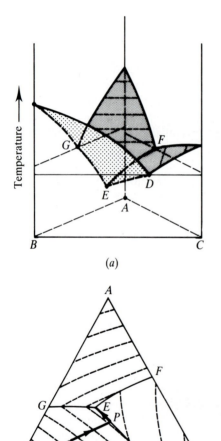

(*a*)

(*b*)

FIGURE 17-8
Hypothetical diagram for a system *ABC* with a ternary eutectic (*E*) and three binary eutectics (*F*, *G,D*). *GE*, *FE*, and *DE* are cotectic lines. Short dashed lines on the liquidus surfaces are temperature contours. Diagram (*a*) is a three-dimensional drawing of the phase diagram and diagram (*b*) is a two-dimensional projection looking down the temperature axis.

away from B (line XP); at P, component C starts to crystallize also, and the liquid composition changes along PE; when the composition reaches E, the third component starts to crystallize, and the composition remains at this point until solidification is complete. Lines like DPE, FE, and GE, representing simultaneous crystallization of two components, are called *cotectic lines*, and point E is a *ternary eutectic*. Note that the course of crystallization of mixture X is down the temperature gradient; the ternary eutectic is the low point of the "temperature surface," and the cotectic lines are the bottoms of valleys. Crystallization in a system of this sort, like crystallization in a binary eutectic, is little affected by removal of early-formed crystals.

A ternary system that forms a series of solid solutions among all its components would be more difficult to portray on a triangular diagram. There could be no cotectic lines outlining fields of crystallization, so that the triangle would be featureless except for contour lines suggesting the shape of liquidus and solidus surfaces (see Figs. 17-2 and 17-7).

We look now at two specific ternary systems of particular interest in chemical petrology.

Forsterite–Silica–Anorthite

This system is a ternary eutectic modified by the incongruent melting of clinoenstatite and by the formation of spinel ($MgAl_2O_4$) from mixtures rich in forsterite and anorthite. In Fig. 17-9 the ternary eutectic is shown as point E. The silica field above the ternary eutectic is modified only by the tridymite–cristobalite transition; as in preceding diagrams, the high-temperature forms of (Fig. 4-1) silica appear rather than quartz because dry melts crystallize at higher temperatures than water-rich natural magmas. Mixtures with compositions near the eutectic crystallize as described in the last section. Thus in mixture A the first crystals to form would be anorthite, the next would be clinoenstatite, and these two would be joined by tridymite when the composition of the remaining liquid reaches E.

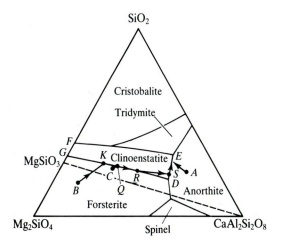

FIGURE 17-9
Phase diagram for the system SiO_2–Mg_2SiO_4–$CaAl_2Si_2O_8$ at 1 bar. (After Bowen, 1928.) Points F (binary eutectic) and G correspond to points F and G in Fig. 17-5. Arrows show crystallization paths for starting fluid compositions represented by points A, B, and C.

The spinel field at the bottom of the diagram is, strictly speaking, part of a four-component system, and cannot be accurately shown on this triangle. But no ordinary rocks have compositions in this part of the diagram, so for present purposes the complication can be disregarded.

The incongruent melting of clinoenstatite is shown by the left-hand edge of the triangle. This edge, in effect, is Fig. 17-5 as seen looking down from above; points *G* and *F* are the same on the two diagrams. The incongruent melting persists even when large amounts of anorthite are present, as is shown by the fact that line *GD* remains on the silica side of the dashed line showing the constant ratio of Mg_2SiO_4 to SiO_2 components in clinoenstatite. Details of crystallization may be illustrated by two examples:

1. From a mixture with composition *B*, forsterite would crystallize first, and the composition of the liquid moves directly away from the forsterite corner. At *K* on *GD*, forsterite becomes unstable and reacts with the liquid to form clinoenstatite, while more clinoenstatite precipitates directly; these processes would make the composition of the liquid change along *KD*. At *D* anorthite starts to crystallize, and the composition of the liquid remains at this point until the supply of liquid is exhausted. The resulting solid would consist of anorthite, clinoenstatite, and the remainder of the forsterite that had failed to react.
2. The mixture *C* would also give forsterite as the first-formed crystals. Here, however, the amount of forsterite would be small when the liquid composition reaches *Q* on *GD*, and reaction with the liquid would quickly convert all of it to clinoenstatite. From this point (*R*) on, clinoenstatite crystallizes directly and the liquid composition changes along *RS*, directly away from the $MgSiO_3$ point. (The point *R* is located by projecting a line from $MgSiO_3$ through *C* to its intersection with *GD*.) At *S*, clinoenstatite is joined by anorthite, and final crystallization together with silica takes place at the eutectic.

This outline assumes that equilibrium is maintained during crystallization. If, on the contrary, early-formed forsterite is removed as fast as it forms, the system may crystallize to give one solid rich in forsterite and another containing silica; for if the reaction between forsterite and liquid is prevented, there is nothing to keep the liquid from ultimately reaching the ternary eutectic, no matter how undersaturated with silica it may have been initially. As with the two-component system shown in Fig. 17-5, this gives a convenient explanation for the reaction rims around olivine grains and for the gravitational segregation of olivine from basaltic magmas.

Diopside–Albite–Anorthite

The deceptively simple diagram for this system (Fig. 17-10) shows only a single cotectic line connecting the binary eutectics diopside–anorthite (*E*) and diopside–albite (*X*). The bottom line of the triangle represents the solid-solution series albite–anorthite; it is a top view, so to speak, of Fig. 17-2. Melts with compositions above the cotectic line give diopside as the first crystals; melts with compositions below the line give plagioclase. If diopside crystallizes first, the melt changes composition

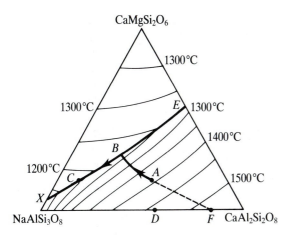

FIGURE 17-10
Phase diagram for the system $CaMgSi_2O_8$–$NaAlSi_3O_8$–$CaAl_2Si_2O_8$ at 1 bar. The light lines are temperature contours. (After Bowen, 1928.)

directly away from the diopside corner. If plagioclase appears first, the cooling history is more complicated.

Consider, for example, the cooling of a mixture represented by point A. The feldspar that appears first will not have a composition 45% albite and 55% anorthite, as one might expect (point D), but a composition richer in anorthite (point F). Formation of these first crystals containing about 80% anorthite causes the liquid to change in composition toward albite; but as cooling progresses and as the crystals take up more and more albite, the direction of change is altered. In other words the composition of the liquid follows a curved path AB, intersecting the cotectic at B. Here diopside begins to crystallize also, and the composition of the melt changes along the cotectic toward X. The last liquid (C) will be used up when the solid plagioclase attains the composition 55% anorthite (D). The precise path of crystallization cannot readily be predicted from this diagram, as it could be from preceding diagrams, but the general outline is well defined.

If equilibrium is not maintained in this system, if in particular early-formed plagioclase is removed, the ultimate composition of the melt is not restricted by the redissolving of the anorthite-rich crystals. The resulting solid might then contain anorthite-rich feldspar in one part and albite-rich feldspar in another part. Early separation of diopside from diopside-rich melts, on the other hand, would not change the character of the crystallizing material.

Compositions in this system are close enough to ordinary basalt to make possible predictions about crystallization of saturated mafic magmas. Note, for example, that no definite "order of crystallization" is to be expected. The first mineral to crystallize in a magma may be either pyroxene or plagioclase, depending on the initial composition, and during most of the cooling the two principal minerals will crystallize simultaneously. The composition of the resulting rock will be uniform if equilibrium is approximately maintained, but segregation of early-formed calcic plagioclase gives the possibility of formation of different rocks from a single original magma. Minor and temporary deviations from equilibrium during cooling may lead to formation of conspicuously zoned plagioclase crystals.

17-6 MAFIC AND INTERMEDIATE MAGMAS

We turn now from artificial systems to the problems of real magmas, beginning with mafic magmas because the application of chemistry to them is somewhat simpler than for more felsic varieties. At first glance the chemical resemblance between the laboratory systems we have considered and actual basaltic lava does not seem very impressive. The mixture shown in Fig. 17-10, for example, has only five oxides (SiO_2, Al_2O_3, MgO, CaO, Na_2O), whereas the composition of basalt includes another five important components (FeO, Fe_2O_3, K_2O, H_2O, TiO_2) plus a host of minor ones. Fortunately, addition of these components to our artificial systems in small amounts would not cause great changes in the conclusions we have drawn—except for H_2O. This oxide can have a marked effect on the crystallization process, particularly by lowering the temperature. The experiments we have been looking at involve dry melts at temperatures ranging up to 1700 or 1800°C, whereas basaltic lavas in nature, because of their content of water and other volatiles, crystallize in the range 850 to 1100°C. Under these conditions some of the solids exhibit different crystal forms: the magnesian pyroxene appears as ordinary enstatite (or hypersthene) rather than clinoenstatite, and silica appears as quartz instead of tridymite or cristobalite. Then water enters the structure of some of the crystallizing solids, giving materials like hornblende or biotite in place of or in addition to pyroxene. Despite these details, the conclusions we have drawn from the experimental work remain useful guides to the chemistry of magmatic crystallization in nature.

Most rocks whose compositions are relatively high in iron and magnesium and low in silica consist dominantly of plagioclase and one or more of the ferromagnesian minerals (olivine, pyroxene, hornblende, sometimes biotite). These are the common rocks gabbro and diorite, and their effusive equivalents basalt and andesite (analyses IV, V, VI in Table 17-1). The plagioclase often shows some variation in composition (zoning), suggesting that it was in process of reacting with the melt at the time consolidation took place. The Mg–Fe minerals too may give evidence of reactions that were incomplete when freezing occurred—not only zoning, but reaction rims of pyroxene around olivine, or hornblende around pyroxene. The reactions involved are chiefly within the systems $CaAl_2Si_2O_8$–$NaAlSi_3O_8$ and CaO–MgO–FeO–SiO_2, which evidently are in large measure independent of one another. The same independence is suggested by the laboratory melts (Figs. 17-2, 17-9, and 17-10): plagioclase and pyroxene may crystallize together as a eutectic, or along a cotectic line or surface, but do not form solid solutions or intermediate compounds.

Thus the order of crystallization of mafic magmas may be summarized by a simple diagram:

Decreasing Temperature $\longrightarrow$

olivine $\longrightarrow$ *pyroxene* $\longrightarrow$ *hornblende* $\longrightarrow$ *biotite*

$\longrightarrow$ (*quartz*, *K-feldspar*)

Ca-plagioclase $\longrightarrow$ *Na-plagioclase*

TABLE 17-1
Average compositions of representative igneous rocks

	I	II	III	IV	V	VI	VII	VIII
					Basalt		Nephe-	
	Alkali	Grano-	Quartz	An-	(tho-	Olivine	line	Peri-
	granite	diorite	diorite	desite	leiite)	basalt	syenite	dotite
SiO_2	73.86	66.68	66.15	58.17	50.83	47.90	55.38	43.54
TiO_2	0.20	0.57	0.62	0.80	2.03	1.65	0.66	0.81
Al_2O_3	13.75	15.66	15.56	17.26	14.07	11.84	21.30	3.99
Fe_2O_3	0.78	1.33	1.36	3.07	2.88	2.32	2.42	2.51
FeO	1.13	2.59	3.42	4.17	9.06	9.80	2.00	9.84
MnO	0.05	0.07	0.08		0.18	0.15	0.19	0.21
MgO	0.26	1.57	1.94	3.23	6.34	14.07	0.57	34.02
CaO	0.72	3.56	4.65	6.93	10.42	9.29	1.98	3.46
Na_2O	3.51	3.84	3.90	3.21	2.23	1.66	8.84	0.56
K_2O	5.13	3.07	1.42	1.61	0.82	0.54	5.34	0.25
H_2O	0.47	0.65	0.69	1.24	0.91	0.59	0.96	0.76
P_2O_5	0.14	0.21	0.21	0.20	0.23	0.19	0.19	0.05

Sources:
Chayes, F.: "The chemical composition of Cenozoic andesites,". *Oregon Dept. Geology and Mineral Resources Bull.*, vol. 65, pp. 1–11, 1969.
Nockolds, S. R.: "Average chemical composition of igneous rocks," *Geol. Soc. America Bull.*, vol. 65, pp. 1007–1032, 1954.

as pointed out by Bowen long ago (1928). The mafic minerals in the upper row, in Bowen's terminology, form a *discontinuous reaction series*, meaning that each substance, as the magma cools, reacts with the melt to form the next mineral in line. The first step in this sequence, from olivine to pyroxene, is easily studied experimentally (Fig. 17-5). In nature the temperatures would be lower than shown in this diagram (because water is present), and the two minerals would contain iron as well as magnesium, but the relationship is the same. The later steps pyroxene–hornblende and hornblende–biotite are more complicated, because these reactions involve water and thus depend on partial pressures of gases as well as silicate compositions. In contrast to this discontinuous sequence, the plagioclase feldspars of the lower line form a *continuous reaction series*, in which crystals react with the liquid continuously until freezing is complete (Fig. 17-2).

Bowen's diagram shows crystallization in the two series proceeding simultaneously as the temperature falls, so that lime-rich plagioclase appears together with olivine or pyroxene, and soda-rich plagioclase together with hornblende or biotite. Crystallization may start in one series or the other, but through most of the freezing process two kinds of crystals are being formed at the same time. The point in the series at which crystallization starts depends on the original composition; the point where it ceases depends on the composition and also on whether equilibrium is maintained as the magma cools.

This representation of magma freezing is too simplified for detailed application to most actual rocks. In many basalts textural relations indicate that calcium-rich monoclinic pyroxene (diopsidic augite) crystallized simultaneously with olivine, rather than by reaction of olivine with the liquid, and that the pyroxene produced at a later stage by conversion of olivine is a different, calcium-poor (orthorhombic) variety. In other words a series of calcium-rich pyroxenes should be added as a third branch to the diagram, a branch that probably converges with the olivine–pyroxene branch at later stages of cooling. For lavas rich in potassium still another branch could be added representing the crystallization of potash feldspars simultaneously with some of the plagioclase.

The progression of minerals in both continuous and discontinuous series shows a general decrease in the ratio of O to Si (16:4 for olivine, 12:4 for pyroxene, 11:4 for amphibole, 10:4 for mica) and a general increase in complexity of the silicate structures (isolated SiO_4 tetrahedra in olivine, chain structures in pyroxene and amphibole, sheet structures in micas). These changes reflect modifications in the structure of the silicate melt as temperature falls: at high temperatures the melt contains many free SiO_4 groups, but these polymerize into more and more complex aggregates as cooling proceeds. One may picture the freezing of a magma as a progressive growth and linking together of silicon–oxygen structures, first in the liquid and then in the minerals that crystallize from it.

The quartz and K-feldspar in parenthesis at the right-hand side of the diagram above are not related as are the others, in that these minerals do not necessarily form by reaction of previously crystallized material with the melt. Quartz and alkali feldspars are rather the minerals that form from the last remaining melt, the "residual liquid" of the mafic magma. How much of this sort of liquid will be present depends on the initial composition and on the effectiveness with which early-formed crystals are removed from the system. Early separation of olivine crystals is particularly effective in concentrating silica in the remaining melt. Much experimental work has shown that almost any melt containing the alkali-metal oxides along with other common oxides of igneous rocks will show enrichment of soda, potash, alumina, and silica in the last liquid to crystallize. Bowen (1928) called a final liquid containing only these four substances "petrogeny's residua system." Its possible compositions can be represented by a triangle whose corners are SiO_2, $KAlSiO_4$, and $NaAlSiO_4$ (Fig. 17-11). This is, so to speak, the ultimate goal of the crystallization process, the liquid that will tend to form even from magmas of quite different initial compositions and histories, provided that removal of crystals from contact with the melt is reasonably effective during consolidation.

Bowen's generalization embodied in the reaction series not only cleared up basic concepts about the freezing of magmas but provided explanations for many petrographic details. The zoning of plagioclase, for example, and reaction rims of pyroxene around olivine, represent failure to maintain complete equilibrium between crystals and melt during cooling. The differentiation sometimes observed in thick sills and flows of mafic rock—the development of an olivine-rich layer just above the basal chilled border, and of interstitial granophyre (a mixture of quartz and alkali feldspar) near the top of the body—is convincingly explained by the settling out of heavy, early-formed olivine crystals, with a resulting enrichment of the remaining magma in silica. The behavior of inclusions of foreign materials in magmas, particularly the partial melting, fragmentation, and disappearance of felsic

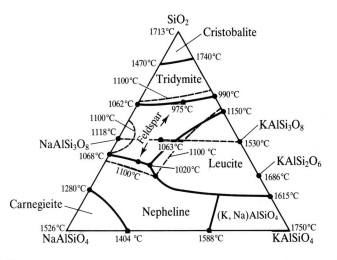

FIGURE 17-11

Approximate phase diagram for the system SiO₂–NaAlSiO₄– KAlSiO₄ at 1 atmosphere pressure with dry melts. The dashed line connecting NaAlSi₃O₈ and KAlSi₃O₈ is not a phase boundary, but is drawn through points having the composition of feldspar. The point of lowest temperature on this line (the minimum melting point of a solid-solution series) is 1063°C. This is a shallow saddle on the temperature surface; from here "valleys" slope down to a low point on the feldspar–nepheline eutectic, and to the temperature minimum on the silica–feldspar cotectic at about 975°C. The general area of lowest temperatures on the diagram is outlined by the 1100°C contour (light dashed lines). Possible directions of crystallization within this area are shown by the arrows, for silica-rich mixtures (upper arrow) and silica-deficient mixtures (lower arrow). (After Schairer, 1950.)

xenoliths in mafic rocks and the changes in composition of mafic xenoliths in granitic bodies, can be accounted for by the partial adjustment of material in the xenoliths to the equilibrium conditions of their new environment. In all these ways and many more, the reaction sequences worked out experimentally by Bowen and his followers have provided explanations for common field and laboratory observations of igneous rocks.

Magmatic Differentiation

A persistent question about the geochemistry of igneous processes is the extent to which a magma can change its composition before or during crystallization. For each separate variety of igneous rock, must we assume a different original magma? Or can we suppose that there are only a few kinds of "original" magma, and that the multiplicity of igneous rocks results from changes in their compositions? Field observations seem to show clearly that some magmas do undergo changes in composition in the course of their history, for it is common to find within a single igneous mass different varieties of rock grading imperceptibly into one another. Furthermore, in successive eruptions from a volcano the lava compositions often show gradual change during the volcano's lifetime, and it is certainly more reasonable to think that the material in a single large magma reservoir beneath is changing than to suppose that the volcano taps a new reservoir with a different kind of magma each month or each year. On the other hand, to suppose that such different and abundant rocks as granite and basalt are produced by changes in a single original magma puts a strain on the imagination. If magmas do alter their composition as they crystallize, how far can the alteration go, and on what scale?

Bowen's reaction series, as noted above, can readily account for some major compositional changes. When a mafic magma solidifies, the minerals in each series undergo progressive changes; if crystals are separated from the magma as they form, the remaining liquid must also change. Rocks resulting from freezing of the last residue of the magma may have compositions quite different from rocks made up of the early crystals that have separated out. This general process of change in the composition of a magma is described as *differentiation by fractional crystallization*. An extreme example is the one cited above—a mafic sill in which early-formed crystals settle into an olivine-rich layer near the base, and a layer containing quartz appears near the top.

The magma in such a sill is said to have differentiated into a part *oversaturated with silica* and a part that is *undersaturated*, the terms referring to the amount of silica in a rock or mineral relative to the amount that ideally could combine with the metallic elements present. Saturated minerals are those that can crystallize from a magma together with quartz, in other words those that contain all the silica their metallic constituents can accommodate; examples are feldspars, pyroxenes, amphiboles, and micas. Undersaturated minerals are those that cannot crystallize at equilibrium with quartz, for example olivine, nepheline, leucite, and melilite. The same terms may be used for magmas (or for rocks): a saturated magma is one from which only saturated minerals can crystallize at equilibrium, an undersaturated magma would give at least some undersaturated minerals, and an oversaturated magma would give quartz together with saturated minerals. Thus differentiation by crystal settling in a mafic sill can produce an undersaturated layer near the bottom and an oversaturated layer near the top.

It is tempting to speculate about the possible large-scale application of this sort of differentiation to problems of igneous geochemistry. Could granitic magma form as a residual liquid resulting from crystal settling in molten basalt—not as thin layers in a sill, but in amounts large enough to form separate intrusive bodies? Such an imagined process could seemingly explain much of the observed diversity of igneous rocks. The familiar sequence of common intrusive rocks often found in a batholith (gabbro, diorite, granodiorite, granite), for example, could represent bodies of magma formed at different stages of the differentiation process, each one having more Si, Na, and K and less Mg, Fe, and Ca than the preceding. Similarly, the succession of different kinds of lava emitted during the lifetime of a volcano is conveniently attributed to slow differentiation taking place in a magma chamber beneath. Certainly some of the variability in igneous rocks can be explained in this manner, but the scale of its operation is unclear. For very large bodies of granitic rocks other processes are doubtlessly involved, in addition to or in place of simple differentiation.

17-7 FELSIC MAGMAS

One possible origin of granitic magma, as we have seen, is the generation of silica-rich fluids by fractional crystallization of mafic magmas. Both field relations of granitic bodies and experimental work on silicate melts suggest, however, that rocks of granitic composition may form in a number of other ways. Granite in the wide sense (to include all varieties from quartz diorite to alkali granite) is an extremely variable rock, and it seems likely that different kinds can be generated in a variety of geologic environments.

Some granitic magma may result from metamorphism, as noted in Chapter 16. Field evidence for such an origin lies in outcrops showing gradational and intertonguing contacts of granite with adjacent metamorphic rocks, showing concordance of structure between layering in the metamorphic rocks and banding in the granite, and showing an abundance of xenoliths within the granite in various stages of breakup and assimilation. Observations like this seem clear evidence that the granite is a product of extreme metamorphism, of the making over of material that was originally sedimentary and volcanic rock. Chemically this is entirely reasonable, since much sedimentary rock, or combinations of sedimentary and volcanic material, can have compositions similar to those of granitic rocks. Genesis of granite by such a process of ultra-high-grade metamorphism is often called *granitization.*

The granitizing process would at first produce small amounts of a silicate liquid with a composition approximating that of "petrogeny's residua system," and if the process continues the liquid would increase in amount and change in composition toward that of a more usual granite. The entire mass need not become liquid; at least locally, rocks resembling granite may form by the making over of metamorphic rock largely in the solid state. But most granite magma that forms in large amounts by the progressive melting or anatexis (Sec. 16-7) of pre-existing rock becomes fluid enough to move as a viscous liquid, or perhaps as a liquid plus residual solids, for considerable distances from its place of origin.

This silica-rich fluid behaves in many ways differently from the mafic fluids we discussed in the last section. In the laboratory, melts of this kind are more difficult to work with than basaltic melts, because dry silica-rich liquids are extremely viscous and freeze to a glass rather than a crystalline solid even when cooled very slowly. In the laboratory study of basaltic melts one normally adds a little water to ensure closer simulation of nature, but the addition is not required for an understanding of the gross features of the crystallization process. By contrast, for granitic melts the addition of water is indispensable to speed up crystallization enough for study.

The composition of many granitic rocks, especially those low in mafic minerals, can be roughly described in terms of three components: SiO_2, $KAlSi_3O_8$, $NaAlSi_3O_8$. We can guess at the general form of a phase diagram for this system by combining the diagrams in Figs. 17-4, 17-6, and 17-7: on a triangular plot (Fig. 17-12) the side quartz–albite should have a simple binary eutectic; the side quartz–K-feldspar also has a eutectic, complicated at low water-vapor pressures by the

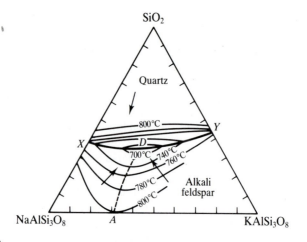

FIGURE 17-12

Phase diagram for the system SiO_2–$NaAlSi_3O_8$– $KAlSi_3O_8$ at a water-vapor pressure of 2000 kg/cm². This diagram is similar to the upper half of Fig. 17-11, but temperatures are lower because of the water vapor. The bottom edge of the triangle is a "top view" of Fig. 17-6, and A is the minimum melting point of the binary system. The dashed line marks the approximate bottom of the temperature trough extending to the low-temperature basin on the quartz–feldspar boundary at D. Arrows show approximate directions of change in composition of various melts as crystallization takes place. (Reprinted by permission from Tuttle and Bowen, 1958.)

incongruent melting of orthoclase; and the side K-feldspar–albite represents a solid-solution series with a minimum melting point. Inside the triangle the simplest possible relationship among the three components would be shown by a cotectic line connecting the two eutectics. The validity of this representation has been demonstrated experimentally.

The diagram resembles the one for diopside–plagioclase (Fig. 17-10), except that the point of lowest temperature is near the middle of the cotectic line rather than at one end. In other words, the minimum melting point between the two alkali feldspars persists even when quartz is added, so that the lowest-melting mixture of all three consists of quartz plus a roughly equimolal mixture of the feldspars. The situation is not far different from a simple ternary eutectic (Fig. 17-8), except that the "cotectic" represented by the minimum melting points is a shallow trough instead of a sharply defined valley. Paths of crystallization, shown by arrows in Fig. 17-12 for various mixtures, all lead into the temperature "basin" near the middle of the cotectic line.

The relationships shown in Fig. 17-12 mean that any magma in this restricted range of compositions should give, on cooling, a residual liquid consisting of roughly equal amounts of SiO_2, $KAlSi_3O_8$, and $NaAlSi_3O_8$, and that this liquid should freeze to form a rock of the same composition. One might expect, then, that many alkali granites, despite differences in texture, would show similar chemical compositions. By plotting many analyses on this same ternary diagram we would find that a large number of granites do in fact have compositions close to the predicted temperature minimum (an example is analysis I, Table 17-1), which is good evidence that many alkali-rich granites are a product of crystallization from a melt.

It follows from the experiments summarized in Fig. 17-12 that a quartz-plus-alkali-feldspar mixture should be both the *last* material to crystallize from a silicate melt and the *first* liquid to form when silicate rock material is heated in the presence of water vapor. A shale, for example, on being heated gradually during metamorphism will show the usual changes in mineral composition (clay minerals to muscovite, biotite, cordierite, garnet, sillimanite, and so on) in the solid state; then if it becomes hot enough for melting to start, the first liquid will have a composition represented by points near the ternary minimum in Fig. 17-12. We may imagine this first liquid being squeezed into pockets and cracks in the metamorphic rock, and there solidifying to form the pods and stringers of quartz plus feldspar (pegmatite) that are often found in areas of high-grade metamorphism. As the liquid increases in amount, it would circulate with greater ease and the mixture of liquid plus remaining solid crystals would grow more and more homogeneous. At any stage the temperature change may be reversed, and the material may freeze to a rock resembling granite if the melting has gone far enough, or to a *migmatite* ("mix rock") if partly altered fragments and bands of recognizably metamorphic rock remain.

Thus anatexis is a possible method of forming granites, especially granites rich in the alkali metals. The melting may occur approximately in place, with little movement of the resulting magma; or one can think of the viscous silicate liquid, perhaps with included crystals and larger solid blocks, as being squeezed upward into regions of lower pressure, or rising upward simply because of its low density relative to the surroundings. The composition of the resulting granitic rock will depend, obviously, on how much of the original solid material has been incorporated into the melt. With rising temperature, more and more of the minerals containing Ca, Mg, and Fe would melt, and the composition of the magma would change from alkali granite to granodiorite to tonalite, perhaps even to diorite. In effect, this is another kind of differentiation, *differentiation by fractional melting*, and field evidence suggests that it is a common mechanism for producing granitic rocks of different compositions.

The predicted changes in magma composition as additional elements from the residual minerals (the *restite*) are added to the initial quartz-plus-alkali-feldspar mixture have been verified experimentally (von Platen, 1965). One important conclusion is that the crystallization of quartz and feldspar is practically unaffected by small amounts of ferromagnesian material. In experimental melts to which biotite had been added, the iron and magnesium separated on cooling as biotite (or biotite and magnetite), but details of appearance of quartz and feldspar were the same whether biotite was present or not.

The addition of anorthite to von Platen's experiments had a more pronounced effect on phase relations. For example, Fig. 17-13 pictures the phase relations for

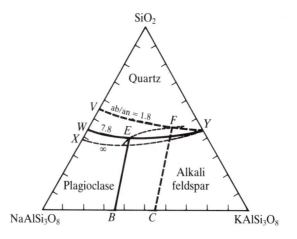

FIGURE 17-13

Phase diagram for the system SiO_2–$KAlSi_3O_8$–$NaAlSi_3O_8$–$CaAl_2Si_2O_8$ (qz–or–ab–an), projected from a tetrahedral diagram onto the SiO_2–$KAlSi_3O_8$–$NaAlSi_3O_8$ (qz–or–ab) plane. Water-vapor pressure 2000 kg/cm². The light dashed line *XY*, the same as *XY* on Fig. 17-12, is the quartz–feldspar boundary for the simple system qz–or–ab, for which the ratio ab/an = ∞. The solid line *WY* is the same boundary when the ratio ab/an = 7.8. Addition of anorthite to the qz–or–ab system causes the alkali-feldspar solid solution to break up into two phases, plagioclase and Na-K-feldspar, whose fields are separated by the boundary *BE*. Hence *E* represents a liquid composition from which three phases crystallize together, rather than the two phases (quartz + solid solution) represented by point *D* in Fig. 17-12. Increasing anorthite causes the plagioclase field to expand, so that the three-phase point moves along the light dotted line, reaching *F* when the ratio ab/an = 1.8. (Reprinted by permission from Von Platen, 1965.)

two albite/anorthite ratios, 1.8 and 7.8, that are projected onto the quartz-plus-alkali-feldspar diagram. Note that as anorthite increases, the range of compositions in which feldspar is the first mineral to crystallize expands at the expense of quartz (line *XY* moves to *WY* and then to *VY* as anorthite increases from zero to an albite/anorthite ratio of 1.8). More striking is the fact that the single-feldspar field of Fig. 17-12 (representing $KAlSi_3O_8$–$NaAlSi_3O_8$ solid solutions) is broken into two parts, one showing initial crystallization of plagioclase and the other initial crystallization of alkali feldspar. In other words the entire field of the diagram now has three subfields, corresponding to the three major mineral constituents of all granitic rocks. The point at which the subfields join is not strictly a ternary eutectic, as it appears to be, because it is on a projection of a line from the tetrahedral phase diagram; it nevertheless represents roughly compositions from which the three minerals will crystallize simultaneously or nearly so. Note that this point shifts toward increasing silica and K-feldspar as the proportion of anorthite in the melt becomes larger (point *E* shifts to *F* as the albite to anorthite ratio falls from 7.8 to 1.8). In a melt with high calcium, therefore, the first mineral to crystallize should be plagioclase over a wide range of compositions (the plagioclase field on the diagram becomes large), and the ultimate residual fluid should be rich in the constituents of quartz and orthoclase. These experimental conclusions fit nicely the common field observations that, in a granodiorite, plagioclase generally has more nearly euhedral crystals than either quartz or orthoclase, and that the felsic dikes (aplite and pegmatite) accompanying a granodiorite usually consist largely of quartz and K-feldspar.

The experimental results are clearly consistent with the origin of granitic rocks either by crystallization–differentiation of a basaltic magma or by progressive melting of many kinds of silicate material. The number and character of possible products in either case depend on how closely equilibrium is maintained. One can imagine, for example, that the early-formed crystals in a cooling melt remain long in contact with the liquid and continue to react with it; or one can think of the crystals as moving out of the melt as fast as they form, so that differences in ultimate compositions between crystalline aggregate and final liquid become exaggerated. Or, when an anatectic melt forms on heating an aggregate of silicate solids (say a mixture of sedimentary and volcanic rocks), the initial melt can remain in place and have its composition progressively modified by continued reaction, or it can be squeezed out as fast as it forms and so retain its original make-up. In nature there will doubtless be all gradations between extremes of separation of early-formed crystals or early-formed melt on the one hand, and extensive reaction on the other—which conveniently explains much of the variety of igneous-rock compositions.

From the experiments described above it appears that silica-rich magmas of widely varying compositions can form either as liquids left when crystals settle out of mafic magmas, or by partial melting of preexisting silicate rocks. For a given granitic body, can its mode of origin be determined from its composition? From its structure? From its geologic situation? These questions are currently the subject of lively controversy among students of granites. We will return to them after a look at one other variable of much importance in the geochemistry of igneous rocks.

17-8 EFFECT OF PRESSURE ON DIFFERENTIATION

The laboratory work considered so far was limited to experiments conducted at pressures of no more than a few kilobars, corresponding to processes that may reasonably be expected to occur in the upper several kilometers of the Earth's crust. How far can the conclusions be projected downward? Would the same processes of differentiation, partial melting, and crystallization operate under the enormous pressures near the base of the crust and in the upper mantle? This hardly seems likely, and experimental work demonstrates that high pressure does indeed have a profound effect on these processes.

Increase in pressure on most silicate minerals in the dry state leads to an increase in melting point, because crystalline silicates in general are more dense than the corresponding liquids (Secs. 8-1 and 16-2), but if the pressure is exerted by water vapor the melting point is commonly lowered. [The lowering by water is partly an example of the general decrease in melting point of one substance by mixing it with another, as predicted by Eqs. (17-5) and (17-6), and partly an effect of water on the structure of silicate liquids.] The same relations with pressure hold for mixtures of silicates: melting temperatures in the dry state are generally increased by pressure, but are decreased if water vapor is present. These relations are nicely illustrated by the curves for melting of granite and basalt in Figs. 1-1 and 16-1: the melting point of dry basalt increases as pressure rises, but the melting point of granite under high water-vapor pressure sharply decreases.

Because the amount of change in melting point with pressure for individual minerals varies greatly from one mineral to another, the composition of eutectic mixtures is often markedly affected by pressure. As an example, Fig. 17-14 shows the effect of high pressures, with and without water vapor, on the simple binary system anorthite–diopside previously illustrated in Fig. 17-1.

Pressure influences not only melting points and eutectic compositions, but the nature of stable mineral phases. If molten basalt, for example, is cooled slowly under a pressure of a few kilobars of water vapor, the crystalline product is not basalt but a rock rich in amphibole. Depending on the composition of the melt and the pressure of water vapor, the amphibole may be accompanied chiefly by plagioclase or chiefly by pyroxene, or it may be practically the only mineral formed. If plagioclase forms at all, it appears late in the crystallization process, not near the beginning as it commonly does at lower pressures. These results mean that basaltic magma may be generated at high pressures by the melting of rocks of widely variable mineral composition: amphibolites, hornblendites, garnet perido-tites. Whether basalt and gabbro in the form we know them, as plagioclase–pyroxene–olivine rocks, are present at lower levels in the crust depends on the amount of water available. Material with the composition of basalt may well exist, at least locally, in the form of widespread amphibole-rich rocks.

In the absence of water vapor, basaltic magma at very high pressures may crystallize to a mixture of garnet and soda-rich pyroxene (omphacite), equivalent in

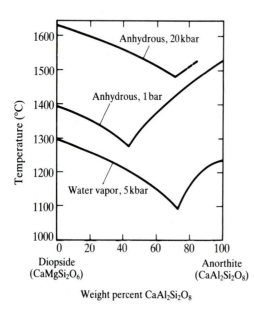

FIGURE 17-14

Effect of pressure on melting in the anorthite–diopside system. The middle curve, for a dry melt under ordinary atmospheric pressure, is the same as Fig. 17-1. The upper (estimated) curve is for a dry melt under a pressure of 20 kbar, and the lower curve is for melting under a water-vapor pressure of 5 kbar. The upper curve is not complete because anorthite at high pressures melts incongruently. (Reprinted by permission from Yoder, 1965.)

composition to the uncommon, unusually dense rock called *eclogite*. [The term "eclogite" is used not only for the high-pressure equivalent of basalt, but also to designate the highest-pressure facies of the blueschist series of metamorphic rocks (Sec. 16-7)—in other words, more generally for any rocks formed under pressure–temperature conditions suitable for the basalt–eclogite transition.] The change from basalt to eclogite may be symbolized

$$NaAlSi_3O_8 + CaAl_2Si_2O_8 + CaMgSi_2O_6 + Mg_2SiO_4 \rightleftharpoons$$

$$\text{plagioclase} \qquad\qquad\qquad \text{diopside} \qquad \text{forsterite}$$

$$NaAlSi_2O_6 + CaMgSi_2O_6 + CaMg_2Al_2Si_3O_{12} + SiO_2 \,.$$

$$\text{omphacite} \qquad\qquad\qquad \text{garnet} \qquad\quad \text{quartz}$$

$$(17\text{-}11)$$

The equation is simplified by omitting minor constituents and by leaving iron out of the formulas of the pyroxenes, olivine, and garnet, but it shows the essential nature of the change: soda enters the heavy pyroxene omphacite rather than feldspar, and garnet replaces the usual plagioclase and olivine. The transition depends on both temperature and pressure, occurring at pressures of 10 to 12 kbar if the temperature is near 500°C, and at 14 to 16 kbar for temperatures near 1000°C. Such pressures correspond to depths of 35 to 55 km, so that formation of eclogite is possible near the base of the crust and in the upper mantle.

If, therefore, a body of basalt magma is generated in the upper mantle, and if the pressure of water vapor is low, its solidification at that depth would follow a very different pattern from the one embodied in Bowen's crystallization sequence. Either omphacite or garnet would crystallize first, depending on the initial composition; eventually it would be joined by the other, and the two would crystallize together during most of the cooling. Since both minerals are complex solid-solutions, there would doubtless be considerable change in the composition of the solids by reaction with the liquid as crystallization proceeds. Extensive differentiation by crystal settling would be possible, but the course of differentiation would be quite different from that at usual pressures.

17-9 ORIGIN OF BASALTIC AND GRANITIC MAGMAS

In preceding paragraphs we have noted the possible changes in a magma that can be produced by fractional crystallization and the variety of magmas that can result from progressive melting, but we have not faced squarely the question of where the major kinds of igneous rock found in nature have their source. Like many of the deeper questions in geology, this one cannot be given an unequivocal answer, but we should be able to make some reasonable guesses.

From a very general, planetary point of view, the rock material of the Earth's crust must have come ultimately from the mantle, and the process of separation must have involved differentiation—either by fractional crystallization of liquid parts of the mantle or by partial melting of solid parts, or by a combination of the two. Probably much of the differentiation was accomplished early in the Earth's history, but certainly the process has continued to be active all during geologic time. We see it operating today most vividly in the upwelling of basalt from the mantle along midocean ridges: here is liquid rock coming out of mantle peridotite and forming oceanic crust as it solidifies. Just where in the mantle the basaltic melt forms is not certain, but probably much of it rises from a layer between depths of roughly 70 and 250 km, called the *asthenosphere*. Here the rock is close to its melting point, and the slowness with which seismic waves are transmitted suggests that pockets and stringers of liquid may be present. Mantle material in this layer is chiefly peridotite containing olivine, two pyroxenes, and garnet (garnet lherzolite), a mineral combination from which experimental work has shown that abundant basalt can be produced by partial melting. Experiments show further that different kinds of basalt—for example, the silica-saturated or nearly saturated basalt of the oceanic floor (tholeitic basalt) and the undersaturated, alkali-metal-rich basalt of many oceanic islands (alkali olivine basalt)—can be reasonably correlated with particular compositions of peridotite or particular depths of melting. It seems well established that much basaltic magma has its origin in localized partial melting of mantle material.

What might cause the local melting? One possible answer is simple release of pressure: solid basalt is denser than the liquid, so if near its melting point would become liquid when pressure decreases. This could account for production of abundant magma along midocean ridges, where pressure on underlying material is released by the spreading apart of the sides of the ridge due to plate motion. Or the upper mantle may be heated locally by rising plumes from far below, forming the basalt of many oceanic islands over "hot spots" (e.g. Hawaii). Another possibility is melting in places where water and other volatiles are concentrated, since the melting point is markedly lowered by a few percent of water; this probably explains abundant basalt magma formation in and near the upper surface of subducting plates, where water is supplied by the hydrothermally altered oceanic crust and sediment being carried down by the moving slab. Sources of heating seem adequate to account for generation of basalt largely out of upper mantle materials.

But what about the granitic magma of the continental crust? Can it also rise directly out of the mantle? This hardly seems likely on a large scale, in view of the extreme scarcity of felsic rocks along midocean ridges and on oceanic islands. Other processes must be involved in the origin of sizable granitic bodies: perhaps the further differentiation of large masses of basaltic melt, or the partial melting of crustal material, or the assimilation of crustal material into mafic or ultramafic magma. Arguments about the relative importance of these processes, and about ways to distinguish granites formed by one process or another, have been loud and persistent among petrologists and geochemists.

One clue about the probable place of origin of much granitic magma comes from plate tectonics. The largest Phanerozoic granite bodies are clearly related to subduction, in that they form the cores of mountain ranges, either present ranges or those of the geologic past, that are situated above subduction zones 100 to 300 km from the point at which the downgoing slab started its descent. In other words, they are vertically above the part of the slab that has moved far enough down into the mantle for melting to commence (Fig. 16-12). The granite can be assumed to result somehow from the melting process, but details of its formation remain speculative. The upper part of the moving slab consists of hydrothermally altered oceanic basalt, probably carrying on its surface a layer of pelagic sediment; the part of the mantle against which it moves (the "mantle wedge") would presumably be chiefly peridotite. Thus a variety of materials is brought together, and melting in such a mixture could follow a number of possible courses, depending on the degree of mixing, the distribution of temperature and pressure, and the amount of water and other volatiles in material at the top of the slab. The first melt, for example, might be basaltic; if fissures provide easy access to the surface, the liquid could rise quickly through the overlying mantle and crust to emerge as a lava flow. Or the liquid could differentiate by crystal settling and thus become more felsic; or if pelagic sediment is abundant, the melt could be more felsic to start with; or a felsic character could be gained by reaction of basalt with crustal material on its way to the surface. There seems no limit to the number of ways in which granitic or andesitic magmas can be imagined to form. Some of the magmas would presumably crystallize in place, some would reach the surface in volcanic eruptions; and the possible variations seem entirely adequate to account for the variety of lavas found among the volcanoes of island arcs and for the multiplicity of forms and compositions often noted among the plutons of a large batholith.

Can the chemistry of a granite or rhyolite provide evidence about the kind of material from which the magma was derived? Some granitic rocks, for example, are relatively rich in Si, Al, and K, others in Ca and Na; could this mean derivation of the former by melting of sedimentary rocks, on the grounds that the first three elements are concentrated in sediments (K and Al in clay minerals and Si in sands), while high Ca and Na would be more likely in melts derived from differentiation of a mafic melt? Trace elements in the former, particularly abnormal amounts of elements with large ionic radii like Cs and Rb, or high values of the ratio $^{87}Sr/^{86}Sr$, would lend weight to the argument for sedimentary origin. In some granitic rocks the principal accessory mineral containing iron is ilmenite ($FeTiO_3$), in others magnetite (Fe_3O_4); the former suggests more reducing conditions when the magma was formed, hence again a sedimentary origin, on the grounds that fine-grained sediments commonly contain a percent or so of organic material. No one of these chemical peculiarities would be convincing in itself, but a combination of these and others could give a strong hint about the history of a felsic rock.

On the basis of such criteria, some petrologists have suggested a broad division of granitic rocks into those of dominantly sedimentary parentage and those generated from the melting or differentiation of igneous materials. The former are referred to as "S-type granites" and the latter "I-type"; another classification suggests "ilmenite-type" and "magnetite-type" for approximately the same grouping. In some places this distinction of granitic types is clearly defined and useful, but as a general rule the chemical characteristics show much overlap and granites of the two kinds are often intimately associated. Perhaps it is best simply to recognize that granitic magma can form in several ways, that assignment of a given body to one or another mode of origin is not always possible, and that various processes can work together to give intermediate types.

One argument for an important role of sediments in the formation of granitic magmas comes from a simple astronomical fact: the Earth is the only known body on which granites in large quantity exist, and also the only known body on which weathering and sedimentation are major surface processes. Why, for example, is granite absent, or nearly so, on the surface of the Moon? Certainly there is abundant basalt, and at times in the past much of the basalt has been liquid; why has fractional crystallization not generated granitic magma? The same question is pertinent for the other terrestrial planets. The abundance of basalt is perhaps less certain, especially for Venus, and the complete absence of granite has not been demonstrated; but what evidence we have suggests basalt as a major component of the planetary surfaces, and there is no topographic sign of large granite intrusions. Could it be that the segregation of elements during weathering and erosion is essential for the formation of most granites—in other words, that the Earth would not have granite as a major kind of rock except for the abundance of air and water at its surface?

SUMMARY

From this long summary of igneous-rock chemistry a few conclusions stand out. Mafic rocks are reasonably accounted for as derivatives of the upper mantle, forming either by the partial melting of solid mantle material or by crystal fractionation from mantle liquids. Felsic magmas do not form in any quantity directly from the mantle, but may be a product of secondary differentiation of basaltic magma which is generated in the mantle. Other felsic magmas may have their origin in the partial melting of previously formed rock material, either sedimentary or igneous or a mixture of both. Large amounts of felsic melt, formed presumably by a combination of the processes of crystal fractionation and partial melting, are generated during the descent of subducting plates to levels where melting of the complex mixture of subducted material and adjacent mantle can take place. The nature of the material from which a given mass of granitic rock was formed is seldom completely clear, but details of its chemistry can often provide at least broad hints about its history.

The discussion has been limited to the commonest of the igneous rocks, those belonging to the series gabbro–diorite–granodiorite–granite and their volcanic equivalents. Igneous rocks not belonging to this series, for the most part rocks that occur only locally and in small masses, would be another fascinating chemical study—the iron-rich rocks formed from magmas differentiating under abnormally reducing conditions, the alkalic rocks undersaturated with silica, the ultramafic rocks, and many more extreme varieties. For these rocks also the processes of fractional crystallization, anatexis, and mixing of melts with solid material would play important roles in their chemical history, but the details are more the province of petrology than of geochemistry.

PROBLEMS

1. Using the phase rule in the form $c - p + 1 = f$ (pressure constant), and remembering that MgO and SiO_2 form the two compounds $MgSiO_3$ (enstatite) and forsterite (Mg_2SiO_4), answer these three questions about each of the following combinations: (*a*) Can the combination exist at equilibrium? (*b*) If it can exist, how many degrees of freedom does it have? (*c*) If it can exist at one temperature, what happens when the temperature is lowered a few degrees?

   (1) Forsterite and liquid containing MgO and SiO_2.
   (2) Forsterite, enstatite, solid silica, and liquid containing MgO and SiO_2.
   (3) Enstatite, solid silica, and liquid containing MgO and SiO_2.
   (4) Liquid containing MgO and SiO_2, with no solid phases present.

2. Using the diagram of Fig. 17-1, describe what happens if (*a*) a melt containing 10% anorthite and 90% diopside is cooled from 1500°C to 1000°C, and (*b*) a solid mixture containing 50% of each is heated from 1000°C to 1500°C. Assume that equilibrium is maintained in both cases. If equilibrium is not maintained, what are the possibilities of forming fractions of different compositions by crystal separation?

3. Forsterite and fayalite form a continuous solid-solution series; pure forsterite melts at 1890°C and pure fayalite at 1205°C. Draw an equilibrium diagram for this system and describe what happens when (*a*) a melt containing 10% fayalite and 90% forsterite is cooled from 1900°C to 1200°C under equilibrium conditions, (*b*) the same mixture is cooled through the same range, but crystals are removed as fast as they form, and (*c*) a mixture of 50% each is heated from 1000°C to 1800°C.

4. The following questions refer to Fig. 17-3:

   (*a*) If a mixture containing 80% silica and 20% nepheline is heated to 1300°C, calculate the composition of each phase present.
   (*b*) Describe the course of crystallization under equilibrium conditions of a mixture containing 40% SiO_2, and of a mixture containing 70% SiO_2.
   (*c*) If equilibrium is not maintained, is it possible for nepheline to crystallize from any mixture containing more than 50% SiO_2? Explain.

5. Using Fig. 17-4, describe how melts of the following compositions would crystallize, first assuming maintenance of equilibrium and second assuming removal of early-formed crystals: (*a*) 10% SiO_2, 90% leucite; (*b*) 30% SiO_2, 70% leucite; (*c*) 50% SiO_2, 50% leucite. Would it be possible to obtain a rock with leucite from a melt containing a greater proportion of silica than is present in orthoclase? If so, under what conditions? If not, why not? What would be the composition of the first liquid obtained from the melting of pure orthoclase? From the melting of an aplite containing 50% orthoclase and 50% quartz?

6. Using Fig. 17-10, describe qualitatively the crystallization of melts with the following compositions, first assuming maintenance of equilibrium and then assuming removal of early-formed crystals: (*a*) 10% diopside, 40% anorthite, 50% albite; (*b*) 70% diopside, 15% anorthite, 15% albite.

7. Using Fig. 17-9, describe the crystallization of mixtures whose compositions lie (*a*) in the anorthite field, (*b*) in the cristobalite field, (*c*) in the clinoenstatite field, (*d*) in the forsterite field near the forsterite corner, and (*e*) in the forsterite field near the clinoenstatite border. In each case, assume first that equilibrium is maintained, and then describe the changes that would follow if the early-formed crystals were removed.

8. The following is a typical section through the Palisades sill in northern New Jersey. Suggest an explanation for the layering.

Fine-grained diabase	40 feet thick
Coarse-grained diabase with interstitial quartz and alkali feldspar	150 feet thick
Medium-grained diabase	800 feet thick
Layer with abundant olivine	15 feet thick
Fine-grained diabase	50 feet thick

9. If a small chunk of granite is engulfed by a partly crystallized basalt flow, what would you expect to happen to the granite?

10. If a granite magma assimilates large amounts of dolomite, could this change the composition of the magma enough to permit nepheline to crystallize from it?

11. A pegmatite dike consists typically of coarse-grained quartz and alkali feldspar, and presumably crystallized from a water-rich liquid with this composition. Where such a dike cuts through an ultramafic rock, the amount of quartz in the dike is often seen to be greatly reduced and may even be replaced by corundum. Suggest an explanation.

12. Calculate the liquidus curves for diopside and anorthite, and the eutectic temperature and liquid composition, for the binary system $CaMgSi_2O_6$–$CaAl_2Si_3O_8$. Use the following data in your calculations:

$$\text{Melting point diopside} = 1665 \text{ K}$$

$$\text{Melting point anorthite} = 1830 \text{ K}$$

$$\text{Heat of fusion diopside} = 142.6 \text{ kJ mol}^{-1}$$

$$\text{Heat of fusion anorthite} = 136 \text{ kJ mol}^{-1}$$

Assume that the enthalpies of fusion are independent of temperature and composition and that activity is equal to mol fraction. Construct a phase diagram similar to Fig. 17-1, showing the results of your calculations in terms of temperature and weight percent anorthite. Explain why the theoretical curves and eutectic temperature and composition differ from the experimental results summarized in Fig. 17-1.

REFERENCES

Barth, T. F. W.: *Theoretical Petrology*, 2d ed., Wiley, New York, 1962.

Bowen, N. L.: *The Evolution of the Igneous Rocks*, Princeton University Press, 1928.

Bowen, N. L., and O. F. Tuttle: "The system $NaAlSi_3O_8$–$KAlSi_3O_8$–H_2O," *Jour. Geology*, vol. 58, pp. 489–511, 1950.

Morse, S. A.: "Alkali feldspars with water at 5 kbar pressure," *Jour. Petrology*, vol. 11, pp. 221–253, 1970.

Schairer, J. F.: "The alkali-feldspar join in the system $NaAlSiO_4$–$KAlSiO_4$–SiO_2," *Jour. Geology*, vol. 58, pp. 512–517, 1950.

Schairer, J. F., and N. L. Bowen: "Temperature–composition diagram of the system leucite-silica at 1 bar," *Geological Society of Finland Bull.*, vol. 20, p. 74, 1948.

Tuttle, O. F., and N. L. Bowen: "Origin of granite in the light of experimental studies in the system $NaAlSi_3O_8$–$KAlSi_3O_8$–SiO_2–H_2O," *Geol. Soc. America Memoir* 74, 1958.

Von Platen, H.: "Kristallisation granitischer Schmelzen," *Beiträge Mineralogie und Petrographie*, vol. 11, pp. 334–381, 1965.

Yoder, H. S., Jr.: "Diopside–anorthite–water at five and ten kilobars and its bearing on explosive volcanism," *Carnegie Institution of Washington Yearbook*, vol. 64, pp. 82–89, 1965.

SUGGESTIONS FOR FURTHER READING

Ernst, W. G.: *Petrologic Phase Equilibria*, W. H. Freeman and Company, San Francisco, 1976. A good reference for phase diagrams of many kinds. Details of experimental determination of phase relations, and computation of phase diagrams from thermodynamic properties.

Hyndman, D. W.: *Petrology of Igneous and Metamorphic Rocks*, 2d ed., McGraw-Hill, New York, 1985. A standard petrology text, with a good discussion of phase diagrams, origin of igneous rocks, and relation of different kinds of igneous rocks to plate tectonics.

Kushiro, I.: "Effect of water on the composition of magmas formed at high pressures," *Jour. Petrology*, vol. 13, pp. 311–334, 1972. Experimental demonstration that silica-rich magmas like andesite and dacite, as well as basalt, can form by partial melting of peridotite if water is present.

Philpotts, A. R.: *Principles of Igneous and Metamorphic Petrology*, Prentice Hall, Englewood Cliffs, New Jersey, 498 pp., 1990. An excellent textbook covering the chemical and physical properties of magmas and the rocks they form.

Wyllie, P. J.: *The Dynamic Earth*, Wiley, New York, 1971. Chapters 5 and 6 cover the chemical relations of crust and mantle, Chapter 8 is an excellent treatment of magma generation based on experimental work, and Chapter 14 relates magma formation to plate tectonics. The book includes a fine historical account of the development of modern ideas about the chemistry of igneous rocks.

Yoder, H. S., Jr., (ed.) *The Evolution of the Igneous Rocks*, Princeton University Press, Princeton, New Jersey, 1979. A book with the same title as N. L. Bowen's classic volume of 1928, and dedicated to his memory. Its eighteen chapters by prominent petrologists and geochemists show how modern ideas on many aspects of igneous-rock chemistry have developed from the foundations that Bowen had laid. Chapter 4 is a good review of phase diagrams by A. Muan.

CHAPTER

18

VOLATILES AND MAGMAS

Most silicate magmas, as we have seen, solidify over a considerable temperature range. The temperature at which crystallization starts, and the range over which it continues, depend in large measure on the amount of water dissolved in the magma. Estimates of the water content of typical magmas, derived both from experimental work on silicate–water systems and from studies of the pressure–temperature conditions of pyroclastic eruptions (Whitney, 1988), are in the range 0.5 to 6 weight percent. The water comes in part from adjacent rocks and in part from the dehydration of minerals like mica, chlorite, and amphibole during anatexis of pre-existing rocks.

Whatever its source, most of the water is concentrated in the residual liquid as crystallization proceeds. The last material to crystallize consists of a fluid made up chiefly of water plus the constituents of alkali feldspar: sodium, potassium, aluminum, and silica. Commonly, but not universally, silica is in excess of the amount necessary to form feldspar, so that crystallization of the residual fluid gives a solid material with the general composition of granite. This kind of water-rich fluid may be produced in small amount by the crystallization of basaltic magma, and in larger amount by crystallization of granitic or granodioritic magma. It may form also by initial melting of heterogeneous rock material during extreme metamorphism. It is a common geologic fluid and a most interesting one, for it often contains, in addition to water plus silica and the elements of feldspar, a variety of minor constituents that form curious kinds of rock and some types of ore deposits. To the details of formation and crystallization of such residual fluids we turn our attention in the following pages.

18-1 KINDS OF EVIDENCE AVAILABLE

Here we proceed on less secure ground than in our discussion of normal igneous rocks. Laboratory data are less extensive, for the good reason that handling corrosive gases at high temperatures is experimentally difficult. Theoretical studies are handicapped by the number of variables that must be considered and by scarcity of data from which reasonable ranges of the variables can be guessed. On the observational side, geologic data are far from satisfactory because the materials we find in the field are not solid equivalents of the residual fluids as such, but only a residue left after much of the fluid has escaped into the atmosphere, groundwater, or ocean. Much of the discussion in this chapter will necessarily be somewhat speculative, based in part on general physicochemical principles, in part on experimental data, and in part on the hints we can pick up from geologic observations that are summarized in the cross section of Fig. 18-1.

We might expect, to begin with, that the residual fluid from a crystallizing magma would show a concentration of substances of low melting points, the so-called *volatiles*, which were originally dissolved in the silicate melt. The general nature of these volatiles can be inferred from the composition of volcanic gases, from tiny fluid inclusions in the minerals of igneous rocks, and from the kind of alteration found in wall rocks around an igneous body. Such evidence suggests that the volatiles may vary a good deal from one body of magma to another, but that they consist predominantly of water with lesser amounts of carbon dioxide, hydrogen chloride, hydrogen fluoride, and compounds of nitrogen, sulfur, and boron. In addition to the volatiles, geologic evidence indicates the presence of compounds of metals: chiefly $NaCl$, with lesser amounts of compounds of potassium, calcium, magnesium, iron, and traces of many others. Some of the minor metals may show strong local concentrations; good examples are copper, lead, and zinc in volcanic sublimates, and beryllium, lithium, uranium, and thorium in some pegmatite dikes. As a generalization, we may conclude that the residual fluid contains material of at least three kinds: substances with particularly stable molecules that are gaseous at magmatic temperatures, compounds that are very soluble in water, and elements whose ions do not fit readily into the structures of growing silicate minerals. These materials, so to speak, are the misfits of the magma, the constituents that for one reason or another cannot readily become a part of the minerals of normal igneous rocks.

To give the discussion a concrete geologic framework, we will use the conventional model of a body of magma cooling in a well-defined "chamber" a few kilometers or tens of kilometers below the Earth's surface. The residual fluid will be thought of as forming in the interstices between the growing crystals, as collecting into pockets and fissures in the nearly solid mass, and as escaping in part into the

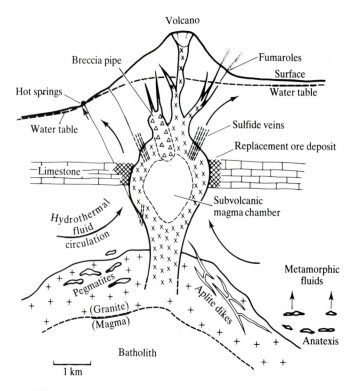

Volcano

Breccia pipe

Fumaroles

Surface

Hot springs

Water table

Water table

Sulfide veins

Replacement ore deposit

Limestone

Subvolcanic
magma chamber

Hydrothermal
fluid
circulation

Metamorphic
fluids

Pegmatites

Aplite dikes

(Granite)
(Magma)

Anatexis

Batholith

1 km

FIGURE 18-1

Schematic cross section of a batholith and subvolcanic magma chamber illustrating the geologic features related to volatiles and magmas discussed in this chapter. Water-rich solutions exsolved from the crystallizing batholith may form pegmatites and aplite dikes, and similar fluids in the shallower subvolcanic magma chamber may form breccia pipes. Sulfide veins, replacement-type ore deposits (skarns), fumaroles, and hot springs probably originate from mixtures of magmatic volatiles, meteoric ground water, and fluids derived from metamorphic reactions.

rocks enclosing the magma body. At best such a model is tremendously over-simplified. It is not realistic at all for incipient melting during metamorphism, where the fluid we are considering would be more "initial" than "residual". But to avoid tedious repetition, we will use a single model, a model that has the merits of mechanical simplicity and demonstrable correspondence with at least one kind of geologic situation where residual fluids have been active.

18-2 PHASE RELATIONS

In imagination, then, we consider a body of magma cooling somewhere below the surface. At first, if temperature and pressure are sufficiently high, the magma consists of a single liquid phase. The potentially volatile materials are dissolved in the liquid, some of them chemically attached to silicon–oxygen groups and some as molecules simply wandering in the spaces between the groups. Crystallization starts, and the system now consists of liquid plus one or more solid phases. (A second liquid phase consisting of sulfides is possible, or at extreme compositions even a second liquid silicate phase, but for present purposes these possibilities will be neglected.) As crystallization progresses, the volume of remaining liquid becomes smaller and smaller, and in this decreasing volume the volatiles are concentrated. The question then arises: Will the volatiles, confined in a smaller and smaller volume of liquid, eventually separate as a gas phase? In other words, will the liquid boil?

This is a trickier question than at first appears. Two factors are operating to change the vapor pressure of the cooling liquid: the vapor pressure tends to decrease because of the cooling, and at the same time it tends to increase because the most volatile constituents are becoming steadily more concentrated. Which factor is more important? Obviously the answer must depend on the amount and nature of the volatiles present. Even if this query could be answered in favor of a net increase in vapor pressure, separation of a vapor phase would not be assured, because boiling occurs only when the vapor pressure exceeds the external pressure. What is the external pressure? Here we run head-on into the inadequacies of our model. Are the walls of the magma chamber impermeable and infinitely strong, so that pressure can build up indefinitely? In this case boiling could be prevented; but such a picture, carried to extremes, is clearly absurd. Are the walls of the chamber impermeable, but limited in their ability to confine gas by the pressure of overlying rock? In this case boiling occurs if the amount of volatile material is large enough to build up a vapor pressure in excess of the lithostatic pressure (for example, roughly 3 kbar at a depth of 10 km), and we may have here a possible mechanism for some explosive volcanic eruptions. On the other hand boiling does not occur if the amount of volatile material is small, so that the maximum vapor pressure reached before crystallization is complete does not exceed the lithostatic pressure. Are the walls of the magma chamber impermeable to a viscous silicate liquid, but freely permeable to gases? With this condition there would be nothing to prevent separation of a gas phase all during the crystallization, for the only pressure to be overcome is hydrostatic pressure of water in fissures above the magma chamber. Thus boiling in the cooling liquid may or may not occur, depending on a variety of factors which cannot be readily evaluated.

If boiling does happen, note that it is a different sort of phenomenon from ordinary boiling. This boiling is in response to *cooling*, and results from an increase in vapor pressure due to confinement of dissolved gases in a smaller and smaller body of liquid. Ordinary boiling follows an increase in vapor pressure due to simple heating. The temperature at which a hypothetical magmatic liquid might boil on cooling is often called its *second boiling point*, and the phenomenon is spoken of as *retrograde* or *resurgent* boiling.

Solubility of Water in Silicate Melts

Experimental data on silicate–water mixtures give a roughly quantitative idea of the temperatures, pressures, and compositions involved in the separation of a gas phase

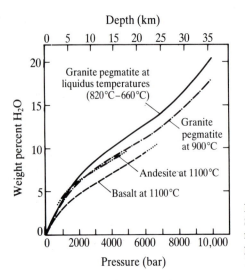

from a magma. Such measurements, summarized in Fig. 18-2, indicate that the amount of water that can be dissolved in a granitic liquid increases indefinitely as the pressure is raised, but that it would not exceed about 10% by weight at temperatures and pressures to be expected in a cooling magma at shallow levels in the crust. The experiments show further that the solubility of water in a silicate melt is not greatly affected by a change in composition even from granite to basalt. From the curves in Fig. 18-2, we can estimate that the residual liquid of an intrusive body at a depth of 10 km would boil when crystallization during cooling has increased its water content to 7 or 8%.

The highest temperature at which retrograde boiling might occur would be the temperature at which crystals begin to appear in a cooling melt, in other words along the liquidus of the silicate–water system. The temperature of the liquidus for the pegmatite shown in the upper curve of Fig. 18-2 is plotted as a function of water vapor pressure in Fig. 18-3. Besides the pegmatite, the diagram includes a melting

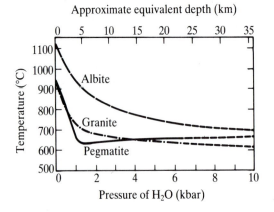

point curve for albite and an incipient melting curve for granite. The effect of water on the melting behavior of these substances is striking. Note, for example, that dry granite does not begin to melt until the temperature rises above 900°C, but in the presence of 1 kbar of water vapor, incipient melting occurs near 700°C.

Separation of a water-rich gas phase from a magma can be brought about by change in pressure as well as by concentration during crystallization. Again a rough quantitative estimate can be made from Fig. 18-2. If, for example, a granitic magma at 800 or 900°C containing 10% dissolved water moves upward from a depth of 20 km to a depth of 3 km, it must lose more than half of its water, since at 1 kbar the solubility is reduced to about 4%.

Water Vapor as a Solvent

We have been describing the vapor that separates as pure water, but of course this is too simple. At the temperatures and pressures we are considering, the water is well above its critical point and is so greatly compressed that its density is comparable to that of ordinary liquid water [Fig. 2-6(a)]. Certainly it would be accompanied by other volatile substances (CO_2, H_2S, HCl, etc.), and in addition, because of its high density, would presumably act as a solvent for some of the rock constituents. Thus we may think of the "vapor" that appears in the experiments (or in nature) as containing in solution appreciable silica and compounds of sodium and potassium, plus smaller amounts of other metals. The dissolved metals might be present as silicates, or more probably in large part combined with volatile constituents to form compounds like NaCl and KCl.

At first sight this seems strange, because we are accustomed to thinking of water vapor as having no appreciable solvent power. We often separate nonvolatile material from water by heating to drive off water vapor, never giving a thought to the possibility that some of the material might dissolve in the vapor and escape. But conditions of the experiments we are considering here differ radically from ordinary laboratory conditions, and a measurable solubility of silica and metal salts in gaseous water at high temperatures and pressures is confirmed by a variety of observations. An accurate description of the phases present in the experiments or in nature would thus include one or more crystalline solid phases, a viscous silicate liquid phase containing a few percent of dissolved water, and a more mobile fluid phase consisting of a supercritical water-rich solution.

Metal salts dissolved in the supercritical fluid would behave somewhat differently from those in low-temperature solutions, because the dielectric constant of the high-temperature water is much lower [Fig. 2-6(b)]. This means that electrolytes would have less tendency to dissociate into ions, and a dissolved salt like NaCl would exist chiefly in the form of undissociated NaCl° rather than the familiar Na^+ and Cl^-. Acids also would be less dissociated: a dissolved volatile like HCl would show reduced acidity, because it would be present mostly as HCl° molecules rather than as ions.

From ordinary experience a lower dielectric constant might also be expected to make metal salts less soluble, but abundant evidence from volcanic emanations and from fluid trapped in tiny cavities (fluid inclusions) in igneous minerals shows that the solution may be fairly concentrated, particularly in sodium chloride. This leads to a puzzling question: dissolved salts are known to raise the critical point of water; can we be certain, then, that the water vapor separating from a cooling magma will remain above its critical point? If it happens, either initially or in the course of later cooling, that the temperature of the gaseous solution drops below its critical point, a further separation of phases becomes possible. In addition to the solid phases we can imagine now three fluid phases: the remaining viscous silicate melt, a water-rich fluid containing abundant dissolved salts, and an even more water-rich vapor. To explore the possibilities more fully, we digress a little to consider the general phenomenon of the critical point in multicomponent systems.

18-3   THE CRITICAL POINT

Carbon dioxide under pressure in a stout-walled glass tube shows a clear separation of liquid and vapor. If the tube is warmed, the meniscus separating the two phases becomes thinner and less curved; at 35°C, quite suddenly, it vanishes altogether. If the tube is then cooled, the meniscus reappears at the same temperature. We say that carbon dioxide has a *critical temperature* of 35°C, meaning that above this temperature two fluid phases cannot exist. The minimum pressure necessary to maintain a liquid at or just below the critical temperature is the *critical pressure*. Above the critical temperature no amount of pressure can cause the appearance of a liquid. The *critical volume* is the specific volume at the critical temperature and critical pressure.

Nomenclature is confusing about the phases just below and just above the critical point. A substance immediately above its critical point cannot properly be described as either a "gas phase" or a "liquid phase", since it has the ability of the former to expand indefinitely and the close molecular spacing of the latter. The term "fluid" is applicable, but this term also refers to ordinary liquids and vapors—to any material capable of flow under infinitely small forces. Some authors restrict the word "gas" to substances above their critical temperatures, and "vapor" to low-density phases below their critical points. More commonly the two terms below the critical point are used interchangeably. For the present discussion a substance above its critical temperature will be referred to loosely as a "gas" or a "fluid," or specifically as a "supercritical fluid" when the meaning is not clear from the context.

Note that a supercritical fluid may have any density, from that of a rarefied gas to one comparable with the density of the low-temperature liquid [Fig. 2-6(a)]. The ability of hot, compressed gases to dissolve nonvolatile substances, referred to in the last section, applies only to supercritical fluids of high density and low dielectric constant. This, of course, requires high confining pressure. Nothing about the supercritical state itself causes greater solubility; a tenuous supercritical gas of low density has just as little solvent power as the vapor of the substance near its boiling point. But high-density supercritical fluids have molecular spacings equivalent to those in a liquid, and they would be expected to show similar dissolving ability. To repeat: it is not primarily high temperature, but high pressure, that makes non-volatile materials soluble in water vapor escaping from a magma.

Effect of Solutes

The critical point of a substance is changed by the addition of solutes, just as its boiling point and freezing point are changed. Experimentally, solutes of greater volatility are found to lower the critical point, whereas less volatile solutes, with a few exceptions, are found to raise it. For example, a 2% solution of NaCl in water (approximately $0.4m$) has a critical point of $399°C$, and a 5% solution one of $424°C$, in contrast to $374°C$ for pure water (Fig. 18-4). Point C on this diagram, at the upper end of the vapor pressure curve, is the critical point for pure water ($374°C$, 221 bar). The vapor-pressure curve does not continue beyond this point, because at higher temperatures and pressures there is no distinction between liquid and gas. The three lighter solid lines in Fig. 18-4 are similar vapor-pressure curves for NaCl solutions of different concentrations; these solutions have lower vapor pressures and higher critical points than pure water. The upper light dashed line extending beyond C is drawn through the critical points of increasingly concentrated NaCl solutions. The field above this line represents a single fluid phase, and the field below the line shows equilibrium between unsaturated solutions and vapor. The lower dashed line is the vapor-pressure curve for saturated solutions, in other words the line representing three-phase equilibrium between solid salt, solution, and vapor.

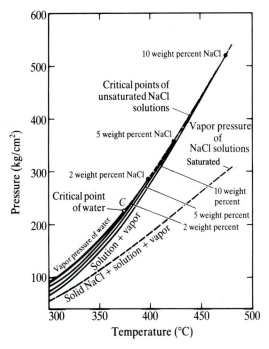

FIGURE 18-4

Vapor-pressure curves and critical-point curve of NaCl solutions. (Reprinted by permission from Ölander and Liander, 1950.)

To show the relations in full detail would require a three-dimensional diagram, with NaCl/H$_2$O ratios as the third variable.

In a partly frozen magma the various solids in equilibrium with residual fluids would be present in large amount. Hence a pertinent question to ask about the experimental data for NaCl is this: Suppose that solid NaCl is present in great excess as the temperature and pressure are raised beyond the limits shown in Fig. 18-4, so that the solution becomes indefinitely more concentrated; would this saturated solution, in continuous contact with the solid, ever reach a critical point? In other words, with solid NaCl present, would we ever find a point where liquid and vapor approach the same composition? From the fact that the critical-point curve and the saturation curve (the two dashed lines) in Fig. 18-4 diverge so sharply, we might guess that the answer would be negative. Fortunately, for this system we have enough experimental data to check our guessing.

Data for the temperature range 350 to 700°C and for pressures up to 1.2 kbar are shown by the heavy solid lines in Fig. 18-5, which are extensions of the two dashed lines in Fig. 18-4. The two curves are extrapolated beyond the experimental data by light lines, which are not drawn to scale. On the left-hand side of the drawing is shown the one-component diagram for pure water (dashed lines); on the right-hand side is the corresponding (not-to-scale) diagram for pure NaCl. Actually these should be in different planes: the drawing is a projection of a three-dimensional figure, with NaCl/H$_2$O compositions as its third axis; the one-

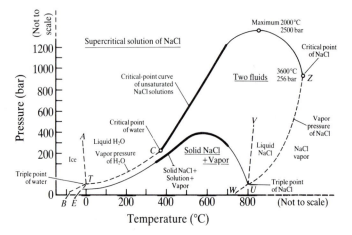

FIGURE 18-5

Pressure–temperature diagram for the system NaCl–H$_2$O. This is a projection of a three-dimensional diagram, with NaCl/H$_2$O ratios as the third axis. Dashed lines show one-component plots for H$_2$O (front face of the three-dimensional figure) and NaCl (back face); these are not drawn to scale. Solid lines (heavy lines drawn to scale from experimental data, light lines hypothetical and not to scale) are curves in space connecting the end faces. Underlined designations of phases refer to regions in space; designations without underlining show phases of pure water and salt on the end planes of the three-dimensional figure. The right-hand side of the figure is diagrammatic only and not to scale. (Reprinted by permission from Sourirajan and Kennedy, 1962; Morey, 1957; and Pitzer, 1984.)

component diagram for H_2O may be considered to lie on the forward face of the figure, and the NaCl diagram on the back face; the two solid lines CZ and TEU are lines in space between the two faces. The upper solid line CZ connects critical points of unsaturated solutions, with solutions rich in water near the left-hand end at C and those rich in NaCl near the critical point of pure NaCl at Z. The lower solid line TEU connects the triple points of the two pure substances and, like the triple points, represents equilibrium between two fluid phases and one solid phase. (This line is the edgewise view of a curved surface in a three-dimensional figure, one side of the surface representing compositions of a liquid solution and the other side compositions of a gaseous solution.) Although experimental data are scanty for the extreme NaCl side of the diagram, the existing data show clearly that the vapor-pressure curve for saturated solutions goes through a maximum and hence does not intersect the critical-point curve. This means that at *any* temperature two fluid phases can coexist. In other words, a critical point is not approached so long as solid NaCl is present.

Substances much less soluble than NaCl behave differently. Silica, for example, raises the critical point only slightly, even when excess solid is present, simply because the solubility is very low at temperatures around 400°C. This situation is shown diagrammatically on the left-hand side of Fig. 18-6. The drawing is not to scale, because the vapor-pressure curve for saturated silica

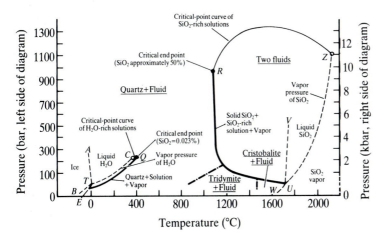

FIGURE 18-6
Pressure–temperature diagram for the system SiO_2–H_2O. This is a projection of a three-dimensional diagram, with SiO_2/H_2O ratios as the third axis. Dashed lines show one-component plots for H_2O (front face of the three-dimensional figure) and SiO_2 (back face); these are not drawn to scale. Solid lines are curves in space between the end faces. The heavy solid line and the dash-dot line are drawn approximately to scale from experimental data; the light solid line on the right-hand side is hypothetical and not to scale; the position of the light solid line on the left-hand side is known from experimental data but is displaced on the diagram for clarity. Underlined designations of phases refer to regions in space; designations without underlining show phases of pure water and silica on the end planes of the three-dimensional figure. (Experimental data from Kennedy et al., 1962; hypothetical reconstruction of the diagram as a whole after Morey, 1957.)

solutions (EQ) is so close to the vapor-pressure curve for pure water that the two would be indistinguishable on any reasonable scale, and the critical points of all silica solutions up to saturation are within a fraction of a degree of the critical point of water. In this system the critical-point curve CQ intersects the vapor-pressure curve at Q, the "critical end point;" in other words, at Q the vapor pressure becomes equal to the critical pressure (or the composition of the "solution" becomes identical with the composition of the "vapor"), and the distinction between liquid and gas vanishes. Hence above the temperature represented by Q only a single fluid phase can exist in contact with solid silica.

The right-hand side of Fig. 18-6 is also of interest. This is the high-temperature end, the end showing mixtures with much silica and only a little water. Again we use a two-dimensional projection of a three-dimensional figure; the one-component curves for pure silica (hypothetical) may be considered to lie on the back face, and the solid lines (ZR and UR) extend out into space toward the H_2O (front) face. The heavy solid line shows experimental data, while the light lines are hypothetical and not to scale. Both the critical point and the melting point of silica are lowered by the addition of water; the sharp reduction of the melting point, from 1720°C to less than 1200°C by a water-vapor pressure of less than 2 kbar, is particularly striking. The experimental work demonstrates the existence of a critical end point (R), where the compositions of liquid and vapor in equilibrium with solid silica become identical. The critical-point curve for silica-rich solutions must extend from this point to the (unknown) critical point of pure silica, shown diagrammatically at Z. In contrast to the NaCl–H_2O system, neither the critical-point curve nor the vapor-pressure curve for saturated solutions is here continuous from one end member to the other. In other words, saturated solutions near both ends of the series have critical points, and over a large range of temperature and pressure only a single fluid phase can exist in contact with solid silica.

In summary, then, critical-point relations in a *saturated* solution of a nonvolatile substance may follow one of two patterns. A solute may become indefinitely more soluble as the temperature is raised, so that a continuous series of solutions exists from pure water to pure solute (for example NaCl); in this case no critical point can be reached as long as excess solute is present, and two fluid phases can exist at any temperature. Or a solute may show only limited solubility, in which case the solution has a critical point slightly higher than the critical point of pure water (for example, SiO_2); at temperatures above this point, and below the corresponding point for limited solubility of water in the molten solute, only one fluid phase can be present in contact with excess solute. Presumably most substances with high solubility, especially substances whose solubility increases rapidly with rising temperature, would show the first type of behavior, but quantitative predictions are not possible from present experimental data.

The Critical Point in Magmas

If the effect of simple salts on the critical point is impossible to predict, the effect of the complex mixture in a magma would seem completely hopeless. Here we must deal not only with many nonvolatile substances, but with a variety of volatile solutes as well. Will such a mixture be above its critical point at some stage of the cooling process (second case above), or will it stay below the critical point at all temperatures (first case)? No answer is possible on theoretical grounds, and the experimental data are not conclusive.

Experimental work, on the one hand, indicates that water and the major constituents of granite have only limited miscibility. In other words, a granite–water mixture behaves like a silica–water mixture, fitting the second case described above—as we might perhaps expect, since the principal constituents of granite are the very insoluble quartz and aluminosilicates. There is the added complication here that the greater number of components makes it possible for a liquid silicate phase to exist in addition to one or more solid phases, but between the gaseous water-rich phase and either a solid–solid or solid–liquid mixture, there is a clean separation. The gas phase can contain only small amounts of silica and the components of feldspar, and in an ideally simple cooling melt it would necessarily be above its critical point.

On the other hand an abundance of NaCl or other soluble salts in the original melt might lead to separation of a concentrated gaseous solution that would be below its critical point (first case above) even in the temperature range of a crystallizing magma, 600 to 700°C. The presence of volatiles like HCl and CO_2 would have an opposite influence on the critical point, so that predictions become mere speculation.

Thus a gamut of possibilities can be envisioned: (1) A water-rich phase may separate from the silicate melt during crystallization, containing so much dissolved salt that it is below its critical point to start with or becomes so as the temperature falls; in this case a maximum of three fluid phases can coexist with the crystallizing silicates. (2) A water-rich phase may separate, either so dilute or mixed with so much volatile material that it remains above its critical point until long after crystallization is complete; in this case the maximum number of fluid phases is two. (3) A water-rich phase may fail to appear, either because the amount of water dissolved originally was so small that it could be accommodated in the crystallization of micas and amphiboles or because excess alkali silicates increased its solubility in the melt; in this case no more than one fluid phase can be present. Which of the three possibilities is most likely depends on the amounts of water, salts, and volatiles in the original melt, and these are generally not determinable. Quite conceivably all three possibilities may be represented in different magma bodies.

Significance of the Critical Point

Why is the existence or nonexistence of the critical state in residual magmatic fluids important? Two reasons may be suggested. The first, now largely of historical interest, is the idea that supercritical fluids may have unique properties, particularly unique ability to dissolve and replace minerals. In the older literature of geology, processes supposedly dependent on supercritical fluids were often given a special name, "pneumatolytic," in contrast with "hydrothermal" processes, which were thought to be limited to subcritical conditions. Modern experimental work has shown that such a distinction is illusory, since there is no discontinuity in the solvent action of a substance on passing through its critical region. Supercritical fluids may indeed be powerful solvents, but as we noted earlier, their solvent ability depends simply on their pressure and temperature and not on their supercritical condition as such.

The second and more important reason for concern with critical phenomena lies in the number of separate fluids that may be imagined to form during the late stages of crystallization of a granitic magma. From observations near granite contacts, several fairly distinct geologic effects are commonly ascribed to residual fluids (Fig. 18-1): formation of pegmatite dikes (coarse-grained quartz–feldspar rock), formation of aplite dikes (fine-to-medium-grained, sugary-textured quartz–feldspar rocks), formation of breccia pipes (consisting of angular rock fragments that host a wide variety of sulfide and silicate mineralization), formation of quartz–sulfide veins, and alteration of wall rocks. The alteration effects may perhaps be due to the same solutions that form either pegmatites or veins, but to think of pegmatites, aplites, and sulfide veins as products of the same solution is not reasonable. An explanation for the radically different compositions and textures becomes easier if two or more different kinds of fluid can be called on, and the most obvious way to obtain different fluids is by a phase separation. The separation of fluid phases, either by simple vaporization or by retrograde boiling, is possible only if the initial fluid is below its critical temperature; the "initial fluid" in this sense can refer either to the original silicate melt (case 2 of a preceding paragraph), or to a separated fluid rich in water and salts (case 1). Thus the possible kinds of explanation we can use for late magmatic phenomena are severely limited by the information we can get, or the guesses we can make, about the presence or absence of supercritical fluids in complex crystallizing melts.

In the following sections we use this discussion of the critical point as a background in considering the origin of pegmatites, aplites, and veins.

18-4 PEGMATITE AND APLITE DIKES

Typical pegmatites and aplites have a similar chemical composition: chiefly quartz and alkali feldspar, often a little biotite or muscovite, less commonly hornblende. The difference between them is textural, pegmatites having crystals much larger than those of ordinary granite and typically showing much variation in texture and composition, aplites having a sugary texture superficially resembling sandstone and remaining monotonously uniform over long distances. Pegmatites are by far the more spectacular of the two, their crystals sometimes attaining enormous size and showing conspicuous segregation into quartz-rich and feldspar-rich zones. Pegmatites are notable for the common occurrence in them of minor black tourmaline and pink garnet; a few contain large amounts of much less common minerals, minerals of a characteristic group of elements which includes lithium, beryllium, niobium, tantalum, tin, rare earth metals, uranium, thorium, tungsten, and zirconium. The word "pegmatite," although generally referring to dikes-with quartz and alkali feldspar as principal constituents, is not restricted to such compositions but can be used for similarly coarse-grained varieties of any igneous rocks. Nepheline–syenite pegmatites, for example, contain chiefly nepheline and feldspar, and gabbro pegmatites consist of coarse-grained plagioclase, pyroxene, and hornblende. In general, pegmatites have a lesser content of mafic minerals and a more sodic plagioclase than the igneous rocks with which they occur.

Pegmatites and aplites are commonly found as dikes cutting intrusive igneous rocks, often extending beyond the contacts into adjacent metamorphic rocks. In addition to dikes with parallel walls, they may form pockets and irregular masses within an intrusive (Fig. 18-1). Pegmatites also are common as pockets and stringers in metamorphic rocks far from intrusive contacts. Aplite and pegmatite are often found together in the same dike, sometimes separated by a sharp contact and sometimes grading into one another.

From their composition and mode of occurrence, these two rocks have long been regarded as material crystallized from the residual liquid of a silicate magma. The coarse texture of pegmatites has commonly been ascribed to crystallization from a residual melt particularly rich in water. This sounds like a reasonable explanation, but it seems to run counter to the experimental results described above showing that the amount of dissolved water in granitic melts is limited to about 10%. How can the postulated water-rich residual fluid evolve if water necessarily boils out of the melt before crystallization has progressed very far?

A possible answer comes from experiments described more than a quarter-century ago by Jahns and Burnham (1969), classical experiments that have been repeated and amplified in much more recent work. In a typical experiment a mixture of quartz, K-feldspar, and Na-feldspar is heated to melting with a measured amount of water slightly less than that necessary for saturation, and allowed to cool slowly. Crystallization from the homogeneous liquid begins with the separation of a fine-grained quartz–feldspar aggregate on the walls of the container. Crystal formation causes increasing concentration of water in the remaining liquid, and eventually conditions for retrograde boiling are reached. A water-rich phase separates in bubbles, and the character of the crystallization changes abruptly. Relatively large, well-formed crystals of quartz and feldspar grow out into the bubbles, becoming larger as the proportion of the water-rich phase increases. Ultimately a crystalline solid is produced showing a gradation from sugary, aplitic-looking material on the outside to coarse-grained material with the texture of pegmatite at the center. The experiment can be modified by changing the original amount of water and by maintaining different pressures during crystallization. If the pressure is high enough at all times to prevent formation of a water-rich phase, the crystalline product has the fine-grained texture of aplite throughout. If the proportion of water is large and if the pressure permits boiling to occur early in the experiment, the relative amount of "aplite" with respect to "pegmatite" is small. By varying pressure and temperature during crystallization, Jahns and Burnham were able to produce zoning and replacement relations like those in natural pegmatites. The resemblance of some of the experimental products to natural rocks is very striking.

The experiments therefore suggest that aplites and pegmatites are two products of the concentration of water in a magma as it cools, aplite forming from the residual viscous silicate liquid and pegmatite chiefly from the less viscous, probably supercritical water-rich fluid that separates by retrograde boiling. The mixtures used by Jahns and Burnham differ from the melts that form natural pegmatites and aplites in at least one important respect, that they do not contain the other solutes—for example, HCl, HF, CO_2, NaCl—which would accompany a residual magma in nature. Nevertheless, the similarity of the experimental mixtures to natural materials is sufficiently close to make it unlikely that these substances have a major influence. Aplites and pegmatites from a magma body may form completely independent dikes, depending on volatile content and pressure–temperature relations at various stages of cooling, or retrograde boiling may occur irregularly during cooling so that the two rocks appear together in outcrop or even intermixed. Pegmatites found in metamorphic rocks at a distance from any intrusive can be regarded as products of the water-rich initial fluid that forms by incipient melting of the original rock. Thus Jahns and Burnham's work provides a mechanism, based solidly on experiment, which explains satisfactorily the principal features of aplites and pegmatites observed in nature.

18-5 QUARTZ–SULFIDE VEIN AND REPLACEMENT DEPOSITS

More difficult to account for is the separation from a magma of the material that ultimately leads to the crystallization of quartz–sulfide veins. The composition of such deposits shows a wide range, from pure quartz to nearly pure mixtures of metallic sulfides. The sulfides may be simple ones like pyrite and galena, or complex sulfosalts like enargite and jamesonite. Vein deposits may contain carbonates, together with quartz or practically to the exclusion of quartz. Metals may occur as native elements (e.g., gold), oxides (e.g., pyrolusite, cassiterite), or carbonates (e.g., rhodochrosite) as well as sulfides. For brevity the deposits will be referred to as "veins," but the same associations of minerals occur also as widespread disseminations and replacement deposits. The most spectacular of the latter deposits are replacement of carbonate strata (skarns) by sulfides and Ca-silicates including wollastonite, garnet, pyroxene, and amphibole.

Many quartz–sulfide veins and areas of replacement ore deposits occur in and adjacent to igneous intrusives, so that a common origin seems clear (Fig. 18-1). On the other hand, many such deposits are also found at long distances from intrusives, some even in areas where no sign of igneous activity is apparent; the lead–zinc deposits of the Mississippi Valley are a good example. Certainly, then, not all quartz–sulfide veins have an origin connected with igneous activity, but enough of them do show an intimate association with intrusive or volcanic rocks to suggest that the cooling of a silicate melt is one possible method of formation.

If we assume that quartz–sulfide veins, like aplites and pegmatites, can be a product of the later stages of magmatic crystallization, what is the relation between the three kinds of material? Because vein and replacement deposits have compositions very different from those of igneous rocks, because their minerals generally show a clear order of crystallization, with the latest ones to form often projecting out as euhedral crystals into open cavities, and because the temperatures of formation, as determined in a variety of ways, are far below the temperatures at which their constituents could be molten, these deposits are almost certainly formed by crystallization out of a hot aqueous solution rather than by freezing of a residual silicate melt. Perhaps the residual liquid of magmatic differentiation separates into two fluids after part of its quartz and feldspar have crystallized to form aplite; when the two fluids are present together, further crystallization of the silica-rich part gives pegmatite; and the remaining water-rich phase, on further cooling, might deposit quartz and sulfides.

Attractive as this hypothesis looks, it is not very successful in explaining geologic relationships. Pegmatites may contain ore minerals, but they are largely nonsulfide minerals of a special group of metals—Li, Be, Sn, U, Th, Nb, Ta—quite different from the usual assemblage in vein and replacement deposits. Even more damaging evidence against the hypothesis is the lack, or at least the great scarcity, of observed transitions between pegmatites and quartz–sulfide veins. Surely, if pegmatites and veins are formed from different fractions of the same original fluid, gradations between the two kinds of material would be fairly common; yet one of the conspicuous observational facts about ore deposits is that pegmatites and quartz–sulfide veins, where they do occur together, show sharp contacts. The field evidence makes it clear that the fluids responsible for pegmatites and quartz–sulfide veins must be sufficiently separated, either in origin or in timing, so that the two kinds of deposits cannot intergrade.

Do phase separations offer any further possibilities, which might take care of this difficulty? In the previous discussion of the critical point, one conceivable set of circumstances was described in which three fluid phases rather than two might be present during the cooling history. This requires that the water-rich phase from the original separation contains so much nonvolatile dissolved material that its critical point is a hundred degrees or more above the critical point of pure water. Such a fluid, when cooled through its critical point, could separate into two phases, both of them water-rich but one of them now a fairly tenuous vapor with far less dissolved salt than the other. One can expand the hypothesis by supposing that a subcritical vapor of this sort might be capable of separating from the original magma at any time, perhaps locally where pressure is drastically reduced, and would not be limited to places where aplite and pegmatite had formed previously; this would account for ore deposits adjacent to intrusives where pegmatites and aplites are scarce or absent. But the hypothesis is little more than free speculation, since there is little good field evidence for the supposed sequence of separations, and no certain proof from experiments that the third fluid would be able to transport the common heavy metals in sufficient amounts to form ore deposits.

Thus the ultimate origin of ore-forming fluids from a cooling magma remains obscure. If, as suggested in the above speculative reconstruction, they separate first as high-temperature, high-density gases, their later history must involve drastic cooling and partial condensation, since evidence seems convincing that actual deposition of most quartz–sulfide ores is from a liquid rather than from a gas (Sec. 19-1). Liquid aqueous solutions with this history, at temperatures in the range 100 through 300°C, are difficult to distinguish from hot underground waters of other origins: heated groundwater containing metals leached from the rocks it has traversed, water derived from volcanic activity, or water set free during progressive metamorphism of sedimentary and igneous rocks. The solutions responsible for ore formation may come from any of these sources and very likely are often mixtures from more than one source.

The problem of distinguishing minerals formed by water with a source in cooling magma from those formed by heated groundwater has been attacked most successfully by study of the isotopes of oxygen (Sec. 10-4). Groundwater is almost entirely meteoric water, meaning that it has its source in rain, and in rainfall the ratio $^{18}O/^{16}O$ is smaller than the ratio in the primary minerals of igneous rocks. This ratio, measured in minerals of veins or of rocks that have been altered by hot solutions, thus serves as a "fingerprint" for identifying the source of the water, or for estimating how much of the water was of meteoric origin (Sec. 11-5).

In many ore deposits so studied, the ratio has turned out to be low, meaning that the water responsible for the deposits was in large part meteoric. This suggests that much postulated "magmatic" water may be groundwater that has been heated and caused to circulate by a magma body below (Fig. 18-1). One can think of the water as part of a convectional system, rising above the hot magma and being replenished by cooler water from all sides. Various mixtures of groundwater and water derived from the magmatic differentiation can be imagined (and have been demonstrated by oxygen-isotope ratios). Then the question arises, does the metal content of quartz–sulfide veins and replacement deposits formed by such waters come chiefly from the magma, or from the surrounding sedimentary or metamorphic rocks? To this question there can be no general answer: very likely in many cases both sources make a contribution to the metals of the ore.

18-6 VOLCANIC GASES AND HOT SPRINGS

All through this long discussion of magmatic volatiles we have faced the difficulty of working with substances for which we have no actual samples. We have based our discussion on the products of the activity of volatiles, the minerals or altered rock they leave behind, but not on samples of the volatiles themselves. It is time that we look briefly at the one place where magmatic volatiles can actually be collected in large amounts: areas where volcanoes are active or have recently been active. Here, obviously, in the gases from fumaroles and the water from hot springs, we find material that has come to the surface as volatile products from a body of magma somewhere below (Fig. 18-1). We note, of course, that the samples we collect are not as pure as could be wished, because the rising volatiles may be altered by contact with rock and have probably mixed with air and heated groundwaters, but at least they can give us useful data about volatile compositions.

The scientific study of fumaroles has a long history, going back to the early

1800's. Much qualitative and semiquantitative information on gases from volcanoes in the Mediterranean, the West Indies, and Iceland was obtained during the 19th century by French, German, and Italian geochemists. The use of modern techniques for collection and analysis of gases dates from the classical work of Day, Jaggar, and Shepherd at Kilauea between 1910 and 1920 (summarized by Shepherd, 1938). In more recent years good samples of volcanic gases have been obtained in many parts of the world, especially Japan, Kamchatka, New Zealand, Iceland, and East Africa. We look at two examples of such studies (Table 18-1), one in Hawaii and one in Ethiopia, both on gas from vents in spatter cones developed on the hardened surface of lava "lakes" in the calderas of basaltic volcanoes—a particularly favorable kind of site for obtaining samples with a minimum of contamination. At both volcanoes, sample temperatures at the time of collection were 1100–1200°C.

The general similarity of gases from these widely separated volcanoes is striking. In most samples water is the chief constituent, CO_2 next, and SO_2 third. The preponderance of these three substances has been noted in high-temperature gases from many other volcanoes. In general, gas compositions reported from volcanoes around the world show much similarity but also perplexing differences, especially in the content of halogen gases (chiefly HCl and HF). Evidence is not convincing as to whether gas composition has any clear relation to kind of lava or kind of eruptive activity.

Some of the observed variation in gas composition may be due to shifting of equilibria as the temperature changes. This possibility can be explored by

TABLE 18-1

Analyses of gases from fumaroles at Kilauea and Erta'Ale (Ethiopia), in mol%

	Kilauea			Erta'Ale		
H_2O	36.18	61.56	67.52	84.8	69.9	79.4
CO_2	47.68	20.93	16.96	7.0	15.8	10.4
CO	1.46	0.59	0.58	0.27	0.68	0.46
COS				0.001	0.01	0.009
SO_2	11.15	11.42	7.91	5.1	10.2	6.5
SO_3	0.42	0.55	2.46			
S_2	0.04	0.25	0.09	0.4	0.0	0.5†
HCl	0.08	0.00	0.20	1.28	1.22	0.42
H_2	0.48	0.32	0.96	0.85	2.11	1.49
N_2	2.41	4.13	3.35	0.10	0.25	0.18
Ar	0.14	0.31	0.66	0.001	0.001	0.001

Sources: For Kilauea, Shepherd, 1938, p. 321; for Erta'Ale, Giggenbach and Le Guern, 1976, p. 26. The three Kilauea samples are Shepherd's Nos. J8, J11, J13; the first two from Erta'Ale are representative individual samples, and the third is an average of 18 samples. Shepherd calculated all chlorine as Cl_2; his numbers are expressed here as HCl, for ease of comparison with the Erta'Ale analyses. The values for S_2 may include some H_2S.

postulating a gas of given composition, in which all possible reactions among the gaseous species have attained equilibrium under given conditions of temperature and pressure, and calculating the effects on various equilibrium reactions that temperature–pressure changes would produce. Of many possible reactions that should be considered, the following list is a sampling:

$$H_2 + CO_2 \rightleftharpoons CO + H_2O \qquad 2H_2 + S_2 \rightleftharpoons H_2S$$
$$4H_2 + 2SO_2 \rightleftharpoons S_2 + 4H_2O \qquad CO_2 + 4H_2 \rightleftharpoons CH_4 + 2H_2O$$
$$N_2 + 3H_2 \rightleftharpoons 2NH_3 \qquad 2CO + S_2 \rightleftharpoons 2COS$$

The calculations follow a pattern we have often used before: standard free energies for the various substances at different temperatures and pressures are first obtained, then from these the equilibrium constants, and from these in turn the proportions of gases present for various assumed overall compositions (Chap. 8). Since each substance takes part in several reactions, all occurring at once, the calculation involves handling a considerable number of simultaneous equations, and is best done with a computer (Chap. 2). Typical results of such calculations are illustrated in Figs. 18-7 and 18-8.

The predicted effects of changing temperature (Fig. 18-7) show reasonable correspondence with analytical results, indicating that reactions among fumarole gas constituents are fast enough to maintain a fairly close approach to equilibrium. Thus the often noted prominence of SO_2 and H_2 at high temperatures, and of H_2S and CO_2 at low temperatures, is nicely accounted for by displacements of equilibrium. Fig. 18-8 shows variations in composition produced by change in oxidation state, notably the increase in SO_2 and SO_3 at the expense of H_2S, and the decrease in H_2 and CO, as the fugacity of oxygen increases.

The analytical data and calculations from them are therefore consistent with a hypothesis that the original gases dissolved in lava are fairly similar from one volcano to another. Some initial differences may exist, but in large part the observed variation is accounted for by adjustment of equilibria to changing temperature and pressure and by admixture of air and meteoric water.

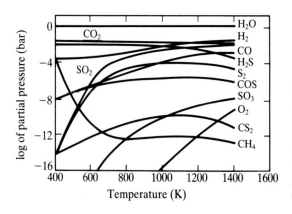

FIGURE 18-7

Calculated change in equilibrium composition of volcanic gas with temperature. Assumed atomic ratios: H/O/C/S = 275.5/142.2/2.680/1.000, from analyses of a typical Kilauea gas. Assumed total pressure = 1 atm. (Reprinted by permission from Heald et al., 1963, p. 547.)

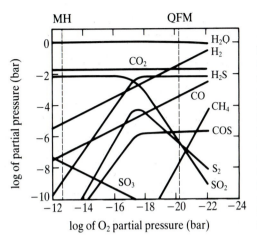

FIGURE 18-8

Change in equilibrium composition of volcanic gas with change in oxidation state. Assumed atomic ratios same as in Fig. 18-7. Temperature 627°C, total pressure 1 atm. The dashed lines are O_2 pressures that might be found in lavas, as indicated by mineral assemblages: MH for the combination magnetite–hematite (sometimes found in rhyolite), and QFM for quartz–fayalite (olivine)–magnetite (possible in basalt). The oxygen fugacities, or partial pressures, of most lavas lie between these extremes. (Reprinted by permission from Heald et al., 1963, p. 549.)

Of particular interest among the gas constituents are the traces of metal compounds. These are commonly not reported in the analyses because their amounts are exceedingly small, but their presence is made obvious by the sublimates that accumulate around the mouths of some fumaroles. Most of the metal compounds are volatile only at magmatic temperatures, hence precipitate readily as the gas cools when it reaches the surface. The most abundant of the metal compounds are generally chlorides and sulfates of Na, K, Ca, Mg, Al, and Fe, oxides of Si and Fe, ammonium chloride, and native sulfur. Minor amounts of many other elements are commonly detectable, of which the most abundant are F, Br, B, P, As, Zn, Cd, Cu, Pb, Mn, and Sn. The list includes several metals commonly found in sulfide ores, hence suggests a relation between volcanic emanations and some kinds of ore deposits.

The hot springs that accompany or replace fumaroles in many volcanic areas are clearly a less reliable source of information about magmatic volatiles, since much of their water is of meteoric origin—simply rainwater that has penetrated deeply enough into the ground to be warmed by the still-hot rocks beneath (Fig. 18-1). This is demonstrated by the topographic position of the springs and the common dependence of flow on rainfall. Still, it is reasonable to suppose that a small part of the spring water has a magmatic source, especially for springs in areas of recent or active volcanism. Such springs, in Yellowstone Park, at Steamboat Springs in Nevada, and in the volcanic areas on the north island of New Zealand and on Iceland, have been the subject of much geochemical study. A compilation of analyses by White, Hem, and Waring (1963) shows evidence of a magmatic component in the waters of many of the hotter springs by abnormally high ratios of Li to Na and of B to Cl, and unusually low I to Cl. Heavy metals are generally present only in traces, but at some springs analyses of solid material suspended in the water or accumulating as muds give evidence that a few metals are precipitating at the present time—notably Sb, As, Ag, and Au (e.g., Krupp and Seward, 1987). In analyses that include concentrations of gases dissolved in the water, the gases found are similar to those from fumaroles: generally CO_2 in greatest abundance, plus minor CO, H_2, N_2, NH_3, H_2S, and Ar in varying amounts.

One other possible way to get information about magmatic volatiles is to heat igneous rocks in the laboratory, in the hope that small amounts of the gases present at the time of consolidation have been trapped in at least three ways: (1) as constituents of minerals, like the water and fluorine in micas and hornblende; (2) as fluid inclusions, tiny cavities within mineral grains filled with liquid or gas or both; (3) by simple adsorption on surfaces or in cracks. Gases held in all these ways should be liberated if the rock is sufficiently heated.

Here again, just as in studies of fumarole gases and hot springs, contamination is a possible problem. It can be minimized by using extreme care to select specimens for heating that show no sign of weathering or hydrothermal alteration, but certainly cannot be altogether eliminated. Results of such experiments show the same assortment of gases as found in fumaroles, and the same variability in proportions from specimen to specimen as that from one fumarole to another. Studies designed specifically to analyze constituents of fluid inclusions have shown gases and dissolved ions generally similar to those in hot-spring waters.

In summary, we possess no really satisfactory method of analyzing volatile materials from cooling magmas, but indirect evidence from three sources— fumarole gases and sublimates, hot springs, and gases from heated rocks—gives a consistent picture of what these volatiles must be. Qualitatively the list of volatile constituents is surprisingly uniform from one volcanic area to another, although there are marked quantitative variations in amounts even among fluids from adjacent sources. The list of fluid substances is consistent also with inferences about the nature of volatiles responsible for contact metamorphic zones (Chap. 16) and for some kinds of ore deposits (Chap. 19).

18-7   RECAPITULATION

The formation of igneous rocks from molten silicate mixtures is a part of geology to which physical chemistry has been broadly and successfully applied. The sequence of crystallization of minerals and the variety of rocks to which this sequence can give rise are strictly chemical problems—conditioned, of course, by the setting under which the crystallization occurs and by mechanical accidents that may disturb the course of cooling. Laboratory experiments with grossly simplified silicate systems show enough resemblance to natural systems that the process of crystallization can be described in terms of a few physicochemical principles. Chief of these principles is the phase rule, which specifies a limitation on the number of solids that can coexist at equilibrium in a system of given composition. Add to the phase rule a few generalities about the melting behavior of mixtures and solutions (which, if desired, can be expressed in the formal language of free energies and entropies of melting), add a few specific data on the melting points of individual compounds, and the main features in the behavior of igneous melts can be summarized neatly and succinctly. We say that we can *explain* the formation of igneous rocks, with an elegance not approached in most parts of geology.

Less successful have been efforts to apply chemical principles to the end stages of magmatic crystallization, particularly to the behavior of volatile substances dissolved in the silicate melt. Chief of the volatile solutes is nearly always water. The water, escaping from the magma in the final stages of crystallization, carries with it much dissolved material, including both volatile and nonvolatile substances. The nature of the volatile materials can be inferred from analyses of volcanic gases, hot springs, fluid inclusions, and gases obtained by heating rocks, but the reliability of such data is limited by possible contamination. Some of the nonvolatile material carried by water escaping from magma can be sampled at fumarole orifices, but fumarole deposits are a poor basis for guessing at the kind of dissolved material in the water under conditions of high temperature and pressure beneath the surface. Materials deposited from such water include aplite and pegmatite dikes and the constituents of some quartz–sulfide vein and replacement deposits. The number of different fluid phases present as these materials were transported and deposited is uncertain, because critical points of the different phases are dependent on the unknown concentrations of volatile and nonvolatile solutes. The multiplicity of unknown variables is simply too great to permit more than guesses about the specific conditions of formation of particular deposits. Chemical principles provide guidance here about general relations, but do not have the detailed applicability that they had for the earlier stages of igneous rock formation.

PROBLEMS

1. What conditions are necessary for a liquid to exhibit a "second boiling point"? Could this phenomenon be demonstrated with a single pure substance? With a mixture of only two substances? With a mixture of three substances?

2. Consider a hypothetical body of granitic magma, with horizontal dimensions 100 by 100 km and a thickness of 10 km, situated in the crust so that its top is 10 km below the surface. (*a*) What would the pressure be at the top of the body due to the weight of overlying rock? (*b*) What would be the pressure at the base of the magma? (*c*) At the level of the top of the body, what would the temperature be if the hot magma were not present and if the temperature were determined only by the normal geothermal gradient (measured values lie between 10 and 50°C/km? (*d*) At approximately what temperature would the last liquid of the magma solidify, assuming that the granite is made up of approximately equal parts of quartz, Na-feldspar, and K-feldspar and that it is under water-vapor pressure equal to the hydrostatic pressure? (*e*) If the magma contains 1% of dissolved water by weight, what weight of water is present? (*f*) Calculate the volume of this weight of water at 700°C and 1 kbar, using the simple gas laws. (*g*) Calculate the volume under the same conditions but using the measured value of density, 0.28 g/cm^3.

3. The critical pressure of water is approximately 220 bar. At what depth in the crust is all water in a supercritical state (i.e., at a supercritical pressure but not necessarily at a supercritical temperature)?

4. What is the geologic evidence for the statement: "Many pegmatites are formed by the residual solutions from crystallization of granitic magmas"? What is the geologic evidence that not all pegmatites have formed in this manner? How does chemical evidence support the conclusion that pegmatites may form in at least two ways?

5. Is the following statement correct? Why or why not? "A pegmatite dike forms by the crystallization of a melt of the same composition."

6. How could you prove that the material of a quartz–sulfide vein is not an igneous rock, i.e., was not formed by crystallization of a melt of roughly the same composition?

7. Would you expect to find H_2S and SO_3 in the same sample of volcanic gas? H_2S and HCl? H_2S and Cl_2? HCl and SO_3? Discuss these combinations by considering possible chemical equilibria. In calculating equilibrium at magmatic temperatures, use the simplifying assumption that ΔH and ΔS remain constant over a temperature range of several hundred degrees, and find free energies from the expression $\Delta G° = \Delta H° - T\Delta S°$.

8. Look up equilibrium data for the reaction

$$SiCl_4(g) + 2H_2O \rightleftharpoons 4HCl + SiO_2(s).$$

Under what conditions (of total pressure, temperature, and HCl/H_2O ratio) would you expect SiO_2 to be deposited from volcanic gas? Under what conditions would the silicon of SiO_2 be volatilized as $SiCl_4$?

9. Given that the free energy change in the reaction

$$PbS(s) + 2HCl \rightleftharpoons PbCl_2(g) + H_2S$$

is about +36.0 kJ at 627°C and +31.4 kJ at 927°C, discuss the effect of temperature, total pressure, and HCl/H_2S ratio on the deposition of PbS from volcanic gas.

REFERENCES

Burnham, C. W., and R. H. Jahns: "A method for determining the solubility of water in silicate melts," *Am. Jour. Sci.*, vol. 260, pp. 721–745, 1962. Solubility of water in melts prepared from a pegmatite and from pure albite, at pressures up to 10 kbar.

Carmichael, I. S. E., F. J. Turner, and J. Verhoogen: *Igneous Petrology*, McGraw-Hill, p. 323, 1974.

Giggenbach, W. F., and F. Le Guern: "The chemistry of magmatic gases from Erta'Ale, Ethiopia," *Geochim. et Cosmochim. Acta*, vol. 40, pp. 25–30, 1976.

Heald, E. F., J. J. Naughton, and I. L. Barnes: "The chemistry of volcanic gases," *Jour. Geophys. Research*, vol. 68, pp. 539–557, 1963. Description of methods of collecting and analyzing volcanic gases, and diagrams showing results of equilibrium calculations based on the analyses.

Jahns, R. H., and C. W. Burnham: "Experimental studies of pegmatite genesis: I. A model for the derivation and crystallization of granitic pegmatites," *Econ. Geology*, vol. 64, pp. 843–864, 1969.

Kennedy, G. C., G. J. Wasserburg, H. C. Heard, and R. C. Newton: "The upper 3-phase region in the system SiO_2–H_2O," *American Jour. Sci.*, vol. 260, pp. 501–521, 1962.

Morey, G. W.: "The solubility of solids in gases," *Econ. Geol.*, vol. 52, pp. 225–251, 1957.

Ölander, A., and H. Liander: "The phase diagram of sodium chloride and steam above the critical point," *Acta Chem. Scand.*, vol. 4, pp. 1437–1445, 1950.

Pitzer, K. S.: "Ionic fluids," *Jour. Phys. Chem.*, vol. 88, pp. 2689–2697, 1984.

Shepherd, E. S.: "Gases in rocks and some related problems," *Am. Jour. Sci.*, vol. 235a, pp. 311–351, 1938. A classical paper, containing analyses of gases obtained by heating specimens of igneous rocks; for comparison, includes a summary of Shepherd's earlier analytical work on gases collected at Kilauea.

Sourirajan, S., and G. C. Kennedy: "The system H_2O–NaCl at elevated temperatures and pressures," *American Jour. Sci.*, vol. 260, pp. 115–141, 1962.

Tuttle, O. F., and N. L. Bowen: "Origin of granite in the light of experimental studies," *Geol. Soc. America Memoir* 74, p. 83, 1958.

SUGGESTIONS FOR FURTHER READING

Carmichael, I. S. E., and H. P. Eugster, (eds): "Thermodynamic modeling of geologic materials: minerals, fluids, and melts," in *Reviews in Mineralogy*, vol. 17, Mineralogical Society of America, 1987. Contains many papers on thermodynamic properties of aqueous species and mineral solubilities in supercritical fluids.

Helgeson, H. C.: "Effects of complex formation in flowing fluids on the hydrothermal solubilities of minerals as a function of fluid pressure and temperature in the critical and supercritical regions of the system H_2O," *Geochimica et Cosmochimica Acta*, vol. 56, pp. 3191–3208, 1992. An excellent summary of predictive thermodynamic calculations on mineral solubilities in supercritical fluids.

Holtz, F., Behrens, H., Dingwell, D. B., and Taylor, R. P.: "Water solubility in aluminosilicate melts of haplogranite composition at 2 kbar," *Chemical Geology*, vol. 96, pp. 289–302, 1992. Solubility of water in melts containing varying proportions of SiO_2, $NaAlSi_3O_8$, and $KAlSi_3O_8$.

Krupp, R. E., and T. M. Seward: "The Rotokawa geothermal system: an active epithermal gold-depositing environment," *Econ. Geology*, vol. 82, pp. 1109–1129, 1987. Muds deposited by hot water in a crater lake contain sulfides of arsenic and antimony with Au, W, Tl, Hg, Ag, and Ga.

Stolper, E. M.: "Water in silicate glasses: an infrared spectrographic study," *Contrib. Mineral. Petrol.*, vol. 81, pp. 1–17, 1982. Demonstration by infrared spectrography that water dissolved in silicate glasses has two forms, OH and H_2O.

Symonds, R. B., M. H. Reed, and W. I. Rose: "Origin, speciation, and fluxes of trace-element gases at Augustine Volcano, Alaska," *Geochim et Cosmochim. Acta*, vol. 56, pp. 633–657, 1992. Ore metals in HCl-rich magmatic gases.

Taylor, H. P.: "$^{18}O/^{16}O$ evidence for meteoric-hydrothermal alteration and ore deposition in the Tonopah, Comstock Lode, and Goldfield mining districts," *Econ. Geology*, vol. 68, pp. 747–764, 1973. A good example of many papers by Taylor and his coworkers on the use of oxygen isotopes in determining the origin of water in vein-forming solutions.

Weissberg, B. G., P. R. L. Browne, and T. M. Seward: "Ore metals in active geothermal systems," in Barnes, H. L. (ed.): *Geochemistry of Hydrothermal Ore Deposits*, Wiley, pp. 738–780, 1979. A review of the geochemistry of solutions and ore minerals in geothermal systems in many parts of the world.

Whitney, J. A.: "The origin of granite: the role and source of water in the evolution of granitic magmas," *Geol. Soc. America Bull.*, vol. 100, pp. 1886–1897, 1988. Review of experimental field studies of possible sources of water in granitic melts and the effects of different amounts of crystallization.

CHAPTER
19

HYDROTHERMAL ORE DEPOSITS

The origin of ore deposits is one of the classic problems of geology, argued at length for more than three centuries. More accurately, it is a large group of unsolved and partially solved problems, for ore deposits take many forms and originate in a variety of ways. Some methods of ore formation are obvious. There is no mystery, for example, about the mechanical process that leads to accumulation of gold in stream placers, or about the chemical reactions that cause iron to precipitate in bogs or aluminum to be concentrated in bauxite. But the problem of origin for the great majority of ore deposits bristles with difficulties, especially so for ores formed at temperatures higher than the normal temperatures of the earth's surface. It is to these deposits that we turn our attention here.

Many of the deposits generated at medium to high temperatures are associated with igneous rocks, and their origin is clearly related to magmatic processes. Some ores can accumulate as direct products of magmatic differentiation: horizons of chromite found in layered mafic intrusions like the Bushveld of South Africa would be an example. More abundant are deposits in which water has played a role—in which the metals have been transported from a distance dissolved in an aqueous fluid, and have somehow been concentrated and precipitated in the place where we now find them. One obvious source of such a fluid is the water-rich residual material from crystallization of magmas that we discussed at length in the last chapter. But this is by no means the only source. Metal-bearing solutions can originate as heated rain (meteoric) or ocean water that has circulated at great depth or has come near an intrusive body, or as seawater originally imprisoned in sediments at the time of their formation (connate water), or as volatile material expelled from rocks during metamorphism. Whatever be their origin, these warm solutions are called *hydrothermal fluids*, and ores that they may deposit are called *hydrothermal ores*.

It is to hydrothermal ores that we devote this chapter. These make up only one kind of ore deposit, but a very important kind that supplies much of the variety of metals on which industrial civilization depends. They are also a kind of deposit to which some of the geochemical principles we have developed in earlier chapters can be usefully applied. We seek here an understanding of the physical and chemical processes responsible for the transport and concentration of metals in the formation of an ore deposit, and then for the movement of metals in surface environments when a deposit has been exposed to weathering and erosion.

19-1 ORE-FORMING SOLUTIONS

What is the nature of the metal-bearing fluid? One hint comes from analyses of hot-spring water and fumarole fluids. In some places these fluids are at present depositing small amounts of ore-metal compounds (Sec. 18-6), and the conclusion seems reasonable that they are similar to solutions that precipitated ore minerals beneath the surface. In hot springs the ore minerals are precipitated from a liquid solution, in fumaroles they crystallize out of a gas. Which of these agents, liquid or gas, plays the chief role in forming an ore deposit? Evidence is strong that most ores are deposited from liquid or supercritical solutions rather than a low-density gas. Especially convincing is the observation that in many deposits the ore minerals have replaced carbonate or silicate minerals, meaning that the carbonates or silicates must have been removed by the ore-forming fluid, and the carrying off of such materials in a gas seems most unlikely. Then in deposits where mineral associations indicate fairly low temperatures of formation, the transport of metals and the constituents of gangue minerals in a gas is clearly impossible.

Volatility of Ore-metal Compounds

Nevertheless, the volatility of metal compounds, especially the chlorides, together with the theoretical likelihood that a water-rich gas would separate at a late stage in the cooling of a magma (Sec. 18-2), makes gaseous transportation of ore metals probable at least in the early high-temperature part of metal concentration. The final deposition of ore may be only the last step in a complex process during which the metal or metals are vaporized, precipitated, dissolved, transported, and reprecipitated. A sequence of events like this can sometimes be observed taking place in the sublimates around fumarole orifices (Sec. 18-6).

In what form would the metals exist in high-temperature gases? Most probably as chlorides, since among all the compounds that might be formed with the constituents of magmatic gases, the chlorides for many of the heavy metals are the most volatile. In whatever combination a metal may exist in a cooling magma (as oxide, sulfide, sulfate, silicate), chlorine or hydrogen chloride in the vapor could form volatile compounds stable enough to hold some of the metal in the gas state at appreciable concentrations. This can be proved by calculating vapor pressures of metal chlorides in equilibrium reactions of the form

$$PbS + 2HCl(g) \rightleftharpoons PbCl_2(g) + H_2S(g). \qquad (19\text{-}1)$$
galena

The calculation is a straightforward application of methods we have used in Chapter 8. Results of such calculations for several common ore metals over a range of conditions that might be expected in gas accompanying a cooling granitic magma at shallow depths are shown in Table 19-1 (the method of calculation is outlined in the table).

What do these numbers mean? More specifically, how high must the volatility of a metal be in order for appreciable amounts of it to be transported in the gaseous form? There is no simple answer to this question, but a rough calculation will indicate the orders of magnitude involved. Imagine a large body of magma measuring 100 by 100 km and 10 km thick, containing 1% of dissolved water. The total weight of water is about 3×10^{18} g, or 10^{17} mols, which would occupy about 10^{16} liters at 800°C and 1000 bar. For a metal whose compounds have a total vapor pressure of 10^{-7} bar at 800°C, and which has an atomic weight of

TABLE 19-1

Maximum vapor pressures (in bars) of metal compounds in magmatic gas, in equilibrium with the most stable solids at 627 and 827°C

	627°C				827°C			
O_2	−14	−17	−20	−23	−8	−11	−14	−17
$FeCl_2$	−3.7	−3.2	−2.9	−2.9	−3.2	−2.7	−2.2	−2.0
$MnCl_2$	−3.9	−3.9	−3.9	−3.6	−3.7	−3.7	−3.7	−3.6
$ZnCl_2$	−0.3	−2.3	−2.3	−2.3	+3.9	−0.1	−1.1	−1.1
$PbCl_2$	+0.9	−1.1	−1.1	−1.1	+2.4	+0.5	−0.4	−0.7
$CuCl^1$	−3.7	−6.7	−6.7	−6.7	+2.8	−3.9	−5.4	−5.4
$AgCl$	−5.8	−6.8	−6.8	−7.4	−3.0	−4.2	−4.8	−5.2
$SnCl_2{}^2$	−3.8	−2.4	−0.9	+0.1	−3.9	−2.5	−1.0	+0.3
MoO_3	−4.6	−7.1	−8.6	−10.1	−1.9	−1.9	−4.2	−5.7
$WO_3 \cdot H_2O$	−12.0	−12.0	−12.0	−12.0	−4.7	−4.7	−4.7	−4.7

Numbers are logarithms of partial pressures expressed in bars.
[1] Cu is present as CuCl and Cu_3Cl_3. The numbers are total copper expressed as CuCl.
[2] The numbers for log $P_{O_2} = -14$ and -8 include a small contribution from $SnCl_4$.
Calculated based on stability of metal compounds that could be present in a magma in contact with a gas having a composition similar to a rough average of many analyses of volcanic gases: water fugacity = 1000 bar, HCl fugacity = 10, HF fugacity = 0.3, and the fugacity of sulfur gases ($H_2S + S_2 + SO_2$) = 10. The fugacities of minor gas components are considered numerically equal to their partial pressures. The numbers for O_2 show the range of oxygen fugacities used, the lowest value at any one temperature representing quartz–fayalite–magnetite equilibrium and the highest value hematite–magnetite equilibrium.

about 100, 10^{16} liters would contain about 1000 tonnes of the metal. For this amount to form an ore deposit, all the gas from the magma would have to precipitate its metal content in one small area, which is not very likely. Certainly, then, 10^{-7} bar is an extreme lower limit of partial pressure, below which the amount of metal transported would be too small to matter. Probably 10^{-6} bar is a more reasonable lower limit for purposes of discussion.

On this basis the numbers in Table 19-1 show that all the metals listed can be present in a gas phase at 827°C in amounts large enough to be important for ore accumulation. The amounts decrease markedly as the temperature falls, but at 627°C all except Cu, Ag, Mo, and W are still above the 10^{-6} bar limit. Possibly some ores are formed by direct precipitation out of such a high-temperature gas phase: deposits of magnetite, chalcopyrite ($CuFeS_2$), or scheelite ($CaWO_4$) immediately adjacent to an intrusive contact would be likely examples. But probably most deposits have a longer history, the metals originally carried as gases becoming part of a liquid solution and being carried in this form to regions of lower temperature or to regions where the solution mixes with fluids from other sources. The calculations, although only estimates, do show clearly that volatility can be an important factor in the high-temperature behavior of common ore metals, and recent data on high-temperature vapor-filled inclusions in igneous rocks support this view.

Metal Complexing in Hydrothermal Solutions

Following the history of any metal from its possible original presence in a magmatic gas to its ultimate deposition from a low-temperature or medium-temperature liquid solution is difficult at best. A possible procedure would be use of the material balance relation of Eq. (11-39) to evaluate fluid transport of metals. Rather than attempt this, we jump to the other end of the temperature scale, to deposits formed at temperatures in the range 50–500°C, where geologic relations and laboratory experiments can give us clearer evidence about the behavior of ore-depositing solutions in liquid form. We turn our attention now to the geochemistry of such moderate-temperature ore-forming solutions without regard to their source, whether it be in a crystallizing igneous melt or in heated groundwater of meteoric or seawater origin, or in water extracted from sedimentary or metamorphic rock.

So we think of liquids that circulate in the cracks and interstices of ordinary rocks, their temperature in the general range of a few hundred degrees and their compositions similar to those of hot-spring waters and waters pumped from geothermal areas. How can such liquids carry metals in solution? Most of the common ore metals—lead, copper, zinc, silver, molybdenum, mercury—occur in ores chiefly as sulfide minerals, but these metal sulfides are among the most insoluble compounds with which a chemist has to deal. To form an ore deposit, the metals must once have been dissolved and must have been somehow present in a solution that also contained sulfur. But the amount of dissolved metal permitted by the solubilities is completely inadequate to form a deposit in geologically reasonable times.

Experimental and theoretical work over the past few decades, however, has shown that the common ore metals can exist in a variety of complex ions and molecules stable enough to keep substantial amounts dissolved, even when sulfide ions are also present in appreciable amounts. One can still argue about which complex is most important for a given metal under particular conditions of temperature and pressure, but the general adequacy of stable complexes to hold the necessary amounts of metal in solution is no longer in doubt. This demonstration, the culmination of work in many laboratories over the world, is a major triumph of geochemistry.

We have touched on this subject before (Sec. 2-5), with a question about zinc sulfide solubility in chloride solutions. If sphalerite stands in contact with a solution containing $0.1m$ Cl^- and $10^{-10}m$ S^{2-} at 250°C, as much as $10^{-4.8}m$ of zinc will dissolve, whereas the concentration calculated from the solubility product of ZnS would be only $10^{-8.9}m$. The 3000-fold difference was attributed to the formation of zinc chloride complexes ($ZnCl$, $ZnCl_2^\circ$, $ZnCl_3-$, $ZnCl_4^{2-}$), according to reactions of the form

$$ZnS + Cl^- \rightleftharpoons ZnCl^+ + S^{2-}. \qquad (19\text{-}2)$$
$$\text{sphalerite}$$

If the effect of ionic strength is included in the calculation, the difference becomes even more pronounced: an assumed ionic strength of 0.5 gives a further increase in total dissolved Zn to about $10^{-4}m$ [Fig. 2-6(a)]. Chloride ion evidently goes far toward overcoming the insolubility of ZnS.

Does it go far enough? Would the increase in solubility enable such a solution to dissolve and transport the metal and sulfide components, and then to precipitate the ore mineral in amounts sufficient to make an ore deposit of reasonable size? This is similar to the question we asked above, as to the minimum concentration of metal in a magmatic gas that would be significant for the formation of ore deposits. We proceed in the same way, using rough numbers to establish a limit of reasonableness. Suppose, for example, that an ore solution carried 10^{-6} mol per liter of Zn (roughly the same as $10^{-6}m$). To deposit 1 tonne of metal would require a minimum of about 10^{10} m^3 of solution, approximately the volume of water carried to the sea each year by the Hudson River (average flow about 300 m^3 sec^{-1}). Such a solution traversing a vein system at a rate of 1 m^3 sec^{-1} could deposit 1 tonne of zinc in 300 years, provided that *all* the dissolved zinc precipitates. Because a medium-sized zinc deposit may contain a million tons of zinc, these estimates of the amount of water and amount of time seem excessive. Thus $10^{-6}m$ can be regarded as an absolute minimum, below which the concentration of metal is too small to be considered. For most purposes a larger figure, say $10^{-5}m$, is a more reasonable minimum.

By this criterion the amount of zinc that could be present in a solution at 250°C together with $10^{-10}m$ S^{2-}, but containing no Cl^-, would be far below the concentration needed to form an ore deposit. If the solution has $0.1m$ NaCl but little else, it could hold barely enough zinc to be of interest. But if additional chloride is present, or enough other ions to give an ionic strength of 0.5 or higher, the amount of Zn in solution would be ample to form an exploitable deposit of sphalerite in a time that seems geologically plausible.

Similar experimental work and similar calculations for many other metals have demonstrated the effectiveness of chloride complexes as a means of increasing solubilities in warm aqueous solutions sufficiently to account for ore formation. And other anions can play a similar role: HS^-, for example, forms stable complexes with Au, Ag, and Hg [$Au(HS)_2^-$, $Ag(HS)_2^-$, $Hg(HS)_2^\circ$]; CO_3^{2-} keeps uranium in solution as $UO_2(CO_3)_2^{2-}$ and $UO_2(CO_3)_3^{4-}$; even OH^- in complexes like $ZnOH^-$ and $Zn(OH)_2^\circ$ can markedly increase solubility. A general reaction for the increased solubility can be written, by analogy with Eq. (19-2):

$$metal\ sulfide + anion \rightarrow metal\text{-}anion\ complex + sulfide\ ion, \qquad (19\text{-}3)$$

where "anion" may respresent Cl^-, HS^-, S^{2-}, OH^-, HCO_3^-, CO_3^{2-}, F^-, and several others. Which complex or complexes will be the most important for a given metal depends on the properties of the metal and composition of the solution, as well as on temperature and pressure.

Theoretical analysis of reactions between potential ore-forming solutions and rocks, using the techniques outlined in Chapters 2, 4, 8, and 9, permit qualitative statements about conditions that favor solubility and about changes in conditions that would lead to deposition of the ore metal or metals. The concentration of a chloride complex, for example, depends on competition for the metal between Cl^- and S^{2-}; if either total sulfide is low or the solution is acidic (which makes S^{2-} low, Figs. 2-4 and 9-7), the amount of metal carried in solution as the complex is high. If such a solution is diluted, or if it moves into a more alkaline environment, much of the metal will precipitate. For a bisulfide complex like $Zn(HS)_3^-$ the relations would be just opposite: high total sulfide and neutral to alkaline conditions would favor high HS^- concentration, hence increased solubility of the metal. The qualitative effect of a change in redox conditions can also be guessed, from the fact that the oxidation state of sulfur in solution is sensitive to oxygen fugacity (Fig. 9-7).

The effect of temperature on the solubility of Fe, Zn, Pb, and Cu sulfides is illustrated by the experimentally determined curves in Fig. 19-1. These isobaric solubility curves show a steep rise between 200 and 500°C, but flatten at higher temperatures because of the increased competition for Cl^- ions between alkali, hydrogen, and base metal cations to form associated species such as NaCl, KCl, HCl, and $PbCl_2$. Experiments of this kind provide one basis for predicting the effect of temperature, as well as pressure, on the transport and deposition of ore metals. For example, arrows in the figure represent the cooling of a hypothetical ore-forming solution that replicates base-metal zoning in the ore deposits called *porphyry coppers*, where Cu is concentrated in the central portion of a deposit and Zn and Pb in the periphery.

Thus the transportation of metals from their source (whether it be a crystallizing magma or not) to the vein or pocket where ore is deposited is largely dependent on the stability of various kinds of complexes. This stability is influenced by changes in temperature and pressure and by changes in fluid composition resulting from irreversible reactions with the rock it flows

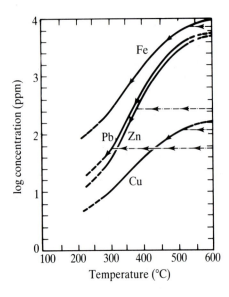

FIGURE 19-1
Experimentally determined (solid and dashed curves) metal solubilities of pyrite, galena, sphalerite, and chalcopyrite at 1 kbar in a $1m$ total chloride solution as a function of temperature. Solution pH is buffered by muscovite, K-feldspar, and quartz equilibrium [Eq. (4-26)]; oxygen and sulfur fugacities are fixed by pyrite–magnetite–pyrrhotite equilibrium (see Fig. 19-8). Arrows and dot–dash lines represent concentrations of a hypothetical ore-forming solution that will produce the Cu–Zn–Pb metal zoning in porphyry copper deposits. For example, the horizontal arrow leading to the solid line marked Zn denotes the initial concentration of Zn in solution; upon cooling, the arrow intersects the solubility curve for Zn, and further cooling will cause sphalerite to precipitate (Reprinted by permission from Hemley and Hunt, 1992.)

through. So important are complexes to an understanding of ore formation that experimental studies of their stability have become a major subject of geochemical inquiry. An impressive body of data has been accumulated on the complex-forming behavior of all the common ore metals in artificial solutions with a variety of compositions and over a big range of temperature and pressure. The data are summarized in tables of equilibrium constants and thermochemical functions (of which Appendix VII is a small sample), and from such numbers metal solubilities can be calculated with much confidence.

Fluid Inclusions

To apply these ideas to actual ore deposits, what we need is an actual sample of the liquid that carried the metals—and it turns out that nature has provided us with such samples. They consist of the tiny specks of fluid—the *fluid inclusions*—that are often visible in great numbers when thin sections of transparent minerals from an ore deposit are examined under high magnification. The fluid of the inclusions was evidently incorporated in the crystals as they grew, and so should represent the material from which the ore minerals were being deposited, or had just been deposited. They may be true samples of the ore-forming solution, and their study has provided some of our best information about this fluid.

The extremely small size of the inclusions—seldom greater than 0.1 mm in diameter, generally less than 0.01 mm—makes the study a real challenge. The great majority of inclusions consist of liquid with a small gas bubble; sometimes more than one liquid is visible, and sometimes recognizable crystals of solid are present (Fig. 19-2). The solids provide some information about the original fluid composition: crystals of halite are by far the most common, and other minerals occasionally

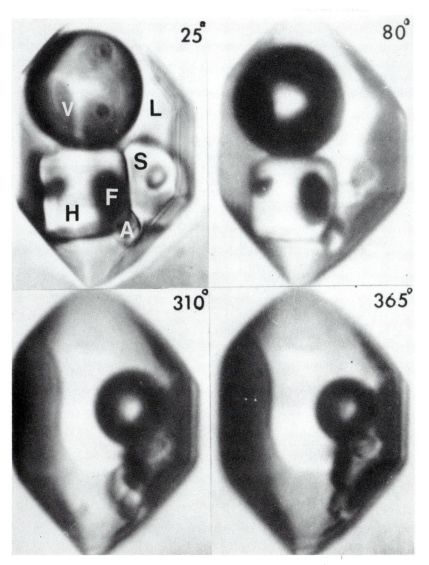

FIGURE 19-2
A multi-phase fluid inclusion in a quartz crystal from the porphyry copper deposit at Bingham Canyon, Utah, showing the changes produced by increasing temperature on a heating stage. The phases visible originally (25°C) are vapor (V), liquid (L), halite (H), sylvite (S), and two phases tentatively identified as anhydrite (A) and hematite (F). Sylvite is gone when the temperature reaches 80°C; most of the halite is gone by 310°C, and all of it has disappeared by 365°C. The hematite and anhydrite remain, but presumably would dissolve at higher temperatures. The inclusion is 35 μm long. [Photo courtesy of Edwin Roedder, reproduced from Roedder (1971, p. 108).]

reported include sylvite, hematite, magnetite, anhydrite, chalcopyrite. If solids are not present, a rough idea about concentrations in the fluid can be obtained by putting the sample on a cooling stage and noting the temperature at which the liquid freezes. Any dissolved solid would lower the freezing point, and on the assumption that the solute is chiefly NaCl, its concentration is readily found from the freezing temperature (a concentration often reported as "weight-percent NaCl equivalent," as in Fig. 19-3). More complete analysis may be accomplished by instrumental techniques, or by crushing the mineral or minerals and dissolving the released fluid in demineralized water.

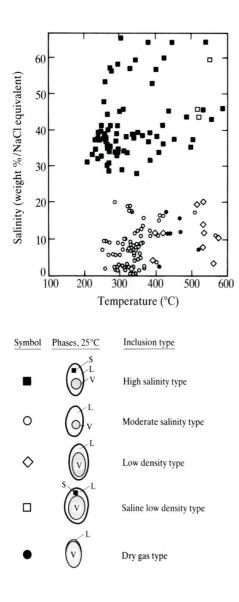

FIGURE 19-3
Salinity of the fluid in inclusions, reported as NaCl equivalent determined from freezing-point measurements, as a function of measured temperatures for liquid–vapor homogenization of inclusions in quartz veins from the Santa Rita porphyry copper deposit, New Mexico. Phases present in the four types of fluid inclusions found in the deposit are schematically shown in the drawings: s = solid, v = vapor, 1 = liquid. (reprinted by permission from Ahmad and Rose, 1980.)

The reliability of such analyses depends on two assumptions: (1) that the inclusions all formed at the same time, corresponding to the time of ore deposition, and (2) that nothing has been added to or taken from the inclusions since they were entrapped. These assumptions are not always justified. An ore deposit after its formation may be invaded by warm solutions, or may be tectonically disturbed; some of its crystals may grow larger and entrap a new set of inclusions, or the existing inclusions may be altered if minute cracks develop and fluid can move in or out. Success in fluid-inclusion study depends in large part on developing skill in distinguishing *primary* inclusions—those trapped during crystal growth—from *secondary* inclusions, those that formed later, usually in microcracks in the crystal.

One of the distinguishing criteria depends on the uniformity of phase relations among the inclusions found in a given thin section. Presumably when the inclusions were trapped at ore-forming temperatures the fluid present was homogeneous, and separation into multiple phases is a result of cooling. Ideally then, inclusions that formed at the same time and cooled through the same temperature range should have the same proportions of the different phases. Very often this expectation is fulfilled: all the inclusions present in a thin section commonly show the same proportional size of the gas bubble compared to the volume of liquid, and this is good evidence for their having been trapped at the same time and not modified later.

Sometimes, however, marked differences appear in the proportion of phases among inclusions within a single ore deposit or within a single vein. One example is shown in Fig. 19-3, where the estimated salinity of each inclusion, as determined by freezing-point measurement, is plotted against the temperature at which the liquid and vapor in the inclusion homogenize to a single phase. The data are from inclusions in vein quartz from the Santa Rita, New Mexico, porphyry copper deposit, a deposit formed in a subvolcanic environment similar to Fig. 18-1. The extreme range in fluid salinity is striking. It could mean that inclusions of different ages are present, and this should be recognized by other criteria. Alternatively the differences may mean that the ore-forming fluid was heterogeneous: perhaps ore deposition occurred at the interface between a hypersaline magmatic fluid and dilute heated groundwater. Another interesting possibility is that the fluid was boiling at the time it was in contact with the growing crystals. In this case one should find a whole series of gas–liquid ratios present, from mostly liquid with a tiny gas bubble to mostly gas with only a trace of liquid—a distribution occasionally observed, which indicates that boiling is not uncommon during ore formation. The data in Fig. 19-3 record the possible diversity of hydrothermal solutions within a single deposit, solutions that are continually modified by mixing, boiling, and reaction with the host rocks.

Some representative analyses of inclusion fluids are listed in Table 19-2. Compositions of the ore-forming fluid may differ widely from one deposit to another, but sodium chloride is nearly always the dominant solute. Its concentration ranges from very small up to more than 50% of the total. Other prominent ions include Ca^{2+}, K^+, and SO_4^{2-}; small amounts of more exotic ones like fluoride and borate may be present, plus some of the ore metals. Of gases in addition to water vapor, CO_2 is the one most commonly reported. Measured pH values are generally

TABLE 19-2
Approximate analyses and densities of fluids trapped in inclusions in minerals of Butte, Montana, copper deposit, in weight percent

H_2O	83.3	63.7	42.6	31.9
CO_2	11.4	0.2	0.2	0.1
NaCl	3.3	35.4	38.7	45.8
KCl	1.1	—	12.7	9.0
Fe_2O_3	0.9	0.4	1.2	0.8
$CaSO_4$	—	0.3	1.5	1.4
unknowns	—	—	3.1	11.0
Density (g/cm^3)	0.29	1.13	1.21	1.30
Wt. % solids	5.3	36.1	57.2	68.0

Source: Roedder (1971).

within one or two units of neutrality. The composition as a whole, except for those containing greater than 50 weight percent NaCl, agrees with what might be expected from analyses of fumarole gases, hot-spring waters, and geothermal brines. The usual high Cl^- concentrations support the conjecture that chloride complexes are responsible for transportation of many ore metals in solution at ore-forming temperatures.

19-2 TEMPERATURES AND PRESSURES OF ORE FORMATION

Temperature and pressure are key variables in any effort to understand the conditions of ore formation, and not surprisingly a number of methods for their determination have been proposed. Some of these methods we have discussed in previous chapters, those based on mineral stability (Chaps. 4, 8, and 16) and on the partitioning of minor elements (Chap. 5) and stable isotopes (Chap. 10) between minerals. Of the two variables, temperature is the one to which ore minerals are most sensitive and to which most attention has been given. But it should be remembered that the variables are closely related, and an accurate determination of one is not possible without some knowledge of the other.

One qualitative method of estimating temperatures and pressures, the oldest, depends simply on repeated observations of mineral sequences in ore deposits related to igneous intrusions. Ore minerals and gangue minerals typically found closest to the intrusive source presumably represent the highest temperatures and pressures, those farthest away the lowest. For example, wolframite [$(Fe,Mn)WO_4$], cassiterite (SnO_2), molybdenite (MoS_2), and pyrrhotite (Fe_7S_8) belong to the high-temperature group; chalcopyrite, bornite (Cu_5FeS_4), and tetrahedrite ($Cu_{12}Sb_4S_{13}$) are typical of intermediate temperatures; argentite (Ag_2S), cinnabar (HgS), stibnite (Sb_2S_3) and marcasite form at lowest temperatures. Obviously the classification cannot be rigid, because some minerals appear in all temperature zones—minerals like galena, sphalerite, pyrite, and native gold. But the sequence of mineral assemblages is sufficiently regular and easy to recognize that it often furnishes a useful first guess at temperature–pressure conditions. This is the basis of Lindgren's (1933) classification of ores as hypothermal (high-temperature, close to the intrusive), mesothermal (intermediate), and epithermal (low-temperature, distant from the source).

Phase relations and zoning patterns of silicates formed by hydrothermal solutions also provide information about temperature variation in an ore deposit. For example, the porphyry copper deposits discussed earlier show a regular zoning of silicates as schematically illustrated in Fig. 19-4(a). Early high-temperature alteration forms K-feldspar and biotite (potassic alteration) in the central portions of the deposit, followed by a later lower-temperature alteration in the upper part characterized by layer silicates such as sericite (muscovite), kaolinite, pyrophyllite, or montmorillonite. Temperatures of formation for these kinds of alteration assemblages can be estimated using phase diagrams like those in Figs. 4-10 and 16-10. A possible example of mineral zoning as related to temperature is shown in Fig. 19-4(b).

Fluid inclusions, besides providing information on ore-fluid composition, give us our best means of estimating the temperatures at which ores were deposited. The technique is simple: a thin-section containing primary inclusions is placed on the heating stage of a microscope, and the temperature at which the inclusions become homogeneous is noted (Fig. 19-3). This *homogenization temperature* is a minimum value for the temperature at which the fluid in each inclusion was caught in the growing crystal. The entrapment could have occurred at a higher temperature, if the pressure on the fluid was greater than that needed to prevent boiling, and thus to obtain the true temperature requires a correction for pressure. Getting a precise value for pressure at the time of ore formation is difficult, but an estimate can be made by reconstructing the geology and guessing at the depth of burial. This estimate permits a rough correction to homogenization temperatures for pressure by use of the constant volume curves for water shown in Fig. 2-6(a). More accurate pressure corrections can be made if needed, but require exact information on the pressure–temperature–density relations for water and mixed electrolyte solutions at ore-forming temperatures. The three variables temperature, pressure, and salt concentration are obviously interrelated: to get a precise value for one requires detailed knowledge of the other two. If additional methods are available, as they often are, for fixing either temperature or pressure, the other can be determined with great accuracy.

Many ways to estimate temperature more precisely have been proposed. Of the additional methods, we look briefly at two: the distribution of minor elements between a pair of co-existing minerals, and the fractionation of sulfur isotopes among sulfide minerals. Both of these possibilities we have mentioned before, the first in Sec. 5-8 with reference to temperatures of metamorphic processes, the second in Sec. 11-5 with reference to general applications of stable-isotope fractionation. Here we apply them specifically to the environment of hydrothermal ore deposition.

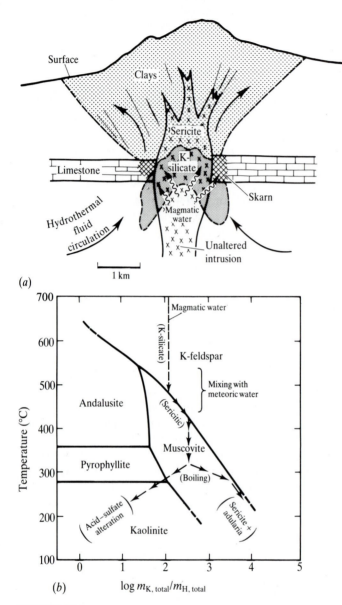

(a)

(b)

FIGURE 19-4

(a) Schematic cross-section of the subvolcanic magma chamber of Fig. 18-1, showing the general distribution of hydrothermal silicates within a porphyry copper deposit. The fine-stippled area denotes the early high-temperature K-silicate or potassic alteration (K-feldspar + biotite) and the coarse-stippled area later, lower-temperature, H^+-metasomatism that produces minerals such as sericite (muscovite), kaolinite, pyrophyllite, and montmorillonite. The solid arrows represent schematically the circulation of meteoric hydrothermal fluids, the wavy arrows the movement of magmatic water. (b) Stability of minerals in the system $K_2O-Al_2O_3-SiO_2-H_2O-HCl$ at 1 kbar in the presence of quartz and a $1m$ total chloride solution as a function of temperature and molal ratio KCl/HCl. The dashed arrows schematically represent the cooling path of magmatic water, a temperature region for mixing of magmatic water and meteoric hydrothermal solutions, and the consequences of boiling as the fluid decompresses. The terms "K-silicate," "sericitic," "sericite + adularia," and "acid-sulfate" refer to different types of alteration products produced by water-rock reactions. (Modified from Hemley and Hunt, 1992.)

Sulfide minerals like galena and sphalerite seldom are pure, but generally contain a variety of other elements in small amounts—for example, cadmium, bismuth, manganese, selenium. If the two minerals form simultaneously, each of the minor elements distributes itself between them, the distribution coefficient for each depending on the relative ease with which it can substitute for the lead, zinc, or sulfur of the dominant sulfides. The coefficient is sensitive to many variables in the milieu of ore formation—temperature, pressure, composition of the ore fluid—but commonly the variation with temperature is largest. This makes analysis of coexisting sulfide minerals a possible means of estimating temperatures.

For example, if galena and sphalerite are in equilibrium with a solution that contains cadmium, the chemical potentials of the component CdS in the three phases will be equal:

$$\mu_{CdS \text{ in galena}} = \mu_{CdS \text{ in sphalerite}} = \mu_{CdS \text{ in solution}}. \tag{19-4}$$

Or the equilibrium between the sulfide minerals can be expressed as

$$CdS_{\text{in sphalerite}} \rightleftharpoons CdS_{\text{in galena}}, \tag{19-5}$$

for which the equilibrium constant is

$$K = \frac{a_{CdS}^{\text{galena}}}{a_{CdS}^{\text{sphalerite}}} = \left(\frac{\lambda_{CdS}^{\text{galena}}}{\lambda_{CdS}^{\text{sphalerite}}} \right) \cdot \left(\frac{X_{CdS}^{\text{galena}}}{X_{CdS}^{\text{sphalerite}}} \right), \tag{19-6}$$

where λ is the activity coefficient and X denotes the mol fraction of CdS in the mineral. The last term in Eq. (19-6) is the distribution coefficient (K_D). It can be expressed as the ratio of mol fractions, or more commonly the ratio of weight percents, of the subscripted component in the two minerals:

$$K_D = \frac{[weight \ percent]_{CdS}^{\text{galena}}}{[weight \ percent]_{CdS}^{\text{sphalerite}}}. \tag{19-7}$$

The variation of equilibrium constant, and thus the distribution coefficient, with temperature for reactions of this sort is proportional to the enthalpy of reaction [ΔH, Eq. (8-61)], which is the difference in partial molal enthalpies of CdS in the two minerals. Similarly, the variation with pressure is proportional to the volume of reaction [ΔV, Eq. (8-71)], where ΔV is the difference in partial molal volumes of CdS in the two minerals. From these formulations it is clear that reactions with a large change in enthalpy make the best geothermometers, those with a large change in volume the best geobarometers (Sec. 8-5).

For the specific case of cadmium distribution between sphalerite and galena, Fig. 19-5 shows experimental results obtained by Bethke and Barton (1971) for log K_D as a function of $1/T$ [Eq. (8-67)]. The figure includes similar plots for the distribution of manganese and selenium between these two minerals, and for selenium between galena and chalcopyrite. Clearly, any of these lines would make a good basis for estimating temperatures of ore formation.

Or would they? Bethke and Barton, having demonstrated the apparently simple and regular variation of distribution coefficients with temperature in

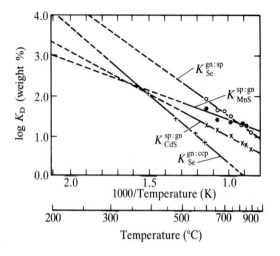

FIGURE 19-5

Variation of distribution coefficients with temperature. For each K_D, the superscript indicates the mineral pair and the subscript the minor element distributed between the two minerals. Abbreviations: gn = galena; sp = sphalerite; ccp = chalcopyrite. (Reprinted by permission from Bethke and Barton, 1971.)

laboratory experiments, go on to point out the many possible pitfalls in applying this method to actual ore deposits. Most obvious are the analytical difficulties: very small amounts of the minor element must be determined accurately, especially in the mineral with the smaller amount. Then the existence of equilibrium during ore formation may be hard to demonstrate, as well as the absence of later alteration of one or both members of the mineral pair. Because of these problems, the authors recommend that ore-formation temperatures obtained in this way be checked by other methods. In a few cases the distribution of minor elements between ore–mineral pairs has given apparently reliable temperature estimates, but the method has often proved disappointing.

Most promising is the determination of temperatures from the distribution of stable isotopes (Sec. 10-4). For application to ore deposition, the isotope ratio most commonly used is $^{34}S/^{32}S$, as measured in a pair of sulfide minerals formed at the same time. The carbon isotope ratio $^{13}C/^{12}C$ can be a measure of temperature in deposits containing carbonate minerals, and oxygen-isotope fractionation between mineral pairs such as quartz–iron oxides, calcite–magnetite, plagioclase–magnetite, and pyroxene–magnetite has also provided useful estimates of ore-deposit temperatures.

Isotope distribution depends on bond strengths in different minerals. The heavy sulfur isotope is favored in minerals with high sulfur bond strengths, and calculations on this basis indicate that the enrichment of ^{34}S should decrease in the order pyrite>sphalerite>chalcopyrite>galena (Fig. 19-6). The enrichment in each mineral is a function of temperature, and the difference in the $^{34}S/^{32}S$ ratio between two minerals that formed at equilibrium with the ore fluid thus provides a measure of temperature. Both theory and laboratory measurements indicate that the enrichment difference, $\Delta^{34}S$ [Eq. (10-33)], is approximately proportional to $10^6/T^2$ for temperatures over about 150°C (Fig. 19-6). Reasons for uncertainty in using this method include the difficulty of establishing simultaneous crystal-

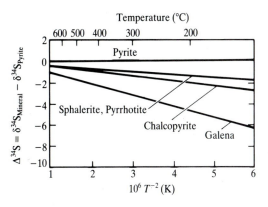

FIGURE 19-6
The temperature dependence of sulfur isotope fractionation between common sulfide minerals. Fractionation is plotted relative to pyrite; thus the curve marked galena is the fractionation of sulfur isotopes between galena and pyrite. (Reprinted by permission from Rye and Ohmoto, 1974.)

lization of the two minerals in a pair, the possibility of isotopic reequilibration with later solutions, and the possible lack of purity of the samples chosen for analysis. Despite such uncertainties, a check of isotopic temperatures with those obtained from fluid inclusions in many deposits has shown satisfactory agreement, and this method of estimating temperatures is often used.

Pressure at the time of ore formation is less easily determined than temperature. What we need for a pressure measurement is some property of ore minerals or their environment that changes more rapidly with pressure than with temperature [in other words, that involves a large change in molal volume rather than molal enthalpy—Eqs. (8-61) and (8-71)]. Some properties of this sort have been explored, but have not provided an accepted method that can be widely applied. Pressure is most commonly estimated, as noted above for fluid inclusions, simply by assigning a probable depth of ore formation from reconstruction of the geology; such an estimate is necessarily imprecise because it involves a further guess as to whether the pressure acting on the ore was lithostatic or hydrostatic or somewhere between. Fortunately the properties used for temperature measurement are not greatly affected by pressure, so that temperatures can be specified within a few tens of degrees even though pressure is known only roughly.

Still another variable that needs consideration for precise temperature–pressure determinations is the composition of the fluid from which the ore minerals are forming. In an ideal world the reconstruction of the environment in which hydrothermal ore has formed would require complete specification of temperature–pressure–composition (T–P–X) conditions, but for geologic purposes a much cruder picture is generally sufficient.

19-3 ESTIMATES OF pH, f_{O_2} AND f_{S_2}

Three other attributes of an ore-forming solution need brief mention: its acidity, its oxidation state represented by the fugacity of oxygen (or Eh or *pe*), and its sulfur concentration expressed as fugacity of sulfur vapor.

The old question of acidity versus alkalinity of the fluids that carry and precipitate ore metals, once a subject of acrimonious debate, can be answered from work in recent years with the simple statement that the pH is probably not far from neutrality, certainly no more than 2 pH units away from the neutral point. One can still argue at length over the minor changes in pH that may have been responsible for ore formation in a particular deposit, but wide-ranging strongly acidic or strongly basic solutions are no longer suggested as potential ore-metal carriers.

Evidence for nearly neutral solutions comes partly from the simple facts that strongly acid solutions in an ore-forming environment would be quickly neutralized by reaction with silicate or carbonate minerals, and strongly basic solutions by reaction with silica (quartz or chalcedony). Many hot-spring waters seem to corroborate this statement, in that they are mostly near the neutral point, but their characteristics are so obviously modified by their immediate surroundings such as boiling, reaction with the atmosphere, and mixing with meteoric ground-water that their properties cannot have great significance. Better evidence is supplied by fluid inclusions, for which the pH measured at room temperature is uniformly not far from neutrality.

"Near neutrality," of course, does not mean a pH near 7 at the temperatures of ore deposition. The dissociation constant of water, as noted in Sec. 2-1, increases with rising temperature to a maximum of $10^{-11.0}$ at 275°C on the boiling point curve, so that concentrations of both H^+ and OH^- in a neutral solution can be as high as $10^{-5.5}m$. In other words, water at first becomes *both* a better acid and a better base as the temperature rises. At still higher temperatures the dissociation of water and electrolytes dissolved in it becomes critically dependent on the density, hence on the pressure, to a greater extent than on temperature. Estimates of the dissociation constant of water at high temperatures and pressures are shown in Fig. 2-1, and as a function of density in Fig. 19-7. Salts like KCl and acids like HCl become practically undissociated when the density falls to 0.3 or 0.4 g cm^{-3}. Under such conditions the ordinary concepts of acids and bases are difficult to apply. A mixture of HCl and H_2O having, say, a density of 0.4 g cm^{-3} at 600°C would be potentially "acid," in the sense that the HCl could react with adjacent rocks and that the HCl would dissociate when the solution cooled; but as long as the density stayed low, the pH would be only a trifle below that of pure water. Thus the kind of question a geochemist asks about the acidity of an ore-forming solution is now framed in terms of the effect of minor changes in pH on the stability of different kinds of complexes, under varying conditions of temperature and pressure.

For hydrothermal conditions where compounds like NaCl, KCl, and HCl are largely undissociated, it is often convenient to represent phase relations among silicate minerals using logarithms of ratios like m_{KCl}/m_{HCl} or $m_{K,total}/m_{H,total}$ as shown in Figs. 4-10, 16-10, and 19-4(*b*), rather than the cation-to-hydrogen activity ratios used in Figs. 4-11 and 4-12. Diagrams of this kind provide a map for estimating changes in temperature and fluid composition from paragenetic relations of gangue minerals in hydrothermal ore deposits [Fig. 19-4(*b*)].

The oxidation state of an ore-forming solution is best expressed as a range of values of the fugacity of oxygen, which can be estimated, at least within broad

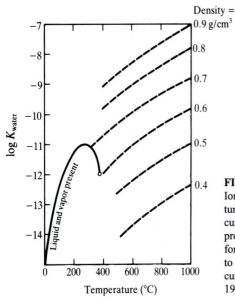

FIGURE 19-7
Ionic dissociation constant of water in high-temperature fluids of various densities. The solid line is the curve of Fig. 2-1, for liquid water under its own vapor pressure. Dashed lines show values of the constant for single-phase fluid water under sufficient pressure to maintain the indicated densities. (*Sources:* Solid curve, Fisher and Barnes, 1972; dashed curves, Quist, 1970.)

limits, from the nature of the ore minerals and gangue minerals present in a given deposit. For example, magnetite is a mineral commonly found in ore deposits and hematite less commonly, so that f_{O_2} probably seldom exceeds the amount fixed by hematite–magnetite equilibrium [Eq. (9-61)]. On the reducing side, a lower limit is suggested by the fact that olivine is not often found associated with ore deposits, so that f_{O_2} should not be lower than the value for quartz–magnetite–fayalite equilibrium [Eq. (9-67)]. Equilibrium values for these reactions, of course, vary with temperature and pressure (Fig. 9-6). For any given deposit other mineral assemblages can be used to specify f_{O_2} limits more precisely. For example, a reaction involving K-feldspar, biotite, and magnetite,

$$\text{KAlSi}_3\text{O}_8 + \text{Fe}_3\text{O}_4 + \text{H}_2\text{O} \rightleftharpoons \text{KFe}_3\text{AlSi}_3\text{O}_{10}(\text{OH})_2 + \tfrac{1}{2}\text{O}_2, \qquad (19\text{-}8)$$

K-feldspar   magnetite   biotite (annite)

has proved useful in estimating oxygen fugacity during high-temperature reactions in porphyry copper deposits.

Still another variable obviously important in describing solutions responsible for deposition of sulfide minerals is the concentration of sulfur. The form that sulfur takes, in or associated with the solution, depends on pH, f_{O_2}, and of course on temperature and pressure. For simplicity the concentration of sulfur is often expressed as the fugacity of the gas molecules S_2 or H_2S, and is commonly represented on phase diagrams either as a major variable plotted against T, pH, f_{O_2}, or a cation activity ratio, or as contours of fugacity on the phase diagram. Examples are shown in Figs. 19-8 and 19-9.

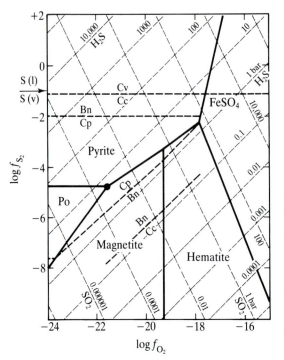

FIGURE 19-8

Relations of S_2 and O_2 fugacities in the Fe–S–O (solid lines) and Cu–Fe–S–O–H (thick dashed lines) systems at 500°C and 1 kbar total pressure. Thin dashed lines indicate SO_2 and H_2S fugacities and the arrow indicates the sulfur condensation point; Po = pyrrhotite, Cp = chalcopyrite, Bn = bornite, Cc = chalcocite, and Cv = covellite. (Reprinted by permission from Hemley et al., 1992.)

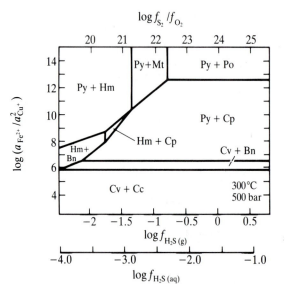

FIGURE 19-9

Theoretical activity–fugacity phase diagram showing phase relations among sulfides and oxides in the Cu–Fe–S–O–H system at 300°C and 500 bar. The lines represent phase boundaries between mineral pairs. Mineral abbrevations are the same as in Fig. 19-8. (Reprinted by permission from Brimhall and Crerar, 1987.)

A complete description of the fluid responsible for a given ore deposit thus includes many variables, for which two-dimensional plots are useful but obviously incomplete. As in so many parts of modern geochemistry, recourse to computer simulation is required for a fuller picture.

19-4   ORIGIN OF HYDROTHERMAL SOLUTIONS

We have looked now in some detail at the composition and principal properties of the hydrothermal solutions responsible for many ore deposits. It is time to return to a question posed earlier: Where do these strange solutions come from, and how do they acquire their content of ore-forming metals?

A century ago the answer would have been simple: the solutions come from a body of magma crystallizing somewhere below the surface. This conclusion is based on the fact that many ore deposits are found in or near masses of granite or other intrusive rock, or in areas where the rocks show evidence of past volcanic activity. Today, of course, we know that the answer is not this easy. There is not one source for hydrothermal solutions, but many sources. Some ore-forming fluids do indeed come from crystallizing magma, but others originate as groundwater or connate water or metamorphic water heated by circulating deep in the crust or by close contact with an intrusive body (Fig. 18-1). From study of the ore minerals and gangue minerals of a particular deposit, one can often make a good guess as to which of the various possible sources of fluid was responsible for its formation.

Evidence about source comes largely from isotope ratios, particularly $^{18}O/^{16}O$ and H/D ratios in minerals of an ore deposit. As noted in Sec. 10-4, these ratios have characteristic values in meteoric water and in the minerals of igneous rocks, so that water which has been in contact with crystals in a cooling melt can be readily distinguished from groundwater that is a product of rainfall (Fig. 10-9). Ratios are preserved in crystals formed from the water, so that measurements of oxygen and hydrogen isotopes permit sources of the water to be recognized (Sec. 18-5), and if temperature can be estimated using the methods outlined above, the isotopic composition of the solution can be computed from experimental data on mineral–water fractionation [for example, see Eq. (10-32)].

Many such measurements in a variety of ore deposits, as summarized in Fig. 19-10, show that meteoric water and magmatic water both play important roles in ore deposition, sometimes one to the exclusion of the other and sometimes the two in combination. Even for ore deposits directly associated with igneous rocks, isotopic evidence in many places indicates that the ore-forming solutions were largely of meteoric origin. Presumably groundwater simply moved close to the intrusive, was heated and perhaps mixed with minor amounts of fluid derived from the crystallizing magma. In some deposits, particularly porphyry coppers, the action of two or more different fluids can be detected—solutions of strictly magmatic origin responsible for early high-temperature potassic alteration and ore deposition, then later solutions of largely meteoric ancestry that produced low-temperature alteration characterized by sericite and clay minerals (Fig. 19-4).

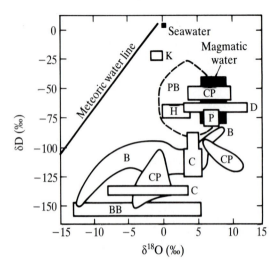

FIGURE 19-10
The range of δD of fluid inclusions plotted against the calculated value of $\delta^{18}O$ of hydrothermal fluids from a number of ore deposits: C = Climax, CP = Casapalca, D = Darwin, P = Providencia, H = Homestake, PB = Pasto Bueno, B = Butte, and BB = BlueBell. Hydrothermal solutions in these deposits represent varying mixtures of meteoric and magmatic waters. Hydrothermal solutions from the Kuroko (K) deposits appear to be mostly of seawater origin. The solid line is the meteoric water line and the box that of primary magmatic water as shown in Fig. 10-9. (Reprinted by permission from Taylor, 1979, p. 260.)

Evidence for a nonigneous source is particularly impressive for large bodies of sulfide ore found in sedimentary rocks far from any known intrusions. Most notable of such ores are the lead–zinc deposits of the Mississippi Valley, ores that have given their name to a kind of deposit found in many parts of the world. Generally, ores of the Mississippi Valley type occur as veins, cavity fillings, and replacement deposits in dolomitized limestone near the margin of a large sedimentary basin; the mineralogy is simple, consisting chiefly of sphalerite and galena in widely varying proportions; temperatures of ore formation, as shown by fluid inclusions, are in the range 70–250°C, with a common average of about 120°C; the chloride concentration of inclusion fluid is generally high, up to $3m$ or more, and pH is near neutral. Although a deeply buried igneous source for the ore-depositing fluids was commonly postulated in the past, an accumulation of recent evidence makes an hypothesis of nonigneous origin much more probable. Fluids derived largely from connate water originally buried with the sediments can be thought of as moving to the margins of the basin under the steadily growing weight of sediments in its central part. For a modern-day analog of such solutions, we can cite the oil-field brines from deep basins (Sec. 15-5), many of which have a high chloride content and abnormal amounts of heavy metals. Details of this hypothesis are still in question, but a largely sedimentary source for the fluids seems well established.

Where do the ore-forming solutions find the metals that they carry? For solutions derived from magma the answer seems easy: ore metals originally disseminated through the molten material would be concentrated in the residual fluids during the process of crystallization. A similar broad generalization is possible about ore metals carried by groundwater: perhaps the metals were originally dispersed in a large volume of solid rock, and were dissolved and concentrated as the warm water circulated through it. The mechanism of concentration is less clear, but appropriate changes of heating, of pH and Eh, and of reaction with wall rocks can be imagined. For either kind of solution, the gleaning of metals from large volumes of rock seems adequate to account for the quantities of metal needed. All ordinary rocks, and magmas derived from them, contain the common ore metals in at least trace concentrations (Appendix III), so that the amounts obtainable are limited only by the size of rock body one imagines to have existed. Not a very satisfying sort of hypothesis, because it leaves so many details uncertain, but for most ore deposits it provides the most reasonable source for the metals that make up the ore.

19-5   WEATHERING OF ORE DEPOSITS

After an ore deposit has formed, uplift and erosion may eventually bring it close to the Earth's surface. How do its minerals respond to the agents of chemical weathering? To answer the question should seemingly require no more than a brief extension of our earlier discussion of weathering processes in Chapters 9 and 13. But the weathering reactions of ore minerals are somewhat more complicated than those of rock-forming minerals, in that most of the elements involved show more than one oxidation state in normal near-surface environments. The reactions, in contrast to those we have been considering in the past several chapters, take place under ordinary temperatures and pressures and are therefore accessible to detailed observation in the field and to experimental study in the laboratory. For this reason, and also because weathering processes often have great economic importance in concentrating the valuable metals of low-grade deposits, the weathering of ores has been the subject of long and intensive study. From the great wealth of accumulated data we select only a few examples to illustrate general principles.

Sulfide Minerals

From our general knowledge of sulfur chemistry, we might expect the sulfide minerals to fall easy prey to chemical weathering. Their sulfur, we could predict, will be oxidized to sulfate ion, SO_4^{2-}, and the metals will be taken into solution or converted into compounds stable under surface conditions (hydroxides, oxides, carbonates, sulfates, silicates). The dissolved metal ions may be removed completely in streams and groundwater (Sec. 6-2), or may be in part carried down into the unoxidized portion of a sulfide deposit and there precipitated by reaction with the sulfide minerals.

A symbolic equation for the oxidation of a simple sulfide is easy to write:

$$PbS + 2O_2 \rightarrow PbSO_4 \qquad (19\text{-}9)$$
$$\text{galena}$$

and

$$ZnS + 2O_2 \rightarrow Zn^{2+} + SO_4^{2-}. \qquad (19\text{-}10)$$
$$\text{sphalerite}$$

But how does the reaction actually take place? Almost certainly it is not a simple matter of collisions between oxygen molecules and solid sulfides, as the equations suggest, because shiny specimens of galena can be kept in contact with ordinary air indefinitely without showing the slightest sign of change. The agency of water in promoting the change is shown by the observation that moist galena surfaces gradually tarnish in the laboratory. Precisely how water acts is not certain, but one possibility is a series of reactions:

$$H_2O + CO_2 \rightarrow H_2CO_3,$$

$$PbS + 2H_2CO_3 \rightarrow Pb^{2+} + 2HCO_3^- + H_2S,$$

$$H_2S + 2O_2 \rightarrow SO_4^{2-} + 2H^+, \tag{19-11}$$

$$Pb^{2+} + SO_4^{2-} \rightarrow PbSO_4,$$

$$H^+ + HCO_3^- \rightarrow H_2CO_3.$$

These reactions added together give Eq. (19-9). According to this mechanism water serves as a catalyst, its catalytic activity consisting of the formation of carbonic acid, which in turn dissolves a tiny amount of PbS or ZnS and permits O_2 to react with dissolved H_2S rather than with the solid sulfide. Reactions of O_2 at ordinary temperatures are slow, but it is known to react more rapidly with molecules and ions in solution than with solids, so that these steps are at least plausible.

But this is far from the only possible mechanism. Sato (1960) has suggested, for example, that water is oxidized in trace amounts to hydrogen peroxide, and that the peroxide is the active agent in oxidizing the sulfide:

$$2H_2O + O_2 \rightarrow 2H_2O_2,$$

and

$$4H_2O_2 + \underset{\text{galena}}{PbS} \rightarrow \underset{\text{anglesite}}{PbSO_4} + 4H_2O. \tag{19-12}$$

The first reaction is known to be slow, and provides a possible general explanation for the slowness of reactions involving molecular O_2. One bit of evidence for this mechanism is the observation that measured Eh values in mine waters are well below the theoretical maximum for the O_2–H_2O couple (Sec. 9-4):

$$2H_2O \rightleftharpoons O_2 + 4H^+ + 4e^- \qquad \text{Eh} = 1.22 - 0.059 \text{ pH}, \tag{19-13}$$

but are close to values for the O_2–H_2O_2 couple:

$$H_2O_2 \rightleftharpoons O_2 + 2H^+ + 2e^- \qquad \text{Eh} = 0.68 - 0.059 \text{ pH}. \tag{19-14}$$

Sato also presents evidence that the oxidation of sulfur in the sulfide does not go directly to sulfate but proceeds by a series of partial oxidations.

Thus the net results of the oxidation of sulfides, regardless of details of the mechanism, are (1) to get the metal ion into solution or into the form of an insoluble compound stable under surface conditions, (2) to convert the sulfur to sulfate ion, and (3) to produce relatively acid solutions. Discussion of such reactions can be refined by use of Eh–pH or $\log f_{O_2}$–pH plots like those described in Chapter 9. Detailed diagrams for many common ore metals have been prepared; some good examples are given in Garrels and Christ (1965, Chap. 7) and Sato (1992).

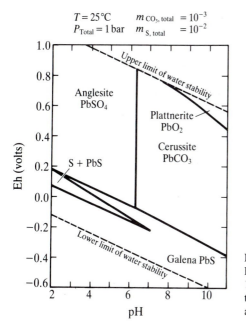

$T = 25\,°C \quad m_{CO_3,\,total} = 10^{-3}$
$P_{Total} = 1\,bar \quad m_{S,\,total} = 10^{-2}$

FIGURE 19-11

Eh–pH diagram for lead minerals at 25°C and 1 bar. Total dissolved carbonate = $10^{-3}m$ and total dissolved sulfur = $10^{-2}m$. (After Garrels and Christ, 1965.)

As one example, we look at an Eh–pH plot for lead shown in Fig. 19-11. Of the four lead minerals shown in the diagram, galena is the most stable (or the least soluble) at low values of Eh, regardless of acidity. When conditions are oxidizing, anglesite is the most stable in an acid environment and cerussite in most alkaline environments. Plattnerite becomes stable only if conditions are highly oxidizing as well as alkaline. These relations, of course, are just what would be expected from the chemistry of lead, and the diagram serves only to express the relations more quantitatively.

Boundaries between fields on the diagram are located by finding Eh and pH values where the Pb^{2+} concentrations on either side are identical. For example, the anglesite–cerussite boundary is drawn through points where the two equilibria

$$\underset{\text{anglesite}}{PbSO_4} \rightleftharpoons Pb^{2+} + SO_4^{2-} \qquad K = 10^{-7.8} \qquad (19\text{-}15)$$

and

$$\underset{\text{cerussite}}{PbCO_3} \rightleftharpoons Pb^{2+} + CO_3^{2-} \qquad K = 10^{-13.1} \qquad (19\text{-}16)$$

are simultaneously satisfied, i.e., where the ratio $a_{SO_4^{2-}}/a_{CO_3^{2-}}$ is equal to $10^{5.3}$. Under the conditions specified on the diagram (total dissolved sulfur = $0.01m$, total dissolved carbonate = $0.001m$), the activity of SO_4^{2-} at moderate and high Eh values can be set equal to the total sulfur, and $a_{CO_3^{2-}}$ on the boundary must therefore be $10^{-2}/10^{5.3}$, or $10^{-7.3}$. From Fig. 2-2, this activity for CO_3^{2-} requires a pH of 6.3. Hence the boundary is a vertical line at pH 6.3.

For boundaries with the PbS field the calculation is more troublesome. The $PbS–PbCO_3$ line is obtained by equating Pb^{2+} activities from the two solubility-product constants, $10^{-27.5}$ and $10^{-13.1}$, from which the ratio $a_{S^{2-}}/a_{CO_3^{2-}}$ must be $10^{-14.4}$. Both sulfide ion and carbonate ion are dependent on pH, and sulfide ion is also determined in part by Eh, so that a formal analytical expression relating Eh to pH is complicated. The simplest procedure is to work out the relation for specific points. Take first the point at the upper end of the line, where the fields of PbS, $PbSO_4$, and $PbCO_3$ come together. From the last paragraph we know that here pH = 6.3 and $a_{CO_3^{2-}} = 10^{-7.3}$. The above ratio tells us that $a_{S^{2-}}$ must be $10^{-14.4} \times 10^{-7.3}$, or $10^{-21.7}$. If $a_{S^{2-}}$ is this low, neither it nor a_{HS^-} can be quantitatively important compared with $a_{SO_4^{2-}}$, so again we set the sulfate activity equal to the total dissolved sulfur, 10^{-2}. From the half-reaction relating S^{2-} to SO_4^{2-},

$$S^{2-} + 4H_2O \rightleftharpoons SO_4^{2-} + 8H^+ + 8e^- \qquad E° = +0.16 \text{ volt}, \qquad (19\text{-}17)$$

we write for Eh, using Eq. (9-19),

$$Eh = E° + \frac{0.059}{n} \log Q = 0.16 + 0.0074 \log \frac{a_{H^+}^8 a_{SO_4^{2-}}}{a_{S^{2-}}}$$

$$= 0.16 + 0.0074 \log \frac{10^{-50.4} \times 10^{-2}}{10^{-21.7}}$$

$$= -0.07 \text{ volt}. \qquad (19\text{-}18)$$

For another point on the line, take pH = 11, high enough that practically all the dissolved carbonate is CO_3^{2-}. If $a_{CO_3^{2-}}$ is set equal to 10^{-3}, then, $a_{S^{2-}} = 10^{-17.4}$, still low enough that $a_{SO_4^{2-}}$ remains effectively equal to the total dissolved sulfur, or 10^{-2}. Repeating the calculation from the half-reaction then gives Eh = -0.38 volt. Other Eh–pH pairs are located similarly.

The diagram does not give directly the activities of Pb^{2+} in equilibrium with the solid phases, but these are easily calculated and may be added as contour lines in the different fields. Over the $PbSO_4$ field, $a_{Pb^{2+}}$ is constant at $10^{-5.8}$, because the sulfate-ion activity is not appreciably affected by either Eh or pH in this range. In the $PbCO_3$ field the Pb^{2+} activity drops off as the pH rises, since $a_{CO_3^{2-}}$ steadily increases; at pH 10.3, for example, where $a_{CO_3^{2-}} = a_{HCO_3^-} = 0.5 \times 10^{-3}$, $a_{Pb^{2+}}$ would be $10^{-9.8}$. Over the PbS field the Pb^{2+} activity would fall as pH rises and as Eh decreases, since both of these changes would cause a_S^{2-} to increase. Different assumptions as to total dissolved sulfur and total carbonate would change the boundaries somewhat, but the alterations are surprisingly minor as long as the concentrations remain within the range commonly found in natural solutions.

Supergene Sulfide Enrichment

When a sulfide deposit is followed downward beneath the oxidized outcrop, the richest ore is often found just at the top of the unoxidized zone. This level in many places corresponds with the regional water table, or with a portion of the water table at some time in the recent geologic past. The rich ore commonly consists of sulfide minerals, which from replacement relations are clearly of a later generation than the primary sulfides found at deeper levels in the deposit. It is a reasonable inference that these secondary sulfides have been formed by downward-moving solutions containing metal ions derived from relatively soluble compounds in the oxidized zone, these metal ions having reacted with and replaced sulfides in the unoxidized rock. This phenomenon, long recognized by students of ore deposits and often of great economic importance, goes by the name of *supergene sulfide enrichment.*

Copper, of all the metals, shows supergene enrichment most commonly and most conspicuously. The simple sulfides covellite (CuS) and especially chalcocite (Cu_2S) are concentrated in thick blankets and lenses near the top of many deposits, often making up the richest part of the ore. Formation of these sulfides is complex in detail, but the general nature of the process is easy to understand. When copper minerals are oxidized near the surface, much of the metal goes into solution, since the possible oxidized compounds are fairly soluble as long as the solutions are slightly acid. The dissolved Cu^{2+}, carried downward by groundwater, comes in contact with primary sulfide minerals below the zone of oxidation. Because the copper sulfides are less soluble than most other common sulfides (Appendix VII), reactions like the following are possible:

$$Cu^{2+} + \underset{\text{sphalerite}}{ZnS} \rightarrow \underset{\text{covellite}}{CuS} + Zn^{2+}, \tag{19-19}$$

$$14Cu^{2+} + 5\underset{\text{pyrite}}{FeS_2} + 12H_2O \rightarrow 7\underset{\text{chalcocite}}{Cu_2S} + 5Fe^{2+} + 3SO_4^{2-} + 24H^+. \tag{19-20}$$

The iron, zinc, and other metals displaced by the copper are carried away in solution, perhaps to be deposited later as limonite, smithsonite, and so on, if the solutions reach an oxidizing and less acid environment.

Silver shares with copper, but to a lesser degree, the properties that make a metal suitable for supergene enrichment: a very insoluble sulfide, and relatively high solubility of compounds it could form with common anions of the oxidized zone. These characteristics are consistent with the well established fact that silver deposits are second only to copper deposits in the prominence of secondary ore minerals just below the zone of oxidation. In contrast to copper and silver, unquestioned examples of notable supergene enrichment are not common for lead, zinc, nickel, cobalt, or mercury.

Oxidation of Uranium Ores

This discussion of oxidation processes could be extended to ores of many other metals. A particularly interesting one is uranium, whose complex and varicolored oxidation products have become well-known because of the search for uranium deposits to provide fuel for nuclear reactors.

The chief primary compound of uranium in vein deposits is the dioxide, UO_2, which occurs in the well crystallized variety uraninite and the microcrystalline form pitchblende. Incipient oxidation and loss of uranium by radioactive decay may increase the oxygen–uranium ratio, so that uraninite and pitchblende seldom show precisely the composition UO_2, often approaching a composition symbolized by U_3O_8. In the zone of weathering, pitchblende and uraninite are converted to one or more of the bright-colored oxidized uranium minerals, such as carnotite $[K_2(UO_2)_2(VO_4)_2 \cdot 3H_2O]$, tyuyamunite $[Ca(UO_2)_2(VO_4)_2 \cdot nH_2O]$, autunite $[Ca(UO_2)_2(PO_4)_2 \cdot nH_2O]$, and rutherfordine (UO_2CO_3). These minerals are slightly soluble, so that their uranium can be carried by surface water or groundwater into reducing environments (a bed of lignite or black shale, for example) and precipitated as pitchblende or coffinite $(USiO_4 \cdot nH_2O)$.

Thus uranium, like iron and copper, is an element showing changes from one oxidation state to another in geologic environments. Details of its chemistry provide an explanation for much of its behavior. Uranium has many oxidation states (+2, +3, +4, +5, and +6), but only +4 and +6 are of geologic interest. In its two lowest oxidation states uranium is such a powerful reducing agent that it can liberate hydrogen from water, and the +5 state in the presence of water is unstable with respect to +4 and +6:

$$2UO_2^+ + 4H^+ \rightleftharpoons UO_2^{2+} + U^{4+} + 2H_2O \qquad \Delta G° = -54.5 \text{ kJ.} \qquad (19\text{-}21)$$

The transition from +4 to +6 has a redox potential within the range for geologic environments:

$$U^{4+} + 2H_2O \rightleftharpoons UO_2^{2+} + 4H^+ + 2e^- \qquad E° = +0.33 \text{ volt,} \qquad (19\text{-}22)$$

so we would expect to find compounds of these two oxidation states in nature.

Concentrations of U^{4+} and UO_2^{2+} (uranyl ion) in solution are limited by the very slight solubility of many of their compounds, for example, the hydroxides (at 25°C and 1 bar):

$$U(OH)_4 \rightleftharpoons U^{4+} + 4OH^- \qquad K = 10^{-50}, \qquad (19\text{-}23)$$

$$UO_2(OH)_2 \rightleftharpoons UO_2OH^+ + OH^- \qquad K = 10^{-14.2}, \qquad (19\text{-}24)$$

$$UO_2OH^+ \rightleftharpoons UO_2^{2+} + OH^- \qquad K = 10^{-8.2}. \qquad (19\text{-}25)$$

These constants mean that in neutral and slightly acid solutions the concentrations of UO_2^{2+} and UO_2OH^+ are appreciable, but U^{4+} is negligibly small. The solubility of U compounds in the +6 state is greatly increased by formation of stable carbonate complexes:

$$UO_2^{2+} + 2CO_3^{2-} \rightarrow UO_2(CO_3)_2^{2-}, \qquad (19\text{-}26)$$

and

$$UO_2(CO_3)_2^{2-} + CO_3^{2-} \rightarrow UO_2(CO_3)_3^{4-}. \qquad (19\text{-}27)$$

Based on these reactions, the general chemistry of uranium in near-surface environments can be readily described. Primary minerals are oxidized to uranyl ion, which is somewhat mobile in weakly acid solutions, and also in neutral and alkaline solutions if CO_3^{2-} is present. From such solutions uranium may be precipitated in the +6 state by a variety of anions, forming the familiar oxidized minerals; or it may be reduced by any one of a number of reducing agents, notably by organic matter, commonly forming UO_2 or one of its hydrates. Processes of these general kinds are taking place, or have taken place recently, in the uranium deposits of the Colorado Plateau and central Wyoming, where oxidized minerals are disseminated in sandstones near the surface, and at lower levels black unoxidized ore is found in sediments containing much organic material.

Such examples demonstrate how simple chemical ideas can help in explaining the reactions that occur when ore deposits are exposed to oxidation. This is an environment in which the agreement of predictions from thermodynamic data with observed mineral associations is especially satisfactory.

19-6 ENVIRONMENTAL EFFECTS OF ORE-DEPOSIT WEATHERING

Concern over the possible presence of toxic metallic constituents in groundwater and surface waters has led in recent years to greatly increased interest in the geochemistry of weathering ore deposits and other metal accumulations. Sulfide deposits containing abundant pyrite or other iron sulfides are generally the most conspicuous ore-related defilers of the environment. Where such a deposit has been opened by mining at or near the surface, or in places where tailings from mining or milling operations have accumulated, the exposure of sulfides to the atmosphere can lead to strongly acid solutions that drain into adjacent streams, carrying with them much iron that coats rocks and vegetation with brown and yellow Fe(III) oxide, plus dissolved ions of more toxic metals (Sec. 6-2 and 13-6). Typical reactions would be

$$2FeS_2 + 7O_2 + 2H_2O \rightarrow 2Fe^{2+} + 4SO_4^{2-} + 4H^+$$

$$2Fe^{2+} + 1/2O_2 + 5H_2O \rightarrow 2Fe(OH)_3 + 4H^+ \qquad (19\text{-}28)$$

$$2CuFeS_2 + 17/2O_2 + 5H_2O \rightarrow 2Cu^{2+} + 2Fe(OH)_3 + 4SO_4^{2-} + 4H^+,$$

where $Fe(OH)_3$ stands for various Fe(III) compounds (oxides, hydroxides, often sulfates) and Cu^{2+} represents other metal ions set free in solution.

A much-studied example of such an environmental menace in northeastern Quebec is described by Blowes and Jambor (1990). A 40-hectare impoundment of tailings from the milling of massive sulfide ore, after weathering for 3 or 4 decades, has developed a layered structure with a zone of complete oxidation a few centimeters thick at its top, an intermediate zone where oxidation is in progress (according to the above reactions), and a lower zone which oxygen has not yet reached. Pore water in the pile shows pH values as low as 2.5 locally, with concentrations of copper up to 60 ppm, of zinc up to nearly 200 ppm, and of cobalt and chromium amounting to several ppm. The tailings have been covered with agricultural limestone to reduce the acidity, and a dense growth of vegetation helps to limit infiltration of rainwater. Surface drainage from the pile goes to a collection pond, where the toxic metals are removed in a treatment plant. Effects on the environment are adequately controlled at present, but careful supervision and monitoring will be necessary for many centuries unless more drastic remedial measures are undertaken.

Uranium ores if exposed to oxidation contribute a different set of toxic metals to groundwater: the uranium itself, which in its oxidized form is readily soluble as uranyl ion, and also the vanadium, molybdenum, and selenium whose minerals are often present in uranium ore. Some of the daughter products of radioactive decay, especially radium, are soluble enough to constitute a further hazard to users of the water. Problems with uranium and its associated elements have been particularly acute with respect to the tailings piles from milling the sandstone ores of the western states: this material, from which the uranium has been largely removed, contains minerals of the other metals in finely divided form, hence unusually soluble because of the large surface area exposed to air and water. The dissolving of toxic metals and the emission of radon gas from the decay of radium have made these tailings piles a hazard to their surroundings and an environmental problem of first magnitude.

A related problem that has received much recent attention from geochemists is the safe disposal of waste material formed when uranium is used for the production of plutonium or the generation of useful energy. In these processes, the nuclei of uranium atoms are split by collisions with neutrons, a reaction that produces enormous energy and also a great variety of radioactive isotopes. The isotopes are part of the debris, the "radioactive waste" that accumulates from the production of nuclear energy. Much of the waste, the "high-level" part, is hot and intensely radioactive, so must be handled by remote control. It becomes a geochemical problem because plans are to put the solidified waste in a mined cavity several hundred meters underground. Here it will eventually be in contact with groundwater, and so in effect will behave much like an ore deposit containing a strange assortment of metals. Some of the metals are isotopic forms of familiar elements like strontium, cesium, zinc, and selenium; others are very unfamiliar, heavy elements with atomic numbers beyond that of uranium—neptunium, plutonium, americium. All are radioactive, and because of their radioactivity will be a hazard to living things if appreciable amounts escape to the surface. So the problem for geochemists is to study the possibility that some of these elements will dissolve in groundwater and be carried to the surface in more than trace quantities. Such a possibility clearly depends partly on the geology of the area in which the cavity containing waste is located and partly on the solubility of the different elements. Can a suitable place be found, and can the waste be put in sufficiently insoluble form, so that safety from escape of radioactive elements in more than minute traces can be guaranteed for as long as the dangerous elements remain active—which means for at least 10,000 years? It is a question that geochemists have been pondering for a long time, and to which the answer is probably "yes"—but still a question about which there is so much controversy that no high-level waste has yet been put underground.

19-7 SUMMARY AND GENERAL REFLECTIONS

Metallic ores come in almost infinite variety, and we have limited discussion here to the most common kind—ores for which geologic evidence tells us that the metals were transported and deposited by solutions at temperatures higher than those at the Earth's surface. Such solutions may originate as residual fluids from the crystallizing of an igneous melt, but recent work, particularly with oxygen isotopes, has demonstrated a variety of other sources for hydrothermal solutions.

The simple fact that ore deposits the world over show a certain regularity in the sequence of their minerals, a sequence that has enabled geologists to classify most deposits in a temperature-depth pattern, is testimony that the deposition of ore takes place according to physicochemical rules that have wide application. The most promising explanation for the observed regularities in ore–mineral associations and sequences has come from a study of stabilities of complex ions and molecules, which offer a means for keeping the metals in solution at appreciable concentrations in the presence of sulfur, and for permitting their crystallization in an orderly sequence in response to changes in temperature, pressure, pH, or oxidation state. The explanation is satisfying as a general logical framework, but many details of the actual mechanism of ore formation at specific deposits remain elusive.

Untouched also are some of the deeper questions about the origin of ores, for example possible connections between ore-forming solutions and plate tectonics. The recent discovery of places along the mid-ocean ridges where hot solutions are emerging and depositing sulfides, particularly sulfides of iron, copper, and zinc, suggests that metals in solution may accompany basalt rising out of the mantle. Do such occurrences provide an explanation for the well-known massive sulfide deposits that are found with mafic and ultramafic rocks, for example the famous copper deposits of Cyprus? Can metals be concentrated by the processes of metamorphism and magma generation thought to take place near the tops of crustal plates as they move downward in subduction zones? Can the apparent rough distribution of ore deposits of different metals in belts parallel to orogenic zones be related to distance from a nearby ocean trench, or to the depth of a subducting plate beneath?

Such broader questions, of course, involve a good deal more than geochemistry. They are fascinating subjects for speculation, and a geochemist who joins the search for answers can play a useful role by insisting that conclusions drawn from geophysics and geology be constrained so as not to violate the physicochemical rules that are known to guide the processes of ore formation.

PROBLEMS

1. Using data from Appendix VII, calculate the solubility of chalcocite in a solution of NaCl due to formation of the chloride complex $CuCl_2^-$. Assume a temperature of 25°C and a pH of 7, and make the calculation for various concentrations of NaCl. Is it possible for significant amounts of copper to be dissolved as $CuCl_2^-$ in a solution that contains sulfide ion? Would the solubility of copper in this form be dependent on pH? Explain.

2. Lead is known to form a stable hydroxy complex ion, $Pb(OH)_3^-$. How alkaline would a solution have to be for appreciable PbS to dissolve to give this ion? Would this be a geologically reasonable method of transporting lead in ore solutions?

3. It has been postulated that metals exist in ore solutions at temperatures above 200°C because sulfur is largely oxidized to SO_2, and that metal sulfides are precipitated below

this temperature because the following equilibrium shifts to the right as temperature falls:

$$SO_2 + 3H_2 \rightleftharpoons 2H_2O + H_2S$$

Make the necessary calculations to evaluate this suggestion, using an arbitrary figure for hydrogen fugacity of 10^{-3} bar.

4. From Fig. 19-7, what is the pH of neutral supercritical water at a temperature of 500°C and a density of 0.5 g cm^{-3}? Assuming that the fluid behaves as a perfect gas, calculate the pressure needed to maintain this density. Do you think the assumption is justified?

5. In the oxidized zone of lead deposits, a crystal of galena is often found surrounded by an inner layer of anglesite and an outer layer of cerussite. Suggest an explanation.

6. The metal cadmium often occurs as a minor constitutent of sphalerite in primary sulfide ores. In partially oxidized ores the sulfide of cadmium (greenockite) is sometimes found as tiny yellow crystals on the surface of sphalerite. Suggest an explanation.

7. Using free energies from Appendix VIII, show how the line separating the PbO_2 and $PbCO_3$ fields in Fig. 19-11 is located.

8. Construct a diagram similar to Fig. 19-11 showing the stability fields of the zinc minerals sphalerite, smithsonite, and zincite. Include "contour" lines showing concentrations of Zn^{2+} in the part of the diagram representing acid, oxidizing conditions. Use any reasonable figures for total dissolved carbonate and total dissolved sulfur.

9. Explain why supergene enrichment is more important for copper than for iron, and more important for silver than for nickel.

10. Using the oxidation potentials in Appendix IX, investigate the possibility of oxidizing and dissolving the metal palladium in the oxidized zone of an ore deposit.

11. Several Fe(II) and Fe(III) sulfate minerals are known [e.g., melanterite, $FeSO_4 \cdot 7H_2O$, and coquimbite, $Fe_2(SO_4)_3 \cdot 9H_2O$]. Under what geologic conditions would you expect these minerals to form?

REFERENCES

Ahmad, S. N., and A. W. Rose: "Fluid inclusions in porphyry and skarn ore at Santa Rita, New Mexico," *Econ. Geol.*, vol. 75, pp. 229–250, 1980.

Bethke, P. M., and P. B. Barton, Jr.: "Distribution of some minor elements between coexisting sulfide minerals," *Econ. Geol.*, vol. 66, p. 159, 1971.

Blowes, D. W., and J. L. Jambor: "The pore-water geochemistry and the mineralogy of the vadose zone of sulfide tailings, Waite Amulet, Quebec, Canada," *Applied Geochemistry*, vol. 5, pp. 327–346, 1990.

Brimhall, G. H., and D. A. Crerar: "Ore fluids: magmatic to supergene," in Carmichael, I. S. E., and H. P. Eugster, (eds): *Thermodyanic Modeling of Geological Materials: Minerals, Fluids, and Melts, Reviews in Mineralogy*, vol. 17, Mineralogical Society of America, pp. 235–282, 1987.

Fisher, J. R., and H. L. Barnes: "The ion-product constant of water to 350°C," *Jour. Phys. Chem.*, vol. 76, pp. 90–99, 1972.

Garrels, R. M., and C. Christ: *Solutions, Minerals, and Equilibria*, Harper and Row, New York, 1965. Chapter 7 has Eh–pH diagrams and partial-pressure diagrams showing relations among the oxidized products of ores of most of the heavy metals.

Hemley, J. J., and J. P. Hunt: "Hydrothermal ore-forming processes in the light of studies in rock-buffered systems: II. Some general geologic applications," *Econ. Geology*, vol. 87, pp. 23–43, 1992.

Hemley, J. J., G. L. Cygan, J. B. Fein, G. R. Robinson, and W. M. d'Angelo: "Hydrothermal ore-forming processes in the light of studies in rock-buffered systems: I. Iron–copper–zinc–lead sulfide solubility relations," *Econ. Geology,* vol. 87, pp. 1–22, 1992. Decribes experimental studies of sulfide solubilities in chloride solutions at 300°C to 700°C and 0.5 to 2 kbar, in solutions buffered by rock-forming minerals.

Quist, A. S.: "The ionization constant of water to 800°C and 4000 bar." *Jour. Phys. Chem.* vol. 74, pp. 3396–3402, 1970.

Roedder, E.: "Fluid inclusion studies on the porphyry-type ore deposits at Bingham Canyon, Utah, Butte, Montana, and Climax, Colorado," *Econ. Geology,* vol. 66, pp. 98–120, 1971.

Roedder, E.: "Fluid inclusions as samples of ore fluids," in Barnes, H. L., (ed.): *Geochemistry of Hydrothermal Ore Deposits*, 2d ed., Wiley, pp. 684–737, 1979.

Rye, R. O., and H. Ohmoto: "Sulfur and carbon isotopes and ore genesis: a review," *Econ. Geology,* vol. 69, pp. 826–842, 1974.

Sato, M.: "Oxidation of sulfide ore bodies. II. Oxidation mechanisms of sulfide minerals at 25°C," *Econ. Geology,* vol. 55, pp. 1202–1231, 1960. An experimental study of details of the mechanisms of oxidation of some common sulfide minerals.

Sato, M.: "Persistency-field Eh–pH diagrams for sulfides and their application to supergene oxidation and enrichment of sulfide ore bodies," *Geochim. et Cosmochim. Acta,* vol. 56, pp. 3133–3156, 1992.

Taylor, H. P., Jr.: "Oxygen and hydrogen isotope relationships in hydrothermal mineral deposits," in Barnes, H. L., (ed.): *Geochemistry of Hydrothermal Ore Deposits,* 2d ed., Wiley, New York, pp. 236–277, 1979. Description of techniques and critical review of the kinds of information obtainable from studies of isotope ratios of hydrogen and oxygen.

SUGGESTIONS FOR FURTHER READING

Guilbert, J. M., and C. F. Park, Jr.: *The Geology of Ore Deposits,* W.H. Freeman, New York, 1986. The physicochemical background of ore formation, and its application to many deposits.

Langmuir, D.: "Uranium solution-mineral equilibria at low temperatures with application to sedimentary ore deposits," *Geochim. et Cosmochim. Acta,* vol. 42, pp. 547–569, 1978. Thermodynamic data applied to movement and deposition of uranium in surface water and groundwater.

Lindgren, W.: *Mineral Deposits,* 4th ed., McGraw-Hill, New York, 1933. Chapter 16 describes Lindgren's classification of ore deposits.

Nash, J. T.: "Fluid-inclusion petrology—data from porphyry copper deposits," *U.S. Geol. Survey Prof. Paper* 907-D, pp. 1–16, 1976. A good review of the use of fluid inclusions for estimating temperatures, pressures, and compositions of ore-forming solutions, especially as applied to one kind of copper deposit.

Ohmoto, H., and R. O. Rye: "Isotopes of sulfur and carbon," in Barnes, H. L., (ed.): *Geochemistry of Hydrothermal Ore Deposits*, 2d ed., Wiley, New York, pp. 509–567, 1979. Data and interpretation of isotope ratios of sulfur and carbon in ore deposits.

Roedder, E.: "Fluid inclusions as samples of ore fluids," in Barnes, H. L., (ed.): *Geochemistry of Hydrothermal Ore Deposits*, 2d ed., Wiley, New York, 1979, pp. 684–737, 1979. An excellent summary of the kinds of information available from detailed study of fluid inclusions in the minerals of ore deposits.

Seward, T. M.: "Thio complexes of gold and the transport of gold in hydrothermal ore solutions," *Geochim. et Cosmochim. Acta,* vol. 37, pp. 379–399, 1973. Experimental data on sulfide, bisulfide, and chloride complexes of gold at 160–300°C.

Sillitoe, R. H.: "The tops and bottoms of porphyry copper deposits," *Econ. Geology,* vol. 68, pp. 799–815, 1973. A model for the development of an important kind of copper deposit, relating such deposits to the combined activity of meteoric and magmatic water, in a granitic body intruded at shallow levels and surmounted by a volcano.

CHAPTER
20

DISTRIBUTION
OF THE
ELEMENTS

The central problem of geochemistry was defined by the Norwegian geochemist V. M. Goldschmidt, early in the 20th century, to be "the determination of the distribution of the elements in materials of the Earth and the reasons for this distribution." In a sense we have been working with this problem from the start of Chapter 1, but our emphasis has been on the role of particular elements in geologic processes as seen from a chemical standpoint. Now we set out to address Goldschmidt's problem more directly.

Basic facts about element distribution are already familiar, just from ordinary experience and from discussions in preceding chapters. Igneous rocks have a characteristic group of elements (Chap. 17), sulfide ores another group (Chap. 19), carbonate sediments another group (Chap. 3), salt deposits still another (Chap. 14). To refine such generalizations, in particular to see how the rarer elements are distributed in various environments, we need a classification, a way of grouping the elements so that their properties may be related to geologic behavior. One such scheme is Mendeleev's periodic table, an overall grouping according to chemical properties that is familiar from elementary chemistry and that we introduced in a geological context back in Chapter 5 (Table 5-3). A few highlights of this table should be recalled.

Elements in each column of the table are closely related in their electronic structures and chemical properties. Resemblances are especially strong among the

534

metallic elements in columns toward the left side of the table and among the nonmetallic elements on the right. On these left and right sides the properties of rare elements, both chemical and geological, can be predicted with considerable accuracy from properties of the better known elements in the same column. The transition elements in the middle of the table, mostly metals, are less regular in their relationships, and prediction of properties is less satisfactory. Two special groups of elements, the actinides and lanthanides, are singled out because they are so very similar among themselves in their properties and in the structures of their atoms. From these relations in Mendeleev's table many useful general statements about the distribution of the rarer elements in geologic materials are possible, but a more specific geochemical classification is also needed.

20-1 GOLDSCHMIDT'S GEOCHEMICAL CLASSIFICATION

One alternative classification was suggested by Goldschmidt in his early work on the rules of element distribution. This grouping of elements was an attempted answer to a hypothetical question: If the Earth at some time in the past was largely molten and if the molten material separated itself on cooling into a metal phase, a sulfide phase, and a silicate phase, how would the elements distribute themselves among these three materials? An answer can be sought from theoretical arguments and also from three kinds of observation: (1) the composition of meteorites, on the assumption that meteorites have an average composition similar to that of the primordial Earth and underwent a similar kind of differentiation; (2) analyses of metal, slag (silicate), and matte (sulfide) phases in metallurgical operations; and (3) the composition of silicate rocks, sulfide ores, and the rare occurrences of native iron found in the Earth's crust.

From a theoretical standpoint, consider first the expected distribution of elements between metallic iron and silicates, in a system with iron in excess. Metals more chemically active (in the sense of a higher free energy of oxidation) than iron would presumably combine with silica to form silicates, and the remaining silica would react so far as possible with iron. Metals less active than iron would have no chance to form silicates but would remain as free metals with the uncombined iron. In other words, the fate of any given metal should depend entirely on the free energy of formation of its silicate (Sec. 7-4). Data on formation energies of silicates are not complete, but to a good approximation we can substitute the free energies of oxides, since in the formation of a silicate the energy of the reaction (with Me standing for any metal)

$$Me + 1/2O_2 \rightleftharpoons MeO \qquad (20\text{-}1)$$

is always much larger than the energy of the reaction

$$MeO + SiO_2 \rightleftharpoons MeSiO_3. \qquad (20\text{-}2)$$

Free energies of formation of representative oxides are given in Table 20-1. From this list we could predict that the elements above iron would go preferentially into

TABLE 20.1
Free energies of formation of oxides per oxygen atom, kJ

	25°C	827°C
CaO	−604.0	−520.8
MgO	−569.4	−481.6
Al_2O_3	−527.4	−442.7
UO_2	−515.9	−447.2
TiO_2	−444.7	−371.9
SiO_2	−428.3	−356.8
MnO	−362.9	−304.3
K_2O	−322.1	−203.2
ZnO	−320.4	−237.9
WO_2	−266.9	−195.0
SnO_2	−259.9	−176.5
FeO	−251.1	−199.7
MoO_3	−222.7	−156.7
CoO	−214.2	−154.9
NiO	−211.7	−140.7
PbO	−188.9	−109.2
Cu_2O	−146.0	−86.9

The numbers are free energies of formation per oxygen atom, in kilojoules. The oxides are arranged in order of decreasing free energy at 25°C. *Source:* Robie et al. (1979).

the oxide (or silicate) phase, and those below iron into the metallic phase. The grouping is not entirely unambiguous, in that two elements (W and Sn) appear to stand above iron at low temperatures and below iron at high temperatures. Similar lists could be drawn up showing the expected distribution of metals between a metal phase and a sulfide phase with iron in excess, and between a sulfide phase and a silicate phase with silica in excess. So many additional assumptions are required, however, that the numerical values are not very helpful.

On the observational side, the compositions of the three phases (metal, sulfide, silicate) in meteorites and in smelter products are well-known. Analyses show that, with a few exceptions, elements above iron in Table 20-1 have higher concentrations in the silicate phases than the metal phases of both materials, and that most of the elements below iron are strongly concentrated in the sulfide phases. Occurrences of elements in silicate rocks, sulfide ores, and native iron agree fairly well with the distributions shown by the meteorite and smelter analyses. The lack of complete agreement is not surprising, as Goldschmidt pointed out, since the conditions of formation of sulfide ore deposits in nature are quite different from the conditions under which sulfides would separate from an artificial melt.

On the basis of such evidence, Goldschmidt suggested that the elements could be usefully grouped into those that preferentially occur with native iron and probably are concentrated in the Earth's iron core (*siderophile elements*); those concentrated in sulfides and therefore characteristic of sulfide ore deposits (*chalcophile elements*); and those that generally occur in or with silicates (*lithophile elements*). For completeness, elements that are prominent in air and other natural gases can be put in a fourth group, the *atmophile elements*. This classification, shown in Table 20-2, is clearly consistent with the data of Table 20-1.

As might be expected, Goldschmidt's classification is closely related to the periodic law. Comparison of Tables 20-2 and 5-3 shows that in general siderophile elements are concentrated in the center of the periodic table, lithophile elements to the left of center, chalcophile elements to the right, and atmophile elements on the extreme right. The classification can also be correlated with electrode potentials: siderophile elements are dominantly noble metals with low electrode potentials, lithophile elements are those with high potentials, and chalcophile elements have an intermediate position.

The grouping of elements shown in Table 20-2 can at best express only tendencies, not quantitative relationships. The different groups overlap, as is shown by the occurrence of many elements in more than one category. Iron, for example, is not only the principal element of the Earth's core but is also common in sulfide deposits and in igneous rocks. Such overlaps are inevitable in a classification that is based partly on distributions in very high-temperature processes and partly on distributions under ordinary surface conditions. Various refinements have been suggested, but there seems little point in trying to make the classification more quantitative or more detailed. It is useful simply as a rough qualitative expression of the geologic behavior of the elements.

TABLE 20-2
Goldschmidt's geochemical classification of the elements

Siderophile	Chalcophile	Lithophile	Atmophile
Fe Co Ni	Cu Ag (Au)[1]	Li Na K Rb Cs	H N (C) (O)
Ru Rh Pd	Zn Cd Hg	Be Mg Ca Sr Ba (Pb)	(F) (Cl) (Br) (I)
Re Os Ir Pt Au	Ga In Tl	B Al Sc Y REE[2]	He Ne Ar Kr Xe
Mo Ge Sn C P	(Ge) (Sn) Pb	(C) Si Ti Zr Hf Th	
(Pb) (As) (W)	As Sb Bi	(P) V Nb Ta	
	S Se Te	O Cr W U	
	(Fe) (Mo) (Re)	(Fe) Mn	
		F Cl Br I	
		(H) (Tl) (Ga) (Ge)	
		(N)	

[1] Parentheses around a symbol indicate that the element belongs primarily in another group, but has some characteristics that relate it to this group. For example, gold is dominantly siderophile, but (Au) appears in the chalcophile group because gold is often found in sulfide veins.
[2] REE = rare-earth elements.

20-2   DISTRIBUTION OF ELEMENTS IN IGNEOUS ROCKS

Point of View

The origin of many igneous rocks, as noted in previous discussions, is a subject of lively debate. Fortunately the various hypotheses of origin lead in general to similar predictions about the distribution of trace elements. To unify the discussion—but not to argue for one hypothesis over another—we start with the "classical" viewpoint that most igneous rocks can be thought of as formed by differentiation of basaltic magma, mafic minerals settling out first to form ultramafic rocks and the remaining melt changing in composition through the series gabbro → diorite → granodiorite → granite → pegmatite. We can speak then of "early-formed minerals" of high melting point, such as olivine and calcic plagioclase, "late-formed minerals" like alkali feldspars and biotite, and "residual fluids" of pegmatitic composition. We can work out semiempirical rules of behavior for various elements during such a differentiation process and compare our predictions with analyses. When this procedure is adopted, it turns out that agreement between predictions and analyses is surprisingly good, and surprisingly uniform from one set of igneous rocks to another.

This agreement, however, is not evidence in favor of the assumed crystallization-differentiation process for the origin of igneous rocks. The same sequence of rock compositions could be formed equally well (but in reverse order) by progressive anatexis of a shale or graywacke, proceeding from the composition of alkali granite through granodiorite, tonalite, and gabbro as more and more mafic material is incorporated into the melt. The same distribution of elements would be expected at comparable stages in this series as in the sequence formed by crystallization differentiation, provided that the composition of the starting material was reasonably close to the crustal average and provided that equilibrium was maintained during the melting. If marked departures from equilibrium occur, during either the fractional crystallization of a mafic magma or the progressive melting of average silicate material, the distribution of minor elements might be considerably different in corresponding parts of the two series. But ordinarily a close enough approach to equilibrium can be assumed so that the rare elements would be expected to distribute themselves according to uniform rules.

Thus the point of view that most igneous rocks have formed by crystallization differentiation, a point of view that underlies much of the following argument, is presented not as a hypothesis to be proved but as a framework to unify the discussion. When we speak of "early-formed minerals," for example, we shall mean minerals that *would* form early during differentiation, but that might equally well form late during metamorphism and anatexis.

Rules of Distribution

The general sequence of minerals that separate during differentiation of a silicate melt at low pressures is already familiar (Chap. 17). Most commonly olivine and calcic plagioclase crystallize first; as the temperature falls, part or all of the olivine reacts with the melt to form pyroxene, which is followed by amphibole and then biotite; the composition of plagioclase meanwhile becomes increasingly sodic; near the end of the crystallization, quartz and potash feldspar appear along with sodic plagioclase. If oxygen fugacity is low this scheme is altered, in that abundant iron remains longer in the melt; this means that ferrous silicates continue to crystallize until a late stage and that the accumulation of silica in the residual fluid is less pronounced. Many other variations are possible: an abundance of water in the melt would emphasize formation of hydrous minerals, abundant chlorine or fluorine or boron would lead to unusual minerals containing these elements, and at higher pressures the dominant early-crystallizing minerals would be garnet and soda-rich pyroxene rather than olivine and plagioclase. But for present purposes we restrict the discussion to the more usual pattern of differentiation, in which major elements drop out in a fairly uniform sequence: most of the magnesium and calcium leave the melt in early stages, and the alkali metals later; iron may be largely concentrated in early minerals or may appear at all stages; aluminum drops out in feldspars all during the differentiation process, and also in micas toward the end. Our objective now is to see how the behavior of the less common elements is related to these changes.

For an element to crystallize in a mineral of its own—a mineral in which it is a major constituent—requires that the element be present in the melt in appreciable amounts. If only a few ions of the element are present, they can be taken up by the crystal structures of the major silicates, either as isomorphous replacements of an abundant element or as random inclusions in the holes of a crystal lattice. *How much* of a given element can be accommodated in the silicate structures depends on the characteristics of its ions. The rare alkali metal rubidium, for example, is so similar to potassium that several tenths of a percent can be accommodated as replacements of Rb^+ for K^+ in micas and feldspars, and no separate rubidium mineral can form. The rare metal zirconium, on the other hand, cannot fit easily into common silicate structures, and even very small amounts of it go into separate crystals of the accessory mineral zircon. Some elements, like beryllium, boron, copper, and uranium, are capable neither of forming their own high-temperature minerals nor of substituting appreciably for common ions in silicate structures, and so are concentrated in the residual solutions that give rise to pegmatites and quartz–sulfide veins. What specific characteristics of an element determine which way it will behave?

The question takes us back to the discussion of isomorphism in Sec. 5-8. On the basis of that earlier treatment, a few rules can be formulated that are generally followed in the isomorphous replacement of one ion by another:

1. A minor element may substitute extensively for a major element if the ionic radii do not differ by more than about 15%.
2. Ions whose charges differ by one unit may substitute for one another, provided their radii are similar and provided the charge difference can be compensated by another substitution. For example, in plagioclase feldspar Na^+ readily substitutes

for Ca^{2+}, and the charge difference is compensated by substitution of Si^{4+} for Al^{3+}.

3. Of two ions that can occupy the same position in a crystal structure, the one that forms the stronger bonds with its neighbors is the one with the smaller radius, higher charge, or both.

4. Substitution of one ion for another may be very limited, even when the size criterion is fulfilled, if the bonds formed differ markedly in covalent character.

Leaving aside the question of bond character for the moment, we find abundant illustrations of the first three rules. We could predict, according to rules 1 and 2 and the table of ionic radii in Appendix VI, that Ba^{2+} would commonly substitute for K^+, Cr^{3+} for Fe^{3+}, Y^{3+} for Ca^{2+}; and these predictions fully accord with analytical data. From the third rule we could generalize that, in an isomorphous pair of compounds, the one containing ions of smaller radius and higher charge would have the higher melting point (because of the stronger bonding) and therefore would appear earlier in a crystallization sequence. Hence a minor element like Li, which substitutes extensively for Mg because of the similarity in their ionic radii, should be concentrated in late-forming Mg minerals rather than early ones because its single charge forms weaker bonds than the double charge on Mg^{2+}; this is borne out by the near absence of Li in olivine and its common presence as a substitute for Mg in the micas of pegmatites. Similarly Rb, having an ion similar in charge to K^+ but somewhat larger, should be enriched in late K minerals rather than early ones, a prediction that agrees with the observed concentration of Rb in the feldspars and micas of pegmatites.

The same rules, it should be noted, are illustrated beautifully by some of the major elements. In the isomorphous series of the olivines, forsterite has a higher melting point than fayalite and is enriched in early crystals (Prob. 17-3), corresponding with the fact that Mg^{2+} is a smaller ion than Fe^{2+}. Early crystals of plagioclase are Ca-rich, late crystals Na-rich (Sec. 17-3), in accordance with the higher charge of the Ca^{2+} ion. The rules are also similar to generalizations we have formulated regarding ion-exchange processes (Sec. 6-3), as they should be, because in both cases we are dealing with the relative strengths of bonds formed between ions and crystal structures.

The rules of substitution according to ionic radius work well as long as we limit discussion to elements in the first three columns of the periodic table, but with the remaining elements agreement between prediction and observation is often much less satisfactory. The ion Cu^+, for example, is similar in size and charge to Na^+, but copper shows no enrichment in sodium minerals. The ion Hg^{2+} closely resembles Ca^{2+} in size and charge, but is not concentrated in calcium minerals. The difficulty goes back to rule 4, the difference in bond character (or polarization): the ions Cu^+ and Hg^{2+} form bonds of markedly less ionic character with the anions of a crystal structure than do Na^+ and Ca^{2+} (last column of Appendix VI). This failure of the rules of substitution is reminiscent of the failure of predictions from simple geometry to account for crystal structures when strongly covalent bonds are involved (Sec. 5-4).

Examples of Minor-Element Distribution

To get a feeling for the effectiveness of the distribution rules in predicting the behavior of minor elements in igneous processes, we look now at some analytical data for typical rock sequences.

Table 20-3 shows average concentrations of major and minor elements in four rock series. The first series consists of worldwide averages for the major igneous rock types, and the other three give analyses for rocks in specific localities. The world-wide averages, of course, do not represent a sequence necessarily related by a differentiation process, but they are pertinent because trace elements would distribute themselves in much the same manner whether rocks form by differentiation or by anatexis during metamorphism. For each of the other three series there is good geological and chemical evidence that the rocks are actually members of a differentiation sequence. The Irish lavas show a fairly typical trend from olivine basalt to rhyolite, with a marked increase in silica and alkali metals and a decrease in iron; the Skaergaard rocks show a notable increase in iron and fairly constant silica until the very end of the differentiation process; the Hawaiian lavas are rich in soda and potash and show only a modest increase in silica. Thus the four sets of analyses enable us to compare the distribution of trace elements in three different kinds of differentiation sequences with the average distribution for igneous rocks in general.

Cursory examination of the data shows that the distribution has a certain amount of regularity, but also much apparently random variation. As so often happens in geochemistry, the numbers seem to follow a vague general rule that is subject to many exceptions. A little reflection will show that this is what we might expect. We are dealing, of course, with a very complicated process. The distribution is affected not only by the various trends of differentiation, but also by details of the way the trace elements enter into individual minerals. Other variables than ionic radius can affect the extent of substitution of one element by another, for example pressure, temperature, and overall rock composition. Furthermore, the steps in differentiation do not necessarily correspond from one set of analyses to another, nor do the extremes necessarily represent equivalent stages in differentiation; we have, in fact, no good way of even defining what we mean by "equivalent stages of differentiation." Recognizing these reasons for lack of complete regularity in the data, we can nevertheless draw useful conclusions from the analyses about general patterns of trace-element behavior.

Trends shown in the table can be visualized with the diagrams in Fig. 20-1. Lines on the figure represent analyses for various elements in the four categories of the world averages: ultramafic rocks, mafic rocks, intermediate rocks, and felsic rocks. These four rock types are shown, from left to right, by four points on each line. The heavy lines represent major elements (K, Ca, Al, Fe, Mg), and the light lines in each column represent possibly related minor elements. Scales are not uniform, so that absolute values cannot be compared from diagram to diagram. The lines simply show trends, the relative increases or decreases of different elements in going from ultramafic to felsic rocks.

TABLE 20-3
Average content of major and minor elements in four rock series[1]

	Ionic radii (Å)	World averages				Tertiary lavas of NE Ireland				Skaergaard intrusive rocks				Hawaiian lavas				
		Ultra-mafic rocks	Mafic rocks	Inter-mediate rocks	Felsic rocks	Oli-vine basalt	Tho-leite basalt	Quartz tra-chyte	Rhyo-lite	Gabbro-picrite	Hyp-olivine gabbro	Ferro-gabbro	Grano-phyre	Picrite-basalt	Basalt	Andes-ine andesite	Oligo-clase andesite	Tra-chyte
Si^{4+}	0.26	19.0	24.0	26.0	32.3	21.1	24.3	30.0	35.2	19.3	21.6	20.3	31.2	22.3	24.0	23.7	24.4	29.0
Al^{3+}	0.54	0.5	8.8	8.9	7.7	7.8	7.5	7.5	6.5	4.6	8.9	7.7	6.7	4.9	6.8	8.7	8.8	9.9
Fe^{3+}	0.64	9.9	8.6	5.9	2.7	1.9	1.8	3.2	0.6	1.9	1.1	2.3	2.6	1.0	1.2	4.1	2.2	3.0
Fe^{2+}	0.78					6.7	6.5	2.2	0.3	8.2	8.1	17.1	3.9	8.1	7.3	4.8	5.7	0.1
Mg^{2+}	0.72	25.9	4.5	2.2	0.6	6.8	3.5	0.5	0.1	16.3	5.8	1.5	0.3	11.4	4.6	2.6	1.9	0.2
Ca^{2+}	1.00	0.7	6.7	4.7	1.6	7.1	7.1	1.5	0.6	4.7	8.1	6.5	2.0	5.5	7.5	4.6	4.5	0.6
Na^+	1.02	0.6	1.9	3.0	2.8	1.5	1.9	3.1	2.0	0.5	1.8	2.1	3.0	1.2	1.5	3.5	4.1	5.1
K^+	1.38	0.03	0.8	2.3	3.3	0.2	0.9	3.5	3.8	0.1	0.2	0.3	2.6	0.3	0.3	1.8	1.8	4.1
P^{5+}	0.17	170	1,400	1,600	700	970	1,260	535	90	90	260	5,000	1,500	1,000	1,140	740	4,060	1,050
Ga^{3+}	0.62	2	18	20	20	33	35	40	43	8	19	20	33	18	23	20	23	25
Cr^{3+}	0.62	2,000	200	50	25	1,600	150	50	<1	1,500	230	<1	11	1,750	470	<1	<1	20
Li^+	0.76	<1	15	20	40	8	18	65	80	2	2	3	20	1	2	15	20	30
Ni^{2+}	0.69	2,000	160	55	8	900	50	65	13	1,000	120	<2	9	950	85	7	13	15
Co^{2+}	0.75	200	45	10	5	140	110	8	2	90	48	20	5	75	35	14	6	2
V^{3+}	0.64	40	200	100	40	630	750	55	2	120	220	4	9	250	280	70	19	<5
Ti^{3+2}	0.67	300	9,000	8,000	2,300	7,000	6,700	3,200	700	9,000	5,000	15,000	5,000	12,000	20,000	16,000	15,000	2,000
Zr^{4+}	0.72	30	100	260	180	180	700	3,000	2,000	30	33	20	1,200	75	100	350	1,250	1,500
Mn^{2+}	0.83	1,500	2,000	1,200	600	1,380	1,360	1,200	130	1,200	700	3,300	800	860	1,010	1,710	1,010	1,170
Sc^{3+}	0.75	5	24	3	3	19	25	25	<10	<10	20	10	<10	<10	10	<10	<10	<10
Cu^{+2}	0.77	20	100	35	20	400	180	50	66	100	67	400	200	150	170	<10	<10	<10
Sr^{2+}	1.18	10	440	800	300	680	1,250	650	170	100	600	400	450	300	800	2,500	3,500	100
Pb^{2+}	1.19	<1	8	15	20	<20	25	43	43	10	No data	20	100	110	<20	No data	100	100
Ba^{2+}	1.35	1	300	650	830	310	1,350	2,000	2,300	10	18	60	1,100	110	120	600	1,000	800
Rb^+	1.52	2	45	100	200	3	90	1,000	930	<20	<20	<20	110	<20	<20	35	55	300

[1] Major elements in weight percent (of the elements, not the oxides), minor elements in parts per million.
2 The elements Ti and Cu may also be present in part as other ions, Ti^{4+} (radius 0.61 Å) and Cu^{2+} (radius 0.73 Å).

Sources:
Ionic radii, Appendix VI.
World averages, Vinogradov (1962).
Irish lavas, Patterson (1952).
Skaergaard rocks, Wager and Mitchell (1951).
Hawaiian lavas, Wager and Mitchell (1953).

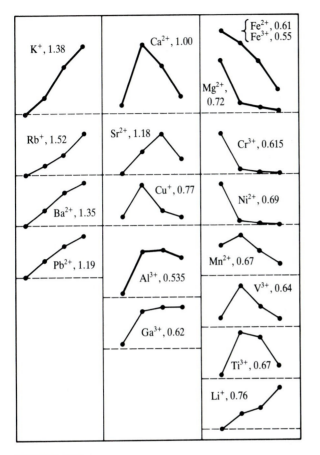

FIGURE 20-1

Trends of minor-element concentration in a differentiation sequence. Data from columns headed "world averages" in Table 20-3. The four points in each line represent concentrations in (from left to right) ultramafic rocks, mafic rocks, intermediate rocks, and felsic rocks. Scales are not uniform; lines show trends only. Heavy lines are trends for major elements, light lines for minor elements. Numbers beside symbols are ionic radii in angstroms.

The first column of Fig. 20-1 gives trends for three large cations that might be expected to substitute for K^+ because of their size. The similarity in their lines is evident, suggesting that Rb^+, Ba^{2+}, and Pb^{2+} do indeed enter potassium minerals in small amounts and hence, like potassium, are concentrated in rocks formed late in the crystallization sequence. The similarity in trend is most marked for Rb^+; because of their double charge and smaller size, Ba^{2+} and Pb^{2+} tend to enter early potassium minerals, hence are not as strongly enriched at the felsic end of the series as is Rb^+. For these elements the rules of substitution work beautifully.

The second column shows the trend for calcium and the related minor element strontium. Here agreement with the rules is less clear. The ion Sr^{2+}, with a radius between those of Ca^{2+} and K^+, can substitute for both, so that its trend is a compromise between the trends for the two major elements. Except for its low abundance in ultramafic rocks, strontium remains fairly constant through the differentiation sequence, because calcium decreases as potassium increases. Copper, with an ionic radius close to that of calcium, has a somewhat similar trend line, but the similarity is probably fortuitous because a chalcophile element like copper forms dominantly covalent bonds with oxygen and cannot readily substitute for an element whose bonds are strongly ionic. Also in the second column are curves for aluminum and gallium, illustrating the great similarity of these two elements and the consequent extensive substitution of Ga^{3+} for Al^{3+} in aluminosilicate minerals.

In the third column of Fig. 20-1 are trend lines for several minor elements with ionic radii similar to those of iron and magnesium. The two major elements show a markedly decreasing abundance as differentiation proceeds, and most of the minor-element trend lines follow this pattern except that some have low concentrations at the ultramafic end of the series. Trends for Cr^{3+} and Ni^{2+} mimic that of Mg^{2+}, as would be expected from the extensive substitution of these metals in early formed Mg minerals (pyroxenes and olivines). Mn^{2+}, V^{3+}, and Ti^{3+} substitute more readily for Fe, and their trend lines are correspondingly similar except at the left-hand ends. Li^{+}, with its single positive charge, substitutes readily for Mg^{2+} only in the latest-formed magnesium minerals (especially micas), hence increases in abundance during the course of differentiation.

Analyses for the three differentiation series in Table 20-3 (northern Ireland, Skaergaard, and Hawaii) show trends similar to those for the world averages. There are a few conspicuous deviations, as might be expected because the rock series have different overall compositions and different degrees of differentiation, but nevertheless the trends displayed in Fig. 20-1 apparently represent a general pattern widely applicable to igneous rocks.

Regularities of Distribution in Igneous Rocks

Important features of this general pattern, as revealed by many analyses of rock series like those in Table 20-3, are the following:

1. Cations with large radii and low electric charges tend to substitute for potassium (radius of K^{+} = 1.38 Å), hence are concentrated in felsic rather than mafic rocks. These elements (Rb, Cs, Ba, Pb, Tl) are often called the *large-ion lithopile* group (LIL). Their abundance in a rock series is a good indication of the extent to which differentiation has sorted out constituents of the original igneous material. The rare-earth elements are often included here, although the substitution of their ions in late-forming minerals is mostly for Ca^{2+} rather than K^{+}.

2. Several cations with smaller radii and mostly with higher charges (U^{4+}, Th^{4+}, B^{3+}, Be^{2+}, Mo^{6+}, W^{6+}, Nb^{5+}, Ta^{5+}, Sn^{4+}, Zr^{4+}) also are concentrated at the felsic end of the series, not because of extensive substitution, but rather because their size and charge make their substitution for any major ions in common silicate minerals difficult. Differentiation segregates them into the late residual solutions,

and if present in appreciable amounts they may form minerals of their own (uraninite, beryl, columbite, zircon, etc.). Because of the small size and high charge, the electric field associated with these ions is unusually strong, and they are often designated *high field-strength elements* (HFSE). Both LIL and HFSE elements, because their ions do not fit readily into the usual positions available in silicate structures, are described as *incompatible* elements.

3. Many elements whose ions have intermediate radii, especially metals of the transition groups, substitute readily for iron and magnesium, hence are abundant in the earlier members of the differentiation sequence. Some (Cr, Ni, Co) are strongly enriched with Mg in ultramafic rocks, others (Mn, V, Ti) have maximum abundance in gabbros and basalts.

4. Among the chalcophile elements, a few substitute to some extent for major cations in silicate structures (Pb^{2+} and Tl^+ for K^+, Zn^{2+} for Fe^{2+} and Mg^{2+}, Bi^{3+} for Ca^{2+}), but for the most part these metals are left to accumulate in the residual solutions that may eventually form sulfide ores. In a magma containing abundant sulfur, part of the chalcophile elements may separate early in differentiation as an immiscible sulfide liquid. The traces of chalcophile metals commonly reported in ordinary igneous rocks may be present largely as tiny sulfide grains rather than as substitutes for major elements. This would explain, for example, the apparently erratic values for copper in the analyses of Table 20-3.

5. Some minor elements are so similar in size and chemical properties to major elements that normal differentiation cannot separate them effectively from their major relatives. Gallium is a good example: it is always present in aluminum minerals, and very seldom becomes sufficiently segregated to appear in a mineral of its own. Other closely related pairs are Rb and K, Hf and Zr, Cd and Zn. The rare members of these pairs are not particularly scarce metals, but they are little known because minerals in which they appear as major constituents are rare or nonexistent.

Thus the principal features of minor-element distribution in igneous rocks can be explained by the simple rules of substitution based on ionic charge and radius, supplemented by empirical statements about the behavior of elements whose bonds are largely covalent. The simple rules, however, have many exceptions, and details of the distribution would require more sophisticated analysis. We have spoken of the distribution, for example, as if it were influenced only by mineral structures in the crystallizing solids; a complete description would have to consider also structures in the silicate liquid with which the solids are in contact. Again, the transition elements show many apparent anomalies in the relative amounts that substitute for Mg^{2+} and the two ions of iron; the anomalies can be in large part clarified by considering effects on the strengths of directed bonds exerted by electric fields in different positions in crystal structures ("crystal-field effects"). Detailed study shows that relationships between major and minor elements in crystal structures are highly complex, so it seems remarkable that the simple rules work as well as they do.

One way to represent the distribution of minor elements is to set up a ratio of concentration in a mineral to concentration in the melt. This ratio, called the *partition coefficient,* is defined as

$$k_{Me} = \frac{[Me]_{mineral}}{[Me]_{melt}},\tag{20-3}$$

where the brackets are some measure of the concentration of a minor element represented by the symbol Me. As an example, elements that are largely incompatible with major elements of silicate minerals have values of k_{Me} much less than one. Analysis of measured concentrations of minor elements in minerals and matrix glasses of extrusive igneous rocks and from laboratory experiments shows that the partition coefficients for many minor elements are dependent on changes in temperature, pressure, and composition, as would be expected on thermodynamic grounds. In most cases these general trends can be predicted using the rules of distribution outlined above.

As an example of the temperature–pressure–composition relationships of partition coefficients, consider the partitioning of Ni between a melt and the mineral enstatite in the system NiO–MgO–SiO_2. At equilibrium we have

$$Ni^{2+} + MgSiO_3 \rightleftharpoons Mg^{2+} + NiSiO_3 \,,\tag{20-4}$$
$$\text{melt} \qquad \text{enstatite} \qquad \text{melt} \qquad \text{Ni-enstatite}$$

and an equilibrium constant of

$$K = \frac{a_{Ni\text{-enstatite}}\, a_{Mg\text{-melt}}}{a_{Ni\text{-melt}}\, a_{Mg\text{-enstatite}}}.\tag{20-5}$$

If activities are expressed in terms of concentration (represented by brackets) and activity coefficients (λ),

$$K = \frac{[Ni]_{enstatite}}{[Ni]_{melt}} \frac{[Mg]_{melt}}{[Mg]_{enstatite}} \left(\frac{\lambda_{Ni\text{-enstatite}}\lambda_{Mg\text{-melt}}}{\lambda_{Ni\text{-melt}}\lambda_{Mg\text{-enstatite}}} \right).\tag{20-6}$$

The first ratio on the right-hand side is the partition coefficient for Ni [Eq. (20-3)]. Its temperature and pressure dependence will be determined, in part, by changes in the equilibrium constant which are governed by the enthalpy and volume of reaction (Sec. 8-5). How the partition coefficient will depend on composition is determined by the activity coefficients: for the melt, activity coefficients reflect melt structure and interaction among Ni and Mg with other melt components, and for enstatite, the coefficients are related to energetics of incorporating Ni and Mg into the mineral structure. All these properties, of course, are functions of temperature and pressure.

Rare-Earth Elements

The distribution of minor elements, as is evident from the simple rules of substitution, leads readily to speculation about the genesis and history of particular igneous rocks and rock associations. The rare-earth elements (REE, Table 5-3) provide a good example, and merit a brief digression.

Despite their name, the REE (Table 20-4) are not especially uncommon in Earth materials. The group as a whole makes up more than 200 ppm of average

TABLE 20-4
Rare-earth (lanthanide) elements

	Ions	Radii (Å) (6-coord)
Lanthanum	La^{3+}	1.03
Cerium	Ce^{3+}	1.01
Praseody- mium	Pr^{3+}	0.99
Neodymium	Nd^{3+}	0.98
(Prome- thium)[1]	Pm^{3+}	0.97
Samarium	Sm^{3+}	0.96
Europium	Eu^{3+}	0.95
	Eu^{2+}	1.17
Gadolinium	Gd^{3+}	0.94
Terbium	Tb^{3+}	0.92
Dysprosium	Dy^{3+}	0.91
Holmium	Ho^{3+}	0.90
Erbium	Er^{3+}	0.89
Thulium	Tm^{3+}	0.88
Ytterbium	Yb^{3+}	0.87
Lutetium	Lu^{3+}	0.86

[1] Promethium does not occur in nature, but radioactive
isotopes may be prepared artificially.
LREE: La–Sm   HREE: Eu–Lu
Source: Shannon (1976).

crust, and some of the individual elements are more abundant than such common metals as lead and copper. From the numbers in Appendix IV it is evident that all of this group, with the single exception of europium, are more concentrated in granite than in basalt, and that the differences are greater for the light elements (La–Sm) than for the heavier ones (Eu–Lu). These simple facts about abundances tell us that the REE are to some extent incompatible with the common ions of igneous magmas, the light ones (LREE) more incompatible than the heavy ones (HREE), and that Eu is something of a maverick.

Elements of this group are remarkable for the similarity in their chemical and physical properties, a similarity which means that they commonly occur together in natural environments and are difficult to separate in the laboratory. The whole assemblage occupies a single spot in Mendeleev's table (Table 5-3), in the third column and sixth period, from which we know immediately that these elements are active metals with a principal oxidation number of +3. The similar properties find a ready explanation in the electron structure of their atoms: all have three valence electrons, and each one differs from the preceding only in the addition of an electron in an f-orbital far down in the electron cloud. As might be expected, the increasing nuclear charge from one element to the next means a steadily decreasing ionic radius through the group (often called the "lanthanide contraction"). The slight decrease in size means slight differences in properties, which are reflected in differences in the distribution of these elements in geologic materials.

The high oxidation number and the large ionic radii (0.86–1.01 Å) mean that the entire group is incompatible in the common minerals of igneous rocks, as we have just guessed from the concentration differences between granite and basalt. The degree of incompatibility varies from mineral to mineral and from one element to another. Certainly the REE would not substitute readily for the smaller Fe^{2+} and Mg^{2+} ions (radii of 0.78 and 0.72 Å, respectively) in the early-crystallizing olivine and orthopyroxene of a mafic melt. Calcium ion is more similar in size (1.00 Å), and some rare-earth concentration is found in calcium materials, but the difference in oxidation number apparently keeps the substitution limited. So the REE by and large remain in the liquid during fractional crystallization, substituting to some extent for LIL elements in the later stages, and at the end may be concentrated enough to form their own minerals (for example, monazite, xenotime, allanite) in alkali granites and pegmatites.

Europium differs from its brethren in that it has an additional stable oxidation number of +2 (ionic radius 1.17 Å), and the relative amounts of Eu^{2+} and Eu^{3+} in a magma depend on its oxidation state. The divalent ion, as might be expected, substitutes readily for Ca^{2+} and even more readily for Sr^{2+} (1.18 Å), so that when plagioclase crystallizes from a mafic magma it often removes a good deal of europium from the liquid. (One other lanthanide, cerium, has an abnormal oxidation number, +4, when conditions are very oxidizing. This peculiarity has little importance in igneous processes, but explains much of cerium's behavior in surface waters and sediments.)

The slight differences in ionic size make the LREE somewhat more incompatible than the HREE, as we noted from abundance figures. Thus in the last residual fluid during crystallization of a magma, or in the earliest melt to form during anatexis, the LREE elements may be more concentrated than their heavier kin. This difference in properties makes analyses of the REE a useful tool in studies of the source and history of the magmas that formed various kinds of igneous rock.

To display analyses showing the distribution of REE in geologic materials, it is common to plot abundances against atomic number. Rather than simple concentrations, however, it is customary to plot ratios—ratios of the amount of each element in a given rock or mineral to the amount in material supposedly representing undifferentiated stuff from which the rock or mineral was derived. In igneous geochemistry the undifferentiated material is commonly taken to be an average of analyses of chondritic meteorites, on the plausible assumption that the Earth and meteorites were formed originally out of the same protoplanetary mixture. A sampling of such plots, showing lanthanide concentrations "normalized" to chondritic concentrations, is shown in Fig. 20-2. In effect, the lines on this figure record for different kinds of igneous rock the changes in REE distribution over the long times since their materials were part of the Earth's primeval substance.

Earth material not greatly differentiated from the original meteoritic composition should plot near the horizontal line representing the ratio "1" in Fig. 20-2. Such material might be a chunk of peridotite carried up from the mantle as a

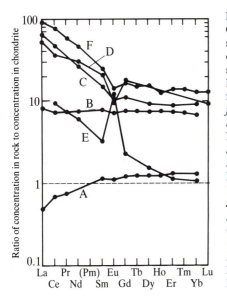

FIGURE 20-2
Chondrite-normalized rare-earth analyses for six rock samples. Each point is the ratio of the concentration of a rare-earth element to the concentration of the same element in an average chondrite meteorite. Rocks and sources: A: *Peridotite* xenolith from lava flow. Jagoutz, E.: *Proc. Lunar and Planetary Conference*, vol. 10, p. 2031, 1979. B: Mid-ocean ridge *basalt*. Langmuir, C. H., J. F. Bender, A. E. Bence, G. N. Hanson, and S. R. Taylor: *Earth and Planetary Science Letters*, vol. 36, p. 133, 1977. C: *Rhyolite*. Jahn, B. M., A. Y. Glikson, J. J. Peucat, and A. H. Hickman: *Geochim. et Cosmochim. Acta,* vol. 45, p. 1633, 1981. D: *Granite*. Shaw, S. E., and R. H. Flood: *Jour. Geophys. Research*, vol. 86, p. 10,530, 1981. E: *Anorthosite*. Simmons, E. C., and G. N. Hanson: *Contrib. Mineralogy and Petrology*, vol. 66, p. 119, 1978. F: North American *shales*. Haskin, L. A., M. A. Haskin, F. A. Frey, and T. R. Wildman, in Ahrens, L. A., ed.: *Origin and Distribution of the Elements*, Pergamon, Oxford, 1968, p. 899.

xenolith in a lava flow (line A). Basalt from the midocean ridges, representing material formed by partial melting of the upper mantle, would be expected to show increased concentrations of all the rare-earth elements, without much separation of LREE and HREE (line B). Further differentiation, leading to marked preferential increases of the light rare earths, is shown by lines C (rhyolite) and D (granite). Line D is noteworthy also because it shows a pronounced negative europium anomaly, presumably meaning that separation of Ca-rich plagioclase has played a role in its history. The frequent strong concentration of europium in plagioclase is strikingly shown by an analysis of anorthosite (line E), a rock consisting almost entirely of this mineral. And finally, line F shows an average analysis of the rare-earth content of North American shales, rocks that represent material formed by repeated differentiation by both igneous and sedimentary processes over long geologic times. It should be noted that Fig. 20-2 shows only a few selected analyses, and that actually curves for any of the various rock types show a great deal of variation. But the general trends of rare-earth behavior during igneous rock formation are well displayed.

This sampling of conclusions that can be drawn from REE distributions is an illustration of the general rule that trace elements are often more sensitive indicators of an igneous rock's history than are its major elements. A rock with an unusual abundance of LIL elements (including LREE), for example, is almost surely the product of extreme differentiation—either the final residual liquid remaining after nearly complete crystallization of a magma, or the initial melt from heating a metamorphic rock, or perhaps the result of segregation of these elements in some previous episode of sedimentation or metamorphism. An igneous rock depleted in LIL elements presumably represents material left after the more easily melted fraction has been removed. Thus the concentrations of the rare-earth elements have proved particularly useful in working out details of igneous rock history.

Summary

The distribution of minor elements in igneous rocks can be explained fairly satisfactorily by assuming the slow crystallization of an orderly sequence of minerals from a melt, usually leading to differentiation. Chalcophile elements may separate early, either in an immiscible sulfide liquid (like the matte of smelting operations) or as sulfide crystals. Most of the trace elements are taken up by the crystallizing silicates, substituting for the major elements in mineral structures largely on the basis of ionic size. Many details of the distribution, both in individual minerals and in rocks formed at different stages of differentiation, can be correlated with such ionic properties as size, charge, and tendency to form covalent bonds. Some elements, like uranium and zirconium, have ions so different from the major elements in size and other properties that little substitution is possible, and hence may form separate minerals even when only small amounts are present. The rare earths are a group of fairly incompatible elements with ions of similar size and charge, and their variations in abundance often provide clues as to the origin and relationships of igneous rocks.

An equally effective explanation for the distribution could be given by assuming that igneous rocks form by progressive melting of material in the mantle, or of average crustal material during extreme metamorphism and anatexis. Emphasis on differentiation of a crystallizing magma in this discussion has been a matter of convenience, and does not imply a choice of one assumed origin over the other.

20-3   DISTRIBUTION IN SEDIMENTARY ROCKS

General

The processes of weathering and sedimentation act like a huge and inefficient sort of chemical separatory process, breaking down the assemblages of elements in igneous rocks and regrouping them into assemblages that are generally simpler. Locally the separation may be very effective, isolating silica in the form of pure quartz sandstone or chert, alumina in bauxite, iron in residual laterite or in sedimentary oxides, carbonates, and silicates, calcium in limestone or gypsum, sodium and potassium in salt deposits. More commonly the breakdown is incomplete, giving only a preponderance of silica in sandstone, of alumina and silica in clays, and of calcium and magnesium in carbonates. In the various operations of this huge separation procedure, what happens to the minor elements? Can they be separated completely from the main elements, or with what main elements are they concentrated?

For detrital sediments the answer is easy. Some trace elements can indeed be separated almost completely from others, simply on the basis of density and resistance of their minerals to solution and abrasion. Placer deposits of gold, platinum, monazite, and zircon come to mind. Minerals that are less resistant to attack often show at least partial concentration: magnetite-rich layers in sandstone are a good example.

More complicated are the chemical processes that may lead to preferential precipitation of minor elements out of solution. Differences in solubility of compounds, adsorption processes, and the activity of organisms all must play a role. In general these processes are not very effective in separating minor elements from major ones. With the exception of phosphates, borates, nitrates, some manganese deposits, and accumulations of copper, vanadium, and uranium with organic matter, the concentrating of rare elements by purely sedimentary processes is not notable.

Explanation of the Distribution

Some rough averages showing the distribution of minor elements in the principal kinds of sedimentary rocks are listed in Table 20-5. One generalization about the distribution stands out immediately: most of the rarer elements are more abundant in shales than in sandstones and limestones. The major exceptions are strontium and

TABLE 20-5
Average concentrations of minor elements in shales, sandstones, and carbonate rocks, in parts per million

	Shales	Sandstones	Carbonates		Shales	Sandstones	Carbonates
*Ba	600	XO	10	Ni	80	2	20
*Li	60	15	5	Pb	20	7	9
*Rb	140	60	3	Sc	15	1	1
*Sr	400	20	610	Th	12	1.7	1.7
				Ti	4,600	1,500	400
Br	5	1	6.2	U	3.5	0.5	2.2
Ce	70	92	11.5	V	130	20	20
Cl	180	10	150	Y	35	40	30
Co	20	0.3	0.1	Zn	90	16	20
Cr	100	35	11	Zr	180	220	19
Cu	50	X	4				
F	600	270	330	°As	10	1	1
Ga	25	12	4	°B	100	35	20
Ge	1.5	0.8	0.2	°Mo	2	0.2	0.4
I	2	1.7	1.2	°P	750	170	400
Mn	850	XO	1,100	°Se	0.6	0.1	0.1

Notes: X means between 1 and 10; XO means between 10 and 100. Ionic potentials greater than 9.5 indicated by (°), less than 2.5 by (*). The order of elements in each group is alphabetical.
Source: For shales, Appendix III; for sandstones and carbonates, Turekian and Wedepohl (1961).

manganese, which are markedly enriched in carbonate sediments, and zirconium and the REE, which are concentrated in sandstone. The enrichment in sandstone is easily explained by the mechanical concentration of the resistant minerals zircon and monazite. The enrichment of strontium in limestones is accounted for by the fact that Sr^{2+} substitutes readily for the very similar ion Ca^{2+}, just as it does in the minerals of igneous rocks. The smaller but appreciable concentrations of manganese, yttrium, and the rare-earth metals in carbonate sediments is probably also attributable to similarities in ionic size.

Now why should most of the minor elements show such a marked preference for the fine-grained sediments? In part the answer may again lie in ionic substitution, this time of elements like manganese and lithium for the magnesium of smectite clays, or of barium for potassium in illite, or of gallium for aluminum in any of the clay minerals. In part the answer rests on reactions of some of the minor elements with organic material, which is generally more abundant in fine-grained sediments than in other kinds. But probably the chief process that leads to enrichment of rare elements in clays and shales is sorption.

From an earlier discussion (Sec. 6-3) we recall the rules of sorption: small ions are more strongly sorbed than big ones, multivalent ions more than univalent, and polarizing ions more than nonpolarizing. We recall also the many exceptions to these rules, particularly the apparent reversal in the order of sorption of the alkali-metal ions, probably because hydration of the ions changes the relative sizes. The order of sorption is somewhat dependent on the nature of the sorbent and on the conditions under which sorption takes place, but in general we might expect a fine-grained sediment to have a sufficient variety of small particles, with a large total surface area and many kinds of sorption sites, so that rare elements of all kinds can be firmly held.

Ionic Potential

An alternative explanation often suggested for the common slight enrichment of minor elements in fine-grained sediments is that some of these elements may be precipitated with clay in the form of hydroxides. The tendency of an ion to precipitate as a hydroxide can then be related, at least theoretically, to a quantity called *ionic potential,* which is defined as the quotient of the positive charge on a simple ion divided by its radius. The name is an unfortunate one because it is easily confused with a very different concept, ionization potential.

The idea of ionic potential is simple. Any positive ion in water is somewhat attracted to the negative ends of the polar water molecules in its vicinity. We can think of this attraction as setting up a competition between the positive ion and hydrogen ion for the oxygen of water:

$$Me-O-H .\qquad(20-7)$$

If the attraction between Me and O is weak compared with the attraction between H and O, the Me ion remains free in solution, merely surrounding itself with a loosely held layer of water molecules; Na^+ is a good example. If the Me-O bond is strong compared with the H-O bond, the Me appropriates one or more O^{2-} ions, forming an anion and leaving H^+ free; an example is the hypothetical ion S^{6+}, which would react with water to form SO_4^{2-}:

$$S^{6+} + 4H_2O \rightleftharpoons SO_4^{2-} + 8H^+. \tag{20-8}$$

If the two bonds are of roughly equal strength, the structure forms an insoluble hydroxide, for example $Zn(OH)_2$. The strength of the Me-O bond will clearly depend on the size and charge of the ion, so that ionic potential should be in some sense a measure of the tendency of an ion to remain free, to form an anion with oxygen, or to precipitate as a hydroxide. An indication of the relationship is shown by the list of elements in Table 20-5: elements marked (*) have low ionic potentials (<2.5), those marked (°) high ionic potentials (>9.5), and unmarked metallic elements have values between 2.5 and 9.5.

As usually happens when ionic properties are expressed as functions of simple geometric quantities, the agreement is excellent for elements on both sides of the periodic table but less satisfactory for those in the middle. Elements in the middle, because of their tendency to form covalent bonds, distort the large O^{2-} anion so that the relative strength of the Me–O and O–H bonds is no longer a simple matter of charge and radius. Attempts have been made to correct the ionic potential for the covalent character of Me–O bonds, but none are very successful. Ionic potential is significant only as a qualitative device for describing the behavior of hydroxides.

With regard to the behavior of minor elements in sediments, the importance of ionic potential is dubious. Elements that form hydroxides of low solubility (which include the vast majority) are indeed somewhat enriched in fine-grained sediments, but there is not the slightest evidence that they are present as hydroxides. Small amounts of such hydroxides would be difficult to detect, but why assume their existence when sorption is a more straightforward explanation? Moreover, as Table 20-5 shows, there are no conspicuous differences in the extent of enrichment of elements with low, intermediate, and high ionic potentials.

Possibly ionic potential has more bearing on the behavior of minor elements in sediments that themselves consist largely of hydroxides, such as bauxite and some sedimentary iron ores. Similarities in ionic potential have been suggested, for example, as an explanation of minor concentrations of beryllium and titanium in bauxite. Even here, however, as long as no separate rare-metal hydroxides are detectable, an explanation in terms of sorption or of replacement of aluminum and iron in their hydroxides seems preferable.

Precipitation, Oxidation, and Reduction

Separation of minor elements from others by direct precipitation of low-solubility compounds is not common. Possible examples are phosphorite (Sec. 14-3), manganese carbonate (Sec. 14-5), and rare evaporite deposits containing minerals of boron, nitrogen, and iodine. Somewhat more common is the precipitation of trace elements following oxidation or reduction.

Manganese is the most important minor element that is often precipitated in sedimentary environments as a result of oxidation (Sec. 14-5). Arsenic and antimony could also be cited here, but the insoluble arsenates and antimonates formed locally by the oxidation of ore deposits make a negligible contribution to sedimentary rocks.

Reduction, usually brought about by the presence of undecayed organic matter, is responsible for the formation of some uranium and vanadium deposits in sediments (for example, the sandstone ores of Colorado and Utah, Sec. 19-5), since compounds of these elements in their lower oxidation states are in general much less soluble than those of the higher oxidation states. It is hardly proper to call this a "sedimentary" process, however, since most such uranium–vanadium deposits show clear evidence that the metals were introduced by solutions well after the sedimentary rocks were formed. Native copper has been observed in modern swamps, under conditions suggesting strongly that it is forming by reduction where dilute copper-bearing solutions encounter organic matter. A similar mechanism is often proposed to account for minor amounts of native copper sometimes found in black shales and associated with organic matter in sandstones, but for such occurrences it is usually difficult to prove that the copper-bearing solutions were contemporaneous with sedimentation. Where reduction takes place in the presence of sulfide ion, many metals can theoretically precipitate as sulfides, but the only metal known to be precipitating in this manner at the present time in ordinary sedimentary environments is iron. Deposits of other metal sulfides in sedimentary rocks (most commonly chalcopyrite, sometimes galena and sphalerite) are often explained by postulating reduction and precipitation during sedimentation, but the evidence is seldom sufficient to eliminate possible introduction of the metals by solutions long after the sediment was formed. How important reduction and precipitation are for the concentration of minor elements during sedimentation remains a much disputed point.

Reactions with Organic Matter

The organic matter of sedimentary rocks is known to be slightly enriched in many of the rarer elements, particularly V, Mo, Ni, Co, As, Cu, Br, I, and locally many others. These elements may form specific minerals, but more commonly are present in the organic matter itself. How the concentration has occurred is seldom entirely clear. In part the metals have doubtlessly been reduced, in the manner just described, from solutions carrying more soluble compounds of higher oxidation states; theoretically, at least, this could happen either during sedimentation or much later. Some of the elements are used by organisms for life processes, and remain when the organisms die and become part of a sediment. Vanadium and nickel, whatever their ultimate source may be, are known to form metal porphyrins, in which they have replaced the magnesium and iron of original porphyrins in living substance. Probably other elements similarly form definite compounds with the organic matter (Sec. 15-2). Simple adsorption on the surface of organic particles may account for some of the metal content. Much remains unknown about the precise mechanisms of concentration, but certainly organic matter has played a role in some important examples of trace-metal enrichment in sedimentary rocks: vanadium and molybdenum in asphalt, uranium and copper in black shales, and germanium in coal are familiar illustrations.

In summary, the major features of the distribution of minor elements in sedimentary rocks can be related, just as in igneous rocks, to ionic size, charge, and bond character. But the influence of these ionic properties is different: substitution of minor-metal ions for major ions in crystal structures is of lesser importance, and sorption of minor-metal ions on the surfaces of particles in fine-grained sediments plays a major role. Most of the minor elements are more abundant in fine-grained detrital sediments than in sandstones or carbonate rocks, probably because of sorption. Other processes helping to determine the distribution of elements in sediments are precipitation following oxidation or reduction and various reactions with organic matter. In general, significant concentrations of minor elements formed by sedimentary processes alone are not common.

20-4 DISTRIBUTION IN METAMORPHIC ROCKS

Information about the distribution of elements during metamorphism and hydro-thermal alteration is meager. In some cases metamorphism produces only small chemical changes. Metamorphism of fine-grained rocks to hornfelses or phyllites causes no detectable change in rare-metal content, unless the rocks have been permeated by solutions during the metamorphic process. At higher grades of metamorphism minor elements redistribute themselves locally among the growing crystals, but again the overall concentrations do not change markedly unless movement of solutions has played a major role. Since some sedimentary rocks have unique assemblages of trace elements, a study of these elements in metamorphic rocks provides a possible way to guess at the nature of the premetamorphic material, but the trace elements seldom give a better basis for guessing than do the more obvious major elements. When metamorphism reaches the ultimate state of partial melting, most of the minor elements go into the melt, and then recrystallize from the melt according to the pattern we have outlined for igneous rocks.

The distribution of major and minor elements in metamorphic rocks is complicated by the flow of reactive fluids through pore networks and fractures. These kinds of fluids and their reactions we have discussed extensively in previous chapters. For example, dehydration and decarbonation reactions release the volatile components H_2O and CO_2 into the pores of the rock. These fluids may flow from their site of origin into other rock types where they are not in equilibrium because of changes in rock composition, temperature, or pressure, or because they mix with fluids of different composition. The resulting reactions dissolve or deposit various elements, thus modifying the composition of both rock and fluid.

During metamorphism the addition or removal of elements by reactive fluids is called *metasomatic mass transfer,* a process we have discussed at length in Chapters 16, 18, and 19. Kinds of metasomatism are commonly specified by using the name of one of the elements that is added to the rock. Thus, H-metasomatism refers to the addition of hydrogen to a rock, as when acid solutions react with feldspar to form sericite or clay minerals, and K-metasomatism means the addition of potassium to a rock, as during the formation of orthoclase in porphyry copper type deposits. Metasomatism can take many forms within the same rock type, as we have seen in low-grade metamorphism of basalts: Na-metasomatism associated with albitization, Ca-metasomatism related to the formation of epidosites, and Mg-metasomatism in the formation of chlorite and smectite. Many examples can be cited for the addition of other major elements such as Si and Fe, or for minor elements like B, Li, Cl, F, S, and Sn. In each example other elements, both major and minor, are added to or removed from the rock during reaction.

Variables that determine the type and extent of element distribution in metamorphic and hydrothermally altered rocks must include the composition of the rock and the fluid, temperature, pressure, rock porosity, rock permeability, fluid-flow potentials, and the rates at which all variables change with time. Thus a general statement concerning major, as well as minor, element distribution is not possible. Generally metamorphism causes only minor changes in element distribution, but examples of extreme changes are shown by the formation of some kinds of ore deposits.

The distribution of elements in various special kinds of geologic materials—soils, evaporites, unusual igneous rocks, sulfide minerals of ore deposits—is a fascinating and productive study, but involves too much detail for our purposes here. In general, the theoretical side of such a study means refinement of the ideas we have outlined in this chapter, particularly the dependence of distribution on such ionic properties as radius, charge, and bond character. Other factors to be considered are the solubilities of elements in silicate and sulfide liquids, aqueous electrolyte solutions, and mixtures of H_2O and CO_2. To fully understand the distribution of elements for a particular rock requires knowledge of the geologic history of all the chemical processes that have acted on the rock. Details of distribution in some geologic environments remain elusive, but the general pattern follows theoretical expectations remarkably well.

PROBLEMS

1. Explain what is meant by the statement that copper is a more chalcophile element than either zirconium or platinum.

2. Vanadium is a more abundant element than boron, yet boron minerals are more common in both igneous and sedimentary rocks. Why?

3. For each of the following ratios, indicate whether you think it would usually be higher in mafic igneous or in felsic igneous rocks, and give reasons for your answer: Rb/Sr, Sr/Ba, B/Mn, Li/Mg, Pb/Rb, Cr/Al, Y/Ca, La/Yb.

4. Of the following ratios, which would you expect to be greater in evaporites than in shales? Why? Ca/Ba, Si/Na, K/Rb, Mg/Mn.

5. "In general, an element is more chalcophile in its lower oxidation states than in its higher states." Is this statement true? Discuss the proposition, using examples to support your opinion.

6. Lithium and cesium are both somewhat concentrated in the late micas of pegmatites, although their ionic radii are very different. Explain.

7. Account for the lack of extensive isomorphous replacement of Na by Li, of Fe^{3+} by Li, of Mg by Nb, of Mn^{2+} by Pt^{2+}.

8. Why are gallium minerals so exceedingly rare, whereas minerals of the much less abundant elements Sn, U, and W are well-known?

9. Account for the fact that the six most abundant elements in seawater are Na, K, Mg, Ca, S, Cl.

10. Why are uranium and thorium more abundant in granites than in ultramafic rocks?

11. Look up the free energies of formation of oxides and sulfides for several metals, and show how these free energies are correlated with the chalcophile or lithophile character of the metals.

12. Most of the rare-earth metals are more abundant in granite than in basalt, but europium is more abundant in basalt. Explain.

REFERENCES

Patterson, E. M.: "A petrochemical study of the Tertiary lavas of northeast Ireland," *Geochim. Cosmochim. Acta*, vol. 2, p. 291, 1952.

Robie, R. A., B. S. Hemingway, and J. R. Fisher: "Thermodynamic properties of minerals and related substances," *U.S. Geological Survey Bull.* 1452, 1979.

Shannon, R, D.: "Revised effective ionic radii," *Acta Crystallographica*, vol. A32, pp. 751–769, 1976.

Turekian, K. K., and K. H. Wedepohl: "Distribution of the elements in some major units of the Earth's crust," *Geol. Soc. America Bull.*, vol. 72, pp. 175–192, 1961.

Vinogradov, A. P.: "Sredniye soderzhaniya khimicheskikh elementov v glavnikh tipakh gornikh porod zemnoi kory," *Geokhimiya*, vol. 1962, pp. 560–561, 1962.

Wager, L. R., and R. L. Mitchell: "Distribution of trace elements during strong fractionation of a basic magma," *Geochim. Cosmochim. Acta*, vol. 1, p. 199 and Table F, 1951.

Wager, L. R., and R. L. Mitchell: "Trace elements in a suite of Hawaiian lavas," *Geochim. Cosmochim. Acta*, vol. 3, p. 218, 1953.

SUGGESTIONS FOR FURTHER READING

Burns, R. G., and W. S. Fyfe: "Trace element distribution rules and their significance," *Chemical Geology*, vol. 2, pp. 89–104, 1967. Review and critique of the distribution rules, with emphasis on the importance of knowing bonding forces not only in crystalline solids but in the liquid surrounding the solids during crystallization.

Henderson, P.: *Inorganic Geochemistry*, Pergamon Press, Oxford, New York, 1982. Chapter 5 gives an excellent review of element distribution in igneous and metamorphic rocks. The chapter also provides useful information on presenting analytical data, and Table 5-2 gives element partition coefficients for a wide variety of igneous rocks.

Henderson, P., (ed.): *Rare-Earth Element Geochemistry*, Elsevier, New York, 1984. A series of papers on rare-earth distribution, especially good on application of rare-earth data to petrogenetic problems.

Mason, B., and C. B. Chapman: *Principles of Geochemistry*, 4th ed., Wiley, New York, 1982. An excellent critical discussion of the behavior of trace elements in geologic processes, and compilations of analytical data on many terrestrial and extraterrestrial materials.

McCarthy, T. S., and R. A. Hasty: "Trace element distribution patterns and their relationship to crystallization of granitic melts," *Geochim. et Cosmochim. Acta*, vol. 40, pp. 1351–1358, 1976.

Theoretical study of the effects of equilibrium and nonequilibrium crystallization on trace-element distribution.

Pearce, J. A., N. B. W. Harris, and A. G. Tindle: "Trace-element discrimination diagrams for the tectonic interpretation of granitic rocks," *Jour. Petrology,* vol. 25, pp. 956–983, 1984. The use of trace elements in distinguishing granites from different tectonic settings.

Wedepohl, K. H., (ed.): *Handbook of Geochemistry,* Springer Verlag, Berlin, 1969–1974. A four-volume compilation of data on the distribution of the elements in geologic materials, including critical discussion of many of the data.

chapter 21

HISTORICAL
GEOCHEMISTRY

We have kept our eyes pretty well fixed, up to now, on chemical changes that occur in or on the Earth's crust. By ordinary human standards this is the most important part of the universe, the part in which familiar geologic processes are taking place and have taken place for a long time in the past. But from other points of view the crust looks less significant. On a cosmic scale it is no more than the thin skin of a small planet, one of several planets circling a star which, in its turn, is only one of many billions of similar objects in the wide universe. It is time that we raise our eyes briefly from detailed study of fluids, rocks, and organic matter in the crust and consider chemistry on this broader scale. We look now at chemical relations between the crust and the great mass of the planet beneath it, and more sketchily at how this one planet fits into the external system of planets and stars. And as geologists we cannot limit ourselves to the cosmochemistry of the present time, but must ask how the chemistry of the Earth and its neighbors may have changed down the ages.

Inquiry in these new directions, so different from the paths we have followed hitherto, requires new kinds of data. Many of the pertinent facts are drawn from astronomy and geophysics rather than from classical geology and chemistry. In a study of the universe as a whole the boundaries between the familiar sciences break down, and we find ourselves grasping for bits of evidence from any discipline that seems relevant. The data, as might be expected, are far from adequate to answer all the questions that would be of interest. Even more than in the obscure parts of crustal geochemistry, we must here depend on imaginative speculation to fill enormous gaps between well established facts.

In so broad and speculative a field it is hard to follow the pattern of our previous discussions, in which, for the most part, ideas have been traced back to well-known observations and principles from elementary science. If we cast loose from such moorings, cosmochemistry becomes a fabric of everchanging hypotheses that rest on assumptions and mathematical arguments which must be taken as items of faith by all but specialists. There is no harm, of course, in weaving such a fabric of hypotheses untestable by the nonspecialist: cosmochemistry, even in the most elementary terms, is a fine exercise for the imagination. Many popular accounts of the subject, based on imaginative appeal rather than strict examination of hypotheses and assumptions, can be found in the current literature. To avoid duplicating such expositions and to keep the length of the discussion within bounds, we shall pass lightly over the more speculative aspects of the subject and pay particular heed to the nature of the basic data.

21-1   COMPOSITION OF THE EARTH

To start with, we might ask if the material that makes up the Earth is a fair sample of the composition of the universe as a whole. Answering such a question requires an estimate of the overall chemistry of our planet. How do we make an analysis on this kind of a scale? It is obviously not easy. For the crust, we can analyze different kinds of rock, guess at their relative amounts, and so calculate an average composition. But the crust is only a tiny fraction of the Earth's bulk (Fig. 21-1). What can we do about the interior?

Geophysical measurements of the Earth's mass, volume, and moment of inertia tell us that the density of Earth materials increases downward from the surface, reaching about 15 g/cm^3 near the center. Measured speeds of earthquake waves give details about the general distribution of materials in the interior, details that are embodied in the familiar model of a crust averaging 40 km thick under the

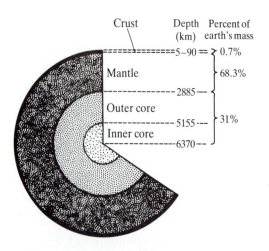

FIGURE 21-1
Schematic section through the Earth showing the thickness and percent mass of the crust, mantle, outer core, and inner core. (Modified from Ernst, 1990.)

continents and 6 km under the oceans, then a mantle extending about halfway to the Earth's center, a liquid core occupying about two-thirds of the remaining distance, and a solid inner core. The mantle, we learn from these measurements, is largely solid, but a solid with most peculiar properties: rigidly solid for short-term disturbances like seismic waves, but flowing slowly like a liquid when acted on by the long-term forces responsible for tectonic activity. As to the chemical nature of these interior materials, seismic waves give us no more than broad hints. For the uppermost part of the mantle we do have some samples that can be directly analyzed: rare exposures of ultramafic rocks brought to the surface by tectonic activity, and xenoliths in some basalt flows and dikes that magma has carried up from the depths where it was generated. But the greater part of the mantle and all of the core remain inaccessible, and their chemistry can be deciphered only indirectly.

Our best guesses about composition come from meteorites. These objects, falling to the Earth from orbits around the Sun, are plausibly interpreted as fragments of a vanished planet, or as residual chunks of the material from which the Earth was originally constructed. On either interpretation the average composition of meteorites should resemble the general make-up of the entire Earth. Obtaining a reasonable average for meteoritic material is troublesome, for meteorites have a wide variety of compositions. At one extreme are the "irons," which are dominantly metallic iron alloyed with a few percent of nickel. At the other extreme are "stones," which consist chiefly of silicates and resemble ultramafic rocks in composition. Because irons are more durable than stones, and more apt to attract attention because they look different from ordinary rocks, museum collections of meteorites commonly have a disproportionate number of irons. From counts of observed falls, however, it seems clear that stones far outweigh irons in actual numbers striking the Earth in recent times. Opinions vary widely as to how meteorite compositions should be weighted to get an average composition, or alternatively what kind of meteorite should be chosen as the best representative of an average composition. The numbers in the fifth column of Table 21-1 are an often quoted guess.

A plausible hypothesis about the chemistry of the Earth's interior, then, can be framed by distributing the metal and silicates of meteorites so as to satisfy the requirements of density and elastic properties deduced from the behavior of seismic waves. This gives the common model of a core consisting largely of iron, molten in its outer part and solid toward the center, surrounded by a mantle made up chiefly of magnesium–iron silicates. Refinements of the model in almost infinite variety have been suggested, in order to secure better fits with seismic data and with the results of experimental work at very high temperatures and pressures. Recent advances in experimental technique make possible laboratory simulation of P–T conditions in all parts of the Earth's interior, and such work has provided details about changes in composition downward in the Earth. For example, the magnesium silicates of the mantle are dominantly olivine and pyroxene just under the crust, but change to denser forms as pressure increases (Fig. 21-2). Spinel appears, then garnet; at still lower levels the crystal structure of olivine changes to a denser form like that of perovskite, in which Si ions are surrounded by 6 oxygens rather than the usual 4, and Mg ions by 12. In the core, the dominance of iron is confirmed by the

TABLE 21-1

Relative abundances of elements in the Earth and the universe (atoms per 10,000 atoms of Si)

	Whole Earth	Crust	Upper mantle	Andesite	Meteorites	Universe	Moon
				Rock-forming elements			
Si	10,000	10,000	10,000	10,000	10,000	10,000	10,000
Al	750	3,050	720	3,540	740	850	740
Fe	11,460	910	1,570	1,030	8,020	9,000	3,000
Mg	9,650	870	13,940	930	9,700	10,750	12,400
Ca	520	920	580	1,340	520	610	430
Na	460	1,230	110	1,080	460	570	20
K	30	670	4	330	30	40	(3)
Mn	70	20	30	20	70	90	(20)
Ti	20	90	20	100	20	20	(20)
Ni	60	10	40		460	590	
Cr		20	50		90	130	(20)
P		30		30	60	100	
				Volatile elements			
H	80	1,460				2.7×10^8	
O	34,000	30,600			34,300	201,000	37,000
N	0.2	1				24,800	
C	70	20				121,000	
S	1,110	20			990	5,150	(130)
F	3	30				8	
Cl	30	4				50	
				Inert gases			
He	3.5×10^{-7}					3.1×10^{-7}	
Ne	12×10^{-7}					86,000	
Ar	5.9×10^{-10}					1,500	
Kr	0.6×10^{-7}					0.51	
Xe	0.05×10^{-7}					0.04	

Sources: Crust, meteorites, and whole Earth: Recalculated from Mason, B., and C. B. Moore: *Principles of Geochemistry*, 4th ed., Wiley, New York, 1982, pp. 18, 46, 52.

Upper mantle: Recalculated from Wyllie, P. J.: *The Dynamic Earth*, Wiley, New York, 1971, p. 114.

Andesite: Recalculated from Gill, J. B.: *Orogenic Andesites and Plate Tectonics*, Springer Verlag, New York, 1981.

Universe: Recalculated from Anders, E., and M. Ebihara: "Solar system abundances of the elements," *Geochim. et Cosmochim. Acta*, vol. 46, pp. 2363–2380, 1982.

Moon: Recalculated from Smith, J. V.: Development of the earth–moon system," in Windley, B. F. (ed) *The Early History of the Earth*, Wiley, New York, 1976. Paraentheses indicate values that Smith regards as less reliable than the others.

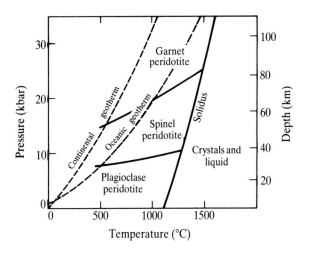

FIGURE 21-2

Generalized phase diagram illustrating the pressure and temperature stability of peridotite mineral facies in the upper mantle and the melting curve (solidus) for peridotite. The diagram shows phase changes for lherzolite, a peridotite consisting of olivine, clinopyroxene, orthopyroxene, and an aluminous phase: plagioclase at low pressures, spinel at intermediate pressures, and garnet at high pressures. These phase boundaries, shown as univariant lines, are sensitive to changes in composition, specifically changes in CaO, Fe_2O_3, Al_2O_3, and Cr_2O_3. The dashed lines show the approximate geotherms for oceanic and continental regions. (Reprinted by permission from Wyllie, P. J.: *The Dynamic Earth*, Wiley, New York, 1971.)

experimental fact that this metal is the only reasonably abundant substance with high pressure properties that correspond with those predicted for the P–T conditions in this part of the Earth. Thus the hypothesis of a "meteorite model" for the Earth's interior is now on a firm basis, much firmer than the mere assumption of a common origin for planets and meteors.

Calculating an overall composition of the Earth based on the meteorite model is a pastime that has attracted many geochemists during the past century. One recent estimate is shown in Table 21-1; another, with different units, is shown in Table 21-2, together with suggested analyses for the crust and mantle. Obviously the numbers in the table cannot be taken too seriously as precise values, but they permit a few qualitative comments. The analyses of the whole Earth and the mantle are very similar, as they must be because the mantle makes up by far the major part of the Earth's bulk (Fig. 21-1). The greater percentage of silica in the crust, and the far greater percentages of aluminum and the alkali metals, suggest some planetary-scale process of differentiation. And the extraordinary similarity between the crustal analysis and the average analysis of andesite indicates that this volcanic rock is a good chemical sample of the entire crust.

To return now to our question about the relation of the Earth's chemistry to that of the universe, we need next an analysis of stars and galaxies—a project that looks even more formidable than finding the composition of the Earth. The chief source of information here is the spectrographic study of stellar atmospheres. The

TABLE 21-2

Estimated compositions of the whole Earth, the mantle, and the continental crust, in weight percent

	Whole Earth	Mantle	Continental crust	Andesite
SiO_2	47.9	45.7	60.2	57.6
TiO_2	0.02	0.09	0.7	0.77
Al_2O_3	3.9	3.4	15.2	17.3
Fe_2O_3			2.5	3.1
FeO	8.9	8.0	3.8	4.3
MnO	0.14	0.14	0.1	0.15
MgO	34.1	38.4	3.1	3.6
CaO	3.2	3.1	5.5	7.2
Na_2O	0.25	0.4	3.0	3.2
K_2O	0.02	0.1(?)	2.9	1.5
P_2O_5				0.21
H_2O				1.0

Sources: Whole Earth: Recalculated from Mason, B., and C. B. Moore: *Principles of Geochemistry*, 4th ed., Wiley, New York, 1982.

Mantle: Ringwood, A. E. *Composition and Petrology of the Earth's Mantle*, McGraw-Hill, New York, 1975.

Continental crust: Ronov, A. B., and A. A. Yaroshevsky: "Chemical composition of the earth's crust," in Hart, P. J., (ed): *The Earth's Crust and Upper Mantle*, Geophysical Monograph 13, Am. Geophys. Union, Washington, DC, 1969.

Andesite: Gill, J. B.: *Orogenic Andesites and Plate Tectonics*, Springer Verlag, New York, 1981. Average of 2500 orogenic andesites.

relatively cool gases surrounding a star absorb light from the incandescent interior, and the absorption produces dark lines in the star's spectrum. Frequencies and intensities of the lines are a great storehouse of information about the composition and state of matter in the absorbing gases. The gases, in turn, are in rough equilibrium with the turbulent matter of the star's interior, and enough is known about the equilibrium relations that an analysis of the atmosphere can be translated into figures for the composition of the star as a whole. Averages for many stars of different kinds, taken together with estimates of interstellar material, give results shown in the sixth column of Table 21-1. No great accuracy can be claimed for such figures, of course, and the estimates for various elements have been often revised as a result of improvements in astronomical techniques.

Comparison of the numbers in columns 1 and 6 of Table 21-1 shows that the Earth is indeed a fair sample of the material that makes up the stars. Some differences appear among the lithophile elements (Table 20-2), but the major discrepancies are with the volatile elements at the bottom of the table. Particularly striking is the contrast in estimated abundances of the inert gases: concentrations of these elements in the Earth are less than concentrations in the stars by factors ranging from 10^7 to 10^{14}. If the Earth's composition ever resembled that of larger bodies in the universe, some process of differentiation during its history has caused tremendous losses of the more volatile elements.

It should be emphasized once more that the numbers in Tables 21-1 and 21-2 are guesses at the compositions of huge masses of material for which, just in the nature of things, no precise analysis is possible. Many similar analyses, often showing large differences in the estimates for particular elements, can be found in the literature. Constructing an analysis of the Earth, or of the universe, or even of accessible parts of the crust, is an operation that requires the fitting together of data from a variety of sources and the weighing of many alternative hypotheses. It becomes a game, like trying to put together the pieces of an enormously complex puzzle, and agreement among the players can hardly be expected. No special virtue, and certainly no great accuracy, may be claimed for the analyses in these two tables. They are simply representative samples of current thinking, and we will find them useful in trying to reconstruct the Earth's past.

21-2 ORIGIN OF THE EARTH

Geochemists, in common with other geologists (and, for that matter, with most of humanity), are fond of speculating about how our planet came to exist. The factual basis for such speculations derived from ordinary observation is extremely meager, since processes of progressive change that can be confidently extrapolated into the distant past are few and questionable. Chemically as well as mechanically the Earth and its sister planets, over the very long term, seem to form a remarkably stable system. The only possibility is to try out various hypotheses of origin, and see which is most successful in predicting a planetary system like the one of which we are a part. We cannot here look into the many hypotheses that have been proposed and discarded, but limit ourselves largely to chemical aspects of the currently most favored lines of speculation.

We might ask first, how can we be sure the Earth actually *had* an origin? Is it not possible (despite philosophical difficulties) that the planets have *always* existed in much their present form? One indication that the Earth cannot be infinitely old comes from astrophysical data about the process of energy production in the sun and stars. Radiation from the sun originates in nuclear reactions by which hydrogen is converted into heavier elements, and the current ratio of hydrogen to other elements shows that the Sun cannot have existed as a star for more than 5 or 6 billion years. The presence in rocks of appreciable amounts of radioactive elements with long half-lives is another indication of finite age, for no known process is capable of producing these elements on or in the Earth today. So the question about origin is not meaningless. Materials that compose the Earth may have existed in some form through an infinite past, but the Earth as a planet revolving around a luminous star must have originated at a definite point in time.

All hypotheses about the Earth's beginning postulate a time when its substance was part of a homogeneous sample of average cosmic matter. Fashionable hypotheses of a few generations ago put the homogeneous sample in the Sun or a similar star. Mechanical difficulties with generating a planetary system from a fully formed star have led more recent speculations back to an earlier hypothesis, in which the Earth's material was once part of a huge cloud of gas and small particles spread thinly over a volume larger than the present orbit of Pluto. The cloud at first was cold, and the nucleus of solid matter that grew into the primordial Earth was formed as an aggregate of cold particles and larger fragments, often called *planetesimals*.

One reason for postulating cold accretion of planetesimals, rather than a fiery beginning for the Earth as a large chunk of incandescent matter torn from the Sun, goes back to the striking difference in amount of volatile elements between cosmic and terrestrial matter (Table 21-1). The Earth's gravitational pull is sufficient at present to hold all gases except hydrogen and helium, so that the deficiency in other volatiles cannot be ascribed to loss from the atmosphere in recent geologic time. If, however, the Earth's substance was once spread over a large volume in the form of planetesimals, the gravitational attraction of the separate chunks would be too weak for retention of most volatile materials. The loss would be greatest for light elements like neon, helium, and hydrogen, in accordance with Table 21-1. A low temperature is indicated, furthermore, by the fact that the inert gases have been lost in much greater amounts than more active volatile materials like water, carbon dioxide, and ammonia. At temperatures of several hundred degrees these active volatile compounds would be just as easily lost as the inert gases, since their molecules would be free to move and their molecular weights are similar. At low temperatures, however, such active compounds as H_2O and CO_2 would be partly retained by the planetesimals as mineral hydrates and carbonates and by sorption on solid surfaces.

When did the accretion of planetesimals to form the Earth take place? An upper limit for the time is set by the calculated age of the Sun, 5 or 6 billion years, since the Sun formed also out of the original cloud, presumably by aggregation of material in the central part. A similar rough maximum age is suggested by the abundance of the lead isotope ^{207}Pb. This isotope is being produced today at a known rate by the radioactive decay of ^{235}U; if *all* the existing ^{207}Pb had been so produced, a time of 5.5×10^9 yr would have been required. Since some ^{207}Pb was probably present in the original Earth, this figure is an extreme maximum. A minimum figure is the age of the oldest rocks, roughly 4.0×10^9 yr. A number between these limits that is probably close to the time of actual accretion of planetesimals comes from analysis of lead isotopes in meteorites (Sec. 10-3): 4.55×10^9 yr. Corroboration is provided by age measurements on samples of rock from the Moon, the oldest of which show ages in the range 4.5 to 4.6 billion years. Acceptance of these ages, of course, requires the commonly used assumption that the Earth, Moon, and meteorites were all formed at about the same time.

21-3 EARLY HISTORY

If the Earth originated by the falling together of planetesimals, the low temperature could not have persisted long. At least two sources of energy were available that must have generated heat in large quantities: the kinetic energy of moving chunks and particles as they came together and collided under the influence of gravity, and the energy of radiation from radioactive elements in the original materials. As long as the Earth was small and most of the planetesimals were far apart, heat from both sources could be radiated away into space. But once the planet had grown to appreciable size, much of the heat would be retained, for rocks are notoriously poor conductors. The temperature of the interior would steadily climb, but how fast and how far the temperature rose remain uncertain.

In the present-day Earth the release of gravitational energy can no longer be a significant source of heat, but some of the heat from the Earth's beginnings still lingers in the deep interior. The stored heat, plus the heat still being generated by radioactive decay, is sufficient to keep the outer part of the core liquid and to maintain in the mantle a huge system of slow convection currents. The convectional system, about which many details are still uncertain, is the driving force of plate tectonics, in that lithospheric plates are carried on the upper horizontal parts of convection cells from divergent centers to subduction zones (Fig. 21-3). Mantle convection, bringing heat from the interior up toward the surface, is a mechanism for cooling the planet; but the motion is so slow, and the conductivity of rock so small, that even after 4.6 billion years much of the original heat remains. How effective is the cooling? Does it compensate for heat being generated by radioactive decay? In other words, is the temperature of the Earth's interior at present increasing or decreasing or remaining about constant?

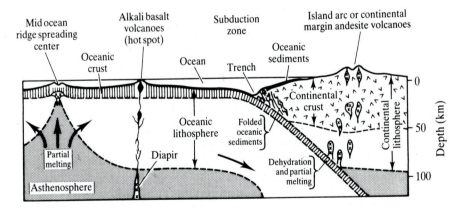

FIGURE 21-3
Schematic diagram of the Earth's crust and lithosphere showing divergent and convergent plate boundaries. Mass movement of the lithosphere is indicated by the large arrows. Partial melting in the asthenosphere produces tholeiitic basalts that rise at the mid-ocean ridges; cooling of this magma produces mid-ocean ridge hydrothermal systems (Fig. 12-3) that react with the oceanic crust. Subduction of the altered oceanic crust and lithosphere beneath the continental crust causes dehydration reactions and partial melting, forming the andesite magmas that rise into the overlying continental crust. Alkali basalt magma, produced deep in the mantle, rises through the lithosphere to form an intraplate hot spot. Magma bodies are represented by the cross pattern.

Available data do not permit a definitive answer. The rate of heat production by the four principal radioactive isotopes is well-known:

$$^{238}\text{U} : \quad 3.0 \text{ joule g}^{-1}\text{yr}^{-1}$$

$$^{235}\text{U} : \quad 19.7 \text{ joule g}^{-1}\text{yr}^{-1}$$

$$^{232}\text{Th} : \quad 0.88 \text{ joule g}^{-1}\text{yr}^{-1}$$

$$^{40}\text{K} : \quad 0.88 \text{ joule g}^{-1}\text{yr}^{-1}$$

But the distribution of the isotopes in various Earth materials is determined only approximately, and the overall heat production depends critically on the precise concentrations. The amount of heat loss from the surface is known with fair accuracy, as is also the conductivity of common rocks, but the amount of heat transferred from the interior by mantle convection and hot-spot plumes remains uncertain. So neither side of the heat-balance equation can be expressed with the accuracy needed to determine whether the Earth is becoming hotter or cooler. It is commonly thought that a rough balance between heat supply and loss exists, and has existed for a long time in the past, but the data on which this conclusion rests are not firm.

If information about the present heat balance is unsatisfactory, our knowledge of the Earth's thermal state in the past is still more speculative. One gross fact gives a hint about conditions at some early period: the Earth has an iron core, and it seems likely that separation of core and mantle required that at one time or another a considerable part of the Earth's interior was hot enough to melt iron. This does not mean that the entire Earth was necessarily molten, because at pressures like those in the interior, in an aggregate of solid metal, sulfide, and silicates, the lowest melting material would be a mixture of iron plus a little iron sulfide. Such a liquid can be imagined to have collected in large bodies and to have moved through interstices in the mass of solid silicates toward the Earth's center. The time required for heating of an originally cold Earth to the point where an iron-plus-minor-sulfide liquid would separate and form a core has been variously estimated; some guesses are as brief as a hundred million years. How much liquid formed remains a point of argument. Calculations regarding the amount of heat that could be generated by infalling planetesimals have suggested to some that a large part of the Earth became molten at one period in its early history; if this happened, formation of the core would have meant separation of immiscible liquids rather than percolation of liquid through a solid framework. Others, estimating a smaller amount of heat, propose that the major separation of core materials had nothing to do with liquids, but resulted from the early settling of heavier, iron-rich chunks of the original swarm of planetesimals toward the center of the swarm as the Earth's material started to accumulate; partial melting at a later time may have made the separation more complete. Information about energy relations in the Earth's earliest history is still too meager for a clear choice between these ideas.

The other major large-scale differentiation process that must, at least in part, go far back in time is the separation of continental material out of the mantle. The overall chemistry of the crust, with granitic rock at higher levels, more mafic rock toward the base, and material of peridotitic composition in the mantle beneath, is strongly reminiscent of the chemical layering observed in large differentiated masses of unquestioned igneous origin, and also of differentiation processes observed in simplified laboratory systems. An easy assumption about the origin of the continental crust, therefore, is an imaginative reconstruction on a planetary scale of a process of crystal settling out of molten material—or, if the mantle was solid, a process of partial melting. The heavier magnesium and iron silicates would drop into, or remain in, the mantle, and a liquid relatively rich in silicon, calcium, and the alkali metals would rise toward the surface. This first liquid differentiate from peridotite would have the composition of basalt, or, if melting was more extensive, a composition tending toward the ultramafic lava komatiite. Only later, after the basalt has been exposed at the surface and has experienced weathering, erosion, sedimentation, deep burial, metamorphism, and remelting, would additional differentiation produce magmas of granitic composition. Thus ultimately granitic crust could form out of mantle peridotite, not directly but in local areas after intermediate steps in which water was involved.

The period of these early differentiation sequences, from the time of the planet's accretion to the formation of the oldest preserved rocks (roughly 4.6 to 4.0 billion years ago) must have been a tumultuous one. The craters that pock the Moon's surface record intense bombardment by meteorites during this time. Presumably the Earth, with its larger gravitational field, would have been pelted with bigger and more numerous projectiles. Unlike the Moon, the Earth's much-eroded surface, continually modified by tectonic processes, retains no vestige of this early bombardment. Conceivably some of the apparent chemical heterogeneity of the Earth's crustal rocks is a heritage from the variety of meteoritic masses that were incorporated into the lithosphere. The odd concentrations of tin ores in some parts of the Earth, and of iron and copper ores in other parts, would find a convenient explanation in such a hypothesis.

Even larger perturbations of the Earth's early surface may have been caused by the Moon. The origin of the Moon is still uncertain, but a currently favored idea envisions our satellite as resulting from collision of the primitive Earth with an object somewhat larger than the planet Mars, at a time when both bodies had differentiated into an iron-rich core and a silicate mantle and when both were partly molten. The Moon would thus consist of a mixture of materials torn from the Earth's mantle and from the colliding object. A fiery origin like this would account for some peculiarities of the Moon's composition, for example a percentage of iron much smaller than that of the Earth, the complete absence of water, and smaller amounts of such relatively volatile elements as sodium and potassium (Table 21-1). Such a catastrophe, added to meteorite impacts, would certainly have ensured a short life for any features of the Earth's surface that might have developed.

The crust that existed during the long meteorite bombardment was almost certainly in large part basalt and ultramafic rock, with some areas of granite. One bit of evidence regarding the nature of this very early crust has come from recent studies of zircon in some of the Earth's most ancient rocks found in western

Australia and northern Canada. These rocks are metamorphosed sediments nearly 4 billion years old, but among the detrital zircons they contain are some that give even older dates, going back as far as 4.3×10^9 years. By 4.0 billion years ago (bya), therefore, weathering and erosion had long been in progress, and some of the rocks that were supplying zircon to the sediments had differentiated from the mantle and solidified at least 300 million years earlier. Because zircon is more common as an accessory mineral in felsic rocks than in mafic rocks, these detrital grains are at least suggestive (but hardly proof) that the very early differentiates included some granitic material.

It seems reasonably certain, then, that by the end of the Earth's first 500 million years its surface was largely covered by differentiated material, most of it basalt. Direct evidence for the presence of some granitic rocks at this time has come from recent age measurements on rocks of the Canadian shield going back to 3.96 bya. The existence of ocean basins, at least small ones, is suggested by the finding of quartzite and a little marble in areas where dates are in the range 3.8 to 4.0 bya. In rocks of slightly younger age, say 3.5 to 3.8 bya, the common Archean assemblage of granite or granite gneiss with greenstones and some sedimentary rock is good evidence for widespread differentiation of crust out of the mantle. Typically the granitic rocks in such areas are in isolated patches, suggesting that early "continents" were small and scattered, probably moved about by slow currents in the mantle beneath. One can imagine that they coalesced and broke apart repeatedly, then toward the end of the Archean (2.5×10^9 years ago) joined into larger aggregates more like the present continents. Details of this early history elude us, but certainly from very early times the Earth's surface was sharply separated, just as it is today, into continental masses of roughly granitic composition and oceanic areas underlain chiefly by basalt.

21-4 LATER GEOCHEMICAL HISTORY OF CRUST AND MANTLE

How soon a plate-tectonic regime like that of the Phanerozoic world was established is a much debated question. Once in operation, it would be the principal mechanism for continued differentiation of mantle material. One can think of basalt derived by partial melting and fractional crystallization from garnet peridotite rising under rift zones, corresponding to the modern mid-ocean ridges, and spreading sideward to form the crust of ocean basins. Hydrothermally altered by reaction with seawater, and moving with the drift of solid plates of the upper mantle, the basalt eventually reaches subduction zones at the edges of small continents (Fig. 21-3). Here it and some of the sediments on its surface descend again into the mantle, and in the upper part of the subducting plates a more complicated sort of differentiation takes place. With rising temperatures at depth, the altered basalt undergoes metamorphism. Dehydration of the hydrothermal minerals formed by reaction with seawater (e.g., smectite, chlorite, talc, actinolite, and epidote) releases water into the surrounding rocks, lowering their melting temperature. As temperature increases further the basalt partly melts, incorporating material from the sediments and perhaps material from the wedge of mantle above the descending plate. Bodies of magma with a variety of compositions are formed and move upward through the mantle wedge and the sedimentary or metamorphic rock above it. Some of them change composition en route by assimilation of adjacent rock, and the molten mixtures rise to form volcanoes at the surface and masses of intrusive rock below the surface. The volcanic rock and intrusives so formed seemingly constitute additions of new material to the continental margins.

This is the sequence of events we visualize in operation at subduction zones in the present world (Fig. 21-3), and very likely a similar sequence had established itself at least as early as the end of the Archean. The plates would doubtless have been smaller and faster-moving because of greater heat flow at that time, but they accomplished the same net result: a gradual segregation of much of the silica, alumina, and alkali metals from the upper mantle into the granitic rock of the continents. We can think of the early islands or small continents of the Archean growing in size, joining together, and ultimately forming continents with dimensions like those we see today.

Subduction, then, can be regarded as a mechanism for addition of material to the continental crust, hence for enlarging the continents, but it can also be imagined in a more destructive role. A downward-moving lithospheric plate, consisting of a thick layer of peridotite surmounted by hydrothermally altered basalt and pelagic sediment, may be partly converted to igneous melt as described above, but at least in part it may continue its descent into deeper and hotter parts of the mantle. This means that sedimentary material derived from erosion of the continental surface, plus a good deal of water and carbon dioxide from the altered basalt, will ultimately be absorbed by peridotite and become again a part of the mantle—which represents a net loss of continental material to the Earth's interior.

Now of the many questions that could be asked about this complicated scenario, the major one in an historical context relates to the balance between the processes of production and destruction of continental crust. Is new crustal material being created faster than other parts of the crust are being carried into the mantle, or have the two processes reached a steady state? Are the continents still growing today, or was most of their growth completed far back in the Precambrian? More accurately, we should ask whether the total mass of continental crust is increasing, because the individual continents have moved about so much during geologic time, joining together and pulling apart in new patterns, that it is misleading to ask about the size of any single one. In the geologic record we find considerable evidence for continuing continental accretion. For example, the Precambrian granites of North America decrease in age from the oldest ones in the Canadian Shield to younger ones near continental margins; and the continent was enlarged in the late Mesozoic by addition of batholiths like those of the Sierra Nevada and the Coast Range of British Columbia formed during subduction at its western edge. Such evidence for particular regions, however, does not prove that the overall mass of the continents has changed greatly, at least in later geologic time.

The question is difficult to approach on theoretical grounds, because too little is known about details of what happens to subducting plates, particularly to the sedimentary layers on their surfaces. How much of the sediment, for example, is actually being carried down with a moving plate into the mantle? Much of it certainly is not, but is scraped off to become part of the contorted layering of the "accretionary wedge" that accumulates at many places where lithospheric plates start their descent. One proof that at least part of the sediment does descend to the lower asthenosphere is provided by a bit of geochemical evidence: island-arc volcanic rock, representing magma that has formed along and risen from a descending plate, in several places has been shown to contain traces of the radioactive isotope ^{10}Be. This isotope, produced by cosmic rays in the upper atmosphere, has a half-life of about 1.5 million years. It is carried by rain into sediments, and because of the short half-life is a good marker for the recent origin of sedimentary material. Its presence in island-arc rocks, therefore, is a sure indication that geologically recent sediments have been involved in production of the magma that fed, or is still feeding, the arc volcanoes.

The observed quantities of ^{10}Be, however, are so small that only a few percent of the sedimentary material on a moving plate need be assumed to take part in magma generation (or to become part of mantle peridotite). This means that our query about the growth or lack of growth of continents is still essentially unanswered.

Another possible source of evidence on the question might lie in observations of changes in the amount and composition of granitic rocks over geologic time. If the global process of differentiation from peridotite-to-basalt-to-granite as described above has gone on without an effective reverse process to counteract it, the amount of granitic rock produced per million years should have reached a maximum in the Precambrian and decreased steadily ever since—because the reservoir of undifferentiated material in the outer part of the Earth would have steadily diminished. In composition, the granitic rocks would presumably have become increasingly felsic, more rich in potassium and silicon and lower in magnesium, as the same material was subjected to repeated differentiation.

Observations of such possible changes give somewhat equivocal results. Regarding amounts of granitic rock, certainly a good case can be made for more plentiful generation during parts of the Precambrian than at any time since; but from the later Precambrian to the present the production of granitic rock has been episodic, with no discernible long-time trends. Granite compositions are not much more informative. Among Precambrian granitoids there is indeed a marked change with decreasing age from rocks rich in Na and Ca to those with dominant K. Later changes are less clear, and attempts to interpret available analytical data have given rise to much argument. One recent effort to detect overall trends in composition on a worldwide scale (Ronov, 1992) seems worthy of mention. This is a compilation of several tens of thousands of analyses, not only of granitic rocks but including also basalts, komatiites, shales, and sandstones, from many parts of the world and of ages from Archean to Holocene. The changes with time shown by the analyses are not large, but over all rock types the direction of trend suggests continuing differentiation of Earth materials on a global scale: a general increase in concentrations of K, Rb, and REE, and a decrease in Mg, Ni, and Cr. Certainly there is a strong suggestion here of directional alteration in the general composition of crustal rocks consistent with long-term separation of felsic material from mantle peridotite over geologic time, hence with continued formation of continental crust.

But this conclusion is not the final word. Skeptics can plausibly maintain that the observed changes are small and their meaning is questionable, that the most impressive fact about overall crustal compositions down through the ages is their near-uniformity. Good arguments can be framed for no appreciable growth of crust since the Archean, or for a decreasing amount. Continued crustal growth does seem the most likely hypothesis on the basis of present data, but it is by no means firmly established.

If the overall chemistry of the solid Earth shows only slight long-term change, this is certainly not true of the surface veneer where weathering and erosion have converted original igneous material into sediments. Some aspects of sedimentation are cyclic, but superposed on the cycles are very real long-term changes in the chemical nature of the sedimentary rocks that have been dominant at different periods in geologic history. Carbonate rocks, for example, are uncommon through much of the Precambrian, but become abundant toward its close and have continued to form in large amounts down to the present; banded iron formation (a rock with conspicuous alternating layers of chert and various iron minerals, most commonly magnetite or hematite) is largely restricted to a period between 2.5 and 1.8×10^9 years ago; placer deposits of easily oxidized minerals like uraninite (Blind River in Ontario and the Witwatersrand in South Africa) are restricted to the Archean, and strongly oxidized sediments ("red beds") become abundant only in the later Proterozoic and Phanerozoic; and evaporites in any quantity are not found until shortly before the Palaeozoic. These changes in rock chemistry are intimately related to long-term changes in the atmosphere and oceans, which are the next subjects on our agenda.

21-5 HISTORY OF THE ATMOSPHERE

The scarcity of the inert gases on Earth compared with their cosmic abundance, as noted in Sec. 21-2, is most plausibly explained by escape of gases from the growing Earth at a time when its materials were still spread through a large volume of space and gravitational attraction was much weaker than it is today. The fact that other gases with similar molecular weights, particularly H_2O, HCl, NH_3, CH_4, CO_2, H_2S, and SO_2, were not lost in the same proportion as the inert gases is good evidence that the original temperature of Earth materials was low, so that these substances could be retained in the planetesimals either as frozen particles or in the form of compounds. As gravitational contraction and radioactive decay caused heating in the newly formed planet, gases would be released from their compounds and accumulate at the surface. Gravity would now be sufficient to retain all except the two lightest gases, hydrogen and helium, provided the surface temperature remained low. How hot the surface became is uncertain; if the Earth at one point was completely molten, as one current hypothesis suggests, these original gases as well as the inert gases would be largely lost. In this case a similar assortment of gases would accumulate gradually as the Earth cooled, from the continuing infall of cold planetesimals (and may still be accumulating, since comets that approach the Earth are known to contain compounds of this sort). Such late additions, while the Earth was still forming and still in a wildly turbulent state, would be added to the Earth's substance and become part of its interior. Thus our present atmosphere is probably in large part of secondary origin, having formed by later "degassing" of the solid body of the planet. Presumably the water of the oceans would also have come to the surface as one of these released gases.

If the atmosphere is largely secondary, how much of it was formed early in the Earth's history and how much has been added later? Are gases still being added, either from the interior or as constituents of meteorites and comets? Was the early atmosphere similar to the present one, or has its composition changed over geologic time?

Some striking evidence regarding such questions has come from simple calculations about the amounts of carbon in various geologic materials. Huge quantities of carbon are locked up in sedimentary rocks, partly as organic material and partly in the carbonate minerals of limestone and dolomite. Practically all of this buried carbon represents atmospheric CO_2 that was added to seawater either directly by dissolution or indirectly by the respiration and decay of organisms. By estimating the total amount of buried carbon, then, we should arrive at a figure indicating how much CO_2 has existed in the air at one time or another. The result is astonishing: carbon in sedimentary rocks exceeds that in the present atmosphere, oceans, and organisms by a factor of about 600! Even if some of the analyses and estimates of volume on which the calculations rest are greatly in error, the figure would still be startlingly large. Beyond any reasonable doubt, the amount of carbon now in the air is only a tiny fraction of the amount that has existed at some time in the geologic past.

If most degassing from the interior took place shortly after the planet was formed, this result means that the atmosphere at some early period was very dense, consisting chiefly of CO_2 at a partial pressure of some 12 bar. Such an extreme is hardly credible, because an atmosphere like this would trap so much of the Sun's radiation that surface temperatures would rise well above the boiling point of water and life as we know it could not have developed. The high pressure of CO_2 would have persisted, and the Earth would be a dead world. Something of the sort may have happened on Venus: this planet, in many ways so similar to the Earth, does indeed have a dense atmosphere with CO_2 as a major constituent and a lifeless surface with temperatures of 300–400°C.

We may owe our existence, then, to the simple fact that the original store of CO_2 in the Earth's planetary material did not escape all at once, as it may have done on Venus, but has been added to the air gradually over geologic time. This suggests another possible extreme: is degassing of the interior approaching an end, so that atmospheric CO_2 will soon be exhausted? We know, of course, that CO_2 is being returned to the air continually by respiration and by organic decay (and increasingly by human activity), but the amount is too small to make up for the carbon that is being steadily removed as precipitated carbonates and organic matter buried with sediments (Sec. 12-4). Rough calculation of the carbon balance indicates that CO_2 in air would fall to a level too low to support plant life within a few centuries, unless some other source of the gas were available. Perhaps, then, we are living near the very end of the history of life on our planet? This hardly seems likely, since the geologic record gives indisputable evidence for the continued existence of multi-cellular organisms for at least 600 million years and of unicellular life for five times that long. During this time the CO_2 content of air cannot have dropped far below its present figure, and it is hardly believable that the present 0.0003 bar has been reached only now after 3 billion years of steady depletion.

So we look for another source of CO_2, and the obvious one is volcanic activity. This gas is one of the most abundant in the emanations from practically all volcanoes, and from hot-spring areas that persist when volcanic activity has ceased (Sec. 18-6). Can we assume that volcanic CO_2 is a manifestation of the degassing of the Earth—that it is rising from the deep interior, and being added to the atmosphere for the first time? Certainly much of it is recycled, coming from decomposition of sediments heated by contact with molten lava and in larger quantity from sediments carried deep down into the mantle by subducting plates. But its abundance in volcanic areas far from subduction zones and areas where sedimentary strata are scarce seems evidence that a part at least does indeed come from the interior. It makes no difference that some of the carbon expelled from volcanoes has the form of CH_4 and CO rather than CO_2, because these gases would be quickly oxidized by atmospheric oxygen and so would contribute to the supply of CO_2. Since geologic evidence shows that volcanoes have been active all during the Earth's history, a mechanism for maintenance of adequate CO_2 in the atmosphere seems assured.

Thus a balance has apparently been established among life processes, sedimentation, and volcanic activity that has kept the partial pressure of CO_2 in air somewhere within range of its present value. How effective has the balance been? How far, in the geologic past, has the pressure strayed from its presently measured 0.0003 bar? This is a question of current active debate (see Prob. 12-2). The continued existence of life certainly sets limits on possible fluctuations of CO_2 pressures, but just where the limits are remains uncertain. A sudden marked increase in CO_2 pressure would have adverse effects on many present-day land organisms, and the resulting decrease in seawater pH would make life difficult for some marine creatures. But the capacity of living things to adjust to gradual changes is enormous, and there seems no reason based on organic processes to suppose that wide deviations may not have occurred at times in the past.

Nor is it easy to pick out inorganic reactions that might be sufficiently sensitive to CO_2 variations to have left a record in the geologic column. Extreme amounts, say a partial pressure greater than 1 bar, can probably be ruled out on the grounds that surface waters would become highly acid, chemical weathering would be greatly speeded, silicates would be largely broken down to free silica, and cations from weathering would accumulate in a very concentrated ocean. Sedimentary rocks of all ages are too uniform to support such a conjecture. On the other hand, the huge amounts of silica deposited with iron minerals in the banded iron formation that is so characteristic of one period in the early to mid Proterozoic has been adduced as evidence for at least a moderately increased CO_2 content of the Precambrian atmosphere (Sec. 12-4). Again, the paleontologic and isotopic evidence for abnormally high temperatures in much of the Cretaceous, as well as the large amounts of limestone among Cretaceous sedimentary rocks, has suggested to many the probability of higher CO_2 in this period also. Quantitative guesses about CO_2 pressures in the geologic past are difficult, but evidence for additions from the interior and for considerable variation in amount is persuasive.

Two inert gases, argon and helium, provide even more direct evidence of the continuing escape of volatiles from the Earth's deep interior. The argon of ordinary air consists chiefly of the isotope ^{40}Ar, whereas the abundant isotope in stellar atmospheres is ^{36}Ar. Because ^{40}Ar is a product of the radioactive decay of ^{40}K, this peculiar fact of distribution suggests that most terrestrial argon has accumulated during geologic time from the decay of potassium in rocks. Such an assumption accounts also for the anomalously large total abundance of argon on the Earth as compared with the other inert gases (Table 21-1). Now if ^{40}Ar comes chiefly from ^{40}K, it is a simple matter to calculate how much ^{40}K must have decayed to give the present amount of argon in air. The result shows that rocks near the surface, in fact all the rocks in the crust, contain only a small fraction of the necessary ^{40}K. Hence argon must have escaped to the surface from the decay of potassium at much deeper levels.

Even more persuasive about continuing escape of gas from the interior is the argument based on helium. The two naturally occurring isotopes of this element, ^3He and ^4He, are minor constituents of air, with a normal ratio of 1.4×10^{-6}. The abundant isotope ^4He is a product of the radioactive decay of uranium and thorium in crustal rocks, and the very minor ^3He comes from a reaction between neutrons and the nuclei of ^6Li. Amounts of the two isotopes and the ratio between them are maintained by a balance between their rates of production and rates of escape from the upper atmosphere. The surprising thing about helium is that the isotope ratio $^3He/^4He$ is consistently larger in volcanic gases and gases liberated from the glassy matrix of volcanic rocks, especially large from rocks (like mid-ocean ridge basalts) that have come directly from the mantle. Largest of all are ratios in volcanic areas situated above mantle-plume hotspots, e.g. Iceland, Yellowstone Park, Hawaii, where the source of gas and rock is deep within the mantle. There seems no doubt that ^3He is rising now from the interior, and presumably has so risen all during the Earth's history.

For atmospheric gases other than He, Ar, and CO_2, no equally direct proof of large-scale emergence from the Earth's interior has been suggested. But if three gases have such an origin, it seems reasonable to suppose that others do also. This is hardly proof, but in the absence of evidence to the contrary, and because the hypothesis is consistent with current ideas about the Earth's origin, it is reasonable to assume that most volatile materials now on the surface, including water, have come from the interior. Very likely the degassing was most rapid early in the Earth's history, but the air and oceans are probably still receiving minor additions in the gases associated with volcanic activity.

The composition of volcanic gases, together with analyses of gases obtained by heating rocks and meteorites, plus spectrographic data on the makeup of stellar atmospheres, gives us a basis for conjecture about the kinds of volatile substances that have escaped from the deep interior. Uncertain as such guesses may be, they tell us at least that the average product of degassing bears little resemblance to the composition of the present atmosphere. Water vapor, hydrogen, carbon oxides, sulfur gases, hydrogen halides, nitrogen or ammonia, possibly hydrocarbons—some such mixture is the primary material that has come out of the Earth and that presumably constituted the primitive atmosphere. How, from such a mixture, could the vastly different atmosphere of our present Earth evolve?

Elimination of the sulfur and halogen gases is no problem, for they would dissolve in surface waters, react with rocks, and eventually appear for the most part as constituents of seawater. The really difficult question is the evolution of large amounts of free oxygen—a highly active gas that is found in comparable amounts nowhere else in the solar system. The answer that comes to mind immediately is the activity of plants, for oxygen is set free in large amounts by the process of photosynthesis in green leaves [Eqs. (12-8) and (12-9)]. This is a good answer, and certainly the correct one regarding the maintenance of the present supply of oxygen in air. But it hardly suffices for the historical question of how the primitive atmosphere of a lifeless Earth was transformed into a mixture that permits higher forms of life to flourish.

The geochemical evolution of the early atmosphere is a subject of great current interest, not only to geologists but to astronomers and biologists as well. If free hydrogen was a prominent gas at the beginning, the earliest atmosphere would have consisted of strongly reduced compounds like methane, ammonia, and hydrogen sulfide. If most of the initial hydrogen escaped during the accumulation of planetesimals, the first atmosphere would have had more free nitrogen and oxides of carbon and sulfur. In either case water would have been a prominent constituent. Water molecules at higher levels in the atmosphere would be decomposed by absorption of ultraviolet radiation from the Sun, producing a little free hydrogen and oxygen; the hydrogen molecules would in part escape from the Earth completely, but the oxygen would be retained. Thus over long periods the gas mixture would slowly become more oxidizing, but free O_2 is used up so rapidly in a variety of processes that its steady-state concentration would remain very small.

Somehow in the primitive ocean, or perhaps in smaller bodies of water on land, or perhaps even in the atmosphere above the water, the first living forms appeared. How this happened remains one of the great unsolved mysteries, and speculations abound. Certainly the essential ingredients of living substance were present on the early Earth—the C, H, O, and N, plus the necessary minor elements—but in the form of simple compounds like methane and ammonia and water. These are very different from the exceedingly complex molecules found in even the most primitive of living creatures, and how the transition occurred from simple to complex remains a matter of speculation. Promising hints have come from experiments set up to simulate supposed early-Earth conditions: in mixtures of the simple gases subjected to shocks like those that can be imagined from lightning discharges or volcanic activity on the early Earth, molecules of organic compounds have been found that could well be intermediate steps on the way toward the still more complex molecules of living things. But the gap between these experimentally produced compounds and actual living substance remains a wide one. In any event, the first organisms were presumably simple anaerobic forms, and only much later did abundant complex plants evolve that could use the energy of sunlight to effect a reduction of atmospheric carbon dioxide to organic compounds and thereby set oxygen free.

That some such sequence of events occurred can hardly be doubted, but the individual steps and the times when major changes in the atmosphere took place are subjects of lively controversy. Seemingly information about atmospheric composition should be obtainable from geologic study of ancient rocks, but the evidence is far from clear. It has long been a speculative possibility that the great increase in abundance and complexity of organisms between the late Precambrian and early Paleozoic might somehow be linked with atmospheric change, perhaps from a CO_2-rich atmosphere to an oxygen atmosphere. Such a change might also help to explain some peculiarities of Precambrian sedimentary rocks. High concentrations of CO_2 in air would make the ocean more acid than at present, and a low concentration of O_2 would permit extensive transportation of iron as Fe^{2+} in surface waters. These are conditions that would favor the deposition of silica in large amounts as chert, together with iron either in reduced form as siderite or pyrite, or where oxygen was not extremely scarce, as magnetite or hematite. Such sediments would eventually become the banded-iron formation that is so characteristic of the middle Precambrian and so important as the world's major source of iron ore. A low pH in seawater would prevent large-scale precipitation of calcium and magnesium carbonates, hence would explain the scarcity of carbonate rocks through much of the Precambrian. A reducing atmosphere would allow the stable existence of uraninite in detrital grains, hence the formation of uranium placers like the famous deposits of the Witwatersrand in South Africa. Extensive "red beds" in the later Precambrian and further in the Phanerozoic suggest an increase in atmospheric oxygen, but the increase need not have been large to make hematite stable. Present levels of O_2 in the air may not have been attained until well into the Paleozoic.

Details of atmospheric evolution remain obscure, but much present thinking inclines to an ultimate origin by gradual degassing of the Earth's interior plus additions from comets and meteors, slow increase in free oxygen accompanying the development of life, and persistence of high CO_2/O_2 ratios until at least the beginning of the Paleozoic era.

21-6 HISTORY OF SEAWATER

The ocean is a complex solution with six major ions (Na^+, Cl^-, SO_4^{2-}, Mg^{2+}, K^+, Ca^{2+}) and a host of minor ones (Sec. 12-3 and Appendix IV). Relative amounts of the major ions are remarkably uniform in different parts of the sea, and the pH remains within narrow limits (generally 8.1 to 8.4 in surface waters, slightly lower at depth). The constancy of composition suggests that strong controls are at work to keep the amounts of different solutes at steady levels. Determining what these controls are and how effectively they have operated over geologic time is an enterprise that has attracted the attention of many geochemists.

Control of concentrations of the major ions, as we have noted previously (Sec. 12-3), can be plausibly ascribed to a balance between rate of supply from rivers, various precipitation and sorption processes, and reaction of seawater with basalt of the oceanic crust. The reaction with basalt is most intense near the mid-ocean ridges, where seawater descends into the basalt, is heated, and rises to the surface again with changed ion ratios, especially increased Ca^{2+}/Mg^{2+}. The extent of such reactions depends on the amount of hot basaltic liquid rising from the mantle under the ridges. Changes in this amount over geologic time have produced some variation in the overall composition of seawater, as reflected in the composition of evaporites formed at different periods (Sec. 14-6). Nevertheless, equilibrium reasoning combined with simple facts about the salts found in marine evaporites can be used to show that major-ion concentrations have not changed greatly during Phanerozoic time. The concentration of Ca^{2+}, for example, is limited by the facts that gypsum would be a common rock accompanying limestone if its concentration had ever been much higher, and that sodium carbonate minerals would be abundant in marine evaporites if it had been much lower (Holland, 1978, 1984). The usual presence of dolomite as a primary mineral at the base of marine evaporite sequences requires that Mg^{2+} cannot have been much below its present value, and the absence or near absence of sepiolite and brucite in evaporites sets an upper limit. Concentrations of the major ions cannot have been much more than twice, or much less than half of, their present values since the late Precambrian. Controls of concentration have been so effective that seawater has not had a very exciting chemical history for much of geologic time.

In earlier periods more radical changes were possible. The Precambrian ocean, as we saw in the last section, was probably in contact with a different kind of atmosphere, and evaporites are scarce among older Precambrian rocks, so that some of our assumptions for evaluating the Phanerozoic ocean are no longer valid. What guesses can be make about changes in seawater composition during this earlier time?

Almost certainly the ocean formed originally by condensation of water vapor escaping from the interior during the early major degassing of the planet. Additional water escaped later, and may be escaping still, but a large part of the ocean's volume was probably present soon after the solid crust became stable. The early ocean, and the streams that flowed into it, must have contained acid gases (CO_2, HCl, SO_2) in abundance, and attack of such gases and such solutions on exposed rock would have been rapid. Much material would have dissolved, and in short order the ocean must have acquired solutes in concentrations comparable to those we find at present.

The composition of the dissolved substances, however, would have been different. The early ocean was most likely slightly acid; some of the acid would be neutralized by reaction with silicate minerals, but in the presence of a CO_2-rich atmosphere and with a supply of acid gases continually renewed by numerous volcanoes, the pH of seawater must have been on the acid side of neutral. The dissolved material, therefore, would have had higher concentrations of Ca^{2+} and

Mg^{2+} than the present ocean, since at low pH these ions can exist together with large amounts of HCO_3^-. Sulfate ion would have been scarce or absent, because sulfur would be largely in reduced form; this means that Ca^{2+} concentrations would not even be limited by precipitation of gypsum. With no more than traces of free oxygen in either air or water, iron could have been present in abundance as Fe^{2+}. Presumably Cl^-, Na^+, and K^+ would have been important ions, just as they are today. Some of the rarer metals whose concentrations at present are kept low by reactions with CO_3^{2-} or SO_4^{2-}, for example Ba^{2+}, Sr^{2+}, and Mn^{2+}, very likely played a more important role in Precambrian seas.

Changes in this early ocean as the carbon dioxide of the atmosphere decreased and oxygen slowly increased are easy to imagine. The pH would gradually rise; Ca^{2+} and Mg^{2+} would precipitate to form the extensive dolomite beds of the later Precambrian; iron would be oxidized and precipitated as hematite; SO_4^{2-} would become an important ion except in stagnant basins where primitive organisms flourished. The overall composition would thus gradually shift toward the proportions that have remained without much variation down to the present day.

Much of this history is speculative, but the general picture of an original slightly acid ocean with abundant dissolved salts of calcium, magnesium, and iron, changing gradually through the Precambrian to the present NaCl-rich and sulfate-rich ocean, accords both with the sedimentary record and with probable changes in the chemistry of the atmosphere and the solid Earth.

A final word about a few elements listed in Appendix IV (notably Cl, Br, S, and B) whose concentrations in seawater seem remarkably high compared with amounts in common rocks. If the figures are used to estimate the total amounts of these elements in the crust and in the ocean, it appears that a sizable fraction of the Earth's supply of each element is dissolved in seawater. The quantity in the sea is far too great to have been supplied by any imaginable process of erosion of ordinary rocks during geologic time. Since the four elements are well-known as constituents of volcanic emanations and since their compounds with common cations are all very soluble, the suggestion is natural that compounds of these elements have been added to seawater directly and in large quantity by volcanic activity rather than by erosion, probably all through the geologic past.

21-7 SOME LARGER PROBLEMS

The pH of seawater, touched on briefly in the last section, is an interesting subject for further rumination. Going back to the discussion of buffers in Chapter 2, we recall that seawater today is protected from major changes in pH by reactions involving carbonic acid and its ions, and to a lesser extent by boric acid and borate ion. This is certainly sufficient to counteract minor additions of acid or alkali, but would it be effective as a long-term control if acid or alkali were added in large quantities? To make the question specific, suppose that the acid gases mentioned in the last section were added to the present sea continuously and in large amounts from volcanic emanations. The carbonate buffer could take up a great deal of these gases: solid carbonates would be dissolved, CO_3^{2-} would be changed to HCO_3^-, and HCO_3^- eventually to H_2CO_3 (Fig. 2-3). The pH would fall slowly as these reactions took place, but even when it dropped to 6.5 or 7 the buffer would still be able to take up H^+.

Eventually, of course, further addition of acid would cause the buffer to break down, and seawater would lose its protection against a drastic lowering of pH. Or would it? A second line of defense can be imagined in the clay minerals. Seawater is in contact with enormous volumes of clay, and ion exchange with the clay minerals could take up huge amounts of H^+. Both the metal ions adsorbed on clay-particle surfaces and interlayer ions like the Na^+ and Ca^{2+} of montmorillonite and the K^+ of illite would be susceptible to replacement by hydrogen ion. This clay buffer system is slower acting and less familiar than the carbonate system, but its capacity is larger, and it may well be the principal long-term regulator of seawater pH—acting, so to speak, behind the scenes while the carbonate system takes care of minor day-to-day fluctuations.

But even ion exchange on clay particles cannot be the ultimate answer. Giving our imagination free reign, we let volcanoes produce enough acid gas to react completely with clay minerals as well as carbonates. Then what would keep seawater from becoming a solution of hydrochloric, hydrobromic, and sulfuric acids? As a third line of defense we turn to silicate minerals other than the clays: feldspars, pyroxenes, amphiboles. These are subject to slow acid attack, breaking down to form silica and kaolinite and contributing their cations to solution. Hence not until our energetic volcanoes had emitted enough acid gas to convert all surface rocks to silica and kaolinite, and to bring all the cations of surface rocks into solution in the ocean, would the acidity of seawater be free to rise.

This series of events has obviously never occurred, and a logical next question might be, why not? For an answer we look into the processes by which a volcano generates acid gas. Here we tread on uncertain ground, but in part at least the gases from a volcano represent volatile material released when sedimentary rock is carried downward into the mantle as part of a subducted plate and is heated sufficiently for melting to start. Reactions of this sort were demonstrated in some of the experiments described in Sec. 16-5: when shales containing pyrite and added NaCl are heated with water vapor under pressure, HCl and H_2S are noted among the products. Evidently the kinds of sedimentary material formed under near-neutral conditions at the Earth's surface are capable of generating acid gases when heated at depth, and the amount of acid gas set free would be limited by the nature of the sediment. In other words we are dealing with a sort of gigantic and complicated equilibrium, in which acid gases and the metal silicates of igneous rocks are the stable forms at high temperature, and hydrogen silicates (for example, kaolinite) plus dissolved salts are the ultimate stable forms at low temperature:

$$2KAlSi_3O_8 + 2HCl + H_2O \rightleftharpoons H_4Al_2Si_2O_9 + 2K^+ + 2Cl^- + 4SiO_2, \qquad (21\text{-}1)$$

feldspar   kaolinite

or, in more general terms:

Low temperature →

metal silicates + acid gases ⇌ hydrogen silicates + dissolved salts   (21-2)

← *High temperature*

Reactions in both directions are slow and incomplete.

Another part of the gas emitted from volcanoes, as we noted above, consists of volatile material out of the deep interior, gases trapped when the Earth was formed and now slowly on their way to the surface. How important a fraction of volcanic gas this primitive material is at the present time remains uncertain, but it is probably not large. We can at least be sure, from geologic evidence, that it has not produced major displacements of the above equilibrium in recent geologic time.

Is there a chance that the buffer system of the ocean might break down in the other direction? Could alkali be added in sufficient amount to convert all carbon dioxide from water and air into carbonate, all boric acid to borate, all clays to montmorillonite and illite? Here we need not consider the active addition of alkali; it is sufficient to imagine simply that processes of weathering are carried far beyond their usual limits. Metal silicates reacting with water give alkaline solutions (Sec. 4-3), and experimental pH's obtained by grinding common silicates under water are reported to rise as high as 11. What restraints are there to an indefinitely rising pH, once the capacity of the common buffers is exhausted? For a time the conversion of silica to an ion of silicic acid would act as a brake. But if weathering proceeded long enough, the free silica of the outer crust would be exhausted, and pH's of solutions would thereafter be determined by the hydrolysis of silicate minerals. Specific prediction is difficult, but there seems little to restrain the pH except the solubility of metal hydroxides. Presumably in nature a final control is neutralization of excess alkali by the acid gases from volcanoes.

The essence of such fanciful arguments has been expressed by Urey (1956) in a symbolic equilibrium much simpler than the generalized equation above:

$$CaSiO_3 + CO_2 \rightleftharpoons CaCO_3 + SiO_2.$$ (21-3)

This equation, in a different context, we have discussed as an example of a simple metamorphic reaction that may occur at granite–limestone contacts (Secs. 16-2 and 16-5). Here we look at the equation from a wider perspective: $CaSiO_3$ stands for any of the common metal silicates, CO_2 represents acid gases either from volcanoes or already present in the atmosphere, $CaCO_3$ is a simple salt, and SiO_2 as the anhydride of silicic acid may stand for the ultimate product of dehydration of the hydrogen silicates. The forward reaction, then, symbolizes the reactions of weathering and dissolution at the Earth's surface—the low-temperature processes by which silicates are attacked, cations are set free or fixed in simple compounds, and alkalinity is increased. The reverse reaction shows the formation of complex silicates and acid gases by the heating of simple sedimentary materials to high temperatures underground. The position of equilibrium in the reaction depends not only on the temperature but on the pressure of CO_2. On a worldwide scale we could read this pressure as the concentration of the constituents of acid gases in the Earth's materials. Thus this simple equation, in rather cryptic form, summarizes a great deal of geochemistry. It expresses not only the ultimate controls on the pH of seawater, but the principal reactions that make up the rock cycle.

It seems appropriate that here at the end of our story we find ourselves looking at one final example of equilibrium. The story began with equilibrium, back in Chapter 1, and we have examined equilibria of many kinds and with varying degrees of complexity. Now in this final discussion, after a long, speculative look at chemical trends down the ages, we hit upon a deceptively simple equation that expresses in the form of an equilibrium much of the subject matter of this and preceding chapters. There are many things we do not know about the Urey equilibrium, and about other equilibria as well. But when problems can be formulated in this manner, we have at least achieved a measure of organization of our thinking and a basis for asking pertinent questions.

REFERENCES AND SUGGESTIONS FOR FURTHER READING

Atreya, S. K., J. B. Pollock, and M. S. Mathews, (eds): *Origin and Evolution of Planetary and Satellite Atmospheres*, University of Arizona Press, 1989. Many articles giving details of the development of atmospheres on planetary bodies of varying sizes and distances from the Sun.

Berner, R. A., A. C. Lasaga, and R. M. Garrels: "The carbonate–silicate geochemical cycle and its effect on atmospheric carbon dioxide over the past 100 million years," *American Jour. Sci.*, vol. 283, pp. 641–683, 1983 and vol. 284, pp. 1125–1192, 1984. An impressive effort to evaluate the sources and sinks of carbon dioxide in the Earth's atmosphere during the Cretaceous and the Cenozoic.

Broecker, W. S.: *How to Build a Habitable Planet*, Eldigio Press, 1985. A lively and entertaining account of the development on a planetary surface of the complex set of conditions needed for life to originate and flourish.

Cloud, P.: "A working model of the primitive earth," *Amer. Jour. Science*, vol. 272, pp. 537–548, 1972. Speculations about early Earth history, with emphasis on the chemical consequences of the development of life.

Engel, A. E. J., S. P. Itson, C. G. Engel, and D. M. Stickney: "Crustal evolution and global tectonics: a petrogenetic view," *Geol. Soc. America Bull.*, vol. 85, pp. 843–858, 1974. Speculations about relation of rock composition to plate tectonics at various times in Earth history.

Ernst, W. G.: *The Dynamic Planet*, Columbia University Press, 1990.

Holland, H. D.: *The Chemistry of the Atmosphere and Oceans*, Wiley, New York, 1978. Compilation of data and discussion of controls of composition.

Holland, H. D.: *The Chemical Evolution of the Atmosphere and Oceans*, Princeton University Press, Princeton, New Jersey, 1984. Changes in composition of the Earth's fluid envelopes through geologic time.

Newsome, H. E., and K. W. W. Sims, "Core formation during early accretion of the earth," *Science*, vol. 252, pp. 926–933, 17 May 1991. Speculations about the distribution of elements during formation of the Earth's core.

Poreda, R. J., P. D. Jenden, I. R. Kaplan, and H. Craig: "Mantle helium in Sacramento basin natural gas wells," *Geochim. et Cosmochim. Acta*, vol. 50, pp. 2847–2853, 1986. Analysis shows that natural gas as well as volcanic gas may have high 3/4 helium ratios, hence contains helium of mantle origin, permitting speculation that some of the methane may also come from the mantle.

Ronov, A. B.: "Chemical evolution of sedimentary and magmatic rocks in the earth's crust," in Schidlowski, M., et al., (eds): *Early Organic Evolution: Implications for Mineral and Energy Resources*, Springer Verlag, Berlin and Heidelberg, 1992.

Rubey, W. W.: "Geologic history of seawater: an attempt to state the problem," *Geol. Soc. America Bull.*, vol. 62, pp. 1111–1147, 1951. A classical and much quoted paper tracing the origin of atmosphere and oceans to volatile materials from the Earth's interior.

Schopf, J. W., and C. Klein, (eds): *The Proterozoic Biosphere*, Cambridge University Press, 1992. The first articles in this massive collection give a good summary of current thinking about the origin and early development of life on the Earth.

Taylor, S. R.: "The origin of the earth," in Brown, G. C., C. J. Hawkesworth, and R. C. L. Wilson, (eds): *Understanding the Earth*, Cambridge University Press, 1992. An excellent brief exposition of current ideas about the early history of the Earth and the origin of the Moon.

CONSTANTS
AND
NUMERICAL
VALUES

e = base of natural logarithms = 2.7183

$\ln x = 2.3026 \log_{10} x$

Gas-law constant, $R = 8.314$ J mol^{-1} deg^{-1}

$\qquad\qquad\qquad\qquad = 1.987$ cal mol^{-1} deg^{-1}

$R \ln x = 19.146 \log x$ J mol^{-1} deg^{-1}

$RT \ln x = 5708.2 \log x$ J mol^{-1} deg^{-1} at 25°C

0°C = 273.15 K   25°C = 298.15 K

Avogadro's number = 6.022×10^{23} molecules per mol

Volume of 1 mol of a perfect gas:

> at 0°C and 1 bar = 22.123 liters (at 1 atm = 22.414 liters)
> at 25°C and 1 bar = 24.15 liters (at 1 atm = 24.47 liters)

Faraday constant = $\mathbf{f}$ = 96,485 J volt^{-1} equivalent^{-1}

$\qquad\qquad\qquad\qquad = 23,061$ cal volt^{-1} equivalent^{-1}

Boltzmann constant = $\mathbf{k}$ = 1.38063×10^{-23} J K^{-1}

1 joule (J) = 0.2390 calorie = 9.868 cm^3-atm = 10 cm^3 bar = 10^7 ergs

1 calorie = 4.184 joules = 41.29 cm^3-atm = 41.84 cm^3 bar

1 bar = 0.987 atm = 10^6 dynes cm^{-2} = 10^5 pascals (Pa)

$\qquad = 14.504$ pounds per square inch (psi)

$\qquad = 1.0197$ kg cm^{-2}

$\qquad = 750.06$ mm Hg = 750.06 torr

1 atm = 1.013 bar = 14.70 psi = 1.033 kg cm^{-2} = 760.0 mm Hg

1 Å = 1 angstrom (ångström) = 10^{-8} cm

Concentration units for solutions:

Molality $= m =$ mols of solute per kg of water

$$= \frac{\text{grams of solute} \times 1000}{\text{gram-molecular weight solute} \times \text{grams water}}$$

Molarity $= M =$ mols of solute per liter of solution

$$= \frac{\text{grams of solute} \times 1000}{\text{gram-molecular weight solute} \times \text{milliliters solution}}$$

$$m = M \times \frac{\text{grams solution}}{\text{grams solution} - \text{total grams solutes}} \times \frac{1}{\text{density solution}}$$

$$\text{Parts per million} = \text{ppm} = \frac{\text{grams solute}}{1,000,000 \text{ grams solution}} = \frac{\text{milligrams solute}}{\text{kilogram solution}}$$

$$\text{Mol fraction} = X = \frac{\text{mols of solute}}{\text{total mols of solution}}$$

$$\text{Equivalent of an acid or base} = \frac{\text{gram-molecular weight}}{\text{number of H's or OH's in formula}}$$

Normality (N) of an acid or base $=$ number of equivalents per liter of solution

Numerical prefixes:

pico (p)	10^{-12}	kilo (k)	10^{3}	
nano (n)	10^{-9}	mega (M)	10^{6}	
micro (μ)	10^{-6}	giga (G)	10^{9}	
milli (m)	10^{-3}	tera (T)	10^{12}	

THE
GEOLOGIC
TIME
SCALE

Eon	Era	Period	Epoch	Duration, millions of years (Ma)	Time since beginning, millions of years (Ma)
Phanerozoic	Cenozoic	Quaternary	Holocene	—	0.01
			Pleistocene	1.6	1.6
		Tertiary	Pliocene	3.7	5.3
			Miocene	18.4	23.7
			Oligocene	12.9	36.6
			Eocene	21.2	57.8
			Paleocene	8.6	66.4
	Mesozoic	Cretaceous		78	144 ± 5
		Jurassic		64	208 ± 18
		Triassic		37	245 ± 20
	Paleozoic	Permian		41	286 ± 12
		Pennsylvanian		34	320 ± 20
		Mississippian		40	360 ± 10
		Devonian		48	408 ± 12
		Silurian		30	438 ± 12
		Ordovician		67	505 ± 32
		Cambrian		65	570 ± 28
Proterozoic	Late			330	900
	Middle			700	1600
	Early			900	2500
Archean	Late			500	3000
	Middle			400	3400
	Early			600(?)	4000

Archean time extends back to the earliest known rocks, about 4000 million years old (4 Ga). The Proterozoic and Archean are called collectively the "Precambrian" or "Prepaleozoic." (*Source:* Geological Society of America.)

APPENDIX
III

LARGE-SCALE DATA ABOUT THE EARTH

Mean radius	6371 km
Volume	1.083×10^{12} km^3
Mass	5.98×10^{27} g
Thicknesses	
Crust	variable; ca. 40 km under continents, 6 km under oceans
Lithosphere	variable; ca. 100 km under continents
Asthenosphere	variable, roughly 300 km
Mantle (Moho to core)	2900 km
Core (radius)	3471 km
Masses	
Crust	0.024×10^{27} g
Mantle	4.02×10^{27} g
Core	1.94×10^{27} g
Density	
Mean for entire earth	5.52 g/cm^3
Mean for crust	2.8 g/cm^3 (2.7 continental, 3.0 oceanic)
Mean for mantle	4.5 g/cm^3
Mean for core	10.7 g/cm^3

APPENDIX
IV

AVERAGE ABUNDANCE OF ELEMENTS IN THE EARTH'S CONTINENTAL CRUST, IN THREE COMMON ROCKS, AND IN SEAWATER
(in parts per million)

	Crust	Granite	Diabase (basalt)	Shale	Seawater
O	466,000	485,000	449,000	495,000	880,000
Si	277,200	339,600	246,100	273,000	2.5
Al	81,300	74,300	79,400	80,000	0.003
Fe	50,000	13,700	77,600	47,200	0.003
Ca	36,300	9,900	78,300	22,100	450
Na	28,300	24,600	16,000	9,600	10,800
K	25,900	45,100	5,300	26,600	390
Mg	20,900	2,400	39,900	15,000	1,290
Ti	4,400	1,500	6,400	4,600	0.0001
H	1,400	400	600	6,300	110,000
P	1,050	390	610	700	0.09
Mn	950	195	1,280	850	0.0002
F	625	700	250	740	1.3

(Continued)

	Crust	Granite	Diabase (basalt)	Shale	Seawater
Ba	425	1,220	160	580	0.015
Sr	375	250	190	300	7.8
S	260	58	123	2,400	900
C	200	200	100	1,000	28
Zr	165	210	105	160	2×10^{-5}
V	135	17	264	130	0.0022
Cl	130	70	200	180	18,800
Cr	100	20	114	90	0.0003
Rb	90	220	21	140	0.12
Ni	75	1	76	68	0.0005
Zn	70	45	86	95	0.0003
Ce	60	170	23	50	2×10^{-6}
Cu	55	13	110	45	0.0002
Y	33	13	25	26	1×10^{-5}
La	30	101	10	24	6×10^{-6}
Nd	28	55	15	24	4×10^{-6}
Co	25	2.4	47	19	1×10^{-6}
Sc	22	2.9	35	13	9×10^{-7}
Li	20	22	15	66	0.18
N	20	59	52	60	670
Nb	20	24	9.5	11	1×10^{-5}
Ga	15	20	16	19	2×10^{-6}
Pb	13	48	7.8	20	3×10^{-6}
B	10	1.7	15	100	4.5
Pr	8.2	19	3.4	6.1	9×10^{-7}
Th	7.2	50	2.4	12	5×10^{-8}
Sm	6.0	8.3	3.6	5.8	8×10^{-7}
Gd	5.4	5	4	5.2	1×10^{-6}
Yb	3.4	1.1	2.1	2.2	2×10^{-6}
Dy	3.0	2.4	4	4.3	2×10^{-6}
Cs	3	1.5	0.9	5	3×10^{-4}
Hf	3	5.2	2.7	2.8	3×10^{-6}
Er	2.8	1.2	2.4	2.7	1×10^{-6}
Be	2.8	3	0.8	3	2×10^{-7}
Br	2.5	0.4	0.4	4	67
Sn	2	3.5	3.2	6.0	6×10^{-7}
Ta	2	1.5	0.5	0.8	2×10^{-6}
U	1.8	3.4	0.6	3.7	0.0032
As	1.8	0.5	1.9	13	0.0017
Ge	1.5	1.1	1.4	1.6	4×10^{-6}
Mo	1.5	6.5	0.6	2.6	0.01
W	1.5	0.4	0.5	1.8	0.0001
Ho	1.2	0.35	0.69	1.2	5×10^{-7}
Eu	1.2	1.3	1.1	1.1	2×10^{-7}
Tb	0.9	0.54	0.65	0.9	2×10^{-7}
Tl	0.5	0.15	0.30	1.0	1×10^{-5}
Lu	0.5	0.19	0.35	0.6	3×10^{-7}
Tm	0.5	0.15	0.30	0.5	3×10^{-7}
I	0.5	<0.03	<0.03	2.2	0.058

(Continued)

	Crust	Granite	Diabase (basalt)	Shale	Seawater
Sb	0.2	0.31	1.0(?)	1.5	0.0002
Cd	0.2	0.03	0.15	0.3	8×10^{-5}
Bi	0.2	0.07	0.05	0.4	4×10^{-9}
In	0.1	0.02	0.07	0.1	1×10^{-7}
Hg	0.08	0.1	0.2	0.4	4×10^{-7}
Ag	0.07	0.05	0.08	0.07	3×10^{-6}
Se	0.05	0.007	0.3	0.6	0.0002
Pd	0.01	0.002	0.025		7×10^{-8}
Pt	0.01	0.0019	0.0012		3×10^{-7}

Te, Re, Ru, Rh, Os, Ir, Au have concentrations of 0.01 ppm or less in rocks and less than 10^{-5} ppm in seawater. Concentrations of inert gases in seawater: He, $6.8\ 10^{-6}$ ppm; Ne, 1.2×10^{-4} ppm; Ar, 4.3×10^{-3} ppm; Kr, 2×10^{-4} ppm; Xe, 5×10^{-5} ppm.

Sources: Concentrations in crust, granite, diabase (basalt), and shale from Mason, B., and C. B. Moore: *Principles of Geochemistry,* 4th ed., Wiley, New York, pp. 46–47 and 176–177, 1982.

Concentrations in seawater from Li, Y.-H.: "Distribution patterns of the elements in the ocean: a synthesis," *Geochimica et Cosmochimica Acta,* vol. 55, pp. 3224–3225, 1991.

Notes: The heading "crust" means the continental crust only, a part of the crust that is assumed to be made up of roughly equal parts of granite and basalt. For the oceanic crust a composition similar to that of average basalt can be assumed. "Granite" includes silica-rich rocks ranging from alkali granite to granodiorite and their volcanic equivalents; the analysis chosen by Mason and Moore as representative is for a granite from Rhode Island. "Diabase" includes the more common varieties of basaltic lava, diabase, and dolerite; the analysis represents a diabase from Virginia. For "shale," Mason and Moore used an earlier compilation by K.K. Turekian and K.H. Wedepohl (*Geol. Soc. America Bull.,* vol. 72, p. 186, 1961), with some updating. "Seawater" is an average analysis of deep Atlantic and deep Pacific water, updated by Y.-H. Li from earlier work by M. Whitfield and D.R. Turner (in Stumm, W., (ed.): *Aquatic Surface Chemistry,* Wiley, pp. 457–493 1987). No great accuracy can be claimed for any of the values, because they depend on subjective judgments about the kinds of material to be included in each category, and because they are subject to change as analytical techniques improve.

APPENDIX

V

SYMBOLS, ATOMIC NUMBERS, AND ATOMIC WEIGHTS OF THE NATURALLY OCCURRING ELEMENTS

Element	Symbol	Atomic number	Atomic weight
Actinium	Ac	89	227.03
Aluminum	Al	13	26.98
Antimony	Sb	51	121.75
Argon	Ar	18	39.95
Arsenic	As	33	74.92
Barium	Ba	56	137.33
Beryllium	Be	4	9.01
Bismuth	Bi	83	208.98
Boron	B	5	10.81
Bromine	Br	35	79.90
Cadmium	Cd	48	112.41
Calcium	Ca	20	40.08
Carbon	C	6	12.01
Cerium	Ce	58	140.12
Cesium	Cs	55	132.91
Chlorine	Cl	17	35.45
Chromium	Cr	24	52.00
Cobalt	Co	27	58.93
Copper	Cu	29	63.55

(Continued)

Element	Symbol	Atomic number	Atomic weight
Dysprosium	Dy	66	162.50
Erbium	Er	68	167.26
Europium	Eu	63	151.96
Fluorine	F	9	19.00
Gadolinium	Gd	64	157.25
Gallium	Ga	31	69.72
Germanium	Ge	32	72.59
Gold	Au	79	196.97
Hafnium	Hf	72	178.49
Helium	He	2	4.003
Holmium	Ho	67	164.93
Hydrogen	H	1	1.008
Indium	In	49	114.82
Iodine	I	53	126.90
Iridium	Ir	77	192.22
Iron	Fe	26	55.85
Krypton	Kr	36	83.80
Lanthanum	La	57	138.91
Lead	Pb	82	207.19
Lithium	Li	3	6.94
Lutetium	Lu	71	174.97
Magnesium	Mg	12	24.31
Manganese	Mn	25	54.94
Mercury	Hg	80	200.59
Molybdenum	Mo	42	95.94
Neodymium	Nd	60	144.24
Neon	Ne	10	20.18
Nickel	Ni	28	58.70
Niobium	Nb	41	92.91
Nitrogen	N	7	14.01
Osmium	Os	76	190.2
Oxygen	O	8	16.00
Palladium	Pd	46	106.4
Phosphorus	P	15	30.97
Platinum	Pt	78	195.09
Polonium	Po	84	209
Potassium	K	19	39.10
Praseodymium	Pr	59	140.91
Protactinium	Pa	91	231.04
Radium	Ra	88	226.03
Radon	Rn	86	222
Rhenium	Re	75	186.21
Rhodium	Rh	45	102.91
Rubidium	Rb	37	85.47
Ruthenium	Ru	44	101.07
Samarium	Sm	62	150.35
Scandium	Sc	21	44.96
Selenium	Se	34	78.96
Silicon	Si	14	28.09

(Continued)

Element	Symbol	Atomic number	Atomic weight
Silver	Ag	47	107.87
Sodium	Na	11	22.99
Strontium	Sr	38	87.62
Sulfur	S	16	32.06
Tantalum	Ta	73	180.95
Tellurium	Te	52	127.60
Terbium	Tb	65	158.93
Thallium	Tl	81	204.37
Thorium	Th	90	232.04
Thulium	Tm	69	168.93
Tin	Sn	50	118.69
Titanium	Ti	22	47.90
Tungsten	W	74	183.85
Uranium	U	92	238.03
Vanadium	V	23	50.94
Xenon	Xe	54	131.30
Ytterbium	Yb	70	173.04
Yttrium	Y	39	88.91
Zinc	Zn	30	65.38
Zirconium	Zr	40	91.22

IONIC RADII AND ELECTRONEGATIVITIES

Element	Ion[1]	Radius for 6-coordination, octahedral (Å)[2]	Commonly occurring coordination numbers[3]	Electro-negativity[4]	Approximate ionic character of bond with oxygen[5]
Aluminum	Al^{3+}	0.535	4, 6	1.5	60
Antimony	Sb^{3+}	0.80 (5)	6		66
	Sb^{5+}	0.60	4, 6	1.9	48
Arsenic	As^{3+}	0.58	4, 6		60
	As^{5+}	0.46	4, 6	2.0	38
Barium	Ba^{2+}	1.35	8–12	0.9	84
Beryllium	Be^{2+}	0.27 (4)	4	1.5	63
Bismuth	Bi^{3+}	1.03	6, 8	1.9	66
Boron	B^{3+}	0.11 (4)	3, 4	2.0	43
Bromine	Br^-	1.96		2.8	
Cadmium	Cd^{2+}	0.95	6, 8	1.7	66
Calcium	Ca^{2+}	1.00	6, 8	1.0	79
Carbon	C^{4+}	0.15 (4)	3, 4, 6	2.5	23
Cerium	Ce^{3+}	1.01	6, 8	1.1	74
Cesium	Cs^+	1.67	12	0.7	89
Chlorine	Cl^-	1.81		3.0	
Chromium	Cr^{3+}	0.615	6	1.6	53
	Cr^{6+}	0.26			23
Cobalt	Co^{2+}	0.745	6	1.8	65
Copper	Cu^+	0.77	6, 8	1.9	71
	Cu^{2+}	0.73	6	2.0	57
Fluorine	F^-	1.33		4.0	

(Continued)

Element	Ion[1]	Radius for 6-coordination, octahedral (Å)[2]	Commonly occurring coordination numbers[3]	Electro-negativity[4]	Approximate ionic character of bond with oxygen[5]
Gallium	Ga^{3+}	0.62	4, 6	1.6	57
Germanium	Ge^{4+}	0.73	4	1.8	49
Gold	Au^+	1.37	8–12	2.4	62
Hafnium	Hf^{4+}	0.71	6	1.3	70
Indium	In^{3+}	0.80	6	1.7	62
Iodine	I^-	2.20		2.5	
	I^{5+}	0.95	6		54
Iron	Fe^{2+}	0.78	6	1.8	69
	Fe^{3+}	0.643	6	1.9	54
Lanthanum	La^{3+}	1.032	8	1.1	77
Lead	Pb^{2+}	1.19	6–10	1.8	72
Lithium	Li^+	0.76	6	1.0	82
Magnesium	Mg^{2+}	0.72	6	1.2	71
Manganese	Mn^{2+}	0.83	6	1.5	72
	Mn^{3+}	0.645	6		51
	Mn^{4+}	0.53	4, 6		38
Mercury	Hg^{2+}	1.02	6, 8	1.9	62
Molybdenum	Mo^{4+}	0.65	6		58
	Mo^{6+}	0.59	4, 6	1.8	47
Nickel	Ni^{2+}	0.69	6	1.8	60
Niobium	Nb^{5+}	0.64	6	1.6	56
Nitrogen	N^{5+}	0.13	3	3.0	9
Oxygen	O^{2-}	1.40		3.5	
Palladium	Pd^{2+}	0.86	6	2.2	61
Phosphorus	P^{5+}	0.17 (4)	4	2.1	35
Potassium	K^+	1.38	8–12	0.8	87
Radium	Ra^{2+}	1.48	8–12	0.9	83
Rare-earth	Ce^{3+}–Sm^{3+}	1.01–0.958	6, 8	1.1–1.2	73–75
metals	Eu^{3+}–Lu^{3+}	0.947–0.861	6	1.2	76
	Eu^{2+}	1.17	8		
Rhenium	Re^{4+}	0.63	6		63
	Re^{7+}	0.53	4, 6		51
Rubidium	Rb^+	1.52	8–12	0.8	87
Scandium	Sc^{3+}	0.745	6	1.3	65
Selenium	Se^{2-}	1.98		2.4	
	Se^{6+}	0.28	4		26
Silicon	Si^{4+}	0.26 (4)	4	1.8	48
Silver	Ag^+	0.94	8, 10	1.9	71
Sodium	Na^+	1.02	6, 8	0.9	83
Strontium	Sr^{2+}	1.18	8	1.0	82
Sulfur	S^{2-}	1.84		2.5	
	S^{6+}	0.12 (4)	4		20
Tantalum	Ta^{5+}	0.64	6	1.5	63
Tellurium	Te^{2-}	2.21		2.1	
	Te^{6+}	0.56	4, 6		36

(Continued)

Element	Ion[1]	Radius for 6-coordination, octahedral (Å)[2]	Commonly occurring coordination numbers[3]	Electro-negativity[4]	Approximate ionic character of bond with oxygen[5]
Thallium	Tl^+	1.50	8–12		79
	Tl^{3+}	0.67	6, 8	1.8	58
Thorium	Th^{4+}	0.94	6, 8	1.3	72
Tin	Sn^{2+}	1.27 (8)	6, 8	1.8	73
	Sn^{4+}	0.69	6	1.9	57
Titanium	Ti^{3+}	0.67	6		60
	Ti^{4+}	0.605	6	1.5	51
Tungsten	W^{6+}	0.60	4, 6	1.7	57
Uranium	U^{4+}	1.00 (8)	6, 8		68
	U^{6+}	0.73	6	1.7	62
Vanadium	V^{3+}	0.64	6	1.6	57
	V^{4+}	0.58	6		45
	V^{5+}	0.54	4, 6		36
Yttrium	Y^{3+}	0.90	6	1.2	74
Zinc	Zn^{2+}	0.74	4, 6	1.7	63
Zirconium	Zr^{4+}	0.72	6	1.4	65

[1] Only ions commonly found in naturally occurring minerals are listed.

[2] *Sources:* Shannon, R. D.: "Revised effective ionic radii and systematic studies of interatomic distances in halides and chalcogenides," *Acta Cryst.,* A32, p. 751–767, 1976, and Shannon, R. D., and C. T. Prewitt: "Effective ionic radii in oxides and fluorides," *Acta Cryst.,* B25, pp. 925–946, 1968. All are radii for 6-coordination except a few for which a different coordination is indicated by a number in parentheses. In general, radii for 4-coordination can be estimated from the 6-coordination radii by subtracting 0.13 Å, and radii for 8-coordination by adding 0.13 Å; for most ions these rules give radii within 0.02 Å of the correct values.

[3] *Source:* Smith, F. G.: *Physical Geochemistry,* Addison-Wesley, Reading, Massachusetts 1963.

[4] *Source:* Pauling, L.: *The Nature of the Chemical Bond,* 3d ed., Copyright © by Cornell University. Used by permission of the publisher, Cornell University Press, N.Y., 1960. The numbers are in arbitrary units, ranging from 0.7 for Cs to 4.0 for F.

[5] *Source:* Smith, *op. cit.,* calculated by Smith from electronegativity values estimated by A. S. Povarennykh, *Dokl. Akad. Nauk SSSR,* vol. 109, pp. 993–996, 1956.

APPENDIX
VII

EQUILIBRIUM
CONSTANTS

TABLE VII-1

Activity solubility products

The numbers below are negative logarithms of activity products at 25°C and 1 bar (pK). For example, the number after PbCl$_2$ is 4.8, which means that

$$\text{solubility product for PbCl}_2 = a_{\text{Pb}^{2+}}a_{\text{Cl}^-}^2 = 10^{-4.8}.$$

The physical state of the compounds is not in every case clear from the literature, but for the most part they are probably finely crystalline precipitates. Because of this uncertainty about physical state and because of differences in experimental methods, values in the literature show considerable disagreement. The numbers cannot be expected to agree precisely with solubility products calculated from the free energies in Appendix VIII, but for most values the discrepancy is small. For values with a discrepancy larger than 0.5, the calculated figure is shown in parenthesis.

For sulfides, additional uncertainty arises from a recent reevaluation of the dissociation constant of HS$^-$ (Schoonen and Barnes, 1988); see Table VII-2. Solubility products calculated using this new figure are shown for some of the sulfides by a second parenthesis.

Chlorides	
CuCl	6.7
PbCl$_2$	4.8
Hg$_2$Cl$_2$	17.9 [Hg$_2^{2+}$][Cl$^-$]^2 Calomel
AgCl	9.7 Cerargyrite
Fluorides	
BaF$_2$	5.8 (6.7)
CaF$_2$	10.4 (9.8) Fluorite
MgF$_2$	8.2 (10.1) Sellaite
SrF$_2$	8.5

(Continued)

Sulfides

Sb_2S_3	90.8 $[Sb^{3+}]^2[S^{2-}]^3$ Stibnite
Bi_2S_3	100 Bismuthinite
CdS	27.0 (28.9) (32.6) Greenockite
Cu_2S	48.5 (47.6) (53.4) Chalcocite
CuS	36.1 (40.8) Covellite
FeS	18.1 (18.8) (26.0) Troilite
PbS	27.5 (33.2) Galena
MnS	13.5 Alabandite
HgS	53.3 (52.7) (58.4) Cinnabar
NiS	26.6 (21.0) (28.1) Millerite
Ag_2S	50.1 (49.2) (54.7) Argentite
ZnS	24.7 (30.3) Sphalerite

Sulfates

$BaSO_4$	10.0 Barite
$CaSO_4$	4.5 Anhydrite
$CaSO_4 \cdot 2H_2O$	4.6 Gypsum
$PbSO_4$	7.8 Anglesite
$SrSO_4$	6.5 Celestite

Carbonates

$BaCO_3$	8.3 Witherite
$CdCO_3$	13.7 (11.2)
$CaCO_3$	8.35 Calcite
$CaCO_3$	8.22 Aragonite
$FeCO_3$	10.7 Siderite
$PbCO_3$	13.1 Cerussite
$MgCO_3$	7.5 (5.2) Magnesite
$MnCO_3$	9.3 (10.6) Rhodochrosite
$NiCO_3$	6.9
$SrCO_3$	9.0 Strontianite
UO_2CO_3	19.6 $[UO_2^{2+}][CO_3^{2-}]$
$ZnCO_3$	10.0 Smithsonite
$Cu_2(OH)_2CO_3$	33.8 $[Cu^{2+}]^2[OH^-]^2[CO_3^{2-}]$ malachite

Phosphates

$AlPO_4 \cdot 2H_2O$	22.1 Variscite
$Ca_3(PO_4)_2$	28.7
$CaHPO_4 \cdot 2H_2O$	6.6
$FePO_4 \cdot 2H_2O$	26.4 Strengite
$Mg_3(PO_4)_2$	25.2
$(UO_2)_3(PO_4)_2$	49.7
$Ca_5(PO_4)_3OH$	57.8 $[Ca^{2+}]^5[PO_4^{3-}]^3[OH^-]$ hydroxylapatite

Sources of data: Smith, R. M., and A. E. Martell: *Critical Stability Constants, vol. 4: Inorganic Complexes,* Plenum Press, New York, 1976. Sillén, L. G.: "Stability constants of metal-ion complexes, Sec. 1: Inorganic ligands," *Chem. Soc. London Spec. Publ.* 17, 1964; and Supplement 1, *Spec. Publ.* 25, 1971. Schoonen, M. A. A., and H. L. Barnes: "An approximation to the second dissociation constant for H_2S," *Geochim. et Cosmochim. Acta,* vol. 52, pp. 649–654, 1988.

TABLE VII-2
Dissociation constants of acids

The numbers are negative logarithms of activity constants at 25°C and 1 bar. For each acid, K_1 is for dissociation of the first H^+, K_2 for dissociation of the second, and K_3 for dissociation of the third. As an example, for H_3PO_4:

$$H_3PO_4 \rightleftharpoons H^+ + HPO_4^- \qquad K_1 = \frac{a_{H^+} a_{H_2PO_4^-}}{a_{H_3PO_4}} = 10^{-2.1}$$

$$H_2PO_4^- \rightleftharpoons H^+ + HPO_4^{2-} \qquad K_2 = \frac{a_{H^+} a_{HPO_4^{2-}}}{a_{H_2PO_4^-}} = 10^{-7.2}$$

$$HPO_4^{2-} \rightleftharpoons H^+ + PO_4^{3-} \qquad K_3 = \frac{a_{H^+} a_{PO_4^{3-}}}{a_{HPO_4^{2-}}} = 10^{-12.4}$$

Acid	Formula	$-\log K_1$	$-\log K_2$	$-\log K_3$
Aluminum hydroxide	H_3AlO_3 (amorph)	12.7		
Arsenious	H_3AsO_3	9.2		
Arsenic	H_3AsO_4	2.2	7.0	11.5
Boric	H_3BO_3	9.2		
Carbonic	H_2CO_3	6.35	10.3	
Hydrofluoric	HF(aq)	3.2		
Water	HOH	14.0		
Phosphoric	H_3PO_4	2.1	7.2	12.4
Hydrosulfuric	H_2S(aq)	7.0	12.9[1]	
Sulfuric	H_2SO_4		2.0	
Hydroselenic	H_2Se(aq)	3.9	15.0	
Selenic	H_2SeO_4		1.9	
Silicic	H_4SiO_4	9.9	11.7	

[1] Determination of this constant is experimentally difficult, and published values are widely divergent. A recent estimate on theoretical grounds is 18.57 (Schoonen and Barnes, 1988). *Sources:* same as for Table VII-1.

TABLE VII-3
Dissociation constants of hydroxides

In the first column, for metals whose oxides are more stable than the hydroxides, formulas are given as oxide + H_2O. The meaning of the dissociation constants is illustrated by the reaction for Fe(III) oxide:

$$1/2 Fe_2O_3 + 3/2 H_2O \rightleftharpoons Fe^{3+} + 3OH^- \qquad K = a_{Fe^{3+}} \cdot a_{OH^-}^3 = 10^{-42.7}.$$

The second column indicates the physical state of the hydroxide, if this is specified in the source. Most hydroxides show a range in values of the dissociation constant, the amorphous material first precipitated dissociating somewhat more than an aged precipitate that has partly crystallized, and often considerably more than the equivalent naturally occurring mineral. Hydroxides for which a form is not specified are most likely precipitates that have been allowed to stand, in other words probably very finely crystalline material. Because of the uncertainty about form, these numbers cannot be expected to agree completely with constants calculated from the free energies in Appendix VIII.

Numerical values in the succeeding columns of the table are negative logarithms of the dissociation (activity) constants at 25°C and 1 bar. The third column gives "total" constants, on the assumption of complete dissociation. Numbers in the next four columns are constants for stepwise dissociation. The column headed K_{aq} shows the extent of solution as undissociated molecules. The last column shows the ability of a hydroxide to dissolve by adding an additional OH^- ion, in other words the extent of its amphoteric character. Cadmium hydroxide will serve as an example:

$$Cd(OH)_2 \rightleftharpoons Cd^{2+} + 2OH^- \qquad K_T = a_{Cd^{2+}} \cdot a_{OH^-}^2 = 10^{-14.4}$$

$$Cd(OH)_2 \rightleftharpoons CdOH^+ + OH^- \qquad K_1 = a_{CdOH^+} \cdot a_{OH^-} = 10^{-10.5}$$

$$CdOH^+ \rightleftharpoons Cd^{2+} + OH^- \qquad K_2 = a_{Cd^{2+}} \cdot a_{OH^-}/a_{CdOH^+} = 10^{-3.9}$$

$$Cd(OH)_2(s) \rightleftharpoons Cd(OH)_2(aq) \qquad K_{aq} = a_{Cd(OH)_2(aq)} = 10^{-6.7}$$

$$Cd(OH)_2 + OH^- \rightleftharpoons Cd(OH)_3^- \qquad K_A = a_{Cd(OH)_3^-}/a_{OH^-} = 10^{-4.1}$$

Hydroxide	Form	$-\log K_T$	$-\log K_1$	$-\log K_2$	$-\log K_3$	$-\log K_4$	$-\log K_{aq}$	$-\log K_A$
NH_4OH	Dissolved	4.7						
$Al(OH)_3$	Amorphous	31.6	12.3	10.3	9.0			-1.1
$Al(OH)_3$	Gibbsite	34.1	14.8	10.3	9.0			1.4
$AlOOH + H_2O$	Boehmite	34.2	14.9	10.3	9.0			1.5
$Cd(OH)_2$		14.4	10.5	3.9			6.7	5.3
$Cr(OH)_3$		29.8	11.6	8.5	11.8			
$Co(OH)_2$		14.9	10.6	4.3			4.0	5.2
$1/2Cu_2O + 1/2H_2O$		14.7						
$Cu(OH)_2$		19.3	13.0	6.3				2.9
$CuO + H_2O$	Tenorite	20.3						
$Fe(OH)_2$		15.1	10.6	4.5			8.4(?)	5.1
$Fe(OH)_3$	Amorphous	38.8	16.5	10.5	11.8			4.4
$1/2Fe_2O_3 + 3/2H_2O$	Hematite	42.7						
$FeOOH + H_2O$	Goethite	41.5						
$PbO + H_2O$	Red	15.3	9.0	6.3			4.4	1.4
$Mg(OH)_2$		11.2	8.6	2.6				
$Mn(OH)_2$		12.8	9.4	3.4				5.1
$HgO + H_2O$	Red	25.4	14.8	10.6			3.6	4.5
$Ni(OH)_2$		15.2	11.1	4.1			7	4
$1/2Ag_2O + 1/2H_2O$		7.7					5.7	3.4
$Th(OH)_4$	Amorphous	44.7			10.3	10.8		5.8
$ThO_2 + 2H_2O$		49.7						
$SnO + H_2O$		26.2	15.8(?)	10.4(?)				
$UO_2 + 2H_2O$		56.2				13.3		3.8
$UO_2(OH)_2$		22.4	14.2	8.2				3.6
$V(OH)_3$		34.4			11.7			
$VO(OH)_2$		23.5	15.2	8.3				
$Zn(OH)_2$	Amorphous	15.5	10.5	5.0			4.4(?)	1.9
$ZnO + H_2O$		16.7						

Sources of data: Smith, R. M., and A. E. Martell: *Critical Stability Constants, vol. 4: Inorganic Complexes,* Plenum Press, New York, 1976. Sillén, L. G.: "Stability constants of metal-ion complexes, Sec. 1: Inorganic ligands," *Chem. Soc. London Spec. Publ.* 17, 1964; and Supplement 1, *Spec. Publ.* 25, 1971.

TABLE VII-4

Equilibrium constants for complex ions and molecules

The numbers are negative logarithms of activity constants at 25°C and 1 bar. For example:

$$AgCl_2^- \rightleftharpoons AgCl(s) + Cl^- \qquad K = a_{Cl^-}/a_{AgCl_2^-} = 10^{+4.4}$$
$$CuCO_3(aq) \rightleftharpoons Cu^{2+} + CO_3^{2-} \qquad K = a_{Cu^{2+}} \cdot a_{CO_3^{2-}}/a_{CuCO_3(aq)} = 10^{-6.8}$$

For many numbers the original data show wide disagreement, so that values in the table should be considered only approximate.

$AlF^{2+} \rightleftharpoons Al^{3+} + F^-$	7.0
$CuCl_2^- \rightleftharpoons CuCl(s) + Cl^-$	−1.2
$CuCl_3^{2-} \rightleftharpoons CuCl_2^- + Cl^-$	0.2
$CuCl^+ \rightleftharpoons Cu^{2+} + Cl^-$	0.4
$CuCO_3(aq) \rightleftharpoons Cu^{2+} + CO_3^{2-}$	6.8
$Cu(CO_3)_2^{2-} \rightleftharpoons CuCO_3(aq) + CO_3^{2-}$	3.2
$FeCl^{2+} \rightleftharpoons Fe^{3+} + Cl^-$	1.5
$FeF^{2+} \rightleftharpoons Fe^{3+} + F^-$	6.0
$PbCl_2(s) \rightleftharpoons PbCl_2(aq)$	3.2
$PbCl_2(aq) \rightleftharpoons PbCl^+ + Cl^-$	0.2
$PbCl^+ \rightleftharpoons Pb^{2+} + Cl^-$	1.4
$HgS_2^{2-} \rightleftharpoons HgS(s) + S^{2-}$	0.6
$AgCl_2^- \rightleftharpoons AgCl(s) + Cl^-$	−4.4
$AgCl(s) \rightleftharpoons AgCl(aq)$	6.4
$SnCl^+ \rightleftharpoons Sn^{2+} + Cl^-$	1.8
$SnF^+ \rightleftharpoons Sn^{2+} + F^-$	Approx. 4.1
$SnF_6^{2-} \rightleftharpoons Sn^{4+} + 6F^-$	Approx. 25
$SnS_3^{2-} \rightleftharpoons SnS_2 + S^{2-}$	5.0
$UO_2(CO_3)_2^{2-} \rightleftharpoons UO_2^{2+} + 2CO_3^{2-}$	14.6
$UO_2(CO_3)_2^{2-} \rightleftharpoons UO_2(CO_3)(s) + CO_3^{2-}$	4.0
$UO_2(CO_3)_3^{4-} \rightleftharpoons UO_2(CO_3)_2^{2-} + CO_3^{2-}$	3.8
$ZnCl^+ \rightleftharpoons Zn^{2+} + Cl^-$	0.4
$ZnF^+ \rightleftharpoons Zn^{2+} + Cl^-$	1.2

Sources of data: Same as for Table VII-3.

TABLE VII-5
Mineral hydrolysis constants

The numbers below are logarithms of equilibrium constants for the hydrolysis of minerals in the system K_2O–Al_2O_3–SiO_2–H_2O at 25°C and 1 bar and at 500°C and 5000 bar. Reactions are written in terms of the aqueous species, K^+, Al^{3+}, $SiO_2(aq)$, and H^+. For example, hydrolysis of kaolinite $[Al_2Si_2O_5(OH)_4]$ is written as

$$Al_2Si_2O_5(OH)_4 + 6H^+ \rightleftharpoons 2Al^{3+} + 2SiO_2(aq) + 5H_2O,$$

and for muscovite

$$KAl_2AlSi_3O_{10}(OH)_2 + 10H^+ \rightleftharpoons K^+ + 3Al^{3+} + 3SiO_2(aq) + 6H_2O.$$

The equilibrium constant for the reaction of muscovite to kaolinite [Eq. (4-27)]

$$KAl_2AlSi_3O_{10}(OH)_2 + 2H^+ + 3H_2O \rightleftharpoons 3Al_2Si_2O_5(OH)_4 + 2K^+$$

is obtained by subtracting 3 times the value of log K for kaolinite from the value of log K for muscovite. Values for other complex reactions can be similarly obtained by adding and subtracting the values of the logarithms of the equilibrium constants given below. These numbers are selected from the extensive tabulation of equilibrium constants reported by Bowers, T. S., K. J. Jackson, and H. C. Helgeson: *Equilibrium Activity Diagrams for Coexisting Minerals and Aqueous Solutions at Pressures and Temperatures to 5 kbar and 600°C,* Springer Verlag, New York, pp. 397, 1984.

Name	Formula	log K 25°C 1 bar	log K 500°C 5000 bars
Amorphous silica	SiO_2	−2.71	−0.69
Andalusite	Al_2SiO_5	16.57	−0.27
Gibbsite	$Al(OH)_3$	7.96	1.50
Kalsilite	$KAlSiO_4$	11.26	4.20
Kaolinite	$Al_2Si_2O_5(OH)_4$	7.43	−0.61
K-feldspar	$KAlSi_3O_8$	0.08	1.40
Kyanite	Al_2SiO_5	16.30	−0.33
Muscovite	$KAl_2AlSi_3O_{10}(OH)_2$	14.56	1.07
Pyrophyllite	$Al_2Si_4O_{10}(OH)_2$	1.06	−2.72
Sillimanite	Al_2SiO_5	16.93	−0.29
Quartz	SiO_2	−4.00	−0.86
Corundum	Al_2O_3	21.38	0.81

APPENDIX
VIII

STANDARD
FREE ENERGIES,
ENTHALPIES,
AND ENTROPIES

Standard state thermodynamic data for geologic substances come from a bewildering array of sources. A variety of experimental and theoretical techniques have been used to obtain the data, and the results are continually modified by new experiments and theoretical algorithms. In the table below, values for standard molal Gibbs free energies, enthalpies, and third-law entropies are those recommended by the National Bureau of Standards for inorganic substances commonly encountered in geochemistry. A full discussion of available data bases can be found in Chapter 12 and Appendix D of Nordstrom, D. K., and J. L. Munoz: *Geochemical Thermodynamics,* The Benjamin/Cummings Publishing Co., 1985.

The second column of the table gives the physical state of each substance, insofar as it is known. Abbreviations: s, solid, form not specified in source; l, liquid; g, gas; aq, dissolved in water at unit activity. The columns headed $\Delta G°$ and $\Delta H°$ give standard free energies and enthalpies of formation from the elements at 25°C and 1 bar, in kilojoules per mol. The column headed $S°$ gives entropies in standard entropy units, joules per mol per degree kelvin. To find the standard free energy change for a reaction, subtract the sum of $\Delta G°$ values for the reactants from the sum of $\Delta G°$ for the products [Eqs. (7-49) and (7-67)]. To find the equilibrium constant for a reaction at 25°C and 1 bar, use the relation [Eq. (8-49)]

$$\log K = -\Delta G°/5709.$$

By using this equation in reverse, approximate free energies of formation for many compounds and ions not given in the table may be calculated from the equilibrium constants in Appendix VII. Numbers in the table are of widely varying accuracy, and are subject to continual revision as new data are reported in the literature.

At temperatures greater than 25°C, approximate values for standard Gibbs free energies of reaction (ΔG_T°) can be found from the equation

$$\Delta G_T^\circ = \Delta H^\circ - 0.001 T \Delta S^\circ,$$

on the assumption that ΔH° and ΔS° are constant (or that $\Delta C_p^\circ \approx 0$). If the heat capacity of the reaction is small, as it is for most solid–solid reactions, this equation will give a close estimate of ΔG_T° for temperatures to a few hundred degrees. To find free energies at higher temperatures, or in general to obtain accurate values for thermodynanic properties of a reaction at elevated temperatures and pressures, requires additional information on the variation of heat capacity and volume of all substances in the reaction [cf. Eqs. (7-41), (7-62), (8-18), and Prob. 8-8].

Formula	Form	ΔG°	ΔH°	S°
	Aluminum			
Al	s	0	0	28.3
Al_2O_3	corundum	−1582.3	−1675.7	50.9
AlOOH	boehmite	−915.9	−990.9	48.5
$Al(OH)_3$	gibbsite	−1155.1	−1293.3	68.5
$Al_2Si_2O_5(OH)_4$	kaolinite	−3799.7	−4119.6	205.0
Al^{3+}	aq	−485	−531	−321.7
$Al(OH)_4^-$	aq	−1305.3	−1502.5	102.9
	Arsenic			
As	metallic	0	0	35.1
As	gas	+261.0	+302.5	174.2
As_4O_6	arsenolite	−1152.4	−1313.9	214.2
As_2O_5	s	−782.3	−924.9	105.4
AsH_3	g	+68.9	+66.4	222.8
As_2S_3	orpiment	−168.6	−169.0	163.6
H_3AsO_3	aq	−639.8	−742.2	195.0
H_3AsO_4	aq	−766.0	−902.5	184.
$H_2AsO_3^-$	aq	−587.1	−714.8	110.5
AsO_4^{3-}	aq	−648.4	−888.1	−162.8
	Barium			
Ba	s	0	0	62.8
BaO	s	−525.1	−553.5	70.4
BaF_2	s	−1156.8	−1207.1	96.4
BaS	s	−456.	−460.	78.2
$BaSO_4$	barite	−1362.2	−1473.2	132.2
$BaCO_3$	witherite	−1137.6	−1216.3	112.1
$BaSiO_3$	s	−1540.2	−1623.6	109.6
Ba^{2+}	aq	−560.8	−537.6	9.6
	Boron			
B	s	0	0	62.8
B_2O_3	s	−1193.7	−1272.8	54.0
H_3BO_3	s	−968.9	−1094.3	88.9
$H_4BO_4^-$	aq	−1153.2	−1344.0	102.5

(Continued)

Formula	Form	$\Delta G°$	$\Delta H°$	$S°$
	Cadmium			
Cd	s	0	0	51.8
CdO	s	−228.4	−258.2	54.8
$Cd(OH)_2$	precipitated	−473.6	−560.7	96.
CdS	greenockite	−156.5	−161.9	64.9
$CdCO_3$	otavite	−669.4	−750.6	92.5
$CdSiO_3$	s	−1105.4	−1189.1	97.5
Cd^{2+}	aq	−77.6	−75.9	−73.2
	Calcium			
Ca	s	0	0	41.4
CaO	s	−604.0	−635.1	39.8
$Ca(OH)_2$	portlandite	−898.5	−986.1	83.4
CaF_2	fluorite	−1167.3	−1219.6	68.9
CaS	s	−477.4	−482.4	56.5
$CaCO_3$	calcite	−1128.8	−1206.9	92.9
$CaCO_3$	aragonite	−1127.8	−1207.1	88.7
$CaMg(CO_3)_2$	dolomite	−2163.4	−2326.3	155.2
$CaSO_4$	anhydrite	−1321.8	−1434.1	106.7
$CaSO_4.2H_2O$	gypsum	−1797.3	−2022.6	194.1
$Ca_3(PO_4)_2$	whitlockite	−3884.7	−4120.8	236.0
$CaSiO_3$	wollastonite	−1549.7	−1634.9	81.9
$CaAl_2Si_2O_8$	anorthite	−4002.3	−4227.9	199.3
$CaMgSi_2O_6$	diopside	−3032.0	−3206.2	142.9
Ca^{2+}	aq	−553.6	−542.8	−53.1
	Carbon			
C	graphite	0	0	5.7
C	diamond	2.9	1.9	2.4
CH_4	g	−50.7	−74.8	186.3
C_2H_6	g	−32.8	−84.7	229.6
C_3H_8	g	−23.5	−104.0	270.0
C_4H_{10}	g	−17.2	−126.1	310.4
C_2H_4	g	+68.2	+52.3	219.6
C_6H_6	l	+129.8	+83.0	269.2
CO	g	−137.2	−110.5	197.7
CO_2	g	−394.4	−393.5	213.7
H_2CO_3	aq	−623.1	−699.7	187.4
HCO_3^-	aq	−586.8	−692.0	91.2
CO_3^{2-}	aq	−527.8	−677.1	−56.9
CN^-	aq	+172.4	+150.6	94.1
	Chlorine			
Cl_2	g	0	0	223.1
HCl	g	−95.3	−92.3	186.9
Cl^-	aq	−131.3	−167.2	56.5
	Chromium			
Cr	s	0	0	23.8
$FeCr_2O_4$	chromite	−1343.8	−1444.7	146.0
CrO_4^{2-}	aq	−727.8	−881.2	50.2
$Cr_2O_7^{2-}$	aq	−1301.1	−1490.3	261.9
	Copper			
Cu	s	0	0	33.2
Cu_2O	cuprite	−146.0	−168.6	93.1

(Continued)

Formula	Form	$\Delta G°$	$\Delta H°$	$S°$
CuO	tenorite	−129.7	−157.3	42.6
$Cu(OH)_2$	s	−357.		
CuCl	s	−119.9	−137.2	86.2
Cu_2S	chalcocite	−86.2	−79.5	120.9
CuS	covellite	−53.6	−53.1	66.5
$Cu_2(OH)_2CO_3$	malachite	−893.6	−1051.4	186.2
Cu^+	aq	+50.0	+71.7	40.6
Cu^{2+}	aq	+65.5	+64.8	−99.6
$CuCl_2^-$	aq	−240.1		
		Fluorine		
F_2	g	0	0	202.8
HF	g	−273.2	−271.1	173.8
HF	aq	−296.8	−320.1	88.7
F^-	aq	−278.8	−332.6	−13.8
		Gold		
Au	s	0	0	47.4
$AuCl_2^-$	aq	−151.1		
$AuCl_4^-$	aq	−235.1	−322.2	266.9
$Au(CN)_2^-$	aq	+285.8	+242.3	172.
		Hydrogen		
H_2	g	0	0	130.7
H^+	aq	0	0	0
		Iron		
Fe	s	0	0	27.3
$Fe_{0.947}O$	wüstite	−245.1	−266.3	57.5
Fe_3O_4	magnetite	−1015.4	−1118.4	146.4
Fe_2O_3	hematite	−742.2	−824.2	87.4
$Fe(OH)_2$	precipitate	−486.5	−569.0	88.
$Fe(OH)_3$	precipitate	−696.5	−823.0	106.7
FeS	troilite	−100.4	−100.0	60.3
Fe_7S_8	pyrrhotite	−748.5	−736.4	485.8
FeS_2	pyrite	−166.9	−178.2	52.9
$FeCO_3$	siderite	−666.7	−740.6	92.9
Fe_2SiO_4	fayalite	−1379.0	−1479.9	145.2
Fe^{2+}	aq	−78.9	−89.1	−137.7
Fe^{3+}	aq	−4.7	−48.5	−315.9
		Lead		
Pb	s	0	0	64.8
Pb	g	+161.9	+195.0	175.4
PbO	s (red)	−188.9	−219.0	66.5
PbO_2	s	−217.3	−277.4	68.6
$Pb(OH)_2$	s	−452.2		
$PbCl_2$	cotunnite	−314.1	−359.4	136.0
PbS	galena	−98.7	−100.4	91.2
$PbSO_4$	anglesite	−813.1	−919.9	148.6
$PbCO_3$	cerussite	−625.5	−699.1	131.0
$PbSiO_3$	s	−1062.1	−1145.7	109.6
Pb^{2+}	aq	−24.4	−1.7	10.5
$Pb(OH)_3^-$	aq	−575.6		

(Continued)

Formula	Form	$\Delta G°$	$\Delta H°$	$S°$
	Magnesium			
Mg	s	0	0	32.7
MgO	periclase	−569.4	−601.7	26.9
$Mg(OH)_2$	brucite	−833.5	−924.5	63.2
MgF_2	sellaite	−1070.2	−1123.4	57.2
MgS	s	−341.8	−346.0	50.3
$MgCO_3$	magnesite	−1012.1	−1095.8	65.7
$MgCO_3.3H_2O$	nesquehonite	−1726.1		
$MgSiO_3$	clinoenstatite	−1462.1	−1549.0	67.7
Mg_2SiO_4	forsterite	−2055.1	−2174.0	95.1
Mg^{2+}	aq	−454.8	−466.9	−138.1
	Manganese			
Mn	s	0	0	32.0
MnO	manganosite	−362.9	−385.2	59.7
Mn_3O_4	hausmannite	−1283.2	−1387.8	155.6
Mn_2O_3	s	−881.1	−959.0	110.5
MnO_2	pyrolusite	−465.1	−520.0	53.1
$Mn(OH)_2$	precipitate	−615.0	−695.4	99.2
MnS	alabandite	−218.4	−214.2	78.2
$MnCO_3$	rhodochrosite	−816.7	−894.1	85.8
$MnSiO_3$	rhodonite	−1240.5	−1320.9	89.1
Mn_2SiO_4	tephroite	−1632.1	−1730.5	163.2
Mn^{2+}	aq	−228.1	−220.8	−73.6
MnO_4^-	aq	−447.2	−541.4	191.2
	Mercury			
Hg	l	0	0	76.0
Hg	g	+31.8	+61.3	175.0
HgO	s, red	−58.5	−90.8	70.3
Hg_2Cl_2	calomel	−210.7	−265.2	192.5
HgS	cinnabar	−50.6	−58.2	82.4
Hg_2^{2+}	aq	+153.5	+172.4	84.5
Hg^{2+}	aq	+164.4	+171.1	−32.2
$HgCl_4^{2-}$	aq	−446.8	−554.0	293.
Hg_2^{2-}	aq	+41.9		
	Molybdenum			
Mo	s	0	0	28.7
MoO_3	s	−668.0	−745.1	77.7
MoS_2	molybdenite	−225.9	−235.1	62.6
$CaMoO_4$	powellite	−1434.6	−1541.4	122.6
MoO_4^{2-}	aq	−836.3	−997.9	27.2
	Nickel			
Ni	s	0	0	29.9
NiO	s	−211.7	−239.7	38.0
$Ni(OH)_2$	s	−447.2	−529.7	88.
NiS	s	−79.5	−82.0	53.0
$NiCO_3$	s	−612.5		
Ni^{2+}	aq	−45.6	−54.0	−128.9
	Nitrogen			
N_2	g	0	0	191.6
N_2O	g	+104.2	+82.1	219.9

(Continued)

Formula	Form	$\Delta G°$	$\Delta H°$	$S°$
NO	g	+86.6	+90.3	210.8
NH_3	g	−16.5	−46.1	192.5
NH_4OH	aq	−263.7	−366.1	181.2
NO_3^-	aq	−108.7	−205.0	146.4
NH_4^+	aq	−79.3	−132.5	113.4
Oxygen				
O_2	g	0	0	205.1
H_2O	l	−237.1	−285.8	69.9
H_2O	g	−228.6	−241.8	188.8
OH^-	aq	−157.2	−230.0	−10.8
Potassium				
K	s	0	0	64.2
KCl	sylvite	−409.1	−436.7	82.6
$KAlSiO_4$	kaliophilite	−2005.3	−2121.3	133.1
$KAlSi_2O_6$	leucite	−2871.4	−3034.2	200.0
$KAlSi_3O_8$	microcline	−3742.9	−3968.1	214.2
$KAl_3Si_3O_{10}(OH)_2$	muscovite	−5608.4	−5984.4	306.3
K^+	aq	−283.3	−252.4	102.5
Silicon				
Si	s	0	0	18.8
SiO_2	α-quartz	−856.6	−910.9	41.8
SiO_2	α-cristobalite	−855.4	−909.5	42.7
SiO_2	α-tridymite	−855.3	−909.1	43.5
SiO_2	glass	−850.7	−903.5	46.9
$SiCl_4$	g	−617.0	−657.0	330.7
SiF_4	g	−1572.7	−1614.9	282.5
SiH_4	g	+56.9	+34.3	204.6
H_4SiO_4	aq	−1316.6	−1468.6	180.
Silver				
Ag	s	0	0	42.6
Ag_2O	s	−11.2	−31.1	121.3
AgCl	cerargyrite	−109.8	−127.1	96.2
Ag_2S	acanthite	−40.7	−32.6	144.0
Ag^+	aq	+77.1	+105.6	72.7
$AgCl_2^-$	aq	−215.4	−245.2	231.4
Sodium				
Na	s	0	0	51.2
NaCl	halite	−384.1	−411.2	72.1
$NaAlSiO_4$	nepheline	−1978.1	−2092.8	124.3
$NaAlSi_2O_6$	jadeite	−2852.1	−3030.9	133.5
$NaAlSi_3O_8$	low albite	−3711.5	−3935.1	207.4
$NaAlSi_2O_6 \cdot H_2O$	analcite	−3082.6	−3300.8	234.3
Na^+	aq	−261.9	−240.1	59.0
Strontium				
Sr	s	0	0	52.3
SrO	s	−561.9	−592.0	54.4
$SrSO_4$	celestite	−1340.9	−1453.1	117.
$SrCO_3$	strontianite	−1140.1	−1220.1	97.1
$SrSiO_3$	c	−1549.7	−1633.9	96.7
Sr^{2+}	aq	−559.5	−545.8	−32.6

(Continued)

Formula	Form	$\Delta G°$	$\Delta H°$	$S°$
	Sulfur			
S	s, orthorhombic	0	0	31.8
S_2	g	+79.3	+128.4	228.2
H_2S	g	−33.6	−20.6	205.8
H_2S	aq	−27.8	−39.7	121.
SO_2	g	−300.2	−296.8	248.2
SO_3	g	−371.1	−395.7	256.8
S^{2-}	aq	+85.8	+33.1	−14.6
HS^-	aq	+12.1	−17.6	62.8
SO_4^{2-}	aq	−744.5	−909.3	20.1
HSO_4^-	aq	−755.9	−887.3	131.8
	Tin			
Sn	s	0	0	51.6
SnO	s	−256.9	−285.8	56.5
SnO_2	cassiterite	−519.6	−580.7	52.3
$Sn(OH)_2$	precipitated	−491.6	−561.1	155.
$SnCl_4$	g	−432.2	−471.5	365.8
SnS	s	−98.3	−100.	77.0
Sn^{2+}	aq	−27.2	−8.8	−17.
Sn^{4+}	aq	+2.5	+30.5	−117.
SnOHCl	aq	−392.0	−453.5	−126.
	Titanium			
Ti	s	0	0	30.6
TiO_2	anatase	−884.5	−939.7	49.9
TiO_2	rutile	−889.5	−944.7	50.3
$TiCl_4$	g	−726.7	−763.2	354.9
	Uranium			
U	g	0	0	50.2
UO_2	uraninite	−1031.7	−1084.9	77.0
UO_3	s	−1145.9	−1223.8	96.1
$UO_2(OH)_2$	s	−1394.8	−1533.9	126.
UF_6	g	−2063.7	−2147.4	377.9
U^{4+}	aq	−531.0	−591.2	−410.
UO_2^{2+}	aq	−953.5	−1019.6	−97.5
	Zinc			
Zn	s	0	0	41.6
Zn	g	+95.1	+130.7	161.0
ZnO	zincite	−318.3	−348.3	43.6
$Zn(OH)_2$	s	−553.5	−641.9	81.2
ZnS	sphalerite	−201.3	−206.0	57.7
$ZnCO_3$	smithsonite	−731.5	−812.8	82.4
Zn_2SiO_4	willemite	−1523.2	−1636.7	131.4
Zn^{2+}	aq	−147.1	−153.9	−112.1
$Zn(OH)_4^{2-}$	aq	−858.5		

Source: Wagman, D. D., W. H. Evans, V. B. Parker, R. H. Schumm, I. Halow, S. M. Bailey, K. L. Churney, and R. L. Nuttall: "NBS tables of chemical thermodynamic properties," *Journal of Physical and Chemical Reference Data,* vol. 11, supplement No. 2, 1982.
 Another excellent source of similar data is Robie, R. A., B. S. Hemingway, and J. R. Fisher: "Thermodynamic properties of minerals and related substances at 298.15 K and 1 bar pressure and at higher temperatures," *U.S. Geol. Survey Bull.* 1452, 1979.

STANDARD ELECTRODE POTENTIALS

The value of E° for each half-reaction is its potential in volts referred to the H_2–H^+ half-reaction, which is assigned the arbitrary value zero. The values are given for 25°C and 1 bar, with all substances at unit activity. Pure substances whose state is not specified in the equations are assumed to be in their standard states at 25°C and 1 bar.

The equation for each couple is written so that the reducing agent is at the left. Potential differences for complete reactions may be obtained by subtracting potentials for the appropriate half-reactions, provided that formulas of oxidizing and reducing agents are identical in the half-reactions and the complete reaction. E° values for half-reactions not shown in the table may be calculated from the free energies in Appendix VIII, by using the equation $E° = \Delta G°/96.5n$, where **n** is the coefficient of e^- in the half-reaction.

Potentials in acid solutions	
$K \rightleftharpoons K^+ + e^-$	−2.93
$Ca \rightleftharpoons Ca^{2+} + 2e^-$	−2.87
$Na \rightleftharpoons Na^+ + e^-$	−2.71
$Mg \rightleftharpoons Mg^{2+} + 2e^-$	−2.37
$Th \rightleftharpoons Th^{4+} + 4e^-$	−1.90
$Al \rightleftharpoons Al^{3+} + 3e^-$	−1.67

(Continued)

Potentials in acid solutions

$U \rightleftharpoons U^{4+} + 4e^-$	-1.38
$Mn \rightleftharpoons Mn^{2+} + 2e^-$	-1.18
$Si + 2H_2O \rightleftharpoons SiO_2 + 4H^+ + 4e^-$	-0.99
$Zn \rightleftharpoons Zn^{2+} + 2e^-$	-0.76
$Cr \rightleftharpoons Cr^{3+} + 3e^-$	-0.74
$Fe \rightleftharpoons Fe^{2+} + 2e^-$	-0.41
$H_2Se \rightleftharpoons Se + 2H^+ + 2e^-$	-0.40
$Co \rightleftharpoons Co^{2+} + 2e^-$	-0.28
$Ni \rightleftharpoons Ni^{2+} + 2e^-$	-0.24
$Sn \rightleftharpoons Sn^{2+} + 2e^-$	-0.14
$Pb \rightleftharpoons Pb^{2+} + 2e^-$	-0.13
$H_2 \rightleftharpoons 2H^+ + 2e^-$	-0.00
$H_2S(aq) \rightleftharpoons S + 2H^+ + 2e^-$	$+0.14$
$Sn^{2+} \rightleftharpoons Sn^{4+} + 2e^-$	$+0.15$
$Cu^+ \rightleftharpoons Cu^{2+} + e^-$	$+0.15$
$S^{2-} + 4H_2O \rightleftharpoons SO_4^{2-} + 8H^+ + 8e^-$	$+0.16$
$H_2SO_3(aq) + H_2O \rightleftharpoons SO_4^{2-} + 4H^+ + 2e^-$	$+0.17$
$Ag + Cl^- \rightleftharpoons AgCl + e^-$	$+0.22$
$As + 2H_2O \rightleftharpoons HAsO_2(aq) + 3H^+ + 3e^-$	$+0.25$
$U^{4+} + 2H_2O \rightleftharpoons UO_2^{2+} + 4H^+ + 2e^-$	$+0.27$
$Cu \rightleftharpoons Cu^{2+} + 2e^-$	$+0.33$
$S + 3H_2O \rightleftharpoons H_2SO_3(aq) + 4H^+ + 4e^-$	$+0.45$
$Cu \rightleftharpoons Cu^+ + e^-$	$+0.52$
$2I^- \rightleftharpoons I_2(s) + 2e^-$	$+0.54$
$3I^- \rightleftharpoons I_3^- + 2e^-$	$+0.54$
$HAsO_2(aq) + 2H_2O \rightleftharpoons H_3AsO_4 + 2H^+ + 2e^-$	$+0.57$
$Pd + 4Cl^- \rightleftharpoons PdCl_4^{2-} + 2e^-$	$+0.59$
$Pt + 4Cl^- \rightleftharpoons PtCl_4^{2-} + 2e^-$	$+0.76$
$Se + 3H_2O \rightleftharpoons H_2SeO_3(aq) + 4H^+ + 4e^-$	$+0.74$
$Fe^{2+} \rightleftharpoons Fe^{3+} + e^-$	$+0.77$
$2Hg \rightleftharpoons Hg_2^{2+} + 2e^-$	$+0.80$
$Ag \rightleftharpoons Ag^+ + e^-$	$+0.80$
$Hg \rightleftharpoons Hg^{2+} + 2e^-$	$+0.85$
$Pd \rightleftharpoons Pd^{2+} + 2e^-$	$+0.95$
$NO(g) + 2H_2O \rightleftharpoons NO_3^- + 4H^+ + 3e^-$	$+0.96$
$Fe^{2+} + 3H_2O \rightleftharpoons Fe(OH)_3 + 3H^+ + e^-$	$+0.97$
$Au + 4Cl^- \rightleftharpoons AuCl_4^- + 3e^-$	$+1.00$
$2Br^- \rightleftharpoons Br_2(l) + 2e^-$	$+1.07$
$2Br^- \rightleftharpoons Br_2(aq) + 2e^-$	$+1.09$
$HgS \rightleftharpoons S + Hg^{2+} + 2e^-$	$+1.11$
$H_2SeO_3(aq) + H_2O \rightleftharpoons SeO_4^{2-} + 4H^+ + 2e^-$	$+1.15$
$\frac{1}{2}I_2(s) + 3H_2O \rightleftharpoons IO_3^- + 6H^+ + 5e^-$	$+1.20$
$2H_2O \rightleftharpoons O_2 + 4H^+ + 4e^-$	$+1.23$

(Continued)

Potentials in acid solutions

$Mn^{2+} + 2H_2O \rightleftharpoons MnO_2(s) + 4H^+ + 2e^-$	+1.23
$2Cr^3 + 7H_2O \rightleftharpoons Cr_2O_7^{2-} + 14H^+ + 6e^-$	+1.35
$2Cl^- \rightleftharpoons Cl_2 + 2e^-$	+1.36
$Pb^{2+} + 2H_2O \rightleftharpoons PbO_2(s) + 4H^+ + 2e^-$	+1.46
$Au \rightleftharpoons Au^{3+} + 3e^-$	+1.50
$Mn^{2+} + 4H_2O \rightleftharpoons MnO_4^- + 8H^+ + 5e^-$	+1.51
$Mn^{2+} \rightleftharpoons Mn^{3+} + e^-$	+1.54
$Au \rightleftharpoons Au^+ + e^-$	+1.69
$Co^{2+} \rightleftharpoons Co^{3+} + e^-$	+1.83
$2F^- \rightleftharpoons F_2(g) + 2e^-$	+2.89

Potentials in basic solution

$Mg + 2OH^- \rightleftharpoons Mg(OH)_2 + 2e^-$	−2.69
$Al + 4OH^- \rightleftharpoons Al(OH)_4^- + 3e^-$	−2.33
$U + 4OH^- \rightleftharpoons UO_2 + 2H_2O + 4e^-$	−2.27
$Mn + 2OH^- \rightleftharpoons Mn(OH)_2 + 2e^-$	−1.56
$Zn + 2OH^- \rightleftharpoons Zn(OH)_2 + 2e^-$	−1.24
$SO_3^{2-} + 2OH^- \rightleftharpoons SO_4^{2-} + H_2O + 2e^-$	−0.93
$Sn(OH)_3^- + 3OH^- \rightleftharpoons Sn(OH)_6^{2-} + 2e^-$	−0.93
$Se^{2-} \rightleftharpoons Se + 2e^-$	−0.92
$Sn + 3OH^- \rightleftharpoons Sn(OH)_3^- + 2e^-$	−0.91
$Fe + 2OH^- \rightleftharpoons Fe(OH)_2 + 2e^-$	−0.89
$H_2 + 2OH^- \rightleftharpoons 2H_2O + 2e^-$	−0.83
$Fe(OH)_2 + OH^- \rightleftharpoons Fe(OH)_3 + e^-$	−0.55
$Pb + 3OH^- \rightleftharpoons Pb(OH)_3^- + 2e^-$	−0.54
$S^{2-} \rightleftharpoons S + 2e^-$	−0.44
$2Cu + 2OH^- \rightleftharpoons Cu_2O + H_2O + 2e^-$	−0.36
$Cr(OH)_3 + 5OH^- \rightleftharpoons CrO_4^{2-} + 4H_2O + 3e^-$	−0.13
$Cu_2O + 2OH^- + H_2O \rightleftharpoons 2Cu(OH)_2 + 2e^-$	−0.09
$Mn(OH)_2 + 2OH^- \rightleftharpoons MnO_2 + 2H_2O + 2e^-$	−0.05
$SeO_3^{2-} + 2OH^- \rightleftharpoons SeO_4^{2-} + H_2O + 2e^-$	+0.05
$Hg + 2OH^- \rightleftharpoons HgO(red) + H_2O + 2e^-$	+0.10
$Mn(OH)_2 + OH^- \rightleftharpoons Mn(OH)_3 + e^-$	+0.10
$Co(OH)_2 + OH^- \rightleftharpoons Co(OH)_3 + e^-$	+0.17
$PbO(red) + 2OH^- \rightleftharpoons PbO_2 + H_2O + 2e^-$	+0.25
$I^- + 6OH^- \rightleftharpoons IO_3^- + 3H_2O + 6e^-$	+0.26
$4OH^- \rightleftharpoons O_2 + 2H_2O + 4e^-$	+0.40

Sources of data: Sillén, L. G.: "Stability constants of metal-ion complexes. Sec. 1: Inorganic ligands," *Chem. Soc. London Spec. Publ.* 17, 1964; and Supplement 1, *Spec. Publ.* 25, 1971. Weast, R. C. *et al.; Handbook of Chemistry and Physics*, 66th ed., CRC Press, Boca Raton, Florida, pp. D-151 to D-158, 1985–86.

APPENDIX
X

BALANCING OXIDATION AND REDUCTION EQUATIONS

Most oxidation–reduction reactions of geologic interest can be balanced by inspection or by simple trial-and-error methods. For the occasional more complicated equation, a set of formal rules is often helpful. With a little practice we soon learn to judge which of the rules may be safely bypassed in a given situation.

1. *Note the pH range in which the reaction occurs.* Usually it suffices to know whether the process takes place in a strongly acid, weakly acid, weakly basic, or strongly basic solution.

2. *Write down the formulas of the substances which are oxidized, the substances which are reduced, and the substances formed from each.* Be sure that the formulas are accurate and that they are appropriate for the pH range. For example, Fe(III) would be written $Fe(OH)_3$ in basic and weakly acid solutions, Fe^{3+} in strong acid; divalent sulfur would be H_2S in acid, HS^- in weakly basic solutions, S^{2-} in strong base; sexivalent chromium would be $Cr_2O_7^{2-}$ in acid and CrO_4^{2-} in base.

3. *If any one of the elements oxidized or reduced does not balance at this stage, put in the appropriate coefficients by inspection.* These coefficients are only tentative. (This step is often not necessary.)

4. *Balance the oxidation process against the reduction process, making sure that the total number of electrons removed from one (or more) kind of atom is balanced by the number added to another kind (or kinds) of atom.* It is often helpful to write the oxidation number of each atom oxidized or reduced above the formula containing it on each side of the equation; then the necessary coefficients must be inserted to make the differences between these numbers identical.

5. *Make sure that all atoms except H and O are balanced.*

6. *Balance the number of + and − charges shown by the formulas so far established by adding H^+ or OH^- to either side of the equation.* Use H^+ if the reaction takes place in an acid solution, OH^- if it takes place in a basic solution.

7. *Balance the number of H atoms by adding H_2O to the side of the equation where H is deficient.*

8. *The equation should now be balanced. Check for balance by counting O atoms on each side.*

For a simple example, consider the oxidation of Cl^- to Cl_2 gas by MnO_2 in acid solution. The only stable form of manganese with a lower oxidation state in acid is Mn^{2+}, so that for step 2 we have

$$Cl^- + MnO_2 \rightleftharpoons Cl_2 + Mn^{2+}.$$

Since one atom of Cl appears on the left and two on the right, we add a coefficient in accordance with step 3:

$$2Cl^- + MnO_2 \rightleftharpoons Cl_2 + Mn^{2+}.$$

These formulas show a gain of 2 electrons by one Mn atom (changing its oxidation state from +4 to +2) and a loss of 2 electrons by two Cl^- ions (changing the oxidation state from −1 to 0). Hence the oxidation and reduction already balance each other, and nothing further is needed to complete step 4. Nor is any change necessary to complete step 5. Now on the right side of the equation we have represented two + charges and on the left side two − charges; to bring these into balance (step 6) we need four H^+ on the left side:

$$4H^+ + 2Cl^- + MnO_2 \rightleftharpoons Cl_2 + Mn^{2+}.$$

Now there are four H atoms on one side and none on the other, so we add the necessary H_2O (step 7):

$$4H^+ + 2Cl^- + MnO_2 \rightleftharpoons Cl_2 + Mn^{2+} + 2H_2O .$$

Inspection shows two O atoms on each side; therefore the equation is balanced (step 8).

For a more complicated example, consider the oxidation of native gold by MnO_2 in an acid solution in the presence of Cl^-. The manganese is again reduced to

Mn^{2+}, and the gold is dissolved as the complex ion $AuCl_4^-$. Hence for step 2:

$$Au + MnO_2 \rightleftharpoons AuCl_4^- + Mn^{2+}.$$

Only one atom of Au and one of Mn are shown on each side, so that step 3 is unnecessary. For the oxidation and reduction, we write oxidation numbers above Au and Mn:

$$\overset{0}{Au} + \overset{+4}{MnO_2} \rightleftharpoons \overset{+3}{AuCl_4^-} + \overset{+2}{Mn^{2+}}.$$

Three electrons are lost by each Au atom, and two are gained by each Mn. To balance this transfer, we need two Au atoms and three Mn (step 4):

$$2Au + 3MnO_2 \rightleftharpoons 2AuCl_4^- + 3Mn^{2+}.$$

Now all atoms (besides O and H) are balanced except Cl. It was originally specified that Cl^- must be present; hence we complete step 5 by adding this ion on the left:

$$8Cl^- + 2Au + 3MnO_2 \rightleftharpoons 2AuCl_4^- + 3Mn^{2+}.$$

To balance charges, we evidently need $12H^+$ on the left (step 6), and then to balance H atoms we must add $6H_2O$ on the right (step 7):

$$12H^+ + 8Cl^- + 2Au + 3MnO_2 \rightleftharpoons 2AuCl_4^- + 3Mn^{2+} + 6H_2O.$$

A count of O atoms shows six on each side, and hence the equation is balanced.

For a final example, take the oxidation of chromite by atmospheric oxygen in the presence of an alkaline solution. Here two elements are oxidized: chromium to chromate ion and iron to ferric hydroxide. Oxygen is reduced to its usual state of -2, but we need not represent this by a separate formula since the reduced oxygen may appear in any of the oxygen-containing formulas among the products. Hence for step 2:

$$FeCr_2O_4 + O_2 \rightleftharpoons Fe(OH)_3 + CrO_4^{2-}.$$

Since only one Cr atom appears on the right, we double the CrO_4^{2-} (step 3). At the same time we may write down the oxidation numbers of the elements oxidized and reduced:

$$\overset{+2}{Fe}\,\overset{+3}{Cr_2}O_4 + \overset{0}{O_2} \rightleftharpoons \overset{+3}{Fe}(OH)_3 + 2\overset{+6}{Cr}\overset{-2}{O_4^{2-}}.$$

[The -2 for O could equally well be written over the O in $Fe(OH)_3$.] For every mol of $FeCr_2O_4$ oxidized, an Fe atom loses 1 electron and two Cr atoms lose 6 electrons, giving a total loss of 7 electrons. Now in the reduction of oxygen each atom must gain 2 electrons; hence it is necessary to double every formula containing Fe or Cr and to show a total of 7 oxygen atoms (step 4):

$$2FeCr_2O_4 + \tfrac{7}{2}O_2 \rightleftharpoons 2Fe(OH)_3 + 4CrO_4^{2-}.$$

(Alternatively one can avoid fractional coefficients by taking $7O_2$ and multiplying the other coefficients by 4.) All atoms except H and O are now accounted for, and step 5 can be skipped. Since an alkaline solution is specified, we balance charges this time with OH^- (step 6):

$$8OH^- + 2FeCr_2O_4 + \tfrac{7}{2}O_2 \rightleftharpoons 2Fe(OH)_3 + 4CrO_4^{2-}.$$

Eight H atoms appear on the left and only 6 on the right; hence a molecule of water must be added (step 7):

$$8OH^- + 2FeCr_2O_4 + \tfrac{7}{2}O_2 \rightleftharpoons 2Fe(OH)_3 + 4CrO_4^{2-} + H_2O.$$

The total of 23 oxygen atoms on each side shows that the equation is balanced.

APPENDIX

XI

ANSWERS TO SELECTED END-OF-CHAPTER PROBLEMS

For questions in the first few chapters, the answer to each question includes an outline of the method by which the answer has been, or could be, obtained. This is to provide help in developing a general approach to the solution of geochemical problems. In later chapters such guidance seems no longer necessary, and the answers are given without preliminaries.

CHAPTER 1

3. (a) The solubility of calcite in pure water as represented by the reaction $CaCO_3 \rightleftharpoons Ca^{2+} + CO_3^{2-}$ is equal to the square root of the equilibrium constant of this reaction:

(1) Solubility of calcite (mols/kg water) = $\sqrt{4.5 \times 10^{-9}} = 6.7 \times 10^{-5}$.

(2) Because the solution is so dilute, mols/kg of water is approximately the same as mols/kg of solution and as mols/liter of solution. The solubility of calcite in g/100 ml solution is therefore approximately

$$\frac{6.7 \times 10^{-5} \text{ mols}}{1000 \text{ ml solution}} \times \frac{100.09 \text{ g}}{\text{mol}} = \frac{6.7 \times 10^{-3} \text{ g}}{1000 \text{ ml solution}} = \frac{6.7 \times 10^{-4} \text{ g}}{100 \text{ ml solution}}.$$

(3) Parts per million of Ca in solution (ppm) =

$$\frac{6.7 \times 10^{-5} \text{ mol}}{1000 \text{ g solution}} \times \frac{40.08 \text{ g}}{\text{mol}} = \frac{2.7 \times 10^{-3} \text{ g}}{1000 \text{ g solution}} = 2.7 \text{ ppm}.$$

(b) Let x represent the solubility of calcite. Then $m_{CO_3^{2-}} = x$ and $m_{Ca^{2+}} = x + 0.05$, so that the solubility product is

$$K = 4.5 \times 10^{-9} = m_{Ca^{2+}} m_{CO_3^{2-}} = (x + 0.05)(x)$$

or

$$x^2 + 0.05x - 4.5 \times 10^{-9} = 0.$$

This equation can be solved using the quadratic formula (see Footnote p. 37), or we can note that $0.05x$ is much greater than x^2, so that for an approximation the latter term can be neglected. The solubility is then

$$x = \frac{4.5 \times 10^{-9}}{0.05} = 9 \times 10^{-8} \text{ (mols calcite dissolved per kg water)}.$$

(c) The reaction for equilibrium among $CaSO_4$, $CaCO_3$, and a solution is obtained by subtracting the solubility reaction for $CaCO_3$ from the reaction for $CaSO_4$ (see Appendix VII-1 for values of solubility products):

$$\begin{array}{ll} CaSO_4 \rightleftharpoons Ca^{2+} + SO_4^{2-} & K_1 = 3.4 \times 10^{-5} \\ -(CaCO_3 \rightleftharpoons Ca^{2+} + CO_3^{2-}) & K_2 = 4.5 \times 10^{-9} \\ \hline CaSO_4 + CO_3^{2-} \rightleftharpoons CaCO_3 + SO_4^{2-} & \end{array}$$

The equilibrium constant for this reaction is

$$K = \frac{m_{SO_4^{2-}}}{m_{CO_3^{2-}}} = \frac{m_{Ca^{2+}} m_{SO_4^{2-}}}{m_{Ca^{2+}} m_{CO_3^{2-}}} = \frac{K_1}{K_2} = \frac{3.4 \times 10^{-5}}{4.5 \times 10^{-9}} = 7,600.$$

4. To calculate the solubility product for Ag_2SO_4 we need values for the molal concentrations of Ag^+ and SO_4^{2-} in the solution and the equilibrium constant relation for the reaction.

$$Ag_2SO_4 \rightleftharpoons 2Ag^+ + SO_4^{2-}$$

$$K = m_{Ag^+}^2 m_{SO_4^{2-}}.$$

First compute the number of mols of Ag_2SO_4 dissolved per kg water,

$$\frac{0.8 \text{ g } Ag_2SO_4}{100 \text{ g water}} \times \frac{mol}{311.8 \text{ g } Ag_2SO_4} = \frac{2.57 \times 10^{-3} \text{ mol } Ag_2SO_4}{100 \text{ g water}}$$

$$= \frac{2.57 \times 10^{-2} \text{ mol } Ag_2SO_4}{kg \text{ water}}.$$

Based on the stoichiometry of the dissolution reaction, the number of mols of Ag_2SO_4 dissolved per kg water is equal to the molality of SO_4^{2-}, which is also equal to twice the molality of Ag^+. Thus $m_{SO_4^{2-}} = 2.57 \times 10^{-2}$ and $m_{Ag^+} = 5.13 \times 10^{-2}$, and the solubility product is

$$K = (5.13 \times 10^{-2})^2 \times (2.57 \times 10^{-2}) = 6.77 \times 10^{-5}.$$

5. From the equilibrium constant relation for the reaction $2 Fe_3O_4 + 0.5 O_2 \rightleftharpoons 3 Fe_2O_3$, we compute the partial pressure of oxygen in equilibrium with magnetite and hematite:

$$K = f_{O_2}^{-1/2} = 5 \times 10^{43}$$

$$f_{O_2} = K^{-2} = (5 \times 10^{43})^{-2} = 4 \times 10^{-88} \text{ bar}.$$

At one bar pressure and 25°C a mol of ideal gas occupies 24.15 liters and contains 6.02×10^{23} molecules (Appendix I). So at 4×10^{-88} bar, one liter would contain

$$(6.02 \times 10^{23} \times 4 \times 10^{-88})/24.15 = 1 \times 10^{-65} \text{ molecules.}$$

This is equivalent to a single molecule in a space approximately equal to the size of the galaxy.

7. First obtain an equilibrium constant for the reaction

$$PbS + Zn^{2+} \rightleftharpoons ZnS + Pb^{2+}$$

by combining the solubility reactions and their equilibrium constants as we did in Prob. 1(c). The constant of equilibrium is $10^{-2.8}$, and the molal ratio of Pb^{2+} to Zn^{2+} in a solution in equilibrium with galena and sphalerite is

$$K = (m_{Pb^{2+}})/(m_{Zn^{2+}}) = 10^{-2.8}.$$

In a solution where $(m_{Zn^{2+}})/(m_{Pb^{2+}}) = 100$, sphalerite would be replaced by galena.

10. From, the solubility products for $CaSO_4$ and $BaSO_4$ we first compute the concentrations of Ca^{2+} and Ba^{2+} as a function of the concentration of SO_4^{2-}:

$$\text{Anhydrite} \quad CaSO_4 \rightleftharpoons Ca^{2+} + SO_4^{2-}$$

$$K = m_{Ca^{2+}} m_{SO_4^{2-}} = 3.4 \times 10^{-5}$$

$$m_{Ca^{2+}} = (3.4 \times 10^{-5})/(m_{SO_4^{2-}})$$

$$\text{Barite} \quad BaSO_4 \rightleftharpoons Ba^{2+} + SO_4^{2-}$$

$$K = m_{Ba^{2+}} m_{SO_4^{2-}} = 1.0 \times 10^{-10}$$

$$m_{Ba^{2+}} = (1.0 \times 10^{-10})/(m_{SO_4^{2-}})$$

Stoichiometric dissolution requires the following mass balance relation:

$$m_{SO_4^{2-}} = m_{Ba^{2+}} + m_{Ca^{2+}}$$

and substituting values for $m_{Ba^{2+}}$ *and* $m_{Ca^{2+}}$ derived above we obtain

$$m_{SO_4^{2-}} = (1.0 \times 10^{-10})/(m_{SO_4^{2-}}) + (3.4 \times 10^{-5})/(m_{SO_4^{2-}})$$

$$(m_{SO_4^{2-}})^2 = 3.4 \times 10^{-5} + 1.0 \times 10^{-10}$$

$$m_{SO_4^{2-}} = 5.8 \times 10^{-3}.$$

Using this value for SO_4^{2-} concentration we obtain the following concentrations for Ca^{2+} and Ba^{2+}: $m_{Ca^{2+}} = 5.8 \times 10^{-3}$ and $m_{Ba^{2+}} = 1.7 \times 10^{-8}$.

CHAPTER 2

1. Large concentrations: Na^+, S^{2-}, HS^-, OH^-. Small concentrations: H_2S, H^+. Add HCl and the concentrations of S^{2-}, HS^-, and OH^- decrease, and H_2S and H^+ increase. Initial solution is alkaline and is buffered by the reaction $S^{2-} + H^+ \rightleftharpoons HS^-$.

3. First compute the molarity of the aqueous species, then the equivalents of $(+)$ and $(-)$ charge:

$$M_{Na^+} = (44.9 \text{ g/L}) \times (22.9 \text{ g/mol})^{-1} = 1.95 \text{ mol/L}$$

$$M_{Ca^{2+}} = (6.6 \text{ g/L}) \times (40.08 \text{ g/mol})^{-1} = 0.167 \text{ mol/L}$$

$$M_{Cl^-} = (81.9 \text{ g/L}) \times (35.45 \text{ g/mol})^{-1} = 2.31 \text{ mol/L}$$

$$M_{SO_4^{2-}} = (1.0 \text{ g/L}) \times (96.06 \text{ g/mol})^{-1} = 0.01 \text{ mol/L}$$

Equivalents of $+$ charge : $1.95 + 0.33 = 2.28$ equiv/liter

Equivalents of $-$ charge : $2.31 + 0.02 = 2.33$ equiv/liter.

Hence solution is acid. Concentration of $H^+ = 0.05m$ is required for electrical neutrality in the solution, and pH = 1.3.

4. The solution is acid, so that $m_{CO_3^{2-}}$ and m_{OH^-} are small relative to $m_{HCO_3^-}$, $m_{H_2CO_3}$, and m_{H^+}. Thus the charge balance constraint of Eq. (2-13) becomes $m_{H^+} = m_{HCO_3^-}$. Because H_2CO_3 is a weak acid, we assume that $m_{H_2CO_3} \approx 10^{-4}$. Using Eq. (2-10) we compute the solution pH:

$$K = 10^{-6.4} = \frac{m_{H^+} m_{HCO_3^-}}{m_{H_2CO_3}} = \frac{m_{H^+}^2}{10^{-4}}.$$

Hence $m_{H^+} = m_{HCO_3^-} = 10^{-5.2}$, the pH = 5.2, and the fraction ionized

$$= \frac{m_{HCO_3^-}}{m_{CO_3, \text{total}}} = 10^{-1.2} = 0.06 = \text{about 6\%. In } 0.01m \text{ } H_2CO_3 \text{ solution, fraction ionized is}$$

0.006, or 0.6%.

5. 120 ppm $SiO_2 = 0.12$ g/kg $= 0.0020$ $m_{SiO_2, \text{total}} = 10^{-2.7} m_{SiO_2, \text{total}}$. Using the mass and charge balance relations,

$$0.0020 = m_{SiO_2, \text{total}} = m_{H_4SiO_4} + m_{H_3SiO_4^-}$$

and

$$m_{H^+} = m_{H_3SiO_4^-} + m_{OH^-},$$

and the assumptions that $m_{H^+} \gg m_{OH^-}$ and $m_{H_4SiO_4} \gg m_{H_3SiO_4^-}$, we obtain $m_{H^+} = m_{H_3SiO_4^-}$ and $m_{H_4SiO_4} = 0.002$. The equilibrium constant for the dissociation of silicic acid is then

$$K = 10^{-9.9} = \frac{m_{H^+} m_{H_3SiO_4^-}}{m_{H_4SiO_4}} = \frac{m_{H^+}^2}{0.002},$$

and the solution pH is approximately 6.3.

6. $I = 0.08$, $\gamma_{Mg^{2+}}$ from Eq. (2.75) is 0.265 and from Eq. (2-76) it is 0.385.

7. The molality of dissolved K_2CO_3 is 20 g/kg $\times$ (138.21 g/mol)$^{-1}$ = 0.15m. From the stoichiometry of Eq. (2-24) we see that $m_{HCO_3^-} = m_{OH^-}$, and because the solution is basic ($m_{H_2CO_3} \approx 0$) the mass balance for total carbonate [Eq. (2-12)] is $m_{CO_3^{2-}} = 0.15 - m_{HCO_3^-}$. Substitute these relations into the equilibrium constant equation for reaction (2-24) [see Eqs. (2-28) and (2-30)],

$$K = 2 \times 10^{-4} = \frac{m_{OH^-} m_{HCO_3^-}}{m_{CO_3^{2-}}} = \frac{m_{OH^-}^2}{0.15 - m_{OH^-}}$$

and solve for m_{OH^-} using the quadratic equation (see footnote, page 37) or by the approximation that $0.15 - m_{OH^-} \approx 0.15$. The pH is 11.7.

12. At pH = 6 we see from Fig. 2-3(a) that $m_{CO_3^{2-}}$ is small relative to the other carbonate species and the mass balance relation of Eq. (2-12) can be written as $m_{HCO_3^-} = 0.1 - m_{H_2CO_3}$. Substitution into Eq. (2-10) gives

$$K = 10^{-6.4} = \frac{10^{-6} \times (0.1 - m_{H_2CO_3})}{m_{H_2CO_3}}$$

or $m_{H_2CO_3} = 0.07$ and $m_{HCO_3^-} = 0.1 - 0.07 = 0.03$. The molality of CO_3^{2-} is computed using these values and Eq. (2-11), $m_{CO_3^{2-}} = 10^{-5.8}$. The product of $m_{Ca^{2+}}$ and $m_{CO_3^{2-}}$ is $10^{-9.8}$, which is less than the solubility product for calcite ($10^{-8.35}$, Appendix VII) so the solution is unsaturated with respect to $CaCO_3$.

13. The addition of 1 ml of $6m$ HCl to the solution in Prob. 12 is equivalent to adding 0.006 mol of H^+ per liter of solution. Because the solution is buffered by the reaction $HCO_3^- + H^+ \rightarrow H_2CO_3$, we obtain the following concentrations:

$$m_{H_2CO_3} = 0.07 + 0.006 = 0.076$$

and

$$m_{HCO_3^-} = 0.03 - 0.006 = 0.024.$$

Using these values in the equilibrium constant relation of Eq. (2-10) gives a pH of 5.9.

CHAPTER 3

4. The equation needed to find the solubility of calcite is

$$\text{calcite} + CO_2 + H_2O \rightleftharpoons Ca^{2+} + 2HCO_3^-, \qquad K = \frac{m_{Ca^{2+}} m^2_{HCO_3^-}}{P_{CO_2}}.$$

The solubility, in mols per kilogram of water, is equal to the concentration of Ca^{2+} (or half the concentration of HCO_3^-), and the problem is to find a value for K so that this concentration can be calculated. The equation can be expressed as the sum of

Calcite $\rightleftharpoons Ca^{2+} + CO_3^{2-}$,	$K_1 = 10^{-8.35}$	[Eq. (2-69)]
$CO_2 + H_2O \rightleftharpoons H_2CO_3$,	$K_2 = 10^{-1.5}$	[Eq. (1-19)]
$H_2CO_3 \rightleftharpoons H^+ + HCO_3^-$,	$K_3 = 10^{-6.4}$	[Eq.(2-8)]
$H^+ + CO_3^{2-} \rightleftharpoons HCO_3^-$,	$K_4 = 10^{+10.3}$	[the reverse of Eq. (2-9)]

and the value of K is then a combination of these four constants:

$$K = 10^{(-8.35 \,-1.5\, -6.4\, +10.3)} = 10^{-5.9}.$$

Because $P_{CO_2} = 0.1$ and $m_{HCO_3^-} = 2m_{Ca^{2+}}$, the expression for K gives $K \times 0.1 = 10^{-6.9} = 4 \times (m_{Ca^{2+}})^3$, from which $m_{Ca^{2+}} = $ solubility of calcite $= 10^{-2.5}$ or 0.003 mol kg^{-1} (or ca. 0.3 g kg^{-1}). The value of pH is found from K_3 above: since $m_{H_2CO_3} = m_{CO_2} \times K_2 = 10^{-1} \times 10^{-1.5} = 10^{-2.5}$; and $m_{HCO_3^-} = 0.006 = 10^{-2.2}$, $K_3 = m_{H^+} \times 10^{-2.2}/10^{-2.5} = 10^{-6.4}$ and pH = 6.7.

5. (a) To obtain the equilibrium constant add the following reactions:

$$MnCO_3 \rightleftharpoons Mn^{2+} + CO_3^{2-} \qquad K_1 = 10^{-9.3}$$

$$H^+ + CO_3^{2-} \rightleftharpoons HCO_3^- \qquad K_2 = 10^{10.3} \qquad \text{[reverse of Eq. (2-9)]}$$

$$H_2CO_3 \rightleftharpoons H^+ + HCO_3^- \qquad K_3 = 10^{-6.4} \qquad \text{[Eq. (2-8)]}$$

where $K = K_1 \times K_2 \times K_3 = 10^{-9.3} \times 10^{10.3} \times 10^{-6.4} = 10^{-5.4}$.

(b) Changing pH has no effect on the equilibrium constant. This constant is a function of temperature, pressure, reaction stoichiometry, and, as we will see in Chapter 8, a function of the standard states chosen for the species in the reaction. It is independent of the composition of minerals and aqueous species in the reaction.

(c) The equilibrium constant relation for the reaction is

$$K = \frac{m_{Mn^{2+}} m_{HCO_3^-}^2}{m_{H_2CO_3}}.$$

If the solution pH is increased from 6 to 8 we see from the third reaction given in (a) above, and from the curves in Fig. 2-3(a), that $m_{H_2CO_3}$ will decrease relative to $m_{HCO_3^-}$. To maintain equilibrium as pH increases from 6 to 8 thus requires the solubility of rhodochrosite (represented by the value of $m_{Mn^{2+}}$) to decrease.

8. Compute the ion activity product (IAP) of Ca^{2+} and F^- in seawater for comparison with the equilibrium constant for fluorite solubility, $CaF_2 \rightleftharpoons Ca^{2+} + 2F^-$, where $K = a_{Ca^{2+}} \cdot a_{F^-}^2$:

$$m_{Ca^{2+}} = \frac{400 \text{ g Ca}}{1000 \text{ kg H}_2\text{O}} \times \frac{1}{40 \text{ g Ca/mol}} = 0.01$$

$$m_{F^-} = \frac{1.3 \text{ g F}}{1000 \text{ kg H}_2\text{O}} \times \frac{1}{19 \text{ g F/mol}} = 0.000068$$

$$\text{IAP} = a_{Ca^{2+}} \cdot a_{F^-}^2 = m_{Ca^{2+}} \gamma_{Ca^2} \cdot m_{F^-}^2 \gamma_{F^-}^2$$

$$\text{IAP} = (0.01)(0.23)(0.000068)^2(0.65)^2 = 10^{-11.34}.$$

The product of ion activities is less than the equilibrium constant of fluorite solubility ($K = 10^{-10.4}$), thus seawater is undersaturated with respect to fluorite. Authigenic fluorine-containing sedimentary minerals: fluorapatite, clays.

9. The equilibrium reaction between calcite and fluorite is obtained by subtracting solubility reactions and logarithms of equilibrium constants for calcite and fluorite given in Appendix VII-1:

$$CaCO_3 \rightleftharpoons Ca^2 + CO_3^{2-} \qquad \log K = -8.35$$

$$\underline{-(CaF_2 \rightleftharpoons Ca^{2+} + 2F^- \qquad \log K = -10.4)}$$

$$CaCO_3 + 2F^- \rightleftharpoons CaF_2 + CO_3^{2-} \qquad \log K = 2.05$$

One way to make predictions about characteristics of solutions in equilibrium with either calcite or fluorite is to use the logarithmic form of the equilibrium constant to make a graph of the relative stabilities of these minerals in terms of the concentrations of CO_3^{2-} and F^-:

$$\log K = 2.05 = \log a_{CO_3^{2-}} - 2 \log a_{F^-}$$

$$\log a_{CO_3^{2-}} = 2.05 + 2 \log a_{F^-}.$$

Assume, as a first approximation, unit activity coefficients for Ca^{2+} and F^-, so that the equilibrium constant equation becomes

$$\log m_{CO_3^{2-}} = 2.05 + 2 \log m_{F^-}.$$

This is an equation for a straight line on a graph with $\log m_{CO_3^{2-}}$ and $\log m_{F^-}$ as axes. The slope of the line is $+2$ and its intercept on the $\log m_{CO_3^{2-}}$ axis is 2.05. Fluids that plot above this line project onto the calcite stability field, and fluids that plot below the line are in the stability field for fluorite. Thus, although fluids with concentrations of $m_{CO_3^{2-}} = m_{F^-} = 0.001$ and of $m_{Ca^{2+}} = 10^{-4}$ are supersaturated with respect to both calcite and fluorite (as determined by their solubility products), the fluid projects onto the calcite stability field in the diagram described above. Thus fluorite is metastable with respect to calcite, and calcite would replace fluorite.

11. First compute the equilibrium constant for equilibrium among strontianite, celestite, and solution:

$$SrCO_3 + SO_4^{2-} \rightleftharpoons SrSO_4 + CO_3^{2-}$$

$$K = (K_{strontianite})/(K_{celestite}) = 10^{-9}/(3 \times 10^{-7}) = 10^{-2.5} = a_{CO_3^{2-}}/a_{SO_4^{2-}}.$$

Then calculate the ion activity product (IAP) for the reaction, using concentrations and activity coefficients given in Table 3-1, and compare this value to the equilibrium constant:

$$IAP = a_{CO_3^{2-}}/a_{SO_4^{2-}}$$

or

$$\log(IAP) = \log a_{CO_3^{2-}} - \log a_{SO_4^{2-}} = -5.079 - (-2.318) = -2.76,$$
$$IAP = 10^{-2.76} = 0.0017 < K = 10^{-2.5} = 0.0032.$$

Thus the ratio of $a_{CO_3^{2-}}/a_{SO_4^{2-}}$ in seawater is less than the ratio required for equilibrium between strontianite and celestite, showing that celestite is stable relative to strontianite if sufficient Sr^{2+} is present in solution.

12. Compare the ion activity product (IAP) to the value of the solubility product for barite:

$$BaSO_4 \rightleftharpoons Ba^{2+} + SO_4^{2-} \qquad K = 10^{-10}$$
$$IAP = a_{Ba^{2+}} \cdot a_{SO_4^{2-}} = m_{Ba^{2+}} \gamma_{Ba^2} \cdot m_{SO_4^{2-}} \cdot \gamma_{SO_4^{2-}}$$

where

$$m_{Ba^{2+}} = \frac{0.015 \text{ g}}{1000 \text{ kg H}_2\text{O}} \times \frac{1}{137.33 \text{ g/mol}} = 1.1 \times 10^{-7}$$
$$IAP = (1.1 \times 10^{-7})(0.23)(0.027)(0.18) = 1.22 \times 10^{-10} > K = 10^{-10}.$$

The calculations suggest that seawater is approximately saturated with respect to barite.

CHAPTER 4

1. Combine reactions to obtain,

$$3K\text{-feldspar} + 2H^+ \rightleftharpoons Muscovite + 2K^+ + 6Quartz \qquad K_1 = 10^{9.69}$$

and

$$\text{Muscovite} + 6H^+ \rightleftharpoons \text{K-feldspar} + 2Al^{3+} + 4H_2O \qquad K_2 = 10^{14.49}.$$

The equilibrium constant equations are

$$\log K_1 = 2 \log a_{K^+} - 2 \log a_{H^+} = 2 \log a_{K^+} + 2pH = 9.69$$

and

$$\log K_2 = 2 \log a_{Al^{3+}} - 6 \log a_{H^+} = 2 \log a_{Al^{3+}} + 6pH = 14.49.$$

For unit activity coefficients of aqueous species these two equations are solved for concentrations of K^+ and Al^{3+} in a solution with a pH of 8.0: $\log m_{K^+} = -3.16$, $\log m_{Al^{3+}} = -16.75$.

3. Following the procedure used for calculating three lines in Fig. 4-8, we combine the reactions in Probs. 4-1 and 4-2 to obtain equilibrium in terms of the three kinds of Al species:

$$\text{Muscovite} + 10H^+ \rightleftharpoons K^+ + 3Al^{3+} + 3\text{Quartz} + 2H_2O \qquad K_1 = 10^{26.57}$$

$$\text{Muscovite} + 7H^+ \rightleftharpoons K^+ + 3Al(OH)^{2+} + 3\text{Quartz} + 3H_2O \qquad K_2 = 10^{12.32}$$

and

$$\text{Muscovite} + 6H_2O \rightleftharpoons K^+ + 3Al(OH)_4^- + 3\text{Quartz} + 2H^+ \qquad K_3 = 10^{-43.24}.$$

For pH's less than approximately 4.75 (see Prob. 4-2), $m_{Al, total} \approx m_{Al^{3+}} = 10^{-5}$ and concentrations of K^+ in equilibrium with muscovite–quartz-solution as a function of pH are represented by the equation for a straight line derived from the logarithmic form of the equilibrium constant for the first reaction:

$$\log m_{K^+} = \log K_1 - 3 \log m_{Al^{3+}} + 10 \log a_{H^+}$$

or

$$\log m_{K^+} = 41.57 - 10pH.$$

Between solution pH's of about 4.75 and 6.17, $m_{Al, total} \approx m_{Al(OH)^{2+}} = 10^{-5}$. Concentrations of K^+ in equilibrium with muscovite–quartz-solution are expressed by a similar equation derived from the equilibrium constant for the second reaction:

$$\log m_{K^+} = 27.32 - 7pH.$$

At solution pH's greater than about 6.17, $m_{Al, total} \approx m_{Al(OH)_4^-} = 10^{-5}$ and concentrations of K^+ are given by an equation derived from the last reaction:

$$\log m_{K^+} = 28.24 + 2pH.$$

5. The completed activity–activity phase diagrams are shown in the figures of pages 10 through 13 of Bowers, Jackson, and Helgeson (1984).

7. To determine what mineral or minerals might be in equilibrium with the analyzed fluid we compute values of $\log (a_{Ca^{2+}}/a_{H^+}^2)$ and $\log (a_{Na^+}/a_{H^+})$ and plot the fluid analysis on

the activity–activity phase diagram. For $\log (a_{Ca^{2+}}/a_{H^+}^2)$ we obtain

$$\log(a_{Ca^{2+}}/a_{H^+}^2) = \log a_{Ca^{2+}} - 2\log a_{H^+}$$
$$= \log m_{Ca^{2+}} + 2pH = -2.8 + 2(6.4) = 10.0.$$

To determine a value for $\log (a_{Na^+}/a_{H^+})$ we need the concentration of free Na^+ ions. First we express the equilibrium constant for NaCl dissociation in terms of m_{Na^+} by substituting the following identities:

$$m_{NaCl} = 1 - m_{Na^+} \quad \text{and} \quad m_{Cl^-} = 1 - m_{NaCl} = m_{Na^+}$$

so that

$$K = \frac{m_{Na^+} m_{Cl^-}}{m_{NaCl}} = \frac{m_{Na^+} m_{Na^+}}{1 - m_{Na^+}} = 2.63.$$

Rearrangement gives

$$m_{Na^+}^2 + K m_{Na^+} - K = 0.$$

This is a quadratic equation of the form $ax^2 + bx + c = 0$, for which the solution is $m_{Na^+} = 0.77$. Using this value, a number is computed for $\log (a_{Na^+}/a_{H^+}) = \log m_{Na^+} + pH = -0.11 + 6.4 = 6.28$. The analyzed fluid composition plots in the stability field of albite.

CHAPTER 6

1. Ions of the alkaline-earth metals, like ions of the alkalis, show a sorption selectivity sequence just opposite to what would be predicted from their ionic sizes alone. The reason is similar: the ions are all hydrated, the smallest (Mg^{2+}) has the greatest attraction for water molecules and hence forms the largest hydrated ion, and so the sorption sequence is reversed. For the transition metals the sequence is determined less by ionic size than by other properties. These ions are sorbed in the form of strongly bonded inner-sphere complexes, and the stability of the complexes depends in large part on electronegativities and the approximate ionic character of bonding with oxygen (Appendix VI). Thus Cu^{2+}, with the highest electronegativity and lowest ionic character of bonding, is the most strongly sorbed.

3. The data summarized in Fig. 6-8 show that selenite (SeO_3^{2-}) forms tightly bonded inner-sphere complexes which are strongly sorbed to the surfaces of iron-oxide minerals in the normal pH range of surface and groundwaters. Selenate ion (SeO_4^{2-}) forms more loosely bonded outer-sphere complexes and shows a much lower degree of sorption under similar pH conditions. Thus selenate ions are more mobile and therefore potentially more toxic to waterfowl.

5. Test the effectiveness of flocculation by the addition of colloids of opposite signs, or of electrolytes with ions of different amounts of charge. For example, Na_2SO_4 should be more effective as a coagulant than NaCl because of the doubly charged SO_4^{2-} ion.

7. Determination of molecular weight by freezing-point depression depends on the uniform dispersion through a solution of solute particles of roughly the same size as water molecules. Colloidal particles are very much larger and have a variety of sizes, hence produce unpredictable effects on the freezing-point.

CHAPTER 7

1.
$$\left(\frac{\partial a}{\partial h}\right)_l = 1$$

$$\left(\frac{\partial a}{\partial l}\right)_h = h$$

$$\left(\frac{\partial p}{\partial h}\right)_l = 2$$

$\left(\frac{\partial l}{\partial h}\right)_p = -1.0$; Substitute $\left(\frac{\partial p}{\partial h}\right)_l = 2$ and $\left(\frac{\partial p}{\partial l}\right)_h = 2$ into the expression for the total derivative of $p = F(h, l)$ and evaluate at $dp = 0:0 = 2\,dh + 2\,dl$. Differentiate this equation with respect to dh at constant p:

$$0 = 2 \times \left(\frac{\partial h}{\partial h}\right)_p + 2 \times \left(\frac{\partial l}{\partial h}\right)_p,$$

and rearrange to get $\left(\frac{\partial l}{\partial h}\right)_p = -1.0.$

$\left(\frac{\partial l}{\partial h}\right)_d = -\frac{h}{l}$; Substitute the identities $\left(\frac{\partial d}{\partial h}\right)_l = -\frac{h}{d}$ and $\left(\frac{\partial d}{\partial l}\right)_h = -\frac{l}{d}$

[obtained by differentiating the equation $d = (h^2 + l^2)^{0.5}$] into the expression for the total derivative of $d = F(h, l)$ and evaluate at constant d.

3. Substitute the relation that volume equals mass divided by density into the expressions for α and β to obtain $\alpha = -\frac{1}{\rho}\left(\frac{\partial \rho}{\partial T}\right)_P$ and $\beta = \frac{1}{\rho}\left(\frac{\partial \rho}{\partial P}\right)_T$. The following relationships demonstrate that α and β are both infinite at the critical point:

$$\beta = -\frac{1}{V}\left(\frac{\partial V}{\partial P}\right)_T = -\frac{1}{V}\left(\frac{1}{\left(\frac{\partial P}{\partial V}\right)_T}\right) = -\frac{1}{V}\left(\frac{1}{0}\right) = \infty.$$

For α we differentiate the expression for the total derivative of $V = F(T,P)$ with respect to P at constant V to obtain an expression for $\left(\frac{\partial V}{\partial T}\right)_P$ and evaluate this equation for the conditions of the critical point,

$$\left(\frac{\partial V}{\partial P}\right)_T = 0 = \left(\frac{\partial V}{\partial T}\right)_P \left(\frac{\partial T}{\partial P}\right)_V + \left(\frac{\partial V}{\partial P}\right)_T \left(\frac{\partial P}{\partial P}\right)_V,$$

$$\left(\frac{\partial V}{\partial T}\right)_P = -\frac{\left(\frac{\partial V}{\partial P}\right)_T}{\left(\frac{\partial T}{\partial P}\right)_V} = -\frac{\left(\frac{\partial P}{\partial T}\right)_V}{\left(\frac{\partial P}{\partial V}\right)_T} = -\frac{\left(\frac{\partial P}{\partial T}\right)_V}{0} = \infty,$$

thus α also becomes infinite at the critical point.

5. $\Delta G_r = (-3,799,700) + 2(-283,300) + 4(-856,600) - 2(-3,742,900)$
   $- 2(0) - (-237,100) = -69,800$ J mol^{-1}.
   $\Delta H_r = (-4,119,600) + 2(-252,400) + 4(-910,900) - 2(-3,968,100)$
   $- 2(0) - (-285,800) = -46,000$ J mol^{-1}.
   $\Delta S_r = 205 + 2(102.5) + 4(41.8) - 2(214.2) - 2(0) - (69.9) = 78.9$ J K^{-1} mol^{-1}.
   To estimate ΔS_r from values for the free energy and enthalpy of reaction use,

$$\Delta S_r = \frac{\Delta H_r - \Delta G_r}{T} = 79.8 \text{ J K}^{-1} \text{ mol}^{-1}.$$

7. Reaction is endothermic and will occur when energy is supplied, for example when temperature increases.

$$\Delta H_r = (-1,206,900) - (-542,800) - (-677,100) = 13,000 \text{ J mol}^{-1}.$$

CHAPTER 8

1. $K = 10^{-11.4}$ at 100°C and $10^{-7.5}$ at 200°C; $f_{H_2}/f_{H_2O} = 10^{-10.4}$ at 100°C and $10^{-6.5}$ at 200°C.
2. $K = 10^{5.8}$ at 25°C and $10^{-1.1}$ at 250°C.
3. Solubility product $= 10^{-28.0}$; calculated solubility $= 10^{-14.0}m$. Hydrolysis of Pb^{2+} and S^{2-} would increase solubility.
4. At 100°C, $K_1 = 10^{-6.2}$ and $K_2 = 10^{-11.1}$, so $m_{S^{2-}}$ is higher.
6. (a) log $K = 0$ and $\Delta G_r^{\circ} = 0$.
   (b) Graph of ΔG_r° as a function of temperature at constant pressure will be a curve with a negative slope and concave downward because:

$$\left(\frac{\partial \Delta G_r^{\circ}}{\partial T}\right)_P = -\Delta S_r^{\circ} < 0 \quad \text{and} \quad \left(\frac{\partial^2 \Delta G_r^{\circ}}{\partial T^2}\right)_P = -\frac{\Delta C_{pr}^{\circ}}{T} < 0.$$

The graph of ΔG_r° as a function of pressure at constant temperature will be a curve with a positive slope and concave downward because:

$$\left(\frac{\partial \Delta G_r^{\circ}}{\partial P}\right)_T = \Delta V_r^{\circ} > 0 \quad \text{and} \quad \left(\frac{\partial^2 \Delta G_r^{\circ}}{\partial P^2}\right)_T = \left(\frac{\partial V}{\partial P}\right)_T < 0.$$

(c) Because log $K = -\log a_{\text{muscovite}} = -\log X_{\text{muscovite}} = -(-0.3) = 0.3$ and the reaction is endothermic, the equilibrium temperature will be greater than 600°C at 2 kbar. The temperature, computed using Eq. (8-67), is about 665°C.

7. (a) The chemical potentials of thermodynamic components are the same in all phases, and the free energy of reaction is equal to zero for any chemical reaction written between the various phases.

(b) The number of components is six (Na_2O–CaO–Al_2O_3–SiO_2–H_2O–HCl) and the number of phases is five, so the degrees of freedom would be three.

(c) pH for 0.1 m NaCl is about 5.6, and for 0.5 m NaCl it is about 4.9.

(d) $\log K = \log a_{Na^+} + pH$, so at constant a_{Na^+} Eq. (8-61) becomes $\left(\dfrac{\partial pH}{\partial T}\right)_P = \dfrac{\Delta H_r^\circ}{2.303RT^2}$. Because ΔH_r° is positive, a decrease in temperature will cause a decrease in pH. For temperatures where ΔH_r° is approximately equal to zero (near 350°C and 100°C), pH will not change appreciably with decreasing temperature. The maximum change in pH with decreasing temperature occurs near the maximum value of ΔH_r° (about 300°C).

CHAPTER 9

1. $2Fe^{2+} + UO_2^{2+} + 4H^+ \rightleftharpoons 2Fe^{3+} + U^{4+} + 2H_2O.$

   $6Cl^- + 2MnO_4^- + 8H^+ \rightleftharpoons 3Cl_2 + 2MnO_2 + 4H_2O.$

   $2V(OH)_3 + O_2 + 6OH^- \rightleftharpoons 2VO_4^{3-} + 6H_2O.$

   $3CH_4 + 4SO_4^{2-} + 8H^+ \rightleftharpoons 3CO_2 + 4S + 10H_2O.$

2. (a) $E^\circ = -1.51$ volt, $\Delta G^\circ = -291$ kJ.   (b) $E^\circ = +0.30$ volt, $\Delta G^\circ = 57.8$ kJ.

   (c) $E^\circ = -0.63$ volt, $\Delta G^\circ = -365$ kJ.   (d) $E^\circ = +0.15$ volt, $\Delta G^\circ = +86.8$ kJ.

   (e) $E^\circ = -1.01$ volt, $\Delta G^\circ = 195$ kJ.

4. 0.096m.

5. +0.77 volt.

7. Decreasing pH: B, A, C, E, D. Decreasing Eh: D, B, A, C, E (positions of A, B, and C uncertain).

CHAPTER 10

2.

$$t = \frac{1}{1.93 \times 10^{-11}} \ln\left(\frac{\left(\dfrac{^{176}Hf}{^{177}Hf}\right) - \left(\dfrac{^{176}Hf_0}{^{177}Hf}\right)}{\left(\dfrac{^{176}Lu}{^{177}Hf}\right)} + 1\right).$$

4. Rearrange Eq. (10-28) to compute $\delta^{18}O_{kaolinite}$ for the two different waters:

$$\delta^{18}O_{kaolinite} = \alpha_{water}^{kaolinite}(\delta^{18}O_{water} + 10^3) - 10^3,$$

for $\delta^{18}O_{water} = -5‰$, $\delta^{18}O_{kaolinite} = 21.9‰$, and for $\delta^{18}O_{water} = -20‰$, $\delta^{18}O_{kaolinite} = 6.5‰$. To compute $\delta D_{kaolinite}$ use the equation $\delta D_{kaolinite} = 7.5\,\delta^{18}O_{kaolinite} - 220$: for water from the equatorial regions, $\delta D_{kaolinite} = -55.8‰$, and for water from the subarctic region, $\delta D_{kaolinite} = -171.3‰$. To calculate the fraction factor for the distribution of hydrogen isotopes between kaolinite and water first compute the value of δD_{water} for each type of water using the equation for the meteoric water line and the values of $\delta^{18}O_{water}$. The equatorial water has a value of $\delta D_{water} = -30\permil$, and the subarctic water a value of $-150\permil$. The fractionation factor is computed using the hydrogen–deuterium analog of Eq. (11-25),

$$\alpha_{water}^{kaolinite} = \frac{R_{kaolinite}}{R_{water}} = \frac{\delta D_{kaolinite} + 1000}{\delta D_{water} + 1000} = 0.97.$$

5. $C \approx 7 \times 10^5$, $T \approx 350°C$.

CHAPTER 11

1. $2.303 \log k_1/k_2 = -E_a/(8.314 \times 308) + E_a/(8.314 \times 298)$.

If $k_1 = 2k_2$: $2.303 \cdot \log 2 = 2.303 \cdot 0.3 = 0.691 = -E_a/2561 + E_a/2478$

$$= E_a(-0.0003905 + 0.0004036)$$
$$= 1.31 \times 10^{-5} \times E_a \text{ and}$$
$$E_a = 52.8 \text{ kJ.}$$

3. $\log K = \log m_{SiO_2} = -1.92$, solubility of quartz $= 10^{-1.92}m = 0.0120 \text{ mol kg}^{-1}$. To dissolve 1 mol requires the reciprocal of this number, or 83.3 kg of water. This amount of water is equal to 83,300/18 or 4630 mols, which at 300°C and 0.5 kbar occupies $4,630 \times 23.17 = 107,000 \text{ cm}^3$. In rock with a porosity of 0.1, this volume of water is contained in $1,070,000 \text{ cm}^3$ or about one m^3 of rock.

5. Rate of flow $= -K(\Delta h/\Delta l) = 10^{-1}(10^4/10^6) = 10^{-3} \text{ cm sec}^{-1}$. Transport for 10 km at this rate requires $10 \times 1000 \times 100/10^{-3} = 10^9$ sec, or $10^9/60 \times 60 \times 24 \times 365 = 31.7$ yr.

CHAPTER 12

4. Concentration of S in seawater $= 900$ ppm (Appendix IV), nearly all in the form of SO_4^{2-}. This is 0.9 parts per thousand or $0.9/32 = 0.028 \text{ mol kg}^{-1}$. An increase in concentration of 5% is $0.028 \times 0.05 = 0.0014 \text{ mol kg}^{-1}$, which requires addition of $0.0014 \times 1.4 \times 10^{21}$ or 1.96×10^{18} mols. This would take $1.96 \times 10^{18}/3 \times 10^{12}$ or about 650,000 yr.

5. Initial decrease in Mg^{2+} and pH: $Mg^{2+} + 2H_2O \rightarrow Mg(OH)_2 + 2H^+$, followed by higher Ca^{2+} and pH: $CaAl_2Si_2O_8 + 3H_2O \rightarrow Ca^{2+} + Al_2Si_2O_5(OH)_4 + 2OH^-$.

CHAPTER 13

4. (a) $CaCO_3 + H_2CO_3 \rightarrow Ca^{2+} + 2HCO_3^-$.
(b) $Ca_3Al_2Si_3O_{12} + 6H_2CO_3 + H_2O \rightarrow 3Ca^{2+} + Al_2Si_2O_5(OH)_4 + H_4SiO_4 + 6HCO_3^-$.
(c) $ZnS + H_2CO_3 \rightarrow Zn^{2+} + HS^- + HCO_3^-$.
(d) $2NaAlSiO_4 + 3H_2O \rightarrow 2Na^+ + Al_2Si_2O_5(OH)_4 + 2OH^-$.

CHAPTER 14

2. (a) $Ca_5(PO_4)_3OH \rightleftharpoons 5Ca^{2+} + 3PO_4^- + OH^-$, $\log K = -57.8$.
(b) $H_2O \rightleftharpoons H^+ + OH^-$, $\log K = -14.0$.

(c) $3CaHPO_4 \rightleftharpoons 3Ca^{2+} + 3HPO_4^{2-}$, $\log K = 10^{-21}$.

(d) $3HPO_4^{2-} \rightleftharpoons 3H^+ + 3PO_4^{3-}$, $\log K = 10^{-37.2}$

Subtract (b), (c), and (d) from (a):

(e) $Ca_5(PO_4)_3OH + 4H^+ \rightleftharpoons 3CaHPO_4 + 2Ca^{2+} + H_2O$, $\log K = +14.4$.

So at equilibrium $a_{Ca^{2+}}/a_{H^+}^2 = 10^{7.2}$

In seawater, $m_{Ca^{2+}} = 450$ ppm $= 0.45$ g kg$^{-1} = 0.011m = 10^{-1.96}$, and $m_{H^+} =$ ca. $10^{-8.2}$. So seawater $m_{Ca^{2+}}/m_{H^+}^2 = 10^{14.4}$, and reaction (e) goes to the left.

3. At pH 6.5 and Eh +0.30 volt, $a_{Fe^{2+}} = 10^{-8.1}$. At pH 8.4 and Eh −0.30 volt, $a_{Fe^{2+}} = 10^{-3.7}$.

4. $MnO_2 + 2Fe(OH)_2 + 2H_2O \rightleftharpoons Mn(OH)_2 + 2Fe(OH)_3$, $\Delta G° = -95.7$ kJ.

$MnO_2 + 2Fe^{2+} + 4H_2O \rightleftharpoons Mn^{2+} + 2Fe(OH)_3 + 2H^+$, $\Delta G° = -49.8$ kJ.

For the second reaction, $\log K = \log a_{Mn^{2+}} + 2\log a_{H^+} - 2\log a_{Fe^{2+}} = 8.7$. If pH = 4, $\log a_{Mn^{2+}} - 2\log a_{Fe^{2+}} = 16.7$. Hence $a_{Mn^{2+}} \gg a_{Fe^{2+}}$ for values of $a_{Fe^{2+}}$ greater than $\approx 10^{-15}$.

7. At Eh 0.4 volt and pH 8.2 in fresh water, the activity of Mn^{2+} in equilibrium with MnO_2 is about 10^{-5}. If the activity coefficient is assumed equal to 1, this means a concentration of $10^{-5}m$ or about 0.5 ppm. In seawater the concentration required for precipitation is four or five times greater than this.

8. (a) At 100°C, $\log a_{H^+} - \log a_{Na^+} = -4.84$, and pH = 4.84.

For 150°C, $\log K_{150°}/K_{100°} = (45,000/2.303 \times 8.314) \times (423 - 373)/(423 \times 373)$

$$= 2,350 \times 3.169 \times 10^{-4} = 0.7447.$$

So $\log K_{150°} = -4.84 - (-0.745) = -4.09$, and pH = 4.09.

(b) $\log K = \log a_{Ca^{2+}} + \log f_{CO_2} + \log a_{H_2O} - 2\log a_{H^+}$.

At 100°C, $9.13 = -1 + \log f_{CO_2} + 0 - 2 \times (-4.84)$, and $f_{CO_2} = 2.8$ bar.

At 150°C, $8.87 = -1 + \log f_{CO_2} + 0 - 2 \times (-4.09)$, and $f_{CO_2} = 49$ bar.

CHAPTER 15

1. $2C_2H_6 \rightleftharpoons C_4H_{10} + H_2$, $\Delta G° = -17.2 - 2(-32.8) = +48.8$ kJ, so reverse is favored.

2. $C_2H_4 + C_2H_6 \rightleftharpoons C_4H_{10}$, $\Delta G° = -52.6$ kJ, so energy not needed.

5. $4CaSO_4 + 3CH_4 + 8H^+ \rightleftharpoons 4S + 4Ca^{2+} + 3CO_2 + 10H_2O$.

15. $3CO_2 + CH_4 \rightleftharpoons 4CO + 2H_2O$.

At 29°C, $\Delta G° = 227.9$ kJ and $\log K = -39.92$.

Assume constant $\Delta H° = +329.7$ kJ and $\Delta S° = +341.0$ J mol^{-1} deg^{-1}.

Then at 100°C, $\Delta G° = \Delta H° - T\Delta S° = 329.7 - 0.341 \times 373 = +202.5$ kJ.

$\Delta G° = 202.5 = -2.303RT \log K = 2.303 \times 0.008314 \times 373 \log K$, and $\log K = -28.35$.

Thus equilibrium is displaced far to the left at both 25 and 100°C.

CHAPTER 16

8. At 630°C and 3 kbar, the slope of the line is about 4500 bar per 120°C, or 37.5 bar deg^{-1}.

$$\Delta S° = \Delta V(dP/dT) = 21.5 \text{ cm}^3 \text{ mol}^{-1} \times 37.5 \text{ bar deg}^{-1} \times 0.1 \text{ J cm}^{-3} \text{ bar}^{-1}$$

$$= 80.6 \text{ J mol}^{-1} \text{ deg}^{-1}.$$

CHAPTER 17

12. Van't Hoff's equation: $\log a_{\text{an in liquid}} - \log a_{\text{an, in solid}} = (\Delta H/2.303R)(1/1830 - 1/T)$. Since a_{an} for the solid is assumed to be unity, this equation is:

$$\log a_{\text{an, liq}} = (136,000/2.303 \times 8.314) \times (1/1830 - 1/T), \text{ from which}$$
$$T = -7103/(\log a_{\text{an}} - 3.881).$$

Similarly, for diopside, $T - 7448/(\log a_{\text{dp}} - 4.473)$.
The calculated eutectic is approximately $1338°C$ at $X_{\text{an}} = 0.3$ and $X_{\text{dp}} = 0.7$.

CHAPTER 18

2. (a) About 2.7 kbar (assuming 2.7 g cm^{-3} for average rock density). (b) 5.4 kbar. (c) 100 to 500°C. (d) About 700°C. (e) 2.7×10^{18} g. (f) 1.21×10^{15} liters. (g) 9.6×10^{15} liters.
3. 0.8 km if pressure assumed lithostatic; 2.2 km if pressure assumed hydrostatic.
9. $\log K = -2.1$ at 627°C and -1.4 at 927°C. Hence deposition of galena is favored by falling temperature and a high $f_{\text{H}_2\text{S}}/f_{\text{HCl}}^2$ ratio. Total pressure has little effect.

CHAPTER 19

1. $Cu_2S + 4Cl^- + H_2O \rightleftharpoons 2CuCl_2- + HS^- + OH^-$; $\Delta G° = +223.2$ kJ.
   $\log K = -39.1 = 2 \log a_{\text{CuCl}_2^-} + \log a_{\text{HS}-} + \log a_{\text{OH}-} - 4 \log a_{\text{Cl}^-}$.
   Let $a_{\text{CuCl}_2^-} = x$ and $a_{\text{HS}-} = 0.5x$: $\log K = -39.1 = 2 \log x + \log 0.5x - 7$
   $-4 \log a_{\text{Cl}-} = 3 \log x - 0.3 - 7 - 4 \log a_{\text{Cl}-}$. Thus $3 \log x = -31.8$, so that $a_{\text{CuCl}_2^-} = 10^{-10.6}$ at $a_{\text{Cl}-} = 1$ and $10^{-9.3}$ at $a_{\text{Cl}-} = 10$. Hence Cu_2S is not appreciably soluble in Cl^- at 25°C, unless pH is an unrealistic 3 or less.
2. For the reaction $PbS + 2OH^- + H_2O \rightleftharpoons Pb(OH)_3^- + HS^-$, $\Delta G° = +86.7$ kJ and $\log K = -15.2$. Even at pH 14, $Pb(OH)_3^-$ is only $10^{-7.6}m$, so solubility is not appreciable. For reaction $PbS + 3OH^- \rightleftharpoons Pb(OH)_3^- + S^{2-}$, $\Delta G° = +80.5$ kJ and $\log K = -14.1$. So at pH 14, $Pb(OH)_3^- = 10^{-7.1}m$, again too small.
3. $SO_2 + 3H_2 \rightleftharpoons 2H_2O + H_2S$, $\Delta G° = -207.4 + 0.0569T$. If $f_{\text{H}_2\text{O}} = 10^3$ bar and $f_{\text{H}_2} = 10^{-3}$ bar, the ratio $f_{\text{H}_2\text{S}}/f_{\text{SO}_2} = 10^{+4.9}$ at 200°C and $10^{-1.9}$ at 400°C.

INDEX